Neuroprotection: Rescue from Neuronal Death in the Brain

Neuroprotection: Rescue from Neuronal Death in the Brain

Editor

Bae Hwan Lee

MDPI • Basel • Beijing • Wuhan • Barcelona • Belgrade • Manchester • Tokyo • Cluj • Tianjin

Editor
Bae Hwan Lee
Yonsei University College of
Medicine
Korea

Editorial Office
MDPI
St. Alban-Anlage 66
4052 Basel, Switzerland

This is a reprint of articles from the Special Issue published online in the open access journal *International Journal of Molecular Sciences* (ISSN 1422-0067) (available at: https://www.mdpi.com/journal/ijms/special_issues/Neuroprotection_Brain).

For citation purposes, cite each article independently as indicated on the article page online and as indicated below:

LastName, A.A.; LastName, B.B.; LastName, C.C. Article Title. *Journal Name* **Year**, *Volume Number*, Page Range.

ISBN 978-3-0365-1994-4 (Hbk)
ISBN 978-3-0365-1995-1 (PDF)

Contents

About the Editor

Bae Hwan Lee is a professor at the Department of Physiology, Yonsei University College of Medicine, Republic of Korea. He received his Ph.D. degree at Korea University in 1993. In 1993, he started his academic career at University of Texas Medical Branch at Galveston, Texas, USA as a post-doc. He became an assistant professor in 1997, an associate professor in 2002, and a professor in 2007 at the Yonsei University College of Medicine. He received Awards for Excellent Papers from Korean Society for Brain and Neural Science in 2008 and from Korean Journal of Physiology & Pharmacology in 2014. He also received Achievement Awards from Korean Society for Brain and Neural Science in 2009 and from Korean Society for Emotion and Sensibility in 2017 for his outstanding service and invaluable contributions to the societies. He is an author of more than 150 scientific publications. His current research focuses on neural injury and functional recovery from cellular levels to behaviors.

Preface to "Neuroprotection: Rescue from Neuronal Death in the Brain"

The brain is located on the top of our body and plays a role as the control tower of our body. Thanks to the brain, we can interact with our environment easily and effectively. In the brain, however, cell death occurs naturally and/or due to various types of injury. In order to maintain the normal functions of the brain, the protection of the brain against cell death is very important.

This Special Issue on the "Neuroprotection: Rescue from Neuronal Death in the Brain" is to present the updated knowledge on the mechanisms of neuroprotection and the developments of therapeutics on cell death in the brain. The research fields covered in this Special Issue include stroke, ischemia, Alzheimer's disease, Parkinson's disease, memory deficit, seizure, oxidative injury, neurotoxic injury, signaling pathways, nanomedicine, trophic factors, antioxidants, phytochemistry, and so on.

This Special Issue brings together a collection of 19 papers which highlight the mechanisms underlying neuroprotection and therapeutics on neuronal death. I would like to thank all the authors for their outstanding contributions and the editorial staff members for their assistance.

Bae Hwan Lee
Editor

International Journal of
Molecular Sciences

Editorial

Neuroprotection: Rescue from Neuronal Death in the Brain

Bae Hwan Lee

Department of Physiology, Brain Korea 21 PLUS Project for Medical Science, Yonsei University College of Medicine, Seoul 03722, Korea; bhlee@yuhs.ac

Citation: Lee, B.H. Neuroprotection: Rescue from Neuronal Death in the Brain. *Int. J. Mol. Sci.* **2021**, *22*, 5525. https://doi.org/10.3390/ijms22115525

Received: 14 May 2021
Accepted: 20 May 2021
Published: 24 May 2021

Publisher's Note: MDPI stays neutral with regard to jurisdictional claims in published maps and institutional affiliations.

The brain plays important roles in mental processing and in controlling other bodily organs. However, the brain tissue is vulnerable to exogenous or endogenous injury. After injury, neuronal death leads to functional deficits and neurological disorders. Therefore, protection from neuronal death is crucial for preserving brain functions.

This Special Issue aimed to understand the mechanisms underlying neuroprotection and to explore potential therapeutic strategies for recovering brain functions from neurological disorders. In this Special Issue, nineteen papers were published, including thirteen original articles and six review papers dealing with interesting subjects which are related to different brain functions and disorders.

Stroke including ischemia is one of the major villains responsible for neuronal death and mortality. A review by Dong et al. [1] summarized the current understanding of the mechanisms of pathogenesis in ischemic stroke and the updated developments of nanomedical therapeutics such as nanoparticles, liposomes, and cell-based nanovesicles. It has been shown that cation-chloride cotransporters (CCCs) are expressed in neurons and contribute to numerous physiological functions. In a review paper, Josiah et al. [2] introduced the recent advances in the functional regulations of the CCCs and their signaling pathways in stroke. Using a rodent model of global cerebral ischemia, Kang et al. [3] observed that amiloride, an inhibitor of the sodium–hydrogen exchanger-1, attenuates hippocampal injury by reducing zinc accumulation in mice, and Hong et al. [4] reported that administration of N-acetyl-L-cysteine, an antioxidant, reduces neuronal death induced by cerebral ischemia and over-activation of transient receptor potential melastatin 2 (TRPM2) channels in rats. Jung et al. [5] described that pyridoxine (a form of vitamin B6) deficiency increases neuronal death after ischemia by enhancing serum homocysteine levels and lipid peroxidation, and by reducing nuclear factor erythroid-2-related factor 2 (Nrf2) levels. Pyridoxine also decreases regenerative potentials by reducing brain-derived neurotrophic factor (BDNF) levels in the hippocampus of gerbils. Markosyan et al. [6] reported that preventive triple gene therapy by adenoviral vectors carrying genes encoding vascular endothelial growth factor (VEGF), glial cell-derived neurotrophic factor (GDNF), and neural cell adhesion molecule (NCAM), or by gene-engineered umbilical cord blood mononuclear cells over-expressing recombinant VEGF, GDNF, and NCAM, has beneficial effects on the recovery from stroke in the brain of rats. Tanioka et al. [7] observed that intra-arterial injection of lipid emulsion during reperfusion shows neuroprotective effects in an in vivo rat model of ischemic reperfusion injury. Using primary astrocyte cultures, Mitroshina et al. [8] demonstrated that ischemic-like conditions decrease network connectivity and the stimulation with ATP under normal conditions increases the number of connections, implying that astrocytes can constitute a functional network in response to a stimulation. Using organotypic hippocampal slices of mice pups, Ort et al. [9] found that neither nimodipine (an L-type voltage-gated calcium channel blocker) alone nor in combination with amiloride (an acid-sensing ion channel inhibitor) showed any improvement in an in vitro ischemia model. The combination of both components dissolved in dimethylsulfoxide (DMSO) even increased cell death, although this effect was not observed in the condition treated with amiloride alone. In this regard, they suggested that DMSO should be used with caution in neuroprotective experiments.

In relation to Alzheimer's disease (AD) and memory functions of the brain, a review paper by Kim et al. [10] focused on the recent progression in the beneficial effects of adiponectin on AD and its potential mechanisms related to the diverse medications of adiponectin in AD treatment. Using a mouse model of AD, Gonzalo-Gobernado et al. [11] observed that intraperitoneal administration of liver growth factor (LGF) restored cognitive deficits and reduced amyloid-β (Aβ) content in the APPswe mice which over-express the Swedish APP mutation, with Aβ plaque deposits in the hippocampus and cerebral cortex. Ye et al. [12] reported that soybean-derived phosphatidylserine, a membrane phospholipid, enhances memory function in memory deficit rats by a trimethyltin, a neurotoxin affecting the brain. In a human study by Piotrowicz et al. [13], BDNF increased after exercise in normoxia and hypoxia athletes but cognitive performance was not improved by acute elevation of BDNF. Indeed, BDNF plays an important role in the growth of neurons and plasticity of neuronal synapses. The roles of BDNF in neuronal plasticity and neuroprotection were highlighted in a review by Colucci-D'Amato et al. [14].

Seizure deteriorates the function of the brain. Huang et al. [15] demonstrated that peroxisome proliferator-activated receptor γ coactivator 1-α (PGC-1α) activated by status epilepticus regulates the VEGF/VEGFR2 (VEGF receptor 2) pathway through mitogen-activated protein kinase kinase (MEK)/extracellular signal-regulated kinase (ERK) signaling and exhibits neuroprotective effects against neuronal death in the hippocampus of rats. In this regard, VEGF plays a crucial role as a neuroprotective factor in the nervous tissues. A review by Silva-Hucha et al. [16] summarized the contribution of VEGF to differences between the oculomotor system and other motor neurons in relation to vulnerability to degeneration.

In the brain, reactive oxygen species (ROS) increase susceptibility of the brain to neuronal damage and functional deficits. Abnormal levels of ROS are regulated by cellular defense mechanisms of antioxidants in the brain. A review paper by Lee et al. [17] presented the roles of various antioxidants in the brain and suggested the potential of antioxidants for the protection of the brain from oxidative insults. Using organic hippocampal slice cultures (OHSCs), Lee et al. [18] found that ascorbic acid (an antioxidant) treatment was effective for survival of neurons in OHSCs after kainic acid (KA) insult, but that its protective effect was different depending on aging of the OHSCs. In Parkinson's disease (PD), oxidative stress induced by ROS is a major cause of the dopaminergic cell death. Mimicking PD-like syndrome to induce cell death through mitochondrial oxidative stress, Haque et al. [19] used neurotoxins including 1-methyl-4-phenylpyridinium ion (MPP$^+$) and hydrogen peroxide (H$_2$O$_2$) in human dopaminergic neuroblastoma cells and found that the pharmacological inhibition by a selective antagonist of G protein-coupled receptor 4 (GPR4) or genetic deletion of GPR4 decreases apoptotic cell death through the modulation of phosphatidylinositol biphosphate (PIP$_2$)-mediated calcium signaling.

As reported in many previous studies and in this Special Issue, mechanisms of neuroprotection are complex. Understanding detailed mechanisms underlying neuroprotection may lead to the development of effective therapeutic strategies for the prevention of neuronal death and for improvement of functional recovery from neural injury or neurodegeneration.

Funding: This work was supported by the Basic Science Research Program through the National Research Foundation of Korea (NRF) funded by the Ministry of Education (MOE) (NRF-2020R1A2C3008481).

Institutional Review Board Statement: Not applicable.

Informed Consent Statement: Not applicable.

Data Availability Statement: Not applicable.

Acknowledgments: I would like to thank all the authors for their outstanding contribution to this Special Issue, the reviewers for their evaluation on the manuscripts, and the IJMS editorial staff members for their assistance.

Conflicts of Interest: The author declares no conflict of interest.

References

1. Dong, X.; Gao, J.; Su, Y.; Wang, Z. Nanomedicine for ischemic stroke. *Int. J. Mol. Sci.* **2020**, *21*, 7600. [CrossRef] [PubMed]
2. Josiah, S.S.; Meor Azlan, N.F.; Zhang, J. Targeting the WNK-SPAK/OSR1 pathway and cation-chloride cotrans-porters for the therapy of stroke. *Int. J. Mol. Sci.* **2021**, *22*, 1232. [CrossRef] [PubMed]
3. Kang, B.S.; Choi, B.Y.; Kho, A.R.; Lee, S.H.; Hong, D.K.; Jeong, J.H.; Kang, D.H.; Park, M.K.; Suh, S.W. An inhibitor of the sodium–hydrogen exchanger-1 (NHE-1), amiloride, reduced zinc accumulation and hippocampal neuronal death after ischemia. *Int. J. Mol. Sci.* **2020**, *21*, 4232. [CrossRef] [PubMed]
4. Hong, D.K.; Kho, A.R.; Lee, S.H.; Jeong, J.H.; Kang, B.S.; Kang, D.H.; Park, M.K.; Park, K.-H.; Lim, M.-S.; Choi, B.Y.; et al. Transient receptor potential melastatin 2 (TRPM2) inhibition by antioxidant, N-acetyl-L-cysteine, reduces global cerebral ischemia-induced neuronal death. *Int. J. Mol. Sci.* **2020**, *21*, 6026. [CrossRef] [PubMed]
5. Jung, H.Y.; Kim, W.; Hahn, K.R.; Kang, M.S.; Kim, T.H.; Kwon, H.J.; Nam, S.M.; Chung, J.Y.; Choi, J.H.; Yoon, Y.S.; et al. Pyridoxine deficiency exacerbates neuronal damage after ischemia by increasing oxidative stress and reduces proliferating cells and neuroblasts in the gerbil hippocampus. *Int. J. Mol. Sci.* **2020**, *21*, 5551. [CrossRef] [PubMed]
6. Markosyan, V.; Safiullov, Z.; Izmailov, A.; Fadeev, F.; Sokolov, M.; Kuznetsov, M.; Trofimov, D.; Kim, E.; Kundakchyan, G.; Gibadullin, A.; et al. Preventive triple gene therapy reduces the negative consequences of ischemia-induced brain injury after modelling stroke in a rat. *Int. J. Mol. Sci.* **2020**, *21*, 6858. [CrossRef]
7. Tanioka, M.; Park, W.K.; Park, J.; Lee, J.E.; Lee, B.H. Lipid emulsion improves functional recovery in an animal model of stroke. *Int. J. Mol. Sci.* **2020**, *21*, 7373. [CrossRef] [PubMed]
8. Mitroshina, E.V.; Krivonosov, M.I.; Burmistrov, D.E.; Savyuk, M.O.; Mishchenko, T.A.; Ivanchenko, M.V.; Vedunova, M.V. Signatures of the consolidated response of astrocytes to ischemic factors in vitro. *Int. J. Mol. Sci.* **2020**, *21*, 7952. [CrossRef] [PubMed]
9. Ort, J.; Kremer, B.; Grüßer, L.; Blaumeiser-Debarry, R.; Clusmann, H.; Coburn, M.; Höllig, A.; Lindauer, U. Failed neuroprotection of combined inhibition of L-type and ASIC1a calcium channels with nimodipine and amiloride. *Int. J. Mol. Sci.* **2020**, *21*, 8921. [CrossRef] [PubMed]
10. Kim, J.Y.; Barua, S.; Jeong, Y.J.; Lee, J.E. Adiponectin: The potential regulator and therapeutic target of obesity and Alzheimer's disease. *Int. J. Mol. Sci.* **2020**, *21*, 6419. [CrossRef]
11. Gonzalo-Gobernado, R.; Perucho, J.; Vallejo-Muñoz, M.; Casarejos, M.J.; Reimers, D.; Jiménez-Escrig, A.; Gómez, A.; Ulzurrun de Asanza, G.M.; Bazán, E. Liver growth factor "LGF" as a therapeutic agent for Alzheimer's disease. *Int. J. Mol. Sci.* **2020**, *21*, 9201. [CrossRef]
12. Ye, M.; Han, B.H.; Kim, J.S.; Kim, K.; Shim, I. Neuroprotective effect of bean phosphatidylserine on TMT-induced memory deficits in a rat model. *Int. J. Mol. Sci.* **2020**, *21*, 4901. [CrossRef] [PubMed]
13. Piotrowicz, Z.; Chalimoniuk, M.; Płoszczyca, K.; Czuba, M.; Langfort, J. Exercise-induced elevated BDNF level does not prevent cognitive impairment due to acute exposure to moderate hypoxia in well-trained athletes. *Int. J. Mol. Sci.* **2020**, *21*, 5569. [CrossRef] [PubMed]
14. Colucci-D'Amato, L.; Speranza, L.; Volpicelli, F. Neurotrophic factor BDNF, physiological functions and therapeutic potential in depression, neurodegeneration and brain cancer. *Int. J. Mol. Sci.* **2020**, *21*, 7777. [CrossRef] [PubMed]
15. Huang, J.-B.; Hsu, S.-P.; Pan, H.-Y.; Chen, S.-D.; Chen, S.-F.; Lin, T.-K.; Liu, X.-P.; Li, J.-H.; Chen, N.-C.; Liou, C.-W.; et al. Peroxisome proliferator-activated receptor γ coactivator 1α activates vascular endothelial growth factor that protects against neuronal cell death following status epilepticus through PI3K/AKT and MEK/ERK signaling. *Int. J. Mol. Sci.* **2020**, *21*, 7247. [CrossRef] [PubMed]
16. Silva-Hucha, S.; Pastor, A.; Morcuende, S. Neuroprotective effect of vascular endothelial growth factor on motoneurons of the oculomotor system. *Int. J. Mol. Sci.* **2021**, *22*, 814. [CrossRef] [PubMed]
17. Lee, K.H.; Cha, M.; Lee, B.H. Neuroprotective effect of antioxidants in the brain. *Int. J. Mol. Sci.* **2020**, *21*, 7152. [CrossRef] [PubMed]
18. Lee, K.H.; Kim, U.J.; Cha, M.; Lee, B.H. Chronic treatment of ascorbic acid leads to age-dependent neuroprotection against oxidative injury in hippocampal slice cultures. *Int. J. Mol. Sci.* **2020**, *22*, 1608. [CrossRef] [PubMed]
19. Haque, E.; Akther, M.; Azam, S.; Choi, D.-K.; Kim, I.-S. GPR4 knockout improves the neurotoxin-induced, caspase-dependent mitochondrial apoptosis of the dopaminergic neuronal cell. *Int. J. Mol. Sci.* **2020**, *21*, 7517. [CrossRef] [PubMed]

International Journal of
Molecular Sciences

Review

Nanomedicine for Ischemic Stroke

Xinyue Dong, Jin Gao, Yujie Su and Zhenjia Wang *

Department of Pharmaceutical Sciences, College of Pharmacy and Pharmaceutical Sciences,
Washington State University, Spokane, WA 99202, USA; xinyue.dong@wsu.edu (X.D.);
jin.gao3@wsu.edu (J.G.); yujie.su@wsu.edu (Y.S.)
* Correspondence: zhenjia.wang@wsu.edu

Received: 29 September 2020; Accepted: 12 October 2020; Published: 14 October 2020

Abstract: Stroke is a severe brain disease leading to disability and death. Ischemic stroke dominates in stroke cases, and there are no effective therapies in clinic, partly due to the challenges in delivering therapeutics to ischemic sites in the brain. This review is focused on the current knowledge of pathogenesis in ischemic stroke, and its potential therapies and diagnosis. Furthermore, we present recent advances in developments of nanoparticle-based therapeutics for improved treatment of ischemic stroke using polymeric NPs, liposomes and cell-derived nanovesicles. We also address several critical questions in ischemic stroke, such as understanding how nanoparticles cross the blood brain barrier and developing in vivo imaging technologies to address this critical question. Finally, we discuss new opportunities in developing novel therapeutics by targeting activated brain endothelium and inflammatory neutrophils to improve the current therapies for ischemic stroke.

Keywords: ischemic stroke; blood brain barrier; nanoparticle-based drug delivery; brain targeting

1. Introduction

Stroke is an unexpected and acute brain disease. It is reported that one of nineteen deaths is related to stroke in the United States, and the mortality rate of stroke is as high as 30% [1]. Stroke is defined by a condition caused by a hemorrhage or occlusion of cerebral blood vessels. Lacking of blood flow in the brain causes dysfunctions of brain cells, oxidative stress, and neurological damage [2]. The symptoms of stroke include numbness, confusion, and aphasia, and those signs are related to injured areas in the brain [3]. Since 87% of strokes are related to ischemia in the brain and 13% are involved with the hemorrhage, this review will focus on discussing how nanotechnology improves therapies and diagnosis of ischemia stroke.

Cerebral ischemia initiates a cascade of pathological processes, eventually causing neuron death. Reperfusion is a clinical method to restore the blood flow in the brain by administration of tissue plasminogen activator (t-PA) or mechanical thrombectomy (MT) for treatment of ischemic stroke [4]. However, reperfusion often leads to tissue damage because oxygen influx of reperfusion generates reactive oxygen species (ROS), initiating inflammatory responses including cytokine production and leukocyte infiltration [5]. Many neuroprotectants have been developed to alleviate reperfusion-induced injury, but none of them were clinically approved. There are several reasons for this failure: (1) ineffective drug delivery into the brain because of blood brain barrier (BBB), (2) drugs with short circulation times, poor stability, and toxicity, (3) difficulty in choosing right drugs and doses due to the heterogenicity of stroke (e.g., disease locations and severity).

Recently, nanotechnology emerges as innovative tools in drug delivery and diagnosis to treat a wide range of diseases, such as cancer and inflammatory disorders [6–13]. In the case of ischemic stroke, nanoparticles could possibly deliver therapeutics across BBB, prolong the drug circulation, and increase the drug accumulation at diseased sites. In addition, nanoparticles can be utilized as innovative diagnostic systems to detect several biomarkers (such as ROS and neurotransmitters) in the brain for early-stage stroke diagnosis. In this review, we firstly describe the physiopathology of ischemic stroke and current limitations in therapy and diagnosis. Then, we discuss how nanotechnology offers opportunities in treating ischemic stroke, highlighting the recent progress on delivering neuroprotective agents, anti-inflammatory drugs, and small interfering RNA (siRNA) to alleviate ischemic injuries. We also review the applications of nanomaterials in stroke diagnosis. Finally, we describe new opportunities in translating nanoparticle-based therapies to clinic in the prevention and treatment of ischemic stroke.

2. Pathology of Stroke and Current Therapies

2.1. Molecular Mechanisms for Pathology in Stroke

Acute ischemic stroke and subsequent reperfusion cause a series of pathophysiological changes including neuronal damage, oxidative stress, local inflammation, and loss of blood brain barrier (BBB) integrity [14]. The cellular process and mechanism of ischemic stroke and the ischemic injury cascade is summarized in Figure 1. Herein, we mainly discuss neuronal damage, oxidative stress, and inflammation responses that contribute to the pathogenesis of ischemic stroke. These three events are intertwined, and each can result in brain damage.

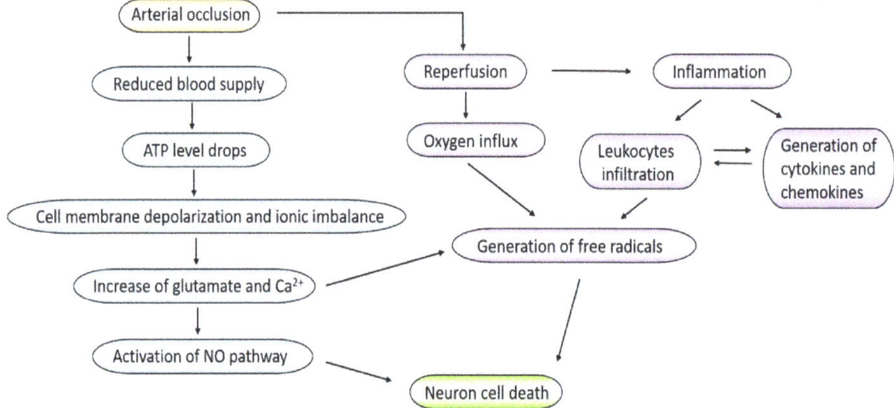

Figure 1. The cellular process and mechanism of ischemic stroke and the ischemic injury cascade.

2.1.1. Neuronal Damage

Neurons are the major components in brain and are particularly vulnerable to ischemia. Studies have shown that acute cerebral ischemia causes the vasculature infarction, thus leading to neuronal death in rodent models [15]. Due to reduced blood supply in the ischemic brain, adenosine triphosphate (ATP) level drops and lactate acidosis leads to the loss of ionic homeostasis in neurons [14]. Failure to maintain the ion gradients results in depolarization of cell membranes and therefore activating a variety of ion channels such as calcium and sodium channels. Those events also cause the excessive release of glutamate, which initiate more loss of neuronal cells [16]. Glutamate is a major excitatory neurotransmitter in nervous system regulating a variety of excitatory synapses, but high levels of extracellular glutamate can induce the cell death in the late stage. Dead neural cells release damage-associated molecular patterns (DAMPs), which activate the innate and adaptive immune system, initiating inflammatory pathways [17].

2.1.2. Oxidative Stress

Oxidative stress is involved with the production of reactive oxygen species (ROS), which activates apoptosis, necrosis, and autophagy pathways during ischemic stroke [18]. Oxidative stress also reduces the bioavailability of nitric oxide (NO) in endothelial cells. Nitric oxide is a potent chemical that inhibits platelet aggregation and leukocyte adhesion [19]. The loss of NO will exacerbate the coagulation, which aggravates the ischemic insult, leading to more blood flow reduction in the brain. In addition, cells under ischemia can increase the levels of calcium, sodium, and adenosine diphosphate (ADP) that produce oxygen species in mitochondria. Reperfusion is a commonly used method to treat ischemic stroke, but this also produces more ROS to cause the secondary damage in the brain. Furthermore, ROS damages lipids, proteins, and nucleic acid via lipid peroxidation, protein oxidation, and DNA fragmentation and leads to further cell death [20].

2.1.3. Inflammation Response

Inflammation response is an important event in ischemic stroke and the subsequent reperfusion. Ischemia and reperfusion activate immune system to respond to tissue injury including activation of endothelium and leukocytes. Specifically, the immune response is involved with the intravascular and parenchymal damage triggered by the interruption of blood supply. In other words, the immune system plays a central role in regulating the outcomes of stroke patients [21]. For the immune response, adhesion molecules on endothelial cells are upregulated during ischemia. For example, P-selectin was highly expressed when the mouse brain was damaged. P-selectin is a major adhesion molecule for the binding of platelets to endothelial cells, and this interaction causes proinflammatory signals [22]. ICAM-1 is another intercellular adhesion molecule and is upregulated during inflammation response. A study showed that knockdown of intercellular adhesion molecule-1 (ICAM-1) gene in mice reduced the infarction of the brain, suggesting that inflammation response plays a major role in pathogenesis of stroke [23].

Innate immune cells are major components during inflammation response because they sense endogenous danger signals through several receptors such as toll-like receptors (TLRs), retinoic acid-inducible gene (RIG)-1-like receptors, nucleotide oligomerization domain (NOD)-like receptors, C-type lectin receptors, and absent in melanoma (AIM2)-like receptors [24]. These receptors activate downstream signaling pathways, such as nuclear factor-κB (NF-κB), mitogen-activated protein kinase, and type 1 interferon pathways. These pathways upregulate proinflammatory cytokines, chemokines, and oxidative metabolites. The products including tumor necrosis factor alpha (TNF-α), interleukin-1β (IL-1β), IL-6, IL-18, and NO activate endothelial cells and astrocytes, and also act on leukocytes for their tissue infiltration in the brain, thus causing more tissue damage. Clinical studies found that in stroke patients, the increase of cytokines appeared in the cerebrospinal fluid and blood and this feature is correlated to patient survival [25]. In addition, the increased infiltration of leukocytes caused by the formation of fibrin and cytokines contributes to further tissue damage. Ischemia and reperfusion activate macrophages in the perivascular space. Activated macrophages release proinflammatory cytokines, which contribute to the expression of adhesion molecules in endothelium, promoting more leukocyte infiltration (especially neutrophils) to cause the BBB damage [26]. The large infiltration of neutrophils in a short period causes the release of proteases that can further destroy the BBB integrity and exacerbate oxidative stress, leading to severe brain damage.

2.2. Blood Brain Barrier in Ischemic Stroke

Blood brain barrier (BBB) plays an important role in ischemia/reperfusion (I/R). During ischemic stroke, there are two phases of BBB breaking. In the early phase of reperfusion (6 h), endothelial transcytosis dominates in increased permeability of BBB, and it may be associated with endothelial vesicle transport [27]. In the later phase, BBB is disrupted because of the loss of integrity of endothelial tight junctions (TJs). Studies showed that the structural changes of TJs occurred at 48–58 h after middle cerebral artery occlusion (MCAO) in mice [28], and these changes were associated with the degradation of TJs by matrix metalloproteinases (MMPs) [29]. MMPs also degraded extracellular matrix proteins of basal membrane of BBB, such as type IV collagen [30].

The immune response also causes the loss of BBB integrity, leading to cell death in stroke [31]. Microglia are macrophages existing in the brain, and they become activated in response to ischemic stroke, which cause the production of ROS, cytokines (e.g., TNF-α, IL-1β, and IL-6) and chemokines (e.g., macrophage inflammatory proteins-1alpha (MIP-1α)/CCL3, monocyte chemoattractant protein-1 (MCP-1)/CCL-2 and chemokine (C-X-C motif) ligand-1 (CXCL-1)). These inflammatory products act on endothelial cells to activate the NF-κB pathway for expression of adhesion molecules, such as vascular cell adhesion protein (VCAM), ICAM-1, and P-selectin. These molecules support leukocyte recruitment to invade brain parenchymal tissues [32], thus increasing the brain permeability [33].

Blood brain barrier maintains the brain homeostasis, but ischemic stroke disrupts this barrier, resulting in increased vascular permeability. As we discussed, the permeability is regulated by transcellular pathways in the early stage and the breaking TJs in the later stage [28]. Therapeutics delivery systems to ischemic stroke lesions can be designed based on its pathogenesis.

2.3. Current Treatment Options for Ischemic Stroke

Treating ischemic stroke is to restore blood flow to the brain as soon as possible, and reperfusion is a major tool to bring the blood back to the brain. The clinical methods include intravenous administration of tissue plasminogen activator (tPA) and mechanical thrombectomy (MT) [34]. Tissue plasminogen activator (tPA) is the gold standard treatment for ischemic stroke. The treatment window using tPA is in 4.5 h when patients show the stroke symptom, and the outcomes dramatically decrease beyond 4.5 h [35]. However, tPA treatment may potentially cause the intracerebral hemorrhage and lead to more mortality [36]. When patients appear with a large artery occlusion (the clot burden is high) or tPA treatment is beyond the best time window, the MT surgery is an alternative strategy to treat ischemic stroke. In MT surgery, a microcatheter is inserted to clotting blood vessels to remove the clots. The treatment procedure for stroke patients is summarized in Figure 2.

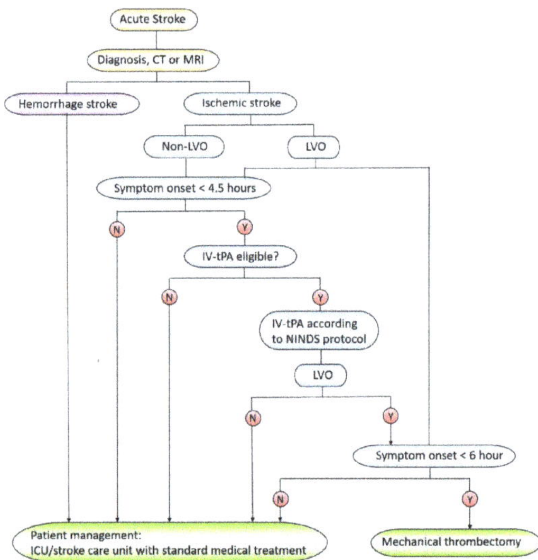

Figure 2. The procedure to treat stroke patients. CT: computed tomography; MRI: magnetic resonance imaging; LVO: large vessel occlusion; IV-tPA: intravenous injection of trans-plasminogen activator; ICU: intensive care unit; NIHSS: National Institutes of Health Stroke Scale; Y: yes; N: no.

Except for the reperfusion, supportive care is also very important. For example, 25% of patients may show neurological deficiency in 24–48 h, and care is needed to prevent further brain damage [37]. Although stroke and its pathogenesis have been investigated for many years, there are no pharmacological agents available in clinic to effectively treat ischemic stroke [38,39].

There are several drugs that have been tested to target inflammatory pathways in ischemic stroke. NXY-059 is a scavenger of free radicals and shows the neuroprotective effect in a pre-clinical stroke model [40]. However, NXY-059 was less effective in Stroke–Acute Ischemic NXY Treatment II (SAINT II) trial [41]. The failure may be due to the high water solubility of NXY-059 that has low permeability to BBB. Therefore, developing new drug delivery approaches are needed to solve the current issues on treating ischemic stroke.

3. Nanomedicine in Stroke Treatment

There are several challenges in treating ischemic stroke. Most therapeutics have short blood circulation and cannot easily cross the blood brain barrier to reach the ischemic tissues. In addition, controlling drug release and developing novel drug delivery vehicles are important factors to improve therapies of ischemic stroke. Nanotechnology is used to design nanoscale materials that can be applied in biomedical applications. Nanoparticles with the sizes of 50–200 nm are generated and can encapsulate drugs to increase drug water solubility and tissue targeting. Nanoparticles are smaller than a cell and larger than small molecules; thus, nanoparticles are excellent carriers to deliver therapeutics (drugs) and control their release for improved therapies in a wide range of diseases [42–44]. Nanoparticle-based drug delivery systems may solve the current challenges in ischemic stroke treatment. In this section, we present several nanoparticle-drug delivery systems used in treating ischemic stroke (Figure 3).

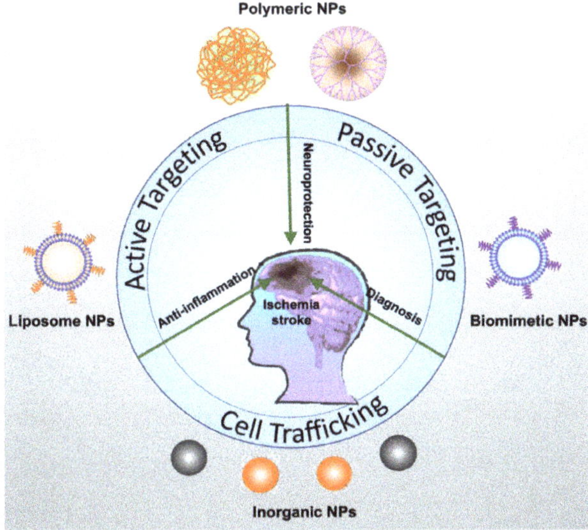

Figure 3. Nanoparticle-based delivery platforms for delivery of neuroprotectants, anti-inflammation reagents, and imaging probes for ischemic stroke therapy and diagnostics. Polymeric NPs include PLGA (poly(lactic-co-glycolic acid) NPs, chitosan NPs, and PAMAN dendrimer. Liposomes are made of a lipid bilayer and are loaded with therapeutics agents. Inorganic NPs include metallic NPs (such as platinum), ceria NPs, and Fe_2O_3 NPs. Biomimetic NPs are new drug delivery platforms generated from cell membrane vesicles. Each type of NPs has been discussed in the manuscript.

3.1. Delivery of Neuroprotectants

Early recanalization (thrombolysis) and neuroprotection are the two major steps for stroke patients [45]. Neuroprotection is an important strategy to protect the neurological damage from ischemic stroke. There are several approaches including reduced immune cell adhesion, blocking of proinflammatory cytokines, decreased lipid peroxidation, and decreased cell apoptosis. Many drugs, such as adenosine [46], have been developed and approved by the FDA to treat myocardial ischemia. However, they failed to be used in treating cerebral I/R injury [47] because they have the short plasma half-life time and lack of permeability to BBB. To address this issue, nanoparticle drug delivery systems have been developed to deliver drugs to cross BBB. The applications of nanoparticle-drug delivery systems have been summarized in Table 1.

Table 1. Studies of nanoparticle-based drug delivery in ischemic stroke.

Nanoparticles	Drugs/agents	Administration	Targeting	Stoke model	Results	Ref.
PEGylated liposomes	Asialo-erythropoietin	i.v.	Ischemic site	MCAO (1 h) * on rats	PEGylated liposomes increased the accumulation of AEPO at ischemia site and ameliorated cerebral I/R injury.	[48]
T7-conjugated liposomes	ZL006	i.v.	TfR over-expressed brain endothelium	MCAO (2 h) * on rats	T7 enhanced the transport of liposomes across the BBB and T7-PLPs/ZL006 reduced infarct volume and neurological deficit.	[49]
SHp and T7 conjugated-PEGylated liposomes	ZL006	i.v.	Ischemic endothelium and neurons	MCAO (2 h) * on rats	Dual targeting peptide enhanced the accumulation of NPs in brain and ameliorated ischemic injury.	[50]
Liposomes	Basic fibroblast growth factor (bFGF)	Intranasal injection	-	MCAO (2 h) * on rats	Liposomes improved bFGF accumulation in brain tissues and the system improved spontaneous locomotor activity of animals.	[51]
PEGylated liposomes	t-PA/fasudil	i.v.	-	MCAO (2 h) * on rats	Treatment of fasudil-lip before t-PA decreased the risk of t-PA-derived cerebral hemorrhage and extended the therapeutic time window of t-PA.	[52]

Table 1. *Cont.*

Nanoparticles	Drugs/agents	Administration	Targeting	Stoke model	Results	Ref.
DSPE-PEG2000 liposome	-	i.v.	Ischemic site	MCAO (20 min) * on mice	Liposomes accumulated in ischemic brain at both early stage (0.5) and late phase (48 h).	[53]
Cationic bovine serum albumin-conjugated PEG-PLA	Tanshinone IIA	i.v.	Brain microvessels	MCAO (2 h) * on rats	CBSA-PEG-TIIA-NPs reduced infarct volume and neurological deficits.	[54]
Cationic bovine serum albumin-conjugated PEG-PLA	Tanshinone IIA	i.v.	Brain microvessels	MCAO (2 h) * on rats	CBSA-PEG-TIIA-NPs suppressed neuronal apoptosis and inflammatory mediators (MMP-9 and COX-2) via MAPK signaling pathways.	[55]
SHp conjugated, red blood cell membrane shelled-polymer	NR2B9C	i.v.	Ischemic neurons	MCAO (2 h) * on rats	SHp-RBC-NP targeted to the ischemic site and ameliorate neuroscores and infarct volume.	[56]
Fas ligand conjugated-PEGylated-lipid NPs	3-n-Butylphthalide	i.v.	Microglia cells in ischaemic region	MCAO (45 min) on mice	By the help of Fas ligand, the NPs reached to ipsilateral region of ischemic brain and delivered the drug to improve brain injury.	[57]
AMD3100-conjugated PEG-T/M-PCL block copolymers (ASNPs)	Glyburide	i.v.	CXCR4 in the ischemic brain	MCAO on mice	ASNPs penetrated the ischemic brain and improve the neurological outcomes after ischemia.	[58]
TfRMAb-conjugated PEG-coated chitosan nanospheres	Z-DEVD-FMK, caspase-3 inhibitor	i.v.	TfR type 1 on the cerebral vasculature	MCAO (2 h) * on mice	Nano-system inhibited caspase activity and had subsequent neuroprotection effect.	[59]
T7-conjugated erythrocyte-coated Mn_3O_4 NPs	-	i.v.	Ischemic endothelium	MCAO (0.5 h) * on rats	The NPs scavenged free radical and restored the oxygen.	[60]

Table 1. *Cont.*

Nanoparticles	Drugs/agents	Administration	Targeting	Stoke model	Results	Ref.
Platelets coated-γ-Fe$_2$O$_3$ magnetic NPs	L-arginine	i.v.	Thrombus	Photochemical-induced thrombosis	L-arginine released at ischemic lesions disrupted the local platelets aggregation and recover blood flow.	[61]
PEGylated-ceria nanoparticles	-	i.v.	-	MCAO on rats	Ceria nanoparticles served as ROS scavenger and reduced ischemic brain damage.	[62]
Angiopep-2 and PEG modified Ceria nanoparticles	Edaravone	i.v.	LRP on BBB	MCAO on rats	E-A/P-CeO2 crossed BBB and delivered Edaravone to eliminate ROS, protected the BBB integrity and ameliorated ischemic injury.	[63]
Platinum nanoparticle (nPt)	-	i.v.	-	MCAO (1 h) * on mice	Platinum nanoparticle reduced ROS production and improved the neurological score.	[64]
Platinum nanoparticle (nPt)	-	i.v.	-	MCAO (1 h) * on mice	nPt protected the ischemic brain via reducing the MMP-9 activity and disruption of neurovascular unit.	[65]
c(RGDyK) peptide-conjugated exosomes	Curcumin	Stereotaxic injection	Ischemic site	MCAO (1 h)* on mice	They targeted the lesion region of the ischemic brain and entered microglia to suppress the inflammatory response and protect ischemic brain.	[66]
R3V6 peptides	Dexamethasone and heme oxygenase-1 siRNA	i.v.	-	MCAO (1 h) * on rats	Dexamethasone enhanced the delivery ability of R3V6 peptides for heme oxygenase-1 (HO-1) gene knockdown and reduced ischemic brain damage.	[67]

Nanoparticles	Drugs/agents	Administration	Targeting	Stoke model	Results	Ref.
Neutrophil membrane-derived nanovesicles	Resolvin D2	i.v.	Inflamed endothelium	MCAO (1 h) * on mice	Nanovesicles specifically targeted the ischemic endothelium and released Resolvin D2 to inhibit neutrophil infiltration and reduce inflammation in ischemic sites.	[68]
Functionalized carbon nanotubes (f-CNT)	Caspase-3 siRNA	Stereotaxic injection	Ischemic neurons	Endothelin Stroke Model	Gene silencing of Caspase-3 by f-CNT in neuronal tissue had neuroprotection effect and improved ischemic insult.	[69]
PAMAM dendrimer amide with basic L-arginine residues	HMGB1 siRNA	i.v.	HMGB1 mRNA in brain	MCAO (1 h) * on rats	HMGB1 siRNA delivered by e-PAM-R reduced neuronal death decreased infarct volume.	[70]

* The times represent the ischemic duration; (-) denotes that the information is not available in the studies.

3.1.1. Liposomes as Delivery Platforms

Liposomes are the first generation of nanocarriers for drug delivery used in academia and pharmaceutical industries [71]. Liposomes are made of biocompatible and biodegradable molecules that can protect drugs from enzymatic degradation and prevent drug clearance in blood.

Ishii and co-workers developed asialo-erythropoietin (AEPO)-modified PEGylated liposomes (AEPO-liposomes) to treat cerebral I/R injury [48]. Asialo-EPO (AEPO) is a neuroprotective agent that binds to EPO receptor (EPOR) on neuronal cells and activates MAPK and PI3K/Akt pathways for improved outcome of cerebral stroke. The results showed that AEPO-liposomes accumulated in the ischemic lesion for more than 24 h after injection, decreased infarct lesions, reduced the neuronal apoptosis, and ameliorated cerebral I/R injury in rats.

ZL006 is another neuroprotectant that can selectively block the ischemia-induced binding between nNOS (neuronal nitric oxide synthase) and PSD-95 (a scaffolding protein) to prevent ischemic damage [72]. However, this drug is limited to cross the BBB. To solve this problem, HAIYPRH (T7), a peptide for targeting transferrin receptor (TfR), was conjugated to PEGylated liposomes (T7-P-LPs) loaded with ZL006 (T7-P-LPs/ZL006), and the liposomes mediated the transport of ZL006 across BBB in the ischemic stroke model [49]. The results in ex vivo fluorescence imaging and particle biodistribution studies suggested that T7 enhanced the transport of liposomes across the BBB. T7-PLPs/ZL006 also significantly reduced infarct volume and prevented neurological deficit in the rat ischemic stroke model compared to free ZL006. The same group developed another dual targeting delivery system using T7 peptide and stroke homing peptide (SHp, CLEVSRKNC)-conjugated liposome (T7&SHp-P-LPs/ZL006) and demonstrated that this system can penetrate BBB to deliver ZL006 to the ischemia brain tissue. Administration of T7&SHp-P-LPs/ZL006 decreased the infarction sizes in the brain and ameliorated neurological damage [50].

Basic fibroblast growth factor (bFGF) has demonstrated the neuroprotective effect in acute stroke; however, bFGF encounters the same issue that it cannot cross the BBB. To overcome this problem, liposomes were developed to deliver bFGF (bFGF-NL) to the brain in a rat model of cerebral I/R. After treatment with bFGF-NL, in three consecutive days, mice showed improved neurological function, reduced infarct volume, and decreased nuclear fragmentation of neurons. The therapeutic effect is associated with PI3K/Akt activation by bFGF-NL [51].

To address the problem of narrow therapeutic time window of tPA (4.5 h), Fukuta et al. combined tPA and liposomal fasudil (fasudil-Lip) and examined whether they can increase the benefit of tPA to treat cerebral I/R. The results showed that the combined therapy increased neuroprotective effects compared to the administration of tPA or fasudil alone. In addition, the study indicated that administration of fasudil-Lip decreased the cerebral hemorrhage and extended the therapeutic time window of tPA compared to treatment with tPA [52].

To determine when drugs should be administered after reperfusion surgery, Al-Ahmady et al. selected a liposomal formulation made from HSPC:CHOL:DSPE-PEG2000 because PEGylated-liposomes have a long blood circulation. In the study, liposomes were intravenously administered at 0.5 or 48 h after reperfusion. Transmission electronic microscopy, histological, and immunostaining were performed, and the results indicated a higher accumulation of liposomes in cerebral ischemic mice than that in healthy mice. The possible mechanism was that the increased accumulation of liposomes in brain in the early phase (0.5 h) was due to transcellular transport. However, at 48 h after stroke, the liposome deposition was associated with both transcellular and paracellular pathways [53]. This is an interesting study, but it is needed to investigate the time course of liposome accumulation in ischemic stroke tissues to determine the therapeutic delivery time to improve the treatment of stroke.

3.1.2. Polymeric Nanoparticles as Delivery Platforms

Biodegradable polymeric NPs have also been developed as carriers for drug delivery to treat neurological diseases because they are not toxic and biocompatible, and they have the sustained-release feature [73]. Liu and co-workers developed tanshinone IIA PEGylated nanoparticles (CBSA-PEG-TIIA-NPs) conjugated with cationic bovine serum albumin (CBSA) to target negatively-charged lumens of brain microvessels [54]. The study demonstrated that NPs delivered tanshinone IIA into the rat brain in the cerebral I/R injury model, significantly suppressing the inflammation response. In addition, NPs reduced the infarcted volume and prevented the neuronal apoptosis. In another study, the same group investigated the mechanism of therapeutic effects using CBSA-PEG-TIIA-NPs. They found that NPs decreased neutrophil infiltration and prevented microglial activation. Furthermore, it is shown that prevention of neuronal apoptosis and suppression of inflammatory mediators (such as MMP-9 and COX-2) were regulated by mitogen-activated protein kinases (MAPK) signal pathways [55].

Nanoparticles were also bioengineered with ROS-responsive features that can control drug release in ischemic brain. In a study [56], Lv et al. designed nanoparticles to deliver NR2B9C (a neuroprotectant agent) to treat ischemic stroke. Nanoparticles (named SHp-RBC-NPs) composed of a red blood cell (RBC) membrane as a shell and a polymer nanoparticle with ROS-responsive boronic ester as a core. To target the ischemic brain, a peptide, SHp (CLEVSRKNC), was conjugated on the surface of nanoparticles. Triggered by high levels of ROS in ischemic region, the nanoparticles can control the release of NR2B9C in ischemic brain tissues. Ex vivo brain fluorescence imaging showed that SHp-RBC-NP can target the ischemic brain to significantly prevent neurological damage and reduce the brain infarction size.

3-n-Butylphthalide (dl-NBP) [74] has demonstrated the therapeutic value to treat ischemic stroke in clinic, but it is challenging to deliver it to the ischemic brain. To solve this problem, PEGylated-lipid nanoparticles (PLNs) conjugated with Fas ligand was developed to target the ischemic region of the brain and deliver 3-n-Butylphthalide (dl-NBP). The results are promising as the nanoparticle formulation alleviates the brain neurological injury [57].

Guo et al. designed AMD3100 (a targeting ligand)-conjugated and size-shrinkable nanoparticles (ASNPs) with the features of protease-responsiveness and brain-targeting. The nanoparticles specifically targeted the CXCR4 (C-X-C chemokine receptor type 4)-enriched ischemic brain tissue and penetrated the ischemic brain. Glyburide, a promising anti-stroke drug, was delivered by the nanoparticles, and this system enhanced the therapeutic outcomes in the ischemic stroke model [58].

Peptide inhibitors for caspases possess neuroprotection effects, but they also cannot cross the BBB. Based on this problem, Karatas et al. developed chitosan nanospheres loaded with N-benzyloxycarbonyl-Asp(OMe)-Glu(OMe)-Val-Asp(OMe)-fluoromethyl ketone (Z-DEVD-FMK), a caspase-3 inhibitor. Anti-mouse transferrin receptor monoclonal antibody (TfRMAb) was conjugated to polyethylene glycol-coated nanospheres because the antibody can selectively bind to TfR type 1 expressed on cerebral vasculature. Nanospheres suppressed the caspase-3 activity and improved neurological repair after ischemic stroke [59].

3.1.3. Biomimetic Nanoparticles as Delivery Platforms

Cell-derived biomimetic carriers have provided new options for drug delivery due to their advanced targeting abilities and better biosafety than artificial carriers. Shi et al. reported a T7 peptide (a brain targeting peptide)-linked erythrocyte membrane nanovesicles loaded with Mn_3O_4 nanoparticles (Mn_3O_4@nanoerythrocyte-T7, MNET), and this complex formulation may scavenge free radicals and change hypoxia environments in the ischemic brain. They found that MNET had a long circulation time and were capable of crossing BBB. In subsequent studies, MNET scavenged free radical and oxygen supply to rescue neurons during ischemic phase, and possibly MNET helped to remove oxygen generated by reperfusion [60].

Platelet membrane-derived nanovesicles were loaded with L-arginine, a nitric oxide (NO) donor, and γ-Fe_2O_3 magnetic nanoparticles (PAMNs) to achieve dual therapeutic and diagnostic purposes. The PAMNs were able to adhere to the damaged cerebral vessel induced by thrombosis and deliver L-arginine when the external magnetic field was applied. Furthermore, the release of L-arginine at ischemic lesions promoted vasodilation and the thrombosis was disrupted to restore the blood flow. This study showed that PAMNs may be a useful tool for both MRI imaging and targeted therapy for ischemic stroke [61].

3.1.4. Inorganic Nanoparticles as ROS Scavengers

Studies have showed that metallic NPs can function as scavengers of ROS. ROS includes superoxide anion, hydrogen peroxide, and hydroxyl radical, and they are generated during ischemic stroke. ROS is involved with oxidative tissue damage that is the mechanism responsible for brain injury in ischemic stroke.

It is reported that ceria nanoparticles can scavenge free radicals by reversibly reacting with oxygen to Ce^{4+} (oxidized) species from Ce^{3+} (reduced) species [62]. Kim et al. firstly reported that ceria nanoparticles reduced ROS production and prevented cell apoptosis, thus protecting the brain from ischemic damage. Although ceria nanoparticles are effective to scavenge ROS, the integrity of BBB is the barrier to prevent their brain deposition. Bao and coworkers developed ceria nanoparticles (E-A/P-CeO_2) modified with Angiopep-2 (ANG) and poly(ethylene glycol), which were able to cross BBB via receptor-mediated transcytosis (Figure 4A). In vitro transmigration assay (Figure 4B) showed that E-A/P-CeO_2 could cross the brain capillary endothelial cells (BCECs) compared to P-CeO_2 [63]. The BBB crossing capability of E-A/P-CeO_2 was also confirmed on healthy rats by determining the ratio of ceria nanoparticles in brain tissue and injected (Figure 4C). Finally, results showed that E-A/P-CeO_2 displayed ROS scavenging ability and decreased the infarct volume on rats (Figure 4D). The results showed that ceria nanoparticles might be promising in treating I/R injury.

Platinum nanoparticles (nPt) are novel ROS scavengers because the large surface area and the high electron density can potentiate their catalytic activity to quench ROS [75]. Takamiya et al. investigated whether nPt (2–3 nm) has the neuroprotection effect against I/R injury in the mouse model of transient middle cerebral artery occlusion (tMCAO) [64]. The results showed that treatment with nPt ameliorated the generation of superoxide via reduction of hydroethidine in the cerebral cortex and decreased the mouse infarct volume. The same group published another work to investigate whether nPt could ameliorate tissue plasminogen activator (tPA)-related ischemic injury since tPA treatment may upregulate the expression of MMP-9, which exacerbates cerebral infarction through low-density lipoprotein receptor-related protein [65]. They found that nPt decreased the MMP-9 activity and ameliorated the disrupted neurovascular unit (NVU) after tMCAO. Those results further strengthened that nPt could be combined with tPA reperfusion treatment to treat ischemic stroke patients.

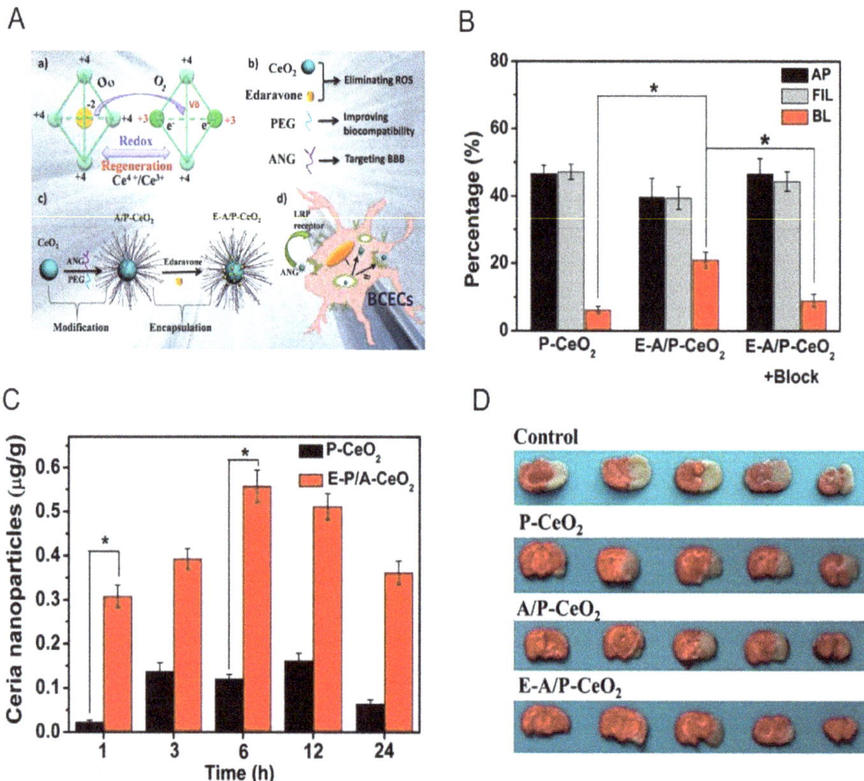

Figure 4. Edaravone-loaded ceria nanoparticles cross the blood brain barrier and protect the brain from ischemic stroke. (**A**) The nanoplatform is comprised of ceria nanoparticle core, PEG shell, and ANG as targeting peptide. (**a**) Transition between cerium(III) and cerium(IV) species in ceria NPs; (**b**) main components of the nanoparticle system; (**c**) synthetic procedure for E-A/P-CeO$_2$; (**d**) receptor-mediated (ANG-LRP) endocytosis of E-A/P-CeO$_2$. (**B**) Transmigrated amount of P-CeO$_2$ and E-A/P-CeO$_2$ in in vitro transmigration model. Free ANG was used as the blocking agent in group 3. (**C**) Concentrations (µg Ce/g brain tissue) of ceria nanoparticles in normal brain tissue. The injection dose was 0.5 mg/kg. (**D**) Representative TTC (triphenyl tetrazolium chloride)-stained brain sections after different treatments within 24 h of stroke. Data represent mean ± standard deviation (SD). * indicates $p < 0.05$. Reproduced with permission [63]. Copyright 2018, American Chemical Society.

3.2. Delivery of Anti-inflammatory Reagents

Inflammation plays a central role in ischemic stroke, thus delivering anti-inflammatory drugs may be a novel strategy to treat ischemic stroke. Exosomes are endogenous vesicles made from cell membrane with cellular targeting features, therefore, they may be drug delivery tools [76–78]. In addition, exosomes have other properties, such as innate stability, biodegradability, low immunogenicity, and ability to cross BBB. Tian et al. used mesenchymal stromal cell (MSC)-derived exosomes conjugated with a c(RGDyK) peptide (called cRGD-Exo) to deliver curcumin to the ischemic brain [66]. Due to the high specific affinity of c(RGDyK) peptide to integrin $\alpha_v\beta_3$, highly expressed on cerebral vascular endothelial cells of ischemic tissues, cRGD-Exo was significantly accumulated in the stroke lesion. Administration of curcumin with exosomes (called cRGD-Exo-cur) suppressed the inflammatory response via the NF-κB pathway and protected the brain from ischemic injury.

Dexamethasone is an anti-inflammatory drug and can reduce proinflammatory cytokines, which prevent cellular damage in the ischemic brain [79]. Lee and co-workers incorporated dexamethasone to the hydrophobic component of a small peptide, R3V6 peptide (with a 3-arginine block and a 6-valine block), to form a stable micelle (named, R3V6-Dexa) used for gene delivery. R3V6-Dexa was loaded with a plasmid DNA, which expressed a heme oxygenase-1 (HO-1) as an antioxidant agent, and the complex of pSV-HO-1/R3V6-Dexa could target ischemic stroke tissues [67]. The authors stereotaxically injected the complex into rats and found that HO-1 expression was significantly increased. In addition, when mice were treated with the pSV-HO-1/R3V6-Dexa complex, the infarction size in the brain was significantly reduced compared to control groups.

Increased leukocyte infiltration is another factor to damage the brain in stroke. Within minutes after ischemia, adhesion molecules, such as P-selectin, are highly expressed on endothelial cells, and proinflammatory signals are rapidly generated [22]. Another strategy is to control the infiltration of neutrophils to alleviate the I/R injury in ischemic stroke. Targeting neutrophils may be a novel approach in treating ischemic stroke, but the methods in anti-neutrophil therapies and neutrophil depletion have failed in clinical studies. This failure may be associated with the complex signaling pathways and numerous receptors involved in neutrophil transmigration. For example, deactivating one receptor, such as anti-ICAM-1 therapy, may not be sufficient because neutrophil infiltration is involved with various types of surface proteins between neutrophils and endothelial cells [80,81]. Neutrophil depletion could be an effective approach, but systemic elimination of neutrophils can make the host vulnerable to bacterial and virus infections [82], and even impair other immune cells, such as natural killer cells [83].

Neutrophil infiltration is regulated by interactions between neutrophils and endothelial cells. Therefore, delivering drugs to the cerebral endothelium may effectively inhibit neutrophil infiltration without interfering neutrophils [77]. Dong et al. reported neutrophil cell membrane-derived nanovesicles (HVs), which can target inflamed endothelium at I/R injury sites and deliver therapeutics to treat the mouse I/R injury [68]. The nanovesicles were generated from differentiated HL-60 cells (neutrophil-like cells) using nitrogen cavitation. The nitrogen cavitation approach was used to disrupt cell membrane to eliminate nuclei and cytosols of cells. Figure 5A shows the scheme of experimental design. The TEM image showed the liposome-like structure of nanovesicles made from HL-60 cells and the size was 200 nm in diameter. In vivo imaging system (IVIS) and confocal microscopy (Figure 5B) showed that HVs (nanovesicles made from differentiated HL-60 cells) were specifically accumulated in the injured half of the brain rather than in the normal brain. To visualize how HVs interacted with brain vasculature, a cranial window was established and intravital microscopy was performed in live mice (Figure 5C). The real-time visualization images strongly indicated that HVs can specifically target ischemic vasculature. To examine the delivery of therapeutics with nanovesicles, Resolvin D2 (RvD2) was loaded into the membrane of nanovesicles because RvD2 is a new lipid mediator to resolve inflammatory responses [84]. After Resolvin D2-loaded nanovesicles (RvD2-HVs) were intravenously injected into mice, reduced neutrophil infiltration was observed using intravital microcopy in live mice, and brain homogenates also confirmed this observation (Figure 5D). Subsequently, the level cytokines, such as TNF-α (Figure 5E), were decreased after RvD2-HVs were administered. The data indicated that RvD2-HVs could alleviate inflammation responses in ischemic stroke. The diminished inflammatory responses reduced the infarction sizes and prevented neurological damage from ischemic stroke (Figure 5F).

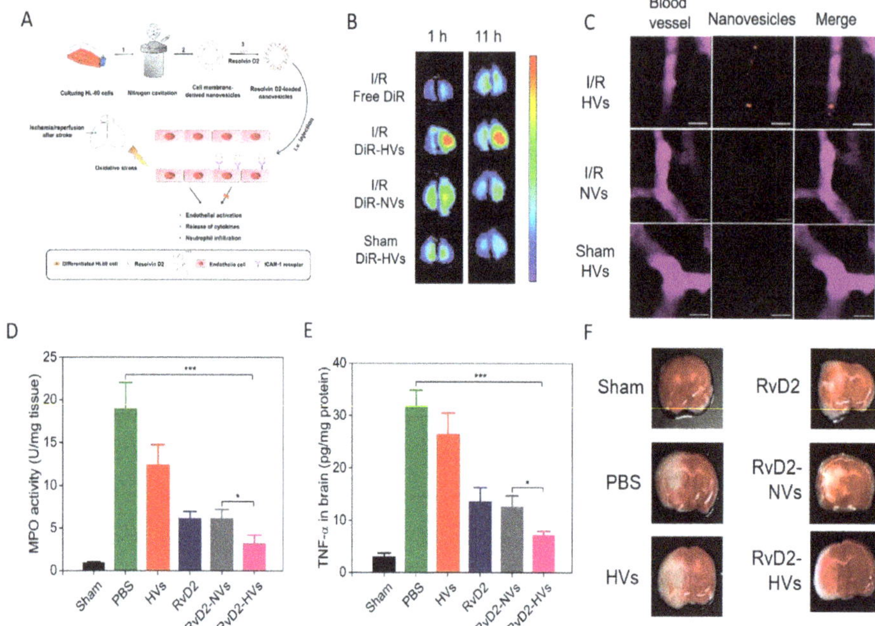

Figure 5. Neutrophil membrane-derived nanovesicles target ischemic endothelium and deliver therapeutics for ischemic stroke treatment. (**A**) A concept of nanovesicles binding to endothelium at ischemic/reperfusion injury sites and alleviating the I/R injury. Resolvin D2-loaded nanovesicles were prepared by: (1) nitrogen cavitation to break the cells; (2) purification; (3) Resolvin D2 loading. (**B**) Nanovesicles (HVs) specifically accumulated in I/R damaged half of brain. (**C**) Intravital microscopy images of HVs (red) specifically bind to I/R endothelium (top panel), instead of normal endothelium (bottom panel). Scale bar = 20 μm. (**D**) Myeloperoxidase (MPO) activity in the injured half of brain after different treatment, which indicates the neutrophil infiltration condition. (**E**) Level of TNF-α in the injured half of brain after different treatment. (**F**) TTC staining of brain sections in different groups. Data represent mean ± standard deviation (SD). * $p < 0.05$ and *** $p < 0.005$. Reproduced with permission [68]. Copyright 2019, American Chemical Society.

3.3. Delivery of siRNA

Small interfering RNAs (siRNAs) can suppress the expression of a specific gene and are also used in ischemic stroke treatment [85]. Caspase-3 activation contributes to brain tissue loss and downstream biochemical events, which lead to programmed cell death in many brain injuries including ischemic stroke [86]. Khuloud et al. investigated whether using carbon nanotube-mediated in vivo RNA interference (RNAi) to silence Caspase-3 could offer a therapeutic opportunity against stroke. Peri-lesional stereotactic administration of functionalized carbon nanotubes (f-CNT) carrying Caspase-3 siRNA (siCas 3) decreased the neurodegeneration, improved behaviors, and reduced ischemic lesion in an endothelin-1 induced stroke model [69].

In another study, a gene delivery vector, PAMAM-Arg, consisting of a polyamidoamine (PAMAM) dendrimer amide grafted with basic L-arginine residues (e-PAM-R), was used to deliver the high mobility group box-1 (HGMB1) siRNA [70]. HMGB1 serves as a danger signal that evokes inflammatory responses in various types of cells. After the e-PAM-R/siRNA complex was administered in the rat MCAO model, HMGB1 levels and neuronal cell death were significantly decreased. In addition, the reduction of infarction sizes in the brain was also observed.

4. Stroke Diagnosis using Nanotechnology

Magnetic resonance imaging (MRI), computed tomography (CT), positron emission tomography (PET), and ultrasound are employed for the diagnosis of stroke. Brain CT and MRI imaging are important to detect the early stage of stroke in patients [87]. CT angiography can discover the arterial dissection in the brain vessels and changes of collateral blood flow. Diffusion-weighted imaging is a type of MRI methods used to detect the early stage of ischemic stroke because the imaging system can localize and measure the ischemic lesion to confirm the infarct sites [88]. I/R injury is associated with vascular inflammation, so fluorescence imaging can be used to investigate the changes of brain vessels using IV injection of fluorescent dyes such as dextran- or BSA-conjugates [89].

Shen's research groups developed a method to track administered stem cells and investigated cell-based therapies in ischemic stroke. Cationic polymersomes formed polymeric vesicles and they were loaded with superparamagnetic iron oxide nanoparticles (SPIONs) and quantum dots (QDs) as imaging agents. Cationic polymersome vesicles were incubated with neural stem cells (NSCs) to label the NSCs, thus NSCs could be tracked after they were injected to the striatum contralateral of the ischemic hemisphere. The migration and location of NSCs were monitored using MRI imaging in six weeks in a rat MCAO model and the optical imaging tracked the cells for four weeks [90]. Another example is that neural stem cells (NSCs) were labeled with an MRI reporter ferritin heavy chain (FTH) and enhanced green fluorescent protein (EGFP) to monitor the stem cells in the long-term for detecting ischemic stroke [91]. Those approaches are promising to understand how stem cells target the ischemic tissues and to develop the cell-based therapies to treat ischemic stroke.

Andreas et al. conducted a clinical phase II pilot trial using ultra small superparamagnetic iron oxide (USPIO)-enhanced MRI for macrophage imaging since USPIO particles have been introduced as a cell-specific MRI contrast agents taken up by macrophages. USPIO contrast agent was infused in ten patients 5 to 6 days after stroke onset, and MRI was performed within 24–36 h or 48–72 h after the infusion. Results showed that USPIO was much better than gadolinium (a clinically used imaging agent). The study indicates that USPIO-enhanced MRI may provide an in vivo tool to track cells in stroke and other CNS pathologies [92].

Theranostics nano-platform is a new way to combine diagnostics and therapeutics in ischemic stroke. In a study, HSP72 antibodies against HSP72, a specific molecular biomarker of the peri-infarct region, were conjugated to liposomes. Rhodamine/gadolinium labelled- and citicoline (CDP-Choline)-loaded liposomes specifically targeted the peri-infarct tissue in cerebral ischemia. MRI results demonstrated that the nanoparticles were accumulated in the injured brain and significantly reduced lesion volumes [93].

Clinically used gadolinium (Gd) chelates as contrast agents have several issues, such as short circulation time, rapid clearance, and low T1 signal. These issues limit the imaging time and resolution [94]. Upconversion nanoparticles doped with Gd ions can greatly enhance the T1 signals. In addition, upconversion nanoparticles can image the deep tissue and possess the photostability. Jing et al. reported synthetic UCNPs of core/shell structure ($NaYF_4$:Yb/Er@$NaGdF_4$) and coated with PEG (PEG-UCNPs). PEG-UCNPs showed the high diagnostic sensitivity to image acute ischemic stroke at a low dosage (5 mg Gd kg^{-1}) compared to the clinical dosage of 108 mg Gd kg^{-1} [95]. This new formulation may be promising in clinic.

5. New Opportunities and Perspectives

Stroke is an acute disease, therefore treating stroke requires the early diagnosis and immediate therapies. This review has highlighted the recent advances in design and engineering of new nanomaterials to target ischemic stroke tissues and new technologies used to image and monitor the disease progression in vivo.

Specifically delivering therapeutics to the injured brain is essential in treating ischemic stroke. While many nanoparticle-based formulations or cell-based platforms have been developed, the fundamental question of how they target ischemic stroke lesion has not been clearly addressed.

For example, BBB is the blood vessel barrier to prevent therapeutics across the blood vessels. Most studies showed the dramatic therapeutic effects of using nanoparticle-based formulations compared to free drugs. The enhanced outcomes claimed that nanoparticles transported therapeutics across BBB, but the direct in vivo experimental data were vague to support this conclusion. Developing advanced in vivo imaging systems [96,97] is needed to visualize the intact brain [68] and to address whether and how nanoparticles cross the BBB.

Brain vasculature during stroke shows the temporal opening, and this disruption of BBB may be a target to guide small molecules across the BBB. Further investigation is needed on the time course of BBB opening to design ideal drug delivery platforms. Although ischemic stroke increases brain vascular permeability, the endothelial gaps are unlikely to allow the efficient transport of nanoparticles because their size is usually larger than endothelial gaps. Therefore, developing new and novel concepts is needed to solve the drug delivery across BBB. Ischemic stroke and reperfusion cause acute inflammatory responses including neutrophil infiltration across blood vessels. Recent studies have shown that rational design of nanoparticles (gold nanoparticles, polymeric nanoparticles, or protein nanoparticles) enables neutrophils to transport nanoparticles across the blood vessel barrier in infection and cancer mouse models [6,98–101]. It is expected that this technology of hijacking neutrophils in vivo may transport nanotherapeutics across BBB for therapies of ischemic stroke.

Ischemic stroke is strongly correlated to inflammatory responses. Inflammatory responses include endothelial activation and neutrophil infiltration, which damages the brain tissues. Targeting activated endothelium using cell membrane-derived nanovesicles has demonstrated the value in delivering therapeutics to treat ischemic stroke [68]. To translate cell-derived nanovesicles, developing new technologies is needed to scale up their production. Recent studies show that nitrogen cavitation methods [77] and other approaches [102] could have the potential to scale up cell-derived nanovesicles for clinical applications. In addition, Dong [68] et al. reported an interesting study to deliver lipid mediators [84] (such as Resolvin D2) to treat ischemic stroke. This study is different from current therapies that mainly deliver anti-inflammatory agents. Anti-inflammatory therapies can cause side effects [103], but Resolvin D2 is a new drug to increase the host immune defense via increased neutrophil apoptosis and macrophage phagocytosis [104]. In the future, it is needed to investigate how to efficiently load lipid mediators in nanovesicles [105] for improved treatment of ischemic stroke. Another direction is the design of new nanoparticles in response to inflammatory environments (such as pH or enzymes) [8] to improve the treatment of ischemic stroke.

Targeting inflammatory neutrophils in situ to block brain neutrophil infiltration is a new opportunity to treat ischemic stroke. A recent study [7] shows that albumin protein-formed nanoparticles loaded with doxorubicin could induce neutrophil apoptosis, thus inhibiting neutrophil infiltration to prevent brain damage in a mouse ischemic stroke model. This is an exciting and new research area to develop novel therapies to solve the lacking pharmacological therapies for ischemic stroke in clinic.

In addition, developing new drugs that can target inflammatory pathways for management of the host injury during ischemic stroke is needed. The pathogenesis of ischemic stroke is complicated, and it is involved with multiple signaling pathways. The molecular mechanisms of ischemic stroke-induced brain injury are needed to be further determined. The timing of administering drugs or nanoparticle-based therapeutics is also very critical. Optimizing therapeutic windows in the future is needed.

Theranostics formulations are interesting and promising in treating ischemic stroke since they combine diagnosis and therapies. Nanoparticle-based platforms are novel constructs because they can contain imaging agents and drugs in single nanoparticle platforms. For instance, formulations with both neuroprotectants and Fe_2O_3 magnetic nanoparticles can achieve the therapy and imaging. In the future, developing more similar drug delivery systems is needed to treat ischemic stroke. However, considering that many inorganic materials do not naturally exist in the body (although iron oxide nanoparticles (IONPs) have been approved by the US food and drug administration (FDA) to treat anemia), fully evaluating the biodistribution and toxicity after systemic administration is required.

6. Conclusions

In summary, we have discussed various nanoparticle-based formulations and cell-based platforms used in treating ischemic stroke. It is essential to understand the pathogenesis of ischemic stroke and how nanoparticles interact with ischemic tissues. The fundamental question on how nanoparticles transport therapeutics across BBB is yet to be addressed, and advances in design and synthesis of nanoparticles and novel in vivo imaging systems (such as intravital microscopy) may address this question. The development of new drugs and novel nanoparticle-based therapeutics will improve the outcomes in treating ischemic stroke patients.

Author Contributions: X.D. was involved in literature review, writing the original draft and editing. J.G. was involved in figure creation and editing. Y.S. was involved in literature review and editing. Z.W. was involved in writing, editing, supervision and funding acquisition. All authors have read and agreed to the published version of the manuscript.

Funding: This work was supported by National Institute of Health, RO1GM116823 and RO1EB027078 to Z.W.

Conflicts of Interest: The authors declare no conflict of interest.

References

1. Virani, S.S.; Alonso, A.; Benjamin, E.J.; Bittencourt, M.S.; Callaway, C.W.; Carson, A.P.; Chamberlain, A.M.; Chang, A.R.; Cheng, S.; Delling, F.N.; et al. Heart Disease and Stroke Statistics-2020 Update: A Report From the American Heart Association. *Circulation* **2020**, *141*, e139–e596. [CrossRef] [PubMed]
2. Malik, R.; Dichgans, M. Challenges and opportunities in stroke genetics. *Cardiovasc. Res.* **2018**, *114*, 1226–1240. [CrossRef] [PubMed]
3. Fernandes, L.F.; Bruch, G.E.; Massensini, A.R.; Frézard, F. Recent Advances in the Therapeutic and Diagnostic Use of Liposomes and Carbon Nanomaterials in Ischemic Stroke. *Front. Neurosci.* **2018**, *12*, 453. [CrossRef] [PubMed]
4. Tsivgoulis, G.; Katsanos, A.H.; Alexandrov, A.V. Reperfusion therapies of acute ischemic stroke: Potentials and failures. *Front. Neurol.* **2014**, *5*, 215. [CrossRef]
5. Granger, D.N.; Kvietys, P.R. Reperfusion injury and reactive oxygen species: The evolution of a concept. *Redox. Biol.* **2015**, *6*, 524–551. [CrossRef]
6. Chu, D.; Dong, X.; Zhao, Q.; Gu, J.; Wang, Z. Photosensitization Priming of Tumor Microenvironments Improves Delivery of Nanotherapeutics via Neutrophil Infiltration. *Adv. Mater.* **2017**, *29*. [CrossRef]
7. Zhang, C.Y.; Dong, X.; Gao, J.; Lin, W.; Liu, Z.; Wang, Z. Nanoparticle-induced neutrophil apoptosis increases survival in sepsis and alleviates neurological damage in stroke. *Sci. Adv.* **2019**, *5*, eaax7964. [CrossRef] [PubMed]
8. Zhang, C.Y.; Gao, J.; Wang, Z. Bioresponsive Nanoparticles Targeted to Infectious Microenvironments for Sepsis Management. *Adv. Mater.* **2018**, *30*, e1803618. [CrossRef]
9. Wang, Z.; Tiruppathi, C.; Cho, J.; Minshall, R.D.; Malik, A.B. Delivery of nanoparticle: Complexed drugs across the vascular endothelial barrier via caveolae. *IUBMB Life* **2011**, *63*, 659–667. [CrossRef]
10. Wang, Z.; Tiruppathi, C.; Minshall, R.D.; Malik, A.B. Size and dynamics of caveolae studied using nanoparticles in living endothelial cells. *ACS Nano* **2009**, *3*, 4110–4116. [CrossRef]
11. Chauhan, V.P.; Jain, R.K. Strategies for advancing cancer nanomedicine. *Nat. Mater.* **2013**, *12*, 958–962. [CrossRef] [PubMed]
12. Zhao, Z.; Ukidve, A.; Kim, J.; Mitragotri, S. Targeting Strategies for Tissue-Specific Drug Delivery. *Cell* **2020**, *181*, 151–167. [CrossRef] [PubMed]
13. Cheng, C.J.; Tietjen, G.T.; Saucier-Sawyer, J.K.; Saltzman, W.M. A holistic approach to targeting disease with polymeric nanoparticles. *Nat. Rev. Drug Discov.* **2015**, *14*, 239–247. [CrossRef]
14. Xing, C.; Arai, K.; Lo, E.H.; Hommel, M. Pathophysiologic cascades in ischemic stroke. *Int. J. Stroke* **2012**, *7*, 378–385. [CrossRef] [PubMed]

15. Baron, J.C.; Yamauchi, H.; Fujioka, M.; Endres, M. Selective neuronal loss in ischemic stroke and cerebrovascular disease. *J. Cereb. Blood Flow Metab.* **2014**, *34*, 2–18. [CrossRef] [PubMed]

16. Nishizawa, Y. Glutamate release and neuronal damage in ischemia. *Life Sci.* **2001**, *69*, 369–381. [CrossRef]

17. Roh, J.S.; Sohn, D.H. Damage-Associated Molecular Patterns in Inflammatory Diseases. *Immune Netw.* **2018**, *18*, e27. [CrossRef]

18. Sun, M.S.; Jin, H.; Sun, X.; Huang, S.; Zhang, F.L.; Guo, Z.N.; Yang, Y. Free Radical Damage in Ischemia-Reperfusion Injury: An Obstacle in Acute Ischemic Stroke after Revascularization Therapy. *Oxid. Med. Cell Longev.* **2018**, *2018*, 3804979. [CrossRef]

19. Lubos, E.; Handy, D.E.; Loscalzo, J. Role of oxidative stress and nitric oxide in atherothrombosis. *Front. Biosci.* **2008**, *13*, 5323–5344. [CrossRef]

20. Nita, M.; Grzybowski, A. The Role of the Reactive Oxygen Species and Oxidative Stress in the Pathomechanism of the Age-Related Ocular Diseases and Other Pathologies of the Anterior and Posterior Eye Segments in Adults. *Oxid. Med. Cell Longev.* **2016**, *2016*, 3164734. [CrossRef]

21. Anrather, J.; Iadecola, C. Inflammation and Stroke: An Overview. *Neurotherapeutics* **2016**, *13*, 661–670. [CrossRef] [PubMed]

22. Iadecola, C.; Anrather, J. The immunology of stroke: From mechanisms to translation. *Nat. Med.* **2011**, *17*, 796–808. [CrossRef]

23. Connolly, E.S., Jr.; Winfree, C.J.; Springer, T.A.; Naka, Y.; Liao, H.; Yan, S.D.; Stern, D.M.; Solomon, R.A.; Gutierrez-Ramos, J.C.; Pinsky, D.J. Cerebral protection in homozygous null ICAM-1 mice after middle cerebral artery occlusion. Role of neutrophil adhesion in the pathogenesis of stroke. *J. Clin. Investig.* **1996**, *97*, 209–216. [CrossRef]

24. Kigerl, K.A.; de Rivero Vaccari, J.P.; Dietrich, W.D.; Popovich, P.G.; Keane, R.W. Pattern recognition receptors and central nervous system repair. *Exp. Neurol.* **2014**, *258*, 5–16. [CrossRef]

25. Chamorro, Á.; Meisel, A.; Planas, A.M.; Urra, X.; van de Beek, D.; Veltkamp, R. The immunology of acute stroke. *Nat. Rev. Neurol.* **2012**, *8*, 401–410. [CrossRef] [PubMed]

26. Jin, R.; Yang, G.; Li, G. Inflammatory mechanisms in ischemic stroke: Role of inflammatory cells. *J. Leukoc. Biol.* **2010**, *87*, 779–789. [CrossRef] [PubMed]

27. Haley, M.J.; Lawrence, C.B. The blood-brain barrier after stroke: Structural studies and the role of transcytotic vesicles. *J. Cereb. Blood Flow Metab.* **2017**, *37*, 456–470. [CrossRef] [PubMed]

28. Knowland, D.; Arac, A.; Sekiguchi, K.J.; Hsu, M.; Lutz, S.E.; Perrino, J.; Steinberg, G.K.; Barres, B.A.; Nimmerjahn, A.; Agalliu, D. Stepwise recruitment of transcellular and paracellular pathways underlies blood-brain barrier breakdown in stroke. *Neuron* **2014**, *82*, 603–617. [CrossRef] [PubMed]

29. Bauer, A.T.; Bürgers, H.F.; Rabie, T.; Marti, H.H. Matrix metalloproteinase-9 mediates hypoxia-induced vascular leakage in the brain via tight junction rearrangement. *J. Cereb. Blood Flow Metab.* **2010**, *30*, 837–848. [CrossRef]

30. Yang, Y.; Rosenberg, G.A. Blood-brain barrier breakdown in acute and chronic cerebrovascular disease. *Stroke* **2011**, *42*, 3323–3328. [CrossRef]

31. da Fonseca, A.C.; Matias, D.; Garcia, C.; Amaral, R.; Geraldo, L.H.; Freitas, C.; Lima, F.R. The impact of microglial activation on blood-brain barrier in brain diseases. *Front. Cell Neurosci.* **2014**, *8*, 362. [CrossRef] [PubMed]

32. Janardhan, V.; Qureshi, A.I. Mechanisms of ischemic brain injury. *Curr. Cardiol. Rep.* **2004**, *6*, 117–123. [CrossRef] [PubMed]

33. Mittal, M.; Siddiqui, M.R.; Tran, K.; Reddy, S.P.; Malik, A.B. Reactive oxygen species in inflammation and tissue injury. *Antioxid Redox Signal.* **2014**, *20*, 1126–1167. [CrossRef] [PubMed]

34. Bhaskar, S.; Stanwell, P.; Cordato, D.; Attia, J.; Levi, C. Reperfusion therapy in acute ischemic stroke: Dawn of a new era? *BMC Neurol.* **2018**, *18*, 8. [CrossRef] [PubMed]

35. Yoo, A.J.; Pulli, B.; Gonzalez, R.G. Imaging-based treatment selection for intravenous and intra-arterial stroke therapies: A comprehensive review. *Expert Rev. Cardiovasc. Ther.* **2011**, *9*, 857–876. [CrossRef]

36. Kurth, T.; Heuschmann, P.U.; Walker, A.M.; Berger, K. Mortality of stroke patients treated with thrombolysis: Analysis of nationwide inpatient sample. *Neurology* **2007**, *68*, 710. [CrossRef]

37. Powers, W.J.; Rabinstein, A.A.; Ackerson, T.; Adeoye, O.M.; Bambakidis, N.C.; Becker, K.; Biller, J.; Brown, M.; Demaerschalk, B.M.; Hoh, B.; et al. Guidelines for the Early Management of Patients With Acute Ischemic Stroke: 2019 Update to the 2018 Guidelines for the Early Management of Acute Ischemic Stroke: A Guideline for Healthcare Professionals From the American Heart Association/American Stroke Association. *Stroke* **2019**, *50*, e344–e418. [CrossRef]

38. Minnerup, J.; Sutherland, B.A.; Buchan, A.M.; Kleinschnitz, C. Neuroprotection for stroke: Current status and future perspectives. *Int J. Mol. Sci.* **2012**, *13*, 11753–11772. [CrossRef]

39. Ginsberg, M.D. Neuroprotection for ischemic stroke: Past, present and future. *Neuropharmacology* **2008**, *55*, 363–389. [CrossRef]

40. Diener, H.C.; Lees, K.R.; Lyden, P.; Grotta, J.; Davalos, A.; Davis, S.M.; Shuaib, A.; Ashwood, T.; Wasiewski, W.; Alderfer, V.; et al. NXY-059 for the treatment of acute stroke: Pooled analysis of the SAINT I and II Trials. *Stroke* **2008**, *39*, 1751–1758. [CrossRef]

41. Shuaib, A.; Lees, K.R.; Lyden, P.; Grotta, J.; Davalos, A.; Davis, S.M.; Diener, H.C.; Ashwood, T.; Wasiewski, W.W.; Emeribe, U. NXY-059 for the treatment of acute ischemic stroke. *N. Engl. J. Med.* **2007**, *357*, 562–571. [CrossRef]

42. Panagiotou, S.; Saha, S. Therapeutic benefits of nanoparticles in stroke. *Front. Neurosci.* **2015**, *9*, 182. [CrossRef]

43. Sarmah, D.; Saraf, J.; Kaur, H.; Pravalika, K.; Tekade, R.K.; Borah, A.; Kalia, K.; Dave, K.R.; Bhattacharya, P. Stroke Management: An Emerging Role of Nanotechnology. *Micromachines* **2017**, *8*, 262. [CrossRef]

44. Kyle, S.; Saha, S. Nanotechnology for the detection and therapy of stroke. *Adv. Healthc. Mater.* **2014**, *3*, 1703–1720. [CrossRef] [PubMed]

45. Shcharbina, N.; Shcharbin, D.; Bryszewska, M. Nanomaterials in stroke treatment: Perspectives. *Stroke* **2013**, *44*, 2351–2355. [CrossRef] [PubMed]

46. Williams-Karnesky, R.L.; Stenzel-Poore, M.P. Adenosine and stroke: Maximizing the therapeutic potential of adenosine as a prophylactic and acute neuroprotectant. *Curr. Neuropharmacol.* **2009**, *7*, 217–227. [CrossRef]

47. Chen, L.; Gao, X. The application of nanoparticles for neuroprotection in acute ischemic stroke. *Ther. Deliv.* **2017**, *8*, 915–928. [CrossRef] [PubMed]

48. Ishii, T.; Asai, T.; Oyama, D.; Fukuta, T.; Yasuda, N.; Shimizu, K.; Minamino, T.; Oku, N. Amelioration of cerebral ischemia-reperfusion injury based on liposomal drug delivery system with asialo-erythropoietin. *J. Control. Release* **2012**, *160*, 81–87. [CrossRef]

49. Wang, Z.; Zhao, Y.; Jiang, Y.; Lv, W.; Wu, L.; Wang, B.; Lv, L.; Xu, Q.; Xin, H. Enhanced anti-ischemic stroke of ZL006 by T7-conjugated PEGylated liposomes drug delivery system. *Sci. Rep.* **2015**, *5*, 12651. [CrossRef] [PubMed]

50. Zhao, Y.; Jiang, Y.; Lv, W.; Wang, Z.; Lv, L.; Wang, B.; Liu, X.; Liu, Y.; Hu, Q.; Sun, W.; et al. Dual targeted nanocarrier for brain ischemic stroke treatment. *J. Control. Release* **2016**, *233*, 64–71. [CrossRef] [PubMed]

51. Zhao, Y.Z.; Lin, M.; Lin, Q.; Yang, W.; Yu, X.C.; Tian, F.R.; Mao, K.L.; Yang, J.J.; Lu, C.T.; Wong, H.L. Intranasal delivery of bFGF with nanoliposomes enhances in vivo neuroprotection and neural injury recovery in a rodent stroke model. *J. Control. Release* **2016**, *224*, 165–175. [CrossRef] [PubMed]

52. Fukuta, T.; Asai, T.; Yanagida, Y.; Namba, M.; Koide, H.; Shimizu, K.; Oku, N. Combination therapy with liposomal neuroprotectants and tissue plasminogen activator for treatment of ischemic stroke. *FASEB J.* **2017**, *31*, 1879–1890. [CrossRef] [PubMed]

53. Al-Ahmady, Z.S.; Jasim, D.; Ahmad, S.S.; Wong, R.; Haley, M.; Coutts, G.; Schiessl, I.; Allan, S.M.; Kostarelos, K. Selective Liposomal Transport through Blood Brain Barrier Disruption in Ischemic Stroke Reveals Two Distinct Therapeutic Opportunities. *ACS Nano* **2019**, *13*, 12470–12486. [CrossRef] [PubMed]

54. Liu, X.; An, C.; Jin, P.; Liu, X.; Wang, L. Protective effects of cationic bovine serum albumin-conjugated PEGylated tanshinone IIA nanoparticles on cerebral ischemia. *Biomaterials* **2013**, *34*, 817–830. [CrossRef] [PubMed]

55. Liu, X.; Ye, M.; An, C.; Pan, L.; Ji, L. The effect of cationic albumin-conjugated PEGylated tanshinone IIA nanoparticles on neuronal signal pathways and neuroprotection in cerebral ischemia. *Biomaterials* **2013**, *34*, 6893–6905. [CrossRef]

56. Lv, W.; Xu, J.; Wang, X.; Li, X.; Xu, Q.; Xin, H. Bioengineered Boronic Ester Modified Dextran Polymer Nanoparticles as Reactive Oxygen Species Responsive Nanocarrier for Ischemic Stroke Treatment. *ACS Nano* **2018**, *12*, 5417–5426. [CrossRef]

57. Lu, Y.M.; Huang, J.Y.; Wang, H.; Lou, X.F.; Liao, M.H.; Hong, L.J.; Tao, R.R.; Ahmed, M.M.; Shan, C.L.; Wang, X.L.; et al. Targeted therapy of brain ischaemia using Fas ligand antibody conjugated PEG-lipid nanoparticles. *Biomaterials* **2014**, *35*, 530–537. [CrossRef] [PubMed]

58. Guo, X.; Deng, G.; Liu, J.; Zou, P.; Du, F.; Liu, F.; Chen, A.T.; Hu, R.; Li, M.; Zhang, S.; et al. Thrombin-Responsive, Brain-Targeting Nanoparticles for Improved Stroke Therapy. *ACS Nano* **2018**, *12*, 8723–8732. [CrossRef]

59. Karatas, H.; Aktas, Y.; Gursoy-Ozdemir, Y.; Bodur, E.; Yemisci, M.; Caban, S.; Vural, A.; Pinarbasli, O.; Capan, Y.; Fernandez-Megia, E.; et al. A nanomedicine transports a peptide caspase-3 inhibitor across the blood-brain barrier and provides neuroprotection. *J. Neurosci.* **2009**, *29*, 13761–13769. [CrossRef]

60. Shi, J.; Yu, W.; Xu, L.; Yin, N.; Liu, W.; Zhang, K.; Liu, J.; Zhang, Z. Bioinspired Nanosponge for Salvaging Ischemic Stroke via Free Radical Scavenging and Self-Adapted Oxygen Regulating. *Nano Lett.* **2020**, *20*, 780–789. [CrossRef]

61. Li, M.; Li, J.; Chen, J.; Liu, Y.; Cheng, X.; Yang, F.; Gu, N. Platelet Membrane Biomimetic Magnetic Nanocarriers for Targeted Delivery and in Situ Generation of Nitric Oxide in Early Ischemic Stroke. *ACS Nano* **2020**, *14*, 2024–2035. [CrossRef] [PubMed]

62. Kim, C.K.; Kim, T.; Choi, I.Y.; Soh, M.; Kim, D.; Kim, Y.J.; Jang, H.; Yang, H.S.; Kim, J.Y.; Park, H.K.; et al. Ceria nanoparticles that can protect against ischemic stroke. *Angew. Chem. Int. Ed. Engl.* **2012**, *51*, 11039–11043. [CrossRef] [PubMed]

63. Bao, Q.; Hu, P.; Xu, Y.; Cheng, T.; Wei, C.; Pan, L.; Shi, J. Simultaneous Blood-Brain Barrier Crossing and Protection for Stroke Treatment Based on Edaravone-Loaded Ceria Nanoparticles. *ACS Nano* **2018**, *12*, 6794–6805. [CrossRef] [PubMed]

64. Takamiya, M.; Miyamoto, Y.; Yamashita, T.; Deguchi, K.; Ohta, Y.; Ikeda, Y.; Matsuura, T.; Abe, K. Neurological and pathological improvements of cerebral infarction in mice with platinum nanoparticles. *J. Neurosci. Res.* **2011**, *89*, 1125–1133. [CrossRef] [PubMed]

65. Takamiya, M.; Miyamoto, Y.; Yamashita, T.; Deguchi, K.; Ohta, Y.; Abe, K. Strong neuroprotection with a novel platinum nanoparticle against ischemic stroke and tissue plasminogen activator-related brain damages in mice. *Neuroscience* **2012**, *221*, 47–55. [CrossRef] [PubMed]

66. Tian, T.; Zhang, H.X.; He, C.P.; Fan, S.; Zhu, Y.L.; Qi, C.; Huang, N.P.; Xiao, Z.D.; Lu, Z.H.; Tannous, B.A.; et al. Surface functionalized exosomes as targeted drug delivery vehicles for cerebral ischemia therapy. *Biomaterials* **2018**, *150*, 137–149. [CrossRef] [PubMed]

67. Lee, J.; Hyun, H.; Kim, J.; Ryu, J.H.; Kim, H.A.; Park, J.H.; Lee, M. Dexamethasone-loaded peptide micelles for delivery of the heme oxygenase-1 gene to ischemic brain. *J. Control. Release* **2012**, *158*, 131–138. [CrossRef]

68. Dong, X.; Gao, J.; Zhang, C.Y.; Hayworth, C.; Frank, M.; Wang, Z. Neutrophil Membrane-Derived Nanovesicles Alleviate Inflammation To Protect Mouse Brain Injury from Ischemic Stroke. *ACS Nano* **2019**, *13*, 1272–1283. [CrossRef]

69. Al-Jamal, K.T.; Gherardini, L.; Bardi, G.; Nunes, A.; Guo, C.; Bussy, C.; Herrero, M.A.; Bianco, A.; Prato, M.; Kostarelos, K.; et al. Functional motor recovery from brain ischemic insult by carbon nanotube-mediated siRNA silencing. *Proc. Natl. Acad. Sci. USA* **2011**, *108*, 10952–10957. [CrossRef]

70. Kim, I.D.; Lim, C.M.; Kim, J.B.; Nam, H.Y.; Nam, K.; Kim, S.W.; Park, J.S.; Lee, J.K. Neuroprotection by biodegradable PAMAM ester (e-PAM-R)-mediated HMGB1 siRNA delivery in primary cortical cultures and in the postischemic brain. *J. Control. Release* **2010**, *142*, 422–430. [CrossRef]

71. Sercombe, L.; Veerati, T.; Moheimani, F.; Wu, S.Y.; Sood, A.K.; Hua, S. Advances and Challenges of Liposome Assisted Drug Delivery. *Front. Pharmacol.* **2015**, *6*, 286. [CrossRef]

72. Zhou, L.; Li, F.; Xu, H.B.; Luo, C.X.; Wu, H.Y.; Zhu, M.M.; Lu, W.; Ji, X.; Zhou, Q.G.; Zhu, D.Y. Treatment of cerebral ischemia by disrupting ischemia-induced interaction of nNOS with PSD-95. *Nat. Med.* **2010**, *16*, 1439–1443. [CrossRef] [PubMed]

73. Masserini, M. Nanoparticles for brain drug delivery. *ISRN Biochem.* **2013**, *2013*, 238428. [CrossRef]

74. Zhang, Y.; Wang, L.; Li, J.; Wang, X.L. 2-(1-Hydroxypentyl)-benzoate increases cerebral blood flow and reduces infarct volume in rats model of transient focal cerebral ischemia. *J. Pharmacol. Exp. Ther.* **2006**, *317*, 973–979. [CrossRef] [PubMed]

75. Watanabe, A.; Kajita, M.; Kim, J.; Kanayama, A.; Takahashi, K.; Mashino, T.; Miyamoto, Y. In vitro free radical scavenging activity of platinum nanoparticles. *Nanotechnology* **2009**, *20*, 455105. [CrossRef] [PubMed]

76. Gao, J.; Chu, D.; Wang, Z. Cell membrane-formed nanovesicles for disease-targeted delivery. *J. Control. Release* **2016**, *224*, 208–216. [CrossRef]

77. Gao, J.; Wang, S.; Wang, Z. High yield, scalable and remotely drug-loaded neutrophil-derived extracellular vesicles (EVs) for anti-inflammation therapy. *Biomaterials* **2017**, *135*, 62–73. [CrossRef] [PubMed]

78. Fang, R.H.; Kroll, A.V.; Gao, W.; Zhang, L. Cell Membrane Coating Nanotechnology. *Adv. Mater.* **2018**, *30*, e1706759. [CrossRef]

79. Tuor, U.I.; Simone, C.S.; Barks, J.D.; Post, M. Dexamethasone prevents cerebral infarction without affecting cerebral blood flow in neonatal rats. *Stroke* **1993**, *24*, 452–457. [CrossRef]

80. Vinten-Johansen, J. Involvement of neutrophils in the pathogenesis of lethal myocardial reperfusion injury. *Cardiovasc. Res.* **2004**, *61*, 481–497. [CrossRef]

81. Chu, D.; Dong, X.; Shi, X.; Zhang, C.; Wang, Z. Neutrophil-Based Drug Delivery Systems. *Adv. Mater.* **2018**, *30*, e1706245. [CrossRef]

82. Fulurija, A.; Ashman, R.B.; Papadimitriou, J.M. Neutrophil depletion increases susceptibility to systemic and vaginal candidiasis in mice, and reveals differences between brain and kidney in mechanisms of host resistance. *Microbiology* **1996**, *142*, 3487–3496. [CrossRef]

83. Jaeger, B.N.; Donadieu, J.; Cognet, C.; Bernat, C.; Ordoñez-Rueda, D.; Barlogis, V.; Mahlaoui, N.; Fenis, A.; Narni-Mancinelli, E.; Beaupain, B.; et al. Neutrophil depletion impairs natural killer cell maturation, function, and homeostasis. *J. Exp. Med.* **2012**, *209*, 565–580. [CrossRef] [PubMed]

84. Serhan, C.N. Pro-resolving lipid mediators are leads for resolution physiology. *Nature* **2014**, *510*, 92–101. [CrossRef]

85. Pardridge, W.M. shRNA and siRNA delivery to the brain. *Adv. Drug Deliv. Rev.* **2007**, *59*, 141–152. [CrossRef] [PubMed]

86. Namura, S.; Zhu, J.; Fink, K.; Endres, M.; Srinivasan, A.; Tomaselli, K.J.; Yuan, J.; Moskowitz, M.A. Activation and cleavage of caspase-3 in apoptosis induced by experimental cerebral ischemia. *J. Neurosci.* **1998**, *18*, 3659–3668. [CrossRef]

87. Birenbaum, D.; Bancroft, L.W.; Felsberg, G.J. Imaging in acute stroke. *West. J. Emerg. Med.* **2011**, *12*, 67–76.

88. Okorie, C.K.; Ogbole, G.I.; Owolabi, M.O.; Ogun, O.; Adeyinka, A.; Ogunniyi, A. Role of Diffusion-weighted Imaging in Acute Stroke Management using Low-field Magnetic Resonance Imaging in Resource-limited Settings. *West. Afr. J. Radiol.* **2015**, *22*, 61–66. [CrossRef] [PubMed]

89. Misgeld, T.; Kerschensteiner, M. In vivo imaging of the diseased nervous system. *Nat. Rev. Neurosci.* **2006**, *7*, 449–463. [CrossRef]

90. Wen, X.; Wang, Y.; Zhang, F.; Zhang, X.; Lu, L.; Shuai, X.; Shen, J. In vivo monitoring of neural stem cells after transplantation in acute cerebral infarction with dual-modal MR imaging and optical imaging. *Biomaterials* **2014**, *35*, 4627–4635. [CrossRef]

91. Zhang, F.; Duan, X.; Lu, L.; Zhang, X.; Chen, M.; Mao, J.; Cao, M.; Shen, J. In Vivo Long-Term Tracking of Neural Stem Cells Transplanted into an Acute Ischemic Stroke model with Reporter Gene-Based Bimodal MR and Optical Imaging. *Cell Transpl.* **2017**, *26*, 1648–1662. [CrossRef] [PubMed]

92. Saleh, A.; Schroeter, M.; Jonkmanns, C.; Hartung, H.P.; Mödder, U.; Jander, S. In vivo MRI of brain inflammation in human ischaemic stroke. *Brain* **2004**, *127*, 1670–1677. [CrossRef] [PubMed]

93. Agulla, J.; Brea, D.; Campos, F.; Sobrino, T.; Argibay, B.; Al-Soufi, W.; Blanco, M.; Castillo, J.; Ramos-Cabrer, P. In vivo theranostics at the peri-infarct region in cerebral ischemia. *Theranostics* **2013**, *4*, 90–105. [CrossRef]

94. Estelrich, J.; Sánchez-Martín, M.J.; Busquets, M.A. Nanoparticles in magnetic resonance imaging: From simple to dual contrast agents. *Int. J. Nanomed.* **2015**, *10*, 1727–1741. [CrossRef]

95. Wang, J.; Zhang, H.; Ni, D.; Fan, W.; Qu, J.; Liu, Y.; Jin, Y.; Cui, Z.; Xu, T.; Wu, Y.; et al. High-Performance Upconversion Nanoprobes for Multimodal MR Imaging of Acute Ischemic Stroke. *Small* **2016**, *12*, 3591–3600. [CrossRef]

96. Wang, Z. Imaging Nanotherapeutics in Inflamed Vasculature by Intravital Microscopy. *Theranostics* **2016**, *6*, 2431–2438. [CrossRef]

97. Wang, Z.; Li, J.; Cho, J.; Malik, A.B. Prevention of vascular inflammation by nanoparticle targeting of adherent neutrophils. *Nat. Nanotechnol.* **2014**, *9*, 204–210. [CrossRef]

98. Chu, D.; Gao, J.; Wang, Z. Neutrophil-Mediated Delivery of Therapeutic Nanoparticles across Blood Vessel Barrier for Treatment of Inflammation and Infection. *ACS Nano* **2015**, *9*, 11800–11811. [CrossRef] [PubMed]

99. Chu, D.; Zhao, Q.; Yu, J.; Zhang, F.; Zhang, H.; Wang, Z. Nanoparticle Targeting of Neutrophils for Improved Cancer Immunotherapy. *Adv. Healthc. Mater.* **2016**, *5*, 1088–1093. [CrossRef]

100. Dong, X.; Chu, D.; Wang, Z. Leukocyte-mediated Delivery of Nanotherapeutics in Inflammatory and Tumor Sites. *Theranostics* **2017**, *7*, 751–763. [CrossRef]

101. Dong, X.; Chu, D.; Wang, Z. Neutrophil-mediated delivery of nanotherapeutics across blood vessel barrier. *Ther. Deliv.* **2018**, *9*, 29–35. [CrossRef] [PubMed]

102. Yurkin, S.T.; Wang, Z. Cell membrane-derived nanoparticles: Emerging clinical opportunities for targeted drug delivery. *Nanomedicine* **2017**, *12*, 2007–2019. [CrossRef] [PubMed]

103. Dinarello, C.A. Anti-inflammatory Agents: Present and Future. *Cell* **2010**, *140*, 935–950. [CrossRef] [PubMed]

104. Spite, M.; Norling, L.V.; Summers, L.; Yang, R.; Cooper, D.; Petasis, N.A.; Flower, R.J.; Perretti, M.; Serhan, C.N. Resolvin D2 is a potent regulator of leukocytes and controls microbial sepsis. *Nature* **2009**, *461*, 1287–1291. [CrossRef]

105. Gc, J.B.; Szlenk, C.T.; Gao, J.; Dong, X.; Wang, Z.; Natesan, S. Molecular Dynamics Simulations Provide Insight into the Loading Efficiency of Proresolving Lipid Mediators Resolvin D1 and D2 in Cell Membrane-Derived Nanovesicles. *Mol. Pharm.* **2020**, *17*, 2155–2164. [CrossRef] [PubMed]

Publisher's Note: MDPI stays neutral with regard to jurisdictional claims in published maps and institutional affiliations.

International Journal of
Molecular Sciences

Review

Targeting the WNK-SPAK/OSR1 Pathway and Cation-Chloride Cotransporters for the Therapy of Stroke

Sunday Solomon Josiah, Nur Farah Meor Azlan and Jinwei Zhang *

Hatherly Laboratories, Institute of Biomedical and Clinical Sciences, Medical School, College of Medicine and Health, University of Exeter, Exeter EX4 4PS, UK; josiahsos2014@gmail.com (S.S.J.); nm503@exeter.ac.uk (N.F.M.A.)
* Correspondence: j.zhang5@exeter.ac.uk; Tel.: +44-(0)1392-72-3828

Abstract: Stroke is one of the major culprits responsible for morbidity and mortality worldwide, and the currently available pharmacological strategies to combat this global disease are scanty. Cation-chloride cotransporters (CCCs) are expressed in several tissues (including neurons) and extensively contribute to the maintenance of numerous physiological functions including chloride homeostasis. Previous studies have implicated two CCCs, the Na^+-K^+-Cl^- and K^+-Cl^- cotransporters (NKCCs and KCCs) in stroke episodes along with their upstream regulators, the with-no-lysine kinase (WNKs) family and STE20/SPS1-related proline/alanine rich kinase (SPAK) or oxidative stress response kinase (OSR1) via a signaling pathway. As the WNK-SPAK/OSR1 pathway reciprocally regulates NKCC and KCC, a growing body of evidence implicates over-activation and altered expression of NKCC1 in stroke pathology whilst stimulation of KCC3 during and even after a stroke event is neuroprotective. Both inhibition of NKCC1 and activation of KCC3 exert neuroprotection through reduction in intracellular chloride levels and thus could be a novel therapeutic strategy. Hence, this review summarizes the current understanding of functional regulations of the CCCs implicated in stroke with particular focus on NKCC1, KCC3, and WNK-SPAK/OSR1 signaling and discusses the current and potential pharmacological treatments for stroke.

Keywords: stroke; electroneutral transport; cation-chloride cotransporters; KCCs; NKCCs; WNK-SPAK/OSR1

Citation: Josiah, S.S.; Meor Azlan, N.F.; Zhang, J. Targeting the WNK-SPAK/OSR1 Pathway and Cation-Chloride Cotransporters for the Therapy of Stroke. *Int. J. Mol. Sci.* **2021**, 22, 1232. https://doi.org/10.3390/ijms22031232

Academic Editor: Bae Hwan Lee
Received: 29 December 2020
Accepted: 24 January 2021
Published: 27 January 2021

Publisher's Note: MDPI stays neutral with regard to jurisdictional claims in published maps and institutional affiliations.

1. Introduction of Cation-Chloride Cotransporter Family

The family of cation-chloride cotransporters (CCCs) comprises the Na^+-K^+-Cl^-, Na^+-Cl^-, and K^+-Cl^- cotransporters (NKCCs, NCC, and KCCs). Identification of these CCCs in several tissues such as red blood cells, epithelia, and neurons have alluded to their extensive contributions to ion and water homeostasis, both at a cellular and trans-epithelial level [1–3]. The identification of the functional properties of most of these transporters dates back to the late 1970s and early 1980s as Cl^--dependent cation fluxes, with red blood cells and Ehrlich ascites tumor cells constituting pivotal model tissues [4–7]. Subsequently, their molecular identities were established about a decade afterwards [8–10]. CCCs are intrinsic membrane proteins that move Na^+, K^+, and Cl^- ions across plasma membranes in a tightly coupled electroneutral manner. They facilitate secondary active transport driven by the gradients generated by the Na^+/K^+-ATPase [11]. The solute carrier family 12 (SLC12) of the CCC family consists of nine members [2]. A group of three Na^+-dependent inward cotransporters comprises of one Na^+-Cl^- cotransporter (NCC)—its sole isoform is found in the kidney and encoded by *SLC12A3* [9] and two Na^+-K^+-Cl^- cotransporters isoforms (NKCC1 and 2)—NKCC1 is ubiquitous whilst NKCC2 is specifically expressed in the kidney and are encoded by *SLC12A2* and *SLC12A1*, respectively [2]. Na^+-independent outward transport of K^+ and Cl^- is facilitated by four K^+-Cl^- cotransporters with distinct functional properties (KCC1 [12], KCC2 [13], KCC3, and KCC4 [14,15]). The KCC isoforms are encoded by *SLC12A4–7* respectively, of which *SLC12A5* (KCC2) is found exclusively in

27

neurons [2]. The additional SLC12 family members, CCC9 and CCC-interacting protein (CIP), are encoded by *SLC12A8* and *SLC12A9*, respectively, and have no physiological role ascribed to them yet [16], though recent genome-wide association studies found novel *SLC12A8* variants may be associated with dyslipidemia [17], and *SLC12A9* may be involved in feather pecking and aggressive behavior [18] (see Table 1).

All proteins in the CCC family have common functional characteristics. These include (1) the coupled transport of one cation (Na^+ and/or K^+) per individually transported anion, hence the appellation of electroneutral cotransporters, (2) chloride is always the transported anion, (3) all cotransporters are modulated by variations in cell volume, (4) changes in the intracellular chloride concentration ($[Cl^-]_i$) influence the modulation of their expression, and (5) the regulation of CCCs activity is achieved through phosphorylation and dephosphorylation processes [19]. The functional and structural characteristics of the Na^+ dependent and Na^+ independent branches clearly distinguish the two. The degree of identity amongst the Na^+ dependent transporters and the Na^+ independent are 50% and 70% respectively. Between the two NKCC isoforms, the degree of identity is 25% [20].

Stroke is one of the major culprits responsible for global death and disability [21]. Currently, there is a paucity of pharmacological strategies to reduce the mental damage as well as the burden triggered by this pathology. Ischemic stroke is the most common type of stroke, which accounts for approximately 85% of the cases of the pathology [22]. Ischemia is the disruption of blood flow and the subsequent depletion of oxygen and glucose. As neuronal components strictly function on aerobic metabolism [23], an ischemia in the brain leads to reduced available ATP levels and ionic imbalance across the neuronal cell membrane [24], which causes irreversible neuronal death, also known as ischemic stroke [25–27]. During this process, an imbalance of excitatory glutamate and inhibitory gamma amino acid butyric acid (GABA) further accelerate neuronal demise [28–30], subsequently leading to the onset of post-stroke seizures [20,31,32]. Neuronal cells excitation is opposed by inhibitory GABA through the activation of $GABA_A$ receptors. The activation of the $GABA_A$ receptors is dependent on the chloride transmembrane gradient [23]. Notably, CCCs are the primary regulators of chloride homeostasis in the brain [2,33]. This role is accomplished through the extrusion of Cl^- via KCC2 and entry of Cl^- via NKCC1 which regulates $[Cl^-]_i$. GABA is inhibitory as a result of lower $[Cl^-]_I$, driven by higher expression of KCC2. Interestingly, immature neurons express less KCC2 and more NKCC1 leading to a higher $[Cl^-]_i$ and excitatory GABA. The switch from excitatory to inhibitory during neurodevelopment, a process termed excitatory-to-inhibitory GABA switch, is generated through reduction in NKCC1 level and increase in KCC2 level [34]. The expanding work on CCC influence on neuronal excitability in physiological conditions especially during development and pathological conditions suggest that they could be a new treatment approach for stroke [27].

In view of this, constant updates on the role of CCCs in stroke and their regulation is highly germane for the development of therapeutic drugs in the management of this pathology. Thus, the aim of this review is to summarize the current understanding of functional regulation of the CCCs and particularly the role of NKCC1 and KCC3 cotransporters in the pathogenesis of stroke. Then, the regulatory role of the with-no-lysine kinase (WNKs) family and STE20/SPS1-related proline/alanine rich kinase (SPAK) or oxidative stress response kinase (OSR1) (WNK-SPAK/OSR1) signaling pathway in stroke will be considered. Lastly, current pharmacological treatments for stroke with respect to potent inhibitors of WNK-SPAK/OSR1 pathway and NKCC1 cotransporter, and activators of KCC3 transporter will be discussed in this review.

Table 1. The solute carrier family 12 (SLC12) of cation-chloride cotransporters in neurological disorders and others. TAL: thick ascending loop of Henle; DCT: distal convoluted tubule; RVI: regulatory volume increase; RVD: regulatory volume decrease; ND: no data (or none). Functional regulation of the cation-chloride cotransporter family.

Encoding Gene (Protein)	Co-Transport Ions	Tissue Distribution	Physiological Functions	Genetic Disorders	References
SLC12A1 (NKCC2)	Na^+, K^+, Cl^-	Kidney-specific (TAL)	NaCl reabsorption in the TAL; regulation of Ca^{2+} excretion; urine concentration	Bartter's syndrome	[2,35–37]
SLC12A2 (NKCC1)	Na^+, K^+, Cl^-	Ubiquitous	Cell volume regulation (RVI); provide ions for secretion	Potential role in human schizophrenia multi-organ system failure, congenital hydrocephalus, hearing, and neurodevelopmental disorder	[2,38–43]
SLC12A3 (NCC)	Na^+, Cl^-	Kidney-specific (DCT)	NaCl reabsorption in the DCT; regulation of Ca^{2+} and K^+ renal excretion;	Gitelman's syndrome	[9,44,45]
SLC12A4 (KCC1)	K^+, Cl^-	Ubiquitous	cell volume regulation (RVD), KCl epithelial Transport	ND	[2,12,45,46]
SLC12A5 (KCC2)	K^+, Cl^-	Neuron-specific	Intraneuronal Cl^- Concentration regulation	Idiopathic generalized epilepsy, developmental apoptosis, neurodevelopmental pathology, Rett syndrome	[2,13,47–51]
SLC12A6 (KCC3)	K^+, Cl^-	Widespread	Volume regulation in the brain; K^+ recycling in the kidney	Anderman's syndrome, Charcot–Marie–Tooth disease, hydrocephalus, sensorimotor neuropathy	[2,14,15,52–57]
SLC12A7 (KCC4)	K^+, Cl^-	Widespread	Participates in acid excretion in alpha intercalated cells of collecting duct	ND	[2,15]
SLC12A8 (CCC9)	Unknown	Widespread	No function ascribed yet	Psoriasis, dyslipidemia	[2,16,17,58,59]
SLC12A9 (CIP)	Unknown	Widespread	No function ascribed yet	May be involved in feather pecking and aggressive behavior	[2,16,18,58]

Undoubtedly, for a cell to function properly it is essential to maintain constant intracellular ionic milieu [53]. Homeostasis of $[Cl^-]_i$ in particular, influences the movement of fluid across epithelia, the polarity of GABA, and more. The electroneutral CCCs are critical determinants of $[Cl^-]_i$ [53,60]. The $[Cl^-]_i$ gradient across the neuronal membrane is crucial for controlling the polarity of GABAergic signaling. $GABA_A$ conducts Cl^- ions. The direction of Cl^- movement through $GABA_A$, which determines whether it is excitatory or inhibitory, is dependent on the $[Cl^-]_i$ gradient. Entry of Cl^- through $GABA_A$ results in the hyperpolarization of neurons and the extrusions of Cl^- through $GABA_A$ depolarizes the neurons [23,61–63]. Changes in expression levels of the CCCs during development reverses the chloride gradient in neurons, generating a switch from an excitatory GABA to an inhibitory GABA [23,64]. NKCCs facilitate Cl^- movement into the cell, while KCCs facilitates Cl^- movement out of the cell. Thus, NKCCs promote an increased participation of $[Cl^-]_i$ in the pathways for regulatory volume increase (RVI), while the KCCs promote decrease in the $[Cl^-]_i$ as one of the regulatory volume decrease (RVD) mechanisms [19,53,65] (Figure 1). These evolutionarily conserved transporters are amongst the most important mediators of ion transport in multicellular organisms, with particular importance in mammalian central nervous system (CNS) regulation of ionic and water homeostasis [66]. As mentioned earlier, the CCCs are involved in several important cellular functions such as trans-epithelial ion transport, cell volume regulation, and maintenance of $[Cl^-]_i$. Their importance in physiological function is evident by the many human Mendelian disorders of the brain and renal phenotype that arise due to mutations in some members of the CCC family and their upstream regulators [19] (Table 1). For instance, reduction in neuronal KCC2 activity results in decreased inhibition and a hyper-excitable network, a feature shared amongst numerous neurological disorders including epilepsy, autism, post-surgical complication, neuropathic pain, and neuropsychiatric disorders [48,53,67,68].

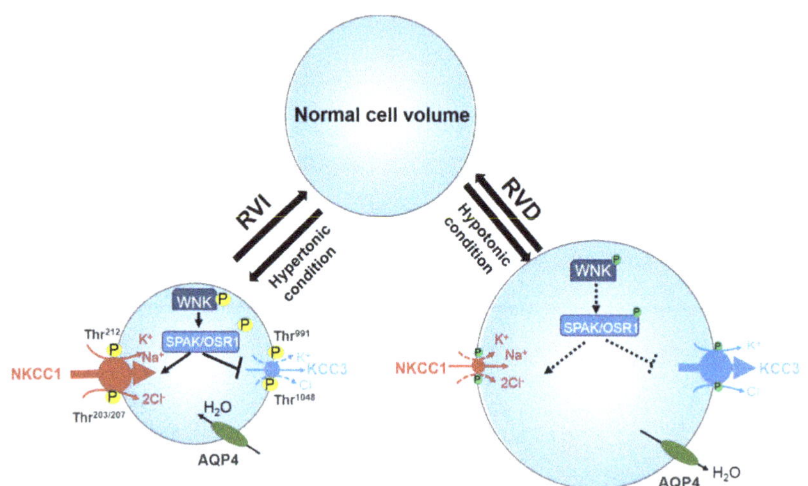

Figure 1. Roles of cation-chloride cotransporters (CCCs) in cell osmoregulation. Intracellular osmolarity changes activate cellular volume regulation. Under hypertonic extracellular conditions of cell shrinkage due to water extrusion from the cell, a counter-response of regulatory volume increase (RVI) restores normal cell volume. In this condition, the WNK-SPAK/OSR1 pathway is activated leading to the phosphorylation of the CCCs. This activates NKCC1 and inhibits KCC, leading to the NKCC1-mediated influx of Na^+, K^+, and $Cl^−$ along with water, thus restoring cell volume. On the contrary, under hypotonic stress conditions of cell swelling, the cell activates a regulatory volume decrease (RVD). The WNK-SPAK/OSR1 pathway remains inactive and NKCC1 and KCCs are dephosphorylated. This stimulates KCC3 but inhibits NKCC1 leading to the efflux K^+ and $Cl^−$ along with water, and cell volume decrease. NKCC1, K^+-$Cl^−$ cotransporters; KCC3, K^+-$Cl^−$ cotransporter 3; WNK, with-no-lysine kinase; SPAK, STE20/SPS1-related proline/alanine rich kinase; OSR1, oxidative stress response kinase; AQP4, aquaporin. Part of figure elements were adapted from Huang et al. [65].

Since CCCs are key players in several important cellular functions and principally responsible for reciprocal cations (Na^+ and K^+) exchange with $Cl^−$ to maintain cellular balance, regulatory mechanisms are crucial to coordinate their activity [53,60,67,69]. Indeed, several kinases and phosphatases regulate their transport activity. However, previous reports have established that members of the WNKs family and their downstream targets, SPAK and OSR1, are the master regulators of CCCs activity [1,53,60,65,70–75].

Though it has been appreciated for some three decades that protein phosphorylation coupled with external osmotic environment are crucial components in regulation of CCC activities [2,54,76–80], knowledge on the enzymes that regulate these signaling networks was sparse then. The WNK family encoded by the genes *WNK1–4* [81], *SPAK*, and *OSR1* play crucial roles in the regulation of cell volume homeostasis through the regulation of intracellular Na^+, K^+, and $Cl^−$ [53,82]. The many roles of the WNK-SPAK/OSR1-CCC pathway which include cell volume homeostasis, epithelial transport, and GABA signaling are associated with an array of pathologies which include essential hypertension, cerebral edema, anemia, and neuropathic pain [1,19,53,60,65,67,71,73]. In response to osmotic stress of low $[Cl^−]_i$, isoforms of WNK are activated through phosphorylation. The WNK isoforms then phosphorylate the related downstream kinases SPAK and/or OSR1 [83,84]. Activated SPAK and/or OSR1 phosphorylates the CCCs, which activates NCC, NKCC1, and NKCC2 but inhibits KCCs through a reciprocal regulatory mechanism [67,72] (Figure 2). The counter regulation of the CCCs coordinates Cl^- movement across the membrane to maintain Cl^- homeostasis and circumvent superfluous energy utilization [65].

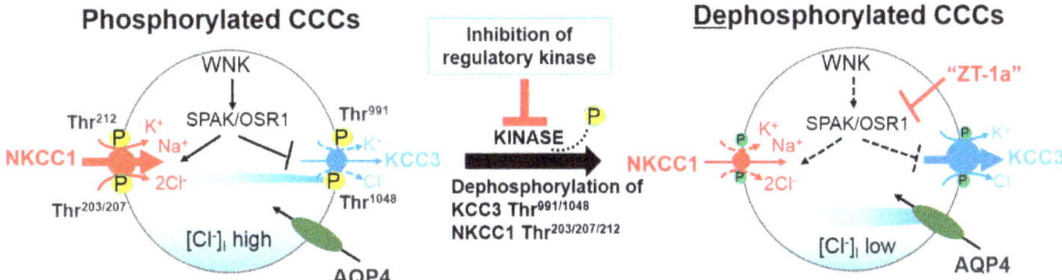

Figure 2. A novel strategy to facilitate cellular Cl^- extrusion by coincident NKCC1 inhibition and KCC3 activation by inhibiting Table 1 kinases. Reversible serine-threonine phosphorylation reciprocally regulates NKCC1 and KCC3. Hypotonic low $[Cl^-]_i$ conditions or a reduction in cell volume activates the WNK-SPAK/OSR1 pathway to promote Cl^- and water influx. This leads to the phosphorylation of NKCC1 and KCC3 and their activation and inhibition respectively. When $[Cl^-]_i$ becomes too high or cell volume increases, WNK-SPAK/OSR1 pathway is inhibited. The cotransporters are dephosphorylated, KCC3 is activated and facilitates $[Cl^-]_i$ and water efflux to restore ion and osmotic homeostasis. CCCs, cation-chloride cotransporters; NKCC1, K^+-Cl^- cotransporters; KCC3, K^+-Cl^- cotransporter 3; WNK, with-no-lysine kinase; SPAK, STE20/SPS1-related proline/alanine rich kinase; OSR1, oxidative stress response kinase; AQP4, aquaporin 4; ZT-1a, specific SPAK inhibitor. Part of figure elements were adapted from Salihu et al. [85].

To maintain cell volume homeostasis, the WNK-SPAK/OSR1 kinase pathway activates NKCC1 and simultaneously inhibits KCCs through phosphorylation. Conversely, dephosphorylation inhibits NKCC1 and activates the KCCs [86]. The major phosphorylation sites of NKCC1 include Thr^{203}, Thr^{207}, and Thr^{212} in the N-terminus whilst the phosphorylation sites of KCC1–4 are located in the C-terminus (Thr^{991} and Thr^{1048} in KCC3 and Thr^{906} and Thr^{1007} in KCC2) [59,72]. Notably, the phosphorylation sites on KCC3, Thr^{991} and Thr^{1048}, are conserved amongst all KCC isoforms in humans [72]. Substitution of these threonine residues that make up the sites of regulated phosphorylation inhibited the phosphorylation and subsequent activation of KCC2 and KCC3 [86–88]. Recently, it was established that the WNK3-SPAK complex is critical for regulated phosphorylation of KCC3 Thr^{991} and Thr^{1048} residues [86] (also see Figures 3 and 4).

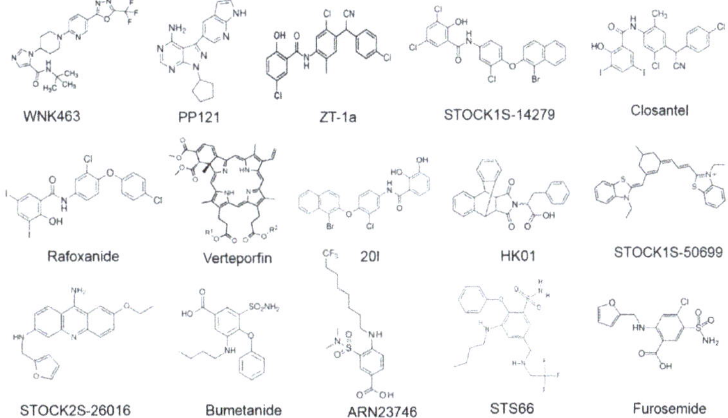

Figure 3. Molecular structures of WNK-SPAK/OSR1-CCCs signaling pathway compounds.

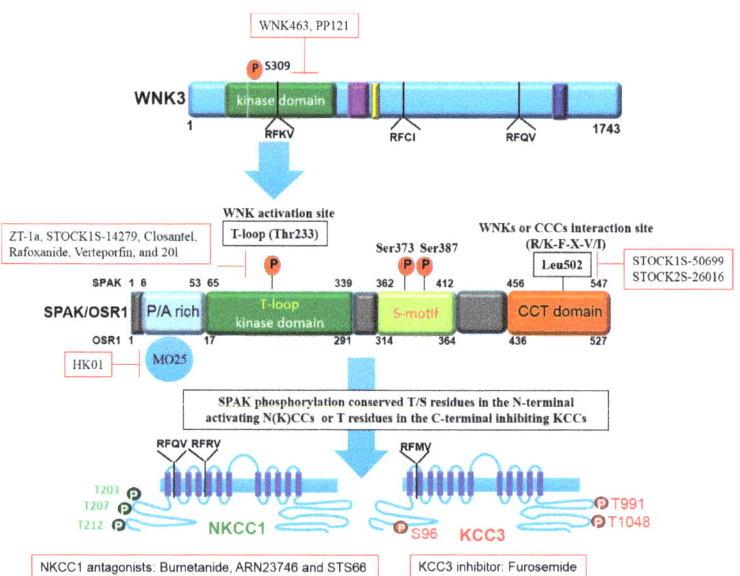

Figure 4. The domain structure of SPAK and the phosphorylation target sites on NKCC1 and KCC3. OSR1 lack the P/A rich (PAPA) domain that is present in SPAK. The figure depicts small molecule inhibitors that target the WNK-SPAK-CCC signaling pathway and their sites of actions. STOCK1S-50699 and STOCK2S-26016 operate through binding to the CCT domain consequently blocking the interaction between SPAK/OSR1 and WNK. STOCK1S-14279, Closantel, Rafoxanide, Verteporfin, and 20l bind the T233E residue on SPAK that is constitutively active or WNK-sensitive. WNK463 and PP121 that inhibit WNKs catalytic activity. HK01, an inhibitor of the mouse protein-25 (M025). Bumetanide, ARN23746 and STS66 are NKCC1 antagonists. Furosemide is a KCC3 inhibitor. ZT-1a is a specific SPAK inhibitor. CCCs, cation-chloride cotransporters; NKCC1, K^+-Cl^- cotransporters; KCC3, K^+-Cl^- cotransporter 3; WNK, with-no-lysine kinase; SPAK, STE20/SPS1-related proline/alanine rich kinase; OSR1, oxidative stress response kinase.

Specific conserved carboxyl-terminal (CCT) domains on SPAK/OSR1 interact with NKCC1 and KCCs [80,86,89]. The Arg-Phe-Xaa-Val/Ile (RFXV/I) domain located in the N-terminal of NKCCs and KCCs is able to recognize the SPAK CCT domain (Figures 3 and 4). Interestingly, a subtype of the KCC2 isoform, KCC2b lacks the RFXV/I motif to facilitate interaction with SPAK. Thus, only KCC2a transport activity decreased when SPAK was overexpressed [53,90]. This interaction of SPAK/OSR1 with both the upstream WNKs and downstream CCCs is crucial for coordinating CCC cellular activity in various osmotic conditions [89,91]. The binding of WNK to SPAK/OSR1 allows for the phosphorylation of residues in the T-loop of the SPAK catalytic domain required for SPAK activation [89,91]. Only once activated is SPAK then able to phosphorylate and inhibit KCC2 and KCC3 at Thr^{1048} and Thr^{1007} respectively and activate NKCC1 at Thr^{203}/Thr^{207}/Thr^{212}. These processes are essential in response to cellular shrinkage and hypertonicity (Figure 1) [53]. In hypertonic conditions, SPAK/OSR1 phosphorylation and activation of NKCC1 is key to achieve RVI [53,67] as an influx of Na^+, K^+, Cl^- through the NKCC1 along with water will allow for cell volume recovery (Figure 1). Under hypotonic extracellular conditions, water enters the cells and causes cell swelling, subsequently triggering a counter-volume regulation response (RVD). The WNK-SPAK/OSR1 pathway in this condition remains inactive and inhibits NKCC1 activity (Figure 1). Furthermore, the dephosphorylation of KCCs mediated by phosphatase stimulates KCC activity and causes efflux of K^+ and Cl^- along with water, decreasing cell volume (Figure 1) [65]. Thus, pharmacological or genetic antagonistic events of WNK-SPAK/OSR1 will lead to a $[Cl^-]_i$ efflux coupled through simultaneous dephosphorylation of NKCC1 and KCCs. This will then mitigate energy failure occasioned by osmotic stress, as evident in some neurological disturbances such as cerebral edema [32,65,67,86,92].

2. Role of NNKCC1 in Stroke

NKCCs play crucial roles in regulating neuronal functions. They are abundantly expressed in neurons throughout the brain and are involved in ion homeostasis maintenance and neuronal excitatory functions [93]. Majorly, they function in regulation and repair of nerve injury through GABAergic signaling [1,94]. However, under specific conditions such as cerebral ischemia, the expression of NKCCs can be altered [94]. Overstimulation of NKCC1 and other major glial ion transporters (such as Na^+/H^+, Na^+/Ca^{2+} and Na^+/HCO_3^- exchangers) can contribute to glial apoptosis, inflammation, demyelination, inflammation, and excitotoxicity [26]. This cascade of events is involved in the development and progression of neurological diseases such as stroke [26]. Studies have demonstrated evidence of increase NKCC1 expression in neurons, a phenotype resembling immature neurons, following an ischemic stroke [95–97]. The altered NKCC1 expression observed post stroke may be responsible for the increased in Na^+ and Cl^- levels in neurons leading to a GABA-mediated depolarization. These events also contribute to a hyper-excitable neuron and cell swelling occasioned by cerebral ischemia [95,98]. Furthermore, disrupted endoplasmic reticulum Ca^{2+} homeostasis [99] and elevated extracellular levels of potassium, glutamate, interleukin-6 [100], interleukin-18 [101], interleukin 1β, and tumor necrotic factor-α [102] which happen during/post cerebral ischemia have been shown to stimulate NKCC1 mRNA gene expression in both neurons and astrocytes. Notably, the elevated extracellular potassium levels seems to be Ca^{2+}-dependent as NKCC1 activation is completely terminated either through the removal of extracellular calcium or using Nifedipine to block L-type voltage-dependent calcium channels [103]. Comparatively, similar effects were seen in the expression of NKCC1 mRNA gene in white and gray matter of mutant and wild-type (WT) mice [104]. In addition, [105] an epigenetic study using quantitative real-time RT-PCR technique on cortical slice culture from rats suggested that DNA methylation/demethylation contribute to the regulation of NKCC1 expression during postnatal development and in response to neuronal injury (ischemia) [105].

Following ischemic stroke, both NKCC1 and KCCs are phosphorylated via the WNK-SPAK/OSR1 signaling pathway, leading to NKCC1 activation and KCC inhibition [22,65]. Other contributors leading to NKCC1 activation following an ischemia include: the WNK-calcium binding protein (Cab39; [106]) as well as antagonists of V1 vasopressin [107], MAPK (p38, ERK, JNK, Raf) pathways, cAMP response element-binding protein (CREB) phosphorylation and the ubiquitous transcription factor; hypoxia inducible factor 1-alpha (HIF-1α). This leads to the stimulation of vascular endothelial growth factor (VEGF) expression and ultimate onset of ischemic stroke [37,108–110]. Studies on the human subacute ischemic stroke brain tissues demonstrate increased NKCC1 mRNA gene expression [106]. The contribution of NKCC1 protein activation to ischemic brain havocs is now evident as genetic deletion of NKCC1 or its upstream regulator WNK3 in mouse transient middle cerebral artery (MCA) occlusion models displayed minimal infarction, edema, and white matter damage [32,104]. Another study demonstrated increased NKCC1 activity in the perilesional cortex of rats challenged with focal cerebral ischemia induced by endothelin-1 (ET-1) [111]. Furthermore, inhibition of NKCC1 has been reported to reduce edema, Na^+ uptake, and ischemic injury in rats subjected to STZ-induced hyperglycemic ischemic stroke [112].

3. Role of KCC3 in Stroke

Here, we recall as stated earlier in Section 2 of this review that the stimulation/inhibition of NKCCs/KCCs pair via protein phosphorylation is through a reciprocal regulatory mechanism [67,72] (Figure 2). NKCC and KCC participations in cell volume regulations via RVI and RVD mechanisms, respectively, have also been earlier highlighted [19,53,65] (Figure 1).

It is only expected that in neuronal functions regulation, activation of KCC3 would play similar physiological roles to those that the inhibition of NKCC1 would. The WNK-Cab39 signaling increased expression of NKCC1 mRNA gene in brain tissues of rats subjected to ischemic stroke. It is proposed to have probable effects on the expression of other cotransporters such as KCC3 [106]. KCC3 expression in the brain requires NKCC1 expression for physiological regulation of cellular homeostasis in the CNS [72,86,87,113,114]. Hence, the roles of WNK-Cab39-KCC signalling in ischemic stroke should be further investigated [106]. In a mouse model study, Lucas et al. [114] demonstrated that alongside inhibited NKCC1, stimulation of KCC3 promoted decreased $[Cl^-]_i$ in the sensory neuron of adult mice. This suggests their involvement in GABAergic/glycinergic transmission as adjudged by its influence on the hyperpolarization of $GABA_A$ equilibrium potential (E_{GABA-A}) resulting in inhibitory GABAergic neurotransmission due to a decrease in $[Cl^-]_i$. Our recent functional kinomics study alluded that regulatory phosphorylation of KCC3 (Thr^{991}/Thr^{1048} residues) is essential for cell volume homeostasis in the mammalian brain [86]. The notion that supports KCC3 physiological role in regulating $[Cl^-]_i$ and consequent influence on GABA polarization state is fascinating and suggests possible relationship between neuronal excitability and cell volume homeostasis [86]. Moreover, this concept behind the physiological function of KCC3 is an indication that it might have a dual role in the regulation of both cell volume and $[Cl^-]_i$ [66] which will be highly relevant in understanding its role in the etiology of stroke.

Furthermore, Byun and Delpire [115] reported that stimulation of KCC3 are involved in cell volume regulation (via RVD) in the nervous system, thereby emphasizing its role in the development and maintenance of myelin and peripheral nerves. The study further established that inhibition of KCC3 by knocking out its expression in mice caused anoxal and periaxonal swelling that ultimately led to neurodegeneration [115]. Another mouse-model study demonstrated that KCC3 gene knockout (KO) in parvalbumin neuron caused peripheral agenesis neuropathy associated with the agenesis of corpus callosum. Similarly, the post-mortem study by Auer and colleagues [116] suggested that neuropathological features observed in the central and peripheral nervous systems (CNS/PNS) could potentially link to genetic defects in axonal KCC3 of CNS/PNS. Indeed, sensory defects in KCC3 knockout ($KCC3^{-/-}$) mice as well as its mutations in humans [63,66,117–119] confirm the fundamental role of the cotransporter in peripheral neurons (also see reviews [53,65,66]). Loss of function mutations function of KCC3 have contributed to the pathogenesis of motor and sensory peripheral neuropathy in adult animals and humans [114,115,120–122]. Manifestations of peripheral neuropathy or fluid-related axonopathy influence cell volume dysregulation [26,115] and may be involved in the pathogenesis of other neurological conditions such as stroke.

4. Role of Regulatory WNK-SPAK/OSR1 Pathway in Stroke

Certainly, the various cellular functional roles of CCCs in the biological system will be compromised without regulatory mechanisms in place. Thus, it is only principally reasonable that the cotransporters actively and continuously maintain their functional integrity through coordinated mechanisms of regulations [53,60,67,69]. Several reports owing to WNK-SPAK/OSR1 kinases as the most involved signaling pathway in the regulation of neuronal Cl^- and cell volume homeostasis do exist [60,65,70,71,123] and these established roles of the WNK-SPAK/OSR1-CCC pathway have alluded to their connection with stroke [19,65,71].

There is a growing body of evidence that the WNK-SPAK/OSR1-CCC pathway is involved in pathogenesis of stroke [53,65,70,106,124]. The established roles of the WNK-SPAK-CCC pathway on GABA signaling and cell volume homeostasis are linked to several neurological diseases such as cerebral stroke [53,60,125]. WNK and SPAK/OSR1 kinases are copiously expressed in the CNS [75]. After an ischemic stroke, both NKCC1 and KCCs are phosphorylated via the WNK-SPAK/OSR1 signaling pathway, leading to activation and inhibition of NKCC1 and KCCs, respectively [22,65]. However, inactivating the WNK-

SPAK-CCC cascade through concurrent inhibition of NKCC-mediated ionic influx and stimulation of the KCC-mediated ion efflux has been shown to reduce cellular swelling in ischemic stroke brains [53,65]. The regulatory role of WNK-SPAK-CCC in cellular ionic homeostasis have also been shown to contribute to post-ischemic stroke infarction and cerebral edema [66]. Thus, inactivation of the WNK-SPAK-CCC cascade would trigger the simultaneous inhibition of NKCC mediated ionic import and stimulation of KCC mediated ionic export to eradicate cellular osmotic imbalance [53,65]. It has also been reported that estradiol increases NKCC1 phosphorylation consequently promoting GABA-mediated depolarization [126]. This occurs through stimulation of SPAK and OSR1 that is transcription dependent [127]. Studies using focal ischemia rat model have shown that estradiol treatment promotes neurogenesis in the subventricular zone of the brain, probably by increased expression of HIF-1α and VEGF [128]. WNK phosphorylate SPAK/OSR1, which in turn, phosphorylate NKCC1 and KCC3 at key regulatory sites [129]. Previous reports have shown that SPAK has a CCT domain to interact with NKCC1 and the KCCs [11,89,91,130]. However, the understanding of their physiological functions in normal and ischemic brains are still elusive [92].

Indeed, WNK isoforms are selectively expressed in the CNS [131] and WNK3 is mostly expressed in the brain [132]. This particular WNK isoform exerts its action on NKCCs and KCCs reciprocally [53,113]. Thus, the reciprocal actions of WNK3 on NKCC1 and the KCCs along with its concurrent expression with cotransporters in GABAergic neurotransmission that undergo dynamic changes in $[Cl^-]_i$, suggest its involvement in regulation of neuronal CCCs [53,133,134]. In fact, Kahle et al. [113] provided a compendium of data that suggested WNK3 as a dynamic regulator of NKCC1 and KCCs physiological activities. Simultaneous expression of WNK3 and NKCC1 in neurons may lead to enhanced phosphorylation of regulatory sites in NKCC1 and a consequent increase in the activity of NKCC1 [32,135]. The target of protein phosphatase 1 (which recognizes the consensus motif: RVNFXD) is a highly conserved RVNFVD sequence that is located in the amino-terminus of NKCC1. The RVNF binding motif overlaps with the SPAK binding motif (RFRV). A slight mutation of this sequence will cause NKCC1 activity to increase [136]. Interestingly, phenotypes of NKCC1 inhibition and KCC activation due to inactive WNK3 signaling pathway are reversed by potential protein phosphatase 1 inhibitors such as calyculin A and cyclosporine A [133,137]. According to Melo et al. [138], WNK3 inhibits the activity of KCC3 by promoting the phosphorylation of Thr^{991} and Thr^{1048} as well as Ser^{96}, a third phospho-site involved in KCC3 regulation (also see Figure 2). Double (KCC3-T991A/T1048A) or triple (KCC3-S96A/T991A/T1048A) alanine mutations of KCC3, activated the cotransporter, which further increased hypotonicity. Thus, the study suggested that the phosphorylation of WNK3 signaling pathway was disabled, subsequently activating KCC3 by cell swelling [138].

Certainly, the upstream WNK3-SPAK/OSR1 pathway regulation of NKCC1 activity coupled with inhibition of KCC3 is implicated in the pathology of ischemic stroke [86]. Previous studies have demonstrated that WNK3 KO mice exhibited a reduction in infarct volume and axonal demyelination coupled with diminished cerebral edema and improved neurological behaviors following cerebral stroke when compared to WT mice with significantly activated WNK3 [32,86,92]. However, it is important to note that WT mice showed better survival and functional outcomes after a brain edema in comparison to mouse models lacking aquaporin-4 (AQP4) [139], a water transport system that allows for bidirectional water flux. As such, further research will need to elucidate the distinction between the role of AQP4 and WNK3 in cerebral edema and stroke. Thus, Begum and colleagues [32] observed stimulation of WNK3 and SPAK kinases in cortical neurons and primary oligodendrocytes cultured from the brain of mice subjected to transient MCA stroke [32]. They further established that cerebral ischemia facilitates hyperphosphorylation of the WNK3-SPAK/OSR1 catalytic T-loop and of NKCC1 stimulatory sites (Thr^{203}/Thr^{207}/Thr^{212}); thus, increased expression of NKCC1 in the brain cells [32]. However, transgenic KO of WNK3, abridged ischemia-mediated SPAK/OSR1-NKCC1 phosphorylation and displayed reduced cerebral edema, axonal demyelination, and infarct volume, as well as improved post-stroke neurological recovery when compared to WT mice [32]. Briefly, the data presented by Begum et al. [32] identify the role of WNK3-SPAK/OSR1-NKCC1 signaling pathway in ischemic neuroglial injury and suggested that obstruction of this pathway could reduce NKCC1 expression in the brain and avert post-stroke neuronal cell death following [32]. Similarly, Zhao et al. [92] demonstrated that KO of the WNK3-SPAK kinase complex in mice instigates decreased expression of NKCC1 and subsequently ameliorated cerebral infarction and edema after MCA stroke. Generally, deletion of the WNK3-SPAK kinase complex significantly produced less cytotoxic edema, less demyelination, and improved post-ischemic stroke neurological outcomes in the transgenic mice [92]. However, it is worth noting that the mechanism(s) of regulations employed by WNK3-SPAK/OSR1-NKCC1 signaling pathway in oligodendrogenesis is still elusive and requires further studies [26]. In addition, we recently demonstrated that WNK3 KO mice exhibit reduced endothelial and perivascular cytotoxic edema of astrocytes following post-ischemic stroke [86]. We further alluded that WNK3-SPAK inhibition confer neuroprotection on mammalian brain through concurrent stimulation of KCC3 activity at Thr^{991} and Thr^{1048} residues and inhibition of NKCC1 activity at Thr^{203}, Thr^{207}, and Thr^{212} residues [86].

In a recent rat model study, Bhuiyan et al. [106] demonstrated that WNK-Cab39-NKCC1 signaling pathway is implicated in ischemia. Furthermore, they suggested that activated WNK-Cab39 pathway increased NKCC1 activity in brain tissues of spontaneous hypertensive rats following subacute ischemic stroke [106]. A more recent similar report by Huang et al. [124] demonstrated that ischemic stroke with hypertension comorbidity further stimulates the WNK-SPAK/OSR1-NKCC1 signaling pathway, which contributes to deteriorated neurological functions/behavior [124]. In fact, the established role of WNK-SPAK/OSR1 signaling pathway in stimulating NKCC1 and inhibiting KCC3, which contribute to the pathogenesis of stroke, are the reasons for our recent pharmacological studies [70,71].

5. Current Pharmacological Treatments for Stroke

We underscored in the earliest section of this review that stroke is one of the major threats to global health. Over the years, stroke had been a chief contributor to mortality and disabled lives across the globe; there are projections that the impact of this disease on global health may be worse in the near future. Presently, there are only few pharmacological strategies available to reduce the health and socio-economic burden triggered by this disease. Thus, there is an urgent need to tackle this disease. In this regard, research on the role of CCCs in the pathogenesis of stroke to inform future drug development is needed. Accordingly, this section of the review highlights current pharmacological approaches in the management of stroke with particular focus on molecular compounds that potentially inhibit SPAK/OSR1 pathway and NKCC1 and stimulate KCC3.

5.1. Inhibitors of WNK-SPAK/OSR1 Pathway

A quick recap of the following crucial information from previous sections, which may contain pharmacological strategies for managing stroke: (1) WNK-SPAK/OSR1 modulates the activities of CCCs through a well-coordinated reciprocal pattern of regulation [53,60,67,69], (2) activation of WNK-SPAK/OSR1 signaling pathway stimulates NKCC1 and inhibits KCC3 expressions [19,53,67,72], and (3) development and progression of stroke have been implicated with phosphorylated WNK-SPAK/OSR1 signaling pathway and subsequent up- and down-regulations of NKCC1 and KCC3 expressions, respectively [22,140]. Therefore, we can safely presume that molecular compounds that act as pharmacological or genetic antagonists of WNK-SPAK/OSR1 kinases are likely potential drug candidates for the treatment of stroke. Recent reports have shown that drugs that are potent blockers of WNK-SPAK/OSR1 signaling pathway reduce phosphorylation of NKCC1 and KCC1 which enables cellular chloride expulsion, subsequently mitigating cerebral edema and other neurological anomalies following ischemia. This then protects against brain damage and enhances post-stroke brain functions [32,65,70,86,92,94,124,141].

The WNK-SPAK/OSR1 pathway constitutes potential therapeutic targets in Cl^- dysregulation [23]. The loop diuretic, bumetanide protects the brain from damage by mediating GABAergic signaling in NKCC1 expression following ischemic injury [22,23,26,142]. Furthermore, treatment with the drug reversed the impact of GABA-mediated depolarization, which may promote functional recovery after stroke via neuron repair/protection as adjudged by its effect of $GABA_A$ receptor antagonist and WNK3 knockout [94,142]. Pharmacologically targeting the WNK-SPAK/OSR1 kinase pathway could be a strategy to restore GABAergic inhibition [23]. Indeed, genetic or pharmacological inhibition of WNK-SPAK/OSR1 activity would lead to cotransporter dephosphorylation: inhibition of NKCC1 and activation KCC3, which would enhance $[Cl^-]_i$ extrusion [67]. Furthermore, enhancement of $[Cl^-]_i$ extrusion in the neurons would facilitate $GABA_A$ receptor-mediated hyperpolarization and thus inhibit neuronal activity through combined NKCC1 inhibition and KCC3 stimulation [19,22,53,65,67] (Figure 2).

Recent studies have indicated that decline in NKCC1 protein expression along with WNK3 knockdown, contributes to lessened post-stroke brain injury and accelerated neurobehavioral recovery [32,86,92]. These reports and more have immensely motivated the development of novel therapeutic strategies that have targeted WNK-SPAK/OSR1 signaling pathways to improve post-stroke physiological functions [70,106,124]. A recent rat model study demonstrated an upregulation of WNK-SPAK/OSR1-NKCC1 signaling pathway in the brains of spontaneously induced-hypertensive rats and subsequently augmented susceptibility to ischemic damage [106]. However, intraperitoneal administration of bumetanide (10 mg/kg) post-reperfusion blocked the WNK-Cab39-NKCC1 signaling pathway and subsequently mitigated post-ischemic infarction and cell swelling and improved neurological functions in animals [106]. Loop diuretics are often used to inhibit NKCCs. Inhibition of NKCC2 promote diuresis in the kidney and reduces pressure due to excess fluid in the lungs. Hence, loop diuretics are a treatment option hypertension and pulmonary edema. Although some loop diuretics inhibit KCCs, they do so very poorly. Researchers have explored the loop diuretics bumetanide and furosemide as novel treatment options for brain disorders [65]. However, a number of unfavorable physiochemical characteristics associated with the use of bumetanide [23,41,65] call for better alternatives in the management of neurological diseases including stroke. Recently, Huang and co-workers [124] reported that a novel NKCC1 inhibitor (STS66) is superior to bumetanide in ameliorating ischemic brain injury following transient MCA occlusion and large-vessel ischemic stroke models. In the study, ischemic injury stimulated WNK-SPAK-NKCC1 cascades in brains of AngiotensinII (AngII)-induced hypertensive mice. However, STS66 treatment completely blocked this pathway and by implication mitigated ischemic infarction, cerebral edema, and neuronal death as well as neurological deficits in both stroke models with hypertension comorbidity [124].

We recently proposed that improved understanding of cooperative interactions among different phospho-sites of cotransporters and the molecular mechanisms involved in their physiological regulations could provide insights to inform potential pharmacological interventions [71]. Recently, we conducted a large-scale phospho-proteomics study with the application of immunoblot and phospho-antibodies immunoprecipitation techniques to investigate the regulatory mechanisms of a broad kinase inhibitor, staurosporine and N-ethylmalemide (NEM), a modulator of both kinase and phosphatase activities on phosphorylation of specific KCC2 and NKCC1 in HEK293 cells and immature cultured hippocampal neurons [71]. Our analyses revealed dephosphorylation of Thr^{203}, Thr^{207}, and Thr^{212} of NKCC1 and Thr^{1007} of KCC2 following application of the two agents. The two compounds resulted in dephosphorylation of sites Thr^{233} and Ser^{373}, phosphorylation sites located within the T-loop and S-loop of SPAK. Hence, the study suggests the inhibitory effect of staurosporine and NEM on WNK-SPAK/OSR1 signaling pathway in the regulation of NKCC1 and KCC2 is in a reciprocal pattern [71]. We are of the opinion that the underlying information from this study will be highly important for future development of integrative therapeutic strategies in the management of neurological diseases such as cerebral stroke.

Importantly, evolving roles of WNK-SPAK/OSR1 signaling in stroke as discussed in this review, points to additional possible applications of WNK-SPAK/OSR1 modulation in neurological diseases. In view of this, a promising strategy could involve exploitation of the unique structure of these kinases to enhance protein specificity [53,70]. Immense efforts to inhibit WNKs or SPAK/OSR1 for the treatment of human diseases such as hypertension have led to the discoveries of small molecule inhibitors. WNK kinase inhibitors include WNK463 [143], PP121 [144], and SPAK inhibitors such as STOCK1S-14279, Closantel [145], Rafoxanide [146], Verteporfin [74], STOCK1S-50699, and STOCK2S-26016, [147], as well as HK01 [148] and 20I [149] (also see Figures 3 and 4). Unfortunately, none of these compounds is an ideal drug candidate for the treatment of brain disorders due to their relatively low penetrability through the blood-brain barrier (BBB). Recently, we employed a scaffold-hybrid strategy in our laboratory to develop a novel compound ZT-1a. ZT-1a is a non-ATP-competitive SPAK blocker, which specifically inhibits this signaling pathway by decreasing SPAK-dependent phosphorylation of NKCC1 and KCCs in cell cultures as well as in vivo mouse and rat brains [70]. In brief, treatment with ZT-1a (2.5–5.0 mg/kg) abated post-stroke related brain injuries and improved neurological features/functions. The data from the study suggests that ZT-1a or related compounds that are CCC modulators could be a therapeutic strategy for neurodegenerative disorders such as cerebral stroke [70]. Hence, we holistically advocate for follow-up with detailed research studies on the development of more WNK-SPAK/OSR1 inhibitors with favorable pharmacokinetic properties for clinical use.

5.2. Inhibitors of NKCC1

It has been established that the recovery process from many neurological disorders including stroke would highly benefit from inhibition of NKCC1 activity [94]. Suppression of NKCC1 activity through bumetanide is neuroprotective and improves post-stroke neurophysiological status [142,150]. Thus, bumetanide has the potential to influence many CNS disorders [94]. Several studies have demonstrated the contribution of NKCC1 in the development and progression of post-stroke edema and cell death, thus targeting NKCC1 could be a potential neuroprotective target [22,26,32,65,111,151]. In fact, a pharmacological study using bumetanide demonstrated a significantly reduced neuronal Na$^+$ overload and cell death. Bumetanide also simultaneously reduced infarct volume and brain edema [104]. Another rat model study showed that bumetanide administered after focal cerebral ischemia in rats (given 7 days post-ischemia, and continued for 21 days), improved behavioral recovery and promoted neurogenesis 4 weeks post-havoc [141]. Low concentrations of bumetanide (2 to 10 μM) are capable of inhibiting NKCCs in vitro with no significant effect on the KCCs; a high concentration has been shown to inhibit the activities of both NKCC1 and KCC2 [152,153]. The expression of NKCC1 is common at the luminal side of endothelial cells of the BBB, thereby allowing easy interaction between the transporter and its inhibitor (bumetanide) when administered intravenously, which subsequently decreased edema in MCA occlusion model of stroke in rats [154]. Bumetanide acts through docking to the binding site at the trans-membrane region of NKCC1. Docking to this region allows for the inhibition of NKCC1 activity and reduced [Cl$^-$]$_i$ in neurons [94]; a more likely mechanism through which the drug confers neuroprotection and neuronal recovery following stroke episodes [94,155].

Simard et al. [156] and Walcott et al. [42] in their respective reviews highlighted the implication of the constitutive expression of NKCC1 and SUR1-regulated NC$_{Ca-ATP}$ (SUR1/TRPM4) channel on the cascade of events that are involved in the pathogenesis of cerebral ischemia and the impact of combinatorial therapy of bumetanide and glibenclamide in ameliorating the havoc. Bhuiyan and colleagues [106] reported that bumetanide downregulated the WNK-Cab39-NKCC1 signaling pathway, consequently reducing the susceptibility of hypertensive rats to ischemic brain damage. Furthermore, in a recent study a synergistic treatment with mild hypothermia (33.5 °C for 30 min) and inhibitor DAPT (50 µM) attenuated the overexpression of NKCC1 mRNA following global cerebral ischemia injury in rats [20].

In animal stroke models, bumetanide administration pre- and post-stroke induction led to the down-regulation of NKCC1 expression. Other observations include a reduction in edema, infarction volume, and ischemic necrotic cell death especially in the early stage of ischemic damage, promotion of neurogenesis, and improved sensorimotor recovery [17,97,109,141,157,158]. In another rat model study, ET-1 was used to induce focal ischemia but post treatment with bumetanide selectively inhibited NKCC1 expression in the cortex and promoted synaptic plasticity in the denervated cervical spinal cord following cerebral ischemia [111]. Similarly, Xu et al. [141] demonstrated that chronic treatment with bumetanide promotes neurogenesis and behavioral recovery after ET-1-induced stroke in rats. In addition, bumetanide (10 µM) was used in another study to block NKCC1 in order to facilitate decreased $[Cl^-]_i$ in hippocampal tissue cultured from rats either during oxygen-glucose deprivation for 120 min or post-exposure. The drug improved neuronal viability during the acute ischemic episode which suggested its critical role in the modulation of transmembrane chloride transport [27].

Indeed, bumetanide appears to be a promising pharmacological inhibitor of NKCC1; it possess some demerits that may limit its application as an anti-stroke drug to some extent [23,94]. Alongside bumetanide, a novel inhibitor STS66 (a prodrug of bumetanide) also exhibits promising potential as a pharmacological inhibition of NKCC1 and has been demonstrated to also reduced ischemic infarction, swelling and neurological deficits in mice model of transient ischemic stroke [124]. Interestingly, STS66 can penetrate BBB more easily and appears to be more efficient in eliciting the aforementioned anti-stroke properties [124], which is one of the various reasons it has been recently proposed as a better therapeutic drug in stroke management when compared with bumetanide [23,65]. A finding contrary to the common hypothesis on the efficacy of bumetanide was recently reported [18]. In this study, post treatment with bumetanide (40 mg/kg) following Intracerebral hemorrhage induction in male Sprague Dawley rats failed to improve behavior or lessen injury neither did the drug normalized ion concentrations after late dosing [18].

In spite of the positive outcomes demonstrated clinically by administration of bumetanide to patients with psychiatric/neurological conditions [159–166]; the drug has exhibited strong diuretic effect resulting from the inhibition of NKCC2 expression in the kidney which may pose serious challenges to issues on drug compliance and health concerns [167–169], thereby limiting the therapeutic applications of bumetanide. Hence, selective inhibition of NKCC1 in lieu of renal NKCC2 may attenuate the diuretic glitches. In this regard, Savardi et al. [170] recently discovered ARN23746, a selective inhibitor of NKCC1 in lieu of NKCC2 and KCC2 in vivo. The reports from the study demonstrated that the pharmacokinetic profile of ARN23746 is better when compared with that of bumetanide in vitro and in vivo. Briefly, the study demonstrated that ARN23746 (10 µM) restored aberrantly high $[Cl^-]_i$ to the physiological level in mature hippocampal neuronal cultures of Ts65Dn mouse model Down syndrome (DS) coupled with rescued cognitive impairment in Ts65Dn with no significant diuretic effect in either the WT or Ts65Dn mice [170]. Furthermore, the researchers demonstrated that intraperitoneal administration of ARN23746 recovered social and repetitive behaviors associated with the main symptoms of autism spectrum disorder (ASD) in valproic acid (VPA) mouse model of ASD. In addition, neither diuretic effect nor overt toxicity of the compound were present in the ARN23746 treated mice [170]. ARN23746 has great potential for further development into a clinically-relevant drug for the treatment of DS, ASD [170], and possibly several other neurological conditions characterized by impaired Cl^- homeostasis including stroke.

5.3. Activator of KCC3

The KCCs, especially KCC2 and KCC3, are popular due to increased findings on human disease-causing mutations [68,121,122]. Hence, the discovery of small molecules that modulate these cotransporters' activities is prioritized within the field. Discovery of such modulators may aid development of therapeutic drugs for the management of KCC-related diseases as well as other pathological conditions including stroke [171]. Currently, the loop diuretics bumetanide and furosemide are the only FDA-approved drugs that modulate the KCCs [172]. In a mouse model study, bumetanide is demonstrated to be involved in the stimulation of KCC3 expression and subsequent extrusion of $[Cl^-]_i$ in the sensory neurons [114]. In addition, Adragna and co-workers [87] in a cell culture study substituted Thr^{991} and Thr^{1048} residues with alanine at the carboxyl terminus of KCC3a protein, which prevented inhibitory phosphorylation at the substituted sites and subsequently triggered increased expression of KCC3a mRNA. Interestingly, the flux condition accompanied a down-regulation of NKCC1 expression, facilitated by the addition of ouabain (0.1 mM), and bumetanide (10 μM) to the flux media [87].

However, Delpire and Weaver [171] recently expressed their concerns for the need to develop modulators of KCC activity to provide insights into KCC modulation as a therapeutic strategy for neurological conditions such as stroke. Unfortunately, these FDA approved drugs (bumetanide and furosemide) are poor inhibitors of KCCs, with a higher potency for NKCC1 or NKCC2 (IC_{50} = 0.5–5.0 μM) in comparison to KCC (IC_{50} = 50–500 mM) [33,171]. In fact, drugs that can act as weak inhibitors might be better alternatives as complete inhibition mimics a loss-of-function, which could presumably be harmful for the nervous system. As the KCC isoforms have different expression patterns and physiological functions, target specificity in the deployed pharmacological approach is also an issue [172]. In a large screening effort targeted against KCC2, Delpire et al. [173] was able to identify inhibitory compounds more potent (3–4x) than the two loop diuretics. However, these compounds are not ideal drug candidates due to the following reasons: (1) non-specificity to KCC2 as they concurrently inhibit KCC3, and (2) poor pharmacokinetic properties [173,174]. Meanwhile, the ability of loop diuretics to reach the CNS/PNS remains obscure [171,172]. Perhaps a better pharmacological approach would be to develop therapeutic compounds that are specific modulators of the KCC3. Fortunately, our recently developed novel molecular compound, ZT-1a, is a SPAK kinase inhibitor that specifically stimulates KCC3 and inhibits NKCC1 by decreasing their SPAK-dependent phosphorylation/signaling pathway in cultured cells and in vivo rat and mouse brains [70] (also see Figure 2). In addition, the systematic administration of ZT-1a ameliorated phosphorylation of co-transporters and cerebral edema following ischemia, protect against brain damage and improve neurological functions after stroke episode [70].

6. Conclusions and Future Directions

The CCCs play crucial roles in regulating neuronal functions. The cotransporters are key mediators of several and important cellular functions such as cell volume regulation, trans-epithelial ion transport, and maintenance of $[Cl^-]_i$. Modulation of NKCC1 and KCC3 expressions by their upstream regulator, WNK-SPAK/OSR1 is implicated in the development and progression of stroke. There are several demonstrations that phosphorylation of NKCC1 and KCC3 via the WNK-SPAK/OSR1 signaling may lead to activation of NKCC1 and inhibition of KCC3 either during or post-stroke episode. In fact, the role of NKCC1 and KCC3 as well as their regulatory proteins in stroke pathogenesis suggests that they are potential targets for the treatment of stroke. The pharmacological strategies that were discussed in this review possess potential therapeutic efficacies for stroke management. Novel compounds must successfully address concerns regarding off-target effects due to the many isoforms and physiological function related to the WNK-SPAK/OSR1-CCC pathway. As advances in stroke therapy may also benefit other neurological impairments, we strongly suggest consistent follow-up actions on currently available pharmacological treatments for stroke through detailed research studies to aid further development of therapeutic drugs with a better pharmacokinetic profile. Hence, we holistically advocate for increased focus on human clinical research on this topic as informed by its paucity to that regards.

Author Contributions: S.S.J. and J.Z. were responsible for writing the whole passage. S.S.J., N.F.M.A., and J.Z. were responsible for checking and revision. All authors have read and agreed to the published version of the manuscript.

Funding: This work was in part supported by a Commonwealth PhD Scholarship (S.S.J.) and the University of Exeter Medical School start-up fund (J.Z.) and NIH Grants R01 NS109358 (J.Z.).

Institutional Review Board Statement: Not applicable.

Informed Consent Statement: Not applicable.

Data Availability Statement: Not applicable.

Conflicts of Interest: The authors declare no competing interests.

References

1. Lu, D.C.-Y.; Hannemann, A.; Wadud, R.; Rees, D.C.; Brewin, J.N.; Low, P.S.; Gibson, J.S. The role of WNK in modulation of KCl cotransport activity in red cells from normal individuals and patients with sickle cell anaemia. *Pflügers Arch. Eur. J. Physiol.* **2019**, *471*, 1539–1549. [CrossRef]
2. Gamba, G. Molecular physiology and pathophysiology of electroneutral cation-chloride cotransporters. *Physiol. Rev.* **2005**, *85*, 423–493. [CrossRef]
3. Meor Azlan, N.; Koeners, M.; Zhang, J. Regulatory control of the Na-Cl co-transporter NCC and its therapeutic potential for hypertension. *Acta Pharm. Sin. B* **2020**. [CrossRef]
4. Lauf, P.; Theg, B. A chloride dependent K$^+$ flux induced by N-ethylmaleimide in genetically low K$^+$ sheep and goat erythrocytes. *Biochem. Biophys. Res. Commun.* **1980**, *92*, 1422–1428. [CrossRef]
5. Dunham, P.B.; Stewart, G.W.; Ellory, J.C. Chloride-activated passive potassium transport in human erythrocytes. *Proc. Natl. Acad. Sci. USA* **1980**, *77*, 1711–1715. [CrossRef]
6. Hoffmann, E.; Sjoholm, C.; Simonsen, L. Anion-Cation Cotransport and Volume Regulation in Ehrlich Ascites Tumor-Cells. *Proc. J. Physiol. Lond.* **1981**, *319*, P94–P95.
7. Gibson, J.S.; Ellory, J.C.; Adragna, N.C.; Lauf, P.K. Pathophysiology of the K$^+$-Cl$^-$ cotransporters: Paths to discovery and overview. In *Physiology and Pathology of Chloride Transporters and Channels in the Nervous System: From Molecules to Disease*; Academic Press: London, UK, 2009; pp. 27–42.
8. Xu, J.-C.; Lytle, C.; Zhu, T.T.; Payne, J.A.; Benz, E.; Forbush, B. Molecular cloning and functional expression of the bumetanide-sensitive Na-K-Cl cotransporter. *Proc. Natl. Acad. Sci. USA* **1994**, *91*, 2201–2205. [CrossRef]
9. Gamba, G.; Saltzberg, S.N.; Lombardi, M.; Miyanoshita, A.; Lytton, J.; Hediger, M.A.; Brenner, B.M.; Hebert, S.C. Primary structure and functional expression of a cDNA encoding the thiazide-sensitive, electroneutral sodium-chloride cotransporter. *Proc. Natl. Acad. Sci. USA* **1993**, *90*, 2749–2753. [CrossRef]
10. Pellegrino, C.M.; Rybicki, A.C.; Musto, S.; Nagel, R.L.; Schwartz, R.S. Molecular identification and expression of erythroid K: Cl cotransporter in human and mouse erythroleukemic cells. *Blood Cells Mol. Dis.* **1998**, *24*, 31–40. [CrossRef]
11. Piechotta, K.; Lu, J.; Delpire, E. Cation chloride cotransporters interact with the stress-related kinases Ste20-related proline-alanine-rich kinase (SPAK) and oxidative stress response 1 (OSR1). *J. Biol. Chem.* **2002**, *277*, 50812–50819. [CrossRef]
12. Gillen, C.M.; Brill, S.; Payne, J.A.; Forbush, B. Molecular cloning and functional expression of the K-Cl cotransporter from rabbit, rat, and human A new member of the cation-chloride cotransporter family. *J. Biol. Chem.* **1996**, *271*, 16237–16244. [CrossRef]
13. Payne, J.A. Functional characterization of the neuronal-specific K-Cl cotransporter: Implications for [K$^+$] oregulation. *Am. J. Physiol. Cell Physiol.* **1997**, *273*, C1516–C1525. [CrossRef]
14. Hiki, K.; D'Andrea, R.J.; Furze, J.; Crawford, J.; Woollatt, E.; Sutherland, G.R.; Vadas, M.A.; Gamble, J.R. Cloning, characterization, and chromosomal location of a novel human K$^+$-Cl$^-$ cotransporter. *J. Biol. Chem.* **1999**, *274*, 10661–10667. [CrossRef]
15. Mercado, A.; Song, L.; George, A.; Delpire, E.; Mount, D. Molecular, functional, and genomic characterization of KCC3 and KCC4. *J. Am. Soc. Nephrol.* **1999**, *10*, 38A.
16. Gagnon, K.B.; Delpire, E. Physiology of SLC12 transporters: Lessons from inherited human genetic mutations and genetically engineered mouse knockouts. *Am. J. Physiol. Cell Physiol.* **2013**, *304*, C693–C714. [CrossRef]
17. Wang, G.; Huang, H.; He, Y.; Ruan, L.; Huang, J. Bumetanide protects focal cerebral ischemia-reperfusion injury in rat. *Int. J. Clin. Exp. Pathol.* **2014**, *7*, 1487.
18. Wilkinson, C.M.; Fedor, B.A.; Aziz, J.R.; Nadeau, C.A.; Brar, P.S.; Clark, J.J.; Colbourne, F. Failure of bumetanide to improve outcome after intracerebral hemorrhage in rat. *PLoS ONE* **2019**, *14*, e0210660. [CrossRef]
19. De los Heros, P.; Pacheco-Alvarez, D.; Gamba, G. Role of WNK kinases in the modulation of cell volume. In *Current Topics in Membranes*; Elsevier: Amsterdam, The Netherlands, 2018; Volume 81, pp. 207–235.
20. Yang, T.; Zhao, K.; Shu, H.; Chen, X.; Cheng, J.; Li, S.; Zhao, Z.; Kuang, Y.; Yu, S. The Nogo receptor inhibits proliferation, migration and axonal extension by transcriptionally regulating WNK1 in PC12 cells. *Neuroreport* **2017**, *28*, 533–539. [CrossRef]

21. Johnson, W.; Onuma, O.; Owolabi, M.; Sachdev, S. Stroke: A global response is needed. *Bull. World Health Organ.* **2016**, *94*, 634. [CrossRef] [PubMed]

22. Martín-Aragón Baudel, M.A.; Poole, A.V.; Darlison, M.G. Chloride co-transporters as possible therapeutic targets for stroke. *J. Neurochem.* **2017**, *140*, 195–209. [CrossRef]

23. Schulte, J.T.; Wierenga, C.J.; Bruining, H. Chloride transporters and GABA polarity in developmental, neurological and psychiatric conditions. *Neurosci. Biobehav. Rev.* **2018**, *90*, 260–271. [CrossRef] [PubMed]

24. Mayor, D.; Tymianski, M. Neurotransmitters in the mediation of cerebral ischemic injury. *Neuropharmacology* **2018**, *134*, 178–188. [CrossRef]

25. Boscia, F.; Begum, G.; Pignataro, G.; Sirabella, R.; Cuomo, O.; Casamassa, A.; Sun, D.; Annunziato, L. Glial Na^+-dependent ion transporters in pathophysiological conditions. *Glia* **2016**, *64*, 1677–1697. [CrossRef]

26. Song, S.; Luo, L.; Sun, B.; Sun, D. Roles of glial ion transporters in brain diseases. *Glia* **2020**, *68*, 472–494. [CrossRef] [PubMed]

27. Zagrean, A.-M.; Grigoras, I.-F.; Iesanu, M.I.; Ionescu, R.-B.; Chitimus, D.M.; Haret, R.M.; Ianosi, B.; Ceanga, M.; Zagrean, L. Neuronal Transmembrane Chloride Transport Has a Time-Dependent Influence on Survival of Hippocampal Cultures to Oxygen-Glucose Deprivation. *Brain Sci.* **2019**, *9*, 360. [CrossRef] [PubMed]

28. Mele, M.; Costa, R.O.; Duarte, C.B. Alterations in GABAA-receptor trafficking and synaptic dysfunction in brain disorders. *Front. Cell. Neurosci.* **2019**, *13*, 77. [CrossRef] [PubMed]

29. Smith, K.R.; Muir, J.; Rao, Y.; Browarski, M.; Gruenig, M.C.; Sheehan, D.F.; Haucke, V.; Kittler, J.T. Stabilization of GABAA receptors at endocytic zones is mediated by an AP2 binding motif within the GABAA receptor β3 subunit. *J. Neurosci.* **2012**, *32*, 2485–2498. [CrossRef]

30. Mielke, J.G.; Wang, Y.T. Insulin exerts neuroprotection by counteracting the decrease in cell-surface GABAA receptors following oxygen–glucose deprivation in cultured cortical neurons. *J. Neurochem.* **2005**, *92*, 103–113. [CrossRef]

31. Fu, C.-Y.; He, X.-Y.; Li, X.-F.; Zhang, X.; Huang, Z.-W.; Li, J.; Chen, M.; Duan, C.-Z. Nefiracetam attenuates pro-inflammatory cytokines and GABA transporter in specific brain regions of rats with post-ischemic seizures. *Cell. Physiol. Biochem.* **2015**, *37*, 2023–2031. [CrossRef]

32. Begum, G.; Yuan, H.; Kahle, K.T.; Li, L.; Wang, S.; Shi, Y.; Shmukler, B.E.; Yang, S.S.; Lin, S.H.; Alper, S.L.; et al. Inhibition of WNK3 Kinase Signaling Reduces Brain Damage and Accelerates Neurological Recovery After Stroke. *Stroke* **2015**, *46*, 1956–1965. [CrossRef]

33. Russell, J.M. Sodium-potassium-chloride cotransport. *Physiol. Rev.* **2000**, *80*, 211–276. [CrossRef] [PubMed]

34. Ben-Ari, Y.; Khalilov, I.; Kahle, K.T.; Cherubini, E. The GABA excitatory/inhibitory shift in brain maturation and neurological disorders. *Neuroscientist* **2012**, *18*, 467–486. [CrossRef]

35. Adachi, M.; Asakura, Y.; SATO, Y.; Tajima, T.; Nakajima, T.; Yamamoto, T.; Fujieda, K. Novel SLC12A1 (NKCC2) mutations in two families with Bartter syndrome type 1. *Endocr. J.* **2007**, *54*, 1003–1007. [CrossRef] [PubMed]

36. Konopacka, A.; Qiu, J.; Yao, S.T.; Greenwood, M.P.; Greenwood, M.; Lancaster, T.; Inoue, W.; de Souza Mecawi, A.; Vechiato, F.M.; de Lima, J.B. Osmoregulation requires brain expression of the renal Na-K-2Cl cotransporter NKCC2. *J. Neurosci.* **2015**, *35*, 5144–5155. [CrossRef] [PubMed]

37. Wallace, B.K.; Jelks, K.A.; O'Donnell, M.E. Ischemia-induced stimulation of cerebral microvascular endothelial cell Na-K-Cl cotransport involves p38 and JNK MAP kinases. *Am. J. Physiol. Cell Physiol.* **2012**, *302*, C505–C517. [CrossRef] [PubMed]

38. Dixon, M.J.; Gazzard, J.; Chaudhry, S.S.; Sampson, N.; Schulte, B.A.; Steel, K.P. Mutation of the Na-K-Cl co-transporter gene Slc12a2 results in deafness in mice. *Hum. Mol. Genet.* **1999**, *8*, 1579–1584. [CrossRef] [PubMed]

39. Nezu, A.; Parvin, M.N.; Turner, R.J. A conserved hydrophobic tetrad near the C terminus of the secretory $Na^+-K^+-2Cl^-$ cotransporter (NKCC1) is required for its correct intracellular processing. *J. Biol. Chem.* **2009**, *284*, 6869–6876. [CrossRef]

40. Orlov, S.N.; Koltsova, S.V.; Kapilevich, L.V.; Gusakova, S.V.; Dulin, N.O. NKCC1 and NKCC2: The pathogenetic role of cation-chloride cotransporters in hypertension. *Genes Dis.* **2015**, *2*, 186–196. [CrossRef]

41. Töllner, K.; Brandt, C.; Töpfer, M.; Brunhofer, G.; Erker, T.; Gabriel, M.; Feit, P.W.; Lindfors, J.; Kaila, K.; Löscher, W. A novel prodrug-based strategy to increase effects of bumetanide in epilepsy. *Ann. Neurol.* **2014**, *75*, 550–562. [CrossRef]

42. Walcott, B.P.; Kahle, K.T.; Simard, J.M. Novel treatment targets for cerebral edema. *Neurotherapeutics* **2012**, *9*, 65–72. [CrossRef]

43. Koumangoye, R.; Bastarache, L.; Delpire, E. NKCC1: Newly Found as a Human Disease-Causing Ion Transporter. *Function* **2021**, *2*, zqaa028. [CrossRef] [PubMed]

44. Blaesse, P.; Airaksinen, M.S.; Rivera, C.; Kaila, K. Cation-chloride cotransporters and neuronal function. *Neuron* **2009**, *61*, 820–838. [CrossRef] [PubMed]

45. Delpire, E.; Mount, D.B. Human and murine phenotypes associated with defects in cation-chloride cotransport. *Annu. Rev. Physiol.* **2002**, *64*, 803–843. [CrossRef] [PubMed]

46. Rust, M.B.; Faulhaber, J.; Budack, M.K.; Pfeffer, C.; Maritzen, T.; Didie, M.; Beck, F.X.; Boettger, T.; Schubert, R.; Ehmke, H.; et al. Neurogenic mechanisms contribute to hypertension in mice with disruption of the K-Cl cotransporter KCC3. *Circ. Res.* **2006**, *98*, 549–556. [CrossRef]

47. Boettger, T.; Hubner, C.A.; Maier, H.; Rust, M.B.; Beck, F.X.; Jentsch, T.J. Deafness and renal tubular acidosis in mice lacking the K-Cl co-transporter Kcc4. *Nature* **2002**, *416*, 874–878. [CrossRef]

48. Kahle, K.T.; Schmouth, J.F.; Lavastre, V.; Latremoliere, A.; Zhang, J.; Andrews, N.; Omura, T.; Laganiere, J.; Rochefort, D.; Hince, P.; et al. Inhibition of the kinase WNK1/HSN2 ameliorates neuropathic pain by restoring GABA inhibition. *Sci. Signal.* **2016**, *9*, ra32. [CrossRef]

49. Mavrovic, M.; Uvarov, P.; Delpire, E.; Vutskits, L.; Kaila, K.; Puskarjov, M. Loss of non-canonical KCC2 functions promotes developmental apoptosis of cortical projection neurons. *EMBO Rep.* **2020**, *21*, e48880. [CrossRef]

50. Hinz, L.; Torrella Barrufet, J.; Heine, V.M. KCC2 expression levels are reduced in post mortem brain tissue of Rett syndrome patients. *Acta Neuropathol. Commun.* **2019**, *7*, 196. [CrossRef]

51. Pisella, L.I.; Gaiarsa, J.L.; Diabira, D.; Zhang, J.; Khalilov, I.; Duan, J.; Kahle, K.T.; Medina, I. Impaired regulation of KCC2 phosphorylation leads to neuronal network dysfunction and neurodevelopmental pathology. *Sci. Signal.* **2019**, *12*, eaay0300. [CrossRef]

52. Garneau, A.P.; Marcoux, A.A.; Frenette-Cotton, R.; Mac-Way, F.; Lavoie, J.L.; Isenring, P. Molecular insights into the normal operation, regulation, and multisystemic roles of K(+)-Cl(-) cotransporter 3 (KCC3). *Am. J. Physiol. Cell Physiol.* **2017**, *313*, C516–C532. [CrossRef]

53. Shekarabi, M.; Zhang, J.; Khanna, A.R.; Ellison, D.H.; Delpire, E.; Kahle, K.T. WNK Kinase Signaling in Ion Homeostasis and Human Disease. *Cell Metab.* **2017**, *25*, 285–299. [CrossRef]

54. Torchia, J.; Lytle, C.; Pon, D.; Forbush, B.; Sen, A. The Na-K-Cl cotransporter of avian salt gland. Phosphorylation in response to cAMP-dependent and calcium-dependent secretogogues. *J. Biol. Chem.* **1992**, *267*, 25444–25450. [CrossRef]

55. Al Shibli, N.; Al-Maawali, A.; Elmanzalawy, A.; Al-Nabhani, M.; Koul, R.; Gabr, A.; Al Murshedi, F. A Novel Splice-Site Variant in SLC12A6 Causes Andermann Syndrome without Agenesis of the Corpus Callosum. *J. Pediatr. Genet.* **2020**, *9*, 293–295. [CrossRef]

56. Jin, S.C.; Furey, C.G.; Zeng, X.; Allocco, A.; Nelson-Williams, C.; Dong, W.; Karimy, J.K.; Wang, K.; Ma, S.; Delpire, E.; et al. SLC12A ion transporter mutations in sporadic and familial human congenital hydrocephalus. *Mol. Genet. Genom. Med.* **2019**, *7*, e892. [CrossRef]

57. Park, J.; Flores, B.R.; Scherer, K.; Kuepper, H.; Rossi, M.; Rupprich, K.; Rautenberg, M.; Deininger, N.; Weichselbaum, A.; Grimm, A.; et al. De novo variants in SLC12A6 cause sporadic early-onset progressive sensorimotor neuropathy. *J. Med. Genet.* **2020**, *57*, 283–288. [CrossRef]

58. Flatman, P.W. Cotransporters, WNKs and hypertension: Important leads from the study of monogenetic disorders of blood pressure regulation. *Clin. Sci.* **2007**, *112*, 203–216. [CrossRef]

59. Richardson, C.; Alessi, D.R. The regulation of salt transport and blood pressure by the WNK-SPAK/OSR1 signalling pathway. *J. Cell Sci.* **2008**, *121*, 3293–3304. [CrossRef]

60. Heubl, M.; Zhang, J.; Pressey, J.C.; Al Awabdh, S.; Renner, M.; Gomez-Castro, F.; Moutkine, I.; Eugene, E.; Russeau, M.; Kahle, K.T.; et al. GABAA receptor dependent synaptic inhibition rapidly tunes KCC2 activity via the Cl(-)-sensitive WNK1 kinase. *Nat. Commun.* **2017**, *8*, 1776. [CrossRef]

61. Rivera, C.; Voipio, J.; Payne, J.A.; Ruusuvuori, E.; Lahtinen, H.; Lamsa, K.; Pirvola, U.; Saarma, M.; Kaila, K. The K$^+$/Cl$^-$ co-transporter KCC2 renders GABA hyperpolarizing during neuronal maturation. *Nature* **1999**, *397*, 251–255. [CrossRef]

62. Delpire, E. Cation-chloride cotransporters in neuronal communication. *Physiology* **2000**, *15*, 309–312. [CrossRef]

63. Kaila, K.; Price, T.J.; Payne, J.A.; Puskarjov, M.; Voipio, J. Cation-chloride cotransporters in neuronal development, plasticity and disease. *Nat. Rev. Neurosci.* **2014**, *15*, 637–654. [CrossRef] [PubMed]

64. Huberfeld, G.; Wittner, L.; Clemenceau, S.; Baulac, M.; Kaila, K.; Miles, R.; Rivera, C. Perturbed chloride homeostasis and GABAergic signaling in human temporal lobe epilepsy. *J. Neurosci.* **2007**, *27*, 9866–9873. [CrossRef] [PubMed]

65. Huang, H.; Song, S.; Banerjee, S.; Jiang, T.; Zhang, J.; Kahle, K.T.; Sun, D.; Zhang, Z. The WNK-SPAK/OSR1 Kinases and the Cation-Chloride Cotransporters as Therapeutic Targets for Neurological Diseases. *Aging Dis.* **2019**, *10*, 626–636. [CrossRef] [PubMed]

66. Kahle, K.T.; Khanna, A.R.; Alper, S.L.; Adragna, N.C.; Lauf, P.K.; Sun, D.; Delpire, E. K-Cl cotransporters, cell volume homeostasis, and neurological disease. *Trends Mol. Med.* **2015**, *21*, 513–523. [CrossRef]

67. Alessi, D.R.; Zhang, J.; Khanna, A.; Hochdorfer, T.; Shang, Y.; Kahle, K.T. The WNK-SPAK/OSR1 pathway: Master regulator of cation-chloride cotransporters. *Sci. Signal.* **2014**, *7*, re3. [CrossRef]

68. Kahle, K.T.; Khanna, A.R.; Duan, J.; Staley, K.J.; Delpire, E.; Poduri, A. The KCC2 Cotransporter and Human Epilepsy: Getting Excited About Inhibition. *Neuroscientist* **2016**, *22*, 555–562. [CrossRef]

69. Hartmann, A.-M.; Nothwang, H.G. Molecular and evolutionary insights into the structural organization of cation chloride cotransporters. *Front. Cell. Neurosci.* **2015**, *8*, 470. [CrossRef]

70. Zhang, J.; Bhuiyan, M.I.H.; Zhang, T.; Karimy, J.K.; Wu, Z.; Fiesler, V.M.; Zhang, J.; Huang, H.; Hasan, M.N.; Skrzypiec, A.E.; et al. Modulation of brain cation-Cl(-) cotransport via the SPAK kinase inhibitor ZT-1a. *Nat. Commun.* **2020**, *11*, 78. [CrossRef]

71. Zhang, J.; Cordshagen, A.; Medina, I.; Nothwang, H.G.; Wisniewski, J.R.; Winklhofer, M.; Hartmann, A.M. Staurosporine and NEM mainly impair WNK-SPAK/OSR1 mediated phosphorylation of KCC2 and NKCC1. *PLoS ONE* **2020**, *15*, e0232967. [CrossRef]

72. De Los Heros, P.; Alessi, D.R.; Gourlay, R.; Campbell, D.G.; Deak, M.; Macartney, T.J.; Kahle, K.T.; Zhang, J. The WNK-regulated SPAK/OSR1 kinases directly phosphorylate and inhibit the K$^+$-Cl$^-$ co-transporters. *Biochem. J.* **2014**, *458*, 559–573. [CrossRef]

73. Brown, A.; Meor Azlan, N.F.; Wu, Z.; Zhang, J. WNK-SPAK/OSR1-NCC kinase signaling pathway as a novel target for the treatment of salt-sensitive hypertension. *Acta Pharmacol. Sin.* **2020**. [CrossRef] [PubMed]
74. AlAmri, M.A.; Kadri, H.; Alderwick, L.J.; Jeeves, M.; Mehellou, Y. The Photosensitising Clinical Agent Verteporfin Is an Inhibitor of SPAK and OSR1 Kinases. *Chembiochem* **2018**, *19*, 2072–2080. [CrossRef] [PubMed]
75. Rinehart, J.; Vazquez, N.; Kahle, K.T.; Hodson, C.A.; Ring, A.M.; Gulcicek, E.E.; Louvi, A.; Bobadilla, N.A.; Gamba, G.; Lifton, R.P. WNK2 kinase is a novel regulator of essential neuronal cation-chloride cotransporters. *J. Biol. Chem.* **2011**, *286*, 30171–30180. [CrossRef] [PubMed]
76. Haas, M.; McBrayer, D.; Lytle, C. [Cl−] i-dependent phosphorylation of the Na-K-Cl cotransport protein of dog tracheal epithelial cells. *J. Biol. Chem.* **1995**, *270*, 28955–28961. [CrossRef] [PubMed]
77. Lytle, C.; Forbush 3rd, B. Regulatory phosphorylation of the secretory Na-K-Cl cotransporter: Modulation by cytoplasmic Cl. *Am. J. Physiol. Cell Physiol.* **1996**, *270*, C437–C448. [CrossRef]
78. Cossins, A.; Weaver, Y.; Lykkeboe, G.; Nielsen, O. Role of protein phosphorylation in control of K flux pathways of trout red blood cells. *Am. J. Physiol. Cell Physiol.* **1994**, *267*, C1641–C1650. [CrossRef]
79. Flatman, P.W.; Adragna, N.C.; Lauf, P.K. Role of protein kinases in regulating sheep erythrocyte K-Cl cotransport. *Am. J. Physiol. Cell Physiol.* **1996**, *271*, C255–C263. [CrossRef]
80. Jennings, M.L.; Schulz, R.K. Okadaic acid inhibition of KCl cotransport. Evidence that protein dephosphorylation is necessary for activation of transport by either cell swelling or N-ethylmaleimide. *J. Gen. Physiol.* **1991**, *97*, 799–817. [CrossRef]
81. McCormick, J.A.; Ellison, D.H. The WNKs: Atypical protein kinases with pleiotropic actions. *Physiol. Rev.* **2011**, *91*, 177–219. [CrossRef]
82. Arroyo, J.P.; Kahle, K.T.; Gamba, G. The SLC12 family of electroneutral cation-coupled chloride cotransporters. *Mol. Asp. Med.* **2013**, *34*, 288–298. [CrossRef]
83. Piala, A.T.; Moon, T.M.; Akella, R.; He, H.; Cobb, M.H.; Goldsmith, E.J. Chloride sensing by WNK1 involves inhibition of autophosphorylation. *Sci. Signal.* **2014**, *7*, ra41. [CrossRef] [PubMed]
84. Maruyama, J.; Kobayashi, Y.; Umeda, T.; Vandewalle, A.; Takeda, K.; Ichijo, H.; Naguro, I. Osmotic stress induces the phosphorylation of WNK4 Ser575 via the p38MAPK-MK pathway. *Sci. Rep.* **2016**, *6*, 18710. [CrossRef] [PubMed]
85. Salihu, S.; Meor Azlan, N.; Josiah, S.; Wu, Z.; Wang, Y.; Zhang, J. Role of the cation-chloride-cotransporters in the circadian system. *Asian J. Pharm. Sci.* **2020**. [CrossRef]
86. Zhang, J.; Gao, G.; Begum, G.; Wang, J.; Khanna, A.R.; Shmukler, B.E.; Daubner, G.M.; de Los Heros, P.; Davies, P.; Varghese, J.; et al. Functional kinomics establishes a critical node of volume-sensitive cation-Cl(-) cotransporter regulation in the mammalian brain. *Sci. Rep.* **2016**, *6*, 35986. [CrossRef] [PubMed]
87. Adragna, N.C.; Ravilla, N.B.; Lauf, P.K.; Begum, G.; Khanna, A.R.; Sun, D.; Kahle, K.T. Regulated phosphorylation of the K-Cl cotransporter KCC3 is a molecular switch of intracellular potassium content and cell volume homeostasis. *Front. Cell Neurosci.* **2015**, *9*, 255. [CrossRef] [PubMed]
88. Rinehart, J.; Maksimova, Y.D.; Tanis, J.E.; Stone, K.L.; Hodson, C.A.; Zhang, J.; Risinger, M.; Pan, W.; Wu, D.; Colangelo, C.M. Sites of regulated phosphorylation that control K-Cl cotransporter activity. *Cell* **2009**, *138*, 525–536. [CrossRef]
89. Thastrup, J.O.; Rafiqi, F.H.; Vitari, A.C.; Pozo-Guisado, E.; Deak, M.; Mehellou, Y.; Alessi, D.R. SPAK/OSR1 regulate NKCC1 and WNK activity: Analysis of WNK isoform interactions and activation by T-loop trans-autophosphorylation. *Biochem. J.* **2012**, *441*, 325–337. [CrossRef]
90. Markkanen, M.; Ludwig, A.; Khirug, S.; Pryazhnikov, E.; Soni, S.; Khiroug, L.; Delpire, E.; Rivera, C.; Airaksinen, M.S.; Uvarov, P. Implications of the N-terminal heterogeneity for the neuronal K-Cl cotransporter KCC2 function. *Brain Res.* **2017**, *1675*, 87–101. [CrossRef]
91. Vitari, A.C.; Thastrup, J.; Rafiqi, F.H.; Deak, M.; Morrice, N.A.; Karlsson, H.K.; Alessi, D.R. Functional interactions of the SPAK/OSR1 kinases with their upstream activator WNK1 and downstream substrate NKCC1. *Biochem. J.* **2006**, *397*, 223–231. [CrossRef]
92. Zhao, H.; Nepomuceno, R.; Gao, X.; Foley, L.M.; Wang, S.; Begum, G.; Zhu, W.; Pigott, V.M.; Falgoust, L.M.; Kahle, K.T.; et al. Deletion of the WNK3-SPAK kinase complex in mice improves radiographic and clinical outcomes in malignant cerebral edema after ischemic stroke. *J. Cereb. Blood Flow Metab.* **2017**, *37*, 550–563. [CrossRef]
93. Cuomo, O.; Vinciguerra, A.; Cerullo, P.; Anzilotti, S.; Brancaccio, P.; Bilo, L.; Scorziello, A.; Molinaro, P.; Di Renzo, G.; Pignataro, G. Ionic homeostasis in brain conditioning. *Front. Neurosci.* **2015**, *9*, 277. [CrossRef] [PubMed]
94. Tao, D.; Liu, F.; Sun, X.; Qu, H.; Zhao, S.; Zhou, Z.; Xiao, T.; Zhao, C.; Zhao, M. Bumetanide: A review of its neuroplasticity and behavioral effects after stroke. *Restor. Neurol. Neurosci.* **2019**, *37*, 397–407. [CrossRef] [PubMed]
95. Jaenisch, N.; Witte, O.W.; Frahm, C. Downregulation of potassium chloride cotransporter KCC2 after transient focal cerebral ischemia. *Stroke* **2010**, *41*, e151–e159. [CrossRef] [PubMed]
96. Wang, T.; Kumada, T.; Morishima, T.; Iwata, S.; Kaneko, T.; Yanagawa, Y.; Yoshida, S.; Fukuda, A. Accumulation of GABAergic neurons, causing a focal ambient GABA gradient, and downregulation of KCC2 are induced during microgyrus formation in a mouse model of polymicrogyria. *Cereb. Cortex* **2014**, *24*, 1088–1101. [CrossRef]
97. Yan, Y.; Dempsey, R.J.; Flemmer, A.; Forbush, B.; Sun, D. Inhibition of Na$^+$–K$^+$–Cl$^-$ cotransporter during focal cerebral ischemia decreases edema and neuronal damage. *Brain Res.* **2003**, *961*, 22–31. [CrossRef]

98. Yan, Y.; Dempsey, R.J.; Sun, D. Na$^+$-K$^+$-Cl$^-$ cotransporter in rat focal cerebral ischemia. *J. Cereb. Blood Flow Metab.* **2001**, *21*, 711–721. [CrossRef]

99. Chen, X.; Kintner, D.B.; Luo, J.; Baba, A.; Matsuda, T.; Sun, D. Endoplasmic reticulum Ca^{2+} dysregulation and endoplasmic reticulum stress following in vitro neuronal ischemia: Role of Na$^+$-K$^+$-Cl$^-$ cotransporter. *J. Neurochem.* **2008**, *106*, 1563–1576. [CrossRef]

100. Chen, H.; Sun, D. The role of Na–K–Cl co–transporter in cerebral ischemia. *Neurol. Res.* **2005**, *27*, 280–286. [CrossRef]

101. Wu, D.; Zhang, G.; Zhao, C.; Yang, Y.; Miao, Z.; Xu, X. Interleukin-18 from neurons and microglia mediates depressive behaviors in mice with post-stroke depression. *Brain Behav. Immun.* **2020**, *88*, 411–420. [CrossRef]

102. Huang, L.-Q.; Zhu, G.-F.; Deng, Y.-Y.; Jiang, W.-Q.; Fang, M.; Chen, C.-B.; Cao, W.; Wen, M.-Y.; Han, Y.-L.; Zeng, H.-K. Hypertonic saline alleviates cerebral edema by inhibiting microglia-derived TNF-α and IL-1β-induced Na-K-Cl Cotransporter up-regulation. *J. Neuroinflamm.* **2014**, *11*, 1–20. [CrossRef]

103. Su, G.; Kintner, D.B.; Flagella, M.; Shull, G.E.; Sun, D. Astrocytes from Na$^+$-K$^+$-Cl$^-$ cotransporter-null mice exhibit absence of swelling and decrease in EAA release. *Am. J. Physiol. Cell Physiol.* **2002**, *282*, C1147–C1160. [CrossRef] [PubMed]

104. Chen, H.; Luo, J.; Kintner, D.B.; Shull, G.E.; Sun, D. Na$^+$-dependent chloride transporter (NKCC1)-null mice exhibit less gray and white matter damage after focal cerebral ischemia. *J. Cereb. Blood Flow Metab.* **2005**, *25*, 54–66. [CrossRef] [PubMed]

105. Lee, H.A.; Hong, S.H.; Kim, J.W.; Jang, I.S. Possible involvement of DNA methylation in NKCC1 gene expression during postnatal development and in response to ischemia. *J. Neurochem.* **2010**, *114*, 520–529. [CrossRef] [PubMed]

106. Bhuiyan, M.I.H.; Song, S.; Yuan, H.; Begum, G.; Kofler, J.; Kahle, K.T.; Yang, S.-S.; Lin, S.-H.; Alper, S.L.; Subramanya, A.R. WNK-Cab39-NKCC1 signaling increases the susceptibility to ischemic brain damage in hypertensive rats. *J. Cereb. Blood Flow Metab.* **2017**, *37*, 2780–2794. [CrossRef] [PubMed]

107. Hertz, L.; Xu, J.; Chen, Y.; Gibbs, M.E.; Du, T. Antagonists of the Vasopressin V1 Receptor and of the β1-Adrenoceptor Inhibit Cytotoxic Brain Edema in Stroke by Effects on Astrocytes-but the Mechanisms Differ. *Curr. Neuropharmacol.* **2014**, *12*, 308–323. [CrossRef] [PubMed]

108. Lu, K.-T.; Huang, T.-C.; Wang, J.-Y.; You, Y.-S.; Chou, J.-L.; Chan, M.W.; Wo, P.Y.; Amstislavskaya, T.G.; Tikhonova, M.A.; Yang, Y.-L. NKCC1 mediates traumatic brain injury-induced hippocampal neurogenesis through CREB phosphorylation and HIF-1α expression. *Pflügers Arch. Eur. J. Physiol.* **2015**, *467*, 1651–1661. [CrossRef] [PubMed]

109. Fu, P.; Tang, R.; Yu, Z.; Huang, S.; Xie, M.; Luo, X.; Wang, W. Bumetanide-induced NKCC1 inhibition attenuates oxygen–glucose deprivation-induced decrease in proliferative activity and cell cycle progression arrest in cultured OPCs via p-38 MAPKs. *Brain Res.* **2015**, *1613*, 110–119. [CrossRef] [PubMed]

110. Yu, Y.; Fu, P.; Yu, Z.; Xie, M.; Wang, W.; Luo, X. NKCC1 inhibition attenuates chronic cerebral hypoperfusion-induced white matter lesions by enhancing progenitor cells of oligodendrocyte proliferation. *J. Mol. Neurosci.* **2018**, *64*, 449–458. [CrossRef]

111. Mu, X.; Wang, H.; Cheng, X.; Yang, L.; Sun, X.; Qu, H.; Zhao, S.; Zhou, Z.; Liu, T.; Xiao, T. Inhibition of Nkcc1 promotes axonal growth and motor recovery in ischemic rats. *Neuroscience* **2017**, *365*, 83–93. [CrossRef] [PubMed]

112. Yuen, N.Y.; Chechneva, O.V.; Chen, Y.-J.; Tsai, Y.-C.; Little, L.K.; Dang, J.; Tancredi, D.J.; Conston, J.; Anderson, S.E.; O'Donnell, M.E. Exacerbated brain edema in a rat streptozotocin model of hyperglycemic ischemic stroke: Evidence for involvement of blood–brain barrier Na–K–Cl cotransport and Na/H exchange. *J. Cereb. Blood Flow Metab.* **2019**, *39*, 1678–1692. [CrossRef]

113. Kahle, K.T.; Rinehart, J.; Lifton, R.P. Phosphoregulation of the Na–K–2Cl and K–Cl cotransporters by the WNK kinases. *Biochim. Biophys. Acta (BBA) Mol. Basis Dis.* **2010**, *1802*, 1150–1158. [CrossRef] [PubMed]

114. Lucas, O.; Hilaire, C.; Delpire, E.; Scamps, F. KCC3-dependent chloride extrusion in adult sensory neurons. *Mol. Cell Neurosci.* **2012**, *50*, 211–220. [CrossRef] [PubMed]

115. Byun, N.; Delpire, E. Axonal and periaxonal swelling precede peripheral neurodegeneration in KCC3 knockout mice. *Neurobiol. Dis.* **2007**, *28*, 39–51. [CrossRef] [PubMed]

116. Auer, R.N.; Laganiere, J.L.; Robitaille, Y.O.; Richardson, J.; Dion, P.A.; Rouleau, G.A.; Shekarabi, M. KCC3 axonopathy: Neuropathological features in the central and peripheral nervous system. *Mod. Pathol.* **2016**, *29*, 962–976. [CrossRef] [PubMed]

117. Dupré, N.; Howard, H.C.; Rouleau, G.A. Hereditary motor and sensory neuropathy with agenesis of the corpus callosum. In *GeneReviews [Internet]*; University of Washington: Seattle, WA, USA, 2014.

118. Uyanik, G.; Elcioglu, N.; Penzien, J.; Gross, C.; Yilmaz, Y.; Olmez, A.; Demir, E.; Wahl, D.; Scheglmann, K.; Winner, B.; et al. Novel truncating and missense mutations of the KCC3 gene associated with Andermann syndrome. *Neurology* **2006**, *66*, 1044–1048. [CrossRef]

119. Delpire, E.; Kahle, K.T. The KCC3 cotransporter as a therapeutic target for peripheral neuropathy. *Expert. Opin. Ther. Targets* **2017**, *21*, 113–116. [CrossRef]

120. Boettger, T.; Rust, M.B.; Maier, H.; Seidenbecher, T.; Schweizer, M.; Keating, D.J.; Faulhaber, J.; Ehmke, H.; Pfeffer, C.; Scheel, O.; et al. Loss of K-Cl co-transporter KCC3 causes deafness, neurodegeneration and reduced seizure threshold. *EMBO J.* **2003**, *22*, 5422–5434. [CrossRef]

121. Howard, H.C.; Mount, D.B.; Rochefort, D.; Byun, N.; Dupre, N.; Lu, J.; Fan, X.; Song, L.; Riviere, J.B.; Prevost, C.; et al. The K-Cl cotransporter KCC3 is mutant in a severe peripheral neuropathy associated with agenesis of the corpus callosum. *Nat. Genet.* **2002**, *32*, 384–392. [CrossRef]

122. Kahle, K.T.; Flores, B.; Bharucha-Goebel, D.; Zhang, J.; Donkervoort, S.; Hegde, M.; Hussain, G.; Duran, D.; Liang, B.; Sun, D.; et al. Peripheral motor neuropathy is associated with defective kinase regulation of the KCC3 cotransporter. *Sci. Signal.* **2016**, *9*, ra77. [CrossRef]

123. Inoue, K.; Furukawa, T.; Kumada, T.; Yamada, J.; Wang, T.; Inoue, R.; Fukuda, A. Taurine inhibits K$^+$-Cl$^-$ cotransporter KCC2 to regulate embryonic Cl$^-$ homeostasis via with-no-lysine (WNK) protein kinase signaling pathway. *J. Biol. Chem.* **2012**, *287*, 20839–20850. [CrossRef]

124. Huang, H.; Bhuiyan, M.I.H.; Jiang, T.; Song, S.; Shankar, S.; Taheri, T.; Li, E.; Schreppel, P.; Hintersteininger, M.; Yang, S.-S. A Novel Na$^+$-K$^+$-Cl$^-$ Cotransporter 1 Inhibitor STS66* Reduces Brain Damage in Mice After Ischemic Stroke. *Stroke* **2019**, *50*, 1021–1025. [CrossRef] [PubMed]

125. Watanabe, M.; Zhang, J.; Mansuri, M.S.; Duan, J.; Karimy, J.K.; Delpire, E.; Alper, S.L.; Lifton, R.P.; Fukuda, A.; Kahle, K.T. Developmentally regulated KCC2 phosphorylation is essential for dynamic GABA-mediated inhibition and survival. *Sci. Signal.* **2019**, *12*, eaaw9315. [CrossRef] [PubMed]

126. Mccarthy, M.M. Estradiol and the developing brain. *Physiol. Rev.* **2008**, *88*, 91–134. [CrossRef] [PubMed]

127. Nugent, B.M.; Valenzuela, C.V.; Simons, T.J.; McCarthy, M.M. Kinases SPAK and OSR1 are upregulated by estradiol and activate NKCC1 in the developing hypothalamus. *J. Neurosci.* **2012**, *32*, 593–598. [CrossRef] [PubMed]

128. Zheng, J.; Zhang, P.; Li, X.; Lei, S.; Li, W.; He, X.; Zhang, J.; Wang, N.; Qi, C.; Chen, X. Post-stroke estradiol treatment enhances neurogenesis in the subventricular zone of rats after permanent focal cerebral ischemia. *Neuroscience* **2013**, *231*, 82–90. [CrossRef]

129. Hannemann, A.; Flatman, P.W. Phosphorylation and transport in the Na-K-2Cl cotransporters, NKCC1 and NKCC2A, compared in HEK-293 cells. *PLoS ONE* **2011**, *6*, e17992. [CrossRef]

130. Zhang, J.; Siew, K.; Macartney, T.; O'Shaughnessy, K.M.; Alessi, D.R. Critical role of the SPAK protein kinase CCT domain in controlling blood pressure. *Hum. Mol. Genet.* **2015**, *24*, 4545–4558. [CrossRef]

131. Shekarabi, M.; Lafreniere, R.G.; Gaudet, R.; Laganiere, J.; Marcinkiewicz, M.M.; Dion, P.A.; Rouleau, G.A. Comparative analysis of the expression profile of Wnk1 and Wnk1/Hsn2 splice variants in developing and adult mouse tissues. *PLoS ONE* **2013**, *8*, e57807. [CrossRef]

132. Shekarabi, M.; Salin-Cantegrel, A.; Laganiere, J.; Gaudet, R.; Dion, P.; Rouleau, G.A. Cellular expression of the K$^+$-Cl$^-$ cotransporter KCC3 in the central nervous system of mouse. *Brain Res.* **2011**, *1374*, 15–26. [CrossRef]

133. De Los Heros, P.; Kahle, K.T.; Rinehart, J.; Bobadilla, N.A.; Vazquez, N.; San Cristobal, P.; Mount, D.B.; Lifton, R.P.; Hebert, S.C.; Gamba, G. WNK3 bypasses the tonicity requirement for K-Cl cotransporter activation via a phosphatase-dependent pathway. *Proc. Natl. Acad. Sci. USA* **2006**, *103*, 1976–1981. [CrossRef]

134. Kahle, K.T.; Rinehart, J.; Ring, A.; Gimenez, I.; Gamba, G.; Hebert, S.C.; Lifton, R.P. WNK protein kinases modulate cellular Cl$^-$ flux by altering the phosphorylation state of the Na-K-Cl and K-Cl cotransporters. *Physiology* **2006**, *21*, 326–335. [CrossRef] [PubMed]

135. Kahle, K.T.; Rinehart, J.; de Los Heros, P.; Louvi, A.; Meade, P.; Vazquez, N.; Hebert, S.C.; Gamba, G.; Gimenez, I.; Lifton, R.P. WNK3 modulates transport of Cl$^-$ in and out of cells: Implications for control of cell volume and neuronal excitability. *Proc. Natl. Acad. Sci. USA* **2005**, *102*, 16783–16788. [CrossRef] [PubMed]

136. Gagnon, K.B.; Delpire, E. Multiple pathways for protein phosphatase 1 (PP1) regulation of Na-K-2Cl cotransporter (NKCC1) function: The N-terminal tail of the Na-K-2Cl cotransporter serves as a regulatory scaffold for Ste20-related proline/alanine-rich kinase (SPAK) AND PP1. *J. Biol. Chem.* **2010**, *285*, 14115–14121. [CrossRef] [PubMed]

137. Dowd, B.F.; Forbush, B. PASK (proline-alanine-rich STE20-related kinase), a regulatory kinase of the Na-K-Cl cotransporter (NKCC1). *J. Biol. Chem.* **2003**, *278*, 27347–27353. [CrossRef] [PubMed]

138. Melo, Z.; de los Heros, P.; Cruz-Rangel, S.; Vazquez, N.; Bobadilla, N.A.; Pasantes-Morales, H.; Alessi, D.R.; Mercado, A.; Gamba, G. N-terminal serine dephosphorylation is required for KCC3 cotransporter full activation by cell swelling. *J. Biol. Chem.* **2013**, *288*, 31468–31476. [CrossRef] [PubMed]

139. Rama Rao, K.V.; Verkman, A.S.; Curtis, K.M.; Norenberg, M.D. Aquaporin-4 deletion in mice reduces encephalopathy and brain edema in experimental acute liver failure. *Neurobiol. Dis.* **2014**, *63*, 222–228. [CrossRef]

140. Zhang, Y.J.; Zheng, H.Q.; Chen, B.Y.; Sun, L.; Ma, M.M.; Wang, G.L.; Guan, Y.Y. WNK1 is required for proliferation induced by hypotonic challenge in rat vascular smooth muscle cells. *Acta Pharmacol. Sin.* **2018**, *39*, 35–47. [CrossRef]

141. Xu, W.; Mu, X.; Wang, H.; Song, C.; Ma, W.; Jolkkonen, J.; Zhao, C. Chloride co-transporter NKCC1 inhibitor bumetanide enhances neurogenesis and behavioral recovery in rats after experimental stroke. *Mol. Neurobiol.* **2017**, *54*, 2406–2414. [CrossRef]

142. Shulga, A.; Magalhães, A.C.; Autio, H.; Plantman, S.; di Lieto, A.; Nykjær, A.; Carlstedt, T.; Risling, M.; Arumäe, U.; Castrén, E. The loop diuretic bumetanide blocks posttraumatic p75NTR upregulation and rescues injured neurons. *J. Neurosci.* **2012**, *32*, 1757–1770. [CrossRef]

143. Yamada, K.; Park, H.M.; Rigel, D.F.; DiPetrillo, K.; Whalen, E.J.; Anisowicz, A.; Beil, M.; Berstler, J.; Brocklehurst, C.E.; Burdick, D.A.; et al. Small-molecule WNK inhibition regulates cardiovascular and renal function. *Nat. Chem. Biol.* **2016**, *12*, 896–898. [CrossRef]

144. Apsel, B.; Blair, J.A.; Gonzalez, B.; Nazif, T.M.; Feldman, M.E.; Aizenstein, B.; Hoffman, R.; Williams, R.L.; Shokat, K.M.; Knight, Z.A. Targeted polypharmacology: Discovery of dual inhibitors of tyrosine and phosphoinositide kinases. *Nat. Chem. Biol.* **2008**, *4*, 691–699. [CrossRef] [PubMed]

145. Kikuchi, E.; Mori, T.; Zeniya, M.; Isobe, K.; Ishigami-Yuasa, M.; Fujii, S.; Kagechika, H.; Ishihara, T.; Mizushima, T.; Sasaki, S.; et al. Discovery of Novel SPAK Inhibitors That Block WNK Kinase Signaling to Cation Chloride Transporters. *J. Am. Soc. Nephrol.* **2015**, *26*, 1525–1536. [CrossRef] [PubMed]

146. AlAmri, M.A.; Kadri, H.; Alderwick, L.J.; Simpkins, N.S.; Mehellou, Y. Rafoxanide and Closantel Inhibit SPAK and OSR1 Kinases by Binding to a Highly Conserved Allosteric Site on Their C-terminal Domains. *ChemMedChem* **2017**, *12*, 639–645. [CrossRef] [PubMed]

147. Mori, T.; Kikuchi, E.; Watanabe, Y.; Fujii, S.; Ishigami-Yuasa, M.; Kagechika, H.; Sohara, E.; Rai, T.; Sasaki, S.; Uchida, S. Chemical library screening for WNK signalling inhibitors using fluorescence correlation spectroscopy. *Biochem. J.* **2013**, *455*, 339–345. [CrossRef]

148. Kadri, H.; Alamri, M.A.; Navratilova, I.H.; Alderwick, L.J.; Simpkins, N.S.; Mehellou, Y. Towards the Development of Small-Molecule MO25 Binders as Potential Indirect SPAK/OSR1 Kinase Inhibitors. *Chembiochem* **2017**, *18*, 460–465. [CrossRef]

149. Fujii, S.; Kikuchi, E.; Watanabe, Y.; Suzuyama, H.; Ishigami-Yuasa, M.; Mori, T.; Isobe, K.; Uchida, S.; Kagechika, H. Structural development of N-(4-phenoxyphenyl) benzamide derivatives as novel SPAK inhibitors blocking WNK kinase signaling. *Bioorganic Med. Chem. Lett.* **2020**, *30*, 127408. [CrossRef]

150. Zhang, J.; Pu, H.; Zhang, H.; Wei, Z.; Jiang, X.; Xu, M.; Zhang, L.; Zhang, W.; Liu, J.; Meng, H. Inhibition of Na^+-K^+-$2Cl^-$ cotransporter attenuates blood-brain-barrier disruption in a mouse model of traumatic brain injury. *Neurochem. Int.* **2017**, *111*, 23–31. [CrossRef]

151. Kahle, K.T.; Simard, J.M.; Staley, K.J.; Nahed, B.V.; Jones, P.S.; Sun, D. Molecular mechanisms of ischemic cerebral edema: Role of electroneutral ion transport. *Physiology* **2009**, *24*, 257–265. [CrossRef]

152. Payne, J.A.; Rivera, C.; Voipio, J.; Kaila, K. Cation–chloride co-transporters in neuronal communication, development and trauma. *Trends Neurosci.* **2003**, *26*, 199–206. [CrossRef]

153. Hamidi, S.; Avoli, M. KCC2 function modulates in vitro ictogenesis. *Neurobiol. Dis.* **2015**, *79*, 51–58. [CrossRef]

154. O'donnell, M.E.; Tran, L.; Lam, T.I.; Liu, X.B.; Anderson, S.E. Bumetanide inhibition of the blood-brain barrier Na-K-Cl cotransporter reduces edema formation in the rat middle cerebral artery occlusion model of stroke. *J. Cereb. Blood Flow Metab.* **2004**, *24*, 1046–1056. [CrossRef] [PubMed]

155. Singh Jaggi, A.; Kaur, A.; Bali, A.; Singh, N. Expanding spectrum of sodium potassium chloride co-transporters in the pathophysiology of diseases. *Curr. Neuropharmacol.* **2015**, *13*, 369–388. [CrossRef] [PubMed]

156. Simard, J.M.; Kahle, K.T.; Gerzanich, V. Molecular mechanisms of microvascular failure in central nervous system injury—synergistic roles of NKCC1 and SUR1/TRPM4: A review. *J. Neurosurg.* **2010**, *113*, 622–629. [CrossRef]

157. Hu, J.-J.; Yang, X.-L.; Luo, W.-D.; Han, S.; Yin, J.; Liu, W.-H.; He, X.-H.; Peng, B.-W. Bumetanide reduce the seizure susceptibility induced by pentylenetetrazol via inhibition of aberrant hippocampal neurogenesis in neonatal rats after hypoxia-ischemia. *Brain Res. Bull.* **2017**, *130*, 188–199. [CrossRef] [PubMed]

158. Glykys, J.; Dzhala, V.; Egawa, K.; Kahle, K.T.; Delpire, E.; Staley, K. Chloride dysregulation, seizures, and cerebral edema: A relationship with therapeutic potential. *Trends Neurosci.* **2017**, *40*, 276–294. [CrossRef] [PubMed]

159. Grandgeorge, M.; Lemonnier, E.; Degrez, C.; Jallot, N. The effect of bumetanide treatment on the sensory behaviours of a young girl with Asperger syndrome. *Case Rep.* **2014**, *2014*, bcr2013202092. [CrossRef]

160. Lemonnier, E.; Lazartigues, A.; Ben-Ari, Y. Treating schizophrenia with the diuretic bumetanide: A case report. *Clin. Neuropharmacol.* **2016**, *39*, 115–117. [CrossRef]

161. Lemonnier, E.; Robin, G.; Degrez, C.; Tyzio, R.; Grandgeorge, M.; Ben-Ari, Y. Treating F ragile X syndrome with the diuretic bumetanide: A case report. *Acta Paediatr.* **2013**, *102*, e288–e290. [CrossRef] [PubMed]

162. Damier, P.; Hammond, C.; Ben-Ari, Y. Bumetanide to treat Parkinson disease: A report of 4 cases. *Clin. Neuropharmacol.* **2016**, *39*, 57–59. [CrossRef]

163. Eftekhari, S.; Mehvari Habibabadi, J.; Najafi Ziarani, M.; Hashemi Fesharaki, S.S.; Gharakhani, M.; Mostafavi, H.; Joghataei, M.T.; Beladimoghadam, N.; Rahimian, E.; Hadjighassem, M.R. Bumetanide reduces seizure frequency in patients with temporal lobe epilepsy. *Epilepsia* **2013**, *54*, e9–e12. [CrossRef]

164. Rahmanzadeh, R.; Eftekhari, S.; Shahbazi, A.; Khodaei Ardakani, M.-r.; Rahmanzade, R.; Mehrabi, S.; Barati, M.; Joghataei, M.T. Effect of bumetanide, a selective NKCC1 inhibitor, on hallucinations of schizophrenic patients; a double-blind randomized clinical trial. *Schizophr Res.* **2017**, *184*, 145–146. [CrossRef] [PubMed]

165. Pressler, R.M.; Boylan, G.B.; Marlow, N.; Blennow, M.; Chiron, C.; Cross, J.H.; de Vries, L.S.; Hallberg, B.; Hellström-Westas, L.; Jullien, V. Bumetanide for the treatment of seizures in newborn babies with hypoxic ischaemic encephalopathy (NEMO): An open-label, dose finding, and feasibility phase 1/2 trial. *Lancet Neurol.* **2015**, *14*, 469–477. [CrossRef]

166. Kahle, K.T.; Barnett, S.M.; Sassower, K.C.; Staley, K.J. Decreased seizure activity in a human neonate treated with bumetanide, an inhibitor of the Na^+-K^+-$2Cl^-$ cotransporter NKCC1. *J. Child Neurol.* **2009**, *24*, 572–576. [CrossRef]

167. Lemonnier, E.; Ben-Ari, Y. The diuretic bumetanide decreases autistic behaviour in five infants treated during 3 months with no side effects. *Acta Paediatr.* **2010**, *99*, 1885–1888. [CrossRef] [PubMed]

168. Lemonnier, É.; Degrez, C.; Phelep, M.; Tyzio, R.; Josse, F.; Grandgeorge, M.; Hadjikhani, N.; Ben-Ari, Y. A randomised controlled trial of bumetanide in the treatment of autism in children. *Transl. Psychiatry* **2012**, *2*, e202. [CrossRef] [PubMed]

169. Lemonnier, E.; Villeneuve, N.; Sonie, S.; Serret, S.; Rosier, A.; Roue, M.; Brosset, P.; Viellard, M.; Bernoux, D.; Rondeau, S. Effects of bumetanide on neurobehavioral function in children and adolescents with autism spectrum disorders. *Transl. Psychiatry* **2017**, *7*, e1056. [CrossRef]
170. Savardi, A.; Borgogno, M.; Narducci, R.; La Sala, G.; Ortega, J.A.; Summa, M.; Armirotti, A.; Bertorelli, R.; Contestabile, A.; De Vivo, M.; et al. Discovery of a Small Molecule Drug Candidate for Selective NKCC1 Inhibition in Brain Disorders. *Chem* **2020**, *6*, 2073–2096. [CrossRef]
171. Delpire, E.; Weaver, C.D. Challenges of Finding Novel Drugs Targeting the K-Cl Cotransporter. *ACS Chem. Neurosci.* **2016**, *7*, 1624–1627. [CrossRef]
172. Flores, B.; Schornak, C.C.; Delpire, E. A role for KCC3 in maintaining cell volume of peripheral nerve fibers. *Neurochem. Int.* **2019**, *123*, 114–124. [CrossRef]
173. Delpire, E.; Days, E.; Lewis, L.M.; Mi, D.; Kim, K.; Lindsley, C.W.; Weaver, C.D. Small-molecule screen identifies inhibitors of the neuronal K-Cl cotransporter KCC2. *Proc. Natl. Acad. Sci. USA* **2009**, *106*, 5383–5388. [CrossRef]
174. Delpire, E.; Baranczak, A.; Waterson, A.G.; Kim, K.; Kett, N.; Morrison, R.D.; Daniels, J.S.; Weaver, C.D.; Lindsley, C.W. Further optimization of the K-Cl cotransporter KCC2 antagonist ML077: Development of a highly selective and more potent in vitro probe. *Bioorg. Med. Chem. Lett.* **2012**, *22*, 4532–4535. [CrossRef] [PubMed]

International Journal of
Molecular Sciences

Review

Adiponectin: The Potential Regulator and Therapeutic Target of Obesity and Alzheimer's Disease

Jong Youl Kim [1,†], Sumit Barua [1,†], Ye Jun Jeong [1] and Jong Eun Lee [1,2,*]

[1] Department of Anatomy, Yonsei University College of Medicine, Seoul 120-752, Korea; jongyoul74@gmail.com (J.Y.K.); drsbarua@gmail.com (S.B.); jyj4453@naver.com (Y.J.J.)

[2] BK21 Plus Project for Medical Sciences, and Brain Research Institute, Yonsei University College of Medicine, Seoul 120-752, Korea

* Correspondence: jelee@yuhs.ac; Tel.: +82-2-2228-1646 (ext. 1659); Fax: +82-2-365-0700

† co-first authors.

Received: 18 June 2020; Accepted: 28 August 2020; Published: 3 September 2020

Abstract: Animal and human mechanistic studies have consistently shown an association between obesity and Alzheimer's disease (AD). AD, a degenerative brain disease, is the most common cause of dementia and is characterized by the presence of extracellular amyloid beta (Aβ) plaques and intracellular neurofibrillary tangles disposition. Some studies have recently demonstrated that Aβ and tau cannot fully explain the pathophysiological development of AD and that metabolic disease factors, such as insulin, adiponectin, and antioxidants, are important for the sporadic onset of nongenetic AD. Obesity prevention and treatment can be an efficacious and safe approach to AD prevention. Adiponectin is a benign adipokine that sensitizes the insulin receptor signaling pathway and suppresses inflammation. It has been shown to be inversely correlated with adipose tissue dysfunction and may enhance the risk of AD because a range of neuroprotection adiponectin mechanisms is related to AD pathology alleviation. In this study, we summarize the recent progress that addresses the beneficial effects and potential mechanisms of adiponectin in AD. Furthermore, we review recent studies on the diverse medications of adiponectin that could possibly be related to AD treatment, with a focus on their association with adiponectin. A better understanding of the neuroprotection roles of adiponectin will help clarify the precise underlying mechanism of AD development and progression.

Keywords: Alzheimer's disease; metabolic disease; adiponectin; insulin; antioxidants

---

1. Introduction

The global prevalence of obesity has increased at an alarming rate over the years. In 2016, it was estimated that nearly 39% of adults aged ≥18 years were overweight worldwide, and 13% were obese [WHO, 2016]. Obesity is generally defined as a body mass index of >30 kg/m^2 and is mainly caused by physical inactivity and westernized dietary habits. Obesity is a major concern because it is a risk factor for a plethora of metabolic diseases that increase mortality rates. Since obesity causes insulin resistance, it is one of the major risk factors for type 2 diabetes (T2DM). After years of obesity-associated hyperinsulinemia, the insulin secretory function in the pancreas could falter and eventually lead to hyperglycemia. A previous study has suggested that obesity and diabetes are linked to Alzheimer's disease (AD) [1]. The pooled effect size for AD in relation to obesity and diabetes was calculated at 1.59 and 1.54 in longitudinal epidemiological studies of body mass, metabolic syndrome (dyslipidemia, hypertension, abdominal obesity, and insulin resistance), diabetes, and glucose and insulin levels [2]. A clinical study has also shown that an increased number of metabolic and vascular risk factors in midlife is critical for amyloid deposition and could lead to the risk of developing AD during old age [3].

Recently, the World Alzheimer Report estimated that the total number of people with AD and dementia is set to triple to 132–152 million cases worldwide by 2050 (World Alzheimer Report 2015). About 60%–70% of dementia cases are caused by AD, a chronic neurodegenerative disease. These dementia cases are often associated with extracellular amyloid beta (Aβ) plaques and intraneuronal deposits of neurofibrillary tangles (NFTs) in the brain. Hyperphosphorylation of tau protein leads to the formation of NFTs, whereas the accumulation of Aβ forms hard, insoluble plaques (Aβ peptide) [4]. For over two decades, these two proteins have been the main target for AD therapeutics; currently, available treatments for AD are symptomatic and do not decelerate or prevent the progression of the disease. However, these therapies demonstrate modest but particularly consistent benefits for cognition, global status, and functional ability. The abovementioned studies have suggested that another approach should be considered for the development of AD treatment based on the relationship between AD, obesity, and T2DM. According to the Mayo Clinic AD Patient Registry, 80% of AD patients had either diabetes or showed impairment in glucose tolerance [5]. Many studies are being conducted to decipher the underlying mechanisms responsible for the association between obesity, T2DM, and AD. In some studies that used a rodent model, a high-fat diet was shown to cause AD pathology due to the accumulation of Aβ peptides and phosphorylated tau proteins, as well as cognitive impairment [6,7].

Adiponectin secreted from the adipose tissue sensitizes the insulin receptor signaling pathway and prevents inflammation. Adiponectin is a protein that modulates a number of metabolic diseases (Figure 1), including diabetes, dyslipidemia, atherosclerosis, and comorbid metabolic dysfunction that occur in cardiovascular diseases such as hypertension. Many studies have indicated that alteration levels of adiponectin in the plasma and cerebrospinal fluid (CSF) correspond to a distinctive condition of mild cognitive impairment (MCI) and AD [8–13]. The reason for the discrepancies among these studies is unclear, but adiponectin may be considered to be a metabolic biomarker for AD.

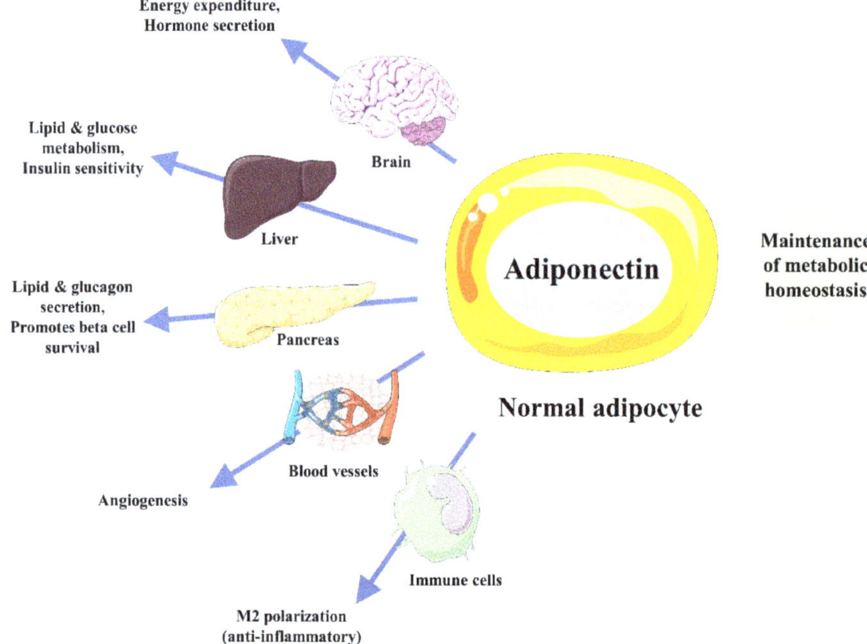

Figure 1. Major mechanisms of adiponectin's actions in the maintenance of metabolic homeostasis.

In this review, we will discuss the potential mechanisms that bridge the relationship between adiponectin and AD. Understanding these mechanisms could narrow down the process of AD's therapeutic window and also help to design novel therapeutic applications against AD and related neurodegenerative diseases.

2. Relationship between Metabolic Disorder and Alzheimer's Disease

The most common age-related neurodegenerative disease is AD, which accounts for the most predominant form (about 60–70%) of dementia, as mentioned by WHO. It is characterized by a gradual loss of learning and memory, particularly episodic memory, and may lead to death within 10 years of onset [14,15]. Limited information is known about the etiology and associated nongenetic risk factors of AD. However, some modifiable events, such as dietary habits, lifestyle, and environmental exposure (toxic chemicals), and the nonmodifiable event of aging are common global factors for AD pathology [16,17]. The prevalence of AD is higher in developed countries than in developing countries, and >40% of AD cases are known to occur in individuals aged ≥80 years.

AD was named after Alois Alzheimer, the first person to describe that extensive neuronal tangles and amyloid plaques, considered to be important hallmarks of the disease, can be found in the AD brain. However, recent studies have shown that AD patients have an insidious loss of neurons and increased reactive gliosis in different parts of the brain, comprising the cerebral cortex and the limbic system, including the amygdala and hippocampus [18]. Cellular changes in the basal ganglia, cerebellum, brain stem, and spinal cord are relatively spared in AD. According to pathological findings of affected AD brain regions, two different types of aggregates are commonly found in the intracellular and extracellular compartments—the intracellular aggregates are known as NFTs and the extracellular aggregates are known as senile plaques that consist of insoluble paired helical filaments of hyperphosphorylated tau protein [19,20]. The major component of the senile plaque is known as Aβ aggregation, which occurs before other pathological events such as NFT formation and neuronal death [21]. Thus, Aβ aggregation is considered the central player of neuronal death, which has also been observed in in-vitro neuronal cells subjected to synthetic Aβ peptide (Aβ1-42/43) treatment [22]. Over the years, the amyloid hypothesis, a process where extracellular Aβ accumulation is followed by the loss of synapses and subsequent neuronal loss, has become the popular concept of AD pathogenesis and has been used by many AD research groups [23].

It has been suggested that metabolic impairment and modification of the AD-related protein levels can contribute to a higher prevalence of AD. Dyslipidemia, hypertension, abdominal obesity, and insulin resistance are the popular hallmarks of metabolic abnormalities, which are collectively known as metabolic syndrome (MetS) [24]. Individuals with higher life expectancy (above 80 years) have a higher prevalence of MetS, which is a potential cause of neurodegenerative diseases, especially cognitive impairment and dementia. Other studies have also found that MetS is not solely responsible for AD progression in elderly people, regardless of their backgrounds. Higher MetS could bridge inappropriate regulation of the central nervous system (CNS) function (especially part of the limbic system, such as the prefrontal cortex, hippocampus, and amygdala) and metabolic regulatory responses that result in functional cognitive impairments [25]. Metabolic products that are produced in peripheral organs, such as estrogen, insulin, cortisol, and leptin, can cross the blood–brain barrier (BBB) and influence cognitive function. However, neuropeptides such as orexin (mice), allastostatin (drosophila), and neuropeptide Y (grass carp) modulate metabolic processes in different animals [26]. In a recent review, Yi et al. suggested that neuropeptide Y receptors in humans can be a promising target for metabolic disorders [27]. Furthermore, food intake and weight gain can be increased by lesions in oxytocin-containing hypothalamic nuclei [28].

Some studies on metabolic diseases have reported that metabolic diseases, such as T2DM and obesity, are associated with AD. Metabolic alteration of substances such as T2DM-related insulin receptors [29] and obesity-related adiponectin [30,31] could alter the process of aging and age-related dementia, such as in AD. In a recent study, Nasoohi et al. suggested the relationship of "type 3 diabetes" to AD pathology, in which the metabolic syndrome consisting of oxidative stress and neuroinflammation leads to brain insulin resistance [32]. This concept suggests that the thioredoxin-interacting protein (TXNIP) is a key regulator of oxidative stress and inflammasome activation, which is associated with impaired insulin function in the brain. Brain insulin resistance is found to be related to the progressive atrophy of the brain regions of early AD progression, which are cingulate cortices, medial temporal lobe, prefrontal gyri, and other regions [33]. Along with the discussion above, brain insulin resistance due to metabolic disorders such as diabetes mellitus, metabolic syndrome, and nonalcoholic fatty liver disease are considered to be a prominent component of AD pathology; the maintenance of brain insulin supply should also be a target of AD therapy for individuals with metabolic disorder [34]. Thus, metabolism and neurological disorders are closely related to each other.

3. Adiponectin as a Modulator of Metabolic Disorder

Adiponectin is secreted with the bioactive molecule leptin from normal adipose tissues (together, they are called adipokines) [35]. It regulates glucose and fatty acid metabolism by increasing insulin sensitivity of peripheral organs [36–38]. However, small amounts of adiponectin, an adipose-tissue-specific protein, can also be synthesized by other cell types. Paradoxically, the adiponectin level decreases with the increase of central adiposity, which causes obesity [39–41]. Obesity causes damage to several organ systems because it regulates MetS, which can be characterized physiologically as excess weight and pathologically as high triglyceride levels and insulin resistance [42]. Moreover, obesity is related to an increase in cognitive decline and AD [43]. Recently, our laboratory has shown that touchscreen-based behavioral testing of high-fat diet mice led to an impairment in cognitive function compared with cognitive function in normal diet mice [44]. MetS, including obesity, cardiovascular disease, type 2 diabetes, and neurodegenerative disorders, is associated with the regulation of adiponectin expression [8,45,46]. Like other hormonal proteins, adiponectin also functions through specific receptors known as adiponectin receptors. Adiponectin receptors have been classified into adiponectin receptor (AdipoR) 1, AdipoR2, and T-cadherin. Through the recruitment of adaptor protein APPL1 by AdipoRs activation, adiponectin signaling regulates a series of signaling pathways [47]. AMP-activated protein kinase (AMPK), peroxisome proliferator-activated receptor-α (PPARα), IkB kinase (IKK)/NF-κB/PTEN, IRS1/2–Akt, and Ras-ERK1/2 signaling are examples of downstream adiponectin signaling [48–52]. Adiponectin receptors are found to be expressed throughout the whole brain. Thundyil et al. suggested that in the cortical neurons, AdipoR1 expression is more prominent than AdipoR2 expression [53]. On the other hand, T-cadherin acts as the coreceptor of a unidentified receptor, through which the adiponectin can transmit the metabolic signals [54].

4. Adiponectin's Potential Role in Alzheimer's Disease

4.1. Adiponectin and Brain

Adiponectin gives a beneficial effect on synaptic regulation and memory in AD (Figure 2). It also promotes synaptic plasticity in AD by improving the hippocampal's long-term potentiation [55]. Adiponectin and its homolog, osmotin, enhance neurite outgrowth and synaptic complexity and improve learning and memory defects in mouse AD models [56–59]. Furthermore, chronic adiponectin deficiency in aged mice leads to AD-like cognitive impairments and pathologies [60]. In a recent clinical study, individuals with higher-than-normal adiponectin levels performed better in cognitive tests, indicating the protective effect of adiponectin against cognitive failure. Furthermore, the study showed that adiponectin could be used to identify the risk of cognitive dysfunction [61]. The upregulation of serum adiponectin expression has been found to be associated with MCI and AD [8]. The adiponectin receptors AdipoR1 and AdipoR2 have approximately 95% homology between human and mice. They are ubiquitously expressed and structurally related in humans and mice, with variable affinity to different isoforms and predominance in some tissues [62]. The expression of AdipoR1 and AdipoR2 is mainly localized to neurons in the hypothalamus, brainstem, and cortex [22], as well as the nucleus basalis of Meynert and the hippocampus, the two main targeted structures in AD [63]. In the hypothalamus and the brainstem, adiponectin is thought to regulate food intake and energy expenditure via AdipoR1-mediated AMPK signaling [23]. However, low levels of adiponectin in CSF may be compensated by the presence of two high-affinity receptors, AdipoR1 and AdipoR2, in the brain [64,65]. Suppression of AdipoR1 can result in metabolic diseases such as obesity and diabetes, which also potentiate spatial learning deficit, memory impairment, and AD pathologies [30]. Hence, studies have evaluated adiponectin and its receptors as therapeutic alternatives for AD.

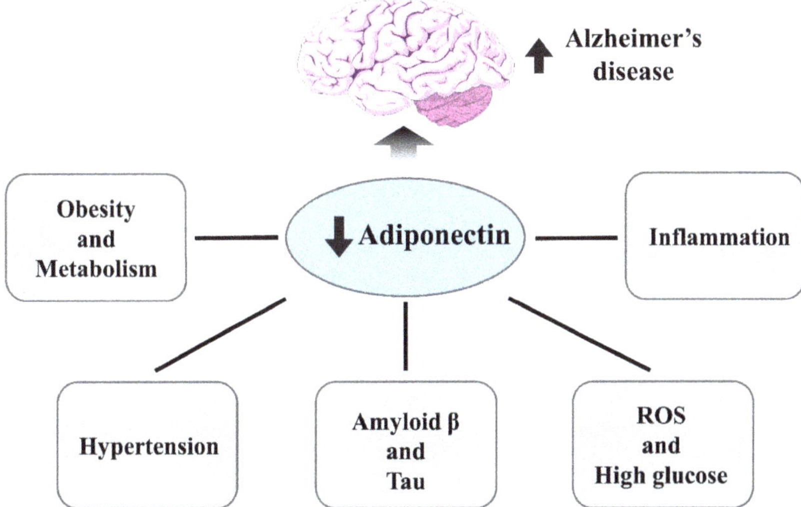

Figure 2. Downregulation of adiponectin involved in the mechanism of Alzheimer's disease exacerbates AD pathology and impairment.

In AD, Aβ has to cross BBB to be transported in the brain, where it is regulated by specific receptors and transporters [66]. Therefore, it is necessary to protect BBB disruption. Adiponectin protects BBB disruption by inhibiting apoptosis of endothelial cells, protecting tight junction integrity via the AdipoR1-mediated NF-κB pathway, and maintaining the balance of Aβ transporters in endothelial cells [67].

4.2. Adiponectin Improves Insulin Signaling

Recently, many studies have provided evidence that insulin signaling dysfunction plays a key role in cognitive decline, such as in MCI and AD [68–70]. It is well known that T2DM is independently associated with cognitive dysfunction and loss of hippocampus volume [71]. Further, insulin signaling prevents Aβ oligomer toxicity [72]. Adiponectin has been found to be beneficial for T2DM because of its ability to enhance insulin sensitivity, and it has been used in T2DM treatment [73]. A few studies have reported the relationship between increased diabetes prevalence and decreased levels of adiponectin [74]. In T2DM patients with low adiponectin, the hippocampus volume is significantly decreased [75]. Lower levels of adiponectin in T2DM have also been associated with lower gray matter volume and reduced cerebral glucose metabolism in the temporal brain regions [76]. In addition, adiponectin-deficient mice have been used as models of insulin resistance and the associated memory pathology [60]. In a rat cognitive-deficient model induced by streptozocin, which is commonly used to induce diabetes, adiponectin attenuated tau hyperphosphorylation and alleviated cognitive function by activating the PI3K/Akt/GSK-3β signaling pathway [77].

4.3. Adiponectin Regulates Glucose/Fatty Acid Metabolism

Deterioration of cerebral glucose metabolism is an important feature in age-related AD and is key to the progression of AD pathogenesis [78–81]. Adiponectin modulates glucose metabolism in hippocampal neurons by increasing glucose uptake, glycolysis, and adenosine triphosphate production rates [82]. Glucose and lactate are considered major energy sources in the brain. However, the amounts of glucose consumption and oxygen utilization in the brain are not the same [83]. In addition, lactate cannot generate energy because of the fast removal of lactate from cells and activated tissues [84]. A study conducted by Dhopeshwarka et al. suggested that there are gaps between glucose consumption and oxygen utilization by the brain and that fatty acids can enter the brain and mitochondria and can be oxidized to produce energy [85]. About 20% (maximum) of the total energy in the brain can be produced from mitochondrial oxidation of fatty acids [86]. Moreover, fatty acids are regarded as key players in the homeostasis of glucose [87]. Fatty acid metabolism has been found to be related to MCI and to adiponectin and its receptors in mice fed a high-fat diet [44,88,89]. Therefore, changes in adiponectin levels can alter the brain metabolism and progression of AD. Adiponectin also activates the AMPK and PPARα pathways through AdipoR1 and AdipoR2, respectively, which reduce hepatic lipogenesis and enhance β-oxidation [90].

4.4. Adiponectin Alleviates Inflammation

One of the key factors for cognitive decline (MCI) and AD is chronic neuroinflammation [91–93]. Amongst renowned anti-inflammatory molecules, adiponectin is considered to be an active contributor to chronic inflammation in obesity and T2DM [91,92,94,95]. Chronic inflammation, which induces AD and metabolic-distress-related pathologies such as neuronal insulin resistance, endoplasmic reticulum stress, synaptotoxicity, and neurodegeneration, is caused by the secretion of proinflammatory cytokines by microglial cell activation [92,96]. In an adiponectin-knockout mouse model, activation of proinflammatory cytokines such as interleukin (IL)-1β, IL-6, and tumor necrotic factor-α has been shown to cause the development of AD-like pathology [60]. Furthermore, adiponectin prevents neuroinflammation by decreasing microglia and regulating the brain macrophage proinflammatory phenotype [97,98]. Therefore, changes in adiponectin levels may be closely related to neuroinflammation in AD.

4.5. Adiponectin Has Protective Effect on Oxidative Stress/Hypoxia

Studies have shown that oxidative stress and hypoxia conditions render an important role in the pathogenesis of age-related neurodegenerative diseases such as AD [78,99,100]. Adiponectin alleviates oxidative stress and oxidative-stress-mediated cytotoxicity [101,102] and has a protective effect in high glucose concentrations in blood [26]. Many of these effects have been reported to occur because of upregulated AMPK signaling [101,103]. Since AMPK is considered the general energy sensor in the brain, inhibition of adiponectin may influence the AMPK pathway, which in turn could affect brain metabolism [104,105]. Similarly, in a hypoxic environment in obese individuals, hypertrophic adipocytes upregulate the expression of hypoxia-inducible factor-1α (HIF-1α) [106]. Upregulation of HIF-1α has been found to inhibit the production of adiponectin. This phenomenon has been confirmed by the expression of high levels of adiponectin mRNA in adipocyte-specific HIF-1α-deficient mice fed a high-fat diet for 7 weeks compared with control mice [107]. With the above study, adiponectin can be considered a modulator of neurocognitive disorders, which might suggest adiponectin as a potential therapeutic target for AD.

4.6. Adiponectin and Neuroprotection/Neurogenesis

Adiponectin has a neuroprotective effect in various conditions. It shows a protective effect in brain injury caused by ischemic stroke and intracerebral hemorrhage [108,109]. Neuroprotective effects of adiponectin have also been demonstrated in a kainite-induced excitotoxicity model [103].

Furthermore, adiponectin plays a role in many deleterious conditions such as Aβ deposition/tau phosphorylation, neuroinflammation, and oxidative stress by protecting neurons and glial cells. Hippocampal neurogenesis is crucial for maintaining cognitive function; however, it is impaired in AD patients [110,111]. In this respect, it seems necessary to pay attention to the neuroproliferative effect of adiponectin in the adult brain. Intracerebroventricular injection of adiponectin has shown neurogenic and proliferative effects in an adiponectin-deficient mice model [55]. An in-vitro and in-vivo study has also indicated that adiponectin stimulates neurogenesis through AdipoR1 [112].

5. Adiponectin-Associated Therapeutic Strategy against AD Induced by Metabolic Diseases

A number of in-vitro, animal, and clinical studies have been conducted to find molecular targets that prevent protein aggregation, oxidative stress, and inflammation for AD treatment. However, a therapeutic target is still unclear. Adiponectin can be considered a protein of interest in the search for new neuroprotective targets for AD. Previous studies have attempted to use adiponectin levels as an AD marker [8–10,13,113] (Table). In addition, numerous therapeutic agents that are being considered new paradigms in AD therapy have been found to be related to adiponectin signaling. These therapeutic agents do not only include adiponectin and AdipoR homologs but also conventional AD drugs, anti-insulin resistance drugs, and cardiovascular drugs.

5.1. Adiponectin as an AD Marker

In some recent studies, the relationship between plasma and CSF adiponectin levels in MCI or AD has been reported. However, there are discrepancies in their results. Some studies have shown decreased adiponectin levels in AD or MCI [113–115], while others have shown increased levels or insignificant changes [9,11–13,116,117]. The reason for the discrepancies among these studies might be because of the ambiguous criteria for classifying AD and MCI patients or failure to exclude other factors that may have affected adiponectin levels. As mentioned earlier, conventional AD medications (acetylcholinesterase inhibitors) can increase serum adiponectin levels. Furthermore, there is also a possibility that increased adiponectin levels may have served as a compensatory mechanism for the progress of AD. Thus, to establish the relationship between adiponectin levels and AD, more controlled studies should be conducted.

5.2. Adiponectin and Adiponectin Receptor Homolog

Osmotin, a protein found in tobacco that structurally and functionally mimics adiponectin, positively modulates the AdipoR1/AMPK/SIRT1 pathway and reduces the AD-related protein Aβ expression (Figure 3) [59]. SIRT1 and AMPK are known for their metabolic activities and cellular energy homeostasis; they positively regulate each [118,119]. Osmotin treatment has been found to inhibit the expression of AD markers, such as amyloid precursor protein, p-tau, and Aβ, in inflammation-induced mouse brains [82,120]. In addition, adiponectin also enhances neurite outgrowth and synaptic complexity via AdipoR1/NgR1 signaling [57].

Adiporon, an agonist of adiponectin receptors that bind AdipoRs, has been known to play a vital role in many neurological diseases. Its ability as an anti-depressive agent and metabolic regulator has been demonstrated in a mouse model of depression, where it also regulated dopaminergic neurons [121]. It has also been shown to modulate fear and intrinsic excitability in the hippocampus [122]. These beneficial neurological effects are possible because of the BBB-penetrating property of this molecule. More recently, Liu et al. have reported that adiporon improves cognitive dysfunction, inhibits Aβ deposition, and restores impaired hippocampal neuron proliferation activation in AD mice by activating the AdipoR1/AMPK pathway [123]. Ultimately, osmotin and adiporon can be effective and realistic therapeutic alternatives for adiponectin-based AD treatment in patients.

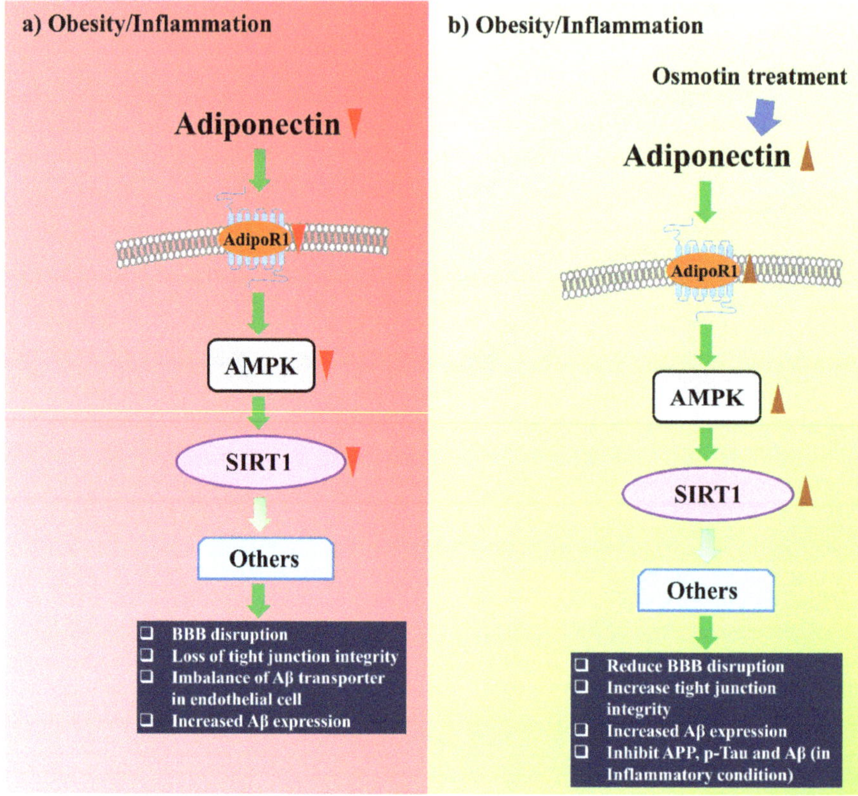

Figure 3. Osmotin treatment positively modulates AD through the AdipoR1/AMPK/SIRT1 pathway. (**a**) Obesity reduces adiponectin and AdipoR1 expressions, which negatively modulate the AMPK/SIRT1 pathway and increase the AD biomarkers. (**b**) Osmotin treatment increases adiponectin and AdipoR1 expressions and reduces AD biomarkers through the AMPK/SIRT1 pathway.

5.3. Adiponectin and Conventional AD Drug

The US Food and Drug Administration (FDA) has approved only two types of medications for the treatment of cognitive dysfunction of AD—acetylcholinesterase inhibitors (AChEI; donepezil and galantamine) and memantine [124]. One study has shown that donepezil increases serum adiponectin levels in AD patients, while another study has shown that galantamine increases serum adiponectin levels in a type 2 diabetes rat model [125]. A recent study has also reported a significant weight loss after AChEI treatment and that AChEI treatment exerts an insulin-sensitizing effect via the activation of IR/PI3K/Akt/GLUT2,4 and Wnt/GSK3β/β-catenin signaling [126]. Although more studies are needed, the beneficial effect of AChEI on AD is thought to be possibly mediated by adiponectin.

Memantine is another FDA-approved medication for AD. Although it cannot be factually claimed that its therapeutic mechanism is directly associated with adiponectin, there are some interesting points that suggest a connection between them. Memantine is an *N*-methyl-D-aspartate (NMDA) receptor antagonist that modulates glutamatergic dysfunction and inhibits excitotoxicity mediated by NMDA [127]. Its therapeutic ability in AD is attributed to this mechanism. Similarly, adiponectin has been shown to have a neuroprotective effect by inhibiting NMDA-mediated excitotoxicity [128]. In addition, adiponectin and memantine play a protective role in glutamate-induced excitotoxicity (shown in both animal/in-vitro models) [129,130]. Furthermore, memantine also has the ability to attenuate insulin resistance and improve brain function in high-fat diet-induced models. Given these similarities, there is a possibility that the comprehensive role of adiponectin in AD could be a hint in conducting research on memantine.

5.4. Adiponectin and Type 2 Diabetes Medications

Several studies have suggested that AD and T2DM are related to each other; thus, T2DM drugs and their potential to be used as AD treatment are drawing attention [131,132]. There are several types of type 2 diabetes drugs, such as biguanide derivatives (metformin), PPARγ agonists (thiazolidinedione [TZD] derivatives), glucagon-like peptide-1 receptor (GLP-1R) agonists, dipeptidyl peptidases 4 (DPP-4) inhibitors, sodium-glucose transport protein 2 inhibitors, and second-generation sulfonylureas [133]. In addition to these drugs' ability to improve cognitive function, some of these drugs are capable of increasing adiponectin levels.

Metformin, the drug used as first-line pharmacologic therapy in T2DM, has been shown to attenuate AD-like pathology in both in-vitro and in-vivo studies [134–136]. An improvement in cognitive function has been observed in many studies on mild cognitive function and dementia patients [60,137,138]. Many studies have shown that metformin increases serum adiponectin levels in various conditions [139–141]. Furthermore, like adiponectin, metformin plays a neuroprotective role by activating AMPK signaling [142,143].

TZD derivatives, including rosiglitazone and pioglitazone, act as PPARγ agonists and are used for T2DM treatment because of their insulin sensitization activity [144,145]. They have also been researched as alternative AD treatments [146,147]. Increased adiponectin by TZD might play a key role in the insulin sensitization activity induced by TZD [132,148]. Thus, TZD's beneficial effect on AD might be related to the increased adiponectin levels.

An insulin-releasing hormone in hyperglycemic conditions, GLP-1, has been found to have a neuroprotective effect [149–151]. GLP-1 receptors are responsible for controlling food intake and body weight and are widely distributed in the brain and special neurons (pyramidal neurons of the hippocampus and neocortex), which means that it plays an important role in neural function and synaptic transmission [149,151–154]. In an AD mice model, Val(8)GLP-1, liraglutide, and exidin-4 (GLP-1 analogs, well known for the upregulation of adiponectin) treatments rescued synaptic plasticity by preventing synaptic degradation, which is also correlated with the increased learning ability of new spatial tasks [155–158]. Several studies have also shown the beneficiary effect of GLP-1 analogs for AD [159–161].

Inhibitors of DPP-4 also increase GLP-1 signaling activation by inhibiting the degradation of GLP-1 [162]. Therefore, they potentially act on AD, like other GLP-1 analogs [163]. They also lead to increased adiponectin and are thought to be alternative medications for AD [164–167]. Accordingly, the GLP-1 functional facilitator, GLP-1 analogs, and DPP-4 inhibitors might be good candidates for AD treatment because of their ability to increase adiponectin.

5.5. Drugs for Cardiovascular Disease and Adiponectin

A recent study has suggested that medications aimed at cardiovascular diseases may also decrease the risk of dementia caused by AD [168]. Several neuroprotective pleiotropic agents have been shown to increase plasma adiponectin levels [169]. In studies with animal models, angiotensin II receptor blockers have shown beneficial effects on cognitive function related to AD [170–172]. Improvements in AD progress and cognitive function have also been observed in human clinical trials [173–175]. Moreover, angiotensin II receptor blockers can reduce neuroinflammation directly or by regulating the infiltration of inflammatory cytokines because of their ability to restore the blood-brain barrier [176–178]. Altogether, angiotensin II receptor blockers may be beneficial in AD therapy because it upregulates adiponectin [179–181].

Angiotensin-converting enzyme inhibitors (ACEIs) regulate the renin–angiotensin system and angiotensin II receptor blockers. Upregulation of blood adiponectin has been observed after treatment with ACEIs [182,183]. Some clinical and meta-analysis studies have also shown that ACEIs reduce the progression and risk of AD [184–187]. This result has also been observed in several animal model studies [188–190]. However, in one meta-analysis study, ACEIs had no effect on cognitive decline [191], and some studies have shown a negative effect of ACEIs in in-vivo AD models [192–194]. Thus, ACEI treatment is controversial for its effect on AD.

Another cardiovascular medication, fibrate, a PPARα agonist, has been shown to increase adiponectin levels [195,196]. Since ligand-activated PPARα decreases tau phosphorylation, Aβ pathology, and neuroinflammation, fibrate might be a good candidate for AD treatment [197].

Statins (such as simvastatin, paravastatin, and atorvastatin) have been reported to increase blood adiponectin levels [198–200]. Statins are well known for their anti-inflammatory properties in diabetic patients [201,202]. According to the above studies, statins might also be useful in the treatment of AD by increasing blood adiponectin levels and reducing neuroinflammation.

6. Conclusions

The conventional approach to developing therapeutics for AD has been focused on the Aβ hypothesis, which insists that Aβ causes AD pathologies. However, this approach has yet to provide a successful treatment or prevention method for AD. Based on the metabolic alterations that occur in the pathological markers of AD, this review suggests that the adipocyte metabolite adiponectin can be considered a therapeutic option for the treatment of AD.

Author Contributions: J.E.L. provided concept, design, and overall supervision of this study. J.Y.K. and S.B. contributed to the writing and drawing. Y.J.J. participated in the discussion and revision. All authors have read and agreed to the published version of the manuscript.

Funding: This study was supported by a grant from the National Research Foundation of Korea (NRF) through a grant funded by the Ministry of Science, ICT, and Future Planning (NRF-2017R1A2B2005350).

Conflicts of Interest: The authors declare no conflict of interest.

References

1. Pugazhenthi, S.; Qin, L.; Reddy, P.H. Common neurodegenerative pathways in obesity, diabetes, and Alzheimer's disease. *Biochim. Biophys. Acta Mol. Basis Dis.* **2017**, *1863*, 1037–1045. [CrossRef] [PubMed]
2. Profenno, L.A.; Porsteinsson, A.P.; Faraone, S.V. Meta-analysis of Alzheimer's disease risk with obesity, diabetes, and related disorders. *Biol. Psychiatry.* **2010**, *67*, 505–512. [CrossRef] [PubMed]
3. Gottesman, R.F.; Schneider, A.L.; Zhou, Y.; Coresh, J.; Green, E.; Gupta, N.; Knopman, D.S.; Mintz, A.; Rahmim, A.; Sharrett, A.R.; et al. Association between midlife vascular risk factors and estimated brain amyloid deposition. *JAMA* **2017**, *317*, 1443–1450. [CrossRef] [PubMed]
4. Mucke, L. Neuroscience: Alzheimer's disease. *Nature* **2009**, *461*, 895–897. [CrossRef] [PubMed]
5. Janson, J.; Laedtke, T.; Parisi, J.E.; O'Brien, P.; Petersen, R.C.; Butler, P.C. Increased risk of type 2 diabetes in Alzheimer disease. *Diabetes* **2004**, *53*, 474–481. [CrossRef]
6. Niu, L.; Han, D.W.; Xu, R.L.; Han, B.; Zhou, X.; Wu, H.W.; Li, S.H.; Qu, C.X.; Liu, M. A high-sugar high-fat diet induced metabolic syndrome shows some symptoms of Alzheimer's disease in rats. *J. Nutr. Health Aging.* **2016**, *20*, 509–513. [CrossRef]
7. Ledreux, A.; Wang, X.; Schultzberg, M.; Granholm, A.C.; Freeman, L.R. Detrimental effects of a high fat/high cholesterol diet on memory and hippocampal markers in aged rats. *Behav. Brain Res.* **2016**, *312*, 294–304. [CrossRef]
8. Une, K.; Takei, Y.A.; Tomita, N.; Asamura, T.; Ohrui, T.; Furukawa, K.; Arai, H. Adiponectin in plasma and cerebrospinal fluid in MCI and Alzheimer's disease. *Eur. J. Neurol.* **2011**, *18*, 1006–1009. [CrossRef]
9. Van Himbergen, T.M.; Beiser, A.S.; Ai, M.; Seshadri, S.; Otokozawa, S.; Au, R.; Thongtang, N.; Wolf, P.A.; Schaefer, E.J. Biomarkers for insulin resistance and inflammation and the risk for all-cause dementia and alzheimer disease: Results from the framingham heart study. *Arch. Neurol.* **2012**, *69*, 594–600. [CrossRef]

10. Waragai, M.; Adame, A.; Trinh, I.; Sekiyama, K.; Takamatsu, Y.; Une, K.; Masliah, E.; Hashimoto, M. Possible involvement of adiponectin, the anti-diabetes molecule, in the pathogenesis of Alzheimer's disease. *J. Alzheimers Dis.* **2016**, *52*, 1453–1459. [CrossRef]

11. Khemka, V.K.; Bagchi, D.; Bandyopadhyay, K.; Bir, A.; Chattopadhyay, M.; Biswas, A.; Basu, D.; Chakrabarti, S. Altered serum levels of adipokines and insulin in probable Alzheimer's disease. *J. Alzheimer Dis. JAD* **2014**, *41*, 525–533. [CrossRef] [PubMed]

12. Ma, J.; Zhang, W.; Wang, H.F.; Wang, Z.X.; Jiang, T.; Tan, M.S.; Yu, J.T.; Tan, L. peripheral blood adipokines and insulin levels in patients with Alzheimer's disease: A replication study and meta-analysis. *Curr. Alzheimer Res.* **2016**, *13*, 223–233. [CrossRef] [PubMed]

13. Wennberg, A.M.; Gustafson, D.; Hagen, C.E.; Roberts, R.O.; Knopman, D.; Jack, C.; Petersen, R.C.; Mielke, M.M. Serum adiponectin levels, neuroimaging, and cognition in the mayo clinic study of aging. *J. Alzheimers Dis.* **2016**, *53*, 573–581. [CrossRef] [PubMed]

14. Albert, M.S.; DeKosky, S.T.; Dickson, D.; Dubois, B.; Feldman, H.H.; Fox, N.C.; Gamst, A.; Holtzman, D.M.; Jagust, W.J.; Petersen, R.C.; et al. The diagnosis of mild cognitive impairment due to Alzheimer's disease: Recommendations from the national institute on aging-Alzheimer's association workgroups on diagnostic guidelines for Alzheimer's disease. *Alzheimers Dement.* **2011**, *7*, 270–279. [CrossRef]

15. Sacks, D.; Baxter, B.; Campbell, B.C.V.; Carpenter, J.S.; Cognard, C.; Dippel, D.; Hirchs, J. Multisociety consensus quality improvement revised consensus statement for endovascular therapy of acute ischemic stroke. *Int. J. Stroke* **2018**, *13*, 612–632. [CrossRef] [PubMed]

16. Alzheimer's, Assotiation. Alzheimer's disease facts and figures. *Alzheimers Dement.* **2016**, *12*, 459–509. [CrossRef]

17. Van Cauwenberghe, C.; van Broeckhoven, C.; Sleegers, K. The genetic landscape of Alzheimer disease: Clinical implications and perspectives. *Genet. Med.* **2016**, *18*, 421–430. [CrossRef]

18. Lee, E.B. Obesity, leptin, and Alzheimer's disease. *Ann. N. Y. Acad. Sci.* **2011**, *1243*, 15–29. [CrossRef]

19. Lee, V.M.; Goedert, M.; Trojanowski, J.Q. Neurodegenerative tauopathies. *Annu. Rev. Neurosci.* **2001**, *24*, 1121–1159. [CrossRef]

20. Mattson, M.P. Pathways towards and away from Alzheimer's disease. *Nature* **2004**, *430*, 631–639. [CrossRef]

21. Morishima, M.; Ihara, Y. Posttranslational modifications of tau in paired helical filaments. *Dementia* **1994**, *5*, 282–288. [CrossRef] [PubMed]

22. Han, X.J.; Hu, Y.Y.; Yang, Z.J.; Jiang, L.P.; Shi, S.L.; Li, Y.R.; Guo, M.Y.; Wu, H.L.; Wan, Y.Y. Amyloid beta-42 induces neuronal apoptosis by targeting mitochondria. *Mol. Med. Rep.* **2017**, *16*, 4521–4528. [CrossRef] [PubMed]

23. Selkoe, D.J.; Hardy, J. The amyloid hypothesis of Alzheimer's disease at 25 years. *EMBO Mol. Med.* **2016**, *8*, 595–608. [CrossRef] [PubMed]

24. Bussler, S.; Penke, M.; Flemming, G.; Elhassan, Y.S.; Kratzsch, J.; Sergeyev, E.; Lipek, T.; Vogel, M.; Spielau, U.; Korner, A.; et al. Novel insights in the metabolic syndrome in childhood and adolescence. *Horm. Res. Paediatr.* **2017**, *88*, 181–193. [CrossRef]

25. Pugazhenthi, S. Metabolic syndrome and the cellular phase of Alzheimer's disease. *Prog. Mol. Biol. Transl. Sci.* **2017**, *146*, 243–258. [CrossRef]

26. Blais, A.; Drouin, G.; Chaumontet, C.; Voisin, T.; Couvelard, A.; Even, P.C.; Couvineau, A. Impact of Orexin-A treatment on food intake, energy metabolism and body weight in mice. *PLoS ONE* **2017**, *12*, e0169908. [CrossRef]

27. Yi, M.; Li, H.; Wu, Z.; Yan, J.; Liu, Q.; Ou, C.; Chen, M. A promising therapeutic target for metabolic diseases: Neuropeptide Y receptors in humans. *Cell Physiol. Biochem.* **2018**, *45*, 88–107. [CrossRef]

28. Lawson, E.A. The effects of oxytocin on eating behaviour and metabolism in humans. *Nat. Rev. Endocrinol.* **2017**, *13*, 700–709. [CrossRef]

29. Frolich, L.; Blum-Degen, D.; Bernstein, H.G.; Engelsberger, S.; Humrich, J.; Laufer, S.; Muschner, D.; Thalheimer, A.; Turk, A.; Hoyer, S.; et al. Brain insulin and insulin receptors in aging and sporadic Alzheimer's disease. *J. Neural. Transm.* **1998**, *105*, 423–438. [CrossRef]

30. Kim, M.W.; Abid, N.B.; Jo, M.H.; Jo, M.G.; Yoon, G.H.; Kim, M.O. Suppression of adiponectin receptor 1 promotes memory dysfunction and Alzheimer's disease-like pathologies. *Sci. Rep.* **2017**, *7*, 12435. [CrossRef]

31. Song, J.; Lee, J.E. Adiponectin as a new paradigm for approaching Alzheimer's disease. *Anat. Cell Biol.* **2013**, *46*, 229–234. [CrossRef]

32. Nasoohi, S.; Parveen, K.; Ishrat, T. Metabolic syndrome, brain insulin resistance, and Alzheimer's disease: Thioredoxin interacting protein (TXNIP) and inflammasome as core amplifiers. *J. Alzheimers Dis.* **2018**, *66*, 857–885. [CrossRef] [PubMed]

33. Willette, A.A.; Xu, G.; Johnson, S.C.; Birdsill, A.C.; Jonaitis, E.M.; Sager, M.A.; Hermann, B.P.; la Rue, A.; Asthana, S.; Bendlin, B.B.; et al. Insulin resistance, brain atrophy, and cognitive performance in late middle-aged adults. *Diabetes Care* **2013**, *36*, 443–449. [CrossRef] [PubMed]

34. De la Monte, S.M. Insulin resistance and neurodegeneration: Progress towards the development of new therapeutics for Alzheimer's disease. *Drugs* **2017**, *77*, 47–65. [CrossRef] [PubMed]

35. Cao, H. Adipocytokines in obesity and metabolic disease. *J. Endocrinol.* **2014**, *220*, T47–T59. [CrossRef]

36. Hotta, K.; Funahashi, T.; Bodkin, N.L.; Ortmeyer, H.K.; Arita, Y.; Hansen, B.C.; Matsuzawa, Y. Circulating concentrations of the adipocyte protein adiponectin are decreased in parallel with reduced insulin sensitivity during the progression to type 2 diabetes in rhesus monkeys. *Diabetes* **2001**, *50*, 1126–1133. [CrossRef]

37. Yamauchi, T.; Kamon, J.; Waki, H.; Terauchi, Y.; Kubota, N.; Hara, K.; Mori, Y.; Ide, T.; Murakami, K.; Tsuboyama-Kasaoka, N.; et al. The fat-derived hormone adiponectin reverses insulin resistance associated with both lipoatrophy and obesity. *Nat. Med.* **2001**, *7*, 941–946. [CrossRef]

38. Yamauchi, T.; Nio, Y.; Maki, T.; Kobayashi, M.; Takazawa, T.; Iwabu, M.; Okada-Iwabu, M.; Kawamoto, S.; Kubota, N.; Kubota, T.; et al. Targeted disruption of AdipoR1 and AdipoR2 causes abrogation of adiponectin binding and metabolic actions. *Nat. Med.* **2007**, *13*, 332–339. [CrossRef]

39. Scherer, P.E.; Williams, S.; Fogliano, M.; Baldini, G.; Lodish, H.F. A novel serum protein similar to C1q, produced exclusively in adipocytes. *J. Biol. Chem.* **1995**, *270*, 26746–26749. [CrossRef]

40. Arita, Y.; Kihara, S.; Ouchi, N.; Takahashi, M.; Maeda, K.; Miyagawa, J.; Hotta, K.; Shimomura, I.; Nakamura, T.; Miyaoka, K.; et al. Paradoxical decrease of an adipose-specific protein, adiponectin, in obesity. *Biochem. Biophys. Res. Commun.* **1999**, *257*, 79–83. [CrossRef]

41. Turer, A.T.; Scherer, P.E. Adiponectin: Mechanistic insights and clinical implications. *Diabetologia* **2012**, *55*, 2319–2326. [CrossRef] [PubMed]

42. Pataky, Z.; Bobbioni-Harsch, E.; Golay, A. Obesity: A complex growing challenge. *Exp. Clin. Endocrinol. Diabetes* **2010**, *118*, 427–433. [CrossRef] [PubMed]

43. Emmerzaal, T.L.; Kiliaan, A.J.; Gustafson, D.R. 2003–2013: A decade of body mass index, Alzheimer's disease, and dementia. *J. Alzheimers Dis.* **2015**, *43*, 739–755. [CrossRef] [PubMed]

44. Lee, S.; Kim, J.Y.; Kim, E.; Seo, K.; Kang, Y.J.; Kim, J.Y.; Kim, C.H.; Song, H.T.; Saksida, L.M.; Lee, J.E.; et al. Assessment of cognitive impairment in a mouse model of high-fat diet-induced metabolic stress with touchscreen-based automated battery system. *Exp. Neurobiol.* **2018**, *27*, 277–286. [CrossRef]

45. Okada-Iwabu, M.; Yamauchi, T.; Iwabu, M.; Honma, T.; Hamagami, K.; Matsuda, K.; Yamaguchi, M.; Tanabe, H.; Kimura-Someya, T.; Shirouzu, M.; et al. A small-molecule AdipoR agonist for type 2 diabetes and short life in obesity. *Nature* **2013**, *503*, 493–499. [CrossRef]

46. Sekiyama, K.; Waragai, M.; Akatsu, H.; Sugama, S.; Takenouchi, T.; Takamatsu, Y.; Fujita, M.; Sekigawa, A.; Rockenstein, E.; Inoue, S.; et al. Disease-modifying effect of adiponectin in model of alpha-synucleinopathies. *Ann. Clin. Transl. Neurol.* **2014**, *1*, 479–489. [CrossRef]

47. Mao, X.; Kikani, C.K.; Riojas, R.A.; Langlais, P.; Wang, L.; Ramos, F.J.; Fang, Q.; Christ-Roberts, C.Y.; Hong, J.Y.; Kim, R.Y.; et al. APPL1 binds to adiponectin receptors and mediates adiponectin signalling and function. *Nat. Cell Biol.* **2006**, *8*, 516–523. [CrossRef]

48. XiaoTian, L.; QiNan, W.; XiaGuang, G.; WuQuan, D.; Bing, C.; ZiWen, L. Exenatide Activates the APPL1-AMPK-PPARalpha axis to prevent diabetic cardiomyocyte apoptosis. *J. Diabetes Res.* **2016**, *2016*, 4219735. [CrossRef]

49. Cheng, K.K.; Lam, K.S.; Wang, Y.; Huang, Y.; Carling, D.; Wu, D.; Wong, C.; Xu, A. Adiponectin-induced endothelial nitric oxide synthase activation and nitric oxide production are mediated by APPL1 in endothelial cells. *Diabetes* **2007**, *56*, 1387–1394. [CrossRef]

50. Chandrasekar, B.; Boylston, W.H.; Venkatachalam, K.; Webster, N.J.; Prabhu, S.D.; Valente, A.J. Adiponectin blocks interleukin-18-mediated endothelial cell death via APPL1-dependent AMP-activated protein kinase (AMPK) activation and IKK/NF-kappaB/PTEN suppression. *J. Biol. Chem.* **2008**, *283*, 24889–24898. [CrossRef]

51. Coope, A.; Milanski, M.; Araujo, E.P.; Tambascia, M.; Saad, M.J.; Geloneze, B.; Velloso, L.A. AdipoR1 mediates the anorexigenic and insulin/leptin-like actions of adiponectin in the hypothalamus. *FEBS Lett.* **2008**, *582*, 1471–1476. [CrossRef] [PubMed]

52. Lee, M.H.; Klein, R.L.; El-Shewy, H.M.; Luttrell, D.K.; Luttrell, L.M. The adiponectin receptors AdipoR1 and AdipoR2 activate ERK1/2 through a Src/Ras-dependent pathway and stimulate cell growth. *Biochemistry* **2008**, *47*, 11682–11692. [CrossRef] [PubMed]

53. Thundyil, J.; Tang, S.C.; Okun, E.; Shah, K.; Karamyan, V.T.; Li, Y.I.; Woodruff, T.M.; Taylor, S.M.; Jo, D.G.; Mattson, M.P.; et al. Evidence that adiponectin receptor 1 activation exacerbates ischemic neuronal death. *Exp. Transl. Stroke. Med.* **2010**, *2*, 15. [CrossRef]

54. Parker-Duffen, J.L.; Nakamura, K.; Silver, M. T-cadherin is essential for adiponectin-mediated revascularization. *J. Biol. Chem.* **2013**, *288*, 24886–24897. [CrossRef] [PubMed]

55. Wang, M.; Jo, J.; Song, J. Adiponectin improves long-term potentiation in the 5XFAD mouse brain. *Sci. Rep.* **2019**, *9*, 8918. [CrossRef]

56. Zhang, D.; Wang, X.; Lu, X.Y. Adiponectin exerts neurotrophic effects on dendritic arborization, spinogenesis, and neurogenesis of the dentate gyrus of male mice. *Endocrinology* **2016**, *157*, 2853–2869. [CrossRef]

57. Yoon, G.; Shah, S.A.; Ali, T.; Kim, M.O. The adiponectin homolog osmotin enhances neurite outgrowth and synaptic complexity via AdipoR1/NgR1 Signaling in Alzheimer's Disease. *Mol. Neurobiol.* **2018**, *55*, 6673–6686. [CrossRef]

58. Ali, T.; Yoon, G.H.; Shah, S.A.; Lee, H.Y.; Kim, M.O. Osmotin attenuates amyloid beta-induced memory impairment, tau phosphorylation and neurodegeneration in the mouse hippocampus. *Sci. Rep.* **2015**, *5*, 11708. [CrossRef]

59. Shah, S.A.; Yoon, G.H.; Chung, S.S.; Abid, M.N.; Kim, T.H.; Lee, H.Y.; Kim, M.O. Novel osmotin inhibits SREBP2 via the AdipoR1/AMPK/SIRT1 pathway to improve Alzheimer's disease neuropathological deficits. *Mol. Psychiatry* **2017**, *22*, 407–416. [CrossRef]

60. Ng, R.C.; Cheng, O.Y.; Jian, M.; Kwan, J.S.; Ho, P.W.; Cheng, K.K.; Yeung, P.K.; Zhou, L.L.; Hoo, R.L.; Chung, S.K.; et al. Chronic adiponectin deficiency leads to Alzheimer's disease-like cognitive impairments and pathologies through AMPK inactivation and cerebral insulin resistance in aged mice. *Mol. Neurodegener.* **2016**, *11*, 71. [CrossRef]

61. Cezaretto, A.; Suemoto, C.K.; Bensenor, I.; Lotufo, P.A.; de Almeida-Pititto, B.; Ferreira, S.R.G.; Group, E.R. Association of adiponectin with cognitive function precedes overt diabetes in the Brazilian Longitudinal Study of Adult Health: ELSA. *Diabetol. Metab. Syndr.* **2018**, *10*, 54. [CrossRef] [PubMed]

62. Yamauchi, T.; Iwabu, M.; Okada-Iwabu, M.; Kadowaki, T. Adiponectin receptors: A review of their structure, function and how they work. *Best Pract. Res. Clin. Endocrinol. Metab.* **2014**, *28*, 15–23. [CrossRef] [PubMed]

63. Thundyil, J.; Pavlovski, D.; Sobey, C.G.; Arumugam, T.V. Adiponectin receptor signalling in the brain. *Br. J. Pharmacol.* **2012**, *165*, 313–327. [CrossRef]

64. Ebinuma, H.; Miida, T.; Yamauchi, T.; Hada, Y.; Hara, K.; Kubota, N.; Kadowaki, T. Improved ELISA for selective measurement of adiponectin multimers and identification of adiponectin in human cerebrospinal fluid. *Clin. Chem.* **2007**, *53*, 1541–1544. [CrossRef]

65. Yamauchi, T.; Kamon, J.; Ito, Y.; Tsuchida, A.; Yokomizo, T.; Kita, S.; Sugiyama, T.; Miyagishi, M.; Hara, K.; Tsunoda, M.; et al. Cloning of adiponectin receptors that mediate antidiabetic metabolic effects. *Nature* **2003**, *423*, 762–769. [CrossRef]

66. Deane, R.; Bell, R.D.; Sagare, A.; Zlokvic, B.V. Clearance of amyloid-β peptide across the blood-brain barrier: Implication for therapies in Alzheimer's disease. *CNS Neurol. Disord. Drug Targets* **2009**, *8*, 16–30. [CrossRef]

67. Varhelyi, Z.P.; Kalman, J.; Olah, Z.; Ivitz, E.V.; Fodor, E.K.; Santha, M.; Datki, Z.L.; Pakaski, M. Adiponectin receptors are less sensitive to stress in a transgenic mouse model of Alzheimer's Disease. *Front. Neurosci.* **2017**, *11*, 199. [CrossRef] [PubMed]

68. Steen, E.; Terry, B.M.; Rivera, E.J.; Cannon, J.L.; Neely, T.R.; Tavares, R.; Xu, X.J.; Wands, J.R.; de la Monte, S.M. Impaired insulin and insulin-like growth factor expression and signaling mechanisms in Alzheimer's disease-Is this type 3 diabetes? *J. Alzheimers Dis.* **2005**, *7*, 63–80. [CrossRef]

69. Arnold, S.E.; Arvanitakis, Z.; Macauley-Rambach, S.L.; Koenig, A.M.; Wang, H.Y.; Ahima, R.S.; Craft, S.; Gandy, S.; Buettner, C.; Stoeckel, L.E.; et al. Brain insulin resistance in type 2 diabetes and Alzheimer disease: Concepts and conundrums. *Nat. Rev. Neurol.* **2018**, *14*, 168–181. [CrossRef] [PubMed]

70. De Felice, F.G. Alzheimer's disease and insulin resistance: Translating basic science into clinical applications. *J. Clin. Investig.* **2013**, *123*, 531–539. [CrossRef]

71. Jack, C.R., Jr.; Petersen, R.C.; Xu, Y.; O'Brien, P.C.; Smith, G.E.; Ivnik, R.J.; Tangalos, E.G.; Kokmen, E. Rate of medial temporal lobe atrophy in typical aging and Alzheimer's disease. *Neurology* **1998**, *51*, 993–999. [CrossRef] [PubMed]

72. De Felice, F.G.; Vieira, M.N.; Bomfim, T.R.; Decker, H.; Velasco, P.T.; Lambert, M.P.; Viola, K.L.; Zhao, W.Q.; Ferreira, S.T.; Klein, W.L. Protection of synapses against Alzheimer's-linked toxins: Insulin signaling prevents the pathogenic binding of Abeta oligomers. *Proc. Natl. Acad. Sci. USA* **2009**, *106*, 1971–1976. [CrossRef] [PubMed]

73. Lopez-Jaramillo, P.; Gomez-Arbelaez, D.; Lopez-Lopez, J.; Lopez-Lopez, C.; Martinez-Ortega, J.; Gomez-Rodriguez, A.; Triana-Cubillos, S. The role of leptin/adiponectin ratio in metabolic syndrome and diabetes. *Horm. Mol. Biol. Clin. Investig.* **2014**, *18*, 37–45. [CrossRef] [PubMed]

74. Berg, A.H.; Combs, T.P.; Du, X.; Brownlee, M.; Scherer, P.E. The adipocyte-secreted protein Acrp30 enhances hepatic insulin action. *Nat. Med.* **2001**, *7*, 947–953. [CrossRef]

75. Masaki, T.; Anan, F.; Shimomura, T.; Fujiki, M.; Saikawa, T.; Yoshimatsu, H. Association between hippocampal volume and serum adiponectin in patients with type 2 diabetes mellitus. *Metabolism* **2012**, *61*, 1197–1200. [CrossRef]

76. Garcia-Casares, N.; Garcia-Arnes, J.A.; Rioja, J.; Ariza, M.J.; Gutierrez, A.; Alfaro, F.; Nabrozidis, A.; Gonzalez-Alegre, P.; Gonzalez-Santos, P. Alzheimer's like brain changes correlate with low adiponectin plasma levels in type 2 diabetic patients. *J. Diabetes Complicat.* **2016**, *30*, 281–286. [CrossRef]

77. Xu, Z.P.; Gan, G.S.; Liu, Y.M.; Xiao, J.S.; Liu, H.X.; Mei, B.; Zhang, J.J. adiponectin attenuates streptozotocin-induced tau hyperphosphorylation and cognitive deficits by rescuing PI3K/Akt/GSK-3beta pathway. *Neurochem. Res.* **2018**, *43*, 316–323. [CrossRef]

78. Butterfield, D.A.; Halliwell, B. Oxidative stress, dysfunctional glucose metabolism and Alzheimer disease. *Nat. Rev. Neurosci.* **2019**, *20*, 148–160. [CrossRef]

79. Cunnane, S.; Nugent, S.; Roy, M.; Courchesne-Loyer, A.; Croteau, E.; Tremblay, S.; Castellano, A.; Pifferi, F.; Bocti, C.; Paquet, N.; et al. Brain fuel metabolism, aging, and Alzheimer's disease. *Nutrition* **2011**, *27*, 3–20. [CrossRef]

80. Jack, C.R., Jr.; Knopman, D.S.; Jagust, W.J.; Shaw, L.M.; Aisen, P.S.; Weiner, M.W.; Petersen, R.C.; Trojanowski, J.Q. Hypothetical model of dynamic biomarkers of the Alzheimer's pathological cascade. *Lancet Neurol.* **2010**, *9*, 119–128. [CrossRef]

81. Small, G.W.; Mazziotta, J.C.; Collins, M.T.; Baxter, L.R.; Phelps, M.E.; Mandelkern, M.A.; Kaplan, A.; La Rue, A.; Adamson, C.F.; Chang, L.; et al. Apolipoprotein E type 4 allele and cerebral glucose metabolism in relatives at risk for familial Alzheimer disease. *JAMA* **1995**, *273*, 942–947. [CrossRef]

82. Cisternas, P.; Martinez, M.; Ahima, R.S.; William Wong, G.; Inestrosa, N.C. Modulation of Glucose Metabolism in Hippocampal Neurons by Adiponectin and Resistin. *Mol. Neurobiol.* **2019**, *56*, 3024–3037. [CrossRef]

83. Dienel, G.A.; Cruz, N.F. Nutrition during brain activation: Does cell-to-cell lactate shuttling contribute significantly to sweet and sour food for thought? *Neurochem. Int.* **2004**, *45*, 321–351. [CrossRef]

84. Dienel, G.A.; Hertz, L. Glucose and lactate metabolism during brain activation. *J. Neurosci. Res.* **2001**, *66*, 824–838. [CrossRef] [PubMed]

85. Dhopeshwarkar, G.A.; Mead, J.F. Fatty acid uptake by the brain. 3. Incorporation of (1-14C)oleic acid into the adult rat brain. *Biochim. Biophys. Acta* **1970**, *210*, 250–256. [CrossRef]

86. Panov, A.; Orynbayeva, Z.; Vavilin, V.; Lyakhovich, V. Fatty acids in energy metabolism of the central nervous system. *Biomed. Res. Int.* **2014**, *2014*, 472459. [CrossRef] [PubMed]

87. Schonfeld, P.; Reiser, G. Why does brain metabolism not favor burning of fatty acids to provide energy? Reflections on disadvantages of the use of free fatty acids as fuel for brain. *J. Cereb. Blood Flow. Metab.* **2013**, *33*, 1493–1499. [CrossRef]

88. Se Oliveira, C.; de Mattos, A.B.; Biz, C.; Oyama, L.M.; Ribeiro, E.B.; do Nascimento, C.M. High-fat diet and glucocorticoid treatment cause hyperglycemia associated with adiponectin receptor alterations. *Lipids Health Dis.* **2011**, *10*, 11. [CrossRef]

89. Yanai, H.; Yoshida, H. Beneficial effects of adiponectin on glucose and lipid metabolism and atherosclerotic progression: Mechanisms and perspectives. *Int. J. Mol. Sci.* **2019**, *20*, 1190. [CrossRef]

90. Yamauchi, T.; Kadowaki, T. Adiponectin receptor as a key player in healthy longevity and obesity-related diseases. *Cell Metab.* **2013**, *17*, 185–196. [CrossRef]

91. Elfeky, M.; Kaede, R.; Okamatsu-Ogura, Y.; Kimura, K. Adiponectin inhibits LPS-induced HMGB1 release through an AMP kinase and heme oxygenase-1-dependent pathway in RAW 264 macrophage cells. *Mediators Inflamm.* **2016**, *2016*, 5701959. [CrossRef] [PubMed]

92. Rizzo, F.R.; Musella, A.; de Vito, F.; Fresegna, D.; Bullitta, S.; Vanni, V.; Guadalupi, L.; Stampanoni Bassi, M.; Buttari, F.; Mandolesi, G.; et al. Tumor necrosis factor and interleukin-1beta modulate synaptic plasticity during neuroinflammation. *Neural. Plast.* **2018**, *2018*, 8430123. [CrossRef] [PubMed]

93. Ferreira, S.T.; Clarke, J.R.; Bomfim, T.R.; De Felice, F.G. Inflammation, defective insulin signaling, and neuronal dysfunction in Alzheimer's disease. *Alzheimers Dement.* **2014**, *10*, S76–S83. [CrossRef] [PubMed]

94. Fruhbeck, G.; Catalan, V.; Rodriguez, A.; Ramirez, B.; Becerril, S.; Salvador, J.; Portincasa, P.; Colina, I.; Gomez-Ambrosi, J. Involvement of the leptin-adiponectin axis in inflammation and oxidative stress in the metabolic syndrome. *Sci. Rep.* **2017**, *7*, 6619. [CrossRef] [PubMed]

95. Yokota, T.; Oritani, K.; Takahashi, I.; Ishikawa, J.; Matsuyama, A.; Ouchi, N.; Kihara, S.; Funahashi, T.; Tenner, A.J.; Tomiyama, Y.; et al. Adiponectin, a new member of the family of soluble defense collagens, negatively regulates the growth of myelomonocytic progenitors and the functions of macrophages. *Blood* **2000**, *96*, 1723–1732. [CrossRef]

96. Lourenco, M.V.; Clarke, J.R.; Frozza, R.L.; Bomfim, T.R.; Forny-Germano, L.; Batista, A.F.; Sathler, L.B.; Brito-Moreira, J.; Amaral, O.B.; Silva, C.A.; et al. TNF-alpha mediates PKR-dependent memory impairment and brain IRS-1 inhibition induced by Alzheimer's beta-amyloid oligomers in mice and monkeys. *Cell Metab.* **2013**, *18*, 831–843. [CrossRef] [PubMed]

97. Nicolas, S.; Cazareth, J.; Zarif, H.; Guyon, A.; Heurteaux, C.; Chabry, J.; Petit-Paitel, A. Globular adiponectin limits microglia pro-inflammatory phenotype through an AdipoR1/NF-kappaB signaling pathway. *Front. Cell Neurosci.* **2017**, *11*, 352. [CrossRef]

98. Chabry, J.; Nicolas, S.; Cazareth, J.; Murris, E.; Guyon, A.; Glaichenhaus, N.; Heurteaux, C.; Petit-Paitel, A. Enriched environment decreases microglia and brain macrophages inflammatory phenotypes through adiponectin-dependent mechanisms: Relevance to depressive-like behavior. *Brain Behav. Immun.* **2015**, *50*, 275–287. [CrossRef]

99. Tonnies, E.; Trushina, E. Oxidative stress, synaptic dysfunction, and Alzheimer's disease. *J. Alzheimers Dis.* **2017**, *57*, 1105–1121. [CrossRef]

100. Snyder, B.; Shell, B.; Cunningham, J.T.; Cunningham, R.L. Chronic intermittent hypoxia induces oxidative stress and inflammation in brain regions associated with early-stage neurodegeneration. *Physiol. Rep.* **2017**, *5*. [CrossRef] [PubMed]

101. Chan, K.H.; Lam, K.S.; Cheng, O.Y.; Kwan, J.S.; Ho, P.W.; Cheng, K.K.; Chung, S.K.; Ho, J.W.; Guo, V.Y.; Xu, A. Adiponectin is protective against oxidative stress induced cytotoxicity in amyloid-β neurotoxicity. *PLoS ONE* **2012**, *7*, e52354. [CrossRef] [PubMed]

102. Park, M.; Youn, B.; Zheng, X.L.; Wu, D.; Xu, A.; Sweeney, G. Globular adiponectin, acting via AdipoR1/APPL1, protects H9c2 cells from hypoxia/reoxygenation-induced apoptosis. *PLoS ONE* **2011**, *6*, e19143. [CrossRef] [PubMed]

103. Qiu, G.; Wan, R.; Hu, J.; Mattson, M.P.; Spangler, E.; Liu, S.; Yau, S.Y.; Lee, T.M.; Gleichmann, M.; Ingram, D.K.; et al. Adiponectin protects rat hippocampal neurons against excitotoxicity. *Age* **2011**, *33*, 155–165. [CrossRef] [PubMed]

104. Ramamurthy, S.; Ronnett, G.V. Developing a head for energy sensing: AMP-activated protein kinase as a multifunctional metabolic sensor in the brain. *J. Physiol.* **2006**, *574*, 85–93. [CrossRef]

105. Kubota, N.; Yano, W.; Kubota, T.; Yamauchi, T.; Itoh, S.; Kumagai, H.; Kozono, H.; Takamoto, I.; Okamoto, S.; Shiuchi, T.; et al. Adiponectin stimulates AMP-activated protein kinase in the hypothalamus and increases food intake. *Cell Metab.* **2007**, *6*, 55–68. [CrossRef]

106. Girgis, C.M.; Cheng, K.; Scott, C.H.; Gunton, J.E. Novel links between HIFs, type 2 diabetes, and metabolic syndrome. *Trends Endocrinol. Metab.* **2012**, *23*, 372–380. [CrossRef]

107. Jiang, C.; Kim, J.H.; Li, F.; Qu, A.; Gavrilova, O.; Shah, Y.M.; Gonzalez, F.J. Hypoxia-inducible factor 1α regulates a SOCS3-STAT3-adiponectin signal transduction pathway in adipocytes. *J. Biol. Chem.* **2013**, *288*, 3844–3857. [CrossRef]

108. Bai, H.; Zhao, L.; Liu, H.; Guo, H.; Guo, W.; Zheng, L.; Liu, X.; Wu, X.; Luo, J.; Li, X.; et al. Adiponectin confers neuroprotection against cerebral ischemia-reperfusion injury through activating the cAMP/PKA-CREB-BDNF signaling. *Brain Res. Bull.* **2018**, *143*, 145–154. [CrossRef]

109. Wang, S.; Li, D.; Huang, C.; Wan, Y.; Wang, J.; Zan, X.; Yang, B. Overexpression of adiponectin alleviates intracerebral hemorrhage-induced brain injury in rats via suppression of oxidative stress. *Neurosci. Lett.* **2018**, *681*, 110–116. [CrossRef]

110. Choi, S.H.; Bylykbashi, E.; Chatila, Z.K.; Lee, S.W.; Pulli, B.; Clemenson, G.D.; Kim, E.; Rompala, A.; Oram, M.K.; Asselin, C.; et al. Combined adult neurogenesis and BDNF mimic exercise effects on cognition in an Alzheimer's mouse model. *Science* **2018**, *361*, eaan8821. [CrossRef]

111. Moreno-Jiménez, E.P.; Flor-García, M.; Terreros-Roncal, J.; Rábano, A.; Cafini, F.; Pallas-Bazarra, N.; Ávila, J.; Llorens-Martín, M. Adult hippocampal neurogenesis is abundant in neurologically healthy subjects and drops sharply in patients with Alzheimer's disease. *Nat. Med.* **2019**, *25*, 554–560. [CrossRef] [PubMed]

112. Song, J.; Kang, S.M.; Kim, E.; Kim, C.H.; Song, H.T.; Lee, J.E. Adiponectin receptor-mediated signaling ameliorates cerebral cell damage and regulates the neurogenesis of neural stem cells at high glucose concentrations: An in vivo and in vitro study. *Cell Death Dis.* **2015**, *6*, e1844. [CrossRef] [PubMed]

113. Teixeira, A.L.; Diniz, B.S.; Campos, A.C.; Miranda, A.S.; Rocha, N.P.; Talib, L.L.; Gattaz, W.F.; Forlenza, O.V. Decreased levels of circulating adiponectin in mild cognitive impairment and Alzheimer's disease. *Neuromol. Med.* **2013**, *15*, 115–121. [CrossRef] [PubMed]

114. Kamogawa, K.; Kohara, K.; Tabara, Y.; Uetani, E.; Nagai, T.; Yamamoto, M.; Igase, M.; Miki, T. Abdominal fat, adipose-derived hormones and mild cognitive impairment: The J-SHIPP study. *Dement. Geriatr. Cogn. Disord.* **2010**, *30*, 432–439. [CrossRef]

115. Gorska-Ciebiada, M.; Saryusz-Wolska, M.; Borkowska, A.; Ciebiada, M.; Loba, J. Adiponectin, leptin and IL-1 β in elderly diabetic patients with mild cognitive impairment. *Metab. Brain Dis.* **2016**, *31*, 257–266. [CrossRef]

116. Kitagawa, K.; Miwa, K.; Okazaki, S.; Sakaguchi, M.; Mochizuki, H. Serum high-molecular-weight adiponectin level and incident dementia in patients with vascular risk factors. *Euro. J. Neurol.* **2016**, *23*, 641–647. [CrossRef]

117. Bigalke, B.; Schreitmüller, B.; Sopova, K.; Paul, A.; Stransky, E.; Gawaz, M.; Stellos, K.; Laske, C. Adipocytokines and CD34 progenitor cells in Alzheimer's disease. *PLoS ONE* **2011**, *6*, e20286. [CrossRef]

118. Hardie, D.G. AMP-activated protein kinase: An energy sensor that regulates all aspects of cell function. *Genes Dev.* **2011**, *25*, 1895–1908. [CrossRef]

119. Ruderman, N.B.; Xu, X.J.; Nelson, L.; Cacicedo, J.M.; Saha, A.K.; Lan, F.; Ido, Y. AMPK and SIRT1: A long-standing partnership? *Am. J. Physiol Endocrinol. Metab.* **2010**, *298*, E751–E760. [CrossRef]

120. Badshah, H.; Ali, T.; Kim, M.O. Osmotin attenuates LPS-induced neuroinflammation and memory impairments via the TLR4/NFkappaB signaling pathway. *Sci. Rep.* **2016**, *6*, 24493. [CrossRef]

121. Nicolas, S.; Debayle, D.; Bechade, C.; Maroteaux, L.; Gay, A.S.; Bayer, P.; Heurteaux, C.; Guyon, A.; Chabry, J. Adiporon, an adiponectin receptor agonist acts as an antidepressant and metabolic regulator in a mouse model of depression. *Transl. Psychiatry* **2018**, *8*, 159. [CrossRef] [PubMed]

122. Zhang, D.; Wang, X.; Wang, B.; Garza, J.C.; Fang, X.; Wang, J.; Scherer, P.E.; Brenner, R.; Zhang, W.; Lu, X.Y. Adiponectin regulates contextual fear extinction and intrinsic excitability of dentate gyrus granule neurons through AdipoR2 receptors. *Mol. Psychiatry* **2017**, *22*, 1044–1055. [CrossRef] [PubMed]

123. Liu, B.; Liu, J.; Wang, J.G.; Liu, C.L.; Yan, H.J. AdipoRon improves cognitive dysfunction of Alzheimer's disease and rescues impaired neural stem cell proliferation through AdipoR1/AMPK pathway. *Exp. Neurol.* **2020**, *327*, 113249. [CrossRef]

124. FDA approves memantine drug for treating AD. *Am. J. Alzheimers Dis. Other Demen.* **2003**, *18*, 329–330.

125. Pákáski, M.; Fehér, A.; Juhász, A.; Drótos, G.; Fazekas, O.C.; Kovács, J.; Janka, Z.; Kálmán, J. Serum adipokine levels modified by donepezil treatment in Alzheimer's disease. *J. Alzheimers Dis.* **2014**, *38*, 371–377. [CrossRef]

126. Ali, M.A.; El-Abhar, H.S.; Kamel, M.A.; Attia, A.S. Antidiabetic Effect of Galantamine: Novel Effect for a Known centrally acting drug. *PLoS ONE* **2015**, *10*, e0134648. [CrossRef]

127. Song, X.; Jensen, M.O.; Jogini, V.; Stein, R.A.; Lee, C.H.; McHaourab, H.S.; Shaw, D.E.; Gouaux, E. Mechanism of NMDA receptor channel block by MK-801 and memantine. *Nature* **2018**, *556*, 515–519. [CrossRef]

128. Liu, B.; Liu, J.; Wang, J.; Sun, F.; Jiang, S.; Hu, F.; Wang, D.; Liu, D.; Liu, C.; Yan, H. Adiponectin protects against cerebral ischemic injury through AdipoR1/AMPK pathways. *Front. Pharmacol.* **2019**, *10*, 597. [CrossRef]

129. Yue, L.; Zhao, L.; Liu, H.; Li, X.; Wang, B.; Guo, H.; Gao, L.; Feng, D.; Qu, Y. Adiponectin Protects against Glutamate-Induced Excitotoxicity via Activating SIRT1-Dependent PGC-1alpha Expression in HT22 Hippocampal Neurons. *Oxid. Med. Cell Longev.* **2016**, *2016*, 2957354. [CrossRef]

130. Shah, S.A.; Lee, H.Y.; Bressan, R.A.; Yun, D.J.; Kim, M.O. Novel osmotin attenuates glutamate-induced synaptic dysfunction and neurodegeneration via the JNK/PI3K/Akt pathway in postnatal rat brain. *Cell Death Dis.* **2014**, *5*, e1026. [CrossRef]

131. Tumminia, A.; Vinciguerra, F.; Parisi, M.; Frittitta, L. Type 2 Diabetes mellitus and Alzheimer's disease: Role of insulin signalling and therapeutic implications. *Int. J. Mol. Sci.* **2018**, *19*, 3306. [CrossRef] [PubMed]

132. Li, J.; Xue, Y.M.; Zhu, B.; Pan, Y.H.; Zhang, Y.; Wang, C.; Li, Y. rosiglitazone elicits an adiponectin-mediated insulin-sensitizing action at the adipose tissue-liver axis in otsuka long-evans tokushima fatty rats. *J. Diabetes Res.* **2018**, *2018*, 4627842. [CrossRef] [PubMed]

133. American Diabetes, A. 9. Pharmacologic approaches to glycemic treatment: Standards of medical care in diabetes-2019. *Diabetes Care* **2019**, *42*, S90–S102. [CrossRef] [PubMed]

134. Kickstein, E.; Krauss, S.; Thornhill, P.; Rutschow, D.; Zeller, R.; Sharkey, J.; Williamson, R.; Fuchs, M.; Kohler, A.; Glossmann, H.; et al. Biguanide metformin acts on tau phosphorylation via mTOR/protein phosphatase 2A (PP2A) signaling. *Proc. Natl. Acad. Sci. USA* **2010**, *107*, 21830–21835. [CrossRef]

135. Gupta, A.; Bisht, B.; Dey, C.S. Peripheral insulin-sensitizer drug metformin ameliorates neuronal insulin resistance and Alzheimer's-like changes. *Neuropharmacology* **2011**, *60*, 910–920. [CrossRef]

136. Li, J.; Deng, J.; Sheng, W.; Zuo, Z. Metformin attenuates Alzheimer's disease-like neuropathology in obese, leptin-resistant mice. *Pharmacol. Biochem. Behav.* **2012**, *101*, 564–574. [CrossRef]

137. Luchsinger, J.A.; Perez, T.; Chang, H.; Mehta, P.; Steffener, J.; Pradabhan, G.; Ichise, M.; Manly, J.; Devanand, D.P.; Bagiella, E. Metformin in amnestic mild cognitive impairment: Results of a pilot randomized placebo controlled clinical trial. *J. Alzheimer Disease JAD* **2016**, *51*, 501–514. [CrossRef]

138. Hsu, C.C.; Wahlqvist, M.L.; Lee, M.S.; Tsai, H.N. Incidence of dementia is increased in type 2 diabetes and reduced by the use of sulfonylureas and metformin. *J. Alzheimer Disease JAD* **2011**, *24*, 485–493. [CrossRef]

139. Abbas, N.A.T.; El Salem, A. Metformin, sitagliptin, and liraglutide modulate serum retinol-binding protein-4 level and adipocytokine production in type 2 diabetes mellitus rat model. *Can. J. Physiol. Pharmacol.* **2018**, *96*, 1226–1231. [CrossRef]

140. Su, J.R.; Lu, Z.H.; Su, Y.; Zhao, N.; Dong, C.L.; Sun, L.; Zhao, S.F.; Li, Y. relationship of serum adiponectin levels and metformin therapy in patients with type 2 diabetes. *Horm. Metab. Res.* **2016**, *48*, 92–98. [CrossRef]

141. Chiang, M.C.; Cheng, Y.C.; Chen, S.J.; Yen, C.H.; Huang, R.N. Metformin activation of AMPK-dependent pathways is neuroprotective in human neural stem cells against Amyloid-beta-induced mitochondrial dysfunction. *Exp. Cell Res.* **2016**, *347*, 322–331. [CrossRef] [PubMed]

142. Kong, W.; Niu, X.; Zeng, T.; Lu, M.; Chen, L. Impact of treatment with metformin on adipocytokines in patients with polycystic ovary syndrome: A meta-analysis. *PLoS ONE* **2015**, *10*, e0140565. [CrossRef]

143. Chung, M.M.; Chen, Y.L.; Pei, D.; Cheng, Y.C.; Sun, B.; Nicol, C.J.; Yen, C.H.; Chen, H.M.; Liang, Y.J.; Chiang, M.C. The neuroprotective role of metformin in advanced glycation end product treated human neural stem cells is AMPK-dependent. *Biochim. Biophys. Acta* **2015**, *1852*, 720–731. [CrossRef] [PubMed]

144. Saltiel, A.R.; Olefsky, J.M. Thiazolidinediones in the treatment of insulin resistance and type II diabetes. *Diabetes* **1996**, *45*, 1661–1669. [CrossRef]

145. Malinowski, J.M.; Bolesta, S. Rosiglitazone in the treatment of type 2 diabetes mellitus: A critical review. *Clin. Ther.* **2000**, *22*, 1151–1168. [CrossRef]

146. Jojo, G.M.; Kuppusamy, G. Scope of new formulation approaches in the repurposing of pioglitazone for the management of Alzheimer's disease. *J. Clin. Pharm. Ther.* **2019**, *44*, 337–348. [CrossRef] [PubMed]

147. Silva-Abreu, M.; Gonzalez-Pizarro, R.; Espinoza, L.C.; Rodriguez-Lagunas, M.J.; Espina, M.; Garcia, M.L.; Calpena, A.C. Thiazolidinedione as an alternative to facilitate oral administration in geriatric patients with Alzheimer's disease. *Eur. J. Pharm. Sci.* **2019**, *129*, 173–180. [CrossRef]

148. Tsuchida, A.; Yamauchi, T.; Takekawa, S.; Hada, Y.; Ito, Y.; Maki, T.; Kadowaki, T. Peroxisome proliferator-activated receptor (PPAR)alpha activation increases adiponectin receptors and reduces obesity-related inflammation in adipose tissue: Comparison of activation of PPARα, PPARγ, and their combination. *Diabetes* **2005**, *54*, 3358–3370. [CrossRef]

149. Holscher, C. Potential role of glucagon-like peptide-1 (GLP-1) in neuroprotection. *CNS Drugs* **2012**, *26*, 871–882. [CrossRef]

150. Kakimura, J.; Kitamura, Y.; Takata, K.; Umeki, M.; Suzuki, S.; Shibagaki, K.; Taniguchi, T.; Nomura, Y.; Gebicke-Haerter, P.J.; Smith, M.A.; et al. Microglial activation and amyloid-beta clearance induced by exogenous heat-shock proteins. *FASEB J.* **2002**, *16*, 601–603. [CrossRef]

151. Perry, T.; Holloway, H.W.; Weerasuriya, A.; Mouton, P.R.; Duffy, K.; Mattison, J.A.; Greig, N.H. Evidence of GLP-1-mediated neuroprotection in an animal model of pyridoxine-induced peripheral sensory neuropathy. *Exp. Neurol.* **2007**, *203*, 293–301. [CrossRef]

152. Baggio, L.L.; Drucker, D.J. Glucagon-like peptide-1 receptors in the brain: Controlling food intake and body weight. *J. Clin. Investig.* **2014**, *124*, 4223–4226. [CrossRef]

153. During, M.J.; Cao, L.; Zuzga, D.S.; Francis, J.S.; Fitzsimons, H.L.; Jiao, X.; Bland, R.J.; Klugmann, M.; Banks, W.A.; Drucker, D.J.; et al. Glucagon-like peptide-1 receptor is involved in learning and neuroprotection. *Nat. Med.* **2003**, *9*, 1173–1179. [CrossRef]

154. Hamilton, A.; Holscher, C. Receptors for the incretin glucagon-like peptide-1 are expressed on neurons in the central nervous system. *Neuroreport* **2009**, *20*, 1161–1166. [CrossRef]

155. Wang, A.; Li, T.; An, P.; Yan, W.; Zheng, H.; Wang, B.; Mu, Y. Exendin-4 upregulates adiponectin level in adipocytes via Sirt1/Foxo-1 signaling pathway. *PLoS ONE* **2017**, *12*, e0169469. [CrossRef] [PubMed]

156. Ryan, D.; Acosta, A. GLP-1 receptor agonists: Nonglycemic clinical effects in weight loss and beyond. *Obesity* **2015**, *23*, 1119–1129. [CrossRef]

157. Gault, V.A.; Holscher, C. GLP-1 agonists facilitate hippocampal LTP and reverse the impairment of LTP induced by beta-amyloid. *Eur. J. Pharmacol.* **2008**, *587*, 112–117. [CrossRef]

158. Han, W.N.; Holscher, C.; Yuan, L.; Yang, W.; Wang, X.H.; Wu, M.N.; Qi, J.S. Liraglutide protects against amyloid-beta protein-induced impairment of spatial learning and memory in rats. *Neurobiol. Aging.* **2013**, *34*, 576–588. [CrossRef]

159. Batista, A.F.; Forny-Germano, L.; Clarke, J.R.; Lyra, E.S.N.M.; Brito-Moreira, J.; Boehnke, S.E.; Winterborn, A.; Coe, B.C.; Lablans, A.; Vital, J.F.; et al. The diabetes drug liraglutide reverses cognitive impairment in mice and attenuates insulin receptor and synaptic pathology in a non-human primate model of Alzheimer's disease. *J. Pathol.* **2018**, *245*, 85–100. [CrossRef]

160. Wicinski, M.; Socha, M.; Malinowski, B.; Wodkiewicz, E.; Walczak, M.; Gorski, K.; Slupski, M.; Pawlak-Osinska, K. Liraglutide and its neuroprotective properties-focus on possible biochemical mechanisms in Alzheimer's disease and cerebral ischemic events. *Int. J. Mol. Sci.* **2019**, *20*, 1050. [CrossRef]

161. Zhang, Y.; Xie, J.Z.; Xu, X.Y.; Hu, J.; Xu, T.; Jin, S.; Yang, S.J.; Wang, J.Z. Liraglutide ameliorates hyperhomocysteinemia-induced alzheimer-like pathology and memory deficits in rats via multi-molecular targeting. *Neurosci. Bull.* **2019**, *35*, 724–734. [CrossRef] [PubMed]

162. Cahn, A.; Cernea, S.; Raz, I. An update on DPP-4 inhibitors in the management of type 2 diabetes. *Expert Opin. Emerg. Drugs* **2016**, *21*, 409–419. [CrossRef] [PubMed]

163. Chen, S.; Zhou, M.; Sun, J.; Guo, A.; Fernando, R.L.; Chen, Y.; Peng, P.; Zhao, G.; Deng, Y. DPP-4 inhibitor improves learning and memory deficits and AD-like neurodegeneration by modulating the GLP-1 signaling. *Neuropharmacology* **2019**, *157*, 107668. [CrossRef] [PubMed]

164. Sahebkar, A.; Ponzo, V.; Bo, S. Effect of dipeptidyl peptidase-4 inhibitors on plasma adiponectin: A systematic review and meta-analysis of randomized controlled trials. *Curr. Med. Chem.* **2016**, *23*, 1356–1369. [CrossRef]

165. Kosaraju, J.; Holsinger, R.M.D.; Guo, L.; Tam, K.Y. Linagliptin, a dipeptidyl peptidase-4 inhibitor, mitigates cognitive deficits and pathology in the 3xTg-AD mouse model of Alzheimer's disease. *Mol. Neurobiol.* **2017**, *54*, 6074–6084. [CrossRef]

166. Isik, A.T.; Soysal, P.; Yay, A.; Usarel, C. The effects of sitagliptin, a DPP-4 inhibitor, on cognitive functions in elderly diabetic patients with or without Alzheimer's disease. *Diabetes Res. Clin. Pract.* **2017**, *123*, 192–198. [CrossRef]

167. Angelopoulou, E.; Piperi, C. DPP-4 inhibitors: A promising therapeutic approach against Alzheimer's disease. *Ann. Transl. Med.* **2018**, *6*, 255. [CrossRef]

168. Claassen, J.A. New cardiovascular targets to prevent late onset Alzheimer disease. *Eur. J. Pharmacol.* **2015**, *763*, 131–134. [CrossRef]

169. Montecucco, F.; Mach, F. Update on therapeutic strategies to increase adiponectin function and secretion in metabolic syndrome. *Diabetes Obes. Metab.* **2009**, *11*, 445–454. [CrossRef]

170. Mogi, M.; Li, J.M.; Tsukuda, K.; Iwanami, J.; Min, L.J.; Sakata, A.; Fujita, T.; Iwai, M.; Horiuchi, M. Telmisartan prevented cognitive decline partly due to PPAR-gamma activation. *Biochem. Biophys. Res. Commun.* **2008**, *375*, 446–449. [CrossRef]

171. Takeda, S.; Sato, N.; Takeuchi, D.; Kurinami, H.; Shinohara, M.; Niisato, K.; Kano, M.; Ogihara, T.; Rakugi, H.; Morishita, R. Angiotensin receptor blocker prevented beta-amyloid-induced cognitive impairment associated with recovery of neurovascular coupling. *Hypertension* **2009**, *54*, 1345–1352. [CrossRef]

172. Ongali, B.; Nicolakakis, N.; Tong, X.K.; Aboulkassim, T.; Papadopoulos, P.; Rosa-Neto, P.; Lecrux, C.; Imboden, H.; Hamel, E. Angiotensin II type 1 receptor blocker losartan prevents and rescues cerebrovascular, neuropathological and cognitive deficits in an Alzheimer's disease model. *Neurobiol. Dis.* **2014**, *68*, 126–136. [CrossRef]

173. Saavedra, J.M. Evidence to consider angiotensin II receptor blockers for the treatment of early Alzheimer's disease. *Cell Mol. Neurobiol.* **2016**, *36*, 259–279. [CrossRef] [PubMed]

174. Li, W.; Zhang, J.W.; Lu, F.; Ma, M.M.; Wang, J.Q.; Suo, A.Q.; Bai, Y.Y.; Liu, H.Q. Effects of telmisartan on the level of Aβ1-42, interleukin-1β, tumor necrosis factor α and cognition in hypertensive patients with Alzheimer's disease. *Zhonghua Yi Xue Za Zhi* **2012**, *92*, 2743–2746.

175. Hsu, C.Y.; Huang, C.C.; Chan, W.L.; Huang, P.H.; Chiang, C.H.; Chen, T.J.; Chung, C.M.; Lin, S.J.; Chen, J.W.; Leu, H.B. Angiotensin-receptor blockers and risk of Alzheimer's disease in hypertension population-A nationwide cohort study. *Circ. J.* **2013**, *77*, 405–410. [CrossRef]

176. Benicky, J.; Sánchez-Lemus, E.; Pavel, J.; Saavedra, J.M. Anti-inflammatory effects of angiotensin receptor blockers in the brain and the periphery. *Cell Mol. Neurobiol.* **2009**, *29*, 781–792. [CrossRef] [PubMed]

177. Panahpour, H.; Nekooeian, A.A.; Dehghani, G.A. Candesartan attenuates ischemic brain edema and protects the blood-brain barrier integrity from ischemia/reperfusion injury in rats. *Iran. Biomed. J.* **2014**, *18*, 232–238. [CrossRef]

178. So, G.; Nakagawa, S.; Morofuji, Y.; Hiu, T.; Hayashi, K.; Tanaka, K.; Suyama, K.; Deli, M.A.; Nagata, I.; Matsuo, T.; et al. Candesartan improves ischemia-induced impairment of the blood-brain barrier in vitro. *Cell Mol. Neurobiol.* **2015**, *35*, 563–572. [CrossRef]

179. Clasen, R.; Schupp, M.; Foryst-Ludwig, A.; Sprang, C.; Clemenz, M.; Krikov, M.; Thone-Reineke, C.; Unger, T.; Kintscher, U. PPARgamma-activating angiotensin type-1 receptor blockers induce adiponectin. *Hypertension* **2005**, *46*, 137–143. [CrossRef]

180. Zorad, S.; Dou, J.T.; Benicky, J.; Hutanu, D.; Tybitanclova, K.; Zhou, J.; Saavedra, J.M. Long-term angiotensin II AT1 receptor inhibition produces adipose tissue hypotrophy accompanied by increased expression of adiponectin and PPARgamma. *Eur. J. Pharmacol.* **2006**, *552*, 112–122. [CrossRef] [PubMed]

181. Makita, S.; Abiko, A.; Naganuma, Y.; Moriai, Y.; Nakamura, M. Potential effects of angiotensin II receptor blockers on glucose tolerance and adiponectin levels in hypertensive patients. *Cardiovasc. Drugs Ther.* **2007**, *21*, 317–318. [CrossRef] [PubMed]

182. Kohlstedt, K.; Gershome, C.; Trouvain, C.; Hofmann, W.K.; Fichtlscherer, S.; Fleming, I. Angiotensin-converting enzyme (ACE) inhibitors modulate cellular retinol-binding protein 1 and adiponectin expression in adipocytes via the ACE-dependent signaling cascade. *Mol. Pharmacol.* **2009**, *75*, 685–692. [CrossRef] [PubMed]

183. Fontana, V.; de Faria, A.P.C.; Oliveira-Paula, G.H.; Silva, P.S.; Biagi, C.; Tanus-Santos, J.E.; Moreno, H. Effects of angiotensin-converting enzyme inhibition on leptin and adiponectin levels in essential hypertension. *Basic. Clin. Pharmacol. Toxicol.* **2014**, *114*, 472–475. [CrossRef] [PubMed]

184. Hajjar, I.M.; Keown, M.; Lewis, P.; Almor, A. Angiotensin converting enzyme inhibitors and cognitive and functional decline in patients with Alzheimer's disease: An observational study. *Am. J. Alzheimer Dis. Other Dement.* **2008**, *23*, 77–83. [CrossRef]

185. Soto, M.E.; van Kan, G.A.; Nourhashemi, F.; Gillette-Guyonnet, S.; Cesari, M.; Cantet, C.; Rolland, Y.; Vellas, B. Angiotensin-converting enzyme inhibitors and Alzheimer's disease progression in older adults: Results from the Réseau sur la Maladie d'Alzheimer Français cohort. *J. Am. Geriatr. Soc.* **2013**, *61*, 1482–1488. [CrossRef]

186. De Oliveira, F.F.; Chen, E.S.; Smith, M.C.; Bertolucci, P.H.F. Pharmacogenetics of angiotensin-converting enzyme inhibitors in patients with Alzheimer's disease dementia. *Curr. Alzheimer Res.* **2018**, *15*, 386–398. [CrossRef]

187. Zhuang, S.; Wang, X.; Wang, H.F.; Li, J.; Wang, H.Y.; Zhang, H.Z.; Xing, C.M. Angiotensin converting enzyme serum activities: Relationship with Alzheimer's disease. *Brain Res.* **2016**, *1650*, 196–202. [CrossRef]

188. AbdAlla, S.; El Hakim, A.; Abdelbaset, A.; Elfaramawy, Y.; Quitterer, U. Inhibition of ACE Retards Tau hyperphosphorylation and signs of neuronal degeneration in aged rats subjected to chronic mild stress. *Biomed. Res. Int.* **2015**, *2015*, 917156. [CrossRef]

189. AbdAlla, S.; Langer, A.; Fu, X.; Quitterer, U. ACE inhibition with captopril retards the development of signs of neurodegeneration in an animal model of Alzheimer's disease. *Int. J. Mol. Sci.* **2013**, *14*, 16917–16942. [CrossRef]

190. Dong, Y.F.; Kataoka, K.; Tokutomi, Y.; Nako, H.; Nakamura, T.; Toyama, K.; Sueta, D.; Koibuchi, N.; Yamamoto, E.; Ogawa, H.; et al. Perindopril, a centrally active angiotensin-converting enzyme inhibitor, prevents cognitive impairment in mouse models of Alzheimer's disease. *FASEB J.* **2011**, *25*, 2911–2920. [CrossRef]

191. Zhuang, S.; Wang, H.F.; Wang, X.; Li, J.; Xing, C.M. The association of renin-angiotensin system blockade use with the risks of cognitive impairment of aging and Alzheimer's disease: A meta-analysis. *J. Clin. Neurosci.* **2016**, *33*, 32–38. [CrossRef]

192. Bernstein, K.E.; Koronyo, Y.; Salumbides, B.C.; Sheyn, J.; Pelissier, L.; Lopes, D.H.J.; Shah, K.H.; Bernstein, E.A.; Fuchs, D.T.; Yu, J.J.Y.; et al. Angiotensin-converting enzyme overexpression in myelomonocytes prevents Alzheimer's-like cognitive decline. *J. Clin. Investig.* **2014**, *124*, 1000–1012. [CrossRef] [PubMed]

193. Liu, S.; Liu, J.; Miura, Y.; Tanabe, C.; Maeda, T.; Terayama, Y.; Turner, A.J.; Zou, K.; Komano, H. Conversion of Aβ43 to Aβ40 by the successive action of angiotensin-converting enzyme 2 and angiotensin-converting enzyme. *J. Neurosci. Res.* **2014**, *92*, 1178–1186. [CrossRef] [PubMed]

194. Zou, K.; Yamaguchi, H.; Akatsu, H.; Sakamoto, T.; Ko, M.; Mizoguchi, K.; Gong, J.-S.; Yu, W.; Yamamoto, T.; Kosaka, K.; et al. Angiotensin-converting enzyme converts amyloid beta-protein 1-42 (Abeta(1-42)) to Abeta(1-40), and its inhibition enhances brain Abeta deposition. *J. Neurosci.* **2007**, *27*, 8628–8635. [CrossRef] [PubMed]

195. Okopien, B.; Buldak, L.; Boldys, A. Fibrates in the management of atherogenic dyslipidemia. *Expert Rev. Cardiovasc. Ther.* **2017**, *15*, 913–921. [CrossRef] [PubMed]

196. Sahebkar, A.; Watts, G.F. Fibrate therapy and circulating adiponectin concentrations: A systematic review and meta-analysis of randomized placebo-controlled trials. *Atherosclerosis* **2013**, *230*, 110–120. [CrossRef]

197. D'Orio, B.; Fracassi, A.; Ceru, M.P.; Moreno, S. Targeting PPARalpha in Alzheimer's Disease. *Curr. Alzheimer Res.* **2018**, *15*, 345–354. [CrossRef]

198. Kai, T.; Arima, S.; Taniyama, Y.; Nakabou, M.; Kanamasa, K. Comparison of the effect of lipophilic and hydrophilic statins on serum adiponectin levels in patients with mild hypertension and dyslipidemia: Kinki Adiponectin Interventional (KAI) Study. *Clin. Exp. Hypertens* **2008**, *30*, 530–540. [CrossRef]

199. Qu, H.Y.; Xiao, Y.W.; Jiang, G.H.; Wang, Z.Y.; Zhang, Y.; Zhang, M. Effect of atorvastatin versus rosuvastatin on levels of serum lipids, inflammatory markers and adiponectin in patients with hypercholesterolemia. *Pharm. Res.* **2009**, *26*, 958–964. [CrossRef]

200. Tsutamoto, T.; Yamaji, M.; Kawahara, C.; Nishiyama, K.; Fujii, M.; Yamamoto, T.; Horie, M. Effect of simvastatin vs. rosuvastatin on adiponectin and haemoglobin A1c levels in patients with non-ischaemic chronic heart failure. *Eur. J. Heart Fail.* **2009**, *11*, 1195–1201. [CrossRef]

201. Guimaraes, E.S.; Cerda, A.; Dorea, E.L.; Bernik, M.M.S.; Gusukuma, M.C.; Pinto, G.A.; Fajardo, C.M.; Hirata, M.H.; Hirata, R.D.C. Effects of short-term add-on ezetimibe to statin treatment on expression of adipokines and inflammatory markers in diabetic and dyslipidemic patients. *Cardiovasc. Ther.* **2017**, *35*. [CrossRef] [PubMed]

202. Hu, Y.; Tong, G.; Xu, W.; Pan, J.; Ryan, K.; Yang, R.; Shuldiner, A.R.; Gong, D.W.; Zhu, D. Anti-inflammatory effects of simvastatin on adipokines in type 2 diabetic patients with carotid atherosclerosis. *Diab. Vasc. Dis. Res.* **2009**, *6*, 262–268. [CrossRef] [PubMed]

International Journal of
Molecular Sciences

Review

Neurotrophic Factor BDNF, Physiological Functions and Therapeutic Potential in Depression, Neurodegeneration and Brain Cancer

Luca Colucci-D'Amato [1,2,*], Luisa Speranza [3] and Floriana Volpicelli [4]

1   Department of Environmental, Biological and Pharmaceutical Sciences and Technologies, University of Campania "Luigi Vanvitelli", 81100 Caserta, Italy
2   InterUniversity Center for Research in Neurosciences (CIRN), University of Campania "Luigi Vanvitelli", 80131 Naples, Italy
3   Department of Neuroscience, Albert Einstein College of Medicine, New York, NY 10461, USA; luisa.speranza@einsteinmed.org
4   Department of Pharmacy, School of Medicine and Surgery, University of Naples Federico II, 80131 Naples, Italy; floriana.volpicelli@unina.it
*   Correspondence: luca.colucci@unicampania.it; Tel.: +39-0823-274577

Received: 29 September 2020; Accepted: 19 October 2020; Published: 21 October 2020

Abstract: Brain-derived neurotrophic factor (BDNF) is one of the most distributed and extensively studied neurotrophins in the mammalian brain. BDNF signals through the tropomycin receptor kinase B (TrkB) and the low affinity p75 neurotrophin receptor (p75NTR). BDNF plays an important role in proper growth, development, and plasticity of glutamatergic and GABAergic synapses and through modulation of neuronal differentiation, it influences serotonergic and dopaminergic neurotransmission. BDNF acts as paracrine and autocrine factor, on both pre-synaptic and post-synaptic target sites. It is crucial in the transformation of synaptic activity into long-term synaptic memories. BDNF is considered an instructive mediator of functional and structural plasticity in the central nervous system (CNS), influencing dendritic spines and, at least in the hippocampus, the adult neurogenesis. Changes in the rate of adult neurogenesis and in spine density can influence several forms of learning and memory and can contribute to depression-like behaviors. The possible roles of BDNF in neuronal plasticity highlighted in this review focus on the effect of antidepressant therapies on BDNF-mediated plasticity. Moreover, we will review data that illustrate the role of BDNF as a potent protective factor that is able to confer protection against neurodegeneration, in particular in Alzheimer's disease. Finally, we will give evidence of how the involvement of BDNF in the pathogenesis of brain glioblastoma has emerged, thus opening new avenues for the treatment of this deadly cancer.

Keywords: BDNF; miRNAs; neurogenesis; synaptic plasticity; depression; neurodegeneration; glioblastoma

---

1. Introduction

The neurotrophin BDNF is one of the most studied and well characterized neurotrophic factors in the CNS. It regulates many different cellular processes involved in the development and maintenance of normal brain function by binding and activating the TrkB, a member of the larger family of Trk receptors. In the brain, BDNF is expressed by glutamatergic neurons [1], glial cells, such as astrocytes isolated from the cortex and hippocampus, but not from the striatum [2], and microglia [3]. During embryogenesis, BDNF–TrkB signaling promotes the differentiation of cortical progenitor cells and later promotes differentiation of cortical progenitor cells into neurons (i.e., neurogenesis) [4]. Several lines of evidence also suggest that the BDNF/TrkB signaling is involved in adult neurogenesis in the hippocampus with differing effects in the dentate gyrus (DG) and subventricular zone (SVZ) [5].

Adult neurogenesis in the dentate gyrus is enhanced by voluntary exercise, exposure to an enriched environment, and chronic antidepressant administration. Recently, it has also been proposed that caloric restriction and intermittent fasting in particular, appears to positively modulate hippocampal neurogenesis and BDNF [6]. The connection between BDNF and the modulation of hippocampal neurogenesis by external stimuli is a topic that has been extensively studied in recent years [7]. It has been demonstrated that voluntary physical exercise, like an enriched environment, increases expression of BDNF in the hippocampus [8], as well as hippocampal neurogenesis [9]. Physical exercise is one particularly effective strategy for increasing circulating levels of BDNF [10,11] and improving brain function [12–14].

In addition, studies also show that BDNF is an important regulator of synaptic transmission and long-term potentiation (LTP) in the hippocampus and in other brain regions. The effects of BDNF on LTP are mediated by the TrkB receptor. Especially in the hippocampus, this neurotrophin is thought to act on both the pre- and post-synaptic compartments, modulating synaptic efficacy, either by changing the pre-synaptic transmitter release, or by increasing post-synaptic transmitter sensitivity [15,16] to induce a long-lasting increase in synaptic plasticity. Additionally, converging data now suggest a role for BDNF in the pathophysiology of brain-associated illnesses. Deficits in BDNF signaling are reported to contribute to the pathogenesis of several major diseases, such as Huntington's disease, Alzheimer's disease (AD), depression, schizophrenia, bipolar, and anxiety disorders. Thus, manipulating the BDNF signaling may present a viable approach to treat a variety of neurological and psychiatric disorders. BDNF protein is also detectable outside of the nervous system in several non-neuronal tissues, such as in endothelial cells [17,18], cardiomyocytes [19], vascular smooth muscle cells [17], leukocytes [20], platelets [21,22], and megakaryocytes [19]. Therefore, it may also be involved in cancer, angiogenesis, reduction of glucose production from the liver [23], and in the uptake of glucose in peripheral tissues (see [24] for review). In addition, BDNF promotes the development of neuromuscular synapses and is required for fiber-type specification, suggesting a potential role as a therapeutic target in muscle diseases [25]. In this review, first we examine the currently known mechanisms of BDNF signaling, information essential for the creation of BDNF-based therapeutics. Next, we focus on the effects of antidepressants on BDNF-mediated plasticity. Additionally, we highlight the function of BDNF as a potent factor capable of conferring protection against neurodegeneration. Finally, we touch on the newly emerging role of BDNF in the pathogenesis of brain gliomas.

2. The Human BDNF Gene: Transcripts and Variants

2.1. BDNF Transcripts

The *BDNF* gene codes for a neurotrophin that is highly expressed in the CNS [26]. At the beginning of 2000s, the only data available about the structure and regulation of the *BDNF* gene were from Timmusk and colleagues, which identified in rats four 5′ exons linked to separate promoters and one 3′ exon encoding the preproBDNF protein [27–29]. These four *BDNF* promoters owned multiple points of *BDNF* mRNA regulation and suggested an activity-dependent regulation [28–31]. Further studies, published in 2007, clarified that *BDNF* has a complex gene structure with 11 different exons in humans, nine different exons in rodents, and nine alternative promoters for both groups [32,33]. Despite this complexity, the coding sequence is located in exon IX in both human and rodents. The latter includes the common sequence that encodes for the proBDNF protein. All other exons are untranslated regions with a start codon present in exons I, VII, VIII, and IX of the human *BDNF* gene. Exon IX is present in all BDNF mRNA isoforms. It is supposed that the nine alternative promoters can regulate the complex spatio-temporal expression of *BDNF* gene and allow BDNF to respond to a greater variety of stimuli. For instance, in human brain tissues, all exons are expressed, but to different degrees and in different brain structures [33]. Human heart tissue, instead, expresses high levels of *BDNF* isoforms containing exon IV and exon IX [33].

Currently, two *BDNF* promoters, promoter I and promoter IV, have been well characterized for their response after the activation of the L-type voltage gated calcium channel (L-VGCC) or the n-methyl-d-aspartate (NMDA) receptor. Activation of L-VGCC and NMDA receptors mediate intracellular Ca^{2+}-signaling and regulate several aspects of brain functions (for review [34,35]). Promoter I is more responsive to neuronal activity and induces activity-dependent expression of BDNF in vitro and in vivo. It contains calcium-responsive elements (CaREs) and cyclic adenosine monophosphate (cAMP)/calcium response element (CRE) [29,36–38]. Deletion of CRE or overexpression of dominant negative of CREB (cAMP-response element-binding protein) significantly impairs rat BDNF promoter I response to neuronal depolarization [38]. Human BDNF promoter I is similar to rat promoter, since an orthologous CRE-like element is also present [39]. However, mutation of this site did not affect human *BDNF* promoter I response to depolarization [39]. Human BDNF promoter I also contains an activator protein 1 (AP1) -like element and an asymmetric E-box-like element [39]. Mutation in E-box-like element reduces human *BDNF* promoter I induction, impairing the response to neuronal depolarization [39].

Another highly characterized *BDNF* promoter is the *BDNF* promoter IV that contributes significantly to activity-dependent *BDNF* transcription. Human and rat *BDNF* promoter IV are similar. In this promoter, three CaREs and three other regulatory elements involved in regulating rat *BDNF* promoter response to NMDA receptor activation have been identified [40]. NMDA receptor activation is capable of triggering *BDNF* exon IV transcription through a protein-signaling cascade requiring extracellular signal-regulated kinase (ERK), Ca^{2+}/calmodulin-dependent protein kinase (CaMK) II/IV, phosphoinositide 3-kinases (PI3K), and phospholipase C (PLC). *BDNF* exon IV expression also seems capable of further stimulating its own expression through TrkB activation [41]. Additionally to the CaREs, two positive regulators have been identified: the NF-kB (nuclear factor kappa-light-chain-enhancer of activated B cells) [42] and NFAT (nuclear factor of activated T-cells) binding sites [30,43]. In contrast to these positive regulators, *BDNF* promoter IV also contains a negative regulatory element, the class B E-box. This is a binding site for a basic helix-loop-helix protein, BHLHB2, a suppressor of the bHLH gene superfamily [43,44]. NMDA treatment is able to remove BHLHB2 binding to the E-box and to increase rat *BDNF* promoter IV activity [43,44]. Disruption of *BDNF* promoter IV in mice significantly reduced the number of parvalbumin GABAergic neurons in the prefrontal cortex and impaired GABAergic activity [45]. These mice displayed depression-like behavior such as anhedonia-like behavior and increased latency to escape in the learned helplessness test [45]. Further evidence suggests a relationship between stress exposure and epigenetic regulation of *BDNF* promoter IV with the development of psychiatric disorders. Specifically, changes in *BDNF* promoter IV methylation levels are implicated in depression [46,47]. Preliminary evidence has demonstrated that patients with major depressive disorder (MDD) present a hypomethylation of the CpG-87 site of the promoter IV region of *BDNF* gene and are less likely to benefit from antidepressants [47,48]. In addition, *BDNF* disruption from promoter IV-derived transcripts impairs fear expression in mice, suggesting that cells expressing *BDNF* from promoter IV critically regulate hippocampal-prefrontal plasticity during fear memory [49,50].

2.2. miRNAs and BDNF

MicroRNAs (miRNAs) are a class of evolutionary conserved small non-coding single-strand RNA molecules, 18–25 nucleotide long, able to bind to 3′ untranslated regions (3′ UTR) of target mRNAs and promote their degradation or suppress their translation into proteins. MiRNAs are expressed abundantly within the nervous system in a tissue-specific manner and are crucial players in several biological processes, including neurogenesis, neuronal maturation, synapse formation, axon guidance, neurite outgrowth and neuronal plasticity [51–53]. Accumulating data indicate that synthesis of BDNF may be affected by miRNAs, indeed, a regulatory negative feedback loop between BDNF and miRNAs exists. That is, while BDNF treatment stimulates neuronal miRNAs expression, miRNAs generally inhibit the expression of BDNF [54]. This negative feedback loop is maintained in a state of equilibrium in normal cells. Alterations in miRNAs or in BDNF contribute to the pathogenic mechanisms involved in neurodegenerative diseases or neuropsychiatric disorders.

A number of recent studies, obtained from high throughput sequencing screening of different brain regions or from neurological disorders, have identified seven miRNAs (miR-15a, miR-206, miR-155-5p, miR-16, miR-103-3p, miR-330-3p, Let-7a-3p) correlated with BDNF [55]. Previous data published by Schratt et al. reported the involvement of miR-134 in BDNF-regulated dendritic spine size in hippocampal neurons. They demonstrate that miR-134 negatively regulates the spine size via repressing the translation of LIM kinase 1, which is known to regulate dendritic structures. BDNF is able to relieve the inhibition of LIM kinase 1 translation, and in this manner contribute to synaptic development, maturation and/or plasticity [56]. Recently, Baby et al. [57] found that miR-134 mediates post-transcriptional regulation of CREB1 and BDNF, as previously described by Gao et al. [58], who demonstrated that mutant mice lacking Sirtuin 1 (SIRT1) catalytic activity shows reduction in both CREB and BDNF proteins and upregulation of miR-134. Thus, higher levels of miR-134 negatively regulate synaptic plasticity [58]. MiR-134-mediated post-transcriptional regulation of CREB1 and BDNF prevents cognitive deficits in chronic unpredicted mild stress model (CUMS) [59]. At the same time, data published by Xin and coworkers [60] demonstrate that miR-202-3p silencing reduces the damage to hippocampal nerve in CUMS rats through the upregulation of BNDF expression. miRNAs could be an effective target also for the treatment of depression. Recent data demonstrate that miR-124-mediated post-transcriptional regulation of CREB1 and BDNF can improve depression-like behavior in a rat model [61]. Instead, miR-153 through the inhibition of activation of the JAK-STAT signaling pathway improves BDNF expression and influences the proliferative ability of hippocampal neurons in autistic mice [62].

In vitro studies have also allowed analysis of the effect of neurotoxins or anesthetic agents on miRNAs expression and in turn, on BDNF expression. For instance, differentiated PC12 cells, treated with 1-methyl-4-phenylpyridinium (MPP), show upregulation of miR-34a, miR-141, and miR-9, suggesting that perturbed expression of them may contribute to Parkinson's disease (PD)-related pathogenic processes, probably by affecting the expression of B-cell lymphoma 2 (BCL2), BDNF, and SIRT1 as potential targets [63].

Instead, studies in vitro on embryonic stem cell-derived neurons demonstrated that inhibition of miR-375 and miR-107 ameliorates ketamine-induced neurotoxicity via inverse regulation of the BDNF gene [64,65].

In summary, understanding the different functions of the various BDNF transcripts, the modulation of the expression of specific exons, and investigating the function of the BDNF-related miRNAs may represent a promising strategy to restore enduring changes in gene expression in response, for example, to environmental insults. This, in turn, might open new therapeutic perspectives for the treatment of neurodegenerative and neuropsychiatric disorders.

2.3. Biology of BDNF

Synthesis and maturation of BDNF is a multistage process, involving the formation of several precursor isoforms. The BDNF protein, discovered in 1982 [66], is a highly conserved protein of 247 amino acids, synthesized and folded in the endoplasmic reticulum as preproBDNF (32–35 kDa). Upon translocation to the Golgi apparatus, the signal sequence of the preregion is rapidly cleaved, and the isoform proBDNF (28–32 kDa) is generated [67]. The proBDNF is further cleaved to reach the mature isoform (mBDNF, 13 kDa) [67,68]. Intracellular proteolytic cleavage of proBDNF may occur by the subtilisin-kexin family of endoproteases such as furin, or in intracellular vesicles by convertases [69,70] (Figure 1).

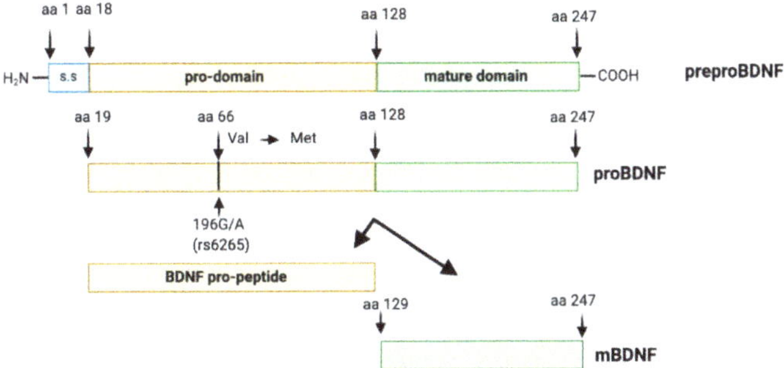

Figure 1. Brain-derived neurotrophic factor (BDNF) protein structure. The preproBDNF consists of three sequences: signal sequence (s.s), pro-domain, and mature domain. The intra- or extracellular cleavage of preproBDNF generates functionally active isoforms: BDNF pro-peptide and mature BDNF (mBDNF), each of which exhibits a characteristic affinity to a specific type of receptor. Arrowheads indicate known protease cleavage sites involved in the processing of mature BDNF. The position of the single nucleotide polymorphism (rs6265, Val66Met) and the substitution of valine (Val) in methionine (Met) at codon (aa) 66 in the human BDNF gene is indicated by an arrow.

Extracellular cleavage of proBDNF is determined by plasmin [71] and matrix metalloproteases 2 and 9 (MMP2 and MMP9) [72,73]. Depending on the cell type, BDNF can be secreted in a constitutive or activity-dependent manner [74]. In neuronal cells, both proBDNF and mBDNF are released following cell membrane depolarization [75–77]. The balance of proBDNF and mBDNF depends on the particular stages of brain development and regions. In the early postnatal period, the concentration of proBDNF is higher and may be considered as an important factor modulating brain function; while mBDNF prevails in adulthood and is important for processes occurring in adulthood, such as neuroprotection and synaptic plasticity [78]. Both proBDNF and mBDNF are active, eliciting opposing effects via the p75 neurotrophin receptor (p75NTR), a member of the tumor necrosis factor (TNF) receptor family and TrkB receptor, respectively. In resting form, both types of receptor are located in the membrane of intracellular vesicles. Stimulation with cAMP, Ca^{2+}, or electrical impulse initiates their transfer and fusion with the cellular membrane [79,80].

The mature domain of proBDNF interacts preferentially with p75NTR, mediating synaptic pruning in the prenatal brain [81]. ProBDNF, through its pro-domain, can also interact with the sortilin receptor or other vacuolar protein sorting 10 protein (Vps10p) (Figure 2). Thus, proBDNF binding to specific receptors triggers signaling pathways, which can determine neuronal fate via promoting their death or survival [82,83]. The proBDNF/p75NTR/sortilin binding complex initiates signaling cascades leading to the activation of c-Jun amino terminal kinase (JNK). This pathway is involved in neuronal apoptosis [82,83]. High levels of p75NTR expression are detected during brain development and post-traumatic recovery [84]. When mature domain of BDNF binds to p75NTR, the RIP2 (serine/threonine-protein kinase 2)/TRAF6 (tumor necrosis factor receptor associated factor 6)-mediated pathway is initiated, which leads to NF-kB activation [82,85]. The activation of NF-kB promotes neuronal survival and maintenance during brain development [85]. In addition, p75NTR interacts also with the Ras homologous (Rho) protein family. This pathway is reported to regulate neuronal growth cone development and motility [85].

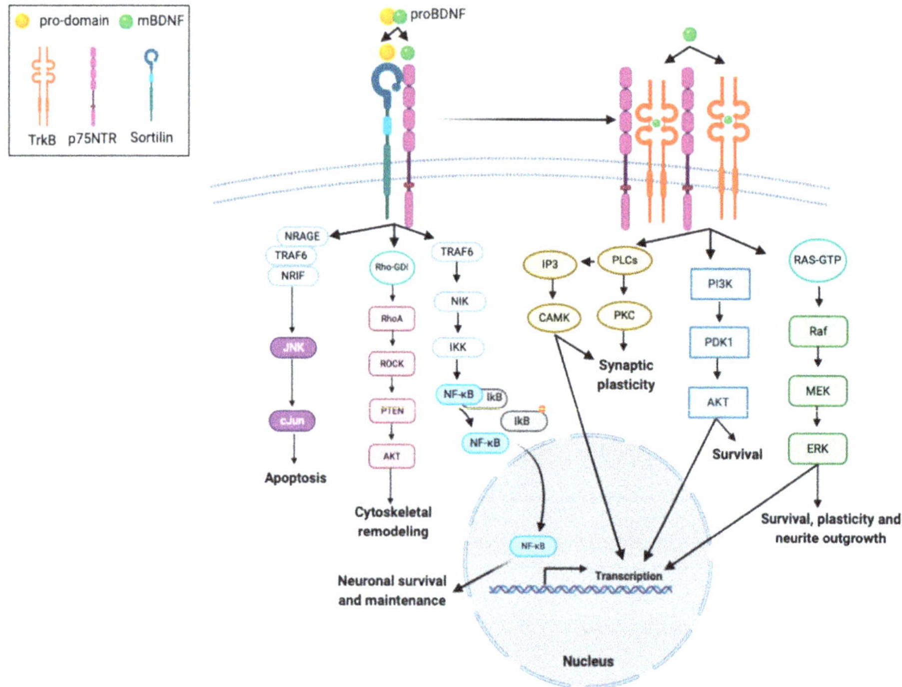

Figure 2. Intracellular signaling cascades activated by interaction of BDNF isoforms with its receptors. proBDNF and mBDNF bind to different receptors, respectively. The mBDNF isoform exhibits highest affinity for the tyrosine kinase B receptor (TrkB) receptor, which when stimulated undergoes homodimerization and autophosphorylation, but also binds the low affinity neurotrophin receptor p75NTR. The interaction between the TrkB receptor and the p75NTR receptor in a complex increases the ligand binding affinity to BDNF. Sortilin is considered a co-receptor for p75NTR. The proBDNF isoform, consisting of two sequences (pro-domain and mature domain), interacts with specific receptors, sortilin and p75NTR, respectively. The binding of proBDNF to a p75NTR/sortilin-complex induces signaling pathways that are specific for proBDNF. The binding of proBDNF in combination with sortilin causes the involvement of neurotrophin receptor-interacting factor (NRIF), tumor necrosis factor receptor-associated factor 6 (TRAF6), and neurotrophin receptor-interacting MAGE homologue (NRAGE) proteins. This pathway activates the JNK-associated pathway that promotes programmed cell death, or the receptor-interacting serine/threonine-protein kinase 2 (RIP2)/TRAF6-mediated pathway is initiated. Multi-subunit IκB kinase (IKK) phosphorylates (orange dot) the inhibitor of kB (IkB) protein, which results in dissociation of IkB from NF-κB. The activated nuclear factor kappa B (NF-kB) is then translocated into the nucleus where it binds to specific sequences of DNA and promotes neuronal survival and maintenance. In addition, p75NTR interacts with the Rho family of proteins, whose activation mediates the activity of Rho-associated protein kinase (ROCK), which subsequently leads to activation of the AKT pathway, involved in cytoskeletal remodeling. The mBDNF/TrkB receptor complex triggers signaling pathways associated with activation of phosphatidylinositol 3-kinase (PI3K), phospholipase C gamma (PLC-γ), and GTP-ases of the Rho family, involved in survival, plasticity and neurite outgrowth, transcription regulation, and synaptic plasticity.

mBDNF binds with the high-affinity TrkB receptor, the receptor dimerizes, and the intracellular tyrosine residues are autophosphorylated [86]. Phosphorylated-TrkB activates several enzymes: PI3K, mitogen-activated protein kinase (MAPK), PLC-γ, and guanosine triphosphate hydrolases (GTP-ases) of the Rho gene family [87–89]. mBDNF-TrkB-signaling pathways regulate multiple events, such as apoptosis and survival of neurons [90–92], dendritic growth [93–96], spine maturation and stabilization, development of synapses [96–98], learning- and memory-processes-dependent synaptic plasticity [99,100].

PI3K/Akt-related pathway exerts antiapoptotic and pro-survival activity and modulates NMDA receptor-dependent synaptic plasticity [101–103]. The PI3K/Akt/mTOR cascade enhances dendritic growth and branching through regulation of protein synthesis and cytoskeleton development [104,105].

The MAPK/Ras-signaling cascade regulates protein synthesis during neuronal differentiation [85] and is also required for the activation of ERK 1/2 and CREB [106,107]. This pathway is crucial not only for early response gene expression (e.g., c-Fos), but also for cytoskeleton protein synthesis (e.g., Arc and cypin) [87], as well as dendritic growth and branching in hippocampal neurons [94,108].

The PLC-γ-dependent pathway evokes activation of CAM kinase and protein kinase C (PKC), which subsequently increases the 1,2-diacylglycerol (DAG) and Ca^{2+} ion concentrations [89]. The PKC-dependent pathway is reported to enhance synaptic plasticity [85] (Figure 2).

In summary, the specific role of BDNF in the regulation of numerous brain physiological processes depends on the interaction of its isoforms with different types of receptors. This, in turn, elicits the activation of signaling pathways that are critical for processes of brain development, synaptic plasticity, and protection and/or regeneration after damage. Perturbation of the BDNF synthesis, resulting in dysfunctions of its signaling cascades, may be responsible for triggering several pathological processes.

2.4. The Human BDNF Variant Val66Met

BDNF level in the peripheral tissues, brain, and blood may be also affected by gene polymorphism. The pro-domain of BDNF is the locus of a functional human BDNF polymorphism (SNPs) Val66Met, also known as rs6265 or G196Apolymorphism [109]. This point mutation causes a substitution of Valine (Val) to Methionine (Met) at codon 66 (Val66Met) in the pro-domain of *BDNF* (Figure 1). The Val66Met polymorphism does not exist in the mouse or other model organisms. Thus, multiple studies aim to mimic the function of BDNF Val66Met in cellular models or in genetically engineered mouse models. The BDNF Val66Met variant was first identified in the late 1990s and in 2002 the first two genetic studies investigating the BDNF Val66Met polymorphism in the pathogenesis of neurodegenerative disease were published [110,111]. The functionality of BDNF Val66Met variant was only confirmed in 2003 [112], where BDNF Val66Met polymorphism was shown to disrupt the episodic memory in humans. In addition, Egan et al. [112] also demonstrated that in hippocampal cultures BDNF Val66Met polymorphism did not alter BDNF expression per se, but the perisomatic localization of BDNF. Then, in 2005 it was discovered that the BDNF Val66Met substitution also disrupts the sortilin-binding site, impairing activity-mediated secretion of BDNF [113]. Likewise, the BDNF Val66Met substitution also disrupts the translin-binding site, which impairs dendritic targeting of BDNF mRNA [114]. Thus, the principle molecular mechanism associated with the BDNF Val66Met polymorphism is the deficient activity-dependent release of BDNF, which consequently impacts the efficiency of BDNF-TrkB signaling [113]. Following the demonstration that this SNP was functionally relevant over the past 18 years, more than 1700 studies have investigated the effects of this polymorphism on brain function in health, as well as in diseases, particularly in neuropsychiatric disorders [115,116]. The BDNF Val66Met polymorphism has been associated with cerebral cortex plasticity [117,118], with gray matter structures [119,120], or white matter integrities and structural networks [121,122]. More specifically, BDNF Val66Met polymorphism is associated with cognitive processes [112,123–127], and cognitive impairment in neurodegenerative disease, such as Parkinson's disease (PD) [128,129] and AD [130,131], and even more with several brain disorders, including MDD and bipolar disorder [132–137], epilepsy [138–140], schizophrenia [125,141–144], aging and dementia [145] and stroke [117,146,147]. Met66, but not Val66, BDNF pro-domain can induce the growth cone retraction in young hippocampal neurons [148]. Although many studies have demonstrated the possible genetic effects of this BDNF polymorphism in diseases or brain function, other articles have failed to replicate the findings. The discrepancies of BDNF Val66Met genetic studies may result from many factors such as environmental factors, ethnicity, age, and sex.

3. Neuroplasticity in MDD: The Effects of Antidepressant Therapies

3.1. Major Depressive Disorder

Major depressive disorder is one of the most prevalent and debilitating psychiatric disorders with high impact on the quality of life and negative effects on mood, behavior, and cognition [149]. Over the past few decades, several mechanisms have been investigated in the pathophysiology of MDD, including altered serotonergic, noradrenergic, dopaminergic, and glutamatergic systems, increased inflammation, hypothalamic-pituitary-adrenal axis abnormalities, vascular changes, and decreased neurogenesis and neuroplasticity. In particular, a decrease in serotonergic neurotransmission is regarded as the main etiopathogenetic mechanism occurring in depressed patients. Thus, the most common drugs used to treat MDD are serotonin reuptake inhibitor (SSRI) that block SERT and thus increase serotonin in the raphe nucleus at post-synapse. Therefore, a misbalance in the serotonin production and/or release is believed to play a central role in determining MDD. This led to finding, by means of genetic, proteomic and pharmacological tools, molecules able to increase the expression of serotonin in neurons by modulating neural genes or proteins [150–152]. Among these molecules, TPH2, the rate-limiting enzyme responsible for brain serotonin biosynthesis, plays a crucial role and is amenable of genetic and pharmacological manipulation [153,154]. Nevertheless, in 1997 Duman and Nestler formulated the neurotrophin hypothesis of depression [155]. This theory is now supported by studies demonstrating a decrease in BDNF mRNA and protein levels in postmortem critical regions, such as the hippocampus, prefrontal cortex and amygdala, in patients with MDD compared to controls.

3.2. BDNF and Neuronal Plasticity

Brain development occurs through coordinated processes of neuro- and gliogenesis, formation of neuronal projections and synaptogenesis, and programmed cell death and elimination of improperly formed connections, together resulting in the formation of the functionally and morphologically adjusted structure of the adult brain [156,157]. Neuroplasticity or brain plasticity is the ability of the nervous system to reorganize its structure, function, and connections in response to extrinsic or intrinsic stimuli [158]. Neuronal plasticity in rodents has been well-documented during the last decades, whereas neuroplasticity in the human brain largely remains indirect, mostly because of methodological limitations as well as ethical constraints. Neuronal plasticity includes different mechanisms excellently reviewed by Castren [159]. One of these is the neurogenesis, i.e., the formation of newborn neurons in proliferative areas. There is solid evidence that neurogenesis occurs in the adult mammalian brain. In rodent adult brains, neurogenesis is mainly restricted to the subventricular zone and the subgranular zone of the dentate gyrus in the hippocampus and olfactory bulbs [156]. An accumulating body of evidence indicates that BDNF is involved in the regulation of migration of neuronal progenitors along the rostral migratory stream and neuronal settlement in the olfactory bulb [160] and also acts during the later stages of neurogenesis [161,162].

Neuronal plasticity is extensively studied during critical periods, a time window during the early phase of brain development, when neuronal circuits are noticeably sensitive to being shaped by external stimuli and experience, producing permanent and large-scale changes to neural circuits. The same circuits can be shaped by experience later in life, but to a lesser degree. After the ending of critical periods, neuronal plasticity and changes in network structure are more restricted. However, recent data indicate that several drugs used for the treatment of neuropsychiatric disorders can directly induce plasticity and reactivate a critical period-like plasticity in the adult brain. The first functional evidence for the role of neurotrophins in plasticity was obtained in the visual cortex. The observation that BDNF synthesis in the visual cortex is regulated by visual stimulation made BDNF the prime candidate for this activity-dependent regulated factor [163–165]. In transgenic mice with early overexpression of

BDNF, an accelerated onset and end of the critical period and precocious maturation of inhibitory circuits was observed. Conversely, mice raised in the dark and resulting in lower levels of BDNF showed a delayed visual plasticity [166,167]. In addition, a disruption in the binding between promoter regions of BDNF exon IV and cAMP response element-binding protein (CREB) results in decreased inhibitory input [168], which impairs the critical period plasticity.

3.3. BDNF and Synaptic Plasticity

Another mechanism involved in neuronal plasticity is the modification of mature neuronal morphology, involving axonal and dendritic arborization and pruning, an increase in spine density, and synaptogenesis [169]. Epigenetic mechanisms involved in the transcriptional regulation of genes also can contribute to synaptic plasticity. Several in vitro and in vivo studies analyzed the effects of BDNF on plasticity. Cazorla et al. proved that 48 h of BDNF stimulation in PC12 cells, transfected with TrkB, increased neurite outgrowth compared to the non-treated cells [170]. Interestingly, BDNF stimulation was able to promote dendritic outgrowth and spine formation [171,172] in primary hippocampal cells grown in B27-deprived medium. This neuroplastic effect is probably achieved through the activation of intracellular-signaling cascades [173,174]. Recent data suggest that intracellular overexpression of BDNF in hippocampal developing neurons induces maturation of excitatory and inhibitory synapses, with respect to exogenous application of BDNF [175]. BDNF mice lacking BDNF die during the second postnatal week [176] and BDNF deficit causes inhibition of dendritic arborization [92, 177] and reduction of expression of genes functionally related to vesicular trafficking and synaptic communication [178]. Instead, heterozygous BDNF mice survive into adulthood and BDNF is required for several forms of LTP, the main mechanism mediating plasticity [179]. At morphological level, these mice display a specific hippocampal volume reduction [180] similar to that observed in heterozygous TrkB mice [181,182], but in contrast to p75NTR-deficient mice [183]. These findings suggest a link between hippocampal volume and BDNF-mediated TrkB signaling [181,182]. Over the last years, BDNF has been extensively studied as an important regulator of synaptic transmission and LTP in the hippocampus and in other brain regions. The effects of BDNF in LTP are mediated by TrkB receptors. In particular, in the hippocampus the neurotrophin is thought to act at pre- and post-synaptic levels, modulating synaptic efficacy either by changes in pre-synaptic transmitter release, or by increased post-synaptic transmitter sensitivity (see e.g., [15,16]) to induce a long-lasting increase in synaptic plasticity. This depends on individual circumstances. Thus, BDNF can be: (i) either, a mediator or a modulator of synaptic plasticity, (ii) both, a neurotransmitter that acts both at pre- and post-synaptic level simultaneously at the same individual synapse. Recent data published from Lin et al. revealed that in CA3 or CA1 regions anterograde BDNF-TrkB signaling is involved in LTP induction, while anterograde and retrograde BDNF-TrkB signaling contributes to LTP maintenance. BDNF in both pre-synaptic and post-synaptic terminals modulate basal neurotransmission and pre-synaptic TrkB, probably regulating pre-synaptic release [184]. In addition, it has also been shown that BDNF regulates the transport of mRNAs along dendrites and their translation at the synapse. These processes occur by modulating the initiation and elongation phases of protein synthesis, and by acting on specific miRNAs [100]. Local protein synthesis responds with rapid and subtle modulation of the proteome to remodel the synaptic regions in response to stimuli [185]. Protein turnover is required for synaptic plasticity, and BDNF-signaling has been also described as a crucial regulator for maintaining the baseline autophagic activity in the brain. BDNF deficiency causes an uncontrolled rise in autophagic degradation [186].

BDNF is one of the most studied synaptic molecules that efficiently modify synaptic strength and can act as a mediator, modulator, or instructor of synaptic plasticity. Specific changes in dendritic spines, as well as in adult hippocampal neurogenesis, can be correlated to several forms of learning and memory. BDNF is one of the most inspiring molecules to better understand the disadvantageous synaptic learning underlying the etiology of depression, accompanied by declines in the rate of adult neurogenesis and in spine densities [181].

3.4. BDNF in Depressed Patients

BDNF protein and TrkB receptor are detectable in several non-neuronal tissues, including endothelial cells [17,18], cardiomyocytes [19], vascular smooth muscle cells [17], leukocytes [20], megakaryocytes [19], and platelets [21,22]. Serum BDNF has been clearly demonstrated to originate from the progenitors of platelets [21]. Platelets are the major source of peripheral BDNF and are important for storing the BDNF secreted from other tissues [187]. Over the last years, there has been a great interest in peripheral BDNF measures in relation to psychiatric illness. It has been studied as biomarker reflecting these disorders [188,189]. However, there is no evidence that serum BDNF is related to brain BDNF and neuroplasticity. Nevertheless, the low serum concentration of BDNF has often been associated with the pathophysiology of MDD [190–192]. An aspect to consider is if the serum BDNF levels are dependent on the release of BDNF from platelets [193]. The significance of the lower BDNF levels in depression is currently unclear. The temporal correlation between serum BDNF levels and the antidepressant effect seems to be indirect: ketamine and electroconvulsive shock treatment increase serum BDNF levels only gradually, while their antidepressant effect appears quickly [194]. There are two studies that directly observed a reduction of BDNF levels in platelets of patients with MDD [195,196]. Another study showed that BDNF levels of platelet were significantly decreased compared to the controls. In this study, the BDNF levels were normalized compared to control with SSRIs treatment [197]. Taken together, these studies strongly suggest that changes in serum BDNF levels reflect altered BDNF release from blood platelets. Thus, given the similarities in the regulation of BDNF synthesis between megakaryocytes and neurons, there may be parallels between the brain, BDNF in serum, and release. Nevertheless, within the CNS a reduction in BDNF and TrkB expression has been reported in the hippocampus and prefrontal cortex of post-mortem brain tissues of suicide victims [198,199]. In addition, several meta-analyses data confirm the association of the Val66Met polymorphism with an increase of susceptibility to develop mood disorders [200–202]. Finally, a recent paper showed that subjects with the Met allele of the BDNF gene are more likely to develop depression [134].

A disruption in serotonin signaling in the brain is also believed to be involved in the pathophysiology of depression. Changes in synaptic serotonin levels and receptor levels are coupled with altered synaptic plasticity and neurogenesis [203,204]. It has been proposed that chronic treatment with conventional antidepressants, such as SSRIs, but not acute administration increases neurogenesis [205–207] and selective SSRIs might reactivate serotonin's ability to mediate developmental plasticity. BDNF acts as a modulator of the 5-HT system and vice versa, acting as the link between the antidepressant drug and the neuroplastic changes. Close molecular connections between serotonin receptors and neurotrophic proteins such as BDNF and intracellular signaling cascades are responsible for cytoskeletal rearrangement [169,208–211]. Thus, dysregulation in 5-HT–BDNF interaction may be responsible for the development of neuropsychiatric and behavioral abnormalities [212].

Understanding the function of the members of the BDNF system in response to the challenges of the environment and the interaction with different 5-HT receptors in health and disease will lead to new classes of drugs that could be used in therapy for psychiatric and neurodegenerative disorders.

3.5. Effect of Antidepressant Therapies on Plasticity BDNF-Mediated

3.5.1. BDNF and Antidepressant Treatments

Multiple lines of evidence suggest that antidepressant treatments increase BDNF mRNA and protein levels in the cerebral cortex and hippocampus (for review see [213,214]). This increase is partly due to a reduction of histone acetylation in the *BDNF* promoter regions. The involvement of BDNF in the efficacy of antidepressant treatments has mainly been demonstrated in rodent models. It has been demonstrated that all pharmacological classes of clinical antidepressants increase TrkB autophosphorylation and signaling in the hippocampus and forebrain, effects observed within hours after the administration of the drug [203,215]. Similar results in BDNF mRNA and TrkB phosphorylation have been observed after acute treatment with ketamine [216–219]. In rodents, injection of BDNF in the hippocampus reduces depression-like behavior [220], in contrast injection of BDNF into the nucleus accumbens or ventral tegmental area promotes depressive effects [221], demonstrating the network-dependent effect of BDNF in mood regulation. Interestingly, conditional knockout of BDNF in forebrain regions increases depressive behavior in females, but not in male mice [222], and blocks the effects of antidepressants desipramine or ketamine [216,223]. Similarly, conditional deletion of TrkB in dentate gyrus or inhibition of TrkB signaling by a dominant-negative TrkB receptor blocks the effects of antidepressants [224,225]. In addition, mice with Val66Met polymorphism are insensitive to antidepressants [226]. Recent evidence demonstrates that the antidepressant effects of GLYX-13, a novel glutamatergic compound that acts as an NMDA modulator with glycine-like partial agonist properties, are blocked by intra-medial prefrontal cortex infusion of an anti-BDNF antibody or in mice with a knock-in of the BDNF Val66Met allele. Pharmacological inhibition of BDNF-TrkB signaling or L-type voltage-dependent Ca^{2+} channels (VDCCs) blocks the antidepressant behavioral actions of GLYX-13 [227].

Taken together, these data suggest that BDNF serves as a transducer, acting as the link between the antidepressant drug and the neuroplastic changes that result in the improvement of depressive symptoms.

3.5.2. Beneficial Effects of Exercise on Plasticity: The Role of BDNF

Several lines of evidence suggest that exercise has beneficial effects on plasticity and BDNF could be a link between plasticity and physical activity. Although it has been proven that exercise in MDD patients reduced depressive symptoms [228–230], neuroplasticity per se has not yet been monitored in these patients. However, voluntary physical exercise, like an enriched environment, increases expression of BDNF in the hippocampus [8], as well as hippocampal neurogenesis [9] and this could improve brain function by enhancing plasticity, cognition, learning, and memory [12–14]. Physical exercise is one particularly effective strategy for increasing circulating levels of BDNF [10,11]. It has repeatedly been demonstrated that an acute bout of aerobic exercise transiently increases both serum and plasma BDNF in an intensity-dependent manner [10,11]. Exercise increases the release of BDNF from the human brain [231,232] suggesting that exercise also mediates central BDNF production in humans. It has been suggested that miR-34a potentially can also mediate changes in BDNF expression and may reflect the decrease in performance after overtraining [233].

Multiple studies suggest that BDNF has a dominant role in mediating the effects of physical activity on cognitive changes [234]. It has been shown that three months of aerobic exercise training increases hippocampal volume in healthy individuals and in patients with schizophrenia by 12% and 16%, respectively [235]. The question whether exercise regulates muscle-derived circulating factors that can pass through the blood–brain barrier and stimulate BDNF production in the brain remains unclear. In 2016, Moon et al. show that the myokine cathepsin B (Ctsb) might be involved in mediating the exercise-induced improvement in hippocampal neurogenesis, memory, and learning [236]. Mice lacking Ctsb showed depression-like symptoms when they were forced to swim [236].

Other papers have demonstrated that exercise induces upregulation in skeletal muscle of PGC1α, a transcriptional co-activator of mitochondrial biogenesis and oxidative metabolism in brown adipose tissue and muscle. In muscle, the increase of PGC1α expression stimulates an upregulation of FNDC5, a membrane protein that is cleaved and secreted into the circulation as the myokine irisin [237]. FNDC5 cross the blood–brain barrier inducing BDNF expression in the hippocampus, in this way BDNF plays a role in neurogenesis and reward-related learning and motivation [238]. Current research has also shown that high intensity exercise increases peripheral lactate and BDNF levels; at the same time lactate infusion at rest can increase peripheral and central BDNF levels. Lactate and BDNF can induce neuroplasticity [239]. In addition, acute elevation of BDNF did not compensate for hypoxia-induced cognition impairment [240].

The identification of exercise-related factors that have a direct or indirect effect on brain function has the potential to highlight novel therapeutic targets for neurodegenerative diseases.

4. The Protective Role of BDNF on Neurodegeneration

Neurodegenerative diseases comprise a wide range of neurological diseases such as AD, PD, Huntington's disease, and amyotrophic lateral sclerosis (ALS), characterized by the deterioration and then the death of selective nuclei of neurons in the brain or the spinal cord. They are chronic and progressive diseases, currently incurable and highly debilitating, causing a tremendous emotional and economic burden on patients, their families, and society. AD, the most frequent among neurodegenerative diseases, accounts for about 70% of dementia cases all over the world, that is about 35 million people. It is estimated to cost more than 480 billion euros each year throughout the world (Sources: OMS, EBC (European Brain Council)). Currently, no pharmacological treatment is available to cure or even significantly slow down the course of neurodegenerative diseases. For these reasons, experimental findings showing that physical exercise, exposure to an enriched environment, metabolic changes and nutritional and/or cognitive intervention, may exert a protective role on neurodegeneration either by delaying the onset and/or curbing the course of the disease, raise hope that these new tools might be useful also in clinical practice. BDNF appears to be crucial or, in some cases even essential, to mediate the neuroprotective effects of the above-mentioned environmental stimuli (Figure 3). In particular, as discussed above, it is well established that BDNF accounts for the hippocampal adult neurogenesis, which, in turn, can be stimulated by a number of conditions such as physical exercise, enriched environment, hormonal balance (i.e., steroid hormones such a cortisol and testosterone) and nutritional intervention (i.e., fasting, low-calorie intake, low-carb diet, selective nutrient intakes), capable of increasing the BDNF level [241].

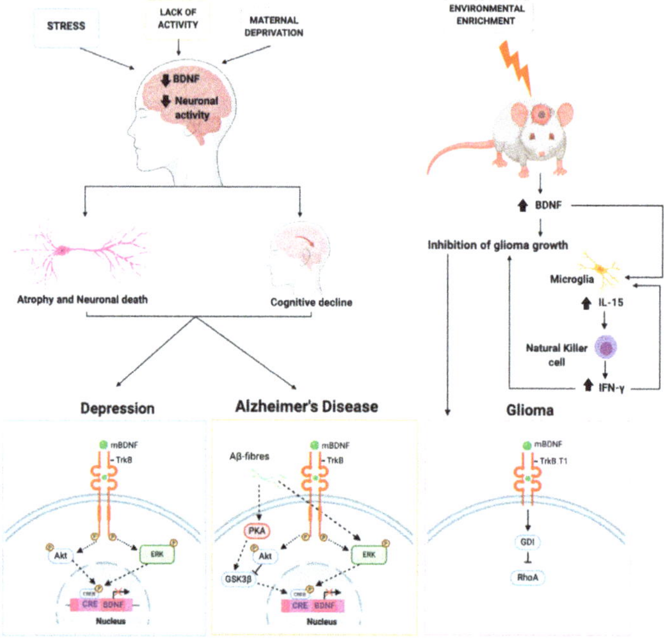

Figure 3. The molecular mechanisms mediated by BDNF involved in depression, Alzheimer's disease, and glioma. External stimuli (stress, maternal deprivation, or lack of activity) causing epigenetic regulation processes can induce a reduction in BDNF expression level and in neuronal activity.

This results in atrophy, neuronal death, and cognitive decline, which may contribute to depression or Alzheimer's disease. In these pathologies the BDNF/TrkB signaling, which activates the downstream Akt and ERK signaling, is altered. Thus, such alterations cause an impairment of CREB signaling resulting in BDNF downregulation. The alteration of phosphorylation (P) inhibits (red X) the transcriptional machinery. In the mouse brain, environmental enrichment induces an increase of BDNF. BDNF, binding the truncated form of TrkB receptor (TrkB.T1), signals directly to the Rho protein dissociation inhibitor (GDI). The latter detaches from TrkB.T1 and binds to the small G protein RhoA, leading to an inhibition of glioma cell migration. BDNF stimulates also indirectly the production of IL-15 in microglia, which in turn stimulates the natural killer cells to produce IFN-γ. IFN-γ contributes to reducing glioma growth.

4.1. The Protective Role of BDNF on Alzheimer's Disease

A reduced level of BDNF has been found in patients affected by neurodegenerative diseases such as Parkinson's, Huntington's and Alzheimer's disease as well as in mild cognitive impairment, the latter being a prodromal stage of AD, characterized by a slight decline of cognitive abilities including memory, thinking and judging skills [242–245]. In some instances, the levels of BDNF even correlate with the severity of the diseases, pointing towards a pathogenetic link between BDNF and AD [246]. Although there are some papers reporting an increase of BDNF in serum or in the post-mortem brain, this might be due either to compensatory mechanisms or its release from immune cells or pharmacological treatments known to raise BDNF (i.e., antidepressants) [241].

Recently, a very complex study explored the role of physical activity in a genetic mouse model of AD. This study provided the most compelling evidence of the relationship between physical activity, adult hippocampal neurogenesis, BDNF, and AD. This study elegantly confirms that adult hippocampal neurogenesis plays a pivotal role in brain resilience to AD. They manipulated with pharmacological and genetic tool neurogenesis as well as BDNF, clearly showing that physical exercise needs neurogenesis to protect the brain from AD and that BDNF is essential for such a protection. In addition, it provides evidence that adult hippocampal neurogenesis can counteract AD memory impairment, only in combination with BDNF, whereas if neurogenesis is experimentally blocked, BDNF does not exert beneficial effects. Finally, pharmacological increase of BDNF further ameliorates AD pathology [247]. Thus, agents that promote both BDNF signaling and neurogenesis might be the key to preventing or curing AD. As far as metabolism is concerned, it has been shown that intermitting fasting, by causing a transition from utilization of carbohydrate and glucose to a fatty acid and ketones source of energy (refer to as "G-K shift") generates a number of beneficial cognitive, metabolic, and cardiovagal effects. BDNF is increased upon intermitting fasting and mediates at least part of these effects. Its increase is stimulated by the ketone body, β-hydroxybutyrate that inhibits histone deacetylases that repress BDNF promoters [248]. Recently, the role of BDNF and neuroprotection in the context of metabolism and fasting has been nicely reviewed by Mattson et al. [6].

4.2. BDNF and Ras-ERK-CREB Signaling in Alzheimer's Disease

BDNF, as also discussed above, causes the activation of the Ras-ERK signaling cascade leading to the phosphorylation of CREB. Such a pathway exerts a well-known trophic and protective role on neuronal cells both in vitro and in vivo in a variety of neurodegenerative models, including AD, PD, and Huntington's diseases. Nevertheless, it has become clear that the Ras-ERK pathway may also foster neurodegeneration or hamper the action of neurotrophic factors when activated by noxious stimuli as occurs for instance in PD and AD [249]. In particular, it has been shown in a number of different cellular models such as primary cortical rat neurons, rat B103 neuroblastoma cells, and A1 mouse mesencephalic cells, that APP and/or Aβ42 oligomer induces the activation of Ras-ERK and GSK-3 signaling, that, in turn, causes hyperphosphorylation of tau and APP at Thr668. The involvement of these molecular events in the pathogenesis AD is corroborated by the finding that activation of Ras-ERK and GSK-3 correlates with Aβ levels in the brain of AD patients [250,251]. Aberrant stimulation of Ras-ERK signaling forces neurons to enter the cell cycle as shown by the expression and nuclear accumulation of cyclin D1 and the subsequent G1/S progression. Since neurons lack functional cell cycle machinery, these events lead to cell death (i.e., mitotic catastrophe) instead of cell division. Interestingly, as clearly shown in the mouse model of familial AD APPswe/PS1ΔE9 mice, although ERK phosphorylation is enhanced compared to the wild type counterpart, it does not result in normal CREB phosphorylation. The impairment in CREB signaling parallels to impairment in a number of cognitive tests [252]. Therefore, in AD, BDNF downregulation is mediated by the impairment of CREB signaling caused by amyloid β [253] (Figure 3).

5. BDNF and Brain Cancer: An Unexpected Role. An Oncogene or a Tumor Suppressor?

The role of BDNF and its cognate receptor TrkB in cancer, including brain cancer, has been recognized for a long time [254]. In many types of cancers, BDNF and/or TrkB have been found expressed or in some cases over-expressed [255]. This is not surprising since growth factors, including neurotrophic factors, and their tyrosine kinase receptors have long been involved in tumors with different cell-dependent mechanisms, fostering proliferation, enhancing anti-apoptotic signaling, and making cells unresponsive to anti-proliferative stimuli [256]. The direct oncogenic activity of TrkB might also be due to the crosstalk with EGF receptors that together with its ligand is well-known to promote cell transformation. BDNF administration not only does phosphorylate TrkB but also EGFR [257]. In line with these observations, it has been recently shown that BDNF produced by glioblastoma (GBM) differentiated cells acts on GBM stem cells, fostering their growth through paracrine signaling [258].

However, recently another study showed that exposing mice to an enriched environment is able to decrease the growth of intracranial glioma, decreasing proliferation and invasion, and improving overall survival. Such an effect is achieved by means of both indirect and direct mechanisms. The former acts via natural killer cells of the innate immune system, whereas the latter utilizes BDNF stimulation of its truncated receptor TrkB.T1 on glioma cancer cells. BDNF binding the TrkB.T1 receptor signals to the Rho protein dissociation inhibitor (RhoGDI), the latter detaches from TrkB.T1 and binds to the small G protein RhoA, leading to its inhibition. The authors found that an enriched environment causes the synthesis of IL-15 and BDNF. When mice bearing the glioma and not housed in enriched environments were infused with BDNF, they reduced tumor size and macrophage infiltration. Thus, showing that at least in part, BDNF accounts for the oncolytic effect elicited by the enriched environment [259]. In a more recent study, the same group delved deeper into the mechanisms, finding that enriched environment changes glioma-associated myeloid cells. BDNF plays a central role by stimulating the production of IL-15 in microglia, which in turn stimulates the natural killer cells to produce IFN-γ. Natural killer cells were responsible for the switch to an oncolytic environment [260] (Figure 3).

Taken together, a scenario emerges where BDNF, acting on different cells is able to reorganize the brain microenvironment in such a way that it becomes resilient to neurodegeneration or oncolytic for tumors. In this regard, although supported by much more preliminary data, it seems that also other compounds might share these properties [261,262].

6. Conclusions

In this review, we discussed the role of BDNF in neurogenesis, differentiation, survival, synaptic plasticity, and transmission to reorganize the brain microenvironment. All BDNF gene products, such as proBDNF, mature BDNF, and even the isolated proBDNF domain, are known to exert functional activity. One of the most important features of BDNF is that it can act as a local, paracrine and/or autocrine factor, on both pre-synaptic and post-synaptic target sites. Here, we presented the contribution of altered BDNF signaling in the pathophysiology of brain diseases, including mental disorders (i.e., depression), neurodegenerative diseases, (i.e., Alzheimer's disease), and brain tumor (i.e., glioblastoma). BDNF is one of the best-studied synaptic molecules that efficiently modify synaptic strength and it can act as a mediator, modulator, or instructor of synaptic plasticity. In neurons, the cellular processes that regulate the amount of both BDNF mRNA and protein, the changes in the efficiency of secretion, and transport of BDNF protein may affect synaptic function and cell survival. BDNF is one of the most inspiring molecules to better understand the disadvantageous synaptic learning underlying the etiology of depression, accompanied by the decline in the rate of adult neurogenesis and in spine densities. Furthermore, BDNF appears to exert a potent role in neuroprotection and/or brain regeneration by modulating signaling pathways such Ras-ERK-CREB, thus rendering neuronal cells resilient to neurodegeneration. Finally, BDNF appears to be crucial in the pathogenesis and development of brain tumors such as glioblastoma by reorganizing its microenvironment. Thus, understanding the physiologic and pathologic BDNF signaling and finding tools to modulate its expression (mRNA and/or protein) is a prerequisite for a potential BDNF-based therapy.

Author Contributions: Original draft preparation, writing and editing, L.C.-D., L.S.; original draft preparation, writing and review, F.V. All authors have read and agreed to the published version of the manuscript.

Funding: This research was funded by program VALERE: VAnviteLli pEr la RicErca to L.C.D.

Conflicts of Interest: The authors declare no conflict of interest.

Abbreviations

AD	Alzheimer's Disease
AP1	Activator Protein 1
BCL2	B-cell lymphoma 2
BDNF	Brain-Derived Neurotrophic Factor
BHLHB2	Helix-loop-helix protein
CaMK	Ca^{2+}/calmodulin-dependent protein kinase
cAMP/CRE	Cyclic adenosine monophosphate calcium response element
CaREs	Calcium-responsive elements
CREB	CREB
Ctsb	Myokine cathepsin B
CUMS	Chronic Unpredicted Mild Stress model
DAG	1,2-diacylglycerol
ERK	Extracellular signal-regulated kinase
GTP-ases	guanosine triphosphate hydrolases
JNK	c-Jun amino terminal kinase
IKK	Multi-subunit IκB kinase
IkB	Inhibitor of kB
L-VGCC	L-type voltage gated calcium channel
LTP	Long-Term Potentiation
MAPK	Mitogen-activated protein kinase
MDD	Major depressive disorder
Met	Methionine
miRNAs	microRNAs
MMP	Matrix Metalloproteases
MPP	1-methyl-4-phenylpyridinium
NF-kB	Nuclear Factor kappa-light-chain-enhancer of activated B cells
NMDA	N-Methyl-D-Aspartate
NRAGE	Neurotrophin receptor-interacting MAGE homologue
NRIF	Neurotrophin receptor interacting factor
PD	Parkinson's disease
p75NTR	p75 neurotrophin receptor
PI3K	Phosphoinositide 3-kinases
PKC	Protein kinase C
PLC	Phospholipase C
Rho	Ras homolog

ROCK	Rho-associated protein kinase
SIRT1	Sirtuin 1
SNPs	Polymorphism
SSRIs	Serotonin reuptake inhibitor
TRAF6	Tumor necrosis factor receptor associated factor 6
TrkB	Tropomycin receptor kinase B
Val	Valine
VDCCs	L-type voltage-dependent Ca^{2+} channels
Vps10p	Vacuolar protein sorting 10 protein

References

1. Andreska, T.; Aufmkolk, S.; Sauer, M.; Blum, R. High abundance of BDNF within glutamatergic presynapses of cultured hippocampal neurons. *Front. Cell Neurosci.* **2014**, *8*, 107. [CrossRef]
2. Clarke, L.E.; Liddelow, S.A.; Chakraborty, C.; Münch, A.E.; Heiman, M.; Barres, B.A. Normal aging induces A1-like astrocyte reactivity. *Proc. Natl. Acad. Sci. USA* **2018**, *115*, E1896–E1905. [CrossRef]
3. Parkhurst, C.N.; Yang, G.; Ninan, I.; Savas, J.N.; Yates, J.R.; Lafaille, J.J.; Hempstead, B.L.; Littman, D.R.; Gan, W.B. Microglia promote learning-dependent synapse formation through brain-derived neurotrophic factor. *Cell* **2013**, *155*, 1596–1609. [CrossRef]
4. Bartkowska, K.; Paquin, A.; Gauthier, A.S.; Kaplan, D.R.; Miller, F.D. Trk signaling regulates neural precursor cell proliferation and differentiation during cortical development. *Development* **2007**, *134*, 4369–4380. [CrossRef]
5. Vilar, M.; Mira, H. Regulation of Neurogenesis by Neurotrophins during Adulthood: Expected and Unexpected Roles. *Front. Neurosci.* **2016**, *10*, 26. [CrossRef]
6. Mattson, M.P.; Moehl, K.; Ghena, N.; Schmaedick, M.; Cheng, A. Intermittent metabolic switching, neuroplasticity and brain health. *Nat. Rev. Neurosci.* **2018**, *19*, 63–80. [CrossRef]
7. Numakawa, T.; Odaka, H.; Adachi, N. Actions of Brain-Derived Neurotrophin Factor in the Neurogenesis and Neuronal Function, and Its Involvement in the Pathophysiology of Brain Diseases. *Int. J. Mol. Sci.* **2018**, *19*, 3650. [CrossRef]
8. Loprinzi, P.D.; Day, S.; Deming, R. Acute Exercise Intensity and Memory Function: Evaluation of the Transient Hypofrontality Hypothesis. *Medicina (Kaunas)* **2019**, *55*, 445. [CrossRef]
9. Fabel, K.; Wolf, S.A.; Ehninger, D.; Babu, H.; Leal-Galicia, P.; Kempermann, G. Additive effects of physical exercise and environmental enrichment on adult hippocampal neurogenesis in mice. *Front. Neurosci.* **2009**, *3*, 50. [CrossRef]
10. Knaepen, K.; Goekint, M.; Heyman, E.M.; Meeusen, R. Neuroplasticity-exercise-induced response of peripheral brain-derived neurotrophic factor: A systematic review of experimental studies in human subjects. *Sports Med.* **2010**, *40*, 765–801. [CrossRef]
11. Szuhany, K.L.; Bugatti, M.; Otto, M.W. A meta-analytic review of the effects of exercise on brain-derived neurotrophic factor. *J. Psychiatr. Res.* **2015**, *60*, 56–64. [CrossRef] [PubMed]
12. van Praag, H.; Shubert, T.; Zhao, C.; Gage, F.H. Exercise enhances learning and hippocampal neurogenesis in aged mice. *J. Neurosci.* **2005**, *25*, 8680–8685. [CrossRef] [PubMed]
13. Tuon, T.; Souza, P.S.; Santos, M.F.; Pereira, F.T.; Pedroso, G.S.; Luciano, T.F.; De Souza, C.T.; Dutra, R.C.; Silveira, P.C.; Pinho, R.A. Physical Training Regulates Mitochondrial Parameters and Neuroinflammatory Mechanisms in an Experimental Model of Parkinson's Disease. *Oxid. Med. Cell. Longev.* **2015**, *2015*, 261809. [CrossRef] [PubMed]
14. Vilela, T.C.; Muller, A.P.; Damiani, A.P.; Macan, T.P.; da Silva, S.; Canteiro, P.B.; de Sena Casagrande, A.; Pedroso, G.D.S.; Nesi, R.T.; de Andrade, V.M.; et al. Strength and Aerobic Exercises Improve Spatial Memory in Aging Rats Through Stimulating Distinct Neuroplasticity Mechanisms. *Mol. Neurobiol.* **2017**, *54*, 7928–7937. [CrossRef]
15. Itami, C.; Kimura, F.; Kohno, T.; Matsuoka, M.; Ichikawa, M.; Tsumoto, T.; Nakamura, S. Brain-derived neurotrophic factor-dependent unmasking of "silent" synapses in the developing mouse barrel cortex. *Proc. Natl. Acad. Sci. USA* **2003**, *100*, 13069–13074. [CrossRef]

16. Edelmann, E.; Lessmann, V.; Brigadski, T. Pre-and postsynaptic twists in BDNF secretion and action in synaptic plasticity. *Neuropharmacology* **2014**, *76 Pt C*, 610–627. [CrossRef]

17. Nakahashi, T.; Fujimura, H.; Altar, C.A.; Li, J.; Kambayashi, J.; Tandon, N.N.; Sun, B. Vascular endothelial cells synthesize and secrete brain-derived neurotrophic factor. *FEBS Lett.* **2000**, *470*, 113–117. [CrossRef]

18. Wang, H.; Ward, N.; Boswell, M.; Katz, D.M. Secretion of brain-derived neurotrophic factor from brain microvascular endothelial cells. *Eur. J. Neurosci.* **2006**, *23*, 1665–1670. [CrossRef]

19. Pius-Sadowska, E.; Machaliński, B. BDNF-A key player in cardiovascular system. *J. Mol. Cell. Cardiol.* **2017**, *110*, 54–60. [CrossRef]

20. Anders, Q.S.; Ferreira, L.V.B.; Rodrigues, L.C.M.; Nakamura-Palacios, E.M. BDNF mRNA Expression in Leukocytes and Frontal Cortex Function in Drug Use Disorder. *Front. Psychiatry* **2020**, *11*, 469. [CrossRef]

21. Chacón-Fernández, P.; Säuberli, K.; Colzani, M.; Moreau, T.; Ghevaert, C.; Barde, Y.A. Brain-derived Neurotrophic Factor in Megakaryocytes. *J. Biol. Chem.* **2016**, *291*, 9872–9881. [CrossRef] [PubMed]

22. Amadio, P.; Sandrini, L.; Ieraci, A.; Tremoli, E.; Barbieri, S.S. Effect of Clotting Duration and Temperature on BDNF Measurement in Human Serum. *Int. J. Mol. Sci.* **2017**, *18*, 1987. [CrossRef] [PubMed]

23. Meek, T.H.; Wisse, B.E.; Thaler, J.P.; Guyenet, S.J.; Matsen, M.E.; Fischer, J.D.; Taborsky, G.J.; Schwartz, M.W.; Morton, G.J. BDNF action in the brain attenuates diabetic hyperglycemia via insulin-independent inhibition of hepatic glucose production. *Diabetes* **2013**, *62*, 1512–1518. [CrossRef] [PubMed]

24. Marosi, K.; Mattson, M.P. BDNF mediates adaptive brain and body responses to energetic challenges. *Trends Endocrinol. Metab.* **2014**, *25*, 89–98. [CrossRef] [PubMed]

25. Delezie, J.; Weihrauch, M.; Maier, G.; Tejero, R.; Ham, D.J.; Gill, J.F.; Karrer-Cardel, B.; Rüegg, M.A.; Tabares, L.; Handschin, C. BDNF is a mediator of glycolytic fiber-type specification in mouse skeletal muscle. *Proc. Natl. Acad. Sci. USA* **2019**, *116*, 16111–16120. [CrossRef]

26. Tettamanti, G.; Cattaneo, A.G.; Gornati, R.; de Eguileor, M.; Bernardini, G.; Binelli, G. Phylogenesis of brain-derived neurotrophic factor (BDNF) in vertebrates. *Gene* **2010**, *450*, 85–93. [CrossRef]

27. Metsis, M.; Timmusk, T.; Arenas, E.; Persson, H. Differential usage of multiple brain-derived neurotrophic factor promoters in the rat brain following neuronal activation. *Proc. Natl. Acad. Sci. USA* **1993**, *90*, 8802–8806. [CrossRef]

28. Timmusk, T.; Lendahl, U.; Funakoshi, H.; Arenas, E.; Persson, H.; Metsis, M. Identification of brain-derived neurotrophic factor promoter regions mediating tissue-specific, axotomy-, and neuronal activity-induced expression in transgenic mice. *J. Cell Biol.* **1995**, *128*, 185–199. [CrossRef]

29. Timmusk, T.; Palm, K.; Metsis, M.; Reintam, T.; Paalme, V.; Saarma, M.; Persson, H. Multiple promoters direct tissue-specific expression of the rat BDNF gene. *Neuron* **1993**, *10*, 475–489. [CrossRef]

30. Vashishta, A.; Habas, A.; Pruunsild, P.; Zheng, J.J.; Timmusk, T.; Hetman, M. Nuclear factor of activated T-cells isoform c4 (NFATc4/NFAT3) as a mediator of antiapoptotic transcription in NMDA receptor-stimulated cortical neurons. *J. Neurosci.* **2009**, *29*, 15331–15340. [CrossRef]

31. Volpicelli, F.; Caiazzo, M.; Greco, D.; Consales, C.; Leone, L.; Perrone-Capano, C.; Colucci D'Amato, L.; di Porzio, U. Bdnf gene is a downstream target of Nurr1 transcription factor in rat midbrain neurons in vitro. *J. Neurochem.* **2007**, *102*, 441–453. [CrossRef] [PubMed]

32. Aid, T.; Kazantseva, A.; Piirsoo, M.; Palm, K.; Timmusk, T. Mouse and rat BDNF gene structure and expression revisited. *J. Neurosci. Res.* **2007**, *85*, 525–535. [CrossRef] [PubMed]

33. Pruunsild, P.; Kazantseva, A.; Aid, T.; Palm, K.; Timmusk, T. Dissecting the human BDNF locus: Bidirectional transcription, complex splicing, and multiple promoters. *Genomics* **2007**, *90*, 397–406. [CrossRef] [PubMed]

34. Paoletti, P.; Bellone, C.; Zhou, Q. NMDA receptor subunit diversity: Impact on receptor properties, synaptic plasticity and disease. *Nat. Rev. Neurosci.* **2013**, *14*, 383–400. [CrossRef] [PubMed]

35. Simms, B.A.; Zamponi, G.W. Neuronal voltage-gated calcium channels: Structure, function, and dysfunction. *Neuron* **2014**, *82*, 24–45. [CrossRef]

36. Shieh, P.B.; Hu, S.C.; Bobb, K.; Timmusk, T.; Ghosh, A. Identification of a signaling pathway involved in calcium regulation of BDNF expression. *Neuron* **1998**, *20*, 727–740. [CrossRef]

37. Tao, X.; Finkbeiner, S.; Arnold, D.B.; Shaywitz, A.J.; Greenberg, M.E. Ca2+ influx regulates BDNF transcription by a CREB family transcription factor-dependent mechanism. *Neuron* **1998**, *20*, 709–726. [CrossRef]

38. Tabuchi, A.; Sakaya, H.; Kisukeda, T.; Fushiki, H.; Tsuda, M. Involvement of an upstream stimulatory factor as well as cAMP-responsive element-binding protein in the activation of brain-derived neurotrophic factor gene promoter I. *J. Biol. Chem.* **2002**, *277*, 35920–35931. [CrossRef]

39. Pruunsild, P.; Sepp, M.; Orav, E.; Koppel, I.; Timmusk, T. Identification of cis-elements and transcription factors regulating neuronal activity-dependent transcription of human BDNF gene. *J. Neurosci.* **2011**, *31*, 3295–3308. [CrossRef]

40. Tao, X.; West, A.E.; Chen, W.G.; Corfas, G.; Greenberg, M.E. A calcium-responsive transcription factor, CaRF, that regulates neuronal activity-dependent expression of BDNF. *Neuron* **2002**, *33*, 383–395. [CrossRef]

41. Zheng, F.; Wang, H. NMDA-mediated and self-induced bdnf exon IV transcriptions are differentially regulated in cultured cortical neurons. *Neurochem. Int.* **2009**, *54*, 385–392. [CrossRef] [PubMed]

42. Lipsky, R.H.; Xu, K.; Zhu, D.; Kelly, C.; Terhakopian, A.; Novelli, A.; Marini, A.M. Nuclear factor kappaB is a critical determinant in N-methyl-D-aspartate receptor-mediated neuroprotection. *J. Neurochem.* **2001**, *78*, 254–264. [CrossRef] [PubMed]

43. Zheng, F.; Zhou, X.; Moon, C.; Wang, H. Regulation of brain-derived neurotrophic factor expression in neurons. *Int. J. Physiol. Pathophysiol. Pharmacol.* **2012**, *4*, 188–200. [PubMed]

44. Jiang, X.; Tian, F.; Du, Y.; Copeland, N.G.; Jenkins, N.A.; Tessarollo, L.; Wu, X.; Pan, H.; Hu, X.Z.; Xu, K.; et al. BHLHB2 controls Bdnf promoter 4 activity and neuronal excitability. *J. Neurosci.* **2008**, *28*, 1118–1130. [CrossRef]

45. Sakata, K.; Woo, N.H.; Martinowich, K.; Greene, J.S.; Schloesser, R.J.; Shen, L.; Lu, B. Critical role of promoter IV-driven BDNF transcription in GABAergic transmission and synaptic plasticity in the prefrontal cortex. *Proc. Natl. Acad. Sci. USA* **2009**, *106*, 5942–5947. [CrossRef]

46. Keller, S.; Sarchiapone, M.; Zarrilli, F.; Videtic, A.; Ferraro, A.; Carli, V.; Sacchetti, S.; Lembo, F.; Angiolillo, A.; Jovanovic, N.; et al. Increased BDNF promoter methylation in the Wernicke area of suicide subjects. *Arch. Gen. Psychiatry* **2010**, *67*, 258–267. [CrossRef]

47. Tadić, A.; Müller-Engling, L.; Schlicht, K.F.; Kotsiari, A.; Dreimüller, N.; Kleimann, A.; Bleich, S.; Lieb, K.; Frieling, H. Methylation of the promoter of brain-derived neurotrophic factor exon IV and antidepressant response in major depression. *Mol. Psychiatry* **2014**, *19*, 281–283. [CrossRef]

48. Lieb, K.; Dreimüller, N.; Wagner, S.; Schlicht, K.; Falter, T.; Neyazi, A.; Müller-Engling, L.; Bleich, S.; Tadić, A.; Frieling, H. BDNF Plasma Levels and BDNF Exon IV Promoter Methylation as Predictors for Antidepressant Treatment Response. *Front. Psychiatry* **2018**, *9*, 511. [CrossRef]

49. Sakata, K.; Duke, S.M. Lack of BDNF expression through promoter IV disturbs expression of monoamine genes in the frontal cortex and hippocampus. *Neuroscience* **2014**, *260*, 265–275. [CrossRef]

50. Hill, J.L.; Hardy, N.F.; Jimenez, D.V.; Maynard, K.R.; Kardian, A.S.; Pollock, C.J.; Schloesser, R.J.; Martinowich, K. Loss of promoter IV-driven BDNF expression impacts oscillatory activity during sleep, sensory information processing and fear regulation. *Transl. Psychiatry* **2016**, *6*, e873. [CrossRef]

51. Volpicelli, F.; Speranza, L.; Pulcrano, S.; De Gregorio, R.; Crispino, M.; De Sanctis, C.; Leopoldo, M.; Lacivita, E.; di Porzio, U.; Bellenchi, G.C.; et al. The microRNA-29a Modulates Serotonin 5-HT7 Receptor Expression and Its Effects on Hippocampal Neuronal Morphology. *Mol. Neurobiol.* **2019**, *56*, 8617–8627. [CrossRef] [PubMed]

52. De Gregorio, R.; Pulcrano, S.; De Sanctis, C.; Volpicelli, F.; Guatteo, E.; von Oerthel, L.; Latagliata, E.C.; Esposito, R.; Piscitelli, R.M.; Perrone-Capano, C.; et al. miR-34b/c Regulates Wnt1 and Enhances Mesencephalic Dopaminergic Neuron Differentiation. *Stem Cell Rep.* **2018**, *10*, 1237–1250. [CrossRef] [PubMed]

53. McNeill, E.; Van Vactor, D. MicroRNAs shape the neuronal landscape. *Neuron* **2012**, *75*, 363–379. [CrossRef] [PubMed]

54. Keifer, J.; Zheng, Z.; Ambigapathy, G. A MicroRNA-BDNF Negative Feedback Signaling Loop in Brain: Implications for Alzheimer's Disease. *Microrna* **2015**, *4*, 101–108. [CrossRef] [PubMed]

55. Khani-Habibabadi, F.; Askari, S.; Zahiri, J.; Javan, M.; Behmanesh, M. Novel BDNF-regulatory microRNAs in neurodegenerative disorders pathogenesis: An in silico study. *Comput. Biol. Chem.* **2019**, *83*, 107153. [CrossRef] [PubMed]

56. Schratt, G.M.; Tuebing, F.; Nigh, E.A.; Kane, C.G.; Sabatini, M.E.; Kiebler, M.; Greenberg, M.E. A brain-specific microRNA regulates dendritic spine development. *Nature* **2006**, *439*, 283–289. [CrossRef]

57. Baby, N.; Alagappan, N.; Dheen, S.T.; Sajikumar, S. MicroRNA-134-5p inhibition rescues long-term plasticity and synaptic tagging/capture in an Aβ(1-42)-induced model of Alzheimer's disease. *Aging Cell* **2020**, *19*, e13046. [CrossRef]

58. Gao, J.; Wang, W.Y.; Mao, Y.W.; Gräff, J.; Guan, J.S.; Pan, L.; Mak, G.; Kim, D.; Su, S.C.; Tsai, L.H. A novel pathway regulates memory and plasticity via SIRT1 and miR-134. *Nature* **2010**, *466*, 1105–1109. [CrossRef]
59. Shen, J.; Xu, L.; Qu, C.; Sun, H.; Zhang, J. Resveratrol prevents cognitive deficits induced by chronic unpredictable mild stress: Sirt1/miR-134 signalling pathway regulates CREB/BDNF expression in hippocampus in vivo and in vitro. *Behav. Brain Res.* **2018**, *349*, 1–7. [CrossRef]
60. Xin, C.; Xia, J.; Liu, Y.; Zhang, Y. MicroRNA-202-3p Targets Brain-Derived Neurotrophic Factor and Is Involved in Depression-Like Behaviors. *Neuropsychiatr. Dis. Treat.* **2020**, *16*, 1073–1083. [CrossRef]
61. Yang, C.R.; Zhang, X.Y.; Liu, Y.; Du, J.Y.; Liang, R.; Yu, M.; Zhang, F.Q.; Mu, X.F.; Li, F.; Zhou, L.; et al. Antidepressant Drugs Correct the Imbalance Between proBDNF/p75NTR/Sortilin and Mature BDNF/TrkB in the Brain of Mice with Chronic Stress. *Neurotox. Res.* **2020**, *37*, 171–182. [CrossRef] [PubMed]
62. You, Y.H.; Qin, Z.Q.; Zhang, H.L.; Yuan, Z.H.; Yu, X. MicroRNA-153 promotes brain-derived neurotrophic factor and hippocampal neuron proliferation to alleviate autism symptoms through inhibition of JAK-STAT pathway by LEPR. *Biosci. Rep.* **2019**, *39*. [CrossRef]
63. Rostamian Delavar, M.; Baghi, M.; Safaeinejad, Z.; Kiani-Esfahani, A.; Ghaedi, K.; Nasr-Esfahani, M.H. Differential expression of miR-34a, miR-141, and miR-9 in MPP+-treated differentiated PC12 cells as a model of Parkinson's disease. *Gene* **2018**, *662*, 54–65. [CrossRef] [PubMed]
64. Zhao, X.; Shu, F.; Wang, X.; Wang, F.; Wu, L.; Li, L.; Lv, H. Inhibition of microRNA-375 ameliorated ketamine-induced neurotoxicity in human embryonic stem cell derived neurons. *Eur. J. Pharmacol.* **2019**, *844*, 56–64. [CrossRef] [PubMed]
65. Jiang, J.D.; Zheng, X.C.; Huang, F.Y.; Gao, F.; You, M.Z.; Zheng, T. MicroRNA-107 regulates anesthesia-induced neural injury in embryonic stem cell derived neurons. *IUBMB Life* **2019**, *71*, 20–27. [CrossRef] [PubMed]
66. Barde, Y.A.; Edgar, D.; Thoenen, H. Purification of a new neurotrophic factor from mammalian brain. *EMBO J.* **1982**, *1*, 549–553. [CrossRef]
67. Foltran, R.B.; Diaz, S.L. BDNF isoforms: A round trip ticket between neurogenesis and serotonin? *J. Neurochem.* **2016**, *138*, 204–221. [CrossRef]
68. Mizui, T.; Ishikawa, Y.; Kumanogoh, H.; Kojima, M. Neurobiological actions by three distinct subtypes of brain-derived neurotrophic factor: Multi-ligand model of growth factor signaling. *Pharmacol. Res.* **2016**, *105*, 93–98. [CrossRef]
69. Lu, J.; Wu, Y.; Sousa, N.; Almeida, O.F. SMAD pathway mediation of BDNF and TGF beta 2 regulation of proliferation and differentiation of hippocampal granule neurons. *Development* **2005**, *132*, 3231–3242. [CrossRef]
70. Kowiański, P.; Lietzau, G.; Czuba, E.; Waśkow, M.; Steliga, A.; Moryś, J. BDNF: A Key Factor with Multipotent Impact on Brain Signaling and Synaptic Plasticity. *Cell Mol. Neurobiol.* **2018**, *38*, 579–593. [CrossRef]
71. Pang, P.T.; Teng, H.K.; Zaitsev, E.; Woo, N.T.; Sakata, K.; Zhen, S.; Teng, K.K.; Yung, W.H.; Hempstead, B.L.; Lu, B. Cleavage of proBDNF by tPA/plasmin is essential for long-term hippocampal plasticity. *Science* **2004**, *306*, 487–491. [CrossRef]
72. Mizoguchi, H.; Nakade, J.; Tachibana, M.; Ibi, D.; Someya, E.; Koike, H.; Kamei, H.; Nabeshima, T.; Itohara, S.; Takuma, K.; et al. Matrix metalloproteinase-9 contributes to kindled seizure development in pentylenetetrazole-treated mice by converting pro-BDNF to mature BDNF in the hippocampus. *J. Neurosci.* **2011**, *31*, 12963–12971. [CrossRef]
73. Vafadari, B.; Salamian, A.; Kaczmarek, L. MMP-9 in translation: From molecule to brain physiology, pathology, and therapy. *J. Neurochem.* **2016**, *139* (Suppl. S2), 91–114. [CrossRef]
74. Mowla, S.J.; Farhadi, H.F.; Pareek, S.; Atwal, J.K.; Morris, S.J.; Seidah, N.G.; Murphy, R.A. Biosynthesis and post-translational processing of the precursor to brain-derived neurotrophic factor. *J. Biol. Chem.* **2001**, *276*, 12660–12666. [CrossRef] [PubMed]
75. Conner, J.M.; Lauterborn, J.C.; Yan, Q.; Gall, C.M.; Varon, S. Distribution of brain-derived neurotrophic factor (BDNF) protein and mRNA in the normal adult rat CNS: Evidence for anterograde axonal transport. *J. Neurosci.* **1997**, *17*, 2295–2313. [CrossRef]
76. Dieni, S.; Matsumoto, T.; Dekkers, M.; Rauskolb, S.; Ionescu, M.S.; Deogracias, R.; Gundelfinger, E.D.; Kojima, M.; Nestel, S.; Frotscher, M.; et al. BDNF and its pro-peptide are stored in presynaptic dense core vesicles in brain neurons. *J. Cell Biol.* **2012**, *196*, 775–788. [CrossRef] [PubMed]
77. Yang, J.; Siao, C.J.; Nagappan, G.; Marinic, T.; Jing, D.; McGrath, K.; Chen, Z.Y.; Mark, W.; Tessarollo, L.; Lee, F.S.; et al. Neuronal release of proBDNF. *Nat. Neurosci.* **2009**, *12*, 113–115. [CrossRef]

78. Yang, J.L.; Lin, Y.T.; Chuang, P.C.; Bohr, V.A.; Mattson, M.P. BDNF and exercise enhance neuronal DNA repair by stimulating CREB-mediated production of apurinic/apyrimidinic endonuclease 1. *Neuromolecular Med.* **2014**, *16*, 161–174. [CrossRef]

79. Du, J.; Feng, L.; Yang, F.; Lu, B. Activity- and Ca(2+)-dependent modulation of surface expression of brain-derived neurotrophic factor receptors in hippocampal neurons. *J. Cell Biol.* **2000**, *150*, 1423–1434. [CrossRef] [PubMed]

80. Meyer-Franke, A.; Wilkinson, G.A.; Kruttgen, A.; Hu, M.; Munro, E.; Hanson, M.G.; Reichardt, L.F.; Barres, B.A. Depolarization and cAMP elevation rapidly recruit TrkB to the plasma membrane of CNS neurons. *Neuron* **1998**, *21*, 681–693. [CrossRef]

81. Deinhardt, K.; Chao, M.V. Shaping neurons: Long and short range effects of mature and proBDNF signalling upon neuronal structure. *Neuropharmacology* **2014**, *76 Pt C*, 603–609. [CrossRef]

82. Anastasia, A.; Deinhardt, K.; Chao, M.V.; Will, N.E.; Irmady, K.; Lee, F.S.; Hempstead, B.L.; Bracken, C. Val66Met polymorphism of BDNF alters prodomain structure to induce neuronal growth cone retraction. *Nat. Commun.* **2013**, *4*, 2490. [CrossRef]

83. Teng, H.K.; Teng, K.K.; Lee, R.; Wright, S.; Tevar, S.; Almeida, R.D.; Kermani, P.; Torkin, R.; Chen, Z.Y.; Lee, F.S.; et al. ProBDNF induces neuronal apoptosis via activation of a receptor complex of p75NTR and sortilin. *J. Neurosci.* **2005**, *25*, 5455–5463. [CrossRef] [PubMed]

84. Roux, P.P.; Colicos, M.A.; Barker, P.A.; Kennedy, T.E. p75 neurotrophin receptor expression is induced in apoptotic neurons after seizure. *J. Neurosci.* **1999**, *19*, 6887–6896. [CrossRef] [PubMed]

85. Reichardt, L.F. Neurotrophin-regulated signalling pathways. *Philos. Trans. R. Soc. Lond. B Biol. Sci.* **2006**, *361*, 1545–1564. [CrossRef]

86. Kaplan, D.R.; Miller, F.D. Neurotrophin signal transduction in the nervous system. *Curr. Opin. Neurobiol.* **2000**, *10*, 381–391. [CrossRef]

87. Gonzalez, A.; Moya-Alvarado, G.; Gonzalez-Billaut, C.; Bronfman, F.C. Cellular and molecular mechanisms regulating neuronal growth by brain-derived neurotrophic factor. *Cytoskeleton (Hoboken)* **2016**, *73*, 612–628. [CrossRef]

88. Huang, E.J.; Reichardt, L.F. Trk receptors: Roles in neuronal signal transduction. *Annu. Rev. Biochem.* **2003**, *72*, 609–642. [CrossRef]

89. Minichiello, L. TrkB signalling pathways in LTP and learning. *Nat. Rev. Neurosci.* **2009**, *10*, 850–860. [CrossRef]

90. Patel, A.V.; Krimm, R.F. BDNF is required for the survival of differentiated geniculate ganglion neurons. *Dev. Biol.* **2010**, *340*, 419–429. [CrossRef]

91. Chen, J.; Shehadah, A.; Pal, A.; Zacharek, A.; Cui, X.; Cui, Y.; Roberts, C.; Lu, M.; Zeitlin, A.; Hariri, R.; et al. Neuroprotective effect of human placenta-derived cell treatment of stroke in rats. *Cell Transplant.* **2013**, *22*, 871–879. [CrossRef] [PubMed]

92. Wu, Y.; Wang, R.; Wang, Y.; Gao, J.; Feng, L.; Yang, Z. Distinct Impacts of Fullerene on Cognitive Functions of Dementia vs. Non-dementia Mice. *Neurotox. Res.* **2019**, *36*, 736–745. [CrossRef] [PubMed]

93. Gorski, J.A.; Zeiler, S.R.; Tamowski, S.; Jones, K.R. Brain-derived neurotrophic factor is required for the maintenance of cortical dendrites. *J. Neurosci.* **2003**, *23*, 6856–6865. [CrossRef] [PubMed]

94. Kwon, M.; Fernández, J.R.; Zegarek, G.F.; Lo, S.B.; Firestein, B.L. BDNF-promoted increases in proximal dendrites occur via CREB-dependent transcriptional regulation of cypin. *J. Neurosci.* **2011**, *31*, 9735–9745. [CrossRef]

95. Orefice, L.L.; Waterhouse, E.G.; Partridge, J.G.; Lalchandani, R.R.; Vicini, S.; Xu, B. Distinct roles for somatically and dendritically synthesized brain-derived neurotrophic factor in morphogenesis of dendritic spines. *J. Neurosci.* **2013**, *33*, 11618–11632. [CrossRef]

96. Zagrebelsky, M.; Tacke, C.; Korte, M. BDNF signaling during the lifetime of dendritic spines. *Cell Tissue Res.* **2020**. [CrossRef]

97. Yoshii, A.; Constantine-Paton, M. Postsynaptic localization of PSD-95 is regulated by all three pathways downstream of TrkB signaling. *Front. Synaptic Neurosci.* **2014**, *6*, 6. [CrossRef]

98. Chen, S.D.; Wu, C.L.; Hwang, W.C.; Yang, D.I. More Insight into BDNF against Neurodegeneration: Anti-Apoptosis, Anti-Oxidation, and Suppression of Autophagy. *Int. J. Mol. Sci.* **2017**, *18*, 545. [CrossRef]

99. Opazo, P.; Watabe, A.M.; Grant, S.G.; O'Dell, T.J. Phosphatidylinositol 3-kinase regulates the induction of long-term potentiation through extracellular signal-related kinase-independent mechanisms. *J. Neurosci.* **2003**, *23*, 3679–3688. [CrossRef]

100. Leal, G.; Bramham, C.R.; Duarte, C.B. BDNF and Hippocampal Synaptic Plasticity. *Vitam. Horm.* **2017**, *104*, 153–195. [CrossRef]

101. Baydyuk, M.; Xu, B. BDNF signaling and survival of striatal neurons. *Front. Cell. Neurosci.* **2014**, *8*, 254. [CrossRef]

102. Waterhouse, E.G.; An, J.J.; Orefice, L.L.; Baydyuk, M.; Liao, G.Y.; Zheng, K.; Lu, B.; Xu, B. BDNF promotes differentiation and maturation of adult-born neurons through GABAergic transmission. *J. Neurosci.* **2012**, *32*, 14318–14330. [CrossRef] [PubMed]

103. Park, H.; Poo, M.M. Neurotrophin regulation of neural circuit development and function. *Nat. Rev. Neurosci.* **2013**, *14*, 7–23. [CrossRef]

104. Jaworski, J.; Spangler, S.; Seeburg, D.P.; Hoogenraad, C.C.; Sheng, M. Control of dendritic arborization by the phosphoinositide-3'-kinase-Akt-mammalian target of rapamycin pathway. *J. Neurosci.* **2005**, *25*, 11300–11312. [CrossRef] [PubMed]

105. Kumar, V.; Zhang, M.X.; Swank, M.W.; Kunz, J.; Wu, G.Y. Regulation of dendritic morphogenesis by Ras-PI3K-Akt-mTOR and Ras-MAPK signaling pathways. *J. Neurosci.* **2005**, *25*, 11288–11299. [CrossRef] [PubMed]

106. Finkbeiner, S.; Tavazoie, S.F.; Maloratsky, A.; Jacobs, K.M.; Harris, K.M.; Greenberg, M.E. CREB: A major mediator of neuronal neurotrophin responses. *Neuron* **1997**, *19*, 1031–1047. [CrossRef]

107. Xing, J.; Kornhauser, J.M.; Xia, Z.; Thiele, E.A.; Greenberg, M.E. Nerve growth factor activates extracellular signal-regulated kinase and p38 mitogen-activated protein kinase pathways to stimulate CREB serine 133 phosphorylation. *Mol. Cell. Biol.* **1998**, *18*, 1946–1955. [CrossRef]

108. Segal, M.; Kreher, U.; Greenberger, V.; Braun, K. Is fragile X mental retardation protein involved in activity-induced plasticity of dendritic spines? *Brain Res.* **2003**, *972*, 9–15. [CrossRef]

109. Petryshen, T.L.; Sabeti, P.C.; Aldinger, K.A.; Fry, B.; Fan, J.B.; Schaffner, S.F.; Waggoner, S.G.; Tahl, A.R.; Sklar, P. Population genetic study of the brain-derived neurotrophic factor (BDNF) gene. *Mol. Psychiatry* **2010**, *15*, 810–815. [CrossRef]

110. Momose, Y.; Murata, M.; Kobayashi, K.; Tachikawa, M.; Nakabayashi, Y.; Kanazawa, I.; Toda, T. Association studies of multiple candidate genes for Parkinson's disease using single nucleotide polymorphisms. *Ann. Neurol.* **2002**, *51*, 133–136. [CrossRef]

111. Ventriglia, M.; Bocchio Chiavetto, L.; Benussi, L.; Binetti, G.; Zanetti, O.; Riva, M.A.; Gennarelli, M. Association between the BDNF 196 A/G polymorphism and sporadic Alzheimer's disease. *Mol. Psychiatry* **2002**, *7*, 136–137. [CrossRef]

112. Egan, M.F.; Kojima, M.; Callicott, J.H.; Goldberg, T.E.; Kolachana, B.S.; Bertolino, A.; Zaitsev, E.; Gold, B.; Goldman, D.; Dean, M.; et al. The BDNF val66met polymorphism affects activity-dependent secretion of BDNF and human memory and hippocampal function. *Cell* **2003**, *112*, 257–269. [CrossRef]

113. Chen, Z.Y.; Patel, P.D.; Sant, G.; Meng, C.X.; Teng, K.K.; Hempstead, B.L.; Lee, F.S. Variant brain-derived neurotrophic factor (BDNF) (Met66) alters the intracellular trafficking and activity-dependent secretion of wild-type BDNF in neurosecretory cells and cortical neurons. *J. Neurosci.* **2004**, *24*, 4401–4411. [CrossRef] [PubMed]

114. Chiaruttini, C.; Vicario, A.; Li, Z.; Baj, G.; Braiuca, P.; Wu, Y.; Lee, F.S.; Gardossi, L.; Baraban, J.M.; Tongiorgi, E. Dendritic trafficking of BDNF mRNA is mediated by translin and blocked by the G196A (Val66Met) mutation. *Proc. Natl. Acad. Sci. USA* **2009**, *106*, 16481–16486. [CrossRef] [PubMed]

115. Tsai, S.J. Salivary gland low-intensity pulsed ultrasound (LIPUS) stimulation as a potential treatment for various BDNF-implicated neuropsychiatric disorders. *Med. Hypotheses* **2020**, *137*, 109560. [CrossRef] [PubMed]

116. Lin, C.C.; Huang, T.L. Brain-derived neurotrophic factor and mental disorders. *Biomed. J.* **2020**, *43*, 134–142. [CrossRef] [PubMed]

117. Di Lazzaro, V.; Pellegrino, G.; Di Pino, G.; Corbetto, M.; Ranieri, F.; Brunelli, N.; Paolucci, M.; Bucossi, S.; Ventriglia, M.C.; Brown, P.; et al. Val66Met BDNF gene polymorphism influences human motor cortex plasticity in acute stroke. *Brain Stimul.* **2015**, *8*, 92–96. [CrossRef]

118. Morin-Moncet, O.; Latulipe-Loiselle, A.; Therrien-Blanchet, J.M.; Theoret, H. BDNF Val66Met polymorphism is associated with altered activity-dependent modulation of short-interval intracortical inhibition in bilateral M1. *PLoS ONE* **2018**, *13*, e0197505. [CrossRef]

119. Hashimoto, T.; Fukui, K.; Takeuchi, H.; Yokota, S.; Kikuchi, Y.; Tomita, H.; Taki, Y.; Kawashima, R. Effects of the BDNF Val66Met Polymorphism on Gray Matter Volume in Typically Developing Children and Adolescents. *Cereb. Cortex* **2016**, *26*, 1795–1803. [CrossRef]

120. Meng, W.D.; Sun, S.J.; Yang, J.; Chu, R.X.; Tu, W.; Liu, Q. Elevated Serum Brain-Derived Neurotrophic Factor (BDNF) but not BDNF Gene Val66Met Polymorphism Is Associated with Autism Spectrum Disorders. *Mol. Neurobiol.* **2017**, *54*, 1167–1172. [CrossRef]

121. Montag, C.; Basten, U.; Stelzel, C.; Fiebach, C.J.; Reuter, M. The BDNF Val66Met polymorphism and anxiety: Support for animal knock-in studies from a genetic association study in humans. *Psychiatry Res.* **2010**, *179*, 86–90. [CrossRef] [PubMed]

122. Park, C.H.; Kim, J.; Namgung, E.; Lee, D.W.; Kim, G.H.; Kim, M.; Kim, N.; Kim, T.D.; Kim, S.; Lyoo, I.K.; et al. The BDNF Val66Met Polymorphism Affects the Vulnerability of the Brain Structural Network. *Front. Hum. Neurosci.* **2017**, *11*, 400. [CrossRef] [PubMed]

123. Hariri, A.R.; Goldberg, T.E.; Mattay, V.S.; Kolachana, B.S.; Callicott, J.H.; Egan, M.F.; Weinberger, D.R. Brain-derived neurotrophic factor val66met polymorphism affects human memory-related hippocampal activity and predicts memory performance. *J. Neurosci.* **2003**, *23*, 6690–6694. [CrossRef] [PubMed]

124. Pezawas, L.; Verchinski, B.A.; Mattay, V.S.; Callicott, J.H.; Kolachana, B.S.; Straub, R.E.; Egan, M.F.; Meyer-Lindenberg, A.; Weinberger, D.R. The brain-derived neurotrophic factor val66met polymorphism and variation in human cortical morphology. *J. Neurosci.* **2004**, *24*, 10099–10102. [CrossRef] [PubMed]

125. Ho, B.C.; Milev, P.; O'Leary, D.S.; Librant, A.; Andreasen, N.C.; Wassink, T.H. Cognitive and magnetic resonance imaging brain morphometric correlates of brain-derived neurotrophic factor Val66Met gene polymorphism in patients with schizophrenia and healthy volunteers. *Arch. Gen. Psychiatry* **2006**, *63*, 731–740. [CrossRef] [PubMed]

126. Montag, C.; Weber, B.; Fliessbach, K.; Elger, C.; Reuter, M. The BDNF Val66Met polymorphism impacts parahippocampal and amygdala volume in healthy humans: Incremental support for a genetic risk factor for depression. *Psychol. Med.* **2009**, *39*, 1831–1839. [CrossRef] [PubMed]

127. Xia, H.; Du, X.; Yin, G.; Zhang, Y.; Li, X.; Cai, J.; Huang, X.; Ning, Y.; Soares, J.C.; Wu, F.; et al. Effects of smoking on cognition and BDNF levels in a male Chinese population: Relationship with BDNF Val66Met polymorphism. *Sci. Rep.* **2019**, *9*, 217. [CrossRef]

128. Altmann, V.; Schumacher-Schuh, A.F.; Rieck, M.; Callegari-Jacques, S.M.; Rieder, C.R.; Hutz, M.H. Val66Met BDNF polymorphism is associated with Parkinson's disease cognitive impairment. *Neurosci. Lett.* **2016**, *615*, 88–91. [CrossRef]

129. Wang, Q.; Liu, J.; Guo, Y.; Dong, G.; Zou, W.; Chen, Z. Association between BDNF G196A (Val66Met) polymorphism and cognitive impairment in patients with Parkinson's disease: A meta-analysis. *Braz. J. Med. Biol. Res.* **2019**, *52*, e8443. [CrossRef]

130. Franzmeier, N.; Ren, J.; Damm, A.; Monté-Rubio, G.; Boada, M.; Ruiz, A.; Ramirez, A.; Jessen, F.; Düzel, E.; Rodríguez Gómez, O.; et al. The BDNF. *Mol. Psychiatry* **2019**. [CrossRef]

131. Yin, Y.; Su, X.; Pan, L.; Li, C. BDNF Val66Met polymorphism and cognitive impairment in Parkinson's disease-a meta-analysis. *Neurol. Sci.* **2019**, *40*, 1901–1907. [CrossRef] [PubMed]

132. Dalby, R.B.; Elfving, B.; Poulsen, P.H.; Foldager, L.; Frandsen, J.; Videbech, P.; Rosenberg, R. Plasma brain-derived neurotrophic factor and prefrontal white matter integrity in late-onset depression and normal aging. *Acta Psychiatr. Scand.* **2013**, *128*, 387–396. [CrossRef] [PubMed]

133. Choi, S.; Han, K.M.; Won, E.; Yoon, B.J.; Lee, M.S.; Ham, B.J. Association of brain-derived neurotrophic factor DNA methylation and reduced white matter integrity in the anterior corona radiata in major depression. *J. Affect. Disord.* **2015**, *172*, 74–80. [CrossRef]

134. Youssef, M.M.; Underwood, M.D.; Huang, Y.Y.; Hsiung, S.C.; Liu, Y.; Simpson, N.R.; Bakalian, M.J.; Rosoklija, G.B.; Dwork, A.J.; Arango, V.; et al. Association of BDNF Val66Met Polymorphism and Brain BDNF Levels with Major Depression and Suicide. *Int. J. Neuropsychopharmacol.* **2018**, *21*, 528–538. [CrossRef]

135. Caldieraro, M.A.; McKee, M.; Leistner-Segal, S.; Vares, E.A.; Kubaski, F.; Spanemberg, L.; Brusius-Facchin, A.C.; Fleck, M.P.; Mischoulon, D. Val66Met polymorphism association with serum BDNF and inflammatory biomarkers in major depression. *World J. Biol. Psychiatry* **2018**, *19*, 402–409. [CrossRef]

136. Rong, C.; Park, C.; Rosenblat, J.D.; Subramaniapillai, M.; Zuckerman, H.; Fus, D.; Lee, Y.L.; Pan, Z.; Brietzke, E.; Mansur, R.B.; et al. Predictors of Response to Ketamine in Treatment Resistant Major Depressive Disorder and Bipolar Disorder. *Int. J. Environ. Res. Public Health* **2018**, *15*, 771. [CrossRef]

137. Mandolini, G.M.; Lazzaretti, M.; Pigoni, A.; Delvecchio, G.; Soares, J.C.; Brambilla, P. The impact of BDNF Val66Met polymorphism on cognition in Bipolar Disorder: A review: Special Section on "Translational and Neuroscience Studies in Affective Disorders"Section Editor, Maria Nobile MD, PhD. This Section of JAD focuses on the relevance of translational and neuroscience studies in providing a better understanding of the neural basis of affective disorders. The main aim is to briefly summaries relevant research findings in clinical neuroscience with particular regards to specific innovative topics in mood and anxiety disorders. *J. Affect. Disord.* **2019**, *243*, 552–558. [CrossRef]

138. Chen, N.C.; Chuang, Y.C.; Huang, C.W.; Lui, C.C.; Lee, C.C.; Hsu, S.W.; Lin, P.H.; Lu, Y.T.; Chang, Y.T.; Hsu, C.W.; et al. Interictal serum brain-derived neurotrophic factor level reflects white matter integrity, epilepsy severity, and cognitive dysfunction in chronic temporal lobe epilepsy. *Epilepsy Behav.* **2016**, *59*, 147–154. [CrossRef] [PubMed]

139. Sidhu, M.K.; Thompson, P.J.; Wandschneider, B.; Foulkes, A.; de Tisi, J.; Stretton, J.; Perona, M.; Thom, M.; Bonelli, S.B.; Burdett, J.; et al. The impact of brain-derived neurotrophic factor Val66Met polymorphism on cognition and functional brain networks in patients with intractable partial epilepsy. *CNS Neurosci. Ther.* **2019**, *25*, 223–232. [CrossRef] [PubMed]

140. Doherty, C.; Hogue, O.; Floden, D.P.; Altemus, J.B.; Najm, I.M.; Eng, C.; Busch, R.M. BDNF and COMT, but not APOE, alleles are associated with psychiatric symptoms in refractory epilepsy. *Epilepsy Behav.* **2019**, *94*, 131–136. [CrossRef]

141. Kheirollahi, M.; Kazemi, E.; Ashouri, S. Brain-Derived Neurotrophic Factor Gene Val66Met Polymorphism and Risk of Schizophrenia: A Meta-analysis of Case-Control Studies. *Cell. Mol. Neurobiol.* **2016**, *36*, 1–10. [CrossRef] [PubMed]

142. Xia, H.; Zhang, G.; Du, X.; Zhang, Y.; Yin, G.; Dai, J.; He, M.X.; Soares, J.C.; Li, X.; Zhang, X.Y. Suicide attempt, clinical correlates, and BDNF Val66Met polymorphism in chronic patients with schizophrenia. *Neuropsychology* **2018**, *32*, 199–205. [CrossRef]

143. Huang, E.; Hettige, N.C.; Zai, G.; Tomasi, J.; Huang, J.; Zai, C.C.; Pivac, N.; Nikolac Perkovic, M.; Tiwari, A.K.; Kennedy, J.L. BDNF Val66Met polymorphism and clinical response to antipsychotic treatment in schizophrenia and schizoaffective disorder patients: A meta-analysis. *Pharmacogenomics J.* **2019**, *19*, 269–276. [CrossRef] [PubMed]

144. Han, M.; Deng, C. BDNF as a pharmacogenetic target for antipsychotic treatment of schizophrenia. *Neurosci. Lett.* **2020**, *726*, 133870. [CrossRef] [PubMed]

145. Brown, A.; Machan, J.T.; Hayes, L.; Zervas, M. Molecular organization and timing of Wnt1 expression define cohorts of midbrain dopamine neuron progenitors in vivo. *J. Comp. Neurol.* **2011**, *519*, 2978–3000. [CrossRef]

146. Ramos-Cejudo, J.; Gutiérrez-Fernández, M.; Otero-Ortega, L.; Rodríguez-Frutos, B.; Fuentes, B.; Vallejo-Cremades, M.T.; Hernanz, T.N.; Cerdán, S.; Díez-Tejedor, E. Brain-derived neurotrophic factor administration mediated oligodendrocyte differentiation and myelin formation in subcortical ischemic stroke. *Stroke* **2015**, *46*, 221–228. [CrossRef]

147. Balkaya, M.; Cho, S. Genetics of stroke recovery: BDNF val66met polymorphism in stroke recovery and its interaction with aging. *Neurobiol. Dis.* **2019**, *126*, 36–46. [CrossRef]

148. Lohia, R.; Salari, R.; Brannigan, G. Sequence specificity despite intrinsic disorder: How a disease-associated Val/Met polymorphism rearranges tertiary interactions in a long disordered protein. *PLoS Comput. Biol.* **2019**, *15*, e1007390. [CrossRef]

149. Rock, P.L.; Roiser, J.P.; Riedel, W.J.; Blackwell, A.D. Cognitive impairment in depression: A systematic review and meta-analysis. *Psychol. Med.* **2014**, *44*, 2029–2040. [CrossRef]

150. Severino, V.; Farina, A.; Colucci-D'Amato, L.; Reccia, M.G.; Volpicelli, F.; Parente, A.; Chambery, A. Secretome profiling of differentiated neural mes-c-myc A1 cell line endowed with stem cell properties. *Biochim. Biophys. Acta* **2013**, *1834*, 2385–2395. [CrossRef]

151. Colucci-D'Amato, L.; Farina, A.; Vissers, J.P.; Chambery, A. Quantitative neuroproteomics: Classical and novel tools for studying neural differentiation and function. *Stem Cell Rev. Rep.* **2011**, *7*, 77–93. [CrossRef]

152. Di Lieto, A.; Leo, D.; Volpicelli, F.; di Porzio, U.; Colucci-D'Amato, L. FLUOXETINE modifies the expression of serotonergic markers in a differentiation-dependent fashion in the mesencephalic neural cell line A1 mes c-myc. *Brain Res.* **2007**, *1143*, 1–10. [CrossRef] [PubMed]

153. Nawa, Y.; Kaneko, H.; Oda, M.; Tsubonoya, M.; Hiroi, T.; Gentile, M.T.; Colucci-D'Amato, L.; Takahashi, R.; Matsui, H. Functional characterization of the neuron-restrictive silencer element in the human tryptophan hydroxylase 2 gene expression. *J. Neurochem.* **2017**, *142*, 827–840. [CrossRef] [PubMed]

154. Gentile, M.T.; Nawa, Y.; Lunardi, G.; Florio, T.; Matsui, H.; Colucci-D'Amato, L. Tryptophan hydroxylase 2 (TPH2) in a neuronal cell line: Modulation by cell differentiation and NRSF/rest activity. *J. Neurochem.* **2012**, *123*, 963–970. [CrossRef] [PubMed]

155. Duman, R.S.; Heninger, G.R.; Nestler, E.J. A molecular and cellular theory of depression. *Arch. Gen. Psychiatry* **1997**, *54*, 597–606. [CrossRef] [PubMed]

156. Colucci-D'Amato, L.; di Porzio, U. Neurogenesis in adult CNS: From denial to opportunities and challenges for therapy. *Bioessays* **2008**, *30*, 135–145. [CrossRef] [PubMed]

157. Colucci-D'Amato, L.; Bonavita, V.; di Porzio, U. The end of the central dogma of neurobiology: Stem cells and neurogenesis in adult CNS. *Neurol. Sci.* **2006**, *27*, 266–270. [CrossRef] [PubMed]

158. Cramer, S.C.; Sur, M.; Dobkin, B.H.; O'Brien, C.; Sanger, T.D.; Trojanowski, J.Q.; Rumsey, J.M.; Hicks, R.; Cameron, J.; Chen, D.; et al. Harnessing neuroplasticity for clinical applications. *Brain* **2011**, *134*, 1591–1609. [CrossRef]

159. Castrén, E. Neuronal network plasticity and recovery from depression. *JAMA Psychiatry* **2013**, *70*, 983–989. [CrossRef]

160. Snapyan, M.; Lemasson, M.; Brill, M.S.; Blais, M.; Massouh, M.; Ninkovic, J.; Gravel, C.; Berthod, F.; Götz, M.; Barker, P.A.; et al. Vasculature guides migrating neuronal precursors in the adult mammalian forebrain via brain-derived neurotrophic factor signaling. *J. Neurosci.* **2009**, *29*, 4172–4188. [CrossRef]

161. Bergami, M.; Rimondini, R.; Santi, S.; Blum, R.; Götz, M.; Canossa, M. Deletion of TrkB in adult progenitors alters newborn neuron integration into hippocampal circuits and increases anxiety-like behavior. *Proc. Natl. Acad. Sci. USA* **2008**, *105*, 15570–15575. [CrossRef] [PubMed]

162. Chan, K.L.; Tong, K.Y.; Yip, S.P. Relationship of serum brain-derived neurotrophic factor (BDNF) and health-related lifestyle in healthy human subjects. *Neurosci. Lett.* **2008**, *447*, 124–128. [CrossRef]

163. Zafra, F.; Lindholm, D.; Castrén, E.; Hartikka, J.; Thoenen, H. Regulation of brain-derived neurotrophic factor and nerve growth factor mRNA in primary cultures of hippocampal neurons and astrocytes. *J. Neurosci.* **1992**, *12*, 4793–4799. [CrossRef] [PubMed]

164. Castrén, E.; Antila, H. Neuronal plasticity and neurotrophic factors in drug responses. *Mol. Psychiatry* **2017**, *22*, 1085–1095. [CrossRef] [PubMed]

165. Castrén, E.; Zafra, F.; Thoenen, H.; Lindholm, D. Light regulates expression of brain-derived neurotrophic factor mRNA in rat visual cortex. *Proc. Natl. Acad. Sci. USA* **1992**, *89*, 9444–9448. [CrossRef] [PubMed]

166. Gianfranceschi, L.; Siciliano, R.; Walls, J.; Morales, B.; Kirkwood, A.; Huang, Z.J.; Tonegawa, S.; Maffei, L. Visual cortex is rescued from the effects of dark rearing by overexpression of BDNF. *Proc. Natl. Acad. Sci. USA* **2003**, *100*, 12486–12491. [CrossRef]

167. Hensch, T.K.; Bilimoria, P.M. Re-opening Windows: Manipulating Critical Periods for Brain Development. *Cerebrum* **2012**, *2012*, 11.

168. Hong, E.J.; McCord, A.E.; Greenberg, M.E. A biological function for the neuronal activity-dependent component of Bdnf transcription in the development of cortical inhibition. *Neuron* **2008**, *60*, 610–624. [CrossRef]

169. Volpicelli, F.; Speranza, L.; di Porzio, U.; Crispino, M.; Perrone-Capano, C. The serotonin receptor 7 and the structural plasticity of brain circuits. *Front. Behav. Neurosci.* **2014**, *8*, 318. [CrossRef]

170. Cazorla, M.; Arrang, J.M.; Prémont, J. Pharmacological characterization of six trkB antibodies reveals a novel class of functional agents for the study of the BDNF receptor. *Br. J. Pharmacol.* **2011**, *162*, 947–960. [CrossRef]

171. Park, S.W.; Nhu, l.H.; Cho, H.Y.; Seo, M.K.; Lee, C.H.; Ly, N.N.; Choi, C.M.; Lee, B.J.; Kim, G.M.; Seol, W.; et al. p11 mediates the BDNF-protective effects in dendritic outgrowth and spine formation in B27-deprived primary hippocampal cells. *J. Affect. Disord.* **2016**, *196*, 1–10. [CrossRef]

172. Katoh-Semba, R.; Asano, T.; Ueda, H.; Morishita, R.; Takeuchi, I.K.; Inaguma, Y.; Kato, K. Riluzole enhances expression of brain-derived neurotrophic factor with consequent proliferation of granule precursor cells in the rat hippocampus. *FASEB J.* **2002**, *16*, 1328–1330. [CrossRef]

173. Cavanaugh, J.E.; Ham, J.; Hetman, M.; Poser, S.; Yan, C.; Xia, Z. Differential regulation of mitogen-activated protein kinases ERK1/2 and ERK5 by neurotrophins, neuronal activity, and cAMP in neurons. *J. Neurosci.* **2001**, *21*, 434–443. [CrossRef]

174. Lin, E.; Cavanaugh, J.E.; Leak, R.K.; Perez, R.G.; Zigmond, M.J. Rapid activation of ERK by 6-hydroxydopamine promotes survival of dopaminergic cells. *J. Neurosci. Res.* **2008**, *86*, 108–117. [CrossRef] [PubMed]

175. Rauti, R.; Cellot, G.; D'Andrea, P.; Colliva, A.; Scaini, D.; Tongiorgi, E.; Ballerini, L. BDNF impact on synaptic dynamics: Extra or intracellular long-term release differently regulates cultured hippocampal synapses. *Mol. Brain* **2020**, *13*, 43. [CrossRef] [PubMed]

176. Ernfors, P.; Lee, K.F.; Jaenisch, R. Mice lacking brain-derived neurotrophic factor develop with sensory deficits. *Nature* **1994**, *368*, 147–150. [CrossRef]

177. Gao, X.; Smith, G.M.; Chen, J. Impaired dendritic development and synaptic formation of postnatal-born dentate gyrus granular neurons in the absence of brain-derived neurotrophic factor signaling. *Exp. Neurol.* **2009**, *215*, 178–190. [CrossRef]

178. Mariga, A.; Zavadil, J.; Ginsberg, S.D.; Chao, M.V. Withdrawal of BDNF from hippocampal cultures leads to changes in genes involved in synaptic function. *Dev. Neurobiol.* **2015**, *75*, 173–192. [CrossRef]

179. Aarse, J.; Herlitze, S.; Manahan-Vaughan, D. The requirement of BDNF for hippocampal synaptic plasticity is experience-dependent. *Hippocampus* **2016**, *26*, 739–751. [CrossRef]

180. Magariños, A.M.; Li, C.J.; Gal Toth, J.; Bath, K.G.; Jing, D.; Lee, F.S.; McEwen, B.S. Effect of brain-derived neurotrophic factor haploinsufficiency on stress-induced remodeling of hippocampal neurons. *Hippocampus* **2011**, *21*, 253–264. [CrossRef]

181. von Bohlen Und Halbach, O.; von Bohlen Und Halbach, V. BDNF effects on dendritic spine morphology and hippocampal function. *Cell Tissue Res.* **2018**, *373*, 729–741. [CrossRef]

182. von Bohlen und Halbach, O.; Minichiello, L.; Unsicker, K. TrkB but not trkC receptors are necessary for postnatal maintenance of hippocampal spines. *Neurobiol. Aging* **2008**, *29*, 1247–1255. [CrossRef]

183. Dokter, M.; Busch, R.; Poser, R.; Vogt, M.A.; von Bohlen Und Halbach, V.; Gass, P.; Unsicker, K.; von Bohlen Und Halbach, O. Implications of p75NTR for dentate gyrus morphology and hippocampus-related behavior revisited. *Brain Struct. Funct.* **2015**, *220*, 1449–1462. [CrossRef]

184. Lin, P.Y.; Kavalali, E.T.; Monteggia, L.M. Genetic Dissection of Presynaptic and Postsynaptic BDNF-TrkB Signaling in Synaptic Efficacy of CA3-CA1 Synapses. *Cell Rep.* **2018**, *24*, 1550–1561. [CrossRef]

185. Eyman, M.; Cefaliello, C.; Mandile, P.; Piscopo, S.; Crispino, M.; Giuditta, A. Training old rats selectively modulates synaptosomal protein synthesis. *J. Neurosci. Res.* **2013**, *91*, 20–29. [CrossRef]

186. Nikoletopoulou, V.; Sidiropoulou, K.; Kallergi, E.; Dalezios, Y.; Tavernarakis, N. Modulation of Autophagy by BDNF Underlies Synaptic Plasticity. *Cell Metab.* **2017**, *26*, 230–242. [CrossRef]

187. Fujimura, H.; Altar, C.A.; Chen, R.; Nakamura, T.; Nakahashi, T.; Kambayashi, J.; Sun, B.; Tandon, N.N. Brain-derived neurotrophic factor is stored in human platelets and released by agonist stimulation. *Thromb. Haemost.* **2002**, *87*, 728–734. [CrossRef]

188. Klein, A.B.; Williamson, R.; Santini, M.A.; Clemmensen, C.; Ettrup, A.; Rios, M.; Knudsen, G.M.; Aznar, S. Blood BDNF concentrations reflect brain-tissue BDNF levels across species. *Int. J. Neuropsychopharmacol.* **2011**, *14*, 347–353. [CrossRef]

189. Mondal, A.C.; Fatima, M. Direct and indirect evidences of BDNF and NGF as key modulators in depression: Role of antidepressants treatment. *Int. J. Neurosci.* **2019**, *129*, 283–296. [CrossRef]

190. Molendijk, M.L.; Spinhoven, P.; Polak, M.; Bus, B.A.; Penninx, B.W.; Elzinga, B.M. Serum BDNF concentrations as peripheral manifestations of depression: Evidence from a systematic review and meta-analyses on 179 associations (N=9484). *Mol. Psychiatry* **2014**, *19*, 791–800. [CrossRef]

191. Polyakova, M.; Stuke, K.; Schuemberg, K.; Mueller, K.; Schoenknecht, P.; Schroeter, M.L. BDNF as a biomarker for successful treatment of mood disorders: A systematic & quantitative meta-analysis. *J. Affect. Disord.* **2015**, *174*, 432–440. [CrossRef] [PubMed]

192. Kishi, T.; Yoshimura, R.; Ikuta, T.; Iwata, N. Brain-Derived Neurotrophic Factor and Major Depressive Disorder: Evidence from Meta-Analyses. *Front. Psychiatry* **2017**, *8*, 308. [CrossRef] [PubMed]

193. Serra-Millàs, M. Are the changes in the peripheral brain-derived neurotrophic factor levels due to platelet activation? *World J. Psychiatry* **2016**, *6*, 84–101. [CrossRef] [PubMed]

194. Allen, A.P.; Naughton, M.; Dowling, J.; Walsh, A.; Ismail, F.; Shorten, G.; Scott, L.; McLoughlin, D.M.; Cryan, J.F.; Dinan, T.G.; et al. Serum BDNF as a peripheral biomarker of treatment-resistant depression and the rapid antidepressant response: A comparison of ketamine and ECT. *J. Affect. Disord.* **2015**, *186*, 306–311. [CrossRef]

195. Lee, B.H.; Kim, Y.K. Reduced platelet BDNF level in patients with major depression. *Prog. Neuropsychopharmacol. Biol. Psychiatry* **2009**, *33*, 849–853. [CrossRef]

196. Pandey, G.N.; Dwivedi, Y.; Rizavi, H.S.; Ren, X.; Zhang, H.; Pavuluri, M.N. Brain-derived neurotrophic factor gene and protein expression in pediatric and adult depressed subjects. *Prog. Neuropsychopharmacol. Biol. Psychiatry* **2010**, *34*, 645–651. [CrossRef]

197. Serra-Millàs, M.; López-Vílchez, I.; Navarro, V.; Galán, A.M.; Escolar, G.; Penadés, R.; Catalán, R.; Fañanás, L.; Arias, B.; Gastó, C. Changes in plasma and platelet BDNF levels induced by S-citalopram in major depression. *Psychopharmacology* **2011**, *216*, 1–8. [CrossRef]

198. Dwivedi, Y. *The Neurobiological Basis of Suicide*; CRC Press: Boca Raton, FL, USA, 2012.

199. Pandey, G.N.; Ren, X.; Rizavi, H.S.; Conley, R.R.; Roberts, R.C.; Dwivedi, Y. Brain-derived neurotrophic factor and tyrosine kinase B receptor signalling in post-mortem brain of teenage suicide victims. *Int. J. Neuropsychopharmacol.* **2008**, *11*, 1047–1061. [CrossRef]

200. Hosang, G.M.; Shiles, C.; Tansey, K.E.; McGuffin, P.; Uher, R. Interaction between stress and the BDNF Val66Met polymorphism in depression: A systematic review and meta-analysis. *BMC Med.* **2014**, *12*, 7. [CrossRef]

201. Li, M.; Du, W.; Shao, F.; Wang, W. Cognitive dysfunction and epigenetic alterations of the BDNF gene are induced by social isolation during early adolescence. *Behav. Brain Res.* **2016**, *313*, 177–183. [CrossRef]

202. Yan, T.; Wang, L.; Kuang, W.; Xu, J.; Li, S.; Chen, J.; Yang, Y. Brain-derived neurotrophic factor Val66Met polymorphism association with antidepressant efficacy: A systematic review and meta-analysis. *Asia Pac. Psychiatry* **2014**, *6*, 241–251. [CrossRef]

203. Castrén, E.; Rantamäki, T. The role of BDNF and its receptors in depression and antidepressant drug action: Reactivation of developmental plasticity. *Dev. Neurobiol.* **2010**, *70*, 289–297. [CrossRef] [PubMed]

204. Crispino, M.; Volpicelli, F.; Perrone-Capano, C. Role of the Serotonin Receptor 7 in Brain Plasticity: From Development to Disease. *Int. J. Mol. Sci.* **2020**, *21*, 505. [CrossRef] [PubMed]

205. Alenina, N.; Klempin, F. The role of serotonin in adult hippocampal neurogenesis. *Behav. Brain Res.* **2015**, *277*, 49–57. [CrossRef] [PubMed]

206. Malberg, J.E.; Eisch, A.J.; Nestler, E.J.; Duman, R.S. Chronic antidepressant treatment increases neurogenesis in adult rat hippocampus. *J. Neurosci.* **2000**, *20*, 9104–9110. [CrossRef] [PubMed]

207. Santarelli, L.; Saxe, M.; Gross, C.; Surget, A.; Battaglia, F.; Dulawa, S.; Weisstaub, N.; Lee, J.; Duman, R.; Arancio, O.; et al. Requirement of hippocampal neurogenesis for the behavioral effects of antidepressants. *Science* **2003**, *301*, 805–809. [CrossRef]

208. Rantamäki, T.; Castrén, E. Targeting TrkB neurotrophin receptor to treat depression. *Expert Opin. Ther. Targets* **2008**, *12*, 705–715. [CrossRef] [PubMed]

209. Speranza, L.; Labus, J.; Volpicelli, F.; Guseva, D.; Lacivita, E.; Leopoldo, M.; Bellenchi, G.C.; di Porzio, U.; Bijata, M.; Perrone-Capano, C.; et al. Serotonin 5-HT7 receptor increases the density of dendritic spines and facilitates synaptogenesis in forebrain neurons. *J. Neurochem.* **2017**, *141*, 647–661. [CrossRef]

210. Speranza, L.; Giuliano, T.; Volpicelli, F.; De Stefano, M.E.; Lombardi, L.; Chambery, A.; Lacivita, E.; Leopoldo, M.; Bellenchi, G.C.; di Porzio, U.; et al. Activation of 5-HT7 receptor stimulates neurite elongation through mTOR, Cdc42 and actin filaments dynamics. *Front. Behav. Neurosci.* **2015**, *9*, 62. [CrossRef]

211. Speranza, L.; Chambery, A.; Di Domenico, M.; Crispino, M.; Severino, V.; Volpicelli, F.; Leopoldo, M.; Bellenchi, G.C.; di Porzio, U.; Perrone-Capano, C. The serotonin receptor 7 promotes neurite outgrowth via ERK and Cdk5 signaling pathways. *Neuropharmacology* **2013**, *67*, 155–167. [CrossRef]

212. Popova, N.K.; Naumenko, V.S. Neuronal and behavioral plasticity: The role of serotonin and BDNF systems tandem. *Expert Opin. Ther. Targets* **2019**, *23*, 227–239. [CrossRef] [PubMed]

213. Castrén, E. Neurotrophins and psychiatric disorders. *Handb. Exp. Pharmacol.* **2014**, *220*, 461–479. [CrossRef] [PubMed]

214. Castrén, E.; Kojima, M. Brain-derived neurotrophic factor in mood disorders and antidepressant treatments. *Neurobiol. Dis.* **2017**, *97*, 119–126. [CrossRef] [PubMed]

215. Castrén, E.; Rantamäki, T. Role of brain-derived neurotrophic factor in the aetiology of depression: Implications for pharmacological treatment. *CNS Drugs* **2010**, *24*, 1–7. [CrossRef]

216. Lepack, A.E.; Fuchikami, M.; Dwyer, J.M.; Banasr, M.; Duman, R.S. BDNF release is required for the behavioral actions of ketamine. *Int. J. Neuropsychopharmacol.* **2014**, *18*. [CrossRef]

217. Abdallah, C.G.; Sanacora, G.; Duman, R.S.; Krystal, J.H. Ketamine and rapid-acting antidepressants: A window into a new neurobiology for mood disorder therapeutics. *Annu. Rev. Med.* **2015**, *66*, 509–523. [CrossRef] [PubMed]

218. Monteggia, L.M.; Zarate, C. Antidepressant actions of ketamine: From molecular mechanisms to clinical practice. *Curr. Opin. Neurobiol.* **2015**, *30*, 139–143. [CrossRef] [PubMed]

219. Björkholm, C.; Monteggia, L.M. BDNF-a key transducer of antidepressant effects. *Neuropharmacology* **2016**, *102*, 72–79. [CrossRef] [PubMed]

220. Hoshaw, B.A.; Malberg, J.E.; Lucki, I. Central administration of IGF-I and BDNF leads to long-lasting antidepressant-like effects. *Brain Res.* **2005**, *1037*, 204–208. [CrossRef]

221. Eisch, A.J.; Bolaños, C.A.; de Wit, J.; Simonak, R.D.; Pudiak, C.M.; Barrot, M.; Verhaagen, J.; Nestler, E.J. Brain-derived neurotrophic factor in the ventral midbrain-nucleus accumbens pathway: A role in depression. *Biol. Psychiatry* **2003**, *54*, 994–1005. [CrossRef]

222. Monteggia, L.M.; Luikart, B.; Barrot, M.; Theobold, D.; Malkovska, I.; Nef, S.; Parada, L.F.; Nestler, E.J. Brain-derived neurotrophic factor conditional knockouts show gender differences in depression-related behaviors. *Biol. Psychiatry* **2007**, *61*, 187–197. [CrossRef] [PubMed]

223. Monteggia, L.M.; Barrot, M.; Powell, C.M.; Berton, O.; Galanis, V.; Gemelli, T.; Meuth, S.; Nagy, A.; Greene, R.W.; Nestler, E.J. Essential role of brain-derived neurotrophic factor in adult hippocampal function. *Proc. Natl. Acad. Sci. USA* **2004**, *101*, 10827–10832. [CrossRef] [PubMed]

224. Saarelainen, T.; Hendolin, P.; Lucas, G.; Koponen, E.; Sairanen, M.; MacDonald, E.; Agerman, K.; Haapasalo, A.; Nawa, H.; Aloyz, R.; et al. Activation of the TrkB neurotrophin receptor is induced by antidepressant drugs and is required for antidepressant-induced behavioral effects. *J. Neurosci.* **2003**, *23*, 349–357. [CrossRef] [PubMed]

225. Li, Y.; Luikart, B.W.; Birnbaum, S.; Chen, J.; Kwon, C.H.; Kernie, S.G.; Bassel-Duby, R.; Parada, L.F. TrkB regulates hippocampal neurogenesis and governs sensitivity to antidepressive treatment. *Neuron* **2008**, *59*, 399–412. [CrossRef]

226. Liu, R.J.; Lee, F.S.; Li, X.Y.; Bambico, F.; Duman, R.S.; Aghajanian, G.K. Brain-derived neurotrophic factor Val66Met allele impairs basal and ketamine-stimulated synaptogenesis in prefrontal cortex. *Biol. Psychiatry* **2012**, *71*, 996–1005. [CrossRef]

227. Kato, T.; Fogaça, M.V.; Deyama, S.; Li, X.Y.; Fukumoto, K.; Duman, R.S. BDNF release and signaling are required for the antidepressant actions of GLYX-13. *Mol. Psychiatry* **2018**, *23*, 2007–2017. [CrossRef]

228. Herring, M.P.; Puetz, T.W.; O'Connor, P.J.; Dishman, R.K. Effect of exercise training on depressive symptoms among patients with a chronic illness: A systematic review and meta-analysis of randomized controlled trials. *Arch. Intern. Med.* **2012**, *172*, 101–111. [CrossRef]

229. Ota, K.T.; Duman, R.S. Environmental and pharmacological modulations of cellular plasticity: Role in the pathophysiology and treatment of depression. *Neurobiol. Dis.* **2013**, *57*, 28–37. [CrossRef]

230. Schuch, F.B.; Vancampfort, D.; Richards, J.; Rosenbaum, S.; Ward, P.B.; Stubbs, B. Exercise as a treatment for depression: A meta-analysis adjusting for publication bias. *J. Psychiatr. Res.* **2016**, *77*, 42–51. [CrossRef]

231. Rasmussen, P.; Brassard, P.; Adser, H.; Pedersen, M.V.; Leick, L.; Hart, E.; Secher, N.H.; Pedersen, B.K.; Pilegaard, H. Evidence for a release of brain-derived neurotrophic factor from the brain during exercise. *Exp. Physiol.* **2009**, *94*, 1062–1069. [CrossRef]

232. Seifert, T.; Brassard, P.; Wissenberg, M.; Rasmussen, P.; Nordby, P.; Stallknecht, B.; Adser, H.; Jakobsen, A.H.; Pilegaard, H.; Nielsen, H.B.; et al. Endurance training enhances BDNF release from the human brain. *Am. J. Physiol. Regul. Integr. Comp. Physiol.* **2010**, *298*, R372–R377. [CrossRef] [PubMed]

233. Xu, L.; Zheng, Y.L.; Yin, X.; Xu, S.J.; Tian, D.; Zhang, C.Y.; Wang, S.; Ma, J.Z. Excessive Treadmill Training Enhances Brain-Specific MicroRNA-34a in the Mouse Hippocampus. *Front. Mol. Neurosci.* **2020**, *13*, 7. [CrossRef] [PubMed]

234. de Assis, G.G.; de Almondes, K.M. Exercise-dependent BDNF as a Modulatory Factor for the Executive Processing of Individuals in Course of Cognitive Decline. A Systematic Review. *Front. Psychol.* **2017**, *8*, 584. [CrossRef] [PubMed]

235. Pajonk, F.G.; Wobrock, T.; Gruber, O.; Scherk, H.; Berner, D.; Kaizl, I.; Kierer, A.; Müller, S.; Oest, M.; Meyer, T.; et al. Hippocampal plasticity in response to exercise in schizophrenia. *Arch. Gen. Psychiatry* **2010**, *67*, 133–143. [CrossRef] [PubMed]

236. Moon, H.Y.; Becke, A.; Berron, D.; Becker, B.; Sah, N.; Benoni, G.; Janke, E.; Lubejko, S.T.; Greig, N.H.; Mattison, J.A.; et al. Running-Induced Systemic Cathepsin B Secretion Is Associated with Memory Function. *Cell Metab.* **2016**, *24*, 332–340. [CrossRef] [PubMed]

237. Boström, P.; Wu, J.; Jedrychowski, M.P.; Korde, A.; Ye, L.; Lo, J.C.; Rasbach, K.A.; Boström, E.A.; Choi, J.H.; Long, J.Z.; et al. A PGC1-α-dependent myokine that drives brown-fat-like development of white fat and thermogenesis. *Nature* **2012**, *481*, 463–468. [CrossRef]

238. Wrann, C.D.; White, J.P.; Salogiannnis, J.; Laznik-Bogoslavski, D.; Wu, J.; Ma, D.; Lin, J.D.; Greenberg, M.E.; Spiegelman, B.M. Exercise induces hippocampal BDNF through a PGC-1α/FNDC5 pathway. *Cell Metab.* **2013**, *18*, 649–659. [CrossRef]

239. Müller, P.; Duderstadt, Y.; Lessmann, V.; Müller, N.G. Lactate and BDNF: Key Mediators of Exercise Induced Neuroplasticity? *J. Clin. Med.* **2020**, *9*, 1136. [CrossRef]

240. Piotrowicz, Z.; Chalimoniuk, M.; Płoszczyca, K.; Czuba, M.; Langfort, J. Exercise-Induced Elevated BDNF Level Does Not Prevent Cognitive Impairment Due to Acute Exposure to Moderate Hypoxia in Well-Trained Athletes. *Int. J. Mol. Sci.* **2020**, *21*, 5569. [CrossRef]

241. Miranda, M.; Morici, J.F.; Zanoni, M.B.; Bekinschtein, P. Brain-Derived Neurotrophic Factor: A Key Molecule for Memory in the Healthy and the Pathological Brain. *Front. Cell Neurosci.* **2019**, *13*, 363. [CrossRef]

242. Hock, C.; Heese, K.; Hulette, C.; Rosenberg, C.; Otten, U. Region-specific neurotrophin imbalances in Alzheimer disease: Decreased levels of brain-derived neurotrophic factor and increased levels of nerve growth factor in hippocampus and cortical areas. *Arch. Neurol.* **2000**, *57*, 846–851. [CrossRef]

243. Zuccato, C.; Marullo, M.; Conforti, P.; MacDonald, M.E.; Tartari, M.; Cattaneo, E. Systematic assessment of BDNF and its receptor levels in human cortices affected by Huntington's disease. *Brain Pathol.* **2008**, *18*, 225–238. [CrossRef] [PubMed]

244. Narisawa-Saito, M.; Wakabayashi, K.; Tsuji, S.; Takahashi, H.; Nawa, H. Regional specificity of alterations in NGF, BDNF and NT-3 levels in Alzheimer's disease. *Neuroreport* **1996**, *7*, 2925–2928. [CrossRef]

245. Hoxha, E.; Lippiello, P.; Zurlo, F.; Balbo, I.; Santamaria, R.; Tempia, F.; Miniaci, M.C. The Emerging Role of Altered Cerebellar Synaptic Processing in Alzheimer's Disease. *Front. Aging Neurosci.* **2018**, *10*, 396. [CrossRef] [PubMed]

246. Peng, S.; Wuu, J.; Mufson, E.J.; Fahnestock, M. Precursor form of brain-derived neurotrophic factor and mature brain-derived neurotrophic factor are decreased in the pre-clinical stages of Alzheimer's disease. *J. Neurochem.* **2005**, *93*, 1412–1421. [CrossRef] [PubMed]

247. Choi, S.H.; Bylykbashi, E.; Chatila, Z.K.; Lee, S.W.; Pulli, B.; Clemenson, G.D.; Kim, E.; Rompala, A.; Oram, M.K.; Asselin, C.; et al. Combined adult neurogenesis and BDNF mimic exercise effects on cognition in an Alzheimer's mouse model. *Science* **2018**, *361*. [CrossRef] [PubMed]

248. Sleiman, S.F.; Henry, J.; Al-Haddad, R.; El Hayek, L.; Abou Haidar, E.; Stringer, T.; Ulja, D.; Karuppagounder, S.S.; Holson, E.B.; Ratan, R.R.; et al. Exercise promotes the expression of brain derived neurotrophic factor (BDNF) through the action of the ketone body β-hydroxybutyrate. *Elife* **2016**, *5*. [CrossRef] [PubMed]

249. Colucci-D'Amato, L.; Perrone-Capano, C.; di Porzio, U. Chronic activation of ERK and neurodegenerative diseases. *Bioessays* **2003**, *25*, 1085–1095. [CrossRef] [PubMed]

250. Kirouac, L.; Rajic, A.J.; Cribbs, D.H.; Padmanabhan, J. Activation of Ras-ERK Signaling and GSK-3 by Amyloid Precursor Protein and Amyloid Beta Facilitates Neurodegeneration in Alzheimer's Disease. *eNeuro* **2017**, *4*. [CrossRef]

251. Nizzari, M.; Barbieri, F.; Gentile, M.T.; Passarella, D.; Caorsi, C.; Diaspro, A.; Taglialatela, M.; Pagano, A.; Colucci-D'Amato, L.; Florio, T.; et al. Amyloid-β protein precursor regulates phosphorylation and cellular compartmentalization of microtubule associated protein tau. *J. Alzheimers Dis.* **2012**, *29*, 211–227. [CrossRef]

252. Bartolotti, N.; Segura, L.; Lazarov, O. Diminished CRE-Induced Plasticity is Linked to Memory Deficits in Familial Alzheimer's Disease Mice. *J. Alzheimers Dis.* **2016**, *50*, 477–489. [CrossRef] [PubMed]

253. Rosa, E.; Fahnestock, M. CREB expression mediates amyloid β-induced basal BDNF downregulation. *Neurobiol. Aging* **2015**, *36*, 2406–2413. [CrossRef] [PubMed]

254. Thiele, C.J.; Li, Z.; McKee, A.E. On Trk–the TrkB signal transduction pathway is an increasingly important target in cancer biology. *Clin. Cancer Res.* **2009**, *15*, 5962–5967. [CrossRef] [PubMed]

255. Xiong, J.; Zhou, L.I.; Lim, Y.; Yang, M.; Zhu, Y.H.; Li, Z.W.; Fu, D.L.; Zhou, X.F. Mature brain-derived neurotrophic factor and its receptor TrkB are upregulated in human glioma tissues. *Oncol. Lett.* **2015**, *10*, 223–227. [CrossRef] [PubMed]

256. Colucci-D'Amato, G.L.; D'Alessio, A.; Califano, D.; Cali, G.; Rizzo, C.; Nitsch, L.; Santelli, G.; de Franciscis, V. Abrogation of nerve growth factor-induced terminal differentiation by ret oncogene involves perturbation of nuclear translocation of ERK. *J. Biol. Chem.* **2000**, *275*, 19306–19314. [CrossRef]

257. Radin, D.P.; Patel, P. BDNF: An Oncogene or Tumor Suppressor? *Anticancer. Res.* **2017**, *37*, 3983–3990. [CrossRef]

258. Wang, X.; Prager, B.C.; Wu, Q.; Kim, L.J.Y.; Gimple, R.C.; Shi, Y.; Yang, K.; Morton, A.R.; Zhou, W.; Zhu, Z.; et al. Reciprocal Signaling between Glioblastoma Stem Cells and Differentiated Tumor Cells Promotes Malignant Progression. *Cell Stem Cell* **2018**, *22*, 514–528. [CrossRef]

259. Garofalo, S.; D'Alessandro, G.; Chece, G.; Brau, F.; Maggi, L.; Rosa, A.; Porzia, A.; Mainiero, F.; Esposito, V.; Lauro, C.; et al. Enriched environment reduces glioma growth through immune and non-immune mechanisms in mice. *Nat. Commun.* **2015**, *6*, 6623. [CrossRef]

260. Garofalo, S.; Porzia, A.; Mainiero, F.; Di Angelantonio, S.; Cortese, B.; Basilico, B.; Pagani, F.; Cignitti, G.; Chece, G.; Maggio, R.; et al. Environmental stimuli shape microglial plasticity in glioma. *Elife* **2017**, *6*. [CrossRef]

261. Colucci-D'Amato, L.; Cimaglia, G. As a potential source of neuroactive compounds to promote and restore neural functions. *J. Tradit. Complement. Med.* **2020**, *10*, 309–314. [CrossRef]

262. Gentile, M.T.; Ciniglia, C.; Reccia, M.G.; Volpicelli, F.; Gatti, M.; Thellung, S.; Florio, T.; Melone, M.A.; Colucci-D'Amato, L. Ruta graveolens L. induces death of glioblastoma cells and neural progenitors, but not of neurons, via ERK 1/2 and AKT activation. *PLoS ONE* **2015**, *10*, e0118864. [CrossRef] [PubMed]

Publisher's Note: MDPI stays neutral with regard to jurisdictional claims in published maps and institutional affiliations.

International Journal of
Molecular Sciences

Review

Neuroprotective Effect of Vascular Endothelial Growth Factor on Motoneurons of the Oculomotor System

Silvia Silva-Hucha, Angel M. Pastor and Sara Morcuende *

Departamento de Fisiología, Facultad de Biología, Universidad de Sevilla, 41012 Sevilla, Spain;
silvia_sh88@hotmail.com (S.S.-H.); ampastor@us.es (A.M.P.)
* Correspondence: smorcuende@us.es; Tel.: +34-954-55-95-49

Abstract: Vascular endothelial growth factor (VEGF) was initially characterized as a potent angiogenic factor based on its activity on the vascular system. However, it is now well established that VEGF also plays a crucial role as a neuroprotective factor in the nervous system. A deficit of VEGF has been related to motoneuronal degeneration, such as that occurring in amyotrophic lateral sclerosis (ALS). Strikingly, motoneurons of the oculomotor system show lesser vulnerability to neurodegeneration in ALS compared to other motoneurons. These motoneurons presented higher amounts of VEGF and its receptor Flk-1 than other brainstem pools. That higher VEGF level could be due to an enhanced retrograde input from their target muscles, but it can also be produced by the motoneurons themselves and act in an autocrine way. By contrast, VEGF's paracrine supply from the vicinity cells, such as glial cells, seems to represent a minor source of VEGF for brainstem motoneurons. In addition, ocular motoneurons experiment an increase in VEGF and Flk-1 level in response to axotomy, not observed in facial or hypoglossal motoneurons. Therefore, in this review, we summarize the differences in VEGF availability that could contribute to the higher resistance of extraocular motoneurons to injury and neurodegenerative diseases.

Keywords: VEGF; oculomotor system; trophic factors; motoneurons; neurodegeneration; axotomy; amyotrophic lateral sclerosis

Citation: Silva-Hucha, S.; Pastor, A.M.; Morcuende, S. Neuroprotective Effect of Vascular Endothelial Growth Factor on Motoneurons of the Oculomotor System. *Int. J. Mol. Sci.* **2021**, *22*, 814. https://doi.org/10.3390/ijms22020814

Received: 10 December 2020
Accepted: 13 January 2021
Published: 15 January 2021

Publisher's Note: MDPI stays neutral with regard to jurisdictional claims in published maps and institutional affiliations.

1. Vascular Endothelial Growth Factor (VEGF)

1.1. History

VEGF was initially described as an angiogenic factor and, consequently, it was named vascular permeability factor (VPF) for its role in inducing vascular permeability in tumor cells [1]. It was not until 1989 when the VEGF protein, whose molecular weight is approximately 45 kDa, was purified and sequenced, and it was definitively assigned the name of vascular endothelial growth factor [2,3].

It is well known that this factor is a highly specific mitogen for vascular endothelial cells, whose family consists of multiple cell signaling proteins involved in angiogenesis, lymphangiogenesis, vasodilation, and vascular leakage, among other functions [4]. In 2001, his essential role in motoneuronal protection was revealed for the first time [5], as will be discussed later in detail.

1.2. VEGF Family

Since the discovery of the first member of the VEGF family, known as VEGF-A, the family has continued to grow and is currently constituted by VEGF-A, VEGF-B [6,7], VEGF-C [8,9], VEGF-D [10,11], VEGF-E [12], VEGF-F [13] and placental growth factor (PlGF [14]).

VEGF-A stands out as the most studied VEGF family members because of its essential roles in neuroprotection. The gene encoding VEGF-A is located on chromosome 6p21.5, giving rise to three different isoforms (VEGF-A 121, VEGF-A 145, VEGF-A 165), which are differentiated by their molecular weight, solubility, biological functions, binding affinities to the components of the extracellular matrix, and tyrosine kinase receptor subtypes (RTKs) [15]. Furthermore, VEGF 165 is the predominant isoform in the central nervous system (CNS), where it acts as a protection factor by promoting the survival of motoneurons [16,17]. This vital role is the one that interests us and the one that we will develop on throughout this review.

On the other hand, VEGF-B is also expressed in the CNS and can regulate adult neurogenesis and even rescuing neurons from apoptosis, but with less vascular effects and worse neuroprotective function on motoneurons than VEGF-A [18]. It has recently been discovered that the presence of VEGF-B is neither necessary nor essential for the survival, maintenance, and development of motoneurons under normal physiological conditions [19,20]. Finally, mention should be made of the members VEGF-C and VEGF-D, which regulate lymphatic angiogenesis, and of VEGF-E, which is virally encoded and specifically expressed in the venom of the habu snake (*Trimeresurus flavoviridis)* [21].

1.3. Functions of VEGF

Throughout this review, we will refer to the VEGF-A 165 isoform, which is the one that exerts direct trophic and neuroprotective effects on many types of neural cells [22], including motoneurons [5,23], astroglia [24], microglia [25], hippocampal, dopaminergic, cortical, cerebellar, sympathetic neurons, and even muscle satellite cells [16,26–32].

Furthermore, this trophic factor plays a fundamental role in stimulating neurogenesis in both developing and adult CNS [16], promoting Schwann cells' proliferation [27]. It also supports synaptic plasticity [33] and favors the growth, survival, differentiation, and migration of neuronal and glial cells [20,34]. Additionally, VEGF guarantees an optimal blood and glucose supply to the brain and spinal cord [5], protecting motoneurons from oxidative stress [35], hypoxia, hypoglycemia [23], and glutamate-mediated excitotoxicity [29,36–39].

VEGF is also a potent inducer of the blood-brain barrier's interruption by increasing its permeability and favoring the supply of oxygen and nutrients to neurons [34,40]. This trophic factor is involved in vasculogenesis during embryological development and promoting angiogenesis in many pathological conditions, such as tumor growth, rheumatoid arthritis, psoriasis, and diabetic retinopathy [4].

1.4. VEGF Expression

Several factors have been found to upregulate VEGF mRNA expression, including tumor necrosis factor (TNF-α), platelet-derived growth factor (PDGF), interleukins, angiopoietins, and erythropoietins [41–45]. Another molecule that regulates the expression of VEGF is nitric oxide, which contributes to the processes of permeabilization of blood vessels and in vasodilation stimulated by VEGF [29,46,47].

However, one of the main and more robust regulators of VEGF expression is hypoxia [48]. VEGF mRNA has a half-life of 30–45 min under normoxic conditions, whereas the mRNA half-life is prolonged in hypoxia [49,50] and cells increase the production of the hypoxia-induced transcription factor 1 (HIF-1), a heterodimer consisting of three subunits (HIF-1α, HIF-1β, and HIF-3) [51,52]. HIF-1α and HIF-1β are produced continuously, but HIF-1α is highly labile in the presence of oxygen, so it degrades under aerobic conditions [53]. When the cell is in a hypoxic environment, HIF-1α persists and translocates to the nucleus, where it associates with HIF-1β and forms the HIF-1α/HIF-1β complex. This complex binds to the hypoxia response element (HRE) [54,55], whose transcriptional activation requires the recruitment of the CREB-binding protein, which is a transcriptional coactivator. Thus, as the 2019 Nobel Laureates in Medicine Kaelin, Ratcliff, and Semenza described, through this mechanism, cells perceive and adapt to changes in oxygen levels, modifying both their metabolism and physiological functions [52,56]. Therefore, transactivation of HRE by the HIF-1α/HIF-1β complex stimulates the gene expression of erythropoietin, glucose transporters, glycolytic enzymes, and VEGF [57,58], among others, the latter being in charge of promoting angiogenesis after binding to its specific receptors [5]. Little is known about the expression and function of HIF-3 [59].

2. VEGF Receptors

The biological activity of the VEGF family is mediated through binding to two classes of receptors: receptors with tyrosine kinase activity and receptors without tyrosine kinase activity. The first group consists of three structurally related receptors characterized by the presence of seven immunoglobulin-like domains in the extracellular region, a single transmembrane region, and an intracellular consensus tyrosine kinase sequence interrupted by a kinase insertion domain. These receptors are VEGR-1 (Flt-1), VEGFR-2 (KDR/Flk-1), and VEGFR-3 (Flt-4). On the other hand, the receptors without kinase activity are neuropilin-1 (NRP-1) and neuropilin-2 (NRP-2), which are also receptors for semaphorins [15,20] (Figure 1).

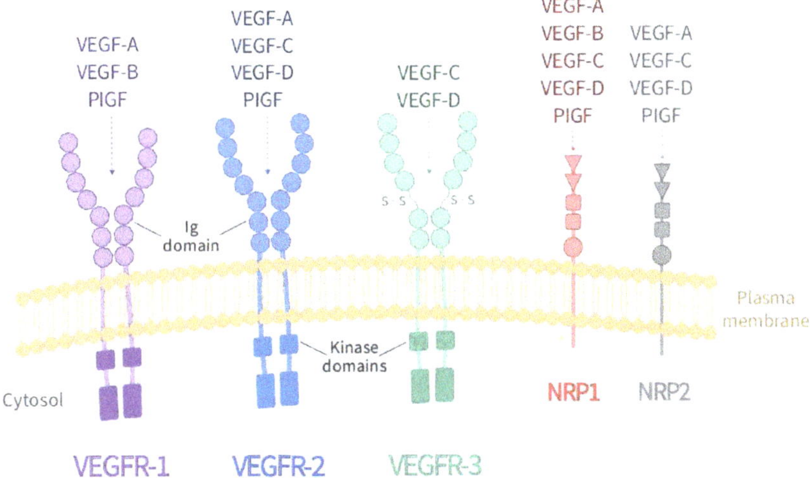

Figure 1. VEGF receptors. The family of VEGF receptors includes three tyrosine kinase receptors (VEGFR-1, VEGFR-2, VEGFR-3) and two non-tyrosine kinase neuropilin receptors (NP-1, NP-2). The different members of the VEGF family bind to the different types of VEGF receptors, as illustrated. The main effect of VEGF-A as a neurotrophic factor is mediated by its binding to VEGFR-2 (Flk-1).

NRP-1 and 2 are expressed in different types of neurons [60,61] and play an essential role in regulating and developing the cardiovascular and nervous systems. Besides, they act as co-receptors for RTKs, presenting and improving VEGF binding to Flk-1 and promoting receptor phosphorylation and neurotrophic factor-mediated signal transduction [62,63].

Many studies indicate that both Flt-1 and Flk-1 activation could produce neuroprotection, but there are differences between the functions of both receptors [5,26,29,30]. Flk-1 predominates in neuronal and Schwann cells and is necessary for endothelial proliferation and migration, while Flt-1 is expressed mainly in vessels, astrocytes, and reactive microglia [29,64]. Besides, it has also been described that Flt-1 acts as a negative regulator for VEGF in endothelial cells, preventing its binding to Flk-1 [4] and that its functions and signaling properties may differ according to the stage of development of the animal, the cell type, and the binding ligand [65].

Unlike Flt-1, the Flk-1 receptor is an important survival promoter for endothelial and CNS cells, being the primary mediator of VEGF functions [4,66]. The main ligand of this Flk-1 receptor is the VEGF-A isoform, being the only one that triggers its autophosphorylation and final glycosylated form [4]. Therefore, the VEGF receptor that we will refer to throughout this review is Flk-1, which is expressed in motoneurons of the human spinal cord [40], mouse [5], rat [38] and is reduced in some patients with amyotrophic lateral sclerosis (ALS) [40]. Flk-1 overexpression in spinal motoneurons of the ALS SOD1 mouse model (with mutations in the gene encoding the antioxidant copper/zinc superoxide dismutase) has been shown to delay both neurodegeneration and disease onset [67], being the primary mediator of the neuroprotective and anti-excitotoxic effects of VEGF on motoneurons [37,38,68]. All these functions are performed by activating the phosphatidylinositol-3 kinase-AKT (PI3-K/Akt) pathway, involved in the processes of cell growth, proliferation, cell survival, and intracellular traffic, among others, in addition to regulating the entry of glucose to the cell through an insulin signaling cascade [69]. Furthermore, VEGF binding to the Flk-1 receptor also exerts a protective effect by suppressing the activation of the mitogen-activated protein kinase p38 (p38MAPK), a determining factor in the cell death pathway [16,30,38,70].

3. Effects of Low Levels of VEGF

The experiments carried out by Oosthuyse et al. in 2001 [5] were the first to suggest that VEGF acted as a neurotrophic factor at the CNS, as the reduction of VEGF function induced a specific degeneration of motoneurons in the adult mice. In these experiments, manipulation of the VEGF gene resulted in homozygous knock-in mice (VEGF$^{\delta/\delta}$), in which the sequence of the hypoxia response element in the VEGF promoter region was removed. Consequently, these mice lost the ability to increase VEGF expression in a hypoxic situation. This alteration caused them severe muscle weakness due to the degeneration of the lower motoneurons, and they became progressively less mobile, with symptoms reminiscent of neuropathological signs of ALS. Although basal VEGF levels in muscles, heart, and fibroblasts were unaffected by removing the hypoxia response element, an overall 40% reduction of the neurotrophic factor was observed in neural tissue.

Those results obtained with the VEGF$^{\delta/\delta}$ mice allowed linking for the first time a low level of VEGF with motoneuronal degeneration. Besides, mice resulting from the crossing of the SOD1 mutant with VEGF$^{\delta/\delta}$ mice exhibited an even more drastic reduction in VEGF levels, thence a more severe degeneration of motoneurons and an earlier onset of symptoms of muscle weakness [17]. All these findings granted VEGF an unexpected neuroprotective role in the degenerative processes that accompany the pathogenesis of motoneurons, supporting the idea that motoneurons seem to be particularly sensitive to a low VEGF support [5,34,71]. Thus, a link was established between low levels of VEGF and neurodegeneration of motoneurons, such as occurs in ALS, and raised great expectation in VEGF as a possible candidate for ALS specific treatment.

VEGF and ALS

ALS is characterized by being an adult neurodegenerative disease that causes progressive degeneration of motoneurons in the lower spinal cord, brainstem, and cortex. Consequently, it triggers astrogliosis, progressive atrophy of the skeletal musculature, and a reduction in voluntary movements, including those of the extremities and respiratory movements [72,73].

The disease affects five out of every 100,000 people worldwide, is progressive, and is generally fatal within 5 years after the onset of symptoms. 95% of cases are sporadic, and only 5% of patients have a family history, with a fifth of these caused by mutations in the SOD1 gene, located on chromosome 21 [15]. There are currently around 180 known mutations in SOD1 which are related to the pathogenesis of the disease [74], with the SOD1^{G93A} mutant mouse model being the most widely used, studied, and well-characterized showing symptoms similar to the disease [75], and to which we refer in this review as the SOD1 model.

It has been shown that the expression of VEGF and its Flk-1 receptor undergo significant downregulation in the motoneurons of the spinal cord of SOD1 mice [76]. These findings correlate with other studies where SOD1 mice were crossed with transgenic mice that overexpressed Flk-1, resulting in a delayed onset of motor impairment and degeneration of motoneurons, and prolonged survival [67]. Moreover, as indicated before, the coincidence of both SOD1 mutation and VEGF$^{\delta/\delta}$ alteration produced an earlier and more severe motoneuronal degeneration [17].

Although the predominant hypothesis is that ALS is a disease of neural origin, some studies indicate that the disease involves a distal axonopathy and denervation of the neuromuscular junctions (NMJs) in the muscles of the extremities in the presymptomatic stage, that is, much before the loss of the motoneuron at the level of the spinal cord [77–79]. Several studies have shown that both the anterograde transport of VEGF and the intracerebroventricular infusion of this factor help protect and preserve the NMJs in a rat model SOD1 [67,80,81]. Likewise, experiments such as gene therapy of VEGF mediated by lentiviral vectors, or transplantation of stem cells that overexpress VEGF, have managed to significantly slow the progression of neurodegeneration, improving motor function and significantly prolonging the survival of motoneurons in the brainstem and the cervical and lumbar spinal cord [82,83]. All these findings give VEGF and its Flk-1 receptor an essential role in the treatment of motoneuronal diseases.

4. Neuroprotective Effect of VEGF

Among the possible pathogenic mechanisms linked to the degenerative neuronal process are oxidative stress, glutamate excitotoxicity, inflammation, mitochondrial and neurofilament dysfunction, protein aggregation, axonal transport abnormalities, and, ultimately, the activation of pathways that trigger apoptosis [84]. Numerous studies show the decisive neuroprotective role that VEGF plays in the CNS. These include experiments where an intracerebroventricular administration of VEGF stimulates neurogenesis in the adult hippocampus, promotes neurites growth, or provides greater protection to motoneurons [85,86]. Furthermore, it has also been shown that the retrograde transport of VEGF, after its intramuscular administration with lentiviral vectors, favors the survival of motoneurons [17,82]. At the same time, the supply of VEGF at the site of a spinal cord injury decreases lesion size, apoptosis levels, and retards neurodegeneration [36,87].

Two main hypotheses have been postulated to explain all of these VEGF effects. The first affirms that this factor promotes the vascular niche necessary for motoneurons to survive, and the second, that the binding of VEGF to Flk-1 promotes cell survival by blocking the process of apoptosis.

Another protective effect of VEGF is also due to the induction of the expression of the GluA2 subunit in AMPA receptors [64], which leads to a reduction in the entry of Ca^{2+} in neurons, which is a relevant mechanism involved in motoneuronal degeneration. All this makes VEGF and its Flk-1 receptor attractive candidates for evaluating its therapeutic potential in neurodegenerative disorders.

4.1. Anti-Apoptotic Effects of VEGF

One of the mechanisms by which the binding of the VEGF-A isoform to the Flk-1 receptor improves and promotes cell survival is by blocking the process of apoptosis through the expression of anti-apoptotic proteins, such as the members of the Bcl-2 family and the generation of neuronal progenitors in the nervous system [16,88,89].

It is well known that the binding of VEGF with Flk-1 directly activates the PI3-K/Akt intracellular signaling pathway. This activation consequently causes an inhibition of the phosphorylation of p38MAPK, which is an essential factor in the cell death pathway [38,64], and an increase of the expression of the anti-apoptotic proteins Bcl-2 and A1, conceding greater protection to motoneurons against excitotoxicity. That effect has been described in diverse models of neurodegeneration, including ALS [38,70].

4.2. Role of Excitotoxicity in Neurodegeneration and VEGF Protection

Glutamate is the major excitatory neurotransmitter in the mammalian CNS and is involved in many aspects of normal brain function. However, an excess in the synaptic transmission of glutamate leads to an over-activation of the different types of receptors for this amino acid, which causes a massive entry of Ca^{2+} in the neurons and triggers the uncontrolled activation of damaging processes that, eventually, produce the destruction of the membrane, neurodegeneration and cell death [90]. Indeed, glutamate-mediated excitotoxicity is considered the primary mechanism leading to a degeneration of motoneurons in various neurodegenerative disorders, including Parkinson's disease, Alzheimer's disease, and ALS [91–93].

Two broad categories of glutamate receptors are known: (i) ionotropic receptors, which are ligand-activated ion channels, and comprise *N*-methyl-D-aspartate (NMDA), α-amino-3-hydroxy-5-isoxazole propionate (AMPA), and kainate receptors; (ii) metabotropic receptors, which are associated to G proteins and coupled to the production of intracellular secondary messengers [94,95]. AMPA ionotropic receptors are heteromeric complexes composed of four subunits, GluA1-GluA4 (formerly GluR1-GluR4), with different combinations. The presence of the GluA2 subunit is known to decrease the permeability of these receptors to Ca^{2+} [95]. Therefore, the hyperactivation of AMPA receptors that lack this GluA2 subunit involves a massive entry of Ca^{2+} into the cell, which produces the activation of phospholipases, proteases, and endonucleases, inducing apoptotic or necrotic cell death, the production of reactive oxygen species (ROS) and a deficiency of mitochondrial function, with the consequent interruption of energy metabolism [90,96]. All of this generates neuronal degeneration, suggesting that the absence of the GluA2 subunit at AMPA receptors is a critical factor for the selective vulnerability of neurons to excitotoxicity.

Motoneurons are particularly susceptible to excitotoxicity due to their low expression of the GluA2 subunit [97,98]. Deficiency of this subunit exacerbates motoneuron degeneration in SOD1 mouse models, whose mutation is implicated in the accumulation of oxidative damage [99]. On top of that, many ALS patients have shown elevated glutamate levels in the cerebrospinal fluid, supporting the excitotoxic hypothesis of degeneration of the motoneurons [100]. Furthermore, glial cells that overexpress the SOD1 mutation are known to adversely affect the viability of spinal motoneurons by producing elevated levels of extracellular glutamate, leading to increased motoneuron degeneration and progressive paralysis [101].

Interestingly, the administration of exogenous recombinant VEGF is capable of preventing both excitotoxic neuronal death, induced by overactivation of AMPA receptors, and the consequent motor disorders [36]. That reduction in glutamate toxicity is mediated by the action of VEGF on PI3-K/Akt and MEK/ERK pathways [102]. Moreover, this neurotrophic factor has been shown, both in vitro and in vivo, to induce an increase in the expression of the GluA2 subunit in AMPA receptors [103], reducing the permeability to Ca^{2+} and, therefore, granting protection to motoneurons against excitotoxicity. Those experiments highlight the relevant role that VEGF plays in reducing excitotoxicity, making this factor an essential piece for the survival of motoneurons.

5. Selective Vulnerability of Motoneurons to Neurodegeneration

As stated above, motoneurons are particularly sensitive to excitotoxic neurodegeneration due, mainly, to their reduced capacity to blockade Ca^{2+} influx. However, a peculiarity of neurodegenerative processes is that specific neuronal populations show superior resistance to degeneration compared to other motor groups. In diseases such as ALS, motoneurons of some motor nuclei offer selective resistance and persist until the last stages of the disease, compared to other motoneurons that degenerate earlier [104]. Motoneurons of the oculomotor system are among those resistant populations [104,105], while motoneurons of the facial, hypoglossal, or trigeminal motor systems are vulnerable populations in the brainstem [106].

Likewise, between the possible differences that mark the selective vulnerability of the different motoneuronal groups is the differential expression of specific proteins. Several studies have shown that ocular motoneurons have a distinct transcriptional profile from other motoneurons in the expression of proteins related to synaptic transmission, including several glutamate and GABA receptor subunits, Ca^{2+}-binding, ubiquitin-dependent proteolysis, mitochondrial function, or immune system processes [107–109].

Moreover, a greater expression of laminins, synaptophysin, and p75 receptor has been detected in muscle fibers of resistant motoneurons [110,111]. Thus, the differential expression of specific proteins and neurotrophic factors on the target muscles seems to influence the selective resistance of the different motor units against neurodegeneration and the deterioration of the NMJs, granting a vital role to the retrograde trophic contribution in motoneuronal survival.

6. Properties of Ocular Motoneurons

The ocular motoneurons present a series of morphological and functional characteristics that differentiate them from the rest of the motoneuronal populations. Several hypotheses have been proposed to explain the greater resistance of these cell populations to neurodegeneration.

Motoneurons of the ocular system show an extensive buffering capability of intracellular Ca^{2+} due to a greater expression of cytosolic Ca^{2+} binding proteins [112–114]. Overexpression of Ca^{2+} binding cytosolic proteins, such as calbindin D-28K (CaBP), calretinin (CR), and parvalbumin (PV), seem to give high protection to motoneurons [115–118]. Likewise, it has been observed that 85–100% of the motoneurons of the primate ocular motor nuclei contain PV, while only 20–30% of the neurons of the trigeminal, facial, and hypoglossal nuclei present it [109]. This PV distribution pattern coincides with the selective vulnerability between the brainstem motor nuclei [104]. Furthermore, additional experiments in SOD1 mice revealed that PV levels were significantly higher in ocular motoneurons compared to hypoglossal motoneurons [108].

Additionally, other studies in ALS models have related a low expression of the neuropeptide calcitonin gene-related peptide (CGRP), with higher resistance of motoneurons. Thus, CGRP could be a factor that promotes neuronal degeneration. Accordingly, motoneurons of the rat oculomotor system show lower expression of CGRP compared to other vulnerable motoneurons, such as facial or spinal ones, which would support these results [119,120]. However, these results have only been demonstrated in rats since a constitutive expression of CGRP has been observed in cats [121].

Notably, the extraocular motoneurons and the EOMs also express a higher proportion of the insulin-like growth factor 2 (IGF-2), which acts as a survival factor for motoneurons, and of its receptor IGF-1R, which mediates its survival effect [122,123]. Moreover, IGF-2 delivery to muscles preserved motoneurons and extended life-span in SOD mice [122]. It has also been shown that receptor $\alpha 1$ of the inhibitory neurotransmitter GABA-A (Gabra1) is preferably present in resistant motoneurons, such as ocular motoneurons, of symptomatic SOD1 mice and patients with end-stage ALS [107,108]. In contrast, vulnerable motoneurons show higher levels of GABA-A receptors $\alpha 2$ (Gabra2), dynein, and peripherin (intermediate neurofilament), which are involved in excitability and retrograde transport, which put these motoneurons at a higher risk [108]. Indeed, dysregulation in the dynein-dynactin or peripherin complex is well known to cause degeneration of the spinal motor neuron in mice due to faulty axonal transport [124,125], which is corroborated by the low levels of dynactin shown by spinal motoneurons in ALS patients [126]. Thus, the lower expression of dynein and peripherin is correlated with a lower vulnerability during neurodegenerative processes since retrograde transport is not affected [127]. Therefore, ocular motoneurons could continue receiving a correct trophic contribution from their target muscles, favoring the maintenance of their NMJs.

Motoneurons present heterogenic neurotrophic dependence [128–130]. It is known that these nerve cells receive NGF, BDNF, and NT-3 from the muscle [131] and that the need for neurotrophic contribution, as well as the expression of the different RTKs, both in a control situation and after inducing a lesion, vary among the diverse populations of motoneurons [120,132–135]. Indeed, the adult rat spinal and cranial motoneurons are known to express the TrkB and TrkC receptors but lack the TrkA receptor [132,136–139].

However, another peculiarity of ocular motoneurons is that they express TrkA both in control and after axotomy in the adult [120,140]. This gives them greater efficiency in their response to NGF [120,141], which acts as a potent survival factor for axotomized neonatal motoneurons [142] and plays an essential synaptotrophic and functional role in axotomized motoneurons of the abducens nuclei [140].

Remarkably, recently it has been shown that motoneurons of the oculomotor nuclei present higher expression of VEGF and Flk-1 in the motoneuronal soma compared to other more vulnerable groups of motoneurons, such as the facial and hypoglossal [143]. Previous studies have also shown weak immunoreactivity for VEGF in hypoglossal and facial motoneurons in control rats [144]. The neuroprotective role of VEGF has been extensively exposed above and, therefore, could also contribute to the extended survival of ocular motoneurons in neurodegenerative diseases.

6.1. VEGF and FLK-1 Expression in the Oculomotor System

A high level of VEGF expression has been broadly related to neuronal survival. The expression of VEGF and Flk-1 is high during the embryonic stages but decreases during the adult state, being restricted to some areas of the adult CNS [145]. A higher basal level of VEGF and its receptor Flk-1 has been detected in ocular motoneurons compared to other brainstem motoneurons that are more vulnerable to neurodegeneration [143]. Thus, these oculomotor neurons could form one of these discrete CNS regions that retain the ability to express VEGF and Flk-1 after development.

Likewise, VEGF decreases the levels of pro-apoptotic proteins caspase-3, caspase-9, and Bax, and induces an increased expression of the anti-apoptotic protein Bcl-2 [146]. Therefore, the higher expression of VEGF observed in oculomotor neurons could yield a greater expression of anti-apoptotic proteins, which may be one of the reasons why these neurons show resistance against neurodegeneration. Furthermore, it has been observed that an increased expression of VEGF leads to a greater expression of its Flk-1 receptor [147], which correlates with the fact that a higher expression of Flk-1 was observed in those motoneurons that in turn expressed more VEGF [143].

Motoneurons are known to be especially susceptible to changes in Flk-1 expression, with a linear relationship between a lower expression of this RTK and a more significant loss of motoneurons. This occurs as a consequence of the blockade of the neuroprotective effect of VEGF, preventing activation of the PI3-K/Akt pathway, phospholipase C, and the p38MAPK protein [38]. These claims are supported by other studies where it was shown that the degeneration of spinal motoneurons could be delayed in transgenic SOD1 mice that overexpressed the Flk-1 receptor, thanks to the survival signals generated after VEGF binding [67]. All these findings give the Flk-1 receptor a key role in selective resistance that specific populations of motoneurons show against excitotoxic processes and neurodegeneration.

6.2. VEGF Sources to Ocular Motoneurons

The high level of VEGF found in the soma of extraocular motoneurons compared to the observed in other brainstem motoneurons [143] could be one more of the reasons for the lower vulnerability of this population to neurodegeneration. But, which is the origin of that higher VEGF content? Several possibilities could be considered: (i) VEGF could be synthesized by the motoneurons themselves and act as an autocrine source for extraocular motoneurons; (ii) it could reach motoneurons from surrounding cells, such as glial cells and endothelial cells on the blood vessels; (iii) VEGF could also come from the target muscles via retrograde to the innervating motoneurons (Figure 2).

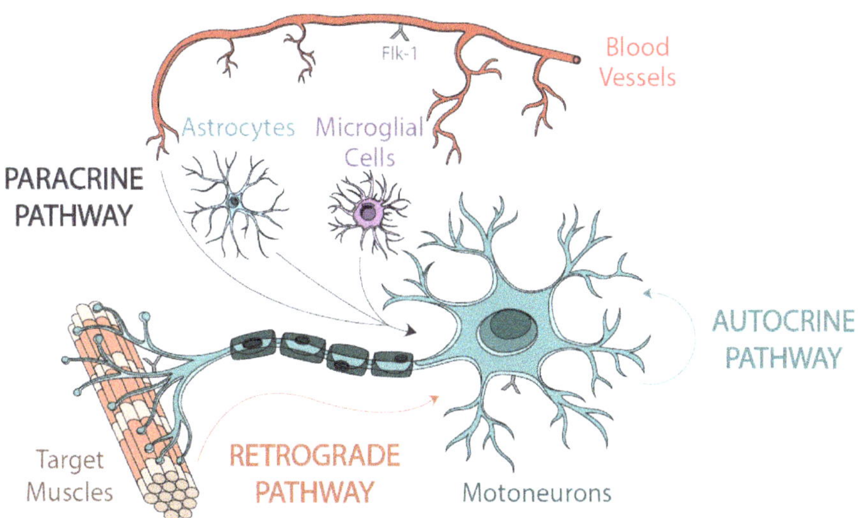

Figure 2. In this scheme, the different pathways of VEGF supply for motoneurons are illustrated. (i) autocrine: self-production of VEGF by the motoneurons themselves; (ii) paracrine: VEGF arriving at the motoneurons from the surrounding cells and blood vessels; and (iii) retrograde: VEGF can also reach the soma of motoneurons from their target muscles.

6.2.1. Via Autocrine

Motoneurons are known to synthesize trophic factors, including neurotrophins such as BDNF, NGF, and NT-3 [148,149]. It is well-known that they also express their receptors, allowing these motoneurons to receive and use the trophic factors as an autocrine source [141,149]. Their production has been shown to vary in response to diverse insults [120,150].

As aforementioned, ocular motoneurons can synthesize VEGF both in control situation and after injury [151], and they express VEGF receptors on their surface [143,151]. Therefore, it is possible that those motoneurons are also acting as an autocrine source of VEGF. Thus, the increased VEGF highlights the autocrine functions of the VEGF, as previously described in the CNS [22,152], this pathway being one of the essential vias of VEGF supply to ocular motoneurons [151].

6.2.2. Via Paracrine

The high level of VEGF located in extraocular motoneurons could also indicate that VEGF is acting as a paracrine factor for the adjacent neurons. The fact that the presence of the Flk-1 receptor is increased in this pool of motoneurons allows them to receive higher amounts of the trophic factor from the neighboring cells. Thus, the upregulation of Flk-1 in the ocular motoneurons highlights the paracrine functions of VEGF [151,152].

Astrocytes are involved in almost all physiological processes that ensure the well-being of neurons [153,154]. Likewise, astrocytes also play a role in neurodegenerative processes since the selective deterioration of the glutamate transporter EAAT2 causes the extracellular accumulation of excitotoxic levels of this amino acid and an increase in the entry of Ca^{2+} in neurons [155]. ROS are believed to induce this oxidative disruption of glutamate transport and promote the spread of this damage, affecting motoneurons. Consequently, glutamate levels increase further, inducing more ROS in motoneurons and triggering a progressive cascade of selective motoneuronal injury, with consequent astrocytic and microglial activation [156]. On the other hand, the cells of the microglia are the specialized macrophages of the CNS [157]. They are an essential component of the inflammatory response to lesions and pathogens [158]. After an injury to the CNS they are the first glial cells to respond, producing pro-inflammatory mediators [159] and promoting the reaction of neurotoxic astrocytes [160].

However, the expression of VEGF driven by the glial cells surrounding brainstem motoneurons is low under control circumstances [151]. A low expression of both mRNA and VEGF protein in glial cells in a control situation was also previously described [161,162], ruling out the possible role of these neural cells as a paracrine source of VEGF to motoneurons at basal conditions. Therefore, astrocytes and microglia do not seem to be contributing to the differential expression of VEGF detected between oculomotor, facial, and hypoglossal motoneurons in a control situation. Nevertheless, glial cells have been reported to modify their VEGF expression under adverse conditions [25,163].

It is well-known that VEGF also acts as a growth factor for vascular endothelial cells forming the blood vessels [2], promoting vascular proliferation and permeability and therefore providing oxygen and nutrients to neurons, which contribute to their wellness. Administration of exogenous VEGF in the brainstem is not accompanied by either angiogenesis or a significantly increased vascular permeability around treated motoneurons [164]. Therefore, it could be assumed that the action of this factor on motoneuron survival was likely due to a direct effect on the motoneurons instead of an indirect effect due to increased blood perfusion. Besides, no differences were observed in the vascularization of these motor nuclei, neither in control nor after an injury [151,164].

Therefore, paracrine actions of VEGF do not seem to be crucial for the differences observed in resistance to degeneration between diverse pools of brainstem motoneurons.

6.2.3. Retrograde Via

Three pairs of extraocular muscles (EOMs) are inserted around the eye, functioning as antagonistic to each other. These are: (i) the medial rectus and lateral rectus muscles, producing eye movements in the horizontal plane; (ii) the superior and inferior rectus muscles, in charge of vertical movements; and (iii) the superior and inferior oblique muscles, which mediate oblique movements [165].

EOMs are anatomically and functionally quite different from other muscles (reviewed in [166]). Most skeletal muscles exclusively have single innervation fibers (SIF), with a single axon forming part of the NMJs and constituting the motor unit, in which a motoneuron innervates 300–2000 muscle fibers (ratio 1: 300–2000). Furthermore, these SIF fibers have a high content of mitochondria and oxidative enzymes, which results in faster contractions. On the other hand, EOMs present a high percentage (20%) of fibers with multiple innervations (MIF), characterized by forming smaller motor units (1:5 ratio), with lower mitochondrial content and fewer oxidative enzymes, a relatively slow, graduated contraction [166,167]. This constitution favors a fine and precise muscular control, modulating the ocular movement and resulting in a more stable vision [168].

It is important to highlight that fast motor units degenerate before slow ones, due, at least in part, to the fact that the motoneurons that supply the slow contraction muscles can compensate the death of its neighboring motoneurons temporarily by generating compensatory axonal branches and reinnervation of the denervated muscle [166]. In contrast, motor-neuronal populations that exclusively present fibers with SIF innervation lose contact with their target muscles much earlier and, consequently, are more vulnerable to neurodegeneration. This resistance has been demonstrated in SOD1 mouse models, where the EOMs remain fully innervated in stages in which the limb muscles show deep denervation [108,169,170]. Therefore, the EOMs can maintain NMJs for a longer time, which leads to a greater retrograde trophic support from the EOMs to the projecting motoneurons.

EOMs have been shown to express VEGF and, therefore, are good candidates to intervene in trophic supply towards motoneurons [151]. Previous studies described the anterograde and retrograde transport of VEGF to neurons, the latter being crucial for maintaining the integrity and functionality of NMJs [82,171]. The importance of trophic supply to ocular motoneurons is also emphasized by the higher expression of BDNF, NGF, and NT-3 found in the EOMs, compared to the buccinator and tongue muscles, target muscles for facial and hypoglossal motoneurons, respectively [172], emphasizing the role of the retrograde pathway as a source of trophic factors.

Although the level of VEGF expression is similar in buccinator, tongue, and EOMs, there is a higher density of Flk-1 receptors in the pre-synaptic terminal of the EOMs compared to the muscles innervated by facial and hypoglossal motoneurons [151]. Previous studies have also shown the presence of the Flk-1 receptor at NMJs level of the abducens motoneurons, projecting by the abducens nerve towards the lateral rectus muscle [173]. These data support the idea that, although extraocular, facial, and hypoglossal muscle fibers were found to be positive for VEGF, not all target muscles appear to be acting to the same extent as the retrograde source of this factor towards the motoneurons that innervate them [151]. In this sense, the VEGF reaching the motoneurons of the ocular motor system through the retrograde pathway may have a more significant influence than the VEGF that comes to the facial or hypoglossal motoneurons.

All this evidence suggests that the retrograde function of VEGF is important and determinant for the survival of brainstem motoneurons.

6.3. Characteristics of the Ocular Motor System after Axotomy

6.3.1. Regulation of Trophic Factors

Upregulation of VEGF expression seems to be a common phenomenon in response to a lesion since any injury to the CNS is known to trigger a hypoxia process involving VEGF expression [144,174]. These findings suggest that the upregulation of endogenous VEGF may be related to its neuroprotective role, so there would be a selective and preferential induction of this neurotrophic factor in some areas of the brain, which would allow greater binding of the ligand to its receptors and would provide greater trophic support [175]. In this sense, it is important to highlight that extraocular motoneurons suffer an increase in the expression of VEGF and Flk-1 in response to axotomy, an increase that was not observed in facial and hypoglossal motoneurons [151].

That increases in VEGF expression has been previously detected in response to ischemia or seizures in both neurons and glial cells of the hippocampus, thalamus, amygdala, and neocortex [175,176]. These results have great functional relevance since VEGF seems to maintain neuromuscular communication even during the denervation processes of the target and are consistent with the neuroprotective role that VEGF exerts on motoneurons, prolonging their survival and improving motor performance [17,82,177,178].

As discussed before, it is well known that the neuroprotective effects that VEGF plays on motoneurons are mediated by the Flk-1 receptor, which is involved in the release of growth factors [64] and mediates trophic functions [37,38,40]. Therefore, this evidence emphasizes the importance of the increased expression of Flk-1 observed in the ocular motoneurons in response to axotomy [151], this being one of the possible keys to the greater resistance shown by this population against neurodegeneration.

Previous studies showed changes in the expression of other trophic factors, such as neurotrophins, or their receptors in motoneurons after axotomy [120,132,179]. It has also been described that axotomized motoneurons experience a decrease in the expression of the protein acetylcholintransferase (ChAT) [120,132,164]. It is important to note that both the immunoreactivity of ChAT and the activity and mRNA of this enzyme are also markedly reduced in cases of ALS [180–182]. In fact, the motoneurons of the anterior horn of the spinal cord suffer a decrease in ChAT activity from early stages, compared to control neurons [183], which suggests that a reduction in ChAT expression is a specific and initial change in disease pathogenesis [184]. It is worth mentioning that the exogenous administration of VEGF at the site of the injury is capable of preventing the loss of the cholinergic phenotype in the axotomized motoneurons, which allows the ocular motoneurons to retain a neurotransmissive phenotype [164]. These studies emphasize the importance of the neuroprotective effect that VEGF has on motoneurons by maintaining its synaptic transmission capacity.

Therefore, all this evidence suggests that facing an injury to the CNS that involves the loss of the muscular target, the ocular motoneurons are capable of expressing a greater amount of VEGF and Flk-1 in their neuronal somas as a compensatory mechanism to keep them protected and in an operational state [151,164]. In summary, the improvement in endogenous VEGF levels highlights autocrine functions, and upregulation of the Flk-1 receptor emphasizes the paracrine functions of VEGF in its neuroprotective effect against degeneration [152].

6.3.2. Administration of Trophic Factors

The axotomy of the oculomotor nerves in the adult, and therefore the loss of synaptic connections with the EOMs target, does not trigger the death of ocular motoneurons [185]. However, alterations in their firing patterns and loss of synaptic inputs are observed, which are reversed by the administration of different neurotrophins [140,186]. Thus, although the survival of adult ocular motoneurons does not depend entirely on the neurotrophic supply, this retrograde signaling is required for the maintenance and regulation of their activity and synaptic properties [187–189].

It is known that the administration of trophic factors can rescue ocular motoneurons from death after axotomy in early postnatal stages [142], as well as reversing the effects of axotomy in adults by promoting the restoration of synaptic coverage and recovery of tonic-phasic triggering [140,186]. Furthermore, neurotrophins have been shown to promote recovery of the cholinergic phenotype after injury [142,190–192]. Therefore, it can be concluded that the ocular motoneurons are characterized by exhibiting a great neurotrophic dependence during the postnatal and adult stages.

In another series of experiments, the recovery of the electrophysiological characteristics of the axotomized motoneurons was observed after the implantation of neural progenitors at the site of the injury [193]. Those neural progenitors expressed NGF, NT-3, and VEGF, revealing for the first time the possible neuroprotective role of VEGF in the oculomotor system. However, it is known that the lack of neurotrophins does not cause motoneuronal degeneration or muscular paralysis, as it does with VEGF deficiency, producing motor alterations similar to that observed during ALS [5,84,194,195]. Therefore, VEGF is probably the most potent of all the neurotrophic factors tested in experimental ALS models [195].

Several experiments have shown that VEGF administration at the injury site can alleviate motoneuronal degeneration in animal models of ALS [36,38,67,82,196]. Recently, the exogenous application of VEGF after axotomy of the abducens nerve prevented the changes observed in axotomized motoneurons, restoring their electrophysiological, morphological properties, and synaptic coverage [173]. These data are correlated with the recovery of ChAT activity observed in axotomized oculomotor neurons due to the administration of VEGF at the site of injury [164].

All these results, together with those that affirm that VEGF administration at the injury supposes a reduction of retrograde axonal degeneration [80,81] and a decrease in injury size and apoptosis levels [87], make this factor an exciting candidate to restore the effects produced by brain damage.

7. Conclusions

In summary, all these data suggest that the higher level of VEGF and its receptor Flk-1 observed in extraocular motoneurons may contribute to their higher resistance shown in adverse conditions, such as excitotoxicity, brain damage, or neurodegenerative diseases, such as ALS. The extraocular motor system presents a series of characteristics that favors the correct contribution of VEGF to ocular motoneurons, even during degeneration and denervation processes, compared to what is observed in other most vulnerable motoneurons. The differential presence of VEGF detected in the soma of the oculomotor and non-oculomotor brainstem motoneurons may be the result of a more generous retrograde trophic contribution of VEGF from the EOMs. Furthermore, the fact that after induction of various types of brain damage, there is an increase in the expression of VEGF and Flk-1 in this specific population of motoneurons further highlights the importance of this neurotrophic factor on motoneuronal survival.

Author Contributions: All authors contributed to the design and writing of this manuscript. All authors have read and agreed to the published version of the manuscript.

Funding: This work was supported by BFU2015-64515-P and PGC2018-094654-B-100 (MCI/AEI/FEDER, UE), and Consejería de Economía, Innovación Ciencia y Empleo, Junta de Andalucía, BIO-297, in Spain.

Conflicts of Interest: The authors declare no conflict of interest.

References

1. Senger, D.R.; Galli, S.J.; Dvorak, A.M.; Perruzzi, C.A.; Harvey, S.V.; Dvorak, H.F. Tumor cells secrete a vascular permeability factor that promotes accumulation of ascites fluid. *Science* **1983**, *219*, 983–985. [CrossRef]
2. Ferrara, N.; Henzel, W.J. Pituitary follicular cells secrete a novel heparin-binding growth factor specific for vascular endothelial cells. *Biochem. Biophys. Res. Commun.* **1989**, *161*, 851–858. [CrossRef]
3. Keck, P.J.; Hauser, S.D.; Krivi, G.; Sanzo, K.; Warren, T.; Feder, J.; Connolly, D.T. Vascular permeability factor, an endothelial cell mitogen related to PDGF. *Science* **1989**, *246*, 1309–1312. [CrossRef]
4. Ferrara, N.; Gerber, H.-P.; LeCouter, J. The biology of VEGF and its receptors. *Nat. Med.* **2003**, *9*, 669–676. [CrossRef]
5. Oosthuyse, B.; Moons, L.; Storkebaum, E.; Beck, H.; Nuyens, D.; Brusselmans, K.; Van Dorpe, J.; Hellings, P.; Gorselink, M.; Heymans, S.; et al. Deletion of the hypoxia-response element in the vascular endothelial growth factor promoter causes motor neuron degeneration. *Nat. Genet.* **2001**, *28*, 131–138. [CrossRef]
6. Grimmond, S.; Lagercrantz, J.; Drinkwater, C.; Silins, G.; Townson, S.; Pollock, P.; Gotley, D.; Carson, E.; Rakar, S.; Nordenskjöld, M.; et al. Cloning and characterization of a novel human gene related to vascular endothelial growth factor. *Genome Res.* **1996**, *6*, 124–131. [CrossRef]
7. Olofsson, B.; Pajusola, K.; Kaipainen, A.; Von Euler, G.; Joukov, V.; Saksela, O.; Orpana, A.; Pettersson, R.F.; Alitalo, K.; Eriksson, U. Vascular endothelial growth factor B, a novel growth factor for endothelial cells. *Proc. Natl. Acad. Sci. USA* **1996**, *93*, 2576–2581. [CrossRef]
8. Joukov, V.; Pajusola, K.; Kaipainen, A.; Chilov, D.; Lahtinen, I.; Kukk, E.; Saksela, O.; Kalkkinen, N.; Alitalo, K. A novel vascular endothelial growth factor, VEGF-C, is a ligand for the Flt4 (VEGFR-3) and KDR (VEGFR-2) receptor tyrosine kinases. *EMBO J.* **1996**, *15*, 290–298. [CrossRef]
9. Lee, J.; Gray, A.; Yuan, J.; Luoh, S.M.; Avraham, H.; Wood, W.I. Vascular endothelial growth factor-related protein: A ligand and specific activator of the tyrosine kinase receptor Flt4. *Proc. Natl. Acad. Sci. USA* **1996**, *93*, 1988–1992. [CrossRef]
10. Orlandini, M.; Marconcini, L.; Ferruzzi, R.; Oliviero, S. Identification of a c-fos-induced gene that is related to the platelet-derived growth factor/vascular endothelial growth factor family. *Proc. Natl. Acad. Sci. USA* **1996**, *93*, 11675–11680. [CrossRef]
11. Yamada, Y.; Nezu, J.I.; Shimane, M.; Hirata, Y. Molecular cloning of a novel vascular endothelial growth factor, VEGF-D. *Genomics* **1997**, *42*, 483–488. [CrossRef]
12. Ogawa, S.; Oku, A.; Sawano, A.; Yamaguchi, S.; Yazaki, Y.; Shibuya, M. A novel type of vascular endothelial growth factor, VEGF-E (NZ-7 VEGF), preferentially utilizes KDR/Flk-1 receptor and carries a potent mitotic activity without heparin-binding domain. *J. Biol. Chem.* **1998**, *273*, 31273–31282. [CrossRef] [PubMed]
13. Yamazaki, Y.; Takani, K.; Atoda, H.; Morita, T. Snake venom vascular endothelial growth factors (VEGFs) exhibit potent activity through their specific recognition of KDR (VEGF Receptor 2). *J. Biol. Chem.* **2003**, *278*, 51985–51988. [CrossRef] [PubMed]
14. Maglione, D.; Guerriero, V.; Viglietto, G.; Delli-Bovi, P.; Persico, M.G. Isolation of a human placenta cDNA coding for a protein related to the vascular permeability factor. *Proc. Natl. Acad. Sci. USA* **1991**, *88*, 9267–9271. [CrossRef] [PubMed]
15. Sathasivam, S. VEGF and ALS. *Neurosci. Res.* **2008**, *62*, 71–77. [CrossRef] [PubMed]
16. Jin, K.; Zhu, Y.; Sun, Y.; Mao, X.O.; Xie, L.; Greenberg, D.A. Vascular endothelial growth factor (VEGF) stimulates neurogenesis in vitro and in vivo. *Proc. Natl. Acad. Sci. USA* **2002**, *99*, 11946–11950. [CrossRef]
17. Lambrechts, D.; Storkebaum, E.; Morimoto, M.; Del-Favero, J.; Desmet, F.; Marklund, S.L.; Wyns, S.; Thijs, V.; Andersson, J.; Van Marion, I.; et al. VEGF is a modifier of amyotrophic lateral sclerosis in mice and humans and protects motoneurons against ischemic death. *Nat. Genet.* **2003**, *34*, 383–394. [CrossRef]
18. Li, X.; Tjwa, M.; Van Hove, I.; Enholm, B.; Neven, E.; Paavonen, K.; Juan, T.D.; Sievers, R.E.; Chorianopoulos, E.; Wada, H.; et al. Reevaluation of the role of VEGF-B suggests a restricted role in the revascularization of the ischemic myocardium. *Arter. Thromb. Vasc. Biol.* **2008**, *28*, 1614–1620. [CrossRef]

19. Poesen, K.; Lambrechts, D.; Van Damme, P.; Dhondt, J.; Bender, F.; Frank, N.; Bogaert, E.; Claes, B.; Heylen, L.; Verheyen, A.; et al. Novel role for vascular endothelial growth factor (VEGF) receptor-1 and its ligand VEGF-B in motor neuron degeneration. *J. Neurosci.* **2008**, *28*, 10451–10459. [CrossRef]
20. Carmeliet, P.; Ruiz de Almodovar, C. VEGF ligands and receptors: Implications in neurodevelopment and neurodegeneration. *Cell. Mol. Life Sci.* **2013**, *70*, 1763–1778. [CrossRef]
21. Shibuya, M. Vascular endothelial growth factor (VEGF) and its receptor (VEGFR) signaling in angiogenesis: A crucial target for anti- and pro-angiogenic therapies. *Genes Cancer* **2011**, *2*, 1097–1105. [CrossRef] [PubMed]
22. Ogunshola, O.O.; Antic, A.; Donoghue, M.J.; Fan, S.Y.; Kim, H.; Stewart, W.B.; Madri, J.A.; Ment, L.R. Paracrine and autocrine functions of neuronal vascular endothelial growth factor (VEGF) in the central nervous system. *J. Biol. Chem.* **2002**, *277*, 11410–11415. [CrossRef] [PubMed]
23. Van Den Bosch, L.; Storkebaum, E.; Vleminckx, V.; Moons, L.; Vanopdenbosch, L.; Scheveneels, W.; Carmeliet, P.; Robberecht, W. Effects of vascular endothelial growth factor (VEGF) on motor neuron degeneration. *Neurobiol. Dis.* **2004**, *17*, 21–28. [CrossRef]
24. Ijichi, A.; Sakuma, S.; Tofilon, P.J. Hypoxia-induced vascular endothelial growth factor expression in normal rat astrocyte cultures. *Glia* **1995**, *14*, 87–93. [CrossRef] [PubMed]
25. Bartholdi, D.; Rubin, B.P.; Schwab, M.E. VEGF mRNA induction correlates with changes in the vascular architecture upon spinal cord damage in the rat. *Eur. J. Neurosci.* **1997**, *9*, 2549–2560. [CrossRef] [PubMed]
26. Silverman, W.F.; Krum, J.M.; Mani, N.; Rosenstein, J.M. Vascular, glial and neuronal effects of vascular endothelial growth factor in mesencephalic explant cultures. *Neuroscience* **1999**, *90*, 1529–1541. [CrossRef]
27. Sondell, M.; Lundborg, G.; Kanje, M. Vascular endothelial growth factor stimulates Schwann cell invasion and neovascularization of acellular nerve grafts. *Brain Res.* **1999**, *846*, 219–228. [CrossRef]
28. Sondell, M.; Lundborg, G.; Kanje, M. Vascular endothelial growth factor has neurotrophic activity and stimulates axonal outgrowth, enhancing cell survival and Schwann cell proliferation in the peripheral nervous system. *J. Neurosci.* **1999**, *19*, 5731–5740. [CrossRef]
29. Sondell, M.; Sundler, F.; Kanje, M. Vascular endothelial growth factor is a neurotrophic factor which stimulates axonal outgrowth through the flk-1 receptor. *Eur. J. Neurosci.* **2000**, *12*, 4243–4254. [CrossRef]
30. Jin, K.L.; Mao, X.O.; Greenberg, D.A. Vascular endothelial growth factor: Direct neuroprotective effect in in vitro ischemia. *Proc. Natl. Acad. Sci. USA* **2000**, *97*, 10242–10247. [CrossRef]
31. Svensson, B.; Peters, M.; König, H.G.; Poppe, M.; Levkau, B.; Rothermundt, M.; Arolt, V.; Kögel, D.; Prehn, J.H.M. Vascular endothelial growth factor protects cultured rat hippocampal neurons against hypoxic injury via an antiexcitotoxic, caspase-independent mechanism. *J. Cereb. Blood Flow Metab.* **2002**, *22*, 1170–1175. [CrossRef]
32. Wick, A.; Wick, W.; Waltenberger, J.; Weller, M.; Dichgans, J.; Schulz, J.B. Neuroprotection by hypoxic preconditioning requires sequential activation of vascular endothelial growth factor receptor and Akt. *J. Neurosci.* **2002**, *22*, 6401–6407. [CrossRef]
33. Licht, T.; Goshen, I.; Avital, A.; Kreisel, T.; Zubedat, S.; Eavri, R.; Segal, M.; Yirmiya, R.; Keshet, E. Reversible modulations of neuronal plasticity by VEGF. *Proc. Natl. Acad. Sci. USA* **2011**, *108*, 5081–5086. [CrossRef]
34. Ruiz de Almodovar, C.; Lambrechts, D.; Mazzone, M.; Carmeliet, P. Role and therapeutic potential of VEGF in the nervous system. *Physiol. Rev.* **2009**, *89*, 607–648. [CrossRef]
35. Li, B.; Xu, W.; Luo, C.; Gozal, D.; Liu, R. VEGF-induced activation of the PI3-K/Akt pathway reduces mutant SOD1-mediated motor neuron cell death. *Mol. Brain Res.* **2003**, *111*, 155–164. [CrossRef]
36. Tovar-y-Romo, L.B.; Zepeda, A.; Tapia, R. Vascular endothelial growth factor prevents paralysis and motoneuron death in a rat model of excitotoxic spinal cord neurodegeneration. *J. Neuropathol. Exp. Neurol.* **2007**, *66*, 913–922. [CrossRef]
37. Tolosa, L.; Mir, M.; Asensio, V.J.; Olmos, G.; Lladó, J. Vascular endothelial growth factor protects spinal cord motoneurons against glutamate-induced excitotoxicity via phosphatidylinositol 3-kinase. *J. Neurochem.* **2008**, *105*, 1080–1090. [CrossRef]
38. Tovar-y-Romo, L.B.; Tapia, R. VEGF protects spinal motor neurons against chronic excitotoxic degeneration in vivo by activation of PI3-K pathway and inhibition of p38MAPK. *J. Neurochem.* **2010**, *115*, 1090–1101. [CrossRef]
39. Ruiz de Almodovar, C.; Fabre, P.J.; Knevels, E.; Coulon, C.; Segura, I.; Haddick, P.C.G.; Aerts, L.; Delattin, N.; Strasser, G.; Oh, W.-J.; et al. VEGF mediates commissural axon chemoattraction through its receptor Flk1. *Neuron* **2011**, *70*, 966–978. [CrossRef]
40. Brockington, A.; Wharton, S.B.; Fernando, M.; Gelsthorpe, C.H.; Baxter, L.; Ince, P.G.; Lewis, C.E.; Shaw, P.J. Expression of vascular endothelial growth factor and its receptors in the central nervous system in amyotrophic lateral sclerosis. *J. Neuropathol. Exp. Neurol.* **2006**, *65*, 26–36. [CrossRef]
41. Folkman, J.; Shing, Y. Angiogenesis. *J. Biol. Chem.* **1992**, *267*, 10931–10934. [CrossRef]
42. Pertovaara, L.; Kaipainen, A.; Mustonen, T.; Orpana, A.; Ferrara, N.; Saksela, O.; Alitalo, K. Vascular endothelial growth factor is induced in response to transforming growth factor-β in fibroblastic and epithelial cells. *J. Biol. Chem.* **1994**, *269*, 6271–6274. [CrossRef]
43. Cohen, T.; Nahari, D.; Cerem, L.W.; Neufeld, G.; Levin, B.Z. Interleukin 6 induces the expression of vascular endothelial growth factor. *J. Biol. Chem.* **1996**, *271*, 736–741. [CrossRef] [PubMed]
44. Ryuto, M.; Ono, M.; Izumi, H.; Yoshida, S.; Weich, H.A.; Kohno, K.; Kuwano, M. Induction of vascular endothelial growth factor by tumor necrosis factor α in human glioma cells: Possible roles of SP-1. *J. Biol. Chem.* **1996**, *271*, 28220–28228. [CrossRef]
45. Deroanne, C.F.; Hajitou, A.; Calberg-Bacq, C.M.; Nusgens, B.V.; Lapière, C.M. Angiogenesis by fibroblast growth factor 4 is mediated through an autocrine up-regulation of vascular endothelial growth factor expression. *Cancer Res.* **1997**, *57*, 5590–5597.

46. Kroll, J.; Waltenberger, J. A novel function of VEGF receptor-2 (KDR): Rapid release of nitric oxide in response to VEGF-A stimulation in endothelial cells. *Biochem. Biophys. Res. Commun.* **1999**, *265*, 636–639. [CrossRef]
47. Shen, B.Q.; Lee, D.Y.; Zioncheck, T.F. Vascular endothelial growth factor governs endothelial nitric-oxide synthase expression via a KDR/Flk-1 receptor and a protein kinase C signaling pathway. *J. Biol. Chem.* **1999**, *274*, 33057–33063. [CrossRef]
48. Richard, D.E.; Berra, E.; Pouysségur, J. Angiogenesis: How a tumor adapts to hypoxia. *Biochem. Biophys. Res. Commun.* **1999**, *266*, 718–722. [CrossRef]
49. Stein, I.; Neeman, M.; Shweiki, D.; Itin, A.; Keshet, E. Stabilization of vascular endothelial growth factor mRNA by hypoxia and hypoglycemia and coregulation with other ischemia-induced genes. *Mol. Cell. Biol.* **1995**, *15*, 5363–5368. [CrossRef]
50. Ikeda, E.; Achen, M.G.; Breier, G.; Risau, W. Hypoxia-induced transcriptional activation and increased mRNA stability of vascular endothelial growth factor in C6 glioma cells. *J. Biol. Chem.* **1995**, *270*, 19761–19766. [CrossRef]
51. Forsythe, J.A.; Jiang, B.H.; Iyer, N.V.; Agani, F.; Leung, S.W.; Koos, R.D.; Semenza, G.L. Activation of vascular endothelial growth factor gene transcription by hypoxia-inducible factor 1. *Mol. Cell. Biol.* **1996**, *16*, 4604–4613. [CrossRef]
52. Semenza, G.L. Hypoxia-inducible factor 1: Master regulator of O_2 homeostasis. *Curr. Opin. Genet. Dev.* **1998**, *8*, 588–594. [CrossRef]
53. Kaelin, W.G. HIF2 inhibitor joins the kidney cancer armamentarium. *J. Clin. Oncol.* **2018**, *36*, 908–910. [CrossRef]
54. Levy, A.P.; Levy, N.S.; Wegner, S.; Goldberg, M.A. Transcriptional regulation of the rat vascular endothelial growth factor gene by hypoxia. *J. Biol. Chem.* **1995**, *270*, 13333–13340. [CrossRef]
55. Liu, Y.; Cox, S.R.; Morita, T.; Kourembanas, S. Hypoxia regulates vascular endothelial growth factor gene expression in endothelial cells: Identification of a 5′ enhancer. *Circ. Res.* **1995**, *77*, 638–643. [CrossRef]
56. Semenza, G.L.; Prabhakar, N.R. The role of hypoxia-inducible factors in carotid body (patho) physiology. *J. Physiol.* **2018**, *596*, 2977–2983. [CrossRef]
57. Ratcliffe, P.J.; Ebert, B.L.; Ferguson, D.J.P.; Firth, J.D.; Gleadle, J.M.; Maxwell, P.H.; Pugh, C.W. State of the art lecture: Regulation of the erythropoietin gene. *Nephrol. Dial. Transplant.* **1995**, *10*, 18–27. [CrossRef]
58. Semenza, G.L. Expression of hypoxia-inducible factor 1: Mechanisms and consequences. *Biochem. Pharmacol.* **2000**, *59*, 47–53. [CrossRef]
59. Wenger, R.H. Mammalian oxygen sensing, signalling and gene regulation. *J. Exp. Biol.* **2000**, *203*, 1253–1263.
60. Kolodkin, A.L.; Levengood, D.V.; Rowe, E.G.; Tai, Y.T.; Giger, R.J.; Ginty, D.D. Neuropilin is a semaphorin III receptor. *Cell* **1997**, *90*, 753–762. [CrossRef]
61. Giger, R.J.; Urquhart, E.R.; Gillespie, S.K.H.; Levengood, D.V.; Ginty, D.D.; Kolodkin, A.L. Neuropilin-2 is a receptor for semaphorin IV: Insight into the structural basis of receptor function and specificity. *Neuron* **1998**, *21*, 1079–1092. [CrossRef]
62. Whitaker, G.B.; Limberg, B.J.; Rosenbaum, J.S. Vascular endothelial growth factor receptor-2 and neuropilin-1 form a receptor complex that is responsible for the differential signaling potency of VEGF165 and VEGF121. *J. Biol. Chem.* **2001**, *276*, 25520–25531. [CrossRef] [PubMed]
63. Geretti, E.; Shimizu, A.; Klagsbrun, M. Neuropilin structure governs VEGF and semaphorin binding and regulates angiogenesis. *Angiogenesis* **2008**, *11*, 31–39. [CrossRef] [PubMed]
64. Lladó, J.; Tolosa, L.; Olmos, G. Cellular and molecular mechanisms involved in the neuroprotective effects of VEGF on motoneurons. *Front. Cell. Neurosci.* **2013**, *7*, 181. [CrossRef] [PubMed]
65. Autiero, M.; Waltenberger, J.; Communi, D.; Kranz, A.; Moons, L.; Lambrechts, D.; Kroll, J.; Plaisance, S.; De Mol, M.; Bono, F.; et al. Role of PlGF in the intra- and intermolecular cross talk between the VEGF receptors Flt1 and Flk1. *Nat. Med.* **2003**, *9*, 936–943. [CrossRef]
66. Zachary, I. Neuroprotective role of vascular endothelial growth factor: Signalling mechanisms, biological function, and therapeutic potential. *NeuroSignals* **2005**, *14*, 207–221. [CrossRef]
67. Storkebaum, E.; Lambrechts, D.; Dewerchin, M.; Moreno-Murciano, M.P.; Appelmans, S.; Oh, H.; Van Damme, P.; Rutten, B.; Man, W.Y.; De Mol, M.; et al. Treatment of motoneuron degeneration by intracerebroventricular delivery of VEGF in a rat model of ALS. *Nat. Neurosci.* **2005**, *8*, 85–92. [CrossRef]
68. Bogaert, E.; Van Damme, P.; Van Den Bosch, L.; Robberecht, W. Vascular endothelial growth factor in amyotrophic lateral sclerosis and other neurodegenerative diseases. *Muscle Nerve* **2006**, *34*, 391–405. [CrossRef]
69. Foster, F.M.; Traer, C.J.; Abraham, S.M.; Fry, M.J. The phosphoinositide (PI) 3-kinase family. *J. Cell Sci.* **2003**, *116*, 3037–3040. [CrossRef]
70. Tolosa, L.; Mir, M.; Olmos, G.; Lladó, J. Vascular endothelial growth factor protects motoneurons from serum deprivation-induced cell death through phosphatidylinositol 3-kinase-mediated p38 mitogen-activated protein kinase inhibition. *Neuroscience* **2009**, *158*, 1348–1355. [CrossRef]
71. Shiote, M.; Nagano, I.; Ilieva, H.; Murakami, T.; Narai, H.; Ohta, Y.; Nagata, T.; Shoji, M.; Abe, K. Reduction of a vascular endothelial growth factor receptor, fetal liver kinase-1, by antisense oligonucleotides induces motor neuron death in rat spinal cord exposed to hypoxia. *Neuroscience* **2005**, *132*, 175–182. [CrossRef] [PubMed]
72. Boillée, S.; Vande Velde, C.; Cleveland, D.W. ALS: A disease of motor neurons and their nonneuronal neighbors. *Neuron* **2006**, *52*, 39–59. [CrossRef] [PubMed]
73. Boillée, S.; Yamanaka, K.; Lobsiger, C.S.; Copeland, N.G.; Jenkins, N.A.; Kassiotis, G.; Kollias, G.; Cleveland, D.W. Onset and progression in inherited ALS determined by motor neurons and microglia. *Science* **2006**, *312*, 1389–1392. [CrossRef] [PubMed]

74. Andersen, P.M.; Sims, K.B.; Xin, W.W.; Kiely, R.; O'Neill, G.; Ravits, J.; Pioro, E.; Harati, Y.; Brower, R.D.; Levine, J.S.; et al. Sixteen novel mutations in the Cu/Zn superoxide dismutase gene in amyotrophic lateral sclerosis: A decade of discoveries, defects and disputes. *Amyotroph. Lateral Scler. Other Mot. Neuron Disord.* **2003**, *4*, 62–73. [CrossRef] [PubMed]
75. Gurney, M.E.; Pu, H.; Chiu, A.Y.; Dal Canto, M.C.; Polchow, C.Y.; Alexander, D.D.; Caliendo, J.; Hentati, A.; Kwon, Y.W.; Deng, H.X. Motor neuron degeneration in mice that express a human Cu,Zn superoxide dismutase mutation. *Science* **1994**, *264*, 1772–1775. [CrossRef] [PubMed]
76. Lu, L.; Zheng, L.; Viera, L.; Suswam, E.; Li, Y.; Li, X.; Estévez, A.G.; King, P.H. Mutant Cu/Zn-superoxide dismutase associated with amyotrophic lateral sclerosis destabilizes vascular endothelial growth factor mRNA and downregulates its expression. *J. Neurosci.* **2007**, *27*, 7929–7938. [CrossRef]
77. Dupuis, L.; Gonzalez de Aguilar, J.L.; Echaniz-Laguna, A.; Eschbach, J.; Rene, F.; Oudart, H.; Halter, B.; Huze, C.; Schaeffer, L.; Bouillaud, F.; et al. Muscle mitochondrial uncoupling dismantles neuromuscular junction and triggers distal degeneration of motor neurons. *PLoS ONE* **2009**, *4*, e5390. [CrossRef]
78. Dadon-Nachum, M.; Melamed, E.; Offen, D. The "dying-back" phenomenon of motor neurons in ALS. *J. Mol. Neurosci.* **2011**, *43*, 470–477. [CrossRef]
79. Rocha, M.C.; Pousinha, P.A.; Correia, A.M.; Sebastião, A.M.; Ribeiro, J.A. Early changes of neuromuscular transmission in the SOD1(G93A) mice model of ALS start long before motor symptoms onset. *PLoS ONE* **2013**, *8*, e73846. [CrossRef]
80. Storkebaum, E.; Carmeliet, P. VEGF: A critical player in neurodegeneration. *J. Clin. Investig.* **2004**, *113*, 14–18. [CrossRef]
81. Storkebaum, E.; Lambrechts, D.; Carmeliet, P. VEGF: Once regarded as a specific angiogenic factor, now implicated in neuroprotection. *Bioessays* **2004**, *26*, 943–954. [CrossRef] [PubMed]
82. Azzouz, M.; Ralph, G.S.; Storkebaum, E.; Walmsley, L.E.; Mitrophanous, K.A.; Kingsman, S.M.; Carmeliet, P.; Mazarakis, N.D. VEGF delivery with retrogradely transported lentivector prolongs survival in a mouse ALS model. *Nature* **2004**, *429*, 413–417. [CrossRef] [PubMed]
83. Hwang, D.H.; Lee, H.J.; Park, I.H.; Seok, J.I.; Kim, B.G.; Joo, I.S.; Kim, S.U. Intrathecal transplantation of human neural stem cells overexpressing VEGF provide behavioral improvement, disease onset delay and survival extension in transgenic ALS mice. *Gene Ther.* **2009**, *16*, 1234–1244. [CrossRef] [PubMed]
84. Gould, T.W.; Oppenheim, R.W. Motor neuron trophic factors: Therapeutic use in ALS? *Brain Res. Rev.* **2011**, *67*, 1–39. [CrossRef]
85. Rosenstein, J.M.; Mani, N.; Khaibullina, A.; Krum, J.M. Neurotrophic effects of vascular endothelial growth factor on organotypic cortical explants and primary cortical neurons. *J. Neurosci.* **2003**, *23*, 11036–11044. [CrossRef]
86. Vande Velde, C.; Cleveland, D.W. VEGF: Multitasking in ALS. *Nat. Neurosci.* **2005**, *8*, 5–7. [CrossRef]
87. Widenfalk, J.; Lipson, A.; Jubran, M.; Hofstetter, C.; Ebendal, T.; Cao, Y.; Olson, L. Vascular endothelial growth factor improves functional outcome and decreases secondary degeneration in experimental spinal cord contusion injury. *Neuroscience* **2003**, *120*, 951–960. [CrossRef]
88. Lu, Y.; Tang, C.; Zhu, L.; Li, J.; Liang, H.; Zhang, J.; Xu, R. The overexpression of TDP-43 protein in the neuron and oligodendrocyte cells causes the progressive motor neuron degeneration in the SOD1 G93A transgenic mouse model of amyotrophic lateral sclerosis. *Int. J. Biol. Sci.* **2016**, *12*, 1140–1149. [CrossRef]
89. Gerber, H.P.; Dixit, V.; Ferrara, N. Vascular endothelial growth factor induces expression of the antiapoptotic proteins Bcl-2 and A1 in vascular endothelial cells. *J. Biol. Chem.* **1998**, *273*, 13313–13316. [CrossRef]
90. Corona, J.C.; Tapia, R. Ca^{2+}-permeable AMPA receptors and intracellular Ca2+ determine motoneuron vulnerability in rat spinal cord in vivo. *Neuropharmacology* **2007**, *52*, 1219–1228. [CrossRef]
91. Tapia, R.; Medina-Ceja, L.; Peña, F. On the relationship between extracellular glutamate, hyperexcitation and neurodegeneration, in vivo. *Neurochem. Int.* **1999**, *34*, 23–31. [CrossRef]
92. Meldrum, B.S. Glutamate and glutamine in the brain glutamate as a neurotransmitter in the brain: Review of physiology and pathology. *J. Nutr.* **2018**, *130*, 1007S–1015S. [CrossRef] [PubMed]
93. Lipton, S.A.; Rosenberg, P.A. Excitatory amino acids as a final common pathway for neurologic disorders. *N. Engl. J. Med.* **1994**, *330*, 613–622. [PubMed]
94. Bettler, B.; Mulle, C. AMPA and kainate receptors. *Neuropharmacology* **1995**, *34*, 123–139. [CrossRef]
95. Heath, P.R.; Shaw, P.J. Update on the glutamatergic neurotransmitter system and the role of excitotoxicity in amyotrophic lateral sclerosis. *Muscle Nerve* **2002**, *26*, 438–458. [CrossRef]
96. Siesjo, B.K. Calcium-mediated processes in neuronal degeneration. *Ann. N. Y. Acad. Sci.* **1994**, *747*, 140–161. [CrossRef] [PubMed]
97. Van Den Bosch, L.; Vandenberghe, W.; Klaassen, H.; Van Houtte, E.; Robberecht, W. Ca2+-permeable AMPA receptors and selective vulnerability of motor neurons. *J. Neurol. Sci.* **2000**, *180*, 29–34. [CrossRef]
98. Van Damme, P.; Van Den Bosch, L.; Van Houtte, E.; Callewaert, G.; Robberecht, W. GluR2-dependent properties of AMPA receptors determine the selective vulnerability of motor neurons to excitotoxicity. *J. Neurophysiol.* **2002**, *88*, 1279–1287. [CrossRef]
99. Van Damme, P.; Braeken, D.; Callewaert, G.; Robberecht, W.; Van Den Bosch, L. GluR2 deficiency accelerates motor neuron degeneration in a mouse model of amyotrophic lateral sclerosis. *J. Neuropathol. Exp. Neurol.* **2005**, *64*, 605–612. [CrossRef]
100. Carriedo, S.G.; Sensi, S.L.; Yin, H.Z.; Weiss, J.H. AMPA exposures induce mitochondrial Ca(2+) overload and ROS generation in spinal motor neurons in vitro. *J. Neurosci.* **2000**, *20*, 240–250. [CrossRef]

101. Rothstein, J.D.; Dykes-Hoberg, M.; Pardo, C.A.; Bristol, L.A.; Jin, L.; Kuncl, R.W.; Kanai, Y.; Hediger, M.A.; Wang, Y.; Schielke, J.P.; et al. Knockout of glutamate transporters reveals a major role for astroglial transport in excitotoxicity and clearance of glutamate. *Neuron* **1996**, *16*, 675–686. [CrossRef]

102. Matsuzaki, H.; Tamatani, M.; Yamaguchi, A.; Namikawa, K.; Kiyama, H.; Vitek, M.P.; Mitsuda, N.; Tohyama, M. Vascular endothelial growth factor rescues hippocampal neurons from glutamate-induced toxicity: Signal transduction cascades. *FASEB J.* **2001**, *15*, 1218–1220. [CrossRef]

103. Bogaert, E.; Van Damme, P.; Poesen, K.; Dhondt, J.; Hersmus, N.; Kiraly, D.; Scheveneels, W.; Robberecht, W.; Van Den Bosch, L. VEGF protects motor neurons against excitotoxicity by upregulation of GluR2. *Neurobiol. Aging* **2010**, *31*, 2185–2191. [CrossRef] [PubMed]

104. Nimchinsky, E.A.; Young, W.G.; Yeung, G.; Shah, R.A.; Gordon, J.W.; Bloom, F.E.; Morrison, J.H.; Hof, P.R. Differential vulnerability of oculomotor, facial, and hypoglossal nuclei in G86R superoxide dismutase transgenic mice. *J. Comp. Neurol.* **2000**, *416*, 112–125. [CrossRef]

105. Haenggeli, C.; Kato, A.C. Differential vulnerability of cranial motoneurons in mouse models with motor neuron degeneration. *Neurosci. Lett.* **2002**, *335*, 39–43. [CrossRef]

106. DePaul, R.; Abbs, J.H.; Caligiuri, M.; Gracco, V.L.; Brooks, B.R. Hypoglossal, trigeminal, and facial motoneuron involvement in amyotrophic lateral sclerosis. *Neurology* **1988**, *38*, 281–283. [CrossRef] [PubMed]

107. Brockington, A.; Ning, K.; Heath, P.R.; Wood, E.; Kirby, J.; Fusi, N.; Lawrence, N.; Wharton, S.B.; Ince, P.G.; Shaw, P.J. Unravelling the enigma of selective vulnerability in neurodegeneration: Motor neurons resistant to degeneration in ALS show distinct gene expression characteristics and decreased susceptibility to excitotoxicity. *Acta Neuropathol.* **2013**, *125*, 95–109. [CrossRef] [PubMed]

108. Comley, L.; Allodi, I.; Nichterwitz, S.; Nizzardo, M.; Simone, C.; Corti, S.; Hedlund, E. Motor neurons with differential vulnerability to degeneration show distinct protein signatures in health and ALS. *Neuroscience* **2015**, *291*, 216–229. [CrossRef]

109. Reiner, A.; Medina, L.; Figueredo-Cardenas, G.; Anfinson, S. Brainstem motoneuron pools that are selectively resistant in amyotrophic lateral sclerosis are preferentially enriched in parvalbumin: Evidence from monkey brainstem for a calcium-mediated mechanism in sporadic ALS. *Exp. Neurol.* **1995**, *131*, 239–250. [CrossRef]

110. Tjust, A.E.; Brannstrom, T.; Pedrosa Domellof, F. Unaffected motor endplate occupancy in eye muscles of ALS G93A mouse model. *Front. Biosci.* **2012**, 1547–1555.

111. Liu, J.X.; Brännström, T.; Andersen, P.M.; Pedrosa-Domellöf, F. Different impact of ALS on laminin isoforms in human extraocular muscles versus limb muscles. *Investig. Ophthalmol. Vis. Sci.* **2011**, *52*, 4842–4852. [CrossRef]

112. Alexianu, M.E.; Ho, B.; Mohamed, A.H.; La Bella, V.; Smith, R.G.; Appel, S.H. The role of calcium-binding proteins in selective motoneuron vulnerability in amyotrophic lateral sclerosis. *Ann. Neurol.* **1994**, *36*, 846–858. [CrossRef] [PubMed]

113. Vanselow, B.K.; Keller, B.U. Calcium dynamics and buffering in oculomotor neurones from mouse that are particularly resistant during amyotrophic lateral sclerosis (ALS)-related motoneurone disease. *J. Physiol.* **2000**, *525 Pt 2*, 433–445. [CrossRef]

114. Laslo, P.; Lipski, J.; Nicholson, L.F.; Miles, G.B.; Funk, G.D. Calcium binding proteins in motoneurons at low and high risk for degeneration in ALS. *Neuroreport* **2000**, *11*, 3305–3308. [CrossRef] [PubMed]

115. Ince, P.; Stout, N.; Shaw, P.; Slade, J.; Hunziker, W.; Heizmann, C.W.; Baimbridge, K.G. Parvalbumin and calbindin D-28k in the human motor system and in motor neuron disease. *Neuropathol. Appl. Neurobiol.* **1993**, *19*, 291–299. [CrossRef] [PubMed]

116. de la Cruz, R.R.; Pastor, A.M.; Martínez-Guijarro, F.J.; López-García, C.; Delgado-García, J.M. Localization of parvalbumin, calretinin, and calbindin D-28k in identified extraocular motoneurons and internuclear neurons of the cat. *J. Comp. Neurol.* **1998**, *390*, 377–391. [CrossRef]

117. Beers, D.R.; Ho, B.K.; Siklós, L.; Alexianu, M.E.; Mosier, D.R.; Habib Mohamed, A.; Otsuka, Y.; Kozovska, M.E.; McAlhany, R.E.; Glenn Smith, R.; et al. Parvalbumin overexpression alters immune-mediated increases in intracellular calcium, and delays disease onset in a transgenic model of familial amyotrophic lateral sclerosis. *J. Neurochem.* **2001**, *79*, 499–509. [CrossRef] [PubMed]

118. Van Den Bosch, L.; Schwaller, B.; Vleminckx, V.; Meijers, B.; Stork, S.; Ruehlicke, T.; Van Houtte, E.; Klaassen, H.; Celio, M.R.; Missiaen, L.; et al. Protective effect of parvalbumin on excitotoxic motor neuron death. *Exp. Neurol.* **2002**, *174*, 150–161. [CrossRef] [PubMed]

119. Fukuoka, T.; Tokunaga, A.; Kondo, E.; Miki, K.; Tachibana, T.; Noguchi, K. Differential regulation of alpha- and beta-CGRP mRNAs within oculomotor, trochlear, abducens, and trigeminal motoneurons in response to axotomy. *Mol. Brain Res.* **1999**, *63*, 304–315. [CrossRef]

120. Morcuende, S.; Matarredona, E.R.; Benítez-Temiño, B.; Muñoz-Hernández, R.; Pastor, A.M.; de la Cruz, R.R. Differential regulation of the expression of neurotrophin receptors in rat extraocular motoneurons after lesion. *J. Comp. Neurol.* **2011**, *519*, 2335–2352. [CrossRef]

121. González-Forero, D.; de la Cruz, R.R.; Delgado-García, J.M.; Álvarez, F.J.; Pastor, A.M. Correlation between CGRP immunoreactivity and firing activity in cat abducens motoneurons. *J. Comp. Neurol.* **2002**, *451*, 201–212. [CrossRef] [PubMed]

122. Allodi, I.; Comley, L.; Nichterwitz, S.; Nizzardo, M.; Simone, C.; Aguila Benitez, J.; Cao, M.; Corti, S.; Hedlund, E. Differential neuronal vulnerability identifies IGF-2 as a protective factor in ALS and SMA. *Sci. Rep.* **2016**, *6*, 25960. [CrossRef] [PubMed]

123. Hedlund, E.; Karlsson, M.; Osborn, T.; Ludwig, W.; Isacson, O. Global gene expression profiling of somatic motor neuron populations with different vulnerability identify molecules and pathways of degeneration and protection. *Brain* **2010**, *133*, 2313–2330. [CrossRef] [PubMed]

124. Hafezparast, M.; Klocke, R.; Ruhrberg, C.; Marquardt, A.; Ahmad-Annuar, A.; Bowen, S.; Lalli, G.; Witherden, A.S.; Hummerich, H.; Nicholson, S.; et al. Mutations in dynein link motor neuron degeneration to defects in retrograde transport. *Science* **2003**, *300*, 808–812. [CrossRef]

125. Millecamps, S.; Robertson, J.; Lariviere, R.; Mallet, J.; Julien, J.P. Defective axonal transport of neurofilament proteins in neurons overexpressing peripherin. *J. Neurochem.* **2006**, *98*, 926–938. [CrossRef]

126. Jiang, Y.M.; Yamamoto, M.; Tanaka, F.; Ishigaki, S.; Katsuno, M.; Adachi, H.; Niwa, J.I.; Doyu, M.; Yoshida, M.; Hashizume, Y.; et al. Gene expressions specifically detected in motor neurons (dynactin 1, early growth response 3, acetyl-CoA transporter, death receptor 5, and cyclin C) differentially correlate to pathologic markers in sporadic amyotrophic lateral sclerosis. *J. Neuropathol. Exp. Neurol.* **2007**, *66*, 617–627. [CrossRef]

127. Collard, J.F.; Côté, F.; Julien, J.P. Defective axonal transport in a transgenic mouse model of amyotrophic lateral sclerosis. *Nature* **1995**, *375*, 61–64. [CrossRef]

128. Oppenheim, R.; Haverkamp, L.; Prevette, D.; McManaman, J.; Appel, S. Reduction of naturally occurring motoneuron death in vivo by a target-derived neurotrophic factor. *Science* **1988**, *240*, 919–922. [CrossRef]

129. Grieshammer, U.; Lewandoski, M.; Prevette, D.; Oppenheim, R.W.; Martin, G.R. Muscle-specific cell ablation conditional upon Cre-mediated DNA recombination in transgenic mice leads to massive spinal and cranial motoneuron loss. *Dev. Biol.* **1998**, *197*, 234–247. [CrossRef]

130. Gould, T.W.; Yonemura, S.; Oppenheim, R.W.; Ohmori, S.; Enomoto, H. The neurotrophic effects of glial cell line-derived neurotrophic factor on spinal motoneurons are restricted to fusimotor subtypes. *J. Neurosci.* **2008**, *28*, 2131–2146. [CrossRef]

131. DiStefano, P.S.; Friedman, B.; Radziejewski, C.; Alexander, C.; Boland, P.; Schick, C.M.; Lindsay, R.M.; Wiegand, S.J. The neurotrophins BDNF, NT-3, and NGF display distinct patterns of retrograde axonal transport in peripheral and central neurons. *Neuron* **1992**, *8*, 983–993. [CrossRef]

132. Koliatsos, V.E.; Crawford, T.O.; Price, D.L. Axotomy induces nerve growth factor receptor immunoreactivity in spinal motor neurons. *Brain Res.* **1991**, *549*, 297–304. [CrossRef]

133. Connor, B.; Young, D.; Lawlor, P.; Gai, W.; Waldvogel, H.; Faull, R.L.M.; Dragunow, M. Trk receptor alterations in Alzheimer's disease. *Mol. Brain Res.* **1996**, *42*, 1–17. [CrossRef]

134. Canals, J.M.; Checa, N.; Marco, S.; Michels, A.; Pérez-Navarro, E.; Alberch, J. The neurotrophin receptors trkA, trkB and trkC are differentially regulated after excitotoxic lesion in rat striatum. *Mol. Brain Res.* **1999**, *69*, 242–248. [CrossRef]

135. Duprey-Díaz, M.V.; Soto, I.; Blagburn, J.M.; Blanco, R.E. Changes in brain-derived neurotrophic factor and trkB receptor in the adult Rana pipiens retina and optic tectum after optic nerve injury. *J. Comp. Neurol.* **2002**, *454*, 456–469. [CrossRef]

136. Koliatsos, V.E.; Clatterbuck, R.E.; Winslow, J.W.; Cayouette, M.H.; Prices, D.L. Evidence that brain-derived neurotrophic factor is a trophic factor for motor neurons in vivo. *Neuron* **1993**, *10*, 359–367. [CrossRef]

137. Merlio, J.P.; Ernfors, P.; Jaber, M.; Persson, H. Molecular cloning of rat trkC and distribution of cells expressing messenger RNAs for members of the trk family in the rat central nervous system. *Neuroscience* **1992**, *51*, 513–532. [CrossRef]

138. Henderson, C.E.; Camu, W.; Mettling, C.; Gouin, A.; Poulsen, K.; Karihaloo, M.; Ruilamas, J.; Evans, T.; McMahon, S.B.; Armanini, M.P.; et al. Neurotrophins promote motor neuron survival and are present in embryonic limb bud. *Nature* **1993**, *363*, 266–270. [CrossRef]

139. Piehl, F.; Frisén, J.; Risling, M.; Hökfelt, T.; Cullheim, S. Increased trkB mRNA expression by axotomized motoneurones. *Neuroreport* **1994**, *5*, 697–700. [CrossRef]

140. Davis-López de Carrizosa, M.A.; Morado-Díaz, C.J.; Morcuende, S.; de la Cruz, R.R.; Pastor, A.M. Nerve growth factor regulates the firing patterns and synaptic composition of motoneurons. *J. Neurosci.* **2010**, *30*, 8308–8319. [CrossRef]

141. Benítez-Temiño, B.; Morcuende, S.; Mentis, G.Z.; de la Cruz, R.R.; Pastor, A.M. Expression of Trk receptors in the oculomotor system of the adult cat. *J. Comp. Neurol.* **2004**, *473*, 538–552. [CrossRef]

142. Morcuende, S.; Muñoz-Hernández, R.; Benítez-Temiño, B.; Pastor, A.M.; de la Cruz, R.R. Neuroprotective effects of NGF, BDNF, NT-3 and GDNF on axotomized extraocular motoneurons in neonatal rats. *Neuroscience* **2013**, *250*, 31–48. [CrossRef] [PubMed]

143. Silva-Hucha, S.; Hernández, R.G.; Benítez-Temiño, B.; Pastor, A.M.; de la Cruz, R.R.; Morcuende, S. Extraocular motoneurons of the adult rat show higher levels of vascular endothelial growth factor and its receptor Flk-1 than other cranial motoneurons. *PLoS ONE* **2017**, *12*, e0178616. [CrossRef] [PubMed]

144. McCloskey, D.P.; Hintz, T.M.; Scharfman, H.E. Modulation of vascular endothelial growth factor (VEGF) expression in motor neurons and its electrophysiological effects. *Brain Res. Bull.* **2008**, *76*, 36–44. [CrossRef] [PubMed]

145. Millauer, B.; Wizigmann-Voos, S.; Schnürch, H.; Martinez, R.; Møller, N.P.H.; Risau, W.; Ullrich, A. High affinity VEGF binding and developmental expression suggest Flk-1 as a major regulator of vasculogenesis and angiogenesis. *Cell* **1993**, *72*, 835–846. [CrossRef]

146. Wang, Y.; Duan, W.; Wang, W.; Wen, D.; Liu, Y.; Liu, Y.; Li, Z.; Hu, H.; Lin, H.; Cui, C.; et al. scAAV9-VEGF prolongs the survival of transgenic ALS mice by promoting activation of M2 microglia and the PI3K/Akt pathway. *Brain Res.* **2016**, *1648*, 1–10. [CrossRef] [PubMed]

147. Pettersson, A.; Nagy, J.A.; Brown, L.F.; Sundberg, C.; Morgan, E.; Jungles, S.; Carter, R.; Krieger, J.E.; Manseau, E.J.; Harvey, V.S.; et al. Heterogeneity of the angiogenic response induced in different normal adult tissues by vascular permeability factor/vascular endothelial growth factor. *Lab. Investig.* **2000**, *80*, 99–115. [CrossRef]

148. Hernández, R.G.; Calvo, P.M.; Blumer, R.; de la Cruz, R.R.; Pastor, A.M. Functional diversity of motoneurons in the oculomotor system. *Proc. Natl. Acad. Sci. USA* **2019**, *116*, 3837–3846. [CrossRef]

149. Benítez-Temiño, B.; Davis-López de Carrizosa, M.A.; Morcuende, S.; Matarredona, E.R.; de la Cruz, R.R.; Pastor, A.M. Functional diversity of neurotrophin actions on the oculomotor system. *Int. J. Mol. Sci.* **2016**, *17*, 2016. [CrossRef]

150. Ernfors, P.; Henschen, A.; Olson, L.; Persson, H. Expression of Nerve Growth Factor Receptor mRNA Is Developmentally Regulated and Increased after Axotomy in Rat Spinal Cord Motoneurons. *Neuron* **1989**, *2*, 1605–1613. [CrossRef]

151. Silva-Hucha, S.; Carrero-Rojas, G.; de Sevilla, M.E.F.; Benítez-Temiño, B.; de Carrizosa, M.A.D.L.; Pastor, A.M.; Morcuende, S. Sources and lesion-induced changes of VEGF expression in brainstem motoneurons. *Brain Struct. Funct.* **2020**, *225*, 1033–1053. [CrossRef]

152. Vijayalakshmi, K.; Ostwal, P.; Sumitha, R.; Shruthi, S.; Varghese, A.M.; Mishra, P.; Manohari, S.G.; Sagar, B.C.; Sathyaprabha, T.N.; Nalini, A.; et al. Role of VEGF and VEGFR2 receptor in reversal of ALS-CSF induced degeneration of NSC-34 motor neuron cell line. *Mol. Neurobiol.* **2015**, *51*, 995–1007. [CrossRef] [PubMed]

153. Falkowska, A.; Gutowska, I.; Goschorska, M.; Nowacki, P.; Chlubek, D.; Baranowska-Bosiacka, I. Energy metabolism of the brain, including the cooperation between astrocytes and neurons, especially in the context of glycogen metabolism. *Int. J. Mol. Sci.* **2015**, *16*, 25959–25981. [CrossRef]

154. Chung, W.S.; Clarke, L.E.; Wang, G.X.; Stafford, B.K.; Sher, A.; Chakraborty, C.; Joung, J.; Foo, L.C.; Thompson, A.; Chen, C.; et al. Astrocytes mediate synapse elimination through MEGF10 and MERTK pathways. *Nature* **2013**, *504*, 394–400. [CrossRef] [PubMed]

155. Taylor, J.P.; Brown, R.H.; Cleveland, D.W. Decoding ALS: From genes to mechanism. *Nature* **2016**, *539*, 197–206. [CrossRef] [PubMed]

156. Barbeito, L.H.; Pehar, M.; Cassina, P.; Vargas, M.R.; Peluffo, H.; Viera, L.; Estévez, A.G.; Beckman, J.S. A role for astrocytes in motor neuron loss in amyotrophic lateral sclerosis. *Brain Res. Rev.* **2004**, *47*, 263–274. [CrossRef]

157. Saijo, K.; Glass, C.K. Microglial cell origin and phenotypes in health and disease. *Nat. Rev. Immunol.* **2011**, *11*, 775–787. [CrossRef]

158. Schafer, D.P.; Lehrman, E.K.; Kautzman, A.G.; Koyama, R.; Mardinly, A.R.; Yamasaki, R.; Ransohoff, R.M.; Greenberg, M.E.; Barres, B.A.; Stevens, B. Microglia sculpt postnatal neural circuits in an activity and complement-dependent manner. *Neuron* **2012**, *74*, 691–705. [CrossRef]

159. Perry, V.H.; Nicoll, J.A.R.; Holmes, C. Microglia in neurodegenerative disease. *Nat. Rev. Neurol.* **2010**, *6*, 193–201. [CrossRef]

160. Kreutzberg, G.W. Microglia: A sensor for pathological events in the CNS. *Trends Neurosci.* **1996**, *19*, 312–318. [CrossRef]

161. Krum, J.M.; Rosenstein, J.M. VEGF mRNA and its receptor flt-1 are expressed in reactive astrocytes following neural grafting and tumor cell implantation in the adult CNS. *Exp. Neurol.* **1998**, *154*, 57–65. [CrossRef] [PubMed]

162. Krum, J.M.; Mani, N.; Rosenstein, J.M. Roles of the endogenous VEGF receptors flt-1 and flk-1 in astroglial and vascular remodeling after brain injury. *Exp. Neurol.* **2008**, *212*, 108–117. [CrossRef] [PubMed]

163. Lennmyr, F.; Ata, K.A.; Funa, K.; Olsoon, Y.; Terént, A. Expression of vascular endothelial growth factor (VEGF) and its receptors (Flt-1 and Flk-1) following permanent and transient occlusion of the middle cerebral artery in the rat. *J. Neuropathol. Exp. Neurol.* **1998**, *57*, 874–882. [CrossRef] [PubMed]

164. Acosta, L.; Morcuende, S.; Silva-Hucha, S.; Pastor, A.M.; de la Cruz, R.R. Vascular endothelial growth factor (VEGF) prevents the downregulation of the cholinergic phenotype in axotomized motoneurons of the adult rat. *Front. Mol. Neurosci.* **2018**, *11*, 241. [CrossRef] [PubMed]

165. Büttner, U.; Büttner-Ennever, J.A. Present concepts of oculomotor organization. *Prog. Brain Res.* **2006**, *151*, 1–42.

166. Nijssen, J.; Comley, L.H.; Hedlund, E. Motor neuron vulnerability and resistance in amyotrophic lateral sclerosis. *Acta Neuropathol.* **2017**, *133*, 863–885. [CrossRef]

167. Zimmermann, L.; Morado-Díaz, C.J.; Carrizosa, M.A.D.L.; de la Cruz, R.R.; May, P.J.; Streicher, J.; Pastor, A.M.; Blumer, R. Axons giving rise to the palisade endings of feline extraocular muscles display motor features. *J. Neurosci.* **2013**, *33*, 2784–2793. [CrossRef]

168. Jacoby, J.; Ko, K.; Weiss, C.; Rushbrook, J.I. Systematic variation in myosin expression along extraocular muscle fibres of the adult rat. *J. Muscle Res. Cell Motil.* **1990**, *11*, 25–40. [CrossRef]

169. Valdez, G.; Tapia, J.C.; Lichtman, J.W.; Fox, M.A.; Sanes, J.R. Shared resistance to aging and als in neuromuscular junctions of specific muscles. *PLoS ONE* **2012**, *7*, e34640. [CrossRef]

170. Tjust, A.E.; Danielsson, A.; Andersen, P.M.; Brännström, T.; Domellöf, F.P. Impact of amyotrophic lateral sclerosis on slow tonic myofiber composition in human extraocular muscles. *Investig. Ophthalmol. Vis. Sci.* **2017**, *58*, 3708–3715. [CrossRef]

171. Heerssen, H.M.; Segal, R.A. Location, location, location: A spatial view of neurotrophin signal transduction. *Trends Neurosci.* **2002**, *25*, 160–165. [CrossRef]

172. Hernández, R.G.; Silva-Hucha, S.; Morcuende, S.; de la Cruz, R.R.; Pastor, A.M.; Benítez-Temiño, B. Extraocular motor system exhibits a higher expression of neurotrophins when compared with other brainstem motor systems. *Front. Neurosci.* **2017**, *11*, 399. [CrossRef] [PubMed]

173. Calvo, P.M.; de la Cruz, R.R.; Pastor, A.M. Synaptic loss and firing alterations in Axotomized Motoneurons are restored by vascular endothelial growth factor (VEGF) and VEGF-B. *Exp. Neurol.* **2018**, *304*, 67–81. [CrossRef] [PubMed]

174. Gao, L.; Zhou, S.; Cai, H.; Gong, Z.; Zang, D. VEGF levels in CSF and serum in mild ALS patients. *J. Neurol. Sci.* **2014**, *346*, 216–220. [CrossRef] [PubMed]

175. Nicoletti, J.N.; Shah, S.K.; McCloskey, D.P.; Goodman, J.H.; Elkady, A.; Atassi, H.; Hylton, D.; Rudge, J.S.; Scharfman, H.E.; Croll, S.D. Vascular endothelial growth factor is up-regulated after status epilepticus and protects against seizure-induced neuronal loss in hippocampus. *Neuroscience* **2008**, *151*, 232–241. [CrossRef]

176. Sun, Y.; Jin, K.; Xie, L.; Childs, J.; Mao, X.O.; Logvinova, A.; Greenberg, D.A. VEGF-induced neuroprotection, neurogenesis, and angiogenesis after focal cerebral ischemia. *J. Clin. Investig.* **2003**, *111*, 1843–1851. [CrossRef]

177. Zheng, C.; Nennesmo, I.; Fadeel, B.; Henter, J.I. Vascular endothelial growth factor prolongs survival in a transgenic mouse model of ALS. *Ann. Neurol.* **2004**, *56*, 564–567. [CrossRef]

178. Krakora, D.; Mulcrone, P.; Meyer, M.; Lewis, C.; Bernau, K.; Gowing, G.; Zimprich, C.; Aebischer, P.; Svendsen, C.N.; Suzuki, M. Synergistic effects of GDNF and VEGF on lifespan and disease progression in a familial ALS rat model. *Mol. Ther.* **2013**, *21*, 1602–1610. [CrossRef]

179. Kobayashi, N.R.; Bedard, A.M.; Hincke, M.T.; Tetzlaff, W. Increased expression of BDNF and trkB mRNA in rat facial motoneurons after axotomy. *Eur. J. Neurosci.* **1996**, *8*, 1018–1029. [CrossRef]

180. Nagata, Y.; Okuya, M.; Watanabe, R.; Honda, M. Regional distribution of cholinergic neurons in human spinal cord transections in the patients with and without motor neuron disease. *Brain Res.* **1982**, *244*, 223–229. [CrossRef]

181. Virgo, L.; de Belleroche, J.; Rossi, M.; Steiner, T.J. Characterisation of the distribution of choline acetyltransferase messenger RNA in human spinal cord and its depletion in motor neurone disease. *J. Neurol. Sci.* **1992**, *112*, 126–132. [CrossRef]

182. Oda, Y.; Imai, S.; Nakanishi, I.; Ichikawa, T.; Deguchi, T. Immunohistochemical study on choline acetyltransferase in the spinal cord of patients with amyotrophic lateral sclerosis. *Pathol. Int.* **1995**, *45*, 933–939. [CrossRef] [PubMed]

183. Kato, T. Choline acetyltransferase activities in single spinal motor neurons from patients with amyotrophic lateral sclerosis. *J. Neurochem.* **1989**, *52*, 636–640. [CrossRef]

184. Oda, Y. Choline acetyltransferase: The structure, distribution and pathologic changes in the central nervous system. *Pathol. Int.* **1999**, *49*, 921–937. [CrossRef] [PubMed]

185. Morcuende, S.; Benítez-Temiño, B.; Pecero, M.L.; Pastor, A.M.; de la Cruz, R.R. Abducens internuclear neurons depend on their target motoneurons for survival during early postnatal development. *Exp. Neurol.* **2005**, *195*, 244–256. [CrossRef]

186. Davis-López de Carrizosa, M.A.; Morado-Díaz, C.J.; Tena, J.J.; Benítez-Temiño, B.; Pecero, M.L.; Morcuende, S.; de la Cruz, R.R.; Pastor, A.M. Complementary actions of BDNF and neurotrophin-3 on the firing patterns and synaptic composition of motoneurons. *J. Neurosci.* **2009**, *29*, 575–587. [CrossRef]

187. de la Cruz, R.R.; Delgado-García, J.M.; Pastor, A.M. Discharge characteristics of axotomized abducens internuclear neurons in the adult cat. *J. Comp. Neurol.* **2000**, *427*, 391–404. [CrossRef]

188. Pastor, A.M.; Delgado-García, J.M.; Martínez-Guijarro, F.J.; López-García, C.; de la Cruz, R.R. Response of abducens internuclear neurons to axotomy in the adult cat. *J. Comp. Neurol.* **2000**, *427*, 370–390. [CrossRef]

189. de la Cruz, R.R.; Benítez-Temiño, B.; Pastor, A.M. Intrinsic determinants of synaptic phenotype: An experimental study of abducens internuclear neurons connecting with anomalous targets. *Neuroscience* **2002**, *112*, 759–771. [CrossRef]

190. Lams, B.E.E.; Isacson, O.; Sofroniew, M.V. Loss of transmitter-associated enzyme staining following axotomy does not indicate death of brainstem cholinergic neurons. *Brain Res.* **1988**, *475*, 401–406. [CrossRef]

191. Matsuura, J.; Ajiki, K.; Ichikawa, T.; Misawa, H. Changes of expression levels of choline acetyltransferase and vesicular acetylcholine transporter mRNAs after transection of the hypoglossal nerve in adult rats. *Neurosci. Lett.* **1997**, *236*, 95–98. [CrossRef]

192. Wang, W.; Salvaterra, P.M.; Loera, S.; Chiu, A.Y. Brain-derived neurotrophic factor spares choline acetyltransferase mRNA following axotomy of motor neurons in vivo. *J. Neurosci. Res.* **1997**, *47*, 134–143. [CrossRef]

193. Morado-Díaz, C.J.; Matarredona, E.R.; Morcuende, S.; Talaverón, R.; Davis-López de Carrizosa, M.A.; de la Cruz, R.R.; Pastor, A.M. Neural progenitor cell implants in the lesioned medial longitudinal fascicle of adult cats regulate synaptic composition and firing properties of abducens internuclear neurons. *J. Neurosci.* **2014**, *34*, 7007–7017. [CrossRef]

194. Lambrechts, D.; Carmeliet, P. VEGF at the neurovascular interface: Therapeutic implications for motor neuron disease. *Biochim. Biophys. Acta* **2006**, *1762*, 1109–1121. [CrossRef] [PubMed]

195. Tovar-y-Romo, L.B.; Ramírez-Jarquín, U.N.; Lazo-Gómez, R.; Tapia, R. Trophic factors as modulators of motor neuron physiology and survival: Implications for ALS therapy. *Front. Cell. Neurosci.* **2014**, *8*, 61. [CrossRef] [PubMed]

196. Dodge, J.C.; Treleaven, C.M.; Fidler, J.A.; Hester, M.; Haidet, A.; Handy, C.; Rao, M.; Eagle, A.; Matthews, J.C.; Taksir, T.V.; et al. AAV4-mediated expression of IGF-1 and VEGF within cellular components of the ventricular system improves survival outcome in familial ALS mice. *Mol. Ther.* **2010**, *18*, 2075–2084. [CrossRef] [PubMed]

International Journal of
Molecular Sciences

Review

Neuroprotective Effect of Antioxidants in the Brain

Kyung Hee Lee [1], Myeounghoon Cha [2] and Bae Hwan Lee [2,3,*]

[1] Department of Dental Hygiene, Division of Health Science, Dongseo University, Busan 47011, Korea; kyhee@dongseo.ac.kr
[2] Department of Physiology, Yonsei University College of Medicine, Seoul 03722, Korea; mhcha@yuhs.ac
[3] Brain Korea 21 PLUS Project for Medical Science, Yonsei University College of Medicine, Seoul 03722, Korea
* Correspondence: bhlee@yuhs.ac; Tel.: +82-2-2228-1711

Received: 31 August 2020; Accepted: 23 September 2020; Published: 28 September 2020

Abstract: The brain is vulnerable to excessive oxidative insults because of its abundant lipid content, high energy requirements, and weak antioxidant capacity. Reactive oxygen species (ROS) increase susceptibility to neuronal damage and functional deficits, via oxidative changes in the brain in neurodegenerative diseases. Overabundance and abnormal levels of ROS and/or overload of metals are regulated by cellular defense mechanisms, intracellular signaling, and physiological functions of antioxidants in the brain. Single and/or complex antioxidant compounds targeting oxidative stress, redox metals, and neuronal cell death have been evaluated in multiple preclinical and clinical trials as a complementary therapeutic strategy for combating oxidative stress associated with neurodegenerative diseases. Herein, we present a general analysis and overview of various antioxidants and suggest potential courses of antioxidant treatments for the neuroprotection of the brain from oxidative injury. This review focuses on enzymatic and non-enzymatic antioxidant mechanisms in the brain and examines the relative advantages and methodological concerns when assessing antioxidant compounds for the treatment of neurodegenerative disorders.

Keywords: neuroprotection; antioxidant; brain; neurodegenerative disease; oxidative stress

---

1. Introduction

Oxidative stress is considered as a detrimental condition for normal brain functioning. Since the brain uses chemically diverse reactive species for signal transmission, it is susceptible to oxidative stress [1]. The human brain consists of more than 86 billion neurons and over 250–300 billion glial +cells, which consume over 20% of the total basal oxygen [2,3]. The mitochondria in the brain use the inspired basal oxygen to reduce O_2 to H_2O to support adenosine triphosphate (ATP) synthesis [4]. The electron transport chain in the mitochondria is not efficient and leakage of reactive oxygen species (ROS) from the mitochondria occurs when there is excess of oxygen [5]. Redox signaling in the brain functions as an intrinsic sensor to oxidative stress, when signals go awry during pathological conditions. Brain tissue has a high rate of oxidative metabolic activity, intense production of reactive oxygen metabolites, relatively low levels of antioxidants, low repair capacity, non-replicating neuronal cells, and a high membrane surface to cytoplasm ratio [6]. The cellular antioxidant system that prevents tissue damage is composed of endogenous and exogenous antioxidants that have the ability to reduce different chemicals. In biological antioxidant defense systems, both endogenous and exogenous antioxidants are classified into enzymatic antioxidants and non-enzymatic antioxidants, including oxidative enzyme inhibitors, antioxidant enzyme cofactors, ROS/RNS scavengers, and transition metal chelators [7,8]. These two main antioxidant systems play an important role in maintaining the balance between pro-oxidant and antioxidant agents in the brain and in mitigating oxidative stress [9]. The present review focuses on enzymatic and non-enzymatic antioxidant mechanisms in the brain and examines the relative advantages of antioxidant compounds for the treatment of various brain diseases. Therefore, this review will describe selective antioxidants shown to have neuroprotective effects that limit neurodegenerative disease.

2. Factors That Contribute to Vulnerability of the Brain to Oxidative Stress

Free radicals or other reactive molecules disturb cellular energy metabolism and cause oxidative stress. Endogenous redox imbalance by pro-oxidant and antioxidant agents occurs as a result of free radicals, which play an important role in oxidative stress, cell death, and tissue damage. Increased free radical production due to excess pro-oxidant mechanisms can react with lipids, proteins, nucleic acids, and other biomolecules which can alter their structure and function. A high concentration of polyunsaturated fatty acids exists in membrane lipids in the brain [10]. These are sources of decomposition reaction in the form of lipid peroxidation, in which a single initiating free radical can precipitate the destruction of adjacent molecules. Polyunsaturated fatty acids especially serve as major biological targets for oxidative damage induced by ROS. Other candidate molecules that serve as biological targets of free radicals are nucleic acids. Breaks in DNA or modified bases can result in aberrant gene expression and cell death [11]. Moreover, free radicals can also oxidize the backbone and side chains of proteins, thereby disrupting the function of enzymes, receptors, neurotransmitters, and structural proteins by oxidative modification [12]. Highly reactive oxygen radicals are produced in the brain as a result of interactions between various transition metals and their reducing equivalents. Transition metals including iron, copper, zinc, and manganese, are associated with increased free radical production by the Fenton reaction [13,14]. Since the brain has abundant lipid content, high energy requirements, and weak antioxidant capacity, it is an easy target for excessive oxidative insults. Specifically, ROS increase susceptibility to neuronal damage and functional decline via brain oxidation in Alzheimer's disease (AD), Parkinson's disease (PD), amyotrophic lateral sclerosis (ALS), cerebrovascular disorders, psychiatric disorders, and other neurodegenerative diseases [6,13,15] (Figure 1).

Considering a detailed perspective on signaling, the rationale underlying the vulnerability of the neurons in the brain to oxidative stress are as follows: In redox signaling, O_2^-/H_2O_2 generation by the reduced form of nicotinamide adenine dinucleotide phosphate (NADPH) oxidase 2 induces the activation of signaling proteins via sulfenic acid formation [16,17]. The brain is dependent on Ca^{2+} signaling for synaptic plasticity, a fundamental brain function, and it expends considerable amounts of ATP to maintain intracellular Ca^{2+} homeostasis. Synaptic terminal glutamate-induced Ca^{2+} transients activate neuronal nitric oxide synthase (nNOS)-mediated $NO^{\cdot}$ generation, and additionally, Ca^{2+} overloaded in the mitochondria induces O_2^-/H_2O_2 generation which can lead to $ONOO^{\cdot}$ production and excitotoxicity [18–20]. Particularly, the mitochondria generate O_2^- in complex I and complex III and the monoamine oxidase enzyme isoform catalyzes H_2O_2 generation during metabolism [21–23]. The human brain consumes over 25% circulating glucose to support neuronal function; however, protein inactivation by the formation of advanced end glycation products (AGE) due to decreased glycolytic rates can cause glucose-induced oxidative stress [24,25]. Neurotransmitters with a catechol group render the brain sensitive to oxidative stress. Redox-active transition metals catalyze the auto-oxidation of dopamine to a semiquinone radical during neurotransmitter oxidation [26–28]. The mature neurons of the brain have abundant and multiple mechanisms that promote long-term neuronal survival and prevent cell death; however, the brain has a comparatively weak endogenous antioxidant defense system relative to other tissues and is susceptible to imbalance in redox homeostasis. For instance, neurons with 50 times lower catalase content than hepatocytes have constrained glutathione peroxidase 4 (GPX4) activity (owing to low glutathione [GSH] content) and a modest antioxidant defense mechanism [29–31]. Microglia monitor neuronal activity for removal of unhealthy cells, neuronal wiring during development, and activity dependent synaptic plasticity. They generate O_2^- via NADPH oxidase 2 (NOX2) within an end-foot process, thus influencing brain oxidation [32,33]. Redox-active transition metals, such as Fe^{2+} and Cu^+, are enriched in the brain and therefore contribute to ferroptosis by catalyzing peroxyl ($ROO^{\cdot}$) and alkoxyl ($RO^{\cdot}$) radical generation [34,35]. Lipid peroxidation, which involves initiation, oxygenation, propagation, and termination, occurs within the neuronal cell membrane [36]. The brain uses nNOS and NOX isoforms to maintain essential functions. However, nNOS/NOX expression is associated with NO˙ generation, which can be spatially co-generated while producing ONOO˙ [37,38]. In summary, these interconnected, myriad factors render the brain vulnerable to oxidative stress. Furthermore, excessive production of reactive species and insufficient activity of antioxidant defense systems have been implicated in the pathogenesis of neurodegeneration.

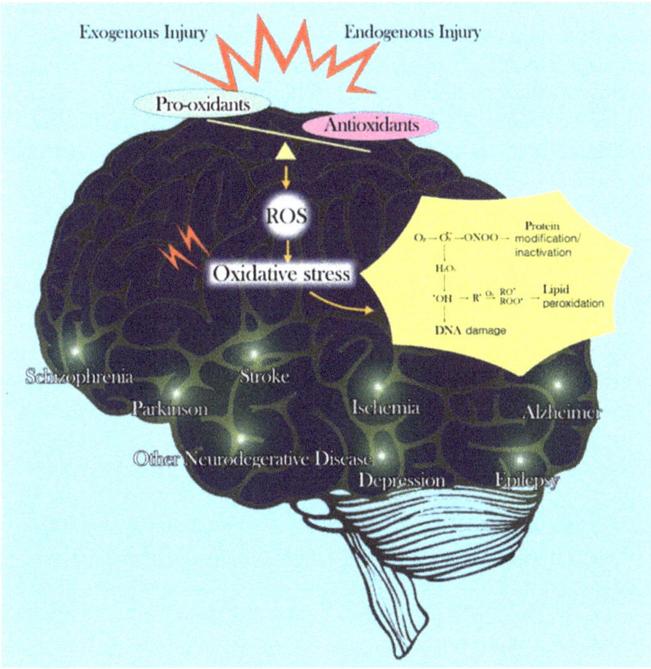

Figure 1. Development of various diseases by the pathophysiology of oxidative stress in the brain. Balance and imbalance between pro-oxidants and antioxidants against reactive oxygen species production induce oxidative stress and are consequently involved in neuronal damage resulting in the neurodegenerative diseases.

3. Role of Antioxidants in the Brain

Antioxidants act in two ways: First, the antioxidant defense systems, which are initiated against oxidative damage, prevent generation of ROS, and block and capture the generated radicals [39]. These systems are present in the aqueous and membrane compartments of cells; they can be enzymatic or non-enzymatic. The first line of defense has been identified as an enzymatic antioxidant system that includes superoxide dismutase, catalase, and glutathione peroxidase. The second line of defense is represented by reduced thiol and non-enzymatic antioxidants, including hydro- and lipo-soluble or metabolic compounds [7]. Second, during the antioxidant repair process, the removal of damaged biomolecules before their aggregation causes alterations in cell metabolism. The intervention by the repair systems consists of repairing oxidatively damaged nucleic acids by specific enzymes, removing oxidized proteins by the proteolytic systems, and repairing oxidized lipids by phospholipases, peroxidases, or acryl transferases [8,39,40].

Antioxidants act to directly scavenge oxidizing radicals and regenerate oxidized biomolecules in organisms. Neuronal cells possess several factors that minimize oxidative damage and a complex antioxidant system consisting of various antioxidant enzymes and non-enzymes. With regard to the mechanism of action of antioxidants, antioxidants respond to different steps of oxidative radical processing and can be considered by taking into account the different steps of lipid peroxidation in cell membranes such as: initiation, propagation, and chain termination [36]. The first step of initiation of lipid peroxidation can be promoted by diverse exogenous factors, endogenous enzyme systems, and the electron transport chain in the mitochondria. The propagation step of peroxidation is initiated by the addition of oxygen to carbon-centered radicals, which are generated at, or near, the diffusion-controlled rate. Propagation is represented by the transfer of a hydrogen atom to the chain carrying peroxyl radicals and occurs at a relatively slow rate. Antioxidants can inhibit lipid peroxidation by removing molecular oxygen or by decreasing the local concentration of molecular oxygen, removing pro-oxidative metal ions, trapping aggressive ROS, scavenging chain initiating radicals, breaking the chain of free radical reactions, and quenching singlet oxygen [7,41]. In its protective role, enzymatic antioxidants are chain-breaking antioxidants which are able to scavenge radical species while non-enzymatic antioxidants are singlet oxygen quenchers, metal chelators, oxidative enzyme inhibitors, peroxide decomposers, and/or ultra violet (UV) radiation absorbers [7]. Additionally, antioxidant enzymes can catalyze the synthesis or regeneration of non-enzymatic antioxidants.

3.1. Enzymatic Antioxidants

3.1.1. Superoxide Dismutase

Superoxide dismutase (SOD), which belongs to the enzymatic defense system, changes the superoxide radical anion to H_2O_2 by oxidative decay. Another function of SOD is to protect dehydratase from free radical superoxide inactivation. There are three forms of SOD in humans. SOD-1, a copper-zinc containing SOD (Cu, Zn-SOD), is present in the cytosol and specifically catalyzes dismutation in a pH-independent medium [42]. SOD-1 protein is a homodimer composed of eight antiparallel beta strands and two metal atoms that catalyze the conversion of toxic O_2^- anions to H_2O_2 and O_2. Specifically, copper minerals play a crucial role in the catalytic function of this enzyme, while zinc is important for structural integrity [43]. SOD-1 plays an important role as a first line of defense by detoxifying superoxide radicals. In addition, the lack of this enzyme exhibited a pronounced susceptibility to paraquat toxicity. SOD-2, a manganese containing SOD (Mn-SOD), is found in the mitochondrial matrix and reduces the superoxide radical anion generated in the electron transport chain [44]. A detailed analysis revealed that the SOD-2 subunit needs one metal atom and that it is functionally disabled in the absence of Mn atoms in the active site. SOD-2 attracts O_2^- and turns positive on the active side. The active metal provides electrons directly to O_2^- then reduces one molecule of O_2^- and one proton to be converted to form H_2O_2 (Figure 2). The amount of SOD-2 is different in each cell type, depending on the number of mitochondria. Specifically, the presence of SOD-2 is essential for the survival of aerobic cells and the development of cellular resistance to oxidative stress. SOD-3, an extracellular SOD (EC-SOD), also contains copper and zinc in its structure, is synthesized inside the cell and secreted into the extracellular matrix [45]. SOD-3 enzyme is not induced by its substrate or other oxidants and is regulated to coordinate with cytokines, rather than as a response of individual cells to oxidants [46].

In neurodegenerative diseases, SOD-1 and SOD-2 genes appear to be disrupted and/or mutated. SOD-1 gene mutations produce diverse cellular changes such as alteration of gene expression, unusual protein interactions, caspase activation, mitochondrial dysfunction, and cytoskeletal abnormalities in familial ALS [47]. Overexpression of SOD-1 protects neurons against the neurotoxic effects of amyloid beta (Aβ), and loss of SOD-1 accelerates aging-related pathologies and reduces lifespan in mice [48–50]. SOD-1 deficiency in an AD model showed that it was associated with augmented Aβ oligomerization and memory impairment mediated by oxidative stress [51]. Overexpression of SOD-2 by treatment with sodium orthovanadate could rescue synaptic failure and neuronal cell death in the hippocampus after kainic acid (KA)-induced oxidative stress; moreover, an increase in SOD-2 protected the neuronal death from oxidative injury [52]. Impaired SOD-2 is a common potential pathogenesis related to oxidative stress in PD and AD [6]. In human amyloid precursor protein (hAPP) transgenic mice with over 50% reduced SOD-2 activity, increased SOD-2 protects the age-related brain against hAPP/Aβ-induced impairments [53]. Extracellularly administered SOD was effective in inhibiting cell death and restoring healthy mitochondrial morphology in a monosodium glutamate-induced excitotoxicity disease model [54]. An increase in SOD-2 can be expected to assist or improve neuronal function and vascular pathology in AD as a therapeutic effect.

3.1.2. Catalase

Catalase (CAT), a heme-containing tetrameric protein is a common antioxidant enzyme naturally produced by the body when exposed to oxygen. H_2O_2 generated inside the cell is enzymatically catabolized in aerobic organisms by catalase and the activity of several peroxidases. CAT is one of the most efficient enzymes that cannot be saturated by H_2O_2 at any concentration. It reacts with H_2O_2 to form water/alcohol and oxygen using Fe as a cofactor [55,56]. CAT protects cells by detoxification of the generated H_2O_2 and plays an important role in the acquisition of tolerance to oxidative stress as an adaptive response [57]. CAT can maintain the concentration of O_2 either for repeated rounds of chemical reduction or for direct interaction with the toxin [58] (Figure 2). Furthermore, inhibition of CAT activity results in enhanced cytotoxicity and increased ROS, indicating an important role of CAT in maintaining the oxidative balance [59]. Specifically, mislocalized CAT is associated with accumulation of H_2O_2 [60] and other ROS in the cells leading to compromised neurological function [61]. Thus, deficiency or malfunction of CAT is postulated to be related to the pathogenesis of many age-associated degenerative diseases.

CAT treatment reduces H_2O_2 levels and improves neuronal survival following Aβ-induced toxicity in neuronal culture [62]. In Aβ toxicity, CAT–SKL (serine-lysine-leucine) treatment reduced the pathology of microglial activation, and rat brains did not show long-term memory impairments via the reduction of H_2O_2 levels [63]. In the PD model, mutant α-synuclein inhibited the expression and activity of CAT and induced low catalase activity and high H_2O_2 production [64]. This antioxidant enzyme plays an important role in maintaining oxidative balance.

3.1.3. Glutathione Peroxidase

Glutathione peroxidase (GPx) catalyzes the reduction of a variety of hydroperoxides (ROOH and H_2O_2) to water or the corresponding alcohols using GSH to protect mammalian cells against oxidative damage [65] (Figure 2). GPx has four identical subunits containing one selenocysteine in each residue for essential enzyme activity, which are classified as selenium-containing GPxs (GPx1–4 and 6) and their non-selenium congeners (GPx5, 7, and 8) [30]. The glutathione redox cycle is mostly active during low levels of oxidant stress, while CAT is more significant in protecting against severe oxidant stress. GPx isoenzymes appear to have antioxidant functions at different locations and cellular compartments, and the levels of expression of each isoform vary depending on the type of tissue, even though their expression is omnipresent. Furthermore, GPx is responsible for the conversion of GSH to oxidized glutathione disulfide (GSSG) and glutathione reductase (GR) reduces GSSG back to GSH [66]. GPx1 and GPx4 are found in most tissues and the predominant forms of GPx are found in brain tissue. GPx1 reduces H_2O_2 and organic hydroperoxides and is expressed both in neurons and astrocytes [67]. GPx4, the phospholipid hydroperoxide glutathione peroxidase, is located in the cytosol and a membrane fraction that can directly reduce phospholipid hydroperoxides, fatty acid hydroperoxides, and cholesterol hydroperoxides [68]. In the brain, mitochondrial and cytosolic GPx4 isoforms have been detected in neurons of the cerebral cortex, hippocampus, and cerebellum. In contrast, GPx4 in glial cells is barely activated under normal physiological conditions, whereas reactive astrocytes upregulate the expression of GPx4 following selective brain injury [31]. Recently, GPx4 has become well known as a key regulatory factor in ferroptosis which is a non-apoptotic and iron-dependent programmed cell death pathway that causes a rapid elevation of oxidative stress [69].

As the function of GPx is related to normal development and cellular metabolism via the regulation of oxidative stress, GPx4 function is potentially the key for cell survival.

GPx activity decreased significantly in PC12 cells during oxygen-glucose deprivation [70], and a two-fold increase in GPx and a three-fold increase in CAT activities were observed during a time course that reflected the temporal increase in nerve growth factor (NGF) in a brain contusion model [71]. The ferroptotic potential of neurons in the forebrain regions of the hippocampus and cerebral cortex is severely affected in AD patients, with increased vulnerability to ferroptosis. The role of GPx4 as a key regulator of ferroptosis was observed in Gpx4BIKO transgenic mice, a mouse model with a conditional deletion of GPx4 in forebrain neurons. Gpx4BIKO transgenic mice showed significant deficits in spatial learning and memory function and exhibited hippocampal neurodegeneration compared to control mice [72]. In human PD tissue biopsies, GPx4 expression was upregulated and ferroptosis-related events were observed [73].

3.1.4. Thioredoxin

The thioredoxin (Trx) system consists of two types of antioxidant oxidoreductase proteins, Trx and thioredoxin reductase (TrxRs) with NADPH as an electron donor. Trx has a conserved active site such as Cys-Pro-Gly-Cys, that has an important function of acting as an active oxidoreductase and electron donor of some peroxiredoxins (Prx) which are crucial for the reduction of peroxides [74]. Trx is an important regulator of cellular function that can respond to redox balance by modulating signaling pathways, transcription factors, and immunological responses for cell survival in many conditions, including neurodegenerative diseases [75]. There are three isoforms in the Trx family such as Trx1 in the cytosol, Trx2 in the mitochondria, and a testis-specific Trx3 in mammalian cells. Trx1 exists in cell compartments such as the nucleus and the plasma membrane or as a secreted protein, depending on its localization and function in different cell types [76]. Trx1 is coupled with Prx 1/2 and methionine sulfoxide reductase and is essential role in the control of growth and apoptosis. TrxR 1 and Trx 1 are observed in neuronal synaptic vesicles. In addition, Trx 1 and 2 expressions are also present mainly in rat brain neurons. Neurons with mitochondrial dysfunction by complex IV inhibition show low levels of Trx and are thus, more vulnerable to H_2O_2. Reduced Trx is a powerful reductase that acts through a disulfide–dithiol exchange mechanism to catalyze the conversion of disulfide bonds into thiols with high efficiency [77]. The disulfides in the oxidized Trx are converted to thiols by the consumption of NADPH [78] (Figure 2). TrxR, a homodimer, catalyzes the reduction of the disulfide at the Trx active site and is encoded by three distinct genes: the cytosolic TrxR (*TrxR1*), mitochondrial TrxR (*TrxR2*), and thioredoxin-glutaredoxin reductase (*TrxR3*). TrxR can directly reduce substrates such as peroxides, including lipid hydroperoxides, H_2O_2, and protein disulfide isomerase [76,79]. TrxR is also involved in the regeneration of other antioxidant molecules such as dehydroascorbate, lipoic acid, and ubiquinone [80–82]. TrxR is very important when other selenoproteins including GPx lose most of their activity and it effectively donates electrons during DNA synthesis [83]. TrxR activation in the rat brain can be maintained at a certain level under severe selenium (Se)-deficient conditions. Both Trx and TrxR are widely expressed in tissues and organs, including the brain. In a central nervous system (CNS) study, it has been suggested that increased expression of Trx and TrxR is closely associated with cell damage due to oxidative stress [84,85]. In addition, NADPH and human TrxR are efficient electron donors to human plasma peroxidase and reduce hyperoxides, even with low GSH capacity [86].

The upregulation of Trx was observed as a neuroprotective effect on retinal ganglion cells against oxidative stress-induced neurodegeneration [87]. Overexpression of Trx1 attenuated endoplasmic reticulum stress by regulating the activation of the molecular mechanism for neuroprotection in an in vitro and in vivo model of PD [88]. In AD brains, TrxR activity was increased in the cerebellum and amygdala [89]. It has been implied that TrxR activation increases a compensatory mechanism during increased oxidative stress that is limited by the substrate Trx and the subsequent neurodegeneration seen in AD. Another evidence for the peripheral response to oxidative stress during neurodegeneration is the reduction of Trx1 and TrxR1 in the plasma and erythrocytes in blood samples from patients with Huntington's disease (HD) [90]. Exogenously administered human recombinant Trx attenuates the generation of ROS involved in cytotoxic mechanisms, ameliorates neuronal damage, and augments neurogenesis following brain ischemia/reperfusion injury in rats [91]. In the ischemic brain, administration of Prx3 and Trx2 shows substantial neuroprotective effects by reducing oxidative stress [92].

Important enzymatic antioxidants such as SOD, CAT, GPx, and Trx remove superoxide and peroxides before they react with metal catalysts to form more reactive species (Figure 2). Peroxidative chain reactions initiated by reactive species that escape enzymatic degradation are terminated via chain-breaking antioxidants, including water-soluble ascorbate, lipid-soluble vitamin E, and ubiquinone. To enhance the antioxidative effect, oxidative stress should be controlled by supplying all the antioxidant nutrients and by minimizing the effect of substances that stimulate reactive oxygen metabolites (ROM) [93].

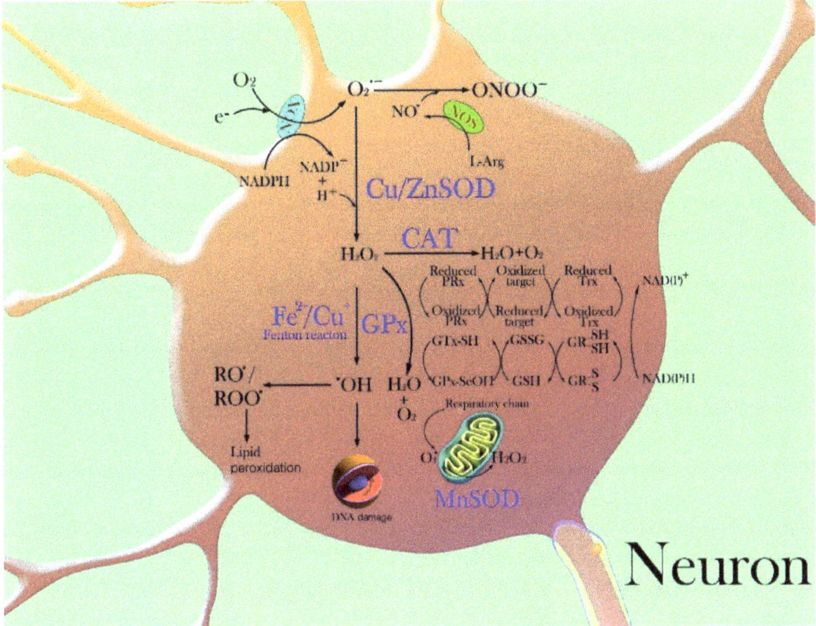

Figure 2. Enzymatic antioxidant defense system against the production of reactive oxygen species (ROS). Various pathways of cell death caused by ROS and its transformation are observed in brain injury. NADPH oxidase and mitochondrial respiratory transport chain are known as major cellular sources of the superoxide radical anion (O_2^-). The superoxide radical anion reacts with nitric oxide (NO) to form the peroxynitrite anion ($ONOO^-$) which mediates oxidative modification of protein residues via an interaction with NO. The superoxide radical is dismuted by the superoxide dismutase enzyme (SOD) to form hydrogen peroxide. In addition, manganese containing superoxide dismutase (Mn-SOD) reduces the superoxide radical anion generated during the electron transport chain in the mitochondrial matrix. Catalase (CAT) and/or glutathione peroxidase (GPx) decomposes hydrogen peroxide to water and oxygen by enzymatic reactions. Hydrogen peroxide is decomposed into reactive hydroxyl radicals by reaction with catalytically active redox metals such as (copper and iron). Hydroxyl radicals can react with oxygen to form peroxyl or alkoxyl radicals which can lead to lipid peroxidation and react with DNA and primarily cause damage to DNA. Peroxiredoxins (Prx) and thioredoxin (Trx) act as redox-regulated proteins to additional redox relay bases.

3.2. Non-Enzymatic Antioxidants

3.2.1. Antioxidant Enzyme Cofactors (Selenium, Coenzyme Q10, Zinc)

Selenium

Selenium (Se) is an enzymatic cofactor and an essential component of selenoaminoacids and selenoproteins. Se has numerous biological functions related to redox signaling, antioxidant defense systems, thyroid hormone metabolism, and humoral and cell-mediated immune responses [94]. Se reacts with redox signals involving one- or two-electron transitions; however, it is less reactive and prefers higher oxidation states. The biological activity of Se varies depending on its concentration: normal growth and development at low concentrations, homeostatic function at moderate concentrations, and induced toxicity at high concentrations [95]. The complex biological functions of Se are found in general body proteins that mainly and effectively act through 25 selenoproteins that have selenocysteine at their active center in humans [96]. Over half of the selenoproteins exhibit antioxidant activity, of which the GPx family and Trx reductase family are well known. Selenoproteins constrain the activation of nuclear factor-κB (NF-κB) via redox signaling, which prohibits a cytokine storm and the formation of reactive oxygen and nitrogen species. Se is essential for brain function, but the Se expression level in the brain is rather poor compared to other tissues [97,98]. The distribution of Se in the human brain enriched in gray matter tended to be higher, while the white matter was found to have reduced Se levels. Specifically, Se level was observed to be highest in the hippocampus, cerebellum, and brainstem in the rat brain [99]. The mechanism of Se neuroprotection is attributed to modulation of Ca^{2+} influx via ion channels, anti-inflammatory effect by abrogation of microglia invasion, and biosynthesis stimulation of antioxidative selenoproteins in the brain [100,101].

Se is essential for the brain and plays an important role in the pathology of neuronal disorders such as AD, PD, ALS, and epilepsy [96]. In Se-deficient transgenic mice, Aβ plaques showed more than a two-fold increase compared to Se-adequate ones [102]. Selenite administration in a rat AD model also showed attenuation of cognitive deficits, oxidative damage and morphological changes in the hippocampus and cerebral cortex [103]. In a paraquat induced PD model, bradykinesia and DNA damage reduction were observed by supplementing Se with selenite in drinking water [104]. Alterations of Se homeostasis in the HD model have been observed and a beneficial effect of selenite supplementation in mice expressing mutant huntingtin has been reported. Additionally, autopsy of the human HD and mouse HD brain showed decreased Se content [105]. Selenite administration prevents secondary pathological events, thus, contributing to the reduction of apoptotic cell death and prevention of neuronal destruction in the cortex and hippocampus of traumatic brain injury [106,107]

Coenzyme Q

Coenzyme Q (CoQ) is a remarkable liposoluble ubiquinone with a long isoprenoid side chain and a main component of the internal mitochondria membrane, Golgi complex membrane, and lysosomal membrane. CoQ is an endogenous free radical scavenger and enzyme cofactor produced by most human cells and is abundant in the brain and intestine in mitochondria, the site of oxidative phosphorylation [108]. It is biosynthesized by all membranes and a component of the mitochondrial respiratory chain, which participates in electron transport [109]. CoQ in either reduced form, hydroquinone or ubiquinol, is a potent lipophilic antioxidant and is also involved in the regeneration and recycling of other antioxidants such as tocopherol and ascorbate [110]. CoQ is an important antioxidant that protects cellular membranes and lipoproteins against the toxic effects of free radicals that are generated during general metabolism [111]. The redox functions of CoQ mainly occur in the mitochondria as electrons shuttle between complexes I and II of the respiratory chain. Dehydrogenases oxidize NADH, NADPH, and dihydroflavine-adenine dinucleotide (FADH2) and transfer protons and electrons to ubiquinone, which is converted to ubiquinol. Finally, the protons transfer to the mitochondrial matrix and the electrons move to the cytochromes. Thus, cytochromes reduce the superoxide radical anion to water with electrons and protons from the matrix. These processes are required to produce ATP [108]. TrxRs within the cytosol induce the reduction of CoQ to ubiquinol and maintain the extra-mitochondrial antioxidant defense [112]. CoQ is synthesized by at least 12 proteins that form a multiprotein complex in the mitochondria, and the pathogenesis of CoQ10 deficiency involves deficient ATP production and excessive ROS formation in humans. However, CoQ10 deficiency is unique among mitochondrial disorders because an effective treatment is available by oral CoQ10 supplementation, to which many patients respond well [113,114]. However, dietary supplementation is still challenging due to the low bioavailability of the compound. Because CoQ10 levels decline with age, it has been studied in a variety of neurodegenerative disorders and aging but with disparate results [115].

In KA-induced oxidative injury in vitro [116], an increased number of surviving CA3 neurons was observed at 0.1 and 1 μM concentrations in CoQ10-treated groups using cresyl violet staining. CoQ10 (0.01, 0.1, and 1 μM) treatment significantly decreased 2,7-dichlorofluorescein fluorescence, and the expression of NQO1 in the CoQ10-treated groups was significantly lower than that in controls, indicating the protective role of CoQ10 in hippocampal neurons against oxidative stress. Rats exposed to intrastriatal CoQ10 showed a larger number of dopaminergic neurons, higher expression of neurogenetic and angiogenetic factors, and less inflammation, and these effects were more prominent than the orally administered CoQ10. Thus, continuous intrastriatal administration of low doses of CoQ10 showed a more effective strategy to prevent dopaminergic neuronal degeneration in a PD rat model induced by 6-hydroxydopamine [117]. In patients with acute ischemic stroke, a significant increase in CoQ10 level was observed with the administration of CoQ10 (300 mg/day group) compared with the placebo group. CoQ10 supplementation attenuated oxidative stress and neuroinflammatory marker levels and improved the neurological outcome scale and antioxidant enzyme activity [118]. In PD patients, the total Unified Parkinson's Disease Rating Scale (UPDRS) scores decreased in the CoQ10-treated group, indicating amelioration of the symptoms and significant improvement in wearing-off in PD [119]. CoQ10 increased a reduced glutathione and SOD levels in patients with pregabalin-treated fibromyalgia. The supplementation of CoQ10 effectively reduced pain sensation, anxiety and brain activity, mitochondrial oxidative stress, and inflammation [120].

Zinc and Essential Metals

Zinc, a redox inactive metal, does not directly interact with ROS but has a crucial role in maintaining redox balance for the antioxidant defense system in various ways in the cell. Zinc increases the activation of antioxidant enzymes such as SOD, GPx, and CAT. It also acts as a direct cofactor of SOD-1 and SOD-3 and as an indirect cofactor for GPx [121]. Zinc inhibits important pro-oxidant enzymes such as NADPH oxidase, inducible nitric oxide synthetase (iNOS), and the reduced form of nicotinamide adenine dinucleotide (NMDA) and regulates oxidant production and metal-induced oxidative damage. Zinc is dynamically associated with sulfur in protein cysteine clusters. It mediates the induction of the zinc-binding protein metallothionein which releases the metal under oxidative conditions and acts as an Se scavenging oxidant. Zinc is involved in the regulation of glutathione metabolism and the overall protein thiol redox status [122]. Zinc competes with redox-active transition metals, iron and copper, for certain binding sites. When zinc binds to these sites, copper and iron are forced to undergo hydrolytic polymerization into unreactive structures, thereby prohibiting the catalysis of free radical formation and the initiation of lipid peroxidation [121–123]. Zinc is mainly expressed in the hippocampus, amygdala, cerebral cortex, thalamus, and olfactory cortex in the brain [124] and is stored as free zinc ions (Zn^{2+}) in the presynaptic glutamatergic neurons. Zinc in synaptic vesicles is released with glutamate and acts as a potent extracellular modulator by interacting with many synaptic receptors during synaptic activity [123].

Co-treatment with zinc and Se significantly decreased mitochondrial dysfunction, ROS levels, and lipid peroxidation levels, while significantly increasing cognitive performance, SOD, glutathione peroxidase, and catalase activity in the mitochondria of the brain in an AD rat model [125]. In a double-blind, placebo-controlled trial of zinc supplementation for premenstrual syndrome, sixty women (18–30 years) were randomly assigned to receive either 30 mg of zinc gluconate and/or placebo for 12 weeks. The zinc-administered group showed beneficial effects on physical and psychological symptoms of premenstrual syndrome, total antioxidant capacity, and brain-derived neurotrophic factor [126].

Other essential metals, such as copper, iron, and magnesium, play an important role in the maintenance of cell homeostasis and preservation of life. They display important structural, regulatory, and catalytic functions in different types of proteins, such as enzymes, receptors, and transporters. Cu^+ and magnesium are the cofactors for enzymes such as COX and/or supper zinc, SOD, and neuronal Cu enrichment predispose to Cu^{2+}-catalyzed Fenton chemistry and H_2O_2-assisted protein oxidation. In particularly, iron continuously shifts between ferrous ion (Fe^{2+}) and ferric ion (Fe^{3+}) states in a redox reaction in the presence of O_2. The constitution of iron in the body is in the form of 65% (Fe^{2+}) ions bound to hemoglobin; less than 10% of ions are expressed with myoglobin (Fe^{2+}), cytochromes (Fe^{2+} or Fe^{3+}), and iron-containing enzymes, and 25% of the ions are bound to iron-storage proteins such as transferrin, ferritin, and hemosiderin [127]. Iron present in the cells in the reduced (Fe^{2+}) and oxidized (Fe^{3+}) states can serve both as an electron donor and electron acceptor. Particularly, the ferrous form of iron can act as a catalyst in the potentiation of oxygen toxicity by generating a wide range of free radical species, including hydroxyl radicals. In addition, an excessive amount of this essential metal induces toxicity, leading to pathological conditions generated by oxidative stress and neurodegeneration [128]. Although homeostasis of iron is essential for physiological functions in the brain, less than 2% of total body iron is present in the brain. Iron contributes to the activity of various enzymes involved in neurotransmitter synthesis and myelination of axons of motor neurons in the brain. Iron accumulation induces features of neurodegenerative disorders including AD, PD, HD, ALS, and neurodegeneration with brain iron accumulation (NBIA). The pathogenesis of neurodegenerative diseases shows a relationship with a dramatic increase in iron content in the brain, which is correlated with the production of ROS [127].

3.2.2. ROS/RNS Scavengers (Vitamin C, E, and A)

Vitamin C

Ascorbate (ascorbic acid, AA), a ubiquitous water-soluble antioxidant and a cofactor for several enzymes, can inhibit the generation of ROS, directly scavenge ROS/RNS, and repair other oxidized scavengers [129]. ROS generation is limited by ascorbate through the inhibition of NOX and nNOS. It also helps in the regeneration of alpha-tocopherol from alpha-tocopheroxyl radical and repair of glutathione. The highest concentration of ascorbate is expressed in the brain and is involved in CNS homeostasis. Endogenous ascorbate exists in two biological forms, the deprotonated ascorbate anion and dehydroascorbate (DHA), which is the product of the two-electron reversible oxidation of ascorbate [129]. The mechanism of ascorbate uptake involves absorption of dietary ascorbate in the intestine by sodium-dependent transporter-1 (SVCT-1) and dissolution in the blood. Ascorbate enters the CNS by slow transport from the plasma to the cerebrospinal fluid across the choroid plexus epithelium. Ascorbate can easily enter the brain through the glucose transporter (GLUT1) when a considerable amount of DHA is present in the blood. Ascorbate or DHA in the cerebrospinal fluid enters the neuron via sodium-dependent transporter-2 (SVCT-2) or GLUT1 transporters [130,131]. Once they enter the neuron, DHA can be reduced to ascorbate or released by GLUT1. Ascorbate produces ascorbate free radicals as one electron donor, which is reduced back to ascorbate within the cells by NADH- and NADPH-dependent reductase. Glial cells obtain ascorbate by reduction of DHA through GLUT1, and ascorbate uptake does not involve SVCT-2, which is different from neurons [131]. DHA goes through GLUT1 slowly to enter the choroid plexus and astrocytes [132–134]. The rapid entry of ascorbate is driven by SVCT-2 through the blood-brain-barrier (BBB) to neurons, and ascorbate modulates SVCT-2 translocation to the plasma membrane, ensuring optimal ascorbate uptake in the neurons [135,136] (Figure 3). The oxidizing and free radical scavenging activity of ascorbate inside the cell is not limited to the aqueous phase, but also includes protection of membranes and other hydrophobic compartments through interaction with vitamin E [137]. Aging decreases the extracellular concentration and the uptake of ascorbate in the brain, which is consistent with an increase in oxidative stress. The neurobiological role of ascorbate in the brain is seen in neuromodulation and neuroprotection [129]. In the role of a neuromodulator, ascorbate release is involved in the uptake or clearance of glutamate from the synapse after its release from the axon terminals. Specifically, L-glutamate promotes ascorbate release as a consequence of removing glutamate from the synapse [138]. Although ascorbate does not act as a classical neurotransmitter, extracellular ascorbate may influence neurotransmission. In particular, ascorbate directly modulates neural excitability through inhibition of T-type Ca^{2+} channels [139], participates in the reduction of extracellular oxidants, which effects the redox status of catecholamines [140], and influences the release of biogenic amines in striatum [141] and pituitary neuropeptides [142]. Another neuroprotective role of ascorbate is the attenuation of neurotoxicity, which results from the scavenging activity [143]. Ascorbate inhibits the oxidative stress triggered by various neurotoxins and protects against ethanol-induced apoptotic neurodegeneration in prenatal rat hippocampal neurons [144]. Oxidative stress in stroke, hypoxia, ischemia, and seizure activity leads to massive glutamate release and subsequent excitotoxicity, a result of over-activation of glutamate receptors [145]. Therefore, ascorbate can protect against glutamate-induced excitotoxicity and neurodegeneration.

Many researchers have reported that neurodegeneration can be reversed or lessened by ascorbate treatment [146–148]. From the viewpoint of the effect of aging, the extracellular concentration and uptake of ascorbate is decreased in the brain, which is consistent with an increase in oxidative stress [149]. Moreover, excess ascorbate intake or deficiency may influence brain aging [150]. Thus, careful maintenance of ascorbate levels in the brain may be important during the life span. There have been extensive reviews on the role of ascorbate in the brain with a focus on neurodegenerative disease [130,131]. Even though ascorbate has a complicated interaction with the neurotransmitter system, ascorbate is considered relevant for use as an antioxidant therapy because neurons are sensitive to ascorbate deficiency and excess oxidant stress. In our previous work [151], we studied the protective effects of AA and DHA on KA-induced oxidative stress using organotypic hippocampal slice cultures. After 12 h of KA treatment, significant delayed neuronal death was detected in the CA3, but not in the CA1 region. Pretreatment with intermediate doses of AA and DHA significantly prevented cell death and reduced ROS levels, as well as mitochondrial dysfunction in the CA3 region. However, pretreatment with high doses of AA or DHA was not effective. Attempting to elevate the brain ascorbate by the systemic administration of high doses of ascorbate is very difficult. The level of AA in the extracellular fluid of the striatum was decreased in a transgenic mouse model of HD; hence, restoring striatal extracellular AA levels with high doses of ascorbate improved behavior [152,153].

Vitamin E

Vitamin E is a major group of lipid-soluble antioxidants called tocopherols and tocotrienols, of which the most biologically active isoform is α-tocopherol [154]. It is a major chain-breaking antioxidant and exists in a low molar ratio compared to unsaturated phospholipids. The most important function of vitamin E is its antioxidant activity, which protects the integrity of cellular membranes from polyunsaturated fatty acid-generated oxygen free radicals and to act as a direct scavenger of superoxide and hydroxyl radicals [154,155]. Based on studies of brain capillary endothelial cells, the mechanism of entry of α-tocopherol into the CNS correlated to α-tocopherol and scavenger receptor class B type 1 (SRB1) levels. α-Tocopherol uptake occurs via a selective high-density lipoprotein (HDL) pathway, which modulates the expression of SRB1 receptor [156–158]. Once α-tocopherol passes through the BBB, it may be directly delivered to specialized astrocytes. Astrocyte-synthesized apolipoprotein E (ApoE) moves through the cerebral spinal fluid transporting α-tocopherol between various cell types in the CNS cell [154,159]. ApoE lipoprotein particles in astrocytes are secreted through membrane proteins and interact with low-density lipoprotein receptor-related protein (LRP) on the neurons [160].

The neurons take up these ApoE particles and distribute them throughout the body, axon, and dendrites to preserve the membrane from lipid peroxidation (Figure 3). The expression of α-tocopherol transfer protein is enhanced by an increase in ROS, and it is useful for combating neuronal damage.

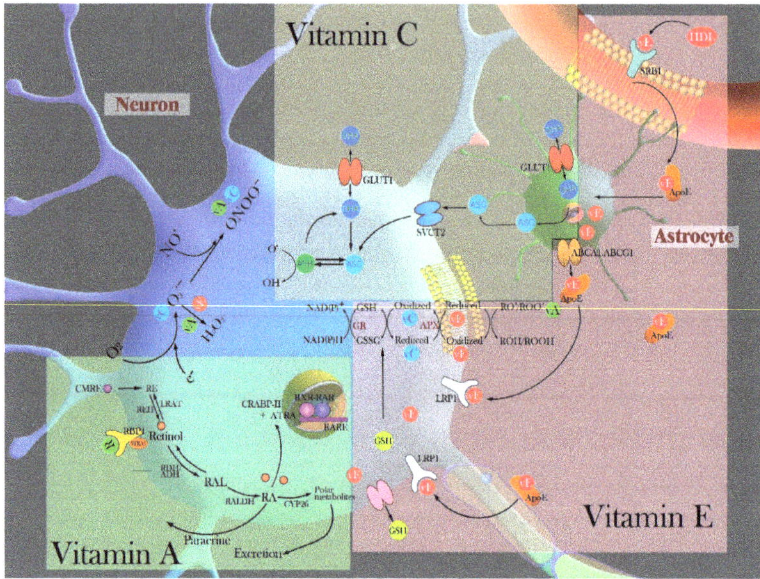

Figure 3. Schematic representation of vitamin uptake and protective mechanisms by exogenous vitamins as antioxidants. **Vitamin C:** During uptake in the CNS, ascorbate passes through the BBB to enter directly through the SVCT-2 and/or possibly DHA through GLUT1s. Moreover, the neuronal uptake of ascorbate occurs through SVCT-2 and DHA via the GLUT1s. In the neuron, DHA can be reduced to ascorbate or released back into the extracellular space by GLUT1. Ascorbate free radicals convert to form DHA and ascorbate. Ascorbate recycles both the ascorbate free radical and DHA by cellular metabolism. Astrocytes contain ascorbate from recycling of DHA which is taken up through GLUT1s. Neurons directly acquire ascorbate via SVCT-2. **Vitamin E:** With respect to vitamin E uptake, HDL particles can pass through SRB1 receptors expressed on endothelial cells. Astrocytes that exist adjacent to the BBB take up vitamin E into the inner cell membrane. Synthesized ApoE lipoproteins take up vitamin E that is left out of an ABC transporter, for transport into neurons through LRP1 requiring vitamin E for maintenance or during conditions of oxidative stress. **Vitamin A:** In cellular retinoid signaling pathways, retinol is metabolized to all-trans-retinoic acid (ATRA). Vitamin A (retinol, ROL) binds to plasma retinol binding protein (RBP4) and circulates; RBP4 protein binds to the membrane receptor STAR6 to promote cellular absorption of retinol from the cells. A chylomicron remnant (CMRE), as a form of circulating vitamin A, can serve as a source of vitamin A for the cells and retinol is esterified and stored by lecithin: retinol acyltransferase (LRAT) and is reversibly oxidized to retinaldehyde (RAL) by retinol dehydrogenase (RDH/ADH). In addition, retinol is further oxidized to RA in an irreversible manner by retinaldehyde dehydrogenase (RALDH). ATRA regulates gene transcription through retinoic acid receptors (RAR) and/or retinoid X receptors (RXRs) which are bound to retinoic acid response elements (RARE) in the nucleus. These representative schematics are modified from [130,154,161]. Abbreviations: VC, vitamin C; VE, vitamin E; VA, vitamin A; ASC, ascorbate; ASF, ascorbate free radical; DHA, dehydroascorbic acid; GLUT1, glucose transports; SVTC-2, sodium-dependent transporters; LRP, lipoprotein receptor-related protein; GSH, glutathione; ApoE, apolipoprotein E; HDL, density lipoprotein; CMRE, chylomicron remnant; CRABP, cellular retinoic acid-binding protein; LRAT, lethicin: retinol acyltransferase; RBP, retinol binding protein; RAL, retinaldehyde; RDH/ADH, retinol dehydrogenase; RALDH, retinaldehyde dehydrogenase; RXR, retinoid X receptors; RAR, retinoic acid receptors; ATRA, all-trans-retinoic acid; BBB, blood–brain barrier.

The antioxidant ability of vitamin E is continuously restored via vitamin E recycling by other antioxidants such as vitamin C, ubiquinols, and thiols. The half-life of vitamin E in the brain tissue is slower than that of other vitamins, and it is also actively retained and protected from auto-oxidation in the brain [162]. The distribution of α-tocopherol is significantly different in the brain. These CNS-regional disparities are suggestive of the specific protective antioxidant effect of α-tocopherol. Specifically, the concentrations of α-tocopherol were relatively higher in the nuclear membranes than in the other membranes of the brain. It has been suggested that α-tocopherol may play a role in nuclear-associated functions in the cerebellum and striatum, wherein preferential accumulation of α-tocopherol in the membrane was most apparent [163]. Vitamin E, similar to other radical scavengers/trappers, influences the flux of lipid hydroperoxide (LOOH), which is derived from both spontaneous and enzymatic formation of lipid peroxyl radicals (LOO$^{\bullet}$) on the cellular membrane [164]. The effects of vitamin E on peroxidation activity appear to involve both the radical scavenging mechanism such as the H atom donor activity and a physical interaction with the polyunsaturated lipid substrate. Tocotrienols, another form of vitamin E that are highly metabolized, show more potent inhibition of the phospholipase A2/lipoxygenases pathway as compared to tocopherols, and have different cellular bioavailability, distribution, and protein interaction in the saturated and unsaturated form of vitamin E in the brain [165]. Deficiency of vitamin E showed increased biochemical and histological markers of oxidative stress, including total glutathione and lipid peroxidation in the CNS.

Therapeutic effects of α-tocopherol by application of vitamin E in neurological lesions caused by neuronal excitotoxicity and the conditional activation of neuroglial cells have also been reported. In KA treatment-induced oxidative stress [166], delayed neuronal death was detected in the hippocampal CA3 region and ROS formation and lipid peroxidation were also increased. Both co-treatment and post-treatment with α-tocopherol (100 μM) or α-tocotrienol (100 μM) significantly increased cell survival and reduced the number of TUNEL-positive cells in the CA3 region. Increased dichlorofluorescein (DCF) fluorescence and thiobarbiturate reactive substance (TBARS) levels were decreased by drug treatment (Figure 4). In AD patients, long-term administration of vitamin D and E alone or in combination could inhibit morphological changes of neurons and improve learning and memory [167]. Vitamin E prevented the memory impairment associated with post-traumatic stress disorder (PTSD)-like behavior in rats. Significant decreases in oxidative stress biomarkers were detected with reduced glutathione/oxidized glutathione (GSH/GSSG) ratio [168]. In PD patients, omega-3 fatty acid and vitamin E co-supplementation had favorable effects on the UPDRS score and increased the total antioxidant capacity (TAC) and GSH concentration compared to placebo [169].

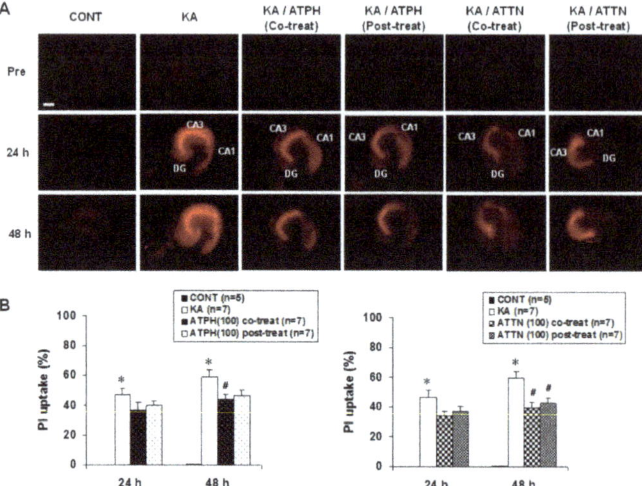

Figure 4. Neuroprotective effect of vitamin E. (**A**): Representative propidium iodide (PI) images. When hippocampal slices were exposed to 5 μM KA for 15 h, PI uptake in the CA3 region was significantly higher than the CA1 region. Co-treatment using ATPH (100 μM) or ATTN (100 μM) with KA significantly reduced PI uptake in the CA3 region compared with KA treatment alone. (**B**): Quantification of PI intensity. * $p < 0.05$, # $p < 0.05$; one-way ANOVA followed by Dunnett's post hoc comparison (* $p < 0.05$ vs. normal, # $p < 0.05$ vs. KA-treated cultures). This present data is a part of our previous research showing neuronal rescue after oxidative stress by alpha-tocopherol and tocotrienol treatment [166]. Abbreviations: ATPH, alpha-tocopherol; ATTN, alpha-tocotrienol; OHSC, organotypic hippocampal slice culture; KA, kainic acid; PI, propidium iodide; DCF, dichlorofluorescein.

Vitamin A

Vitamin A, carotenoids including retinol and beta-carotene, are fat-soluble chemicals synthesized by plants and some microorganisms and have many functions in human growth, development, and health [170,171]. Vitamin A is available in the human diet as pro-vitamin A carotenoids and preformed vitamin A (retinol-alcohol form, retinal-aldehyde form, retinoic acid-carboxylic acid form, and retinyl ester-ester form). The preformed vitamin A from animal-derived food and pro-vitamin A carotenoids from plant-derived foods are converted to all-trans-retinol as vitamin A alcohol by a series of reactions in the intestine. Carotenoids, mainly via dietary intake, can function directly as antioxidants by quenching ROS through energy transfer [172]. Vitamin A deprivation was investigated in the deficiency of cognitive function in adult mice and rats, which highlights the importance of adequate vitamin A status by the retinoid signaling pathways [161]. Carotenoids are classified into pro-vitamin A carotenoids such as β-carotene and β-cryptoxanthin, which are capable of converting to retinal, and non-pro-vitamin A carotenoids such as lycopene and lutein, which cannot be converted to retinal [173]. Carotenoids act through several pathways and interact with free radicals in the plasma, mitochondria, and nuclear membranes of cells via electron transfer, hydrogen abstraction, and physical quenching [174]. Carotenoids indirectly react with cell signaling cascades, including the nuclear factor elytroid 2 (NF-E2)-related factor 2 (Nrf2), NF-κB, or mitogen-activated protein kinase (MAPK) [175,176]. The antioxidant action of carotenoids involves singlet oxygen quenching and trapping of peroxyl radicals. Retinol-binding proteins observed in the BBB regulate the access of retinol into the brain [177]. High concentrations of retinol and carotenoids have been observed in the postmortem human frontal lobe cortex [178]. β-carotene, a precursor of retinol and retinoic acid, is reported to be a potent free radical quencher, singlet oxygen scavenger, and lipid antioxidant in tissues and plasma. Therefore, β-carotene acts in a hydrophobic environment such as the lipid core of the membranes and is used faster than α-tocopherol, implying that β-carotene is more favorable than α-tocopherol to quench lipophilic radicals in the membrane. The most efficient synergistic inhibition during oxidative stress was observed with a combination treatment of α-tocopherol and ascorbic acid [179].

Retinoids, compounds structurally related to vitamin A, are considered vitamin A derivatives that contribute to regular cellular morphogenesis, proliferation, and differentiation. Retinoids are involved in normal signaling cascades in modulating brain functions [180]. Retinoids modulate the availability of glucocorticosteroids in the brain, an important biological mechanism that can be explored in many stress-related pathologies to prevent alterations in the plasticity of the hippocampus [181]. Retinol metabolic pathways have shown that retinol can be stored intracellularly as retinyl esters and metabolized into all-trans-retinoic acid (ATRA) as a bioactive derivative. ATRA induces cellular differentiation and growth by reacting to retinoic acid receptors (RARs). Cellular retinol-binding proteins (CRBP-I and II) and cellular retinoic acid-binding proteins (CRABP-I, II) are distributed in the adult CNS. Furthermore, CRBP-I distribution parallels that of ATRA with expression in the meninges, hippocampus, amygdala, and olfactory bulb [182] (Figure 3). Under oxidative stress conditions such as metal exposure and production and accumulation of ROS, retinoids protect the cells against this imbalance through multiple mechanisms, including interference with ROS production, scavenging free radicals directly, upregulation of antioxidant enzymes, and signaling pathways involved in defense system such as Nrf2 signaling [183]. It has also been observed that retinoic acid has a protective effect on neuronal apoptosis and oxidative damage by reducing glutathione [184] and restoring SOD-1 and SOD-2 in the hippocampal cells [185]. The role of retinoid signal transduction in the control of dopaminergic neurotransmission was observed in the presence of high levels of retinoic acid-synthesizing enzymes [186] and RAR, which may play a critical role in controlling the survival, adaptation, and homeostatic regulation of the dopaminergic system [187]. Retinoid signaling play a physiological role in synaptic plasticity and learning and memory behaviors [188].

Retinoic acid supplementation upregulated μ-type opioid receptor 1 (MOR1) and its signaling and alleviated dyskinetic movements, which is a known consequence of prolonged administration of L-DOPA, in Pitx3$^{ak/ak}$ mice [189]. Moreover, RA triggered the neuroprotective effect on DA neurons in MPTP-treated mice model of PD. Administration of RA-loaded polymeric nanoparticle significantly reduced the loss of DA neuron in the substantia nigra as well as their neuronal fiber/axonal innervations in the striatum [190]. Moreover, oral administration of lycopene (5–20 mg/kg), a carotenoid with unique pharmacological properties, attenuated oxidative stress in mice with PD, which was induced with intraperitoneal injection of 1-methyl-4-phenyl-1,2,3,6-tetrahydropyridine (MPTP). Lycopene supplementation also inhibits apoptosis in PD mice by decreasing Bax and caspases, and contrarily increasing Bcl-2 [191]. Dietary supplementation with astaxanthin, another carotenoid family member, significantly decreased intracellular ROS accumulation in a hippocampal neuronal cell line after exposure to glutamate and induced antioxidant mediators, such as heme oxygenase-1 (HO-1) and nuclear Nrf2 expression in vitro [192]. In multiple sclerosis (MS) patients, vitamin A supplementation had a significant effect in the treatment group for fatigue and depression. In addition, when a synthetic retinoid was tested, it showed reduction in inflammation, Aβ burden, and tau phosphorylation with associated cognitive benefits in AD patients [193,194].

Aside from elucidating the efficiency of endogenous antioxidants, several studies have increasingly accredited the role of various exogenous antioxidants. Limited clinical studies are reported herein and are illustrated in Table 1. Table 1 also shows several examples of previously mentioned exogenous antioxidant reactions that have worked in clinical settings. Although the administration of exogenous antioxidants showed positive effects, there are some inconsistencies in clinical trials. The clinical trials performed multivariable analyses with various factors, such as sample size, replication or validation studies using the same agent and outcome measures, assessment of vitamins, different conditions of endogenous antioxidants, long-term monitoring, etc., which caused variability because of inconsistencies. In general, several common antioxidants and their clinical effects are described in Table 1.

Table 1. Clinical trials of antioxidant in neurodegenerative diseases.

Antioxidant	Number of Patients	Follow up Period	Dosage	Route	Effects	Disease	Reference
CoQ10	609 40	60 month 96 week	2400 mg/day 300 mg/day	Oral Oral	No Y	Huntington Parkinson	[119,195]
Selenium	7540	7 year 6 month	200 µg/day 1000 µg/day	Oral I.V.	No Y	Alzheimer Traumatic brain injury	[107,196]
Zinc	43	12 week	20 mg/day	Oral	Y	Depression with multiple sclerosis	[197]
Vitamin A	50 101	28 day 6 month	30 mg/day 25,000 UI/day	Oral Oral	Y Y	Alzheimer Multiple sclerosis	[191,198]
Vitamin C	12 60	6 day 3 month	2 g/day 500 mg/day	Oral Oral	Y Y	Hyperoxia Trauma surgery	[147,199]
Vitamin E	7540 60 50	7 year 3 month 12 month	400 UI/day 400 UI/day 45 UI/day	Oral Oral Oral	No Y Y	Alzheimer Parkinson Parkinson	[168,196,200]

3.2.3. Nrf2 Antioxidant System

The transcription factor Nrf2 is characterized as a regulator of redox homeostasis and antioxidant defense mechanisms. This protective pathway also encompasses the activation of a detoxification network such as oxidation/reduction factors (Phase I), metabolizing enzymes (Phase II), efflux transporters (Phase III), and free radical scavengers [201]. Oxidation/reduction factors consisting of nearly 500 genes encoding proteins, including redox balancing factors, stress response proteins, detoxifying enzymes, and metabolic enzymes such as NAD(P)H quinone oxidoreductase (NQO1), HO-1, SOD, GST, GSR, GSH-Px, carbonyl reductase (CR), and glutamate-cysteine ligase (GCL) play important roles in antioxidant and pro-survival effects and detoxification of xenobiotics. Nrf2 is generally targeted for ubiquitin-mediated degradation by its endogenous inhibitor Keap1, but oxidative modification of Keap1 inhibits the Nrf2 degradation process during conditions of redox imbalance [202]. To maintain cellular redox homeostasis, basal Nrf2 accumulation increases to mediate the normal expression of antioxidant response element (ARE)-dependent genes in the nucleus. The mitochondrial membrane directly interacts with Nrf2 which can respond to mitochondrial oxidative stressors that can collapse cellular bioenergetics leading to cell apoptosis [203]. Thus, the Keap1/Nrf2 system is another good homeostatic regulator of intrinsic cellular antioxidant defense and mitochondrial health. The suppression of Nrf2 activity increases the susceptibility of the brain to the damaging effects of oxidative stress and inflammatory stimuli [204]. Nrf2 activation promotes neuroprotective effects in both in vitro and in vivo neurodegenerative models. Nrf2 activity diminishes with age, and consequently, the effect of antioxidant enzyme activity decreases. Nrf2/ARE system impairment leads to higher susceptibility to oxidative injury, abnormal protein aggregation and neurodegeneration in the brain. Many studies have demonstrated the importance of the Nrf2/ARE pathway in the pathogenesis and control of neurological disorders, including PD, AD, ischemia, and other neurodegenerative diseases.

After middle cerebral artery occlusion (MCAO) in rats, Keap1 levels are decreased and this loss is correlated with an increase in Nrf2 and its downstream proteins, such as thioredoxins, GSH synthases, and HO-1 [205]. Stroke models of Nrf2-deficient mice exhibit higher levels of ROS than the wild-type littermates, which supports the natural compensatory mechanism of Nrf2 [206]. In AD transgenic mice, Nrf2-deficiency brains presented increased marker of oxidative stress and exhibited deficits in spatial learning and memory [204]. Nrf2 knockout and wild-type mice can be administered MPTP with doses ranging from 20 to 60 mg/kg for an animal model of PD. Nrf2$^{-/-}$ mice exhibited increased sensitivity to the dopaminergic toxins MPTP and 6-OHDA [207]. Nrf2 deficiency increased the MPTP sensitivity by 30 mg/kg administration of MPTP, but astrocytic Nrf2 overexpression showed the amelioration of MPTP toxicity in ARE-hPAP mice [208]. Astrocytic Nrf2 reduced the MPTP neurotoxicity by β-lapachone treatment in PD models [209].

4. Conclusions

Brains are metabolically active and have a high demand for large amounts of ATP to maintain their physiological function. Additionally, the brain has a relatively low level of antioxidants, low repair capacity, non-replicating nature of neuronal cells, and a high ratio of membrane surface to cytoplasm. ROS production is largely through oxidative phosphorylation and increased free radicals play a central role in neurological disorders by the imbalance of pro-oxidant and antioxidant agents in the brain. This underpins the importance of targeting antioxidant systems to counteract the oxidative stress and associated brain diseases. Indeed, the antioxidant system is important for the rescue of neuronal cells from oxidative stress and preservation of the right redox balance in the brain tissue by promoting antioxidative defenses for neutralizing ROS and by blocking transcription. Currently, there is growing research interest in the development of new/combination exogenous supplementation of antioxidants, retention of the functional integrity of intrinsic antioxidant systems for preventing harmful CNS disorders, and identification of novel approaches to therapy to prevent and/or reduce brain injury. Thus, high levels and variable metabolites in the brain indicate an increased requirement for antioxidant defense systems as enzymatic and non-enzymatic molecules. Redox biology fulfills important physiological functions that extend beyond its role on oxidative stress, such as cell signaling [210,211]. Perhaps, ROS generation may be necessary for cell function. Although their role is still ill-defined, the antioxidant systems may contribute to the controlled release of ROS for normal brain function.

Author Contributions: All authors contributed substantially to the preparation of this study. K.H.L. designed the study and drafted the manuscript. M.C. assisted with drafting of the manuscript and preparation of the figures. B.H.L. supervised the entire project and prepared the manuscript draft. All authors have read and agreed to the published version of the manuscript.

Funding: This work was supported by the Basic Science Research Program through the National Research Foundation of Korea (NRF) funded by the Ministry of Education (MOE) (NRF-2016R1D1A3B2008194).

Conflicts of Interest: The authors declare no conflict of interest.

Abbreviations

AD	Alzheimer disease
ALS	amyotrophic lateral sclerosis
ApoE	apolipoprotein E
ARE	antioxidant response element
ATP	adenosine triphosphate
ATPH	alpha-tocopherol
ATRA	all-trans-retinoic acid
ATTN	alpha-tocotrienol
Aβ	beta-amyloid
CA3	cornu ammonis 3
CAT	catalase
CNS	central nervous system
CoQ	coenzyme Q
COX	cyclooxygenase
CR	carbonyl reductase
CRABP	cellular retinoic acid-binding proteins
CRBP	cellular retinol binding proteins
DCF	dichlorofluorescein
DHA	dehydroascorbate
EC-SOD	extracellular superoxide dismutase
FADH2	dihydroflavine-adenine dinucleotide
Fe^{2+}	ferrous ion
Fe^{3+}	ferric ion

GCL	glutamate-cysteine ligase
GLUT1	glucose transports
GPx	glutathione peroxidase
GR	glutathione reductase
GSH	glutathione
GSSG	glutathione disulfide
hAPP	human amyloid precursor protein
HD	Huntington's disease
HDL	high density lipoprotein
HO-1	heme oxygenase-1
iNOS	inducible nitric oxide synthase
KA	kainic acid
LRP	lipoprotein receptor-related protein
MAPK	mitogen-activated protein kinase
MCAO	middle cerebral artery occlusion
Mn-SOD	manganese containing superoxide dismutase
MOR	μ-type opioid receptor
MPTP	1-methyl-4-phenyl-1,2,3,6-tetrahydropyridine
NADH	reduced form of nicotinamide adenine dinucleotide
NADPH	reduced form of nicotinamide adenine dinucleotide phosphate
NBIA	neurodegeneration brain iron accumulation
NF-κB	nuclear factor-κB
nNOS	neuronal nitric oxide synthase
NOX	NADPH oxidase
NQO1	NADPH quinone oxidoreductase
Nrf2	nuclear factor elytroid 2-related factor 2
6-OHDA	6-hydroxydopamine
OHSC	organotypic hippocampal slice culture
PD	Parkinson disease
PI	propidium iodide
PTSD	post-traumatic stress disorder
RAR	retinoic acid receptors
RNS	reactive nitrogen species
ROS	reactive oxygen species
Se	selenium
SOD	superoxide dismutase
SRB1	scavenger receptor class B type 1
SVCT	sodium-vitamin C co-transporters
TAC	total antioxidant capacity
TBARS	thiobarbiturate reactive substance
Trx	thioredoxin
TrxR	thioredoxin reductase
UV	ultra violet

References

1. Cobley, J.N.; Fiorello, M.L.; Bailey, D.M. 13 reasons why the brain is susceptible to oxidative stress. *Redox Biol.* **2018**, *15*, 490–503. [CrossRef] [PubMed]
2. Magistretti, P.J.; Allaman, I. A cellular perspective on brain energy metabolism and functional imaging. *Neuron* **2015**, *86*, 883–901. [CrossRef] [PubMed]
3. Nedergaard, M.; Ransom, B.; Goldman, S.A. New roles for astrocytes: Redefining the functional architecture of the brain. *Trends Neurosci.* **2003**, *26*, 523–530. [CrossRef] [PubMed]
4. Bailey, D.M.; Bärtsch, P.; Knauth, M.; Baumgartner, R.W. Emerging concepts in acute mountain sickness and high-altitude cerebral edema: From the molecular to the morphological. *Cell. Mol. Life Sci.* **2009**, *66*, 3583–3594. [CrossRef] [PubMed]

5. Turrens, J.F. Mitochondrial formation of reactive oxygen species. *J. Physiol.* **2003**, *552*, 335–344. [CrossRef] [PubMed]

6. Niedzielska, E.; Smaga, I.; Gawlik, M.; Moniczewski, A.; Stankowicz, P.; Pera, J.; Filip, M. Oxidative stress in neurodegenerative diseases. *Mol. Neurobiol.* **2016**, *53*, 4094–4125. [CrossRef] [PubMed]

7. Pisoschi, A.M.; Pop, A. The role of antioxidants in the chemistry of oxidative stress: A review. *Eur. J. Med. Chem.* **2015**, *97*, 55–74. [CrossRef] [PubMed]

8. Poljsak, B.; Šuput, D.; Milisav, I. Achieving the balance between ROS and antioxidants: When to use the synthetic antioxidants. *Oxid. Med. Cell. Longev.* **2013**, *2013*, 956792. [CrossRef]

9. Jiang, T.; Sun, Q.; Chen, S. Oxidative stress: A major pathogenesis and potential therapeutic target of antioxidative agents in Parkinson's disease and Alzheimer's disease. *Prog. Neurobiol.* **2016**, *147*, 1–19. [CrossRef]

10. Watson, B.D. Evaluation of the concomitance of lipid peroxidation in experimental models of cerebral ischemia and stroke. In *Progress in Brain Research*; Elsevier: Amsterdam, The Netherlands, 1993; Volume 96, pp. 69–95.

11. Choi, B.H. Oxygen, antioxidants and brain dysfunction. *Yonsei Med. J.* **1993**, *34*, 1–10. [CrossRef]

12. Poon, H.F.; Calabrese, V.; Scapagnini, G.; Butterfield, D.A. Free radicals: Key to brain aging and heme oxygenase as a cellular response to oxidative stress. *J. Gerontol. A Biol. Sci. Med. Sci.* **2004**, *59*, M478–M493. [CrossRef] [PubMed]

13. Rao, A.; Balachandran, B. Role of oxidative stress and antioxidants in neurodegenerative diseases. *Nutr. Neurosci.* **2002**, *5*, 291–309. [CrossRef] [PubMed]

14. Packer, L. Oxidative stress, Antioxidants, Aging and Disease. In *Oxidative Stress and Aging*; Springer: Berlin/Heidelberg, Germany, 1995; pp. 1–14.

15. Li, J.; Li, W.; Jiang, Z.G.; Ghanbari, H.A. Oxidative stress and neurodegenerative disorders. *Int. J. Mol. Sci.* **2013**, *14*, 24438–24475. [CrossRef] [PubMed]

16. Conway, M.E.; Lee, C. The redox switch that regulates molecular chaperones. *Biomol. Concepts* **2015**, *6*, 269–284. [CrossRef]

17. Dickinson, B.C.; Peltier, J.; Stone, D.; Schaffer, D.V.; Chang, C.J. Nox2 redox signaling maintains essential cell populations in the brain. *Nat. Chem. Biol.* **2011**, *7*, 106–112. [CrossRef]

18. Brown, G.C. Nitric oxide and mitochondrial respiration. *Biochim. Biophys. Acta* **1999**, *1411*, 351–369. [CrossRef]

19. Carballal, S.; Bartesaghi, S.; Radi, R. Kinetic and mechanistic considerations to assess the biological fate of peroxynitrite. *Biochim. Biophys. Acta* **2014**, *1840*, 768–780. [CrossRef]

20. Lipton, S.A.; Choi, Y.B.; Pan, Z.H.; Lei, S.Z.; Chen, H.S.; Sucher, N.J.; Loscalzo, J.; Singel, D.J.; Stamler, J.S. A redox-based mechanism for the neuroprotective and neurodestructive effects of nitric oxide and related nitroso-compounds. *Nature* **1993**, *364*, 626–632. [CrossRef]

21. Sena, L.A.; Chandel, N.S. Physiological roles of mitochondrial reactive oxygen species. *Mol. Cell* **2012**, *48*, 158–167. [CrossRef]

22. Brand, M.D.; Goncalves, R.L.; Orr, A.L.; Vargas, L.; Gerencser, A.A.; Jensen, M.B.; Wang, Y.T.; Melov, S.; Turk, C.N.; Matzen, J.T. Suppressors of superoxide-H_2O_2 production at site IQ of mitochondrial complex I protect against stem cell hyperplasia and ischemia-reperfusion injury. *Cell Metab.* **2016**, *24*, 582–592. [CrossRef]

23. Goncalves, R.L.; Rothschild, D.E.; Quinlan, C.L.; Scott, G.K.; Benz, C.C.; Brand, M.D. Sources of superoxide/H_2O_2 during mitochondrial proline oxidation. *Redox Biol.* **2014**, *2*, 901–909. [CrossRef]

24. Bélanger, M.; Allaman, I.; Magistretti, P.J. Brain energy metabolism: Focus on astrocyte-neuron metabolic cooperation. *Cell Metab.* **2011**, *14*, 724–738. [CrossRef] [PubMed]

25. Pun, P.B.L.; Logan, A.; Darley-Usmar, V.; Chacko, B.; Johnson, M.S.; Huang, G.W.; Rogatti, S.; Prime, T.A.; Methner, C.; Krieg, T. A mitochondria-targeted mass spectrometry probe to detect glyoxals: Implications for diabetes. *Free Radic. Biol. Med.* **2014**, *67*, 437–450. [CrossRef] [PubMed]

26. Przedborski, S. The two-century journey of Parkinson disease research. *Nat. Rev. Neurosci.* **2017**, *18*, 251. [CrossRef] [PubMed]

27. Cohen, G.; Heikkila, R.E. The generation of hydrogen peroxide, superoxide radical, and hydroxyl radical by 6-hydroxydopamine, dialuric acid, and related cytotoxic agents. *J. Biol. Chem.* **1974**, *249*, 2447–2452.

28. Miller, D.M.; Buettner, G.R.; Aust, S.D. Transition metals as catalysts of "autoxidation" reactions. *Free Radic. Biol. Med.* **1990**, *8*, 95–108. [CrossRef]

29. Patra, R.; Swarup, D.; Dwivedi, S. Antioxidant effects of α tocopherol, ascorbic acid and L-methionine on lead induced oxidative stress to the liver, kidney and brain in rats. *Toxicology* **2001**, *162*, 81–88. [CrossRef]

30. Brigelius-Flohé, R.; Maiorino, M. Glutathione peroxidases. *Biochim. Biophys. Acta* **2013**, *1830*, 3289–3303. [CrossRef]

31. Savaskan, N.E.; Borchert, A.; Bräuer, A.U.; Kuhn, H. Role for glutathione peroxidase-4 in brain development and neuronal apoptosis: Specific induction of enzyme expression in reactive astrocytes following brain injury. *Free Radic. Biol. Med.* **2007**, *43*, 191–201. [CrossRef]

32. Hong, S.; Dissing-Olesen, L.; Stevens, B. New insights on the role of microglia in synaptic pruning in health and disease. *Curr. Opin. Neurobiol.* **2016**, *36*, 128–134. [CrossRef]

33. Bedard, K.; Krause, K.-H. The NOX family of ROS-generating NADPH oxidases: Physiology and pathophysiology. *Physiol. Rev.* **2007**, *87*, 245–313. [CrossRef] [PubMed]

34. Cheng, Z.; Li, Y. What is responsible for the initiating chemistry of iron-mediated lipid peroxidation: An update. *Chem. Rev.* **2007**, *107*, 748–766. [CrossRef] [PubMed]

35. Halliwell, B. Oxidative stress and neurodegeneration: Where are we now? *J. Neurochem.* **2006**, *97*, 1634–1658. [CrossRef] [PubMed]

36. Yin, H.; Xu, L.; Porter, N.A. Free Radical Lipid Peroxidation: Mechanisms and Analysis. *Chem. Rev.* **2011**, *111*, 5944–5972. [CrossRef] [PubMed]

37. Ferrer-Sueta, G.; Radi, R. Chemical biology of peroxynitrite: Kinetics, diffusion, and radicals. *ACS Chem. Biol.* **2009**, *4*, 161–177. [CrossRef]

38. Thomas, D.D. Breathing new life into nitric oxide signaling: A brief overview of the interplay between oxygen and nitric oxide. *Redox Biol.* **2015**, *5*, 225–233. [CrossRef]

39. Lobo, V.; Patil, A.; Phatak, A.; Chandra, N. Free radicals, antioxidants and functional foods: Impact on human health. *Pharmacogn. Rev.* **2010**, *4*, 118. [CrossRef]

40. Cadet, J.; Davies, K.J. Oxidative DNA damage & repair: An introduction. *Free Radic. Biol. Med.* **2017**, *107*, 2–12.

41. Gutteridge, J. Biological origin of free radicals, and mechanisms of antioxidant protection. *Chem. Biol. Interact.* **1994**, *91*, 133–140. [CrossRef]

42. Ellerby, L.M.; Cabelli, D.E.; Graden, J.A.; Valentine, J.S. Copper–zinc superoxide dismutase: Why not pH-dependent? *J. Am. Chem. Soc.* **1996**, *118*, 6556–6561. [CrossRef]

43. Banci, L.; Benedetto, M.; Bertini, I.; Del Conte, R.; Piccioli, M.; Viezzoli, M.S. Solution structure of reduced monomeric Q133M2 copper, zinc superoxide dismutase (SOD). Why is SOD a dimeric enzyme? *Biochemistry* **1998**, *37*, 11780–11791. [CrossRef] [PubMed]

44. Guan, Y.; Hickey, M.J.; Borgstahl, G.E.; Hallewell, R.A.; Lepock, J.R.; O'Connor, D.; Hsieh, Y.; Nick, H.S.; Silverman, D.N.; Tainer, J.A. Crystal structure of Y34F mutant human mitochondrial manganese superoxide dismutase and the functional role of tyrosine 34. *Biochemistry* **1998**, *37*, 4722–4730. [CrossRef] [PubMed]

45. Adachi, T.; Wang, X.L. Association of extracellular-superoxide dismutase phenotype with the endothelial constitutive nitric oxide synthase polymorphism. *FEBS Lett.* **1998**, *433*, 166–168. [CrossRef]

46. Matés, J.M.; Sánchez-Jiménez, F. Antioxidant enzymes and their implications in pathophysiologic processes. *Front. Biosci.* **1999**, *4*, D339–D345. [CrossRef] [PubMed]

47. Liu, J.; Lillo, C.; Jonsson, P.A.; Velde, C.V.; Ward, C.M.; Miller, T.M.; Subramaniam, J.R.; Rothstein, J.D.; Marklund, S.; Andersen, P.M. Toxicity of familial ALS-linked SOD1 mutants from selective recruitment to spinal mitochondria. *Neuron* **2004**, *43*, 5–17. [CrossRef]

48. Li, Q.X.; Mok, S.S.; Laughton, K.M.; McLean, C.A.; Volitakis, I.; Cherny, R.A.; Cheung, N.S.; White, A.R.; Masters, C.L. Overexpression of Aβ is associated with acceleration of onset of motor impairment and superoxide dismutase 1 aggregation in an amyotrophic lateral sclerosis mouse model. *Aging Cell* **2006**, *5*, 153–165. [CrossRef]

49. Iadecola, C.; Zhang, F.; Niwa, K.; Eckman, C.; Turner, S.K.; Fischer, E.; Younkin, S.; Borchelt, D.R.; Hsiao, K.K.; Carlson, G.A. SOD1 rescues cerebral endothelial dysfunction in mice overexpressing amyloid precursor protein. *Nat. Neurosci.* **1999**, *2*, 157–161. [CrossRef]

50. De Haan, J.B.; Newman, J.D.; Kola, I. Cu/Zn superoxide dismutase mRNA and enzyme activity, and susceptibility to lipid peroxidation, increases with aging in murine brains. *Brain Res. Mol. Brain Res.* **1992**, *13*, 179–187. [CrossRef]

51. Murakami, K.; Murata, N.; Noda, Y.; Tahara, S.; Kaneko, T.; Kinoshita, N.; Hatsuta, H.; Murayama, S.; Barnham, K.J.; Irie, K. SOD1 (copper/zinc superoxide dismutase) deficiency drives amyloid β protein oligomerization and memory loss in mouse model of Alzheimer disease. *J. Biol. Chem.* **2011**, *286*, 44557–44568. [CrossRef]

52. Kim, U.J.; Lee, B.H.; Lee, K.H. Neuroprotective effects of a protein tyrosine phosphatase inhibitor against hippocampal excitotoxic injury. *Brain Res.* **2019**, *1719*, 133–139. [CrossRef]

53. Esposito, L.; Raber, J.; Kekonius, L.; Yan, F.; Yu, G.-Q.; Bien-Ly, N.; Puoliväli, J.; Scearce-Levie, K.; Masliah, E.; Mucke, L. Reduction in mitochondrial superoxide dismutase modulates Alzheimer's disease-like pathology and accelerates the onset of behavioral changes in human amyloid precursor protein transgenic mice. *J. Neurosci.* **2006**, *26*, 5167–5179. [CrossRef] [PubMed]

54. Liao, R.; Wood, T.R.; Nance, E. Superoxide dismutase reduces monosodium glutamate-induced injury in an organotypic whole hemisphere brain slice model of excitotoxicity. *J. Biol. Eng.* **2020**, *14*, 1–12.

55. Hunt, C.R.; Sim, J.E.; Sullivan, S.J.; Featherstone, T.; Golden, W.; Von Kapp-Herr, C.; Hock, R.A.; Gomez, R.A.; Parsian, A.J.; Spitz, D.R. Genomic instability and catalase gene amplification induced by chronic exposure to oxidative stress. *Cancer Res.* **1998**, *58*, 3986–3992. [PubMed]

56. Fita, I.; Rossmann, M.G. The active center of catalase. *J. Mol. Biol.* **1985**, *185*, 21–37. [CrossRef]

57. Usui, S.; Komeima, K.; Lee, S.Y.; Jo, Y.-J.; Ueno, S.; Rogers, B.S.; Wu, Z.; Shen, J.; Lu, L.; Oveson, B.C. Increased expression of catalase and superoxide dismutase 2 reduces cone cell death in retinitis pigmentosa. *Mol. Ther.* **2009**, *17*, 778–786. [CrossRef]

58. Speranza, M.J.; Bagley, A.; Lynch, R. Cells enriched for catalase are sensitized to the toxicities of bleomycin, adriamycin, and paraquat. *J. Biol. Chem.* **1993**, *268*, 19039–19043.

59. Terlecky, S.R.; Koepke, J.I.; Walton, P.A. Peroxisomes and aging. *Biochim. Biophy. Acta* **2006**, *1763*, 1749–1754. [CrossRef]

60. Sheikh, F.G.; Pahan, K.; Khan, M.; Barbosa, E.; Singh, I. Abnormality in catalase import into peroxisomes leads to severe neurological disorder. *Proc. Natl. Acad. Sci. USA* **1998**, *95*, 2961–2966. [CrossRef]

61. Baxter, P.S.; Bell, K.F.; Hasel, P.; Kaindl, A.M.; Fricker, M.; Thomson, D.; Cregan, S.P.; Gillingwater, T.H.; Hardingham, G.E. Synaptic NMDA receptor activity is coupled to the transcriptional control of the glutathione system. *Nat. Commun.* **2015**, *6*, 1–13. [CrossRef]

62. Zhang, Z.; Rydel, R.E.; Drzewiecki, G.J.; Fuson, K.; Wright, S.; Wogulis, M.; Audia, J.E.; May, P.C.; Hyslop, P.A. Amyloid β-mediated oxidative and metabolic stress in rat cortical neurons: No direct evidence for a role for H_2O_2 generation. *J. Neurochem.* **1996**, *67*, 1595–1606. [CrossRef]

63. Nell, H.J.; Au, J.L.; Giordano, C.R.; Terlecky, S.R.; Walton, P.A.; Whitehead, S.N.; Cechetto, D.F. Targeted antioxidant, catalase–SKL, reduces beta-amyloid toxicity in the rat brain. *Brain Pathol.* **2017**, *27*, 86–94. [CrossRef] [PubMed]

64. Yakunin, E.; Kisos, H.; Kulik, W.; Grigoletto, J.; Wanders, R.J.; Sharon, R. The regulation of catalase activity by PPAR γ is affected by α-synuclein. *Ann. Clin. Transl. Neurol.* **2014**, *1*, 145–159. [CrossRef] [PubMed]

65. Ursini, F.; Maiorino, M.; Brigelius-Flohé, R.; Aumann, K.D.; Roveri, A.; Schomburg, D.; Flohé, L. Diversity of glutathione peroxidases. *Methods Enzymol.* **1995**, *252*, 38–53. [PubMed]

66. Kemp, M.; Go, Y.M.; Jones, D.P. Nonequilibrium thermodynamics of thiol/disulfide redox systems: A perspective on redox systems biology. *Free Radic. Biol. Med.* **2008**, *44*, 921–937. [CrossRef] [PubMed]

67. Taylor, J.M.; Ali, U.; Iannello, R.C.; Hertzog, P.; Crack, P.J. Diminished Akt phosphorylation in neurons lacking glutathione peroxidase-1 (Gpx1) leads to increased susceptibility to oxidative stress-induced cell death. *J. Neurochem.* **2005**, *92*, 283–293. [CrossRef] [PubMed]

68. Cardoso, B.R.; Hare, D.J.; Bush, A.I.; Roberts, B.R. Glutathione peroxidase 4: A new player in neurodegeneration? *Mol. Psychiatry* **2017**, *22*, 328–335. [CrossRef]

69. Dixon, S.J.; Lemberg, K.M.; Lamprecht, M.R.; Skouta, R.; Zaitsev, E.M.; Gleason, C.E.; Patel, D.N.; Bauer, A.J.; Cantley, A.M.; Yang, W.S. Ferroptosis: An iron-dependent form of nonapoptotic cell death. *Cell* **2012**, *149*, 1060–1072. [CrossRef]

70. Chen, Z.H.; Yoshida, Y.; Saito, Y.; Niki, E. Adaptation to hydrogen peroxide enhances PC12 cell tolerance against oxidative damage. *Neurosci. Lett.* **2005**, *383*, 256–259. [CrossRef]

71. Goss, J.R.; Taffe, K.M.; Kochanek, P.M.; DeKosky, S.T. The antioxidant enzymes glutathione peroxidase and catalase increase following traumatic brain injury in the rat. *Exp. Neurol.* **1997**, *146*, 291–294. [CrossRef]

72. Hambright, W.S.; Fonseca, R.S.; Chen, L.; Na, R.; Ran, Q. Ablation of ferroptosis regulator glutathione peroxidase 4 in forebrain neurons promotes cognitive impairment and neurodegeneration. *Redox Biol.* **2017**, *12*, 8–17. [CrossRef]

73. Do Van, B.; Gouel, F.; Jonneaux, A.; Timmerman, K.; Gelé, P.; Pétrault, M.; Bastide, M.; Laloux, C.; Moreau, C.; Bordet, R. Ferroptosis, a newly characterized form of cell death in Parkinson's disease that is regulated by PKC. *Neurobiol. Dis.* **2016**, *94*, 169–178. [CrossRef] [PubMed]

74. Chae, H.Z.; Kim, H.J.; Kang, S.W.; Rhee, S.G. Characterization of three isoforms of mammalian peroxiredoxin that reduce peroxides in the presence of thioredoxin. *Diabetes Res. Clin. Pract.* **1999**, *45*, 101–112. [CrossRef]

75. Burke-Gaffney, A.; Callister, M.E.; Nakamura, H. Thioredoxin: Friend or foe in human disease? *Trends Pharmacol. Sci.* **2005**, *26*, 398–404. [CrossRef] [PubMed]

76. Silva-Adaya, D.; Gonsebatt, M.E.; Guevara, J. Thioredoxin system regulation in the central nervous system: Experimental models and clinical evidence. *Oxid. Med. Cell. Longev.* **2014**, *2014*, 590808. [CrossRef]

77. Ren, X.; Zou, L.; Zhang, X.; Branco, V.; Wang, J.; Carvalho, C.; Holmgren, A.; Lu, J. Redox signaling mediated by thioredoxin and glutathione systems in the central nervous system. *Antioxid. Redox Signal.* **2017**, *27*, 989–1010. [CrossRef]

78. Lundström, J.; Holmgren, A. Protein disulfide-isomerase is a substrate for thioredoxin reductase and has thioredoxin-like activity. *J. Biol. Chem.* **1990**, *265*, 9114–9120.

79. Soerensen, J.; Jakupoglu, C.; Beck, H.; Förster, H.; Schmidt, J.; Schmahl, W.; Schweizer, U.; Conrad, M.; Brielmeier, M. The role of thioredoxin reductases in brain development. *PLoS ONE* **2008**, *3*, e1813. [CrossRef]

80. May, J.M.; Cobb, C.E.; Mendiratta, S.; Hill, K.E.; Burk, R.F. Reduction of the ascorbyl free radical to ascorbate by thioredoxin reductase. *J. Biol. Chem.* **1998**, *273*, 23039–23045. [CrossRef]

81. Arnér, E.S.; Nordberg, J.; Holmgren, A. Efficient reduction of lipoamide and lipoic acid by mammalian thioredoxin reductase. *Biochem. Biophys. Res. Commun.* **1996**, *225*, 268–274. [CrossRef]

82. Xia, L.; Björnstedt, M.; Nordman, T.; Eriksson, L.C.; Olsson, J.M. Reduction of ubiquinone by lipoamide dehydrogenase: An antioxidant regenerating pathway. *Eur. J. Biochem.* **2001**, *268*, 1486–1490. [CrossRef]

83. Lu, J.; Zhong, L.; Lönn, M.E.; Burk, R.F.; Hill, K.E.; Holmgren, A. Penultimate selenocysteine residue replaced by cysteine in thioredoxin reductase from selenium-deficient rat liver. *FASEB J.* **2009**, *23*, 2394–2402. [CrossRef] [PubMed]

84. Takagi, Y.; Hattori, I.; Nozaki, K.; Mitsui, A.; Ishikawa, M.; Hashimoto, N.; Yodoi, J. Excitotoxic hippocampal injury is attenuated in thioredoxin transgenic mice. *J. Cereb. Blood Flow Metab.* **2000**, *20*, 829–833. [CrossRef] [PubMed]

85. Takagi, Y.; Mitsui, A.; Nishiyama, A.; Nozaki, K.; Sono, H.; Gon, Y.; Hashimoto, N.; Yodoi, J. Overexpression of thioredoxin in transgenic mice attenuates focal ischemic brain damage. *Proc. Natl. Acad. Sci. USA* **1999**, *96*, 4131–4136. [CrossRef]

86. Björnstedt, M.; Xue, J.; Huang, W.; Akesson, B.; Holmgren, A. The thioredoxin and glutaredoxin systems are efficient electron donors to human plasma glutathione peroxidase. *J. Biol. Chem.* **1994**, *269*, 29382–29384. [PubMed]

87. Munemasa, Y.; Kim, S.H.; Ahn, J.H.; Kwong, J.M.; Caprioli, J.; Piri, N. Protective effect of thioredoxins 1 and 2 in retinal ganglion cells after optic nerve transection and oxidative stress. *Investig. Ophthalmol. Vis. Sci.* **2008**, *49*, 3535–3543. [CrossRef] [PubMed]

88. Zeng, X.S.; Jia, J.J.; Kwon, Y.; Wang, S.D.; Bai, J. The role of thioredoxin-1 in suppression of endoplasmic reticulum stress in Parkinson disease. *Free Radic. Biol. Med.* **2014**, *67*, 10–18. [CrossRef]

89. Lovell, M.A.; Xie, C.; Gabbita, S.P.; Markesbery, W.R. Decreased thioredoxin and increased thioredoxin reductase levels in Alzheimer's disease brain. *Free Radic. Biol. Med.* **2000**, *28*, 418–427. [CrossRef]

90. Sánchez-López, F.; Tasset, I.; Agüera, E.; Feijóo, M.; Fernández-Bolaños, R.; Sánchez, F.M.; Ruiz, M.C.; Cruz, A.H.; Gascón, F.; Túnez, I. Oxidative stress and inflammation biomarkers in the blood of patients with Huntington's disease. *Neurol. Res.* **2012**, *34*, 721–724. [CrossRef]

91. Zhou, F.; Liu, P.P.; Ying, G.Y.; Zhu, X.D.; Shen, H.; Chen, G. Effects of thioredoxin-1 on neurogenesis after brain ischemia/reperfusion injury. *CNS Neurosci. Ther.* **2013**, *19*, 204. [CrossRef]

92. Hwang, I.K.; Yoo, K.-Y.; Kim, D.W.; Lee, C.H.; Choi, J.H.; Kwon, Y.-G.; Kim, Y.-M.; Choi, S.Y.; Won, M.-H. Changes in the expression of mitochondrial peroxiredoxin and thioredoxin in neurons and glia and their protective effects in experimental cerebral ischemic damage. *Free Radic. Biol. Med.* **2010**, *48*, 1242–1251. [CrossRef]

93. Miller, J.; Brzezinska-Slebodzinska, E.; Madsen, F. Oxidative stress, antioxidants, and animal function. *J. Dairy Sci.* **1993**, *76*, 2812–2823. [CrossRef]

94. Hardy, G.; Hardy, I.; Manzanares, W. Selenium supplementation in the critically ill. *Nutr. Clin. Pract.* **2012**, *27*, 21–33. [CrossRef] [PubMed]

95. Hamilton, S.J. Review of selenium toxicity in the aquatic food chain. *Sci. Total Environ.* **2004**, *326*, 1–31. [CrossRef] [PubMed]

96. Solovyev, N.D. Importance of selenium and selenoprotein for brain function: From antioxidant protection to neuronal signalling. *J. Inorg. Biochem.* **2015**, *153*, 1–12. [CrossRef]

97. Santamaría, A.; Vázquez-Román, B.; La cruz, V.P.D.; González-Cortés, C.; Trejo-Solís, M.C.; Galván-Arzate, S.; Jara-Prado, A.; Guevara-Fonseca, J.; Ali, S.F. Selenium reduces the proapoptotic signaling associated to NF-κB pathway and stimulates glutathione peroxidase activity during excitotoxic damage produced by quinolinate in rat corpus striatum. *Synapse* **2005**, *58*, 258–266. [CrossRef] [PubMed]

98. Zachara, B.A.; Pawluk, H.; Bloch-Boguslawska, E.; Śliwka, K.M.; Korenkiewicz, J.; Skok, Ź.; Ryć, K. Tissue level, distribution and total body selenium content in healthy and diseased humans in Poland. *Arch. Environ. Health* **2001**, *56*, 461–466. [CrossRef]

99. Kühbacher, M.; Bartel, J.; Hoppe, B.; Alber, D.; Bukalis, G.; Bräuer, A.U.; Behne, D.; Kyriakopoulos, A. The brain selenoproteome: Priorities in the hierarchy and different levels of selenium homeostasis in the brain of selenium-deficient rats. *J. Neurochem.* **2009**, *110*, 133–142. [CrossRef]

100. Dalla Puppa, L.; Savaskan, N.E.; Braeuer, A.U.; Behne, D.; Kyriakopoulos, A. The role of selenite on microglial migration. *Ann. N. Y. Acad. Sci.* **2007**, *1096*, 179–183. [CrossRef]

101. Uğuz, A.C.; Nazıroğlu, M. Effects of selenium on calcium signaling and apoptosis in rat dorsal root ganglion neurons induced by oxidative stress. *Neurochem. Res.* **2012**, *37*, 1631–1638. [CrossRef]

102. Haratake, M.; Yoshida, S.; Mandai, M.; Fuchigami, T.; Nakayama, M. Elevated amyloid-β plaque deposition in dietary selenium-deficient Tg2576 transgenic mice. *Metallomics* **2013**, *5*, 479–483. [CrossRef]

103. Ishrat, T.; Parveen, K.; Khan, M.M.; Khuwaja, G.; Khan, M.B.; Yousuf, S.; Ahmad, A.; Shrivastav, P.; Islam, F. Selenium prevents cognitive decline and oxidative damage in rat model of streptozotocin-induced experimental dementia of Alzheimer's type. *Brain Res.* **2009**, *1281*, 117–127. [CrossRef] [PubMed]

104. Ellwanger, J.H.; Molz, P.; Dallemole, D.R.; dos Santos, A.P.; Müller, T.E.; Cappelletti, L.; da Silva, M.G.; Franke, S.I.R.; Prá, D.; Henriques, J.A.P. Selenium reduces bradykinesia and DNA damage in a rat model of Parkinson's disease. *Nutrition* **2015**, *31*, 359–365. [CrossRef] [PubMed]

105. Lu, Z.; Marks, E.; Chen, J.; Moline, J.; Barrows, L.; Raisbeck, M.; Volitakis, I.; Cherny, R.A.; Chopra, V.; Bush, A.I. Altered selenium status in Huntington's disease: Neuroprotection by selenite in the N171-82Q mouse model. *Neurobiol. Dis.* **2014**, *71*, 34–42. [CrossRef] [PubMed]

106. Yeo, J.E.; Kang, S.K. Selenium effectively inhibits ROS-mediated apoptotic neural precursor cell death in vitro and in vivo in traumatic brain injury. *Biochim. Biophys. Acta* **2007**, *1772*, 1199–1210. [CrossRef]

107. Khalili, H.; Ahl, R.; Cao, Y.; Paydar, S.; Sjölin, G.; Niakan, A.; Dabiri, G.; Mohseni, S. Early selenium treatment for traumatic brain injury: Does it improve survival and functional outcome? *Injury* **2017**, *48*, 1922–1926. [CrossRef] [PubMed]

108. Nohl, H.; Kozlov, A.V.; Staniek, K.; Gille, L. The multiple functions of coenzyme Q. *Bioorg. Chem.* **2001**, *29*, 1–13. [CrossRef] [PubMed]

109. Ernster, L.; Dallner, G. Biochemical, physiological and medical aspects of ubiquinone function. *Biochim. Biophys. Acta* **1995**, *1271*, 195–204. [CrossRef]

110. Bhagavan, H.N.; Chopra, R.K. Coenzyme Q10: Absorption, tissue uptake, metabolism and pharmacokinetics. *Free Radic. Res.* **2006**, *40*, 445–453. [CrossRef]

111. Hargreaves, I.P. Ubiquinone: Cholesterol's reclusive cousin. *Ann. Clin. Biochem.* **2003**, *40*, 207–218. [CrossRef]

112. Xia, L.; Nordman, T.; Olsson, J.M.; Damdimopoulos, A.; Björkhem-Bergman, L.; Nalvarte, I.; Eriksson, L.C.; Arnér, E.S.; Spyrou, G.; Björnstedt, M. The mammalian cytosolic selenoenzyme thioredoxin reductase reduces ubiquinone a novel mechanism for defense against oxidative stress. *J. Biol. Chem.* **2003**, *278*, 2141–2146. [CrossRef]

113. Montini, G.; Malaventura, C.; Salviati, L. Early coenzyme Q10 supplementation in primary coenzyme Q10 deficiency. *N. Engl. J. Med.* **2008**, *358*, 2849–2850. [CrossRef] [PubMed]

114. Salviati, L.; Sacconi, S.; Murer, L.; Zacchello, G.; Franceschini, L.; Laverda, A.M.; Basso, G.; Quinzii, C.; Angelini, C.; Hirano, M.; et al. Infantile encephalomyopathy and nephropathy with CoQ10 deficiency: A CoQ10-responsive condition. *Neurology* **2005**, *65*, 606–608. [CrossRef]

115. Hargreaves, I.; Mantle, D. Supplementation with selenium and coenzyme Q10 in critically ill patients. *Br. J. Hosp. Med.* **2019**, *80*, 589–593. [CrossRef] [PubMed]

116. Won, R.; Lee, K.H.; Lee, B.H. Coenzyme Q10 protects neurons against neurotoxicity in hippocampal slice culture. *Neuroreport* **2011**, *22*, 721–726. [CrossRef] [PubMed]

117. Park, H.W.; Park, C.G.; Park, M.; Lee, S.H.; Park, H.R.; Lim, J.; Paek, S.H.; Choy, Y.B. Intrastriatal administration of coenzyme Q10 enhances neuroprotection in a Parkinson's disease rat model. *Sci. Rep.* **2020**, *10*, 1–12. [CrossRef] [PubMed]

118. Ramezani, M.; Sahraei, Z.; Simani, L.; Heydari, K.; Shahidi, F. Coenzyme Q10 supplementation in acute ischemic stroke: Is it beneficial in short-term administration? *Nutr. Neurosci.* **2020**, *23*, 640–645. [CrossRef] [PubMed]

119. Yoritaka, A.; Kawajiri, S.; Yamamoto, Y.; Nakahara, T.; Ando, M.; Hashimoto, K.; Nagase, M.; Saito, Y.; Hattori, N. Randomized, double-blind, placebo-controlled pilot trial of reduced coenzyme Q10 for Parkinson's disease. *Parkinsonism Relat. Disord.* **2015**, *21*, 911–916. [CrossRef] [PubMed]

120. Sawaddiruk, P.; Apaijai, N.; Paiboonworachat, S.; Kaewchur, T.; Kasitanon, N.; Jaiwongkam, T.; Kerdphoo, S.; Chattipakorn, N.; Chattipakorn, S.C. Coenzyme Q10 supplementation alleviates pain in pregabalin-treated fibromyalgia patients via reducing brain activity and mitochondrial dysfunction. *Free Radic. Res.* **2019**, *53*, 901–909. [CrossRef] [PubMed]

121. Kloubert, V.; Rink, L. Zinc as a micronutrient and its preventive role of oxidative damage in cells. *Food Funct.* **2015**, *6*, 3195–3204. [CrossRef] [PubMed]

122. Oteiza, P.I. Zinc and the modulation of redox homeostasis. *Free Radic. Biol. Med.* **2012**, *53*, 1748–1759. [CrossRef] [PubMed]

123. Kawahara, M.; Tanaka, K.-i.; Kato-Negishi, M. Zinc, carnosine, and neurodegenerative diseases. *Nutrients* **2018**, *10*, 147. [CrossRef]

124. Frederickson, C.J.; Suh, S.W.; Silva, D.; Frederickson, C.J.; Thompson, R.B. Importance of zinc in the central nervous system: The zinc-containing neuron. *J. Nutr.* **2000**, *130*, 1471S–1483S. [CrossRef] [PubMed]

125. Farbood, Y.; Sarkaki, A.; Mahdavinia, M.; Ghadiri, A.; Teimoori, A.; Seif, F.; Dehghani, M.A.; Navabi, S.P. Protective effects of co-administration of zinc and selenium against streptozotocin-induced Alzheimer's disease: Behavioral, mitochondrial oxidative stress, and GPR39 expression alterations in rats. *Neurotox. Res.* **2020**, *38*, 398–407. [CrossRef] [PubMed]

126. Jafari, F.; Amani, R.; Tarrahi, M.J. Effect of zinc supplementation on physical and psychological symptoms, biomarkers of inflammation, oxidative stress, and brain-derived neurotrophic factor in young women with premenstrual syndrome: A randomized, double-blind, placebo-controlled trial. *Biol. Trace Elem. Res.* **2020**, *194*, 89–95. [CrossRef] [PubMed]

127. Carocci, A.; Catalano, A.; Sinicropi, M.S.; Genchi, G. Oxidative stress and neurodegeneration: The involvement of iron. *Biometals* **2018**, *31*, 715–735. [CrossRef] [PubMed]

128. Farina, M.; Avila, D.S.; Da Rocha, J.B.T.; Aschner, M. Metals, oxidative stress and neurodegeneration: A focus on iron, manganese and mercury. *Neurochem. Int.* **2013**, *62*, 575–594. [CrossRef] [PubMed]

129. Ballaz, S.J.; Rebec, G.V. Neurobiology of vitamin C: Expanding the focus from antioxidant to endogenous neuromodulator. *Pharmacol. Res.* **2019**, *146*, 104321. [CrossRef]

130. Moretti, M.; Fraga, D.B.; Rodrigues, A.L.S. Preventive and therapeutic potential of ascorbic acid in neurodegenerative diseases. *CNS Neurosci. Ther.* **2017**, *23*, 921–929. [CrossRef]

131. Harrison, F.E.; May, J.M. Vitamin C function in the brain: Vital role of the ascorbate transporter SVCT2. *Free Radic. Biol. Med.* **2009**, *46*, 719–730. [CrossRef]

132. García-Krauss, A.; Ferrada, L.; Astuya, A.; Salazar, K.; Cisternas, P.; Martínez, F.; Ramírez, E.; Nualart, F. Dehydroascorbic acid promotes cell death in neurons under oxidative stress: A protective role for astrocytes. *Mol. Neurobiol.* **2016**, *53*, 5847–5863. [CrossRef]

133. Astuya, A.; Caprile, T.; Castro, M.; Salazar, K.; García, M.d.l.A.; Reinicke, K.; Rodríguez, F.; Vera, J.C.; Millán, C.; Ulloa, V. Vitamin C uptake and recycling among normal and tumor cells from the central nervous system. *J. Neurosci. Res.* **2005**, *79*, 146–156. [CrossRef]

134. Ulloa, V.; García-Robles, M.; Martínez, F.; Salazar, K.; Reinicke, K.; Pérez, F.; Godoy, D.F.; Godoy, A.S.; Nualart, F. Human choroid plexus papilloma cells efficiently transport glucose and vitamin C. *J. Neurochem.* **2013**, *127*, 403–414. [CrossRef] [PubMed]

135. Ziylan, Y.Z.; Diler, A.S.; Lefauconnier, J.-M.; Bourre, J.-M. Evidence for ascorbic acid transport system in rat brain capillaries. *Int. J. Neurosci.* **2006**, *116*, 25–38. [CrossRef]

136. Covarrubias-Pinto, A.; Acuna, A.; Boncompain, G.; Papic, E.; Burgos, P.; Perez, F.; Castro, M. Ascorbic acid increases SVCT2 localization at the plasma membrane by accelerating its trafficking from early secretory compartments and through the endocytic-recycling pathway. *Free Radic. Biol. Med.* **2018**, *120*, 181–191. [CrossRef] [PubMed]

137. Getoff, N. Vitamin C: Electron emission, free radicals and biological versatility. *In Vivo* **2013**, *27*, 565–570. [PubMed]

138. Lane, D.J.; Lawen, A. Ascorbate and plasma membrane electron transport—Enzymes vs efflux. *Free Radic. Biol. Med.* **2009**, *47*, 485–495. [CrossRef] [PubMed]

139. Nelson, M.T.; Joksovic, P.M.; Su, P.; Kang, H.-W.; Van Deusen, A.; Baumgart, J.P.; David, L.S.; Snutch, T.P.; Barrett, P.Q.; Lee, J.-H. Molecular mechanisms of subtype-specific inhibition of neuronal T-type calcium channels by ascorbate. *J. Neurosci.* **2007**, *27*, 12577–12583. [CrossRef]

140. Smythies, J. Redox aspects of signaling by catecholamines and their metabolites. *Antioxid. Redox Signal.* **2000**, *2*, 575–583. [CrossRef]

141. Sandstrom, M.I.; Rebec, G.V. Extracellular ascorbate modulates glutamate dynamics: Role of behavioral activation. *BMC Neurosci.* **2007**, *8*, 1–6. [CrossRef]

142. Karanth, S.; Wen, H.Y.; Walczewska, A.; Mastronardi, C.A.; McCann, S.M. Ascorbic acid stimulates gonadotropin release by autocrine action by means of NO. *Proc. Natl. Acad. Sci. USA* **2001**, *98*, 11783–11788. [CrossRef]

143. Harrison, F.; Dawes, S.; Meredith, M.; Babaev, V.; Li, L.; May, J. Low vitamin C and increased oxidative stress and cell death in mice that lack the sodium-dependent vitamin C transporter SVCT2. *Free Radic. Biol. Med.* **2010**, *49*, 821–829. [CrossRef] [PubMed]

144. Naseer, M.; Ullah, N.; Ullah, I.; Koh, P.; Lee, H.; Park, M.; Kim, M. Vitamin C protects against ethanol and PTZ-induced apoptotic neurodegeneration in prenatal rat hippocampal neurons. *Synapse* **2011**, *65*, 562–571. [CrossRef] [PubMed]

145. Choi, D.W. Excitotoxic cell death. *J. Neurobiol.* **1992**, *23*, 1261–1276. [CrossRef] [PubMed]

146. Barak, O.F.; Caljkusic, K.; Hoiland, R.L.; Ainslie, P.N.; Thom, S.R.; Yang, M.; Jovanov, P.; Dujic, Z. Differential influence of vitamin C on the peripheral and cerebral circulation after diving and exposure to hyperoxia. *Am. J. Physiol. Regul. Integr. Comp. Physiol.* **2018**, *315*, R759–R767. [CrossRef]

147. Jain, S.K.; Dar, M.Y.; Kumar, S.; Yadav, A.; Kearns, S.R. Role of anti-oxidant (Vitamin-C) in post-operative pain relief in foot and ankle trauma surgery: A prospective randomized trial. *Foot Ankle Surg.* **2019**, *25*, 542–545. [CrossRef]

148. Kim, E.J.; Park, Y.G.; Baik, E.J.; Jung, S.J.; Won, R.; Nahm, T.S.; Lee, B.H. Dehydroascorbic acid prevents oxidative cell death through a glutathione pathway in primary astrocytes. *J. Neurosci. Res.* **2005**, *79*, 670–679. [CrossRef]

149. Siqueira, I.R.; Elsner, V.R.; Leite, M.C.; Vanzella, C.; dos Santos Moysés, F.; Spindler, C.; Godinho, G.; Battú, C.; Wofchuk, S.; Souza, D.O. Ascorbate uptake is decreased in the hippocampus of ageing rats. *Neurochem. Int.* **2011**, *58*, 527–532. [CrossRef]

150. Naziroglu, M.; Butterworth, P.J.; Sonmez, T.T. Dietary vitamin C and E modulates antioxidant levels in blood, brain, liver, muscle, and testes in diabetic aged rats. *Int. J. Vitam Nutr. Res.* **2011**, *81*, 347. [CrossRef]

151. Kim, E.J.; Won, R.; Sohn, J.-H.; Chung, M.-A.; Nam, T.S.; Lee, H.-J.; Lee, B.H. Anti-oxidant effect of ascorbic and dehydroascorbic acids in hippocampal slice culture. *Biochem. Biophys. Res. Commun.* **2008**, *366*, 8–14. [CrossRef]

152. Rebec, G.V.; Barton, S.J.; Ennis, M.D. Dysregulation of ascorbate release in the striatum of behaving mice expressing the Huntington's disease gene. *J. Neurosci.* **2002**, *22*, RC202. [CrossRef]

153. Rebec, G.V.; Barton, S.J.; Marseilles, A.M.; Collins, K. Ascorbate treatment attenuates the Huntington behavioral phenotype in mice. *Neuroreport* **2003**, *14*, 1263–1265. [CrossRef] [PubMed]

154. Lee, P.; Ulatowski, L.M. Vitamin E: Mechanism of transport and regulation in the CNS. *IUBMB Life* **2019**, *71*, 424–429. [CrossRef] [PubMed]

155. Atkinson, J.; Epand, R.F.; Epand, R.M. Tocopherols and tocotrienols in membranes: A critical review. *Free Radic. Biol. Med.* **2008**, *44*, 739–764. [CrossRef] [PubMed]

156. Goti, D.; Hrzenjak, A.; Levak-Frank, S.; Frank, S.; Van Der Westhuyzen, D.R.; Malle, E.; Sattler, W. Scavenger receptor class B, type I is expressed in porcine brain capillary endothelial cells and contributes to selective uptake of HDL-associated vitamin E. *J. Neurochem.* **2001**, *76*, 498–508. [CrossRef] [PubMed]

157. Balazs, Z.; Panzenboeck, U.; Hammer, A.; Sovic, A.; Quehenberger, O.; Malle, E.; Sattler, W. Uptake and transport of high-density lipoprotein (HDL) and HDL-associated α-tocopherol by an in vitro blood–brain barrier model. *J. Neurochem.* **2004**, *89*, 939–950. [CrossRef]

158. Acton, S.; Rigotti, A.; Landschulz, K.T.; Xu, S.; Hobbs, H.H.; Krieger, M. Identification of scavenger receptor SR-BI as a high density lipoprotein receptor. *Science* **1996**, *271*, 518–520. [CrossRef]

159. Vance, J.E.; Hayashi, H. Formation and function of apolipoprotein E-containing lipoproteins in the nervous system. *Biochim. Biophys. Acta* **2010**, *1801*, 806–818. [CrossRef]

160. Hayashi, H.; Campenot, R.B.; Vance, D.E.; Vance, J.E. Apolipoprotein E-containing lipoproteins protect neurons from apoptosis via a signaling pathway involving low-density lipoprotein receptor-related protein-1. *J. Neurosci.* **2007**, *27*, 1933–1941. [CrossRef]

161. Lane, M.A.; Bailey, S.J. Role of retinoid signalling in the adult brain. *Prog. Neurobiol.* **2005**, *75*, 275–293. [CrossRef]

162. Ambrogini, P.; Torquato, P.; Bartolini, D.; Albertini, M.C.; Lattanzi, D.; Di Palma, M.; Marinelli, R.; Betti, M.; Minelli, A.; Cuppini, R. Excitotoxicity, neuroinflammation and oxidant stress as molecular bases of epileptogenesis and epilepsy-derived neurodegeneration: The role of vitamin E. *Biochim. Biophys. Acta* **2019**, *1865*, 1098–1112. [CrossRef]

163. Mohn, E.S.; Kuchan, M.J.; Erdman, J.W.; Neuringer, M.; Matthan, N.R.; Chen, C.-Y.O.; Johnson, E.J. The subcellular distribution of alpha-tocopherol in the adult primate brain and its relationship with membrane arachidonic acid and its oxidation products. *Antioxidants* **2017**, *6*, 97. [CrossRef] [PubMed]

164. Shah, R.; Shchepinov, M.S.; Pratt, D.A. Resolving the role of lipoxygenases in the initiation and execution of ferroptosis. *ACS Cent. Sci.* **2018**, *4*, 387–396. [CrossRef] [PubMed]

165. Sen, C.K.; Khanna, S.; Rink, C.; Roy, S. Tocotrienols: The Emerging Face of Natural Vitamin E. In *Vitam. Horm*; Academic Press: Cambridge, MA, USA, 2007; Volume 76, pp. 203–261.

166. Jung, N.Y.; Lee, K.H.; Won, R.; Lee, B.H. Neuroprotective effects of α-tocotrienol on kainic acid-induced neurotoxicity in organotypic hippocampal slice cultures. *Int. J. Mol. Sci.* **2013**, *14*, 18256–18268. [CrossRef]

167. Mehrabadi, S.; Sadr, S.S. Administration of Vitamin D3 and E supplements reduces neuronal loss and oxidative stress in a model of rats with Alzheimer's disease. *Neurol. Res.* **2020**, *38*, 1–7. [CrossRef] [PubMed]

168. Ahmed, M.; Alzoubi, K.H.; Khabour, O.F. Vitamin E prevents the cognitive impairments in post-traumatic stress disorder rat model: Behavioral and molecular study. *Psychopharmacology (Berlin)* **2020**, *237*, 599–607. [CrossRef]

169. Taghizadeh, M.; Tamtaji, O.R.; Dadgostar, E.; Kakhaki, R.D.; Bahmani, F.; Abolhassani, J.; Aarabi, M.H.; Kouchaki, E.; Memarzadeh, M.R.; Asemi, Z. The effects of omega-3 fatty acids and vitamin E co-supplementation on clinical and metabolic status in patients with Parkinson's disease: A randomized, double-blind, placebo-controlled trial. *Neurochem. Int.* **2017**, *108*, 183–189. [CrossRef]

170. Shannon, S.R.; Moise, A.R.; Trainor, P.A. New insights and changing paradigms in the regulation of vitamin A metabolism in development. *Wiley Interdiscip. Rev. Dev. Biol.* **2017**, *6*, e264. [CrossRef]

171. Duester, G. Retinoic acid synthesis and signaling during early organogenesis. *Cell* **2008**, *134*, 921–931. [CrossRef]

172. Mueller, L.; Boehm, V. Antioxidant activity of β-carotene compounds in different in vitro assays. *Molecules* **2011**, *16*, 1055–1069. [CrossRef]

173. Honarvar, N.M.; Saedisomeolia, A.; Abdolahi, M.; Shayeganrad, A.; Sangsari, G.T.; Rad, B.H.; Muench, G. Molecular anti-inflammatory mechanisms of retinoids and carotenoids in Alzheimer's disease: A review of current evidence. *J. Mol. Neurosci.* **2017**, *61*, 289–304. [CrossRef]

174. Woodall, A.A.; Lee, S.W.-M.; Weesie, R.J.; Jackson, M.J.; Britton, G. Oxidation of carotenoids by free radicals: Relationship between structure and reactivity. *Biochim. Biophys. Acta* **1997**, *1336*, 33–42. [CrossRef]

175. Ben-Dor, A.; Steiner, M.; Gheber, L.; Danilenko, M.; Dubi, N.; Linnewiel, K.; Zick, A.; Sharoni, Y.; Levy, J. Carotenoids activate the antioxidant response element transcription system. *Mol. Cancer Ther.* **2005**, *4*, 177–186. [PubMed]

176. Palozza, P.; Serini, S.; Torsello, A.; Nicuolo, F.D.; Piccioni, E.; Ubaldi, V.; Pioli, C.; Wolf, F.I.; Calviello, G. Nutrient-gene interactions-b-carotene regulates NF-kB DNA-binding activity by a redox mechanism in human leukemia and colon adenocarcinoma cells. *J. Nutr.* **2003**, *133*, 381–388. [CrossRef] [PubMed]

177. MacDonald, P.N.; Bok, D.; Ong, D.E. Localization of cellular retinol-binding protein and retinol-binding protein in cells comprising the blood-brain barrier of rat and human. *Proc. Natl. Acad. Sci. USA* **1990**, *87*, 4265–4269. [CrossRef]

178. Craft, N.; Haitema, T.; Garnett, K.; Fitch, K.; Dorey, C. Carotenoid, tocopherol, and retinol concentrations in elderly human brain. *Exp. Anim.* **2004**, *21*, 22.

179. de Oliveira, B.F.; Veloso, C.A.; Nogueira-Machado, J.A.; de Moraes, E.N.; dos Santos, R.R.; Cintra, M.T.G.; Chaves, M.M. Ascorbic acid, alpha-tocopherol, and beta-carotene reduce oxidative stress and proinflammatory cytokines in mononuclear cells of Alzheimer's disease patients. *Nutr. Neurosci.* **2012**, *15*, 244–251. [CrossRef]

180. Shudo, K.; Fukasawa, H.; Nakagomi, M.; Yamagata, N. Towards retinoid therapy for Alzheimer's disease. *Curr. Alzheimer Res.* **2009**, *6*, 302–311. [CrossRef]

181. Bonhomme, D.; Minni, A.M.; Alfos, S.; Roux, P.; Richard, E.; Higueret, P.; Moisan, M.-P.; Pallet, V.; Touyarot, K. Vitamin A status regulates glucocorticoid availability in Wistar rats: Consequences on cognitive functions and hippocampal neurogenesis? *Front. Behav. Neurosci.* **2014**, *8*, 20. [CrossRef]

182. Zetterström, R.H.; Lindqvist, E.; De Urquiza, A.M.; Tomac, A.; Eriksson, U.; Perlmann, T.; Olson, L. Role of retinoids in the CNS: Differential expression of retinoid binding proteins and receptors and evidence for presence of retinoic acid. *Eur. J. Neurosci.* **1999**, *11*, 407–416. [CrossRef]

183. Sodhi, R.K.; Singh, N. Retinoids as potential targets for Alzheimer's disease. *Pharmacol. Biochem. Behav.* **2014**, *120*, 117–123. [CrossRef]

184. Ahlemeyer, B.; Krieglstein, J. Inhibition of glutathione depletion by retinoic acid and tocopherol protects cultured neurons from staurosporine-induced oxidative stress and apoptosis. *Neurochem. Int.* **2000**, *36*, 1–5. [CrossRef]

185. Ahlemeyer, B.; Bauerbach, E.; Plath, M.; Steuber, M.; Heers, C.; Tegtmeier, F.; Krieglstein, J. Retinoic acid reduces apoptosis and oxidative stress by preservation of SOD protein level. *Free Radic. Biol. Med.* **2001**, *30*, 1067–1077. [CrossRef]

186. McCaffery, P.; Dräger, U.C. High levels of a retinoic acid-generating dehydrogenase in the meso-telencephalic dopamine system. *Proc. Natl. Acad. Sci. USA* **1994**, *91*, 7772–7776. [CrossRef] [PubMed]

187. Lévesque, D.; Rouillard, C. Nur77 and retinoid X receptors: Crucial factors in dopamine-related neuroadaptation. *Trends Neurosci.* **2007**, *30*, 22–30. [CrossRef]

188. Crandall, J.; Sakai, Y.; Zhang, J.; Koul, O.; Mineur, Y.; Crusio, W.E.; McCaffery, P. 13-cis-retinoic acid suppresses hippocampal cell division and hippocampal-dependent learning in mice. *Proc. Natl. Acad. Sci. USA* **2004**, *101*, 5111–5116. [CrossRef]

189. Pan, J.; Yu, J.; Sun, L.; Xie, C.; Chang, L.; Wu, J.; Hawes, S.; Saez–Atienzar, S.; Zheng, W.; Kung, J. ALDH1A1 regulates postsynaptic μ–opioid receptor expression in dorsal striatal projection neurons and mitigates dyskinesia through transsynaptic retinoic acid signaling. *Sci. Rep.* **2019**, *9*, 1–14. [CrossRef]

190. Esteves, M.; Cristóvão, A.C.; Saraiva, T.; Rocha, S.M.; Baltazar, G.; Ferreira, L.; Bernardino, L. Retinoic acid-loaded polymeric nanoparticles induce neuroprotection in a mouse model for Parkinson's disease. *Front. Aging Neurosci.* **2015**, *7*, 20. [CrossRef]

191. Prema, A.; Janakiraman, U.; Manivasagam, T.; Thenmozhi, A.J. Neuroprotective effect of lycopene against MPTP induced experimental Parkinson's disease in mice. *Neurosci. Lett.* **2015**, *599*, 12–19. [CrossRef]

192. Wen, X.; Huang, A.; Hu, J.; Zhong, Z.; Liu, Y.; Li, Z.; Pan, X.; Liu, Z. Neuroprotective effect of astaxanthin against glutamate-induced cytotoxicity in HT22 cells: Involvement of the Akt/GSK-3β pathway. *Neuroscience* **2015**, *303*, 558–568. [CrossRef]

193. Bitarafan, S.; Saboor-Yaraghi, A.; Sahraian, M.-A.; Soltani, D.; Nafissi, S.; Togha, M.; Moghadam, N.B.; Roostaei, T.; Honarvar, N.M.; Harirchian, M.-H. Effect of vitamin A supplementation on fatigue and depression in multiple sclerosis patients: A double-blind placebo-controlled clinical trial. *Iran. J. Allergy Asthma Immunol.* **2016**, *15*, 13–19.

194. Corbett, A.; Pickett, J.; Burns, A.; Corcoran, J.; Dunnett, S.B.; Edison, P.; Hagan, J.J.; Holmes, C.; Jones, E.; Katona, C. Drug repositioning for Alzheimer's disease. *Nat. Rev. Drug Discov.* **2012**, *11*, 833–846. [CrossRef] [PubMed]

195. McGarry, A.; McDermott, M.; Kieburtz, K.; de Blieck, E.A.; Beal, F.; Marder, K.; Ross, C.; Shoulson, I.; Gilbert, P.; Mallonee, W.M. A randomized, double-blind, placebo-controlled trial of coenzyme Q10 in Huntington disease. *Neurology* **2017**, *88*, 152–159. [CrossRef] [PubMed]

196. Kryscio, R.J.; Abner, E.L.; Caban-Holt, A.; Lovell, M.; Goodman, P.; Darke, A.K.; Yee, M.; Crowley, J.; Schmitt, F.A. Association of antioxidant supplement use and dementia in the prevention of Alzheimer's disease by vitamin E and selenium trial (PREADViSE). *JAMA Neurol.* **2017**, *74*, 567–573. [CrossRef] [PubMed]

197. Salari, S.; Khomand, P.; Arasteh, M.; Yousefzamani, B.; Hassanzadeh, K. Zinc sulphate: A reasonable choice for depression management in patients with multiple sclerosis: A randomized, double-blind, placebo-controlled clinical trial. *Pharmacol. Rep.* **2015**, *67*, 606–609. [CrossRef]

198. Endres, K.; Fahrenholz, F.; Lotz, J.; Hiemke, C.; Teipel, S.; Lieb, K.; Tüscher, O.; Fellgiebel, A. Increased CSF APPs-α levels in patients with Alzheimer disease treated with acitretin. *Neurology* **2014**, *83*, 1930–1935. [CrossRef]

199. Thorsen, E.; Haave, H.; Hofsø, D.; Ulvik, R.J. Exposure to hyperoxia in diving and hyperbaric medicine–effects on blood cell counts and serum ferritin. *Undersea Hyperb. Med.* **2001**, *28*, 57–62.

200. Group, P.S. Effects of tocopherol and deprenyl on the progression of disability in early Parkinson's disease. *N. Engl. J. Med.* **1993**, *328*, 176–183.

201. Zhang, D.; Hannink, D.D. Distinct Cysteine residues in keap1 are required for keap1-dependent ubiquitination of Nrf2 and for stabilization of Nrf2 by chemopreventive agents and oxidative stress. *Mol. Cell. Biol.* **2003**, *23*, 8137–8151. [CrossRef]

202. Suzuki, T.; Motohashi, H.; Yamamoto, M. Toward clinical application of the Keap1–Nrf2 pathway. *Trends Pharmacol. Sci.* **2013**, *34*, 340–346. [CrossRef]

203. Strom, J.; Xu, B.; Tian, X.; Chen, Q.M. Nrf2 protects mitochondrial decay by oxidative stress. *FASEB J.* **2016**, *30*, 66–80. [CrossRef]

204. Rojo, A.I.; Pajares, M.; Rada, P.; Nuñez, A.; Nevado-Holgado, A.J.; Killik, R.; Van Leuven, F.; Ribe, E.; Lovestone, S.; Yamamoto, M.; et al. NRF$_2$ deficiency replicates transcriptomic changes in Alzheimer's patients and worsens APP and TAU pathology. *Redox Biol.* **2017**, *13*, 444–451. [CrossRef] [PubMed]

205. Tanaka, N.; Ikeda, Y.; Ohta, Y.; Deguchi, K.; Tian, F.; Shang, J.; Matsuura, T.; Abe, K. Expression of Keap1–Nrf2 system and antioxidative proteins in mouse brain after transient middle cerebral artery occlusion. *Brain Res.* **2011**, *1370*, 246–253. [CrossRef]

206. Zhao, X.; Sun, G.; Zhang, J.; Strong, R.; Dash, P.K.; Kan, Y.W.; Grotta, J.C.; Aronowski, J. Transcription factor Nrf2 protects the brain from damage produced by intracerebral hemorrhage. *Stroke* **2007**, *38*, 3280–3286. [CrossRef] [PubMed]

207. Burton, N.C.; Kensler, T.W.; Guilarte, T.R. In vivo modulation of the Parkinsonian phenotype by Nrf2. *Neurotoxicology* **2006**, *27*, 1094–1100. [CrossRef] [PubMed]

208. Chen, P.C.; Vargas, M.R.; Pani, A.K.; Smeyne, R.J.; Johnson, D.A.; Kan, Y.W.; Johnson, J.A. Nrf2-mediated neuroprotection in the MPTP mouse model of Parkinson's disease: Critical role for the astrocyte. *Proc. Natl. Acad. Sci. USA* **2009**, *106*, 2933–2938. [CrossRef] [PubMed]

209. Park, J.S.; Leem, Y.H.; Park, J.E.; Kim, D.Y.; Kim, H.S. Neuroprotective effect of β-lapachone in MPTP-induced Parkinson's disease mouse model: Involvement of astroglial p-AMPK/Nrf2/HO-1 signaling pathways. *Biomol. Ther.* **2019**, *27*, 178. [CrossRef] [PubMed]

210. Rhee, S.G. Cell signaling. H2O2, a necessary evil for cell signaling. *Science* **2006**, *312*, 1882–1883. [CrossRef]
211. Schieber, M.; Chandel, N.S. ROS function in redox signaling and oxidative stress. *Curr. Biol.* **2014**, *24*, R453–R462. [CrossRef]

International Journal of
Molecular Sciences

Article

An Inhibitor of the Sodium–Hydrogen Exchanger-1 (NHE-1), Amiloride, Reduced Zinc Accumulation and Hippocampal Neuronal Death after Ischemia

Beom Seok Kang [1], Bo Young Choi [1], A Ra Kho [1], Song Hee Lee [1], Dae Ki Hong [1], Jeong Hyun Jeong [1], Dong Hyeon Kang [2], Min Kyu Park [1] and Sang Won Suh [1,*]

[1]   Department of Physiology, College of Medicine, Hallym University, Chuncheon 24252, Korea;
      ttiger1993@gmail.com (B.S.K.); bychoi@hallym.ac.kr (B.Y.C.); rnlduadkfk136@hallym.ac.kr (A.R.K.);
      sshlee@hallym.ac.kr (S.H.L.); zxnm01220@gmail.com (D.K.H.); jd1422@hanmail.net (J.H.J.);
      bagmingyu50@gmail.com (M.K.P.)
[2]   Department of Medical Science, College of Medicine, Hallym University, Chuncheon 24252, Korea;
      ehdgus6312@gmail.com
*   Correspondence: swsuh@hallym.ac.kr; Tel.: +82-10-8573-6364

Received: 10 May 2020; Accepted: 12 June 2020; Published: 14 June 2020

Abstract: Acidosis in the brain plays an important role in neuronal injury and is a common feature of several neurological diseases. It has been reported that the sodium–hydrogen exchanger-1 (NHE-1) is a key mediator of acidosis-induced neuronal injury. It modulates the concentration of intra- and extra-cellular sodium and hydrogen ions. During the ischemic state, excessive sodium ions enter neurons and inappropriately activate the sodium–calcium exchanger (NCX). Zinc can also enter neurons through voltage-gated calcium channels and NCX. Here, we tested the hypothesis that zinc enters the intracellular space through NCX and the subsequent zinc accumulation induces neuronal cell death after global cerebral ischemia (GCI). Thus, we conducted the present study to confirm whether inhibition of NHE-1 by amiloride attenuates zinc accumulation and subsequent hippocampus neuronal death following GCI. Mice were subjected to GCI by bilateral common carotid artery (BCCA) occlusion for 30 min, followed by restoration of blood flow and resuscitation. Amiloride (10 mg/kg, intraperitoneally (*i.p.*)) was immediately injected, which reduced zinc accumulation and neuronal death after GCI. Therefore, the present study demonstrates that amiloride attenuates GCI-induced neuronal injury, likely via the prevention of intracellular zinc accumulation. Consequently, we suggest that amiloride may have a high therapeutic potential for the prevention of GCI-induced neuronal death.

Keywords: global cerebral ischemia; amiloride; sodium–hydrogen exchanger-1; zinc; neuronal death; neuroprotection

---

1. Introduction

Ischemic stroke is one of the most severe cerebral pathological conditions and can manifest via a number of clinical symptoms such as problems in cognition, dizziness, or loss of vision on one side of the visual field [1]. In particular, the development of ischemic conditions in the brain is very dangerous because only a momentary lack of adequate blood flow to the brain can lead to oxygen deprivation, insufficient nutrient provision, and potentially irreversible neural injury [2–4]. There are two basic types of ischemic injury: global and focal ischemia. Global ischemia impacts wide areas of brain tissue at once due to the blockage of blood flow to an entire region of the brain, while focal ischemia is limited to a specific region of the brain tissue and is due to the more local disruption of cerebral blood flow. Ischemia-induced brain damage can be recovered by early reperfusion, but this reperfusion process can also initiate independent cascades of cell death pathways such as zinc release, microglial activation, and blood–brain barrier (BBB) disruption [5]. If the interruption of blood circulation happens for an extended period prior to the restoration of circulation, brain damage can be permanent.

Furthermore, after a long period of ischemia, secondary brain damage occurs when blood flow is initiated, which is known as "reperfusion injury". The sudden recovery of blood supply leads to mitochondrial dysfunction, producing excessive reactive oxygen species (ROS) and, finally, cell death [6]. Under healthy conditions, superoxide production via neuronal nicotinamide adenine dinucleotide phosphate (NADPH) oxidase is thought to play a part in normal physiological processes such as long-term potentiation and intracellular signaling [7,8]. However, under ischemic conditions, mitochondrial dysfunction and excessive ROS generation predominate and promote pathological responses such as leukocyte invasion and the disruption of the BBB [9,10]. Also, tissue pH is typically reduced to 6.0–6.5 during ischemia, falling even lower during severe ischemic conditions [11–13]. As a result, this study suggests that the increase in NADPH activity after ischemia may also promote cell death by acidifying brain cells.

Zinc is one of the most essential transition metals in our body, especially in the brain. Furthermore, zinc regulates physiological functions that control DNA synthesis, cell division, and signal transduction. Most of the zinc is present in a protein-bound form in neuronal cytoplasm. Free or chelatable zinc is localized within the vesicles of synaptic terminals [14,15]. Taken together, these lines of circumstantial evidence suggest that zinc plays a key role in maintaining cellular homeostasis. However, previous studies have demonstrated that excessive neuronal zinc accumulation occurs after traumatic brain injury, ischemia, hyperglycemia, and epilepsy [16–19]. Under these various pathological conditions, ROS changes the protein-mediated sequestration of zinc and, thus, increases intracellular free zinc levels which increases ROS production. If this condition is maintained for long periods, it can lead to neuronal death [18]. Several neurological injuries, accelerate zinc release from synaptic vesicles, and zinc then moves into neurons via multiple classes of membrane-bound channels [20,21].

Sodium–hydrogen exchangers (NHEs) are membrane transporters that mediate hydrogen efflux into cells. Sodium–hydrogen exchanger-1 (NHE-1) is a ubiquitous and essential membrane ion transporter that mediates the electroneutral exchange of hydrogen and sodium to regulate intracellular pH [22]. Other NHEs, namely, NHE-2–5, indicate more distinct cell-type- and tissue-dependent expressions than NHE-1 and play key roles in regulating transcellular sodium and potassium ion absorption [23]. The NHE-6 expression is localized to early endosomes [24]. The NHE-7 and -8 isoforms have not been localized in the brain, and the NHE-9 isoform is localized to late recycling endosomes [25]. Here, we focused on NHE-1 which is expressed in high abundance in the brain.

Amiloride, also known as thiazide, is related to other loop diuretic agents and blocks sodium hydrogen exchanger-1 [26]. Amiloride, an inhibitor of NHE-1, has demonstrated neuroprotective effects in various neuropathological conditions involving brain injuries such as ischemia, dementia, and epilepsy [27,28]. Previous studies have suggested that injection of amiloride can reduce cerebral-hypoxia-induced neuronal death after seizure and spinal cord injury [29–31]. Under the acidic conditions found in the post-ischemic state, as NHE-1 becomes activated, hydrogen is released into the extracellular space and sodium moves into the intracellular compartment. Thus, intracellular sodium is increased and the sodium–calcium exchanger (NCX) becomes activated, so that sodium is released into the extracellular space driving calcium entry, leading to a calcium overload state and, ultimately, resulting in neuronal death. Considering this, we hypothesized that not only calcium, but also zinc may enter via NCX and that NHE-1 blockade by amiloride decreases zinc accumulation and neuronal death through reduced NCX activity. Therefore, we investigated whether the administration of amiloride (10 mg/kg, intraperitoneally (*i.p.*)) reduces zinc accumulation, neuronal degeneration, oxidative stress, and microtubule damage after global cerebral ischemia. As a result, the present study proved that the administration of amiloride reduces ischemia-induced neuronal degeneration.

2. Results

2.1. Amiloride Reduced Global Cerebral Ischemia-Induced Hippocampus Neuronal Death after 24 Hour Post-Insult

To investigate whether amiloride has neuroprotective effects after global cerebral ischemia (GCI)-induced hippocampus neuronal death, experimental mice were immediately intraperitoneally injected with amiloride (10 mg/kg) after GCI. The mice were sacrificed 24 h after GCI with or without amiloride treatment. After insult, a histological evaluation to detect degenerating neurons was conducted in the hippocampal subiculum (Sub), cornus ammonis 1 (CA1), CA2, and CA3 regions. Fluoro-Jade B (FJB) staining showed widespread neuronal degeneration in the Sub, CA1, CA2, and CA3 regions of the hippocampus ($p < 0.05$) (Figure 1A). This staining protocol is a sensitive and selective marker of degenerating neurons. The number of degenerating neurons was increased in the GCI-induced group compared with the sham-operated group. When compared with the GCI-vehicle-treated groups, the amiloride-injected groups showed a dramatically reduced number of degenerating hippocampal neurons. Figure 1B shows the quantified FJB (+) neurons in the Sub, CA1, CA2, and CA3 regions. Amiloride-administered groups displayed a reduction of FJB (+) neurons of approximately 64% in the Sub (GCI-vehicle, 76.9 ± 8.4; GCI-amiloride, 27.2 ± 10.5), 84% in the CA1 (GCI-vehicle, 126.2 ± 18.5; GCI-amiloride, 19.5 ± 17.6), 83% in the CA2 (GCI-vehicle, 125.3 ± 22; GCI-amiloride, 20.7 ± 17.9), and 50% in the CA3 (GCI-vehicle, 220.6 ± 30.3; GCI-amiloride, 109.5 ± 29) regions compared with the vehicle-treated groups.

2.2. Amiloride Reduced Global Cerebral Ischemia-Induced Hippocampal Zinc Accumulation after 24 Hour Post-Insult

To estimate GCI-induced zinc accumulation, brain sections were histologically evaluated by N-(6-methoxy-8-quinolyl)-para-toluenesulfonamide (TSQ) staining 24 h after GCI. Zinc accumulation is known to advance the neuronal NADPH oxidase activity and ROS responsible for neuronal death. Under normal conditions, zinc levels are controlled by zinc transporters and zinc-binding proteins [32]. However, under conditions such as ischemia, traumatic brain injury, and seizure, neuronal death occurs in part by the destruction of zinc homeostasis. Thus, we performed TSQ fluorescence staining to confirm whether amiloride can reduce zinc accumulation in the brain hippocampal CA1 region. The intensity of TSQ staining in the CA1 region was reduced in the amiloride-treated group compared with the GCI-induced group ($p < 0.05$) (Figure 1C,D). Amiloride-administered groups displayed an approximately 44% reduction of TSQ (+) neurons in the CA1 region (GCI-vehicle, 36.9 ± 5.9; GCI-amiloride, 20.6 ± 4.1) compared with the vehicle-treated groups.

2.3. Amiloride Reduced Global Cerebral Ischemia-Induced Astrocyte and Microglial Activation 24 Hour Post-Insult

Multiple studies have previously reported that ischemia induces astrocyte and microglial activation in the brain hippocampal region [3,18,33], and it is well known that inflammation contributes to the severity of astrocyte and microglial activation. In addition, activated astrocytes are potentially harmful, because they can produce nitric oxide synthase (NOS) and neurotoxic nitric oxide (NO). To test whether amiloride affects GCI-induced glial activation, we performed immunofluorescence staining using the marker glial fibrillary acidic protein (GFAP) and ionized calcium-binding adaptor molecule 1 (Iba-1). We confirmed an assessment of the microglia number, morphology, and intensity and astrocyte intensity [34]. Global cerebral ischemia triggers astroglia and microglial activation which is considered to have macrophage-like activity. However, activated astrocytes were reduced by approximately 45% in the GCI–amiloride groups compared with the GCI-vehicle groups in the CA1 region (GCI-vehicle, 33.2 ± 3.3; GCI–amiloride, 18.2 ± 1.8). Further, activated microglia were reduced by 32% in the GCI–amiloride groups in the CA1 region (GCI-vehicle, 5.8 ± 0.3; GCI–amiloride, 3.9 ± 0.3) (Figure 2B,D).

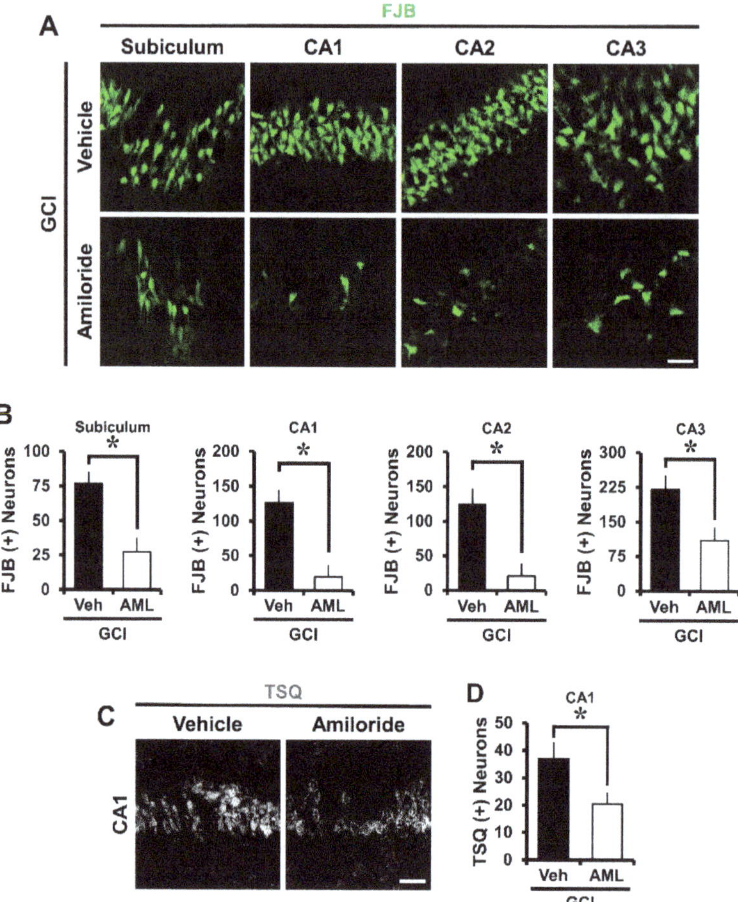

Figure 1. Amiloride treatment decreased the number of degenerating neurons and zinc accumulation after global cerebral ischemia (GCI). GCI-induced hippocampus neuronal death was confirmed in the subiculum (Sub), cornus ammonis 1 (CA1), CA2, and CA3 regions after ischemic insult. Zinc accumulation was confirmed in the CA1 region after ischemic insult. (**A**) Fluorescent images show degenerated neurons in the Sub, CA1, CA2, and CA3 regions. Intraperitoneal post-treatment with amiloride (10 mg/kg) reduced neuronal death in the Sub, CA1, CA2, and CA3 regions at 24 h after ischemia. Scale bar = 20 µm. (**B**) Bar graph displaying the quantification of degenerating neurons in the hippocampal regions. The number of FJB (+) neurons was decreased in the amiloride-injected (10 mg/kg) group in the Sub, CA1, CA2, and CA3 regions compared with the vehicle-treated group (GCI-vehicle, $n = 8$; GCI-amiloride, $n = 8$). (**C**) Representative images show *N*-(6-methoxy-8-quinolyl)-para-toluenesulfonamide (TSQ) (+) neurons in the CA1 region. Scale bar = 20 µm. (**D**) The bar graph indicates the TSQ (+) neurons in the hippocampal CA1 region (GCI-vehicle, $n = 8$; GCI-amiloride, $n = 10$). Data are mean ± S.E.M. * Considerably different from the vehicle-treated group, $p < 0.05$. (Mann–Whitney U test (B) Sub: $z = 2.626$, $p = 0.007$; CA1: $z = 2.838$, $p = 0.003$; CA2: $z = 2.836$, $p = 0.003$; CA3: $z = 2.205$, $p = 0.028$; (D) CA1: $z = 2.134$, $p = 0.034$).

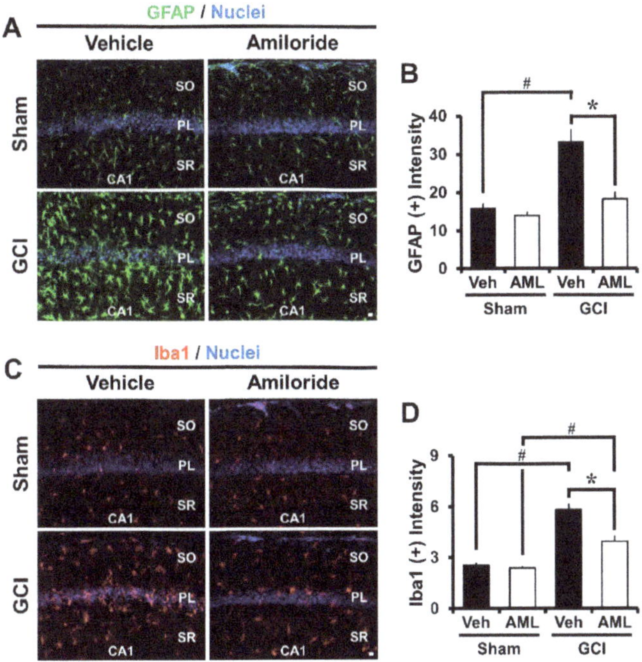

Figure 2. Amiloride treatment reduced astrocyte and microglial activation. GCI induces an inflammatory response by promoting astrocyte and microglia activation in the damaged brain. This figure shows astrocyte and microglia activation in the CA1 region 24 h after GCI. (**A**,**C**) show glial fibrillary acidic protein (GFAP) and Ionized calcium-binding adaptor molecule-1 (Iba-1) activation and (**B**,**D**) show GFAP and Iba-1 quantification in the CA1 region from sham-operated or GCI-induced mice. It was increased in the GCI-induced group compared with the sham-operated group. However, the amiloride-treated group showed reduced astrocyte and microglia activation after GCI. Scale bar = 20 μm. (Sham-vehicle, n = 5; sham-amiloride, n = 5; GCI-vehicle, n = 6; GCI-amiloride, n = 8). Data are mean ± SEM. * Considerably different from the vehicle-treated group, $p < 0.05$; # sham versus vehicle-operated group; sham versus vehicle-treated group, $p < 0.05$. (Kruskal–Wallis test (B) Chi square = 14.612, df = 3, $p = 0.002$; (D) Chi square = 18.606, df = 3, $p < 0.001$) (SO: stratum oriens; PL: pyramidal cell layer; SR: stratum lacunosum-moleculare).

2.4. Amiloride Reduces Global Cerebral Ischemia-Induced Oxidative Damage after 24 Hour Post-Insult

We estimated oxidative stress by using 4-hydroxynonenal (4HNE) staining. The brain was immunohistochemically stained with a 4HNE antibody 24 h after global cerebral ischemia induction to discover whether hippocampal neurons had experienced oxidative stress. The sham-vehicle and amiloride-injected groups showed no difference in 4HNE fluorescence signals. However, the 4HNE fluorescence signal of the GCI-vehicle groups increased in the hippocampal Sub, CA1, CA2, and CA3 regions. The amiloride-treated group showed a significant reduction of 4HNE intensity compared with the vehicle-treated groups (Figure 3A). Oxidative damage was reduced in the GCI-amiloride groups compared with the GCI-vehicle groups by approximately 42% in the Sub (GCI-vehicle, 21.8 ± 2.4; GCI-amiloride, 12.6 ± 0.5), 48% in the CA1 (GCI-vehicle, 23.5 ± 2.3; GCI-amiloride, 12.2 ± 1.5), 36% in the CA2 (GCI-vehicle, 21.5 ± 2.4; GCI-amiloride, 13.7 ± 1.4), and 48% in the CA3 (GCI-vehicle, 31.2 ± 3.3; GCI-amiloride, 16.1 ± 2.2) regions (Figure 3B).

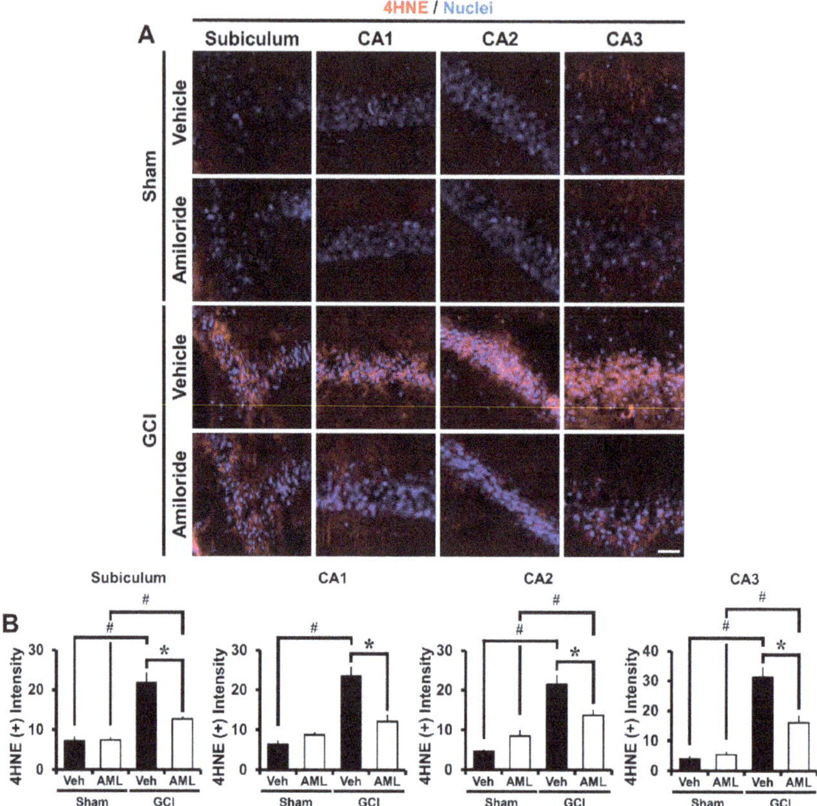

Figure 3. Amiloride reduced oxidative injury after GCI. Oxidative injury was detected by 4-hydroxynonenal (4HNE, red color) staining from the hippocampal Sub, CA1, CA2, and CA3 regions 24 h after GCI. (**A**) Sham-operated groups showed minimal 4HNE fluorescence signals in the hippocampus. Amiloride-treated groups showed reduced immunoreactive fluorescence intensity for 4HNE in the hippocampus compared with the vehicle-treated group after GCI. Scale bar = 20 μm. (**B**) The bar graph presents the 4HNE fluorescence intensity in the Sub, CA1, CA2, and CA3 regions. The fluorescence intensity showed a significant difference among groups (sham-vehicle, $n = 6$; sham-amiloride, $n = 5$; GCI-vehicle, $n = 8$; GCI-amiloride, $n = 8$). Data are mean ± S.E.M. * Considerably different from the vehicle-treated group, $p < 0.05$; # sham versus vehicle-operated group, sham versus vehicle-treated group, $p < 0.05$. (Kruskal–Wallis test (B) Sub: Chi square = 22.444, df = 3, $p < 0.001$; CA1: Chi square = 17.896, df = 3, $p < 0.001$; CA2: Chi square = 20.967, df = 3, $p < 0.001$; CA3: Chi square = 20.986, df = 3, $p < 0.001$).

2.5. Amiloride Reduced Global Cerebral Ischemia-Induced Microtubule Damage after 24 Hour Post-Insult

To evaluate whether GCI-induced microtubule damage occurred, brain sections were histologically processed with antibodies against microtubule-associated protein 2 (MAP2) 24 h after GCI. The GCI-vehicle groups showed a significant reduction in MAP2 immunoreactivity (IR) in the hippocampus and cortex compared with the GCI-amiloride group, indicating a loss of microtubules. Amiloride injection reduced GCI-induced microtubule damage compared with the vehicle-treated group (Figure 4A). Microtubule intensity was increased in the GCI-amiloride groups compared with the GCI-vehicle groups by approximately 42% in the Sub (GCI-vehicle, 41.9 ± 6.3; GCI-amiloride, 72.5 ± 7.2), 60% in the CA1 (GCI-vehicle, 43.3 ± 8.4; GCI-amiloride, 109.7 ± 12.5), 43% in the CA2

(GCI-vehicle, 56.2 ± 11; GCI-amiloride, 99.6 ± 10.1), and 59% in the CA3 (GCI-vehicle, 23.2 ± 5.9; GCI-amiloride, 57.2 ± 10.6) regions (Figure 4B).

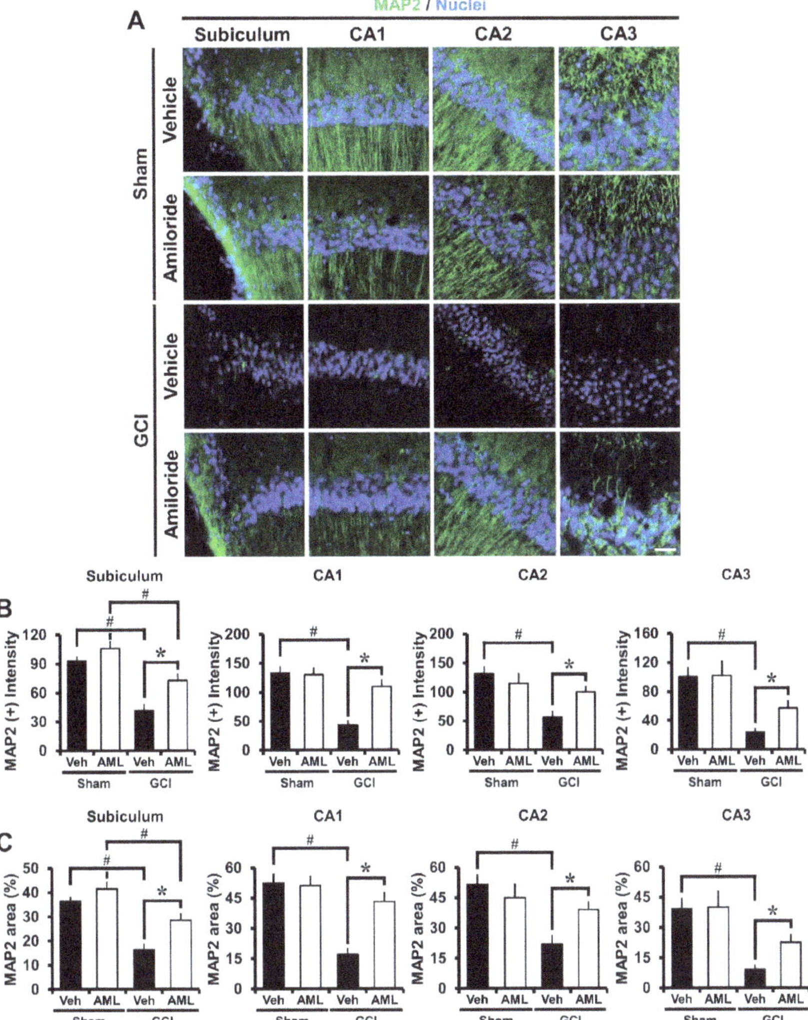

Figure 4. Microtubule damage was detected by microtubule-associated protein 2 (MAP2, green color) staining at the hippocampal Sub, CA1, CA2, and CA3 regions 24 h after GCI. (**A**) Sham-operated groups showed MAP2 fluorescence signals in the hippocampus. Amiloride-administered groups showed a reduced microtubule loss in the hippocampal regions compared with the vehicle-treated group. Scale bar = 20 μm. (**B**) The bar graph indicates the MAP2 fluorescence intensity in the hippocampus. (**C**) The bar graph indicates the MAP2 percent area in the hippocampus (sham-vehicle, $n = 6$; sham-amiloride, $n = 5$; GCI-vehicle, $n = 8$; GCI-amiloride, $n = 8$). Data are mean ± S.E.M. * Considerably different from the vehicle-treated group, $p < 0.05$; # sham versus vehicle-operated group, sham versus vehicle-treated group, $p < 0.05$. (Kruskal–Wallis test (B) Sub: Chi square = 18.901, df = 3, $p < 0.001$; CA1: Chi square = 15.166, df = 3, $p < 0.002$; CA2: Chi square = 15.054, df = 3, $p < 0.002$; CA3: Chi square = 17.137, df = 3, $p < 0.001$).

2.6. Amiloride Prevented Global Cerebral Ischemia-Induced Blood–Brain Barrier Disruption after 24 Hour Post-Insult

To verify the degree of blood–brain barrier (BBB) disruption, we stained brain sections to evaluate extravasation of serum immunoglobulin G (IgG) by using immunohistochemistry as described before [35,36]. In sham-operated brain sections, leakage of IgG was not detected. However, in ischemia-induced mice, we observed excessive extravascular IgG leakage in the hippocampus (Figure 5A). Figure 5B shows a bar graph of the scale of IgG extravasation from the damaged BBB in the hippocampus. IgG leakage was reduced by 24% in the GCI-amiloride group compared with the GCI-vehicle group (GCI-vehicle, 1.37 ± 0.09; GCI-amiloride, 1.04 ± 0.07) (Figure 5B).

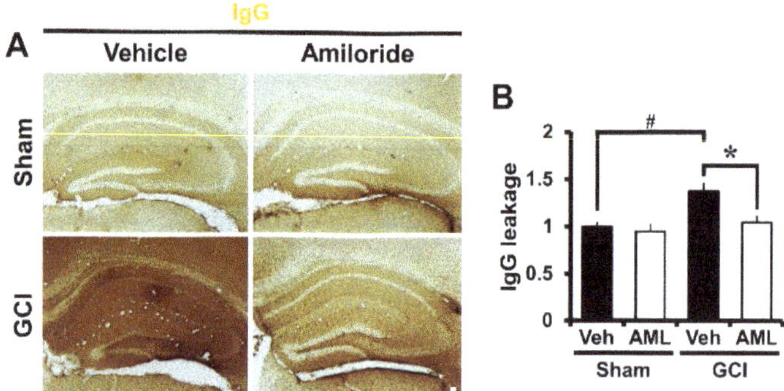

Figure 5. GCI-induced blood–brain barrier (BBB) disruption was decreased by amiloride administration. Brain sections were stained with antibodies against IgG to detect BBB disruption. (**A**) Indicates magnification (4×) of a microscopic image of IgG staining in the hippocampus in each group. These images indicate that BBB disruption occurred after GCI. The GCI-amiloride group had decreased leakage of serum IgG in the hippocampus compared with the GCI-vehicle group. Scale bar = 100 μm. (**B**) Bar graph shows the quantification of IgG serum extravasation in the hippocampus (sham-to-GCI ratio, sham-vehicle, $n = 6$; sham-amiloride, $n = 5$; GCI-vehicle, $n = 7$; GCI-amiloride, $n = 8$). Data are mean ± SEM. * Considerably different from the vehicle-treated group, $p < 0.05$; # sham versus vehicle-operated group; $p < 0.05$. (Kruskal–Wallis test (B) Chi square = 11.126, df = 3, $p = 0.011$).

2.7. Amiloride Improves Global Cerebral Ischemia-Induced Survival of Hippocampal Neurons at 3 Days Post-Insult

To investigate whether amiloride promotes neuronal survival after GCI-induced hippocampus, amiloride (10 mg/kg) was immediately injected to the intraperitoneal space after termination of the blood reperfusion process. Both vehicle- and amiloride-administrated cohorts were sacrificed at 3 days following ischemic insult. After the insult, a histological evaluation using NeuN staining to detect and quantify newly generated neurons was conducted in the hippocampal Sub, CA1, CA2, and CA3 regions. NeuN positive neurons were widespread in the hippocampal regions we examined ($p < 0.05$) and increased in number rapidly (Figure 6A). Figure 6B shows the counted NeuN (+) neurons in the hippocampal regions. Amiloride-administered groups displayed an increase of NeuN (+) neurons of 26% in the Sub (GCI-vehicle, 115.1 ± 4.0; GCI-amiloride, 156.4 ± 10.8), 17% in the CA1 (GCI-vehicle, 177.7 ± 3.5; GCI-amiloride, 213.4 ± 11.1), 17% in the CA2 (GCI-vehicle, 219.5 ± 5.9; GCI-amiloride, 265 ± 10.9), and 17% in the CA3 (GCI-vehicle, 251.2 ± 8.2; GCI-amiloride, 304.5 ± 13.2) regions compared with the vehicle-treated groups.

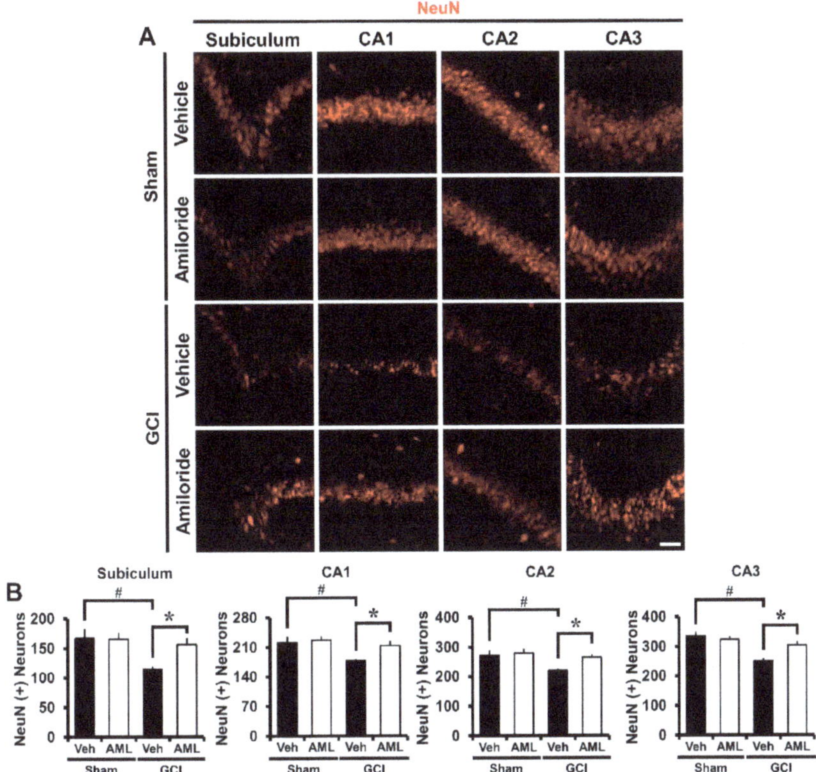

Figure 6. Amiloride treatment improves neuronal survival after GCI. The presence of live neurons after ischemic insult was confirmed in the Sub, CA1, CA2, and CA3 regions. (**A**) Fluorescent images show surviving neurons in the Sub, CA1, CA2, and CA3 regions. Intraperitoneal post-treatment with amiloride (10 mg/kg) increased the number of live neurons in the Sub, CA1, CA2, and CA3 regions at 3 days post-ischemia. Scale bar = 20 μm. (**B**) Bar graph displaying the quantification of surviving neurons in hippocampal subregions. The number of NeuN (+) neurons was increased in the amiloride-injected (10 mg/kg) group in the Sub, CA1, CA2, and CA3 regions compared with the vehicle-treated group (GCI-vehicle, $n = 8$; GCI-amiloride, $n = 6$). Data are mean ± SEM. * Considerably different from the vehicle-treated group, $p < 0.05$; # sham versus vehicle-operated group, $p < 0.05$. (Kruskal–Wallis test (B) Sub: Chi square = 14.214, df = 3, $p = 0.003$; CA1: Chi square = 13.422, df = 3, $p = 0.004$; CA2: Chi square = 14.249, df = 3, $p = 0.003$; CA3: Chi square = 16.158, df = 3, $p = 0.001$).

3. Discussion

Previous studies have demonstrated that amiloride showed neuroprotective effects in two different stroke models; middle cerebral artery occlusion (MCAO) in rat and transient forebrain ischemia in gerbil [27,37]. However, no studies have been performed with the global cerebral ischemia model with mice. The present study investigated whether amiloride administration has potential therapeutic effects for GCI-induced hippocampus neuronal damage and zinc accumulation via inhibition of NHE-1 in mice. Consequently, we found that amiloride significantly reduced zinc accumulation, neuronal degeneration, oxidative damage, microtubule damage, astrocyte and microglial activation, and BBB disruption.

Under ischemic conditions, blood flow to the brain is rapidly and dramatically reduced. This phenomenon leads to a lack of oxygen and other substrates to the nervous tissue.

Thus, the extracellular concentration of glucose is quickly reduced [38–40]. As a result, the physiological glycolysis process was disturbed, and glucose 6-phosphate altered ribulose 5-phosphate by the pentose phosphate pathway (PPP). While glucose is modified to ribulose 5-phosphate, hydrogen is released and triggers NHE-1 [41,42]. When the intracellular hydrogen ion concentration increases, NHE-1 is activated in the cell membrane. NHE-1 activation contributes to neuronal electron exchange via sodium and hydrogen ion exchange across the cell membrane [43]. Previous studies have shown that NHE-1 inhibitors reduced the activity of NCX [27,44–47]. Furthermore, additional studies demonstrated that zinc enters intracellularly through NCX [48,49]. Thus, we can speculate that NHE-1 inhibition may indirectly inhibit zinc influx into neurons. Also, ischemia-induced ROS formation degrades zinc-binding proteins in the intracellular space, thus significantly increasing free zinc levels within the intracellular space [18]. As this cycle becomes sustained, the accumulation of excessive amounts of zinc contributes to neuronal cell death [3,18,27,50] (Figure 7).

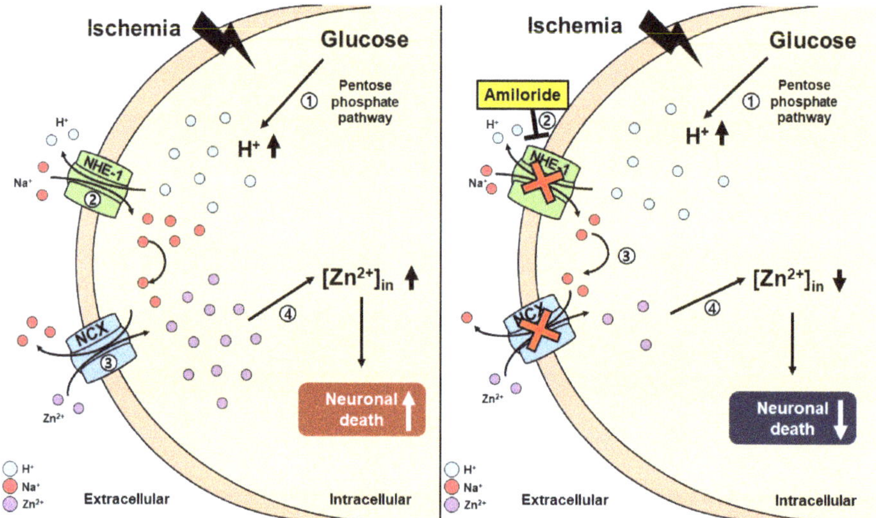

Figure 7. This schematic illustration shows amiloride action via inhibition and downregulation of the sodium–hydrogen exchanger-1 (NHE-1) channel. (**A**) [1] Global cerebral ischemia insult results in increasing levels of intracellular hydrogen. [2] Hydrogen ions are moved to the extracellular space via NHE-1. Sodium ions are moved to the intracellular space via NHE-1. [3] When sodium is overloaded in the intracellular space, it is released into the extracellular space through the sodium–calcium exchanger (NCX). Extracellular zinc enters the cell through NCX. [4] Intracellular zinc accumulation occurs, leading to neuronal death. (**B**) However, [1] after global cerebral ischemia, [2] amiloride administration inhibits intracellular sodium accumulation via the NHE-1 channel. [3] Sodium does not enter the intracellular space and becomes NCX inactivated. [4] Because of this agent's mechanism, zinc accumulation is reduced, resulting in reduced neuronal death after global cerebral ischemic insult.

Amiloride has previously been described as a diuretic and as a non-specific inhibitor for NHE-1. However, it has been used as an NHE-1 inhibitor in several studies [26,51]. Additionally, amiloride has been known to cross the blood–brain barrier and has neuroprotective effects against global cerebral ischemia. Several studies have demonstrated that amiloride has neuroprotective properties using a dose of 10 mg/kg [27,52–54]. Following this logic, we hypothesized that GCI-induced hippocampal damage can be protected by reducing intracellular zinc accumulation through inhibition of NHE-1 [27,55].

The histological evaluation, we performed further supports our hypothesis. The number of hippocampal degenerating neurons was estimated using FJB staining. The number of FJB fluorescence-signal-positive neurons in the hippocampal Sub, CA1, CA2, and CA3 regions was significantly reduced in the amiloride-administered group. Intraneuronal free zinc accumulation was displayed by TSQ staining. The number of the TSQ-positive neurons were significantly reduced in the amiloride-administered groups compared with the vehicle groups. These results demonstrated that the intraneuronal free zinc accumulation was reduced by inhibiting NHE-1 channels using amiloride. We confirmed that under ischemic conditions, zinc accumulation increases, and neuronal death occur. However, blocking NHE-1 by injection of amiloride reduces zinc accumulation and, subsequently, neuronal death was reduced. Additionally, we confirmed the presence of live neurons at 3 days following ischemic insult and stained for NeuN to identify neurons that survived the insult. The number of NeuN positive neurons in the hippocampal Sub, CA1, CA2 and CA3 regions were greater in the amiloride-administered group. So, we concluded that the administration of amiloride improved the survival of hippocampal neurons after 3 days post-insult.

Astrocytes and microglia play important roles in the brain. During neurological disorders such as ischemia, neuroinflammation, and neurodegenerative disease, astrocytes and microglia are over activated. Under ischemic conditions, activated astrocytes, together with reactive microglia, release several pro-inflammatory factors such as tumor necrosis factor-α (TNF-α), ROS, NO, and interleukin-1β (IL-1β), which exacerbate tissue damage [18,56–58]. Previous studies have demonstrated that NHE-1 expression occurs in astrocyte and microglial cells. In addition, NHE-1 activity caused by ischemia activates astrocyte and microglial cells, leading to neuronal death [18,59]. So, we thought that the administration of amiloride, an NHE-1 inhibitor, would reduce the activation of astrocytes and microglia, and thus reduce neuronal death. The present study verified astrocyte activation by GFAP and microglia activation by Iba-1 immunofluorescence staining in the hippocampal CA1 region. We found that amiloride administration reduced reactive astrocyte and microglial activation after GCI.

Ischemic damage leaded to microtubule damage and ROS production. In the present study, we found that amiloride administration reduced microtubule damage after GCI. ROS formation is caused via multiple intracellular signaling cascades, such as iron-associated free radical formation, depletion of antioxidant enzymes, and an increase in the breakdown of lipids and fatty acids after GCI [60]. Because of this, the zinc accumulation described above occurs and microtubules are damaged [18,61]. In addition, several previous studies have suggested that peroxynitrite (PN) toxicity is mediated by intracellular zinc release [62,63]. Peroxynitrite is produced by a combination of nitric oxide and superoxide. It has been reported that PN are endogenous reactive nitrogen species formed when superoxide radicals, or oxygen reacts with nitric oxide formed by inducible nitric oxide synthase (iNOS). Peroxynitrite can induce cytoplasmic free zinc release, mitochondria dysfunction, and lead to BBB disruption, and finally neuronal death, in several types of brain injuries [64–66]. To test whether ROS activation was reduced by amiloride, 4HNE staining was performed in the hippocampal Sub, CA1, CA2, and CA3 regions. The 4HNE fluorescence signal was significantly increased in the GCI-vehicle group. For the amiloride-administered group after GCI, the 4HNE fluorescence signal was significantly decreased in hippocampal regions compared with the GCI-vehicle group. Amiloride, which inhibits NHE-1 and decreases NCX activity, reduces ROS formation by regulating the intracellular ion balance.

Finally, we evaluated the BBB balance after GCI. Abnormal intracellular zinc accumulation and vesicular zinc release may mediate BBB disruption after brain insults such as ischemia, multiple sclerosis, and traumatic brain injury [67]. Thus, administration of amiloride may decrease zinc accumulation and reduces BBB disruption after GCI. After GCI insult, the BBB was destroyed, leading to extravasation of plasma components, such as erythrocytes, leukocytes, and several immunoglobulins. BBB disruption triggers neurodegenerative processes and produces neurotoxic substrates, which causes brain dysfunction that is deleterious to synapse function and disturbs neural transmission [68,69]. So, to evaluate the effects of amiloride on BBB disruption, we conducted IgG staining. As a result, we found that the IgG staining intensity was reduced in the amiloride-administered group compared with the vehicle group after GCI. This result indicates that the neuroprotective provided by amiloride administration might be mediated via protection against BBB disruption, which is one of the main mechanisms associated with GCI-induced neuronal death [70].

Taken together, we conclude that inhibition of NHE-1 by amiloride reduces neuronal death and zinc accumulation. We found that amiloride reduced various deleterious features associated with GCI, such as neurodegeneration, zinc accumulation, oxidative damage, microtubule damage, glial activation, and BBB disruption, which strongly indicates that amiloride administration has neuroprotective effects by decreasing ROS production and zinc accumulation in hippocampal neurons through the inhibition of NHE-1. Therefore, the present study suggests that amiloride can be a potential therapeutic tool to prevent ischemia-induced neuronal death.

4. Materials and Methods

4.1. Ethics Statement

The present study was performed in accordance with the protocols of the Guidelines for the Use and Care of Laboratory Animals, allowed by the National Institutes of Health. Animal studies were conducted in accordance with the guidelines of the Committee on Animal Use for Study and Education at Hallym University (protocol # Hallym-2018-32; Data of approval: July 19,2018). We sacrificed mice under isoflurane anesthesia to minimize any pain and suffering.

4.2. Experimental Animals

The present study used 2–3 month old adult male C57BL/6J mice (20–25 g, DBL Co., Chungcheongbuk-do, Eumseong-gun, Korea). The mice were housed three to four mice per cage under conditions of sustained temperature (20 ± 2 °C) and humidity (55% ± 5%). Animal room lights were managed automatically, turned on and off in a 12 h cycle (on at 6:00 a.m. and off at 6:00 p.m.).

4.3. Global Cerebral Ischemia Surgery

Male C57BL/6J mice (aged 2–3 months, weight 20–25 g) from DBL (Chungcheongbuk-do, Eumseong-gun, Korea) were used as controls. The mice were anesthetized with 2% isoflurane in a 75:25 mixture of oxygen and nitrous oxide. Core temperature was maintained at 36.7–37.5 °C with a homeothermic blanket control unit (Harvard Apparatus, Holliston, MA, USA). Bilateral common carotid arteries (BCCAs) were exposed through a midline neck incision. The BCCAs were loosely encircled with a 4/O silk suture before the occlusion. Aneurysmal clips were used to occlude BCCAs. Mice were subjected to common carotid artery occlusion for 30 min while anesthetized with 1% isoflurane [71–73]. The aneurysmal clips were removed, and the BCCAs were inspected for normal recovery of blood flow after the end of the 30 min ischemic period. Anesthetics were discontinued following the suture of the skin incision. When mice confirmed spontaneous respiration, they were returned to a recovery room retained at 37 °C. Sham-operated animals received the same neck skin incision under isoflurane anesthesia without BCCA occlusion.

4.4. Amiloride Administration

To confirm the effect of amiloride on GCI-induced neuronal death, the experimental groups were divided into four groups: sham (vehicle, amiloride) and global cerebral ischemia (vehicle, amiloride). The amiloride-treated groups were administrated amiloride (10 mg/kg, i.p.) dissolved in 0.9% normal saline. After GCI induction, we immediately injected the amiloride into the intraperitoneal space. The vehicle group was given 0.9% normal saline instead of amiloride. All experimental groups were sacrificed 24 h after GCI.

4.5. Brain Sample Preparation

Mice were sacrificed at 24 h or 3 days after GCI using urethane (1.5 g/kg, i.p.) to deeply anesthetize them. After anesthetizing, mice were perfused transcardially with 0.9% saline, followed by 4%

paraformaldehyde (PFA). Afterwards, harvested brains were post-fixed for approximately 1 h in 4% PFA. After fixation in PFA, brains were moved into a 30% sucrose solution overnight for cryoprotection. After the brain sank to the bottom of the sucrose solution, they were frozen in the freezing medium for 10 min and then cut with cryostats at 30 μm thicknesses. Brain slices were kept in storage solution until used for immunohistochemistry and immunofluorescence staining.

4.6. Confirmation of Hippocampal Neuronal Death

To confirm neuronal death after GCI, brain sections (30 μm) were put on gelatin-coated slides (Fisher Scientific, Pittsburgh, PA, USA). To detect degenerating neurons, brain slices were stained by the FJB staining method [74,75]. Firstly, a slide with a brain section was soaked in a 100% ethanol solution for 3 min, a 70% ethanol solution for 1 min, distilled water for 1 min, and then in 0.06% potassium permanganate for 15 min. Next, the slides were put into 0.001% FJB (Histo-Chem Inc., Jefferson, AR, USA) solution for 30 min and washed three times for 10 min in distilled water. After washing, slides were dried by a gentle air flow (Labtech, Co., Ltd., Namyangju, Korea), dehydrated in xylene for 2 min, and then mounted with DPX (Sigma-Aldrich Co., St. Louis, MO, USA). Slides were checked under a fluorescence microscope (Olympus, Japan) via blue (450–490 nm) excitation light. We chose about six to eight coronal brain sections that were collected from each mouse. A blinded observer counted the FJB-positive cells. The FJB-positive cells were counted and evaluated in the hippocampal Sub, CA1, CA2, and CA3 regions from the bilateral hemisphere. The total number of FJB-positive cells from the hippocampal region was used for statistical analysis. A blinded observer counted the FJB-positive cells. The FJB-positive cells were counted and evaluated in the hippocampal Sub, CA1, CA2, and CA3 regions from the bilateral hemisphere. The total number of FJB-positive cells from the hippocampal region was used for statistical analysis.

4.7. Confirmation of Hippocampal Zinc Translocation

Intracellular free zinc was verified using TSQ staining [76]. Mice were sacrificed 24 h after amiloride (10 mg/kg, i.p.) administration and the fresh frozen, but not fixed, brains were coronally sectioned at 10 μm thicknesses in a −15 °C cryostat, then mounted on gelatin-coated slides and dried. Five evenly spaced sections were chosen from the hippocampal region of each brain and soaked in a solution of 4.5 mmol/L TSQ (Enzo Life Science, Enzo Biochem, Inc, Farmingdale, New York, NY, USA, ENZ-52153) for 1 min, then washed for 1 min in 0.9% saline. Each sample was photographed with a microscope under 360 nm UV light and a 500 nm long-pass filter. We used the Image J (National Institute of Health, Bethesda, Rockville, MD, USA) program to measure zinc intensity and evaluated the mean gray value.

4.8. Evaluation of Hippocampal Oxidative Stress

To analyze oxidative damage induced by the lipid peroxidation product from the brain sections, 4HNE was detected by immunofluorescence staining. 4HNE antibodies (Alpha Diagnostic Intl. Inc., San Antonio, TX, USA) for immunohistochemical staining were used as in previous studies [77,78]. Brain sections were soaked in a monoclonal mouse anti-4HNE serum (diluted 1:500, Alpha Diagnostic Intl. Inc., San Antonio, TX, USA) with the PBS containing 0.3% TritonX-100 overnight in a 4 °C incubator. After overnight incubation, brain sections were washed three times for 10 min with 0.01 M PBS, and then the brain sections were also soaked in a solution of Alexa-Fluor-594-conjugated donkey anti-mouse IgG secondary antibody (diluted 1:250, Invitrogen, Grand Island, NY, USA) for 2 h at room temperature (RT). The brain sections were raised on gelatin-coated slides for analysis under a microscope. We used the Image J (NIH, Bethesda, Rockville, MD, USA) program to measure the oxidative injury and measured the mean gray value [79].

4.9. Evaluation of Hippocampal Microtubule Damage

To analyze microtubule damage from the brain sections, MAP2 was detected by immunofluorescence staining. MAP2 antibodies (Alpha Diagnostic Intl. Inc., San Antonio, TX, USA) for immunohistochemical staining were used as in a previous study [80]. Brain sections were soaked in a polyclonal rabbit anti-MAP2 serum (diluted 1:200, Alpha Diagnostic Intl. Inc., San Antonio, TX, USA) with PBS containing 0.3% TritonX-100 overnight in a 4 °C incubator. After overnight incubation, we washed the sections three times for 10 min with 0.01 M PBS, and then the brain sections were soaked in a solution of Alexa-Fluor-488-conjugated donkey anti-rabbit IgG secondary antibody (diluted 1:250, Invitrogen, Grand Island, NY, USA) for 2 h at RT. The brain sections were raised on gelatin-coated slides for analysis under a microscope. We used the Image J (NIH, Bethesda, Rockville, MD, USA) program to measure the microtubule damage and measured the mean gray value.

4.10. Evaluation of Hippocampal Astrocytes and Microglia

To analyze astrocyte and microglial activation, we performed Iba-1 and GFAP staining. Staining was used with a mixture of goat antibody to mouse Iba-1 (diluted 1:500, Abcam, Cambridge, UK) and rabbit antibody to mouse GFAP (diluted 1:1000, Abcam, Cambridge, UK). Following incubation in 0.01 M PBS containing 0.3% TritonX-100, we left it overnight in a 4 °C incubator. After overnight incubation, we washed the sections three times for 10 min with 0.01 M PBS. Then, the sections were soaked in a secondary antibody (Alexa-Fluor-488-conjugated donkey anti-rabbit IgG secondary antibody and Alexa-Fluor-594-conjugated donkey anti-goat IgG secondary antibody, both diluted 1:250, Invitrogen, Grand Island, NY, USA) for 2 h at RT. The brain sections were raised on gelatin-coated slides for analysis under a microscope. We used the Image J (NIH, Bethesda, Rockville, MD, USA) program to measure the astrocytes. In the case of microglia, five brain sections were scored with the same area (20× magnification) of the hippocampal CA1 region. The functional standards of microglia cells were their number, morphology, and intensity of microglia activation. Iba-1-immunoreactive cell score of 0: no cells are present; 1:1–9 cells; 2:10–20 cells; and 3:>20 cells with continuous processers per 100 μm^2. Morphology score of 0:no activated morphology (amoeboid morphology with enlarged soma and thickened processes); 1:1–45% of microglia activation; 2:45–90% of microglia activation; and 3: >90% of microglia with the activated morphology. The intensity of microglial activation was measured using the Image J (NIH, Bethesda, Rockville, MD, USA) program. After the measurements, an intensity score of 1:0–19% expression; 2:20–29% expression; and 3:>29% expression. Therefore, the total score summed up the three scores depending on the categories, ranging from 0 to 9 [18,81,82].

4.11. Evaluation of BBB Disruption

To analyze the putative breakdown of the BBB, we used immunohistochemistry to find serum IgG leakage [83]. To detect IgG-like immunoreactivity, the ABC immunoperoxidase protocol was used [84]. Mouse brains were fixed by transcardiac perfusion with 0.9% normal saline, followed by 4% paraformaldehyde. We used anti-mouse IgG (diluted 1:250, Burlingame, Vector, CA, USA) which can discover leakages of IgG when the BBB is damaged. After washing in 0.01 M PBS, brain sections were deeply soaked in the ABC complex mixture (Vector, Burlingame, CA, USA) for 2 h at RT. The immunoreactivity was visualized with 0.06% 3,3'-diaminobenzidine (DAB ager, Sigma–Aldrich Co., St. Louis, MO, USA) in 0.1 M PBS buffer. Leaked IgG extravasations were detected using a bright-field microscope.

4.12. Evaluation of Live Hippocampal Neurons

To assess the number of live neurons present in a sample, NeuN was detected by immunofluorescence staining. NeuN antibodies (diluted 1:500, EMD Millipore, Billerica, MA, USA) for immunohistochemical staining were used as in previous studies. Brain sections were soaked in a monoclonal rabbit anti-NeuN serum with PBS containing 0.3% TritonX-100 overnight in a 4 °C incubator. After overnight incubation, brain sections were washed three times for 10 min each with 0.01 M PBS, and then the brain sections were also soaked in a solution of Alexa-Fluor-594-conjugated donkey anti-rabbit IgG secondary antibody (diluted 1:250, Invitrogen, Grand Island, NY, USA) for 2 h at room temperature (RT). The brain sections were raised on gelatin-coated slides for analysis under a microscope. A blinded observer counted the NeuN-positive cells. NeuN-positive cells were counted and evaluated in the hippocampal Sub, CA1, CA2, and CA3 regions from each hemisphere. The total number of NeuN-positive cells from the hippocampal region was used for statistical analysis [18].

4.13. Statistical Analysis

We conducted nonparametric testing to confirm any statistical significance between the experimental groups. Data were analyzed using the Bonferroni post-hoc test, and the Kruskal–Wallis test was employed to compare among the four groups. For comparison across two groups, data were analyzed using the Mann–Whitey U test. Data are displayed as the mean $\pm$ SEM. Statistical significance is described as $p < 0.05$.

5. Conclusions

This present study supports the hypothesis that the administration of amiloride protects against hippocampal neuronal death and zinc accumulation after GCI. In addition, inhibition of NHE-1 by amiloride may have considerable therapeutic potential for the prevention of GCI.

Author Contributions: B.S.K. researched and collected the data and reviewed and edited the manuscript. B.Y.C. reviewed and edited the manuscript. A.R.K., S.H.L., D.K.H., J.H.J., D.H.K., and M.K.P. researched and collected the data. S.W.S. contributed to the discussion and wrote, reviewed, and edited the manuscript. All authors have read and agree to the published version of the manuscript.

Funding: This study was supported by funding from the National Research Foundation of Korea (NRF) (NRF-2019R1A2C4004912) to B.Y.C. Additionally, this work was supported by the Brain Research Program through the NRF, funded by the Ministry of Science, Information, and Communication Technology and Future Planning (NRF-2017M3C7A1028937 and 2020R1A2C2008480) to S.W.S.

Conflicts of Interest: The authors declare no conflict of interest.

References

1. Wan, L.; Cheng, Y.; Luo, Z.; Guo, H.; Zhao, W.; Gu, Q.; Yang, X.; Xu, J.; Bei, W.; Guo, J. Neuroprotection, learning and memory improvement of a standardized extract from Renshen Shouwu against neuronal injury and vascular dementia in rats with brain ischemia. *J. Ethnopharmacol.* **2015**, *165*, 118–126. [CrossRef] [PubMed]

2. Lee, R.H.C.; Lee, M.H.H.; Wu, C.Y.C.; Couto, E.S.A.; Possoit, H.E.; Hsieh, T.H.; Minagar, A.; Lin, H.W. Cerebral ischemia and neuroregeneration. *Neural Regen. Res.* **2018**, *13*, 373–385. [PubMed]

3. Hong, D.K.; Choi, B.Y.; Kho, A.R.; Lee, S.H.; Jeong, J.H.; Kang, B.S.; Kang, D.H.; Park, K.H.; Suh, S.W. Carvacrol attenuates hippocampal neuronal death after global cerebral ischemia via inhibition of transient receptor potential melastatin 7. *Cells* **2018**, *7*, 231. [CrossRef] [PubMed]

4. Ding, Q.; Liao, S.J.; Yu, J. Axon guidance factor netrin-1 and its receptors regulate angiogenesis after cerebral ischemia. *Neurosci. Bull.* **2014**, *30*, 683–691. [CrossRef] [PubMed]

5. Dirnagl, U.; Lindauer, U.; Them, A.; Schreiber, S.; Pfister, H.W.; Koedel, U.; Reszka, R.; Freyer, D.; Villringer, A. Global cerebral ischemia in the rat: Online monitoring of oxygen free radical production using chemiluminescence in vivo. *J. Cereb. Blood Flow Metab.* **1995**, *15*, 929–940. [CrossRef] [PubMed]

6. White, B.C.; Sullivan, J.M.; DeGracia, D.J.; O'Neil, B.J.; Neumar, R.W.; Grossman, L.I.; Rafols, J.A.; Krause, G.S. Brain ischemia and reperfusion: Molecular mechanisms of neuronal injury. *J. Neurol. Sci.* **2000**, *179*, 1–33. [CrossRef]

7. Rhee, S.G. Cell signaling. H_2O_2, a necessary evil for cell signaling. *Science* **2006**, *312*, 1882–1883. [CrossRef]

8. Klann, E. Cell-permeable scavengers of superoxide prevent long-term potentiation in hippocampal area CA1. *J. Neurophysiol.* **1998**, *80*, 452–457. [CrossRef]

9.   Kalogeris, T.; Baines, C.P.; Krenz, M.; Korthuis, R.J. Cell biology of ischemia/reperfusion injury. *Int Rev. Cell Mol. Biol.* **2012**, *298*, 229–317.

10. Pan, J.; Konstas, A.A.; Bateman, B.; Ortolano, G.A.; Pile-Spellman, J. Reperfusion injury following cerebral ischemia: Pathophysiology, MR imaging, and potential therapies. *Neuroradiology* **2007**, *49*, 93–102. [CrossRef]

11. Rehncrona, S. Brain acidosis. *Ann. Emerg. Med.* **1985**, *14*, 770–776. [CrossRef]

12. Nedergaard, M.; Kraig, R.P.; Tanabe, J.; Pulsinelli, W.A. Dynamics of interstitial and intracellular pH in evolving brain infarct. *Am. J. Physiol.* **1991**, *260*, R581–R588. [CrossRef] [PubMed]

13. Siesjo, B.K.; Katsura, K.; Kristian, T. Acidosis-related damage. *Adv. Neurol.* **1996**, *71*, 209–233. [PubMed]

14. Cull, R.E. Role of axonal transport in maintaining central synaptic connections. *Exp. Brain Res.* **1975**, *24*, 97–101. [CrossRef] [PubMed]

15. Takeda, A.; Kodama, Y.; Ohnuma, M.; Okada, S. Zinc transport from the striatum and substantia nigra. *Brain Res. Bull.* **1998**, *47*, 103–106. [CrossRef]

16. Choi, D.W. Calcium-mediated neurotoxicity: Relationship to specific channel types and role in ischemic damage. *Trends Neurosci.* **1988**, *11*, 465–469. [CrossRef]

17. Chuah, M.I.; Tennent, R.; Jacobs, I. Response of olfactory Schwann cells to intranasal zinc sulfate irrigation. *J. Neurosci. Res.* **1995**, *42*, 470–478. [CrossRef]

18. Kho, A.R.; Choi, B.Y.; Lee, S.H.; Hong, D.K.; Lee, S.H.; Jeong, J.H.; Park, K.H.; Song, H.K.; Choi, H.C.; Suh, S.W. Effects of protocatechuic acid (PCA) on global cerebral ischemia-induced hippocampal neuronal death. *Int. J. Mol. Sci.* **2018**, *19*, 1420. [CrossRef]

19. Kim, J.H.; Jang, B.G.; Choi, B.Y.; Kwon, L.M.; Sohn, M.; Song, H.K.; Suh, S.W. Zinc chelation reduces hippocampal neurogenesis after pilocarpine-induced seizure. *PLoS ONE* **2012**, *7*, e48543. [CrossRef]

20. Inoue, K.; Branigan, D.; Xiong, Z.G. Zinc-induced neurotoxicity mediated by transient receptor potential melastatin 7 channels. *J. Biol. Chem.* **2010**, *285*, 7430–7439. [CrossRef]

21. Choi, B.Y.; Lee, B.E.; Kim, J.H.; Kim, H.J.; Sohn, M.; Song, H.K.; Chung, T.N.; Suh, S.W. Colchicine induced intraneuronal free zinc accumulation and dentate granule cell degeneration. *Metallomics* **2014**, *6*, 1513–1520. [CrossRef] [PubMed]

22. Yao, H.; Ma, E.; Gu, X.Q.; Haddad, G.G. Intracellular pH regulation of CA1 neurons in Na$^+$/H$^+$ isoform 1 mutant mice. *J. Clin. Invest.* **1999**, *104*, 637–645. [CrossRef] [PubMed]

23. Noel, J.; Pouyssegur, J. Hormonal regulation, pharmacology, and membrane sorting of vertebrate Na$^+$/H$^+$ exchanger isoforms. *Am. J. Physiol.* **1995**, *268*, C283–C296. [CrossRef] [PubMed]

24. Ohgaki, R.; Matsushita, M.; Kanazawa, H.; Ogihara, S.; Hoekstra, D.; van Ijzendoorn, S.C. The Na$^+$/H$^+$ exchanger NHE6 in the endosomal recycling system is involved in the development of apical bile canalicular surface domains in HepG2 cells. *Mol. Biol. Cell* **2010**, *21*, 1293–1304. [CrossRef] [PubMed]

25. Nakamura, N.; Tanaka, S.; Teko, Y.; Mitsui, K.; Kanazawa, H. Four Na$^+$/H$^+$ exchanger isoforms are distributed to Golgi and post-Golgi compartments and arc involved in organelle pH regulation. *J. Biol. Chem.* **2005**, *280*, 1561–1572. [CrossRef] [PubMed]

26. Kleyman, T.R.; Cragoe, E.J., Jr. Amiloride and its analogs as tools in the study of ion transport. *J. Membr. Biol.* **1988**, *105*, 1–21. [CrossRef]

27. Hwang, I.K.; Yoo, K.Y.; An, S.J.; Li, H.; Lee, C.H.; Choi, J.H.; Lee, J.Y.; Lee, B.H.; Kim, Y.M.; Kwon, Y.G.; et al. Late expression of Na$^+$/H$^+$ exchanger 1 (NHE1) and neuroprotective effects of NHE inhibitor in the gerbil hippocampal CA1 region induced by transient ischemia. *Exp. Neurol.* **2008**, *212*, 314–323. [CrossRef]

28. Verma, V.; Bali, A.; Singh, N.; Jaggi, A.S. Implications of sodium hydrogen exchangers in various brain diseases. *J. Basic Clin. Physiol. Pharmacol.* **2015**, *26*, 417–426. [CrossRef] [PubMed]

29. Tai, K.K.; Truong, D.D. Amiloride but not memantine reduces neurodegeneration, seizures and myoclonic jerks in rats with cardiac arrest-induced global cerebral hypoxia and reperfusion. *PLoS ONE* **2013**, *8*, e60309. [CrossRef]

30. Ou-Yang, T.P.; Zhu, G.M.; Ding, Y.X.; Yang, F.; Sun, X.L.; Jiang, W. The effects of amiloride on seizure activity, cognitive deficits and seizure-induced neurogenesis in a novel rat model of febrile seizures. *Neurochem. Res.* **2016**, *41*, 933–942. [CrossRef]

31. Imai, T.; Katoh, H.; Suyama, K.; Kuroiwa, M.; Yanagisawa, S.; Watanabe, M. Amiloride promotes oligodendrocyte survival and remyelination after spinal cord injury in rats. *J. Clin. Med.* **2018**, *7*, 46. [CrossRef] [PubMed]

32. Foster, M.; Samman, S. Zinc and redox signaling: Perturbations associated with cardiovascular disease and diabetes mellitus. *Antioxid. Redox Signal.* **2010**, *13*, 1549–1573. [CrossRef] [PubMed]

33. Kho, A.R.; Choi, B.Y.; Lee, S.H.; Hong, D.K.; Jeong, J.H.; Kang, B.S.; Kang, D.H.; Park, K.H.; Park, J.B.; Suh, S.W. The effects of sodium dichloroacetate on mitochondrial dysfunction and neuronal death following hypoglycemia-induced injury. *Cells* **2019**, *8*, 405. [CrossRef] [PubMed]

34. Ali, S.M.; Dunn, E.; Oostveen, J.A.; Hall, E.D.; Carter, D.B. Induction of apolipoprotein E mRNA in the hippocampus of the gerbil after transient global ischemia. *Brain Res. Mol. Brain Res.* **1996**, *38*, 37–44. [CrossRef]

35. Hoane, M.R.; Kaplan, S.A.; Ellis, A.L. The effects of nicotinamide on apoptosis and blood-brain barrier breakdown following traumatic brain injury. *Brain Res.* **2006**, *1125*, 185–193. [CrossRef]

36. Tang, X.N.; Berman, A.E.; Swanson, R.A.; Yenari, M.A. Digitally quantifying cerebral hemorrhage using Photoshop and Image J. *J. Neurosci. Methods* **2010**, *190*, 240–243. [CrossRef]

37. Li, W.; Ward, R.; Dong, G.; Ergul, A.; O'Connor, P. Neurovascular protection in voltage-gated proton channel Hv1 knock-out rats after ischemic stroke: Interaction with Na^+/H^+ exchanger-1 antagonism. *Physiol. Rep.* **2019**, *7*, e14142. [CrossRef]

38. Sun, D.; Nguyen, N.; DeGrado, T.R.; Schwaiger, M.; Brosius, F.C., 3rd. Ischemia induces translocation of the insulin-responsive glucose transporter GLUT4 to the plasma membrane of cardiac myocytes. *Circulation* **1994**, *89*, 793–798. [CrossRef]

39. Tian, R.; Abel, E.D. Responses of GLUT4-deficient hearts to ischemia underscore the importance of glycolysis. *Circulation* **2001**, *103*, 2961–2966. [CrossRef]

40. Young, L.H.; Renfu, Y.; Russell, R.; Hu, X.; Caplan, M.; Ren, J.; Shulman, G.I.; Sinusas, A.J. Low-flow ischemia leads to translocation of canine heart GLUT-4 and GLUT-1 glucose transporters to the sarcolemma in vivo. *Circulation* **1997**, *95*, 415–422. [CrossRef]

41. Wang, Y.P.; Zhou, L.S.; Zhao, Y.Z.; Wang, S.W.; Chen, L.L.; Liu, L.X.; Ling, Z.Q.; Hu, F.J.; Sun, Y.P.; Zhang, J.Y.; et al. Regulation of G6PD acetylation by SIRT2 and KAT9 modulates NADPH homeostasis and cell survival during oxidative stress. *EMBO J.* **2014**, *33*, 1304–1320. [CrossRef] [PubMed]

42. Kruger, N.J.; von Schaewen, A. The oxidative pentose phosphate pathway: Structure and organisation. *Curr. Opin. Plant. Biol.* **2003**, *6*, 236–246. [CrossRef]

43. Orlowski, J.; Grinstein, S. Diversity of the mammalian sodium/proton exchanger SLC9 gene family. *Pflügers Arch.* **2004**, *447*, 549–565. [CrossRef] [PubMed]

44. Avkiran, M. Protection of the ischaemic myocardium by Na^+/H^+ exchange inhibitors: Potential mechanisms of action. *Basic Res. Cardiol.* **2001**, *96*, 306–311. [CrossRef] [PubMed]

45. Ferrazzano, P.; Shi, Y.; Manhas, N.; Wang, Y.; Hutchinson, B.; Chen, X.; Chanana, V.; Gerdts, J.; Meyerand, M.E.; Sun, D. Inhibiting the Na^+/H^+ exchanger reduces reperfusion injury: A small animal MRI study. *Front. Biosci. (Elite Ed.)* **2011**, *3*, 81–88. [CrossRef] [PubMed]

46. Yeves, A.M.; Ennis, I.L. Na^+/H^+ exchanger and cardiac hypertrophy. *Hipertens. Riesgo. Vasc.* **2020**, *37*, 22–32. [CrossRef]

47. Sensi, S.L.; Canzoniero, L.M.; Yu, S.P.; Ying, H.S.; Koh, J.Y.; Kerchner, G.A.; Choi, D.W. Measurement of intracellular free zinc in living cortical neurons: Routes of entry. *J. Neurosci.* **1997**, *17*, 9554–9564. [CrossRef]

48. Ohana, E.; Segal, D.; Palty, R.; Ton-That, D.; Moran, A.; Sensi, S.L.; Weiss, J.H.; Hershfinkel, M.; Sekler, I. A sodium zinc exchange mechanism is mediating extrusion of zinc in mammalian cells. *J. Biol. Chem.* **2004**, *279*, 4278–4284. [CrossRef]

49. Sekler, I.; Sensi, S.L.; Hershfinkel, M.; Silverman, W.F. Mechanism and regulation of cellular zinc transport. *Mol. Med.* **2007**, *13*, 337–343. [CrossRef]

50. Lee, B.K.; Jung, Y.S. The Na^+/H^+ exchanger-1 inhibitor cariporide prevents glutamate-induced necrotic neuronal death by inhibiting mitochondrial Ca^{2+} overload. *J. Neurosci Res.* **2012**, *90*, 860–869. [CrossRef]

51. Masereel, B.; Pochet, L.; Laeckmann, D. An overview of inhibitors of Na^+/H^+ exchanger. *Eur. J. Med. Chem.* **2003**, *38*, 547–554. [CrossRef]

52. Durham-Lee, J.C.; Mokkapati, V.U.; Johnson, K.M.; Nesic, O. Amiloride improves locomotor recovery after spinal cord injury. *J. Neurotrauma* **2011**, *28*, 1319–1326. [CrossRef] [PubMed]

53. Xiong, Z.G.; Zhu, X.M.; Chu, X.P.; Minami, M.; Hey, J.; Wei, W.L.; MacDonald, J.F.; Wemmie, J.A.; Price, M.P.; Welsh, M.J.; et al. Neuroprotection in ischemia: Blocking calcium-permeable acid-sensing ion channels. *Cell* **2004**, *118*, 687–698. [CrossRef] [PubMed]

54. Arias, R.L.; Sung, M.L.; Vasylyev, D.; Zhang, M.Y.; Albinson, K.; Kubek, K.; Kagan, N.; Beyer, C.; Lin, Q.; Dwyer, J.M.; et al. Amiloride is neuroprotective in an MPTP model of Parkinson's disease. *Neurobiol. Dis.* **2008**, *31*, 334–341. [CrossRef]

55. Lazarewicz, J.W.; Rybkowski, W.; Sadowski, M.; Ziembowicz, A.; Alaraj, M.; Wegiel, J.; Wisniewski, H.M. N-methyl-D-aspartate receptor-mediated, calcium-induced calcium release in rat dentate gyrus/CA4 in vivo. *J. Neurosci. Res.* **1998**, *51*, 76–84. [CrossRef]

56. Kreutzberg, G.W. Microglia: A sensor for pathological events in the CNS. *Trends Neurosci.* **1996**, *19*, 312–318. [CrossRef]

57. Jin, R.; Yang, G.; Li, G. Inflammatory mechanisms in ischemic stroke: Role of inflammatory cells. *J. Leukoc. Biol.* **2010**, *87*, 779–789. [CrossRef]

58. Yenari, M.A.; Kauppinen, T.M.; Swanson, R.A. Microglial activation in stroke: Therapeutic targets. *Neurotherapeutics* **2010**, *7*, 378–391. [CrossRef]

59. Shi, Y.; Chanana, V.; Watters, J.J.; Ferrazzano, P.; Sun, D. Role of sodium/hydrogen exchanger isoform 1 in microglial activation and proinflammatory responses in ischemic brains. *J. Neurochem.* **2011**, *119*, 124–135. [CrossRef]

60. Hua, J.S.; Li, L.P.; Zhu, X.M. Effects of moxibustion pretreating on SOD and MDA in the rat of global brain ischemia. *J. Tradit. Chin. Med.* **2008**, *28*, 289–292. [CrossRef]

61. Chen, H.; Yoshioka, H.; Kim, G.S.; Jung, J.E.; Okami, N.; Sakata, H.; Maier, C.M.; Narasimhan, P.; Goeders, C.E.; Chan, P.H. Oxidative stress in ischemic brain damage: Mechanisms of cell death and potential molecular targets for neuroprotection. *Antioxid. Redox Signal.* **2011**, *14*, 1505–1517. [CrossRef] [PubMed]

62. Choi, B.Y.; Jung, J.W.; Suh, S.W. The emerging role of zinc in the pathogenesis of multiple sclerosis. *Int. J. Mol. Sci.* **2017**, *18*, 2070. [CrossRef]

63. Li, S.; Vana, A.C.; Ribeiro, R.; Zhang, Y. Distinct role of nitric oxide and peroxynitrite in mediating oligodendrocyte toxicity in culture and in experimental autoimmune encephalomyelitis. *Neuroscience* **2011**, *184*, 107–119. [CrossRef] [PubMed]

64. Zhang, Y.; Wang, H.; Li, J.; Dong, L.; Xu, P.; Chen, W.; Neve, R.L.; Volpe, J.J.; Rosenberg, P.A. Intracellular zinc release and ERK phosphorylation are required upstream of 12-lipoxygenase activation in peroxynitrite toxicity to mature rat oligodendrocytes. *J. Biol. Chem.* **2006**, *281*, 9460–9470. [CrossRef] [PubMed]

65. Khan, M.; Dhammu, T.S.; Sakakima, H.; Shunmugavel, A.; Gilg, A.G.; Singh, A.K.; Singh, I. The inhibitory effect of S-nitrosoglutathione on blood-brain barrier disruption and peroxynitrite formation in a rat model of experimental stroke. *J. Neurochem.* **2012**, *123*, 86–97. [CrossRef]

66. Torreilles, F.; Salman-Tabcheh, S.; Guerin, M.; Torreilles, J. Neurodegenerative disorders: The role of peroxynitrite. *Brain Res. Brain Res. Rev.* **1999**, *30*, 153–163. [CrossRef]

67. Choi, B.Y.; Lee, S.H.; Choi, H.C.; Lee, S.K.; Yoon, H.S.; Park, J.B.; Chung, W.S.; Suh, S.W. Alcohol dependence treating agent, acamprosate, prevents traumatic brain injury-induced neuron death through vesicular zinc depletion. *Transl. Res.* **2019**, *207*, 1–18. [CrossRef]

68. Abbott, N.J.; Patabendige, A.A.; Dolman, D.E.; Yusof, S.R.; Begley, D.J. Structure and function of the blood-brain barrier. *Neurobiol. Dis.* **2010**, *37*, 13–25. [CrossRef]

69. Dalkara, T.; Gursoy-Ozdemir, Y.; Yemisci, M. Brain microvascular pericytes in health and disease. *Acta Neuropathol.* **2011**, *122*, 1–9. [CrossRef]

70. Woodruff, T.M.; Thundyil, J.; Tang, S.C.; Sobey, C.G.; Taylor, S.M.; Arumugam, T.V. Pathophysiology, treatment, and animal and cellular models of human ischemic stroke. *Mol. Neurodegener.* **2011**, *6*, 11. [CrossRef]

71. Won, S.J.; Yoo, B.H.; Brennan, A.M.; Shin, B.S.; Kauppinen, T.M.; Berman, A.E.; Swanson, R.A.; Suh, S.W. EAAC1 gene deletion alters zinc homeostasis and exacerbates neuronal injury after transient cerebral ischemia. *J. Neurosci.* **2010**, *30*, 15409–15418. [CrossRef]

72. Jang, B.G.; Won, S.J.; Kim, J.H.; Choi, B.Y.; Lee, M.W.; Sohn, M.; Song, H.K.; Suh, S.W. EAAC1 gene deletion alters zinc homeostasis and enhances cortical neuronal injury after transient cerebral ischemia in mice. *J. Trace Elem. Med. Biol.* **2012**, *26*, 85–88. [CrossRef] [PubMed]

73. Suh, S.W.; Shin, B.S.; Ma, H.; Van Hoecke, M.; Brennan, A.M.; Yenari, M.A.; Swanson, R.A. Glucose and NADPH oxidase drive neuronal superoxide formation in stroke. *Ann. Neurol.* **2008**, *64*, 654–663. [CrossRef]

74. Schmued, L.C.; Hopkins, K.J. Fluoro-Jade B: A high affinity fluorescent marker for the localization of neuronal degeneration. *Brain Res.* **2000**, *874*, 123–130. [CrossRef]

75. Suh, S.W.; Aoyama, K.; Chen, Y.; Garnier, P.; Matsumori, Y.; Gum, E.; Liu, J.; Swanson, R.A. Hypoglycemic neuronal death and cognitive impairment are prevented by poly(ADP-ribose) polymerase inhibitors administered after hypoglycemia. *J. Neurosci.* **2003**, *23*, 10681–10690. [CrossRef] [PubMed]

76. Frederickson, C.J.; Kasarskis, E.J.; Ringo, D.; Frederickson, R.E. A quinoline fluorescence method for visualizing and assaying the histochemically reactive zinc (bouton zinc) in the brain. *J. Neurosci. Methods* **1987**, *20*, 91–103. [CrossRef]

77. Suh, S.W.; Gum, E.T.; Hamby, A.M.; Chan, P.H.; Swanson, R.A. Hypoglycemic neuronal death is triggered by glucose reperfusion and activation of neuronal NADPH oxidase. *J. Clin. Invest.* **2007**, *117*, 910–918. [CrossRef] [PubMed]

78. Kim, J.H.; Jang, B.G.; Choi, B.Y.; Kim, H.S.; Sohn, M.; Chung, T.N.; Choi, H.C.; Song, H.K.; Suh, S.W. Post-treatment of an NADPH oxidase inhibitor prevents seizure-induced neuronal death. *Brain Res.* **2013**, *1499*, 163–172. [CrossRef] [PubMed]

79. Kho, A.R.; Choi, B.Y.; Kim, J.H.; Lee, S.H.; Hong, D.K.; Lee, S.H.; Jeong, J.H.; Sohn, M.; Suh, S.W. Prevention of hypoglycemia-induced hippocampal neuronal death by N-acetyl-L-cysteine (NAC). *Amino Acids* **2017**, *49*, 367–378. [CrossRef] [PubMed]

80. Lee, S.H.; Choi, B.Y.; Lee, S.H.; Kho, A.R.; Jeong, J.H.; Hong, D.K.; Suh, S.W. Administration of protocatechuic acid reduces traumatic brain injury-induced neuronal death. *Int. J. Mol. Sci.* **2017**, *18*, 2510. [CrossRef]

81. Kauppinen, T.M.; Swanson, R.A. Poly(ADP-ribose) polymerase-1 promotes microglial activation, proliferation, and matrix metalloproteinase-9-mediated neuron death. *J. Immunol.* **2005**, *174*, 2288–2296. [CrossRef] [PubMed]

82. Kauppinen, T.M.; Higashi, Y.; Suh, S.W.; Escartin, C.; Nagasawa, K.; Swanson, R.A. Zinc triggers microglial activation. *J. Neurosci.* **2008**, *28*, 5827–5835. [CrossRef] [PubMed]

83. Ruth, R.E.; Feinerman, G.S. Foreign and endogenous serum protein extravasation during harmaline tremors or kainic acid seizures in the rat: A comparison. *Acta Neuropathol.* **1988**, *76*, 380–387. [CrossRef] [PubMed]

84. Hsu, S.M.; Raine, L.; Fanger, H. Use of avidin-biotin-peroxidase complex (ABC) in immunoperoxidase techniques: A comparison between ABC and unlabeled antibody (PAP) procedures. *J. Histochem. Cytochem.* **1981**, *29*, 577–580. [CrossRef]

International Journal of
Molecular Sciences

Article

Transient Receptor Potential Melastatin 2 (TRPM2) Inhibition by Antioxidant, *N*-Acetyl-L-Cysteine, Reduces Global Cerebral Ischemia-Induced Neuronal Death

Dae Ki Hong [1], A Ra Kho [1], Song Hee Lee [1], Jeong Hyun Jeong [1], Beom Seok Kang [1], Dong Hyeon Kang [2], Min Kyu Park [1], Kyoung-Ha Park [3], Man-Sup Lim [4], Bo Young Choi [1,*] and Sang Won Suh [1,*]

[1] Department of Physiology, College of Medicine, Hallym University, Chuncheon 24252, Korea; zxnm01220@gmail.com (D.K.H.); rnlduadkfk136@hallym.ac.kr (A.R.K.); sshlee@hallym.ac.kr (S.H.L.); jd1422@hanmail.net (J.H.J.); ttiger1993@gmail.com (B.S.K.); bagmingyu50@gmail.com (M.K.P.)

[2] Department of Medical Science, College of Medicine, Hallym University, Chuncheon 24252, Korea; ehdgus6312@gmail.com

[3] Division of Cardiovascular Diseases, Hallym University Medical Center, Anyang 14068, Korea; pkhmd@naver.com

[4] Department of Medical Education, College of Medicine, Hallym University, Chuncheon 24252, Korea; ellemes@hallym.ac.kr

* Correspondence: bychoi@hallym.ac.kr (B.Y.C.); swsuh@hallym.ac.kr (S.W.S.); Tel.: +82-10-8888-0646 (B.Y.C.); +82-10-8573-6364 (S.W.S.)

Received: 9 July 2020; Accepted: 20 August 2020; Published: 21 August 2020

Abstract: A variety of pathogenic mechanisms, such as cytoplasmic calcium/zinc influx, reactive oxygen species production, and ionic imbalance, have been suggested to play a role in cerebral ischemia induced neurodegeneration. During the ischemic state that occurs after stroke or heart attack, it is observed that vesicular zinc can be released into the synaptic cleft, and then translocated into the cytoplasm via various cation channels. Transient receptor potential melastatin 2 (TRPM2) is highly distributed in the central nervous system and has high sensitivity to oxidative damage. Several previous studies have shown that TRPM2 channel activation contributes to neuroinflammation and neurodegeneration cascades. Therefore, we examined whether anti-oxidant treatment, such as with *N*-acetyl-L-cysteine (NAC), provides neuroprotection via regulation of TRPM2, following global cerebral ischemia (GCI). Experimental animals were then immediately injected with NAC (150 mg/kg/day) for 3 and 7 days, before sacrifice. We demonstrated that NAC administration reduced activation of GCI-induced neuronal death cascades, such as lipid peroxidation, microglia and astroglia activation, free zinc accumulation, and TRPM2 over-activation. Therefore, modulation of the TRPM2 channel can be a potential therapeutic target to prevent ischemia-induced neuronal death.

Keywords: global cerebral ischemia; *N*-acetyl-L-cysteine; transient receptor potential melastatin 2; zinc; neurodegeneration

---

1. Introduction

Recent decades have seen a dramatic increase in the attention and focus on the study of brain disease. In particular, cerebral ischemia is a global problem and is the leading cause of brain disease and morbidity [1,2]. There are many reasons why cerebral ischemia is so widespread, in part due to the diverse number of individual pathologies that converge on the emergence of the ischemic state; the largest risk factor for development of cerebral ischemia is occlusion of blood vessels to the brain.

Upon the occlusion of these blood vessels, cerebral blood flow is considerably reduced to regions of the brain and essential supplies of oxygen, nutrients, glucose, and many other cellular substrates are reduced [3]. Cerebral ischemia is divided into two sub-categories. The first, focal cerebral ischemia, involves transiently or permanently reduced blood flow and nutrient supply to particular brain areas that are directly fed by the occluded vasculature. Damage to the brain after focal ischemia is thus focused on the affected areas and injury occurs locally [4]. Global cerebral ischemia (GCI) is similar to focal cerebral ischemia, although the mechanisms of induction and patterns of damage to specific brain regions are different. This brain injury occurs due to several reasons, but is best described by a sudden and near-complete stop in cerebral blood-flow, such as is seen under cardiac arrest-induced cessation of blood circulation [5].

Once the ischemic insult has occurred, abnormal physiological changes to brain function have already began to occur. Under the ischemic condition, disruption of brain homeostasis is disturbed and cell death mechanisms are initiated. This cerebral injury leads to an alteration in ionic balance within the intra-extracellular space [6], reactive oxygen species (ROS) production [7], microglia [8] and astrocyte [9] activation, zinc accumulation [10], and neuronal cell degeneration [11–13] in the brain. Blockage of blood flow to the brain by ischemic insult causes tissue infarction, dysfunction of ionic balance, and intracellular calcium overload [14]. In a previous study, excessive release of calcium ions from synaptic vesicles after brain ischemia was shown to have deleterious effects on neurons and glia [15]. Overloaded intracellular calcium ions trigger glutamate release, proteolysis, mitochondrial dysfunction, and other deleterious cascades [16]. As discussed above, calcium overload contributes to ischemic injury-induced cell death mechanisms. According to a recent report, the cation zinc has an essential role in brain ischemia, in addition to the calcium ion [17].

Zinc is the most widely distributed metal ion in the central nervous system, and is especially abundant in the hippocampus. Zinc has many important roles in enzymes and transcription factors, and its concentration in neurons and glia is controlled by binding proteins, known as metallothionein, as well as several transporters and channels [17–19]. In addition, zinc regulates neural synaptic transmission, innate and adaptive immunity, signal transduction, and cell proliferation, particularly for DNA synthesis [20,21]. In the pre-synaptic vesicles, chelatable zinc is abundant. It is released into the synaptic cleft, moved into pre-synaptic neurons via several transporters, and subsequently undergoes reuptake from the synaptic cleft under normal physiological conditions [22,23]. Under abnormal physiological conditions, such as neurological disorders, zinc homeostasis is disturbed and dysregulated. In the case of neurological disorders, vesicular zinc is excessively released and largely accumulates within post-synaptic neurons. In addition to releasing zinc from synaptic vesicles, zinc that is loosely bound by proteins can be divided into the categories of chelatable free zinc and proteins [12,24,25]. Pathological zinc accumulation can influence cellular dysfunction in several ways. In a previous study, it was demonstrated that zinc-treated cultured neurons produce ROS from their mitochondria [18] and trigger microglial activation [26]. In addition, previous studies by our group have observed that excessive zinc accumulation within brain hippocampal regions can lead to neurodegeneration after cerebral insult, such as ischemia [10,27], head trauma [25], hypoglycemia [28], and cultured neurons [29].

N-acetyl-L-cysteine (NAC), which contains an acetylated cysteine residue, acts as a powerful antioxidant by supplementation of cysteine and promotion of glutathione biosynthesis [30,31]. Injury-induced oxidative damage plays a primary role in central nervous system dysfunction and contributes to neurodegenerative cascades [32,33]. To protect against brain injury-induced oxidative damage, our group has previously used NAC to treat hypoglycemia and traumatic brain injury. NAC stabilized glutathione concentrations, reduced degenerating neurons, and decreased oxidative damage within hippocampal regions [24,34]. Hippocampal neurodegeneration can be prevented or reduced when oxidative damage and other deleterious cascades are prevented by regulating physiologic concentrations of glutathione via providing ample cysteine, which is supplied from NAC [35]. Moreover, NAC has binding sites for trace metals that include copper, magnesium, and zinc.

These studies demonstrate that NAC can neutralize toxically accumulated trace metals occurring in neuropathies [24,36].

Transient receptor potential melastatin 2 (TRPM2) is one of the non-selective cationic channels, and the activity of this channel is triggered by ROS production promoted by deleterious cell death mechanisms in several neurological disorders. Recently, the TRPM2 channel has been reported to be included in both physiological and pathological conditions [37,38]. Several previous studies have shown that the TRPM2 channel is expressed in the central nervous system (CNS), which is largely expressed within the hippocampus, stratum, and cortex [39]. These studies found that several divalent cations, for example, calcium and magnesium, moved into the intracellular space via TRPM2, and that its activity was regulated by external stimuli such as ROS, hydrogen peroxide, and tumor necrosis factor-α (TNF-α) [40–42]. In addition, TRPM2 is distributed in both neurons and glia, and is involved in pathophysiological conditions of nitric oxide (NO) and ROS production signaling [38,43]. Following several neurological injuries, excessive calcium influx via TRPM2 can trigger release of diverse cytokines and thus contributes to the inflammatory response [44]. Several brain diseases, such as stroke or head trauma, lead to excitotoxicity, which is caused by the inappropriate activation of voltage-dependent calcium channels and excitatory amino acid release from the dendritic and presynaptic space into the extracellular cleft. Previous studies verified that in a rodent disease model of ischemia, TRPM2 mRNA levels were significantly increased. Interestingly, a recent study demonstrated that TRPM2 deficiency in mice decreased cytosolic zinc concentrations and reduced zinc influx from the extracellular space [45]. These findings suggest that, during several brain diseases, TRPM2 is activated and this contributes to cell death mechanisms [44,46].

In the present study, we tested the hypothesis that if ischemia-induced oxidative damage leads to TRPM2 activation, inhibition of this channel may have therapeutic effects through blockade of zinc influx-induced cell damage, which is supported by evidence that NAC administration decreased brain damage in several other neurological diseases [24,34]. Following this logic, we hypothesized that if the TRPM2 level was increased by global cerebral ischemia-induced cell death cascades, regulation of TRPM2 as a potential target for preventing ischemia-induced injury could ameliorate a diverse number of cerebral pathologies associated with ischemia. In addition, we speculated that the zinc ion is closely related to TRPM2 and is excessively moved into the intracellular space via this channel. To test our hypothesis, we used a GCI brain disease model in adult rats and conducted several types of histological evaluations and behavioral assessments.

2. Results

2.1. Global Cerebral Ischemia-Induced Neuronal Death and Zinc Accumulation Is Attenuated by Post-Administration of NAC

Timeline showing the experimental design for 3 days and 7 days (Figure 1A). Whole GCI conducted experimental animals were monitored for arterial blood pressure and electroencephalography continuously during the before-isoelectric, isoelectric, and after-blood reperfusion condition (Figure 1B–D).

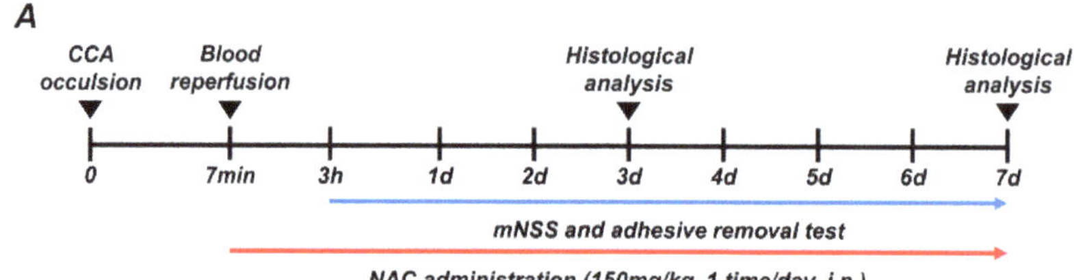

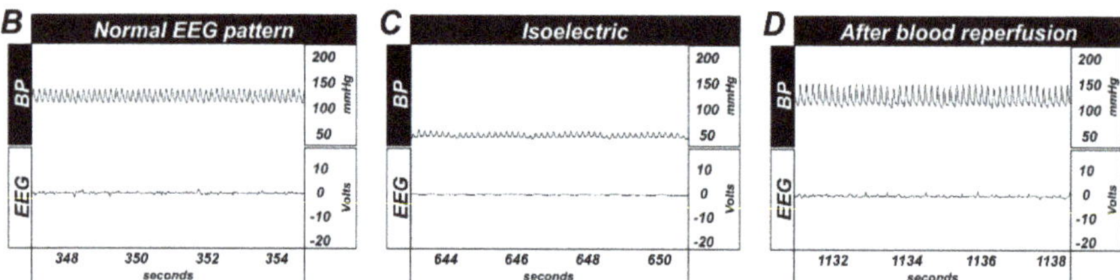

Figure 1. (**A**) Timeline representing the experimental design for 3 and 7 days following global cerebral ischemia (GCI). Experimental animals were given vehicle and *N*-acetyl-ʟ-cysteine (NAC) (dosage: 150 mg/kg) once per day for 3 and 7 days. During the 7 days of the experiment, animals were subjected to behavioral outcome assessment by adhesive removal test and modified neurological severity score measurement. Global cerebral ischemia (GCI) induces electroencephalograph (EEG) and blood pressure changes, and timeline of experiment. (**B**) Normal blood pressure, resting EEG pattern. (**C**) Blood pressure decreased up to 40 (diastolic)–50 (systolic) mmHg, and EEG displayed isoelectric pattern. (**D**) Blood pressure and EEG pattern completely normalized after blood reperfusion.

To determine whether GCI-induced neuronal death and zinc accumulation was attenuated by NAC post-administration, we used FJB (Fluoro-Jade B) and TSQ (*N*-(6-methoxy-8-quinolyl)-para-toluenesulfonamide staining at 3 days after GCI. The degenerating neurons were strongly present in the GCI–vehicle groups. When compared with vehicle-treated groups, NAC-treated groups showed significantly reduced numbers of degenerating neurons (Figure 2A). Quantified numbers of degenerating neurons are displayed using bar graphs (vehicle: subiculum, 168.48 ± 19.63; CA1, 140.03 ± 16.81; CA2, 132.75 ± 14.96; NAC: subiculum, 52.17 ± 20.09; CA1, 54.97 ± 19.74; CA2, 54.98 ± 19.24; Figure 2B). Consequently, GCI-induced zinc accumulation in post-synaptic neurons was observed in vehicle-treated groups. However, NAC administration after GCI reduced zinc accumulation in hippocampal areas (Figure 2C). Representative bar graph shows number of quantified TSQ-positive cells in both GCI–vehicle and GCI–NAC groups (vehicle: subiculum, 84.6 ± 4.29; CA1, 76.28 ± 5.32; CA2, 76.53 ± 3.18; NAC: subiculum, 37.66 ± 2.18; CA1, 32.63 ± 10.52; CA2, 27.77 ± 4.77; Figure 2D).

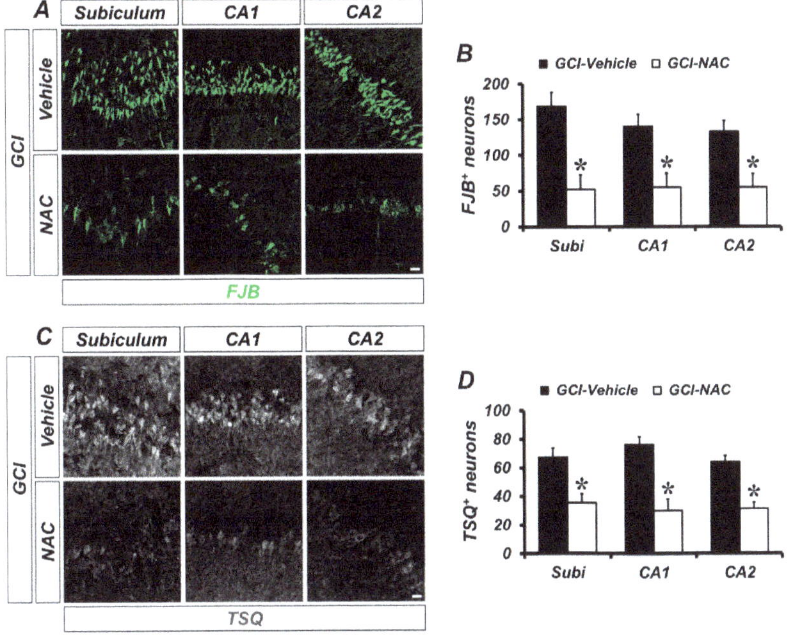

Figure 2. Number of GCI-induced degenerating neurons and zinc accumulation was attenuated by NAC treatment. (**A**) Brain sections were stained with fluorescence dye for detection of degenerating neurons, Fluoro-Jade B (FJB, green signal) in the hippocampal subiculum, cornus ammonis (CA) 1 and CA2 regions. Brain tissues that suffered GCI condition have numerous degenerating neurons in hippocampal regions and administration of NAC reduces the number of FJB-positive neurons in the same regions. Scale bar = 20 µm. (**B**) Quantification of FJB-positive neurons counted in each hippocampal region. Data are mean ± standard error of mean (SEM), $n = 6$ each group, * $p < 0.05$ versus vehicle-treated group (Mann–Whitney U test, Subiculum: z = 2.562, $p = 0.010$; CA1: z = 2.402, $p = 0.016$; CA2: z = 2.082, $p = 0.037$). (**C**) Fluorescence microscopic images of free zinc stained by N-(6-methoxy-8-quinolyl)-para-toluenesulfonamide (TSQ) staining. Bright blue fluorescence signal means ischemic condition-induced excessive zinc release and translocation into hippocampal neurons. The NAC-administered group had a remarkably reduced number of TSQ-positive neurons. Scale bar = 20 µm. (**D**) Quantification of TSQ fluorescence signal-positive neurons counted from the hippocampus. Data are mean ± SEM, $n = 4$–5 each group, * $p < 0.05$ versus vehicle-treated group (Mann–Whitney U test, Subiculum: z = 2.205, $p = 0.027$; CA1: z = 2.449, $p = 0.014$; CA2: z = 2.449, $p = 0.014$).

2.2. NAC Post-Treatment Regulates Global Cerebral Ischemia-Induced Loss of Neuronal Glutathione and Lipid Peroxidation

One strategy to solve the problem of ischemic injury is to focus on reducing oxidative damage. We speculated that post-administration of NAC can supply cysteine to produce the powerful anti-oxidant glutathione and influence neuronal oxidative damage. Representative images show glutathione (GS-NEM) levels (Figure 3A). Sham groups show a wide distribution of glutathione within neurons, but the GCI–vehicle group shows a loss of glutathione after injury, particularly oxidative damage. NAC treatment after GCI elevates glutathione concentration in damaged neurons. Based on the above results, we confirmed whether NAC administration directly protects against ROS-induced neuronal oxidative damage by enhancing glutathione levels. ROS triggers lipid peroxidation, and 4-hydroxy-2-nonenal (4-HNE) is a histological marker for lipid peroxidation. GCI led to an increase in ROS, and 4-HNE-positive fluorescence signal was largely increased in the hippocampal subiculum, CA1, and CA2 regions (Figure 3B). Increased lipid peroxidation by GCI-induced ROS production was significantly attenuated following post-treatment of NAC (mean gray value, GCI–vehicle: subiculum, 25.93 ± 2.08; CA1, 28.89 ± 1.69; CA2, 26.36 ± 2.41; GCI–NAC: subiculum, 17.20 ± 0.97; CA1, 16.34 ± 0.99; CA2, 16.90 ± 0.31; Figure 3C). Sham groups had no differences in 4-HNE signal (sham–vehicle: subiculum, 11.12 ± 0.37; CA1, 9.76 ± 0.64; CA2, 11.12 ± 0.97; sham NAC: subiculum, 10.10 ± 0.92; CA1, 8.65 ± 0.43; CA2, 10.89 ± 1.21; Figure 3C).

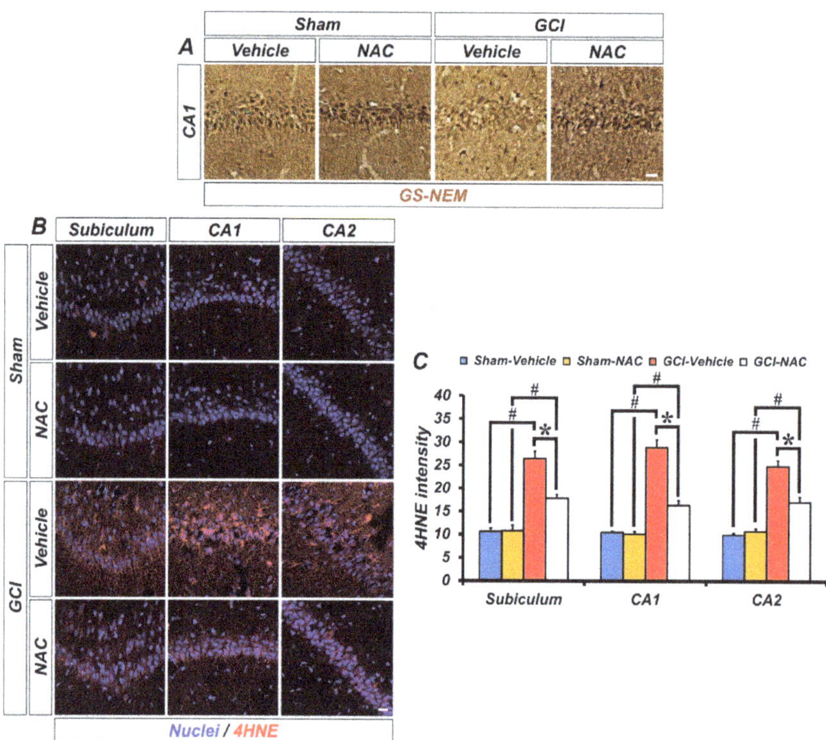

Figure 3. GCI-induced decreasing concentration of glutathione and increasing oxidative damage was restored by anti-oxidant, NAC. (**A**) Representative histological images show concentration of glutathione marker GS-NEM in the hippocampal CA1. Neuronal glutathione level was remarkably reduced after GCI and was restored to normal levels. Scale bar = 20 μm. (**B**) 4-hydroxy-2-nonenal (4-HNE) immunofluorescence staining was used as a marker for lipid peroxidation. Sham–vehicle and NAC groups show minimized 4-HNE-positive signal (red color) in hippocampal regions. It was remarkably increased in the GCI–vehicle group and NAC administration after GCI attenuates oxidative damage-induced lipid peroxidation. Scale bar = 20 μm. (**C**) Bar graph means intensity of 4-HNE signal in the hippocampal regions. Data are mean ± SEM, n = 4–7 each group, * p < 0.05 versus vehicle-treated group; # p < 0.05 versus sham-operated group (Kruskal–Wallis test followed by Bonferroni post-hoc test, Subiculum: chi square = 18.688, df = 3, p < 0.001; CA1: chi square = 18.783, df = 3, p < 0.001; CA2: chi square = 18.229, df = 3, p < 0.001).

2.3. NAC Attenuates TRPM2 Activation in Hippocampal Neurons and Glial Cells

GCI activates TRPM2 channels in the brain, which is localized to glial cells in addition to hippocampal neurons. To demonstrate whether TRPM2 is located within neurons or glial cells in our experimental setting, we conducted immuno-fluorescence staining. Figure 4A shows distribution of TRPM2 channels in the hippocampal pyramidal layer CA1 region, where it was co-localized with neuronal nuclei (NeuN). There was no difference in TRPM2-positive fluorescence signals in sham groups (mean gray value, sham–vehicle: 11.08 ± 0.58; sham NAC: 11.09 ± 0.62; Figure 4B). However, the ischemic condition contributes to activation of TRPM2 in the brain and this is highly localized to the hippocampal pyramidal CA1 layer, which is also known to be especially vulnerable to ischemic injury. The TRPM2-positive signal was dramatically increased in the GCI–vehicle group and significantly decreased in the NAC post-treatment group (GCI–vehicle: 29.55 ± 1.15; GCI–NAC: 18.25 ± 1.76, Figure 4B).

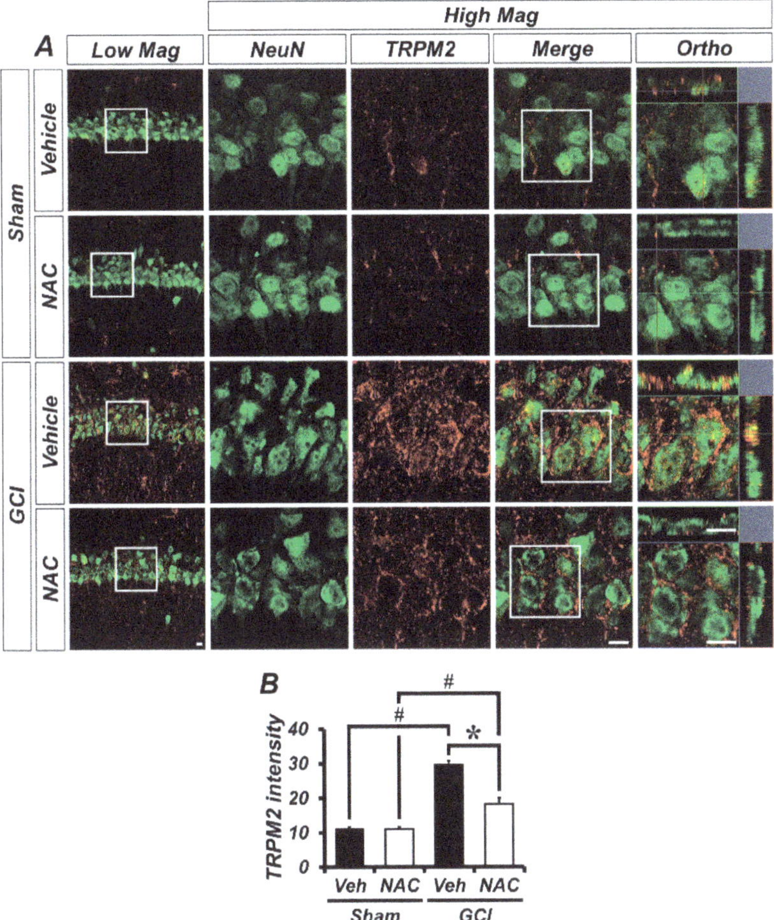

Figure 4. GCI-induced transient receptor potential melastatin 2 (TRPM2) activation was attenuated by NAC treatment. (**A**) Representative histological images of TRPM2-positive signal in the hippocampal CA1. The neuronal TRPM2 level was remarkably increased in the GCI–vehicle group and NAC administration after GCI reduced TRPM2 activation. Scale bar = 20 µm. (**B**) Bar graph shows intensity of TRPM2-positive signal (red color) within the hippocampal pyramidal layer. Data are mean ± SEM, *n* = 5 each group, * $p < 0.05$ versus vehicle-treated group; # $p < 0.05$ versus sham-operated group (Kruskal–Wallis test followed by Bonferroni post-hoc test: chi square = 16.097, df = 3, $p = 0.001$).

To explore the localization of TRPM2 in non-neuronal cells, we conducted glial cell immunostaining. Sham groups showed that Iba-1-positive microglia were distributed throughout the hippocampus in an inactivated state. However, GCI triggers microglial activation, which was significantly decreased by NAC post-treatment (Figure 5A). The bar graph shows quantified microglia intensity (mean gray value, sham–vehicle, 7.59 ± 0.70; sham NAC, 7.82 ± 0.54; GCI–vehicle, 48.35 ± 4.60; GCI–NAC, 23.59 ± 1.01; Figure 5B). Figure 5C shows co-localized TRPM2 with microglia in GCI–vehicle groups. Similarly, astrocytes were distributed throughout the hippocampus, and there were no differences in sham groups. GCI triggers astrocyte activation, which was significantly reduced by NAC post-treatment (Figure 5D). The bar graph shows quantified astrocyte activation (mean gray value, sham–vehicle, 12.52 ± 1.29; sham NAC, 11.88 ± 1.96; GCI–vehicle, 61.46 ± 5.21; GCI–NAC, 36.75 ± 4.58; Figure 5E). Figure 5F shows co-localized TRPM2 with astrocyte in GCI–vehicle groups.

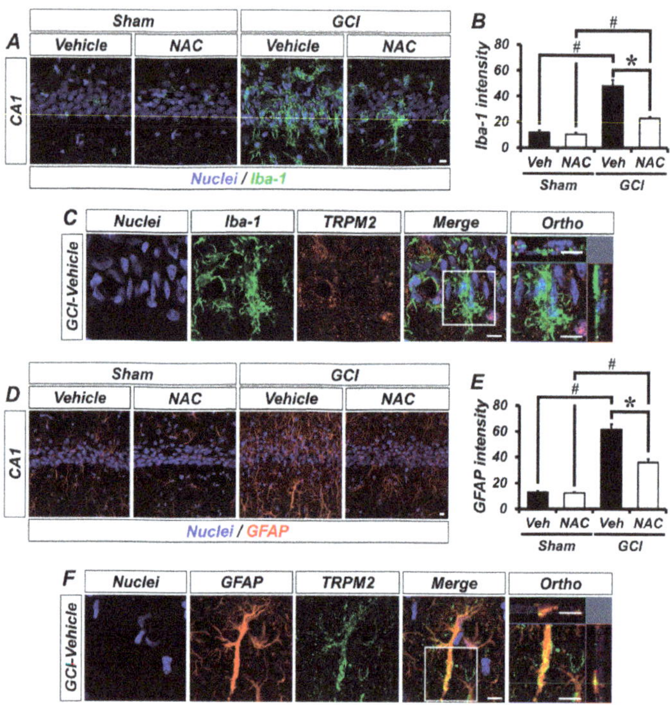

Figure 5. GCI-induced glial cell activation was attenuated by NAC treatment. (**A**) Microglial activation was estimated by Iba-1 immuno-staining. Sham groups (vehicle and NAC) had no difference in Iba-1-positive signal (green color). GCI triggers excessive microglial activation in the hippocampal pyramidal layer, and this was attenuated by NAC administration for 3 days after onset of GCI. Scale bar = 10 µm. (**B**) Bar graph representing intensity of Iba-1-positive signals from hippocampal pyramidal layer administered with vehicle and NAC for 3 days after sham and GCI surgery. Data are mean ± SEM, n = 5–7 each group, * $p < 0.05$ versus vehicle-treated group; # $p < 0.05$ versus sham-operated group (Kruskal–Wallis test followed by Bonferroni post-hoc test: chi square = 20.248, df = 3, $p < 0.001$). (**C**) Distribution of TRPM2 in microglial cells (merged image). (**D**) Astrocyte activation was evaluated by GFAP immuno-staining. Sham groups had no difference in GFAP-positive signal (red color). Astrocyte activation was stimulated by GCI, and reduced by NAC administration. Scale bar = 10 µm. (**E**) Bar graph displaying intensity of GFAP-positive signals from the hippocampal pyramidal layer in vehicle and NAC-treated groups after sham and GCI surgery. Data are mean ± SEM, n = 5–6 each group, * $p < 0.05$ versus vehicle-treated group; # $p < 0.05$ versus sham-operated group (Kruskal–Wallis test followed by Bonferroni post-hoc test: chi square = 18.747, df = 3, $p < 0.001$). (**F**) Distribution of TRPM2 in astrocyte following GCI (merged image).

2.4. NAC Restores Global Cerebral Ischemia-Induced Sensorimotor Deficit, Neurologic Decline, and Neurodegeneration

To investigate the effects of NAC post-administration on GCI-induced sensorimotor deficit and loss of neurological function, experimental animals were analyzed using the adhesive removal test and modified neurological severity score (mNSS) methods. First, to evaluate sensorimotor deficit after ischemic damage we used the adhesive removal test method. The GCI–vehicle-treated groups spent longer periods of time recognizing adhesive tapes on palms than the GCI–NAC-treated groups after ischemic insult. Sham–vehicle and NAC groups had no difference in removal time during the test days. These results showed that GCI-induced sensorimotor deficit was restored by NAC treatment (Figure 6A). Next, we also assessed whether NAC administration restored GCI-induced neurological decline. Because the sham groups had no impairment, the mNSS count was zero. Following GCI, neurological impairment lead to high scores in the vehicle group and NAC administration attenuated this neurologic decline (Figure 6B). In addition, we determined whether post-administration of NAC provides neuroprotective effects on GCI-induced neurodegeneration following 7 days of consecutive treatment (Figure 6C). Sham groups show numerous neuronal nuclei (NeuN)-positive cells in hippocampal subiculum, CA1, and CA2 regions. There were no significant differences in the number of NeuN-positive cells between the sham–vehicle and NAC group (sham–vehicle: subiculum, 170.5 ± 5.5; CA1, 180.5 ± 5.5; CA2, 210 ± 11, sham-NAC: subiculum, 177 ± 7; CA1, 179 ± 10; CA2, 205 ± 7; Figure 6D). In contrast, the number of NeuN-positive cells was dramatically decreased in the GCI–vehicle group. However, NAC administration significantly restored the number of NeuN-positive cells in the hippocampal regions (GCI–vehicle: subiculum, 79.95 ± 7.69; CA1, 62.74 ± 11.15; CA2, 66.82 ± 4.91, GCI–NAC: subiculum, 110.47 ± 6.42; CA1, 108.12 ± 9.87; CA2, 110.31 ± 7.81; Figure 6D).

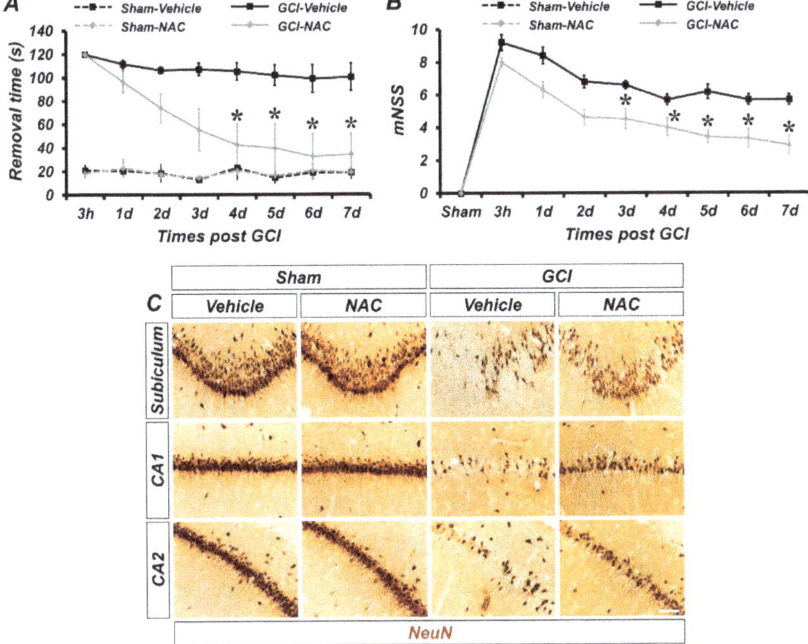

Figure 6. *Cont.*

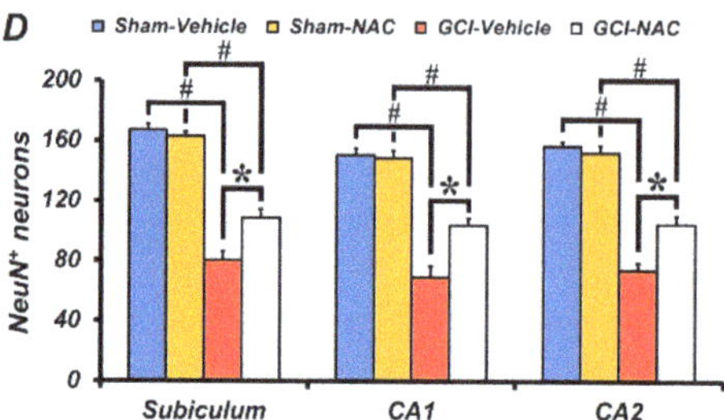

Figure 6. Post administration of NAC reduces GCI-induced behavioral outcome deficits and delays neuronal loss. (**A**) The adhesive removal test to assess sensorimotor deficit was conducted at both 3 h and 7 days after sham and GCI surgery. Sham groups showed no difference in time to remove the adhesive tape during the test. GCI–vehicle group took significantly longer to remove the tape. Post treatment of NAC after GCI reduces removal time (time dependently). Data are mean ± SEM, n = 5–8 each group, * $p < 0.05$ versus vehicle-treated group; # $p < 0.05$ versus sham-operated group (repeated measure test followed by ANOVA, time: F = 6.365, $p < 0.001$; group: F = 33.354, $p < 0.001$; time * group: F = 3.061, $p < 0.001$). (**B**) The mNSS was conducted at 3 h and 7 consecutive days after sham and GCI surgery. Sham groups indicate a 0 score which means all tasks were performed completely without defect. Under the ischemic condition, the vehicle group showed a failure of mNSS sessions. Post administration, NAC improved mNSS procedures and reduced failure of tasks. Data are mean ± SEM, n = 6–8 each group, * $p < 0.05$ versus the vehicle-treated group (repeated measure test followed by ANOVA, time: F = 35.621, $p < 0.001$; group: F = 17.576, $p = 0.001$; time * group: F = 2.322, $p = 0.032$). (**C**) Immunohistochemistry images show NeuN-positive neurons in the hippocampal subiculum, CA1, and CA2 regions. Scale bar = 50 µm. (**D**) Bar graph shows analysis of NeuN-positive cells in each hippocampal region. Post treatment of NAC restores GCI-induced delayed neuronal loss in the following 7 days. Data are mean ± SEM, n = 5–6 each group, * $p < 0.05$ versus vehicle-treated group; # $p < 0.05$ versus sham-operated group (Kruskal–Wallis test followed by Bonferroni post-hoc test, Subiculum: chi square = 17.792, df = 3, $p < 0.001$; CA1: chi square = 17.958, df = 3, $p < 0.001$; CA2: chi square = 17.735, df = 3, $p < 0.001$).

3. Discussion

In the present study, we verified whether NAC administration attenuates GCI-induced neuronal death via inhibition of TRPM2 channels. Our findings indicate that NAC reduces accumulation of intracellular free zinc and TRPM2 over-activation by maintaining a sufficient supply of glutathione production. Cysteine is the rate-limiting substrate required for the biosynthesis of glutathione and the cysteine residue present in NAC can serve as an immediate source of cellular cysteine to maintain high levels of intracellular glutathione, which in turn protects against oxidative damage arising from free radicals. In addition, daily administration of NAC for 3 days after GCI also clearly attenuated several GCI-induced cell death cascades involving neurodegeneration, intracellular free zinc accumulation, oxidative damage-induced lipid peroxidation, loss of glutathione, TRPM2 over-activation, and glial activation. Furthermore, post-treatment of NAC for 7 days prevented GCI-induced delayed hippocampal neuron death, sensorimotor impairment, and neurological deficits. Taken together, these results suggest that reduction of intracellular free zinc and restoration of neuronal GSH concentration by NAC administration attenuates GCI-induced neurodegenerative cascades and promotes behavioral impairments. Therefore, NAC may be an outstanding therapeutic agent for preventing GCI-induced neuronal death.

In general, ionic gradients are actively maintained within finely tuned ranges via several selective and non-selective ion channels throughout the central nervous system, and this process is fundamental to preserving healthy brain function. However, during many neurological injuries that include ischemic stroke [12] and traumatic brain injury [25] damage leads to an ionic imbalance, which in turn allows for the initiation of multiple neurodegenerative cascades. Our previous studies have focused on these mechanisms that follow ionic imbalance in several neurological diseases, such as global ischemia [13,47], hypoglycemia [48], and traumatic brain injury [49]. In particular, we have focused on the abnormal phenomenon of excessive release of vesicular zinc ions following brain injuries and its subsequent cellular pathology. Chelatable zinc that is liberated in this manner moves into post-synaptic neurons through non-selective cation channels, TRPM2 [50], and once chelatable zinc has accumulated within post-synaptic neurons it acts as a neurotoxin [51]. TRPM2 is one of the non-selective cation channels that regulates physiological and pathological processes in the central nervous system via modulation of numerous signaling pathways [38]. In addition, brain disease-induced ROS production triggers TRPM2 activation, leading to calcium influx from the extracellular space, and finally triggers cell death cascades that include PARP-1/PARG and caspase-dependent cell-death pathways [50,52,53]. In this present study, we focused on attenuating the effects of excessively released free zinc ions from synaptic vesicles after GCI that translocate into post-synaptic neurons via the TRPM2 channel. Hence, we hypothesized and verified that by reducing ROS production through supplementation of cysteine by NAC administration to maximize glutathione production, one can reduce hippocampal neurodegeneration through inhibition of zinc accumulation and TRPM2 activation.

N-acetyl-L-cysteine (NAC) is a derivative of cysteine and acts as a scavenger for elimination of free radicals [54]. In the present study, we focused on NAC as a cysteine donor to improve glutathione synthesis, and also as a chelator of heavy metal ions. Excessively released and accumulated free zinc ions after ischemic stroke have a deleterious effect on the nervous system. Previous studies determined that NAC can bind heavy metals (nickel, copper, zinc) because NAC has a thiol group (-SH), which allows it to bind cationic metal ions [55]. In addition, NAC can restore neuronal glutathione concentrations and protect against brain injuries. Kho et al. recently demonstrated that hypoglycemic brain disease-induced hippocampal neurodegeneration, lipid peroxidation, and zinc accumulation were restored to pre-insult levels by NAC administration [24]. Furthermore, Choi et al. recently verified that genetic deletion of the cysteine-related transporter, excitatory amino acid carrier 1 (EAAC1), exacerbates traumatic brain injury (TBI)-induced neurological damage, compared with wild type controls. These cell death cascades were attenuated by NAC treatment in both wild type and EAAC1 gene deletion animals [34]. In light of the abovementioned results, we hypothesized that NAC treatment after GCI may chelate zinc ions, and reduce accumulation of zinc-induced neurotoxicity and downregulation of the TRPM2 channel by reducing ROS production.

The post-global ischemic neuro-inflammatory response contributes to formation of the glial scar and exacerbates neuronal damage. Both microglia and astrocytes play an essential role in maintaining ionic gradients within the extracellular space and in supporting neurotransmission in neighboring neurons. In general, inactivated (resting) microglia and astrocytes regulate the maintenance of central nervous system homeostasis [56]. However, neurodegenerative diseases, such as head trauma, multiple sclerosis, epilepsy and stroke, trigger gliosis which is a non-specific alteration of glial cells and this can lead to glial scar formation if unchecked [57]. Lakhan et al. demonstrated that stroke-induced excitotoxicity and oxidative damage triggers activation of microglia and astrocytes, and these cells secrete several cytokines, chemokines, and matrix metalloproteases [58]. In addition, Kauppinen et al. verified that the zinc-overloaded condition alters the morphology of resting microglia to an activated form in cultured microglia, which was attenuated by treatment with the extracellular zinc chelator, CaEDTA, during the ischemic condition [59]. Barreto et al. reported that astrocyte viability was important for the restoration of neurons following ischemic damage. Furthermore, they asserted that although reactive astrocytes trigger the production of pro-inflammatory cytokines, they also perform functions for maintaining neuronal survival [9].

In the present study, our findings indicate that TRPM2 activation in neurons and glial cells were found after GCI, but were significantly correlated with astrocytes. Astrocytes have a dual nature when activated that can both alleviate or exacerbate the effects of brain damage depending on a sensitive mix of cellular factors. Accordingly, Barreto et al. assert that any solution that adequately addresses the problem of stroke-induced neurodegenerative must involve astrocytes [9]. No details about the interaction of TRPM2 and astrocytes has been reported, but we assumed that there must be some ischemia-related cascades between TRPM2 channels and astrocytes, and began to test this hypothesis. In the case of our research, TRPM2 activation in microglia was only mildly observed compared with astrocytes. However, microglia have a strong influence on CNS diseases, and it is also necessary to consider them for understanding injury-induced neuroinflammation and neurodegeneration in several neurological conditions. Malko et al. reported that calcium influx via TRPM2 after external stimuli triggers microglial activation and generates pro-inflammatory mediators from activated microglial cells, which further exacerbate damage. Furthermore, the TRPM2 activation process was closely related with NADPH oxidase-mediated ROS production, and PARP-1 (poly ADP-ribose polymerase-1) and ADPR (ADP-ribose) activation [60]. As mentioned in the introduction, oxidative damage following ischemic stroke is an essential factor for understanding brain damage, and is closely related with regulation of cation channel activity.

Based on our study and other previous studies, we speculated that a possible neuronal death cascade could be triggered by zinc influx via TRPM2 (Figure 7). In Figure 7A, schematic drawings describe the process of GCI-induced zinc accumulation, ROS production, and TRPM2 activation. Predicted steps of zinc-related neuronal death are as follows. (1) GCI induces vesicular zinc release from the pre-synaptic terminal to the synaptic cleft [10,13], where it translocated into post-synaptic neurons via the non-selective cation channel, TRPM2 [50,61]. In the post-synaptic neurons, translocated zinc contributes several deleterious cascades such as apoptosis [62,63], necrosis [63], as well as ROS-producing signaling. (2) Following ischemia zinc aggravates ROS production via neuronal NADPH oxidase activation [64]. Synaptically released zinc after GCI stimulates protein kinase C (PKC), and PKC contributes NADPH oxidase activation [65]. (3) A major activator of TRPM2 is ADP-ribose (ADPR), which is released upon oxidative stress including ROS [66,67]. (4) ROS-induced TRPM2 activation triggers increased secondary zinc influx [61]. (5) Secondary translocation of zinc promotes increased ROS production [59,68]. (6) Zinc and ROS induce neuronal death. In Figure 7B, the predicted mechanism of NAC on TRPM2 is shown, as follows. (7) GCI induces zinc translocation from the synaptic terminal into neurons through TRPM. Zinc increases ROS production via NADPH oxidase activation. (8) NAC-supplied cysteine enters into post-synaptic neurons through EAAC1. EAAC1 acts as a cysteine carrier in the brain. (9) Cysteine is converted to glutathione, which has an anti-oxidative effect. ROS production is attenuated by glutathione. (10) Reduction of ROS production decreases TRPM2 activation. (11) Reduced TRPM2 induces diminished secondary zinc influx. Zinc influx reduction contributes neuroprotection. (12) NAC administration induced reduced ROS production and diminished zinc influx, which finally attenuates neuronal death. Thus, as discussed above, steps (1)–(6) shown in the Figure 7A have been demonstrated by our and other group's previous studies. However, steps (7)–(12) in the Figure 7B have been verified in this study; it is shown in Figures 2–6 that NAC treatment reduced GCI-induced zinc accumulation, oxidative stress, TRPM2 activation and neuronal death.

Taken together, the present study demonstrates the efficacy of NAC in preventing GCI-induced neuronal death, sensorimotor deficits, and neurological pathology via several signaling cascades. NAC inhibits GCI-induced zinc accumulation and influx via TRPM2 into postsynaptic neurons. Therefore, we speculate that inhibition of zinc influx and TRPM2 activation by NAC may have a potential for the treatment of ischemic stroke.

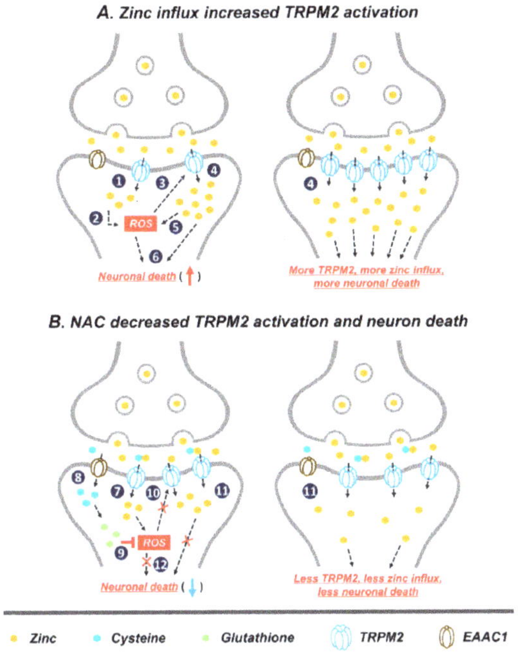

A. Zinc influx increased TRPM2 activation

More TRPM2, more zinc influx, more neuronal death

Neuronal death

B. NAC decreased TRPM2 activation and neuron death

Less TRPM2, less zinc influx, less neuronal death

Neuronal death

Zinc Cysteine Glutathione TRPM2 EAAC1

Figure 7. Predicted reactions of NAC post treatment on GCI-induced neuronal death. Representative schematic illustration indicates NAC action via reduction of reactive oxygen species (ROS) production and subsequent decrease in TRPM2 activation. (**A**) (1) Released zinc after GCI moves into the post-synaptic neurons through TRPM2. (2) Translocated zinc contributes to ROS production. (3) Oxidative damage including ROS activated TRPM2. (4) ROS-induced TRPM2 activation increased secondary zinc influx. (5) Increased translocated of zinc contributes to greater ROS production. (6) Zinc and ROS induces neuronal death. (1)–(6) Image shows the hypothesized cascade produced upon GCI-induced zinc influx and TRPM2 activation. (**B**) (7) Zinc translocated into the post-synaptic neurons via TRPM2 after GCI. (8) Cysteine supplied by NAC administration moves into post-synaptic neurons through excitatory amino acid carrier 1 (EAAC1). (9) Cysteine is converted to glutathione which has an anti-oxidant effect and inhibits ROS production. (10) Inhibition of ROS production leads to reduced TRPM2 activation. (11) Reduced TRPM2 activation leads to less secondary zinc influx. (12) Reduction of ROS production and secondary zinc influx reduces neuronal death. (7)–(12) Process shows possible neuroprotective cascade of NAC after GCI.

4. Materials and Methods

4.1. Ethics Statement and Experimental Animals

All animal care protocols and experimental procedures were approved by the Committee on Animal Use for Research at Hallym University (protocol # Hallym 2019-70, 21 February 2020). To test our hypothesis, experimental animal groups were divided as follows: sham (Vehicle, NAC) and GCI (Vehicle, NAC). All rats were housed in a consistently controlled environment (temperature: 22 ± 2 °C, humidity: 55 ± 5%, light regulation: 12-h light/dark cycle) and given water and standard feed by Purina (Gyeonggi-do, Korea) ad libitum. To minimize any suffering of experimental animals, we used 2~3% isoflurane anesthesia during all procedures. Additionally, this manuscript was written in accordance with the standards put forth in ARRIVE (Animals in Research: Reporting In Vivo Experiments) [69].

4.2. Global Cerebral Ischemia Surgery

The experimental disease model of global cerebral ischemia was performed as previously described and reported [70]. To explain again in detail, rats were deeply anesthetized with 2~3% isoflurane, which was ventilated using mixed 70% nitrous oxide and 30% oxygen and kept at a body temperature of 37 ± 1 °C using a homeothermic monitoring system (Harvard Apparatus, Holliston, MA, USA). To consistently monitor systemic arterial blood pressure and to remove blood as needed, a catheter filled with 10 units of heparin was inserted into the femoral artery. Then, common carotid arteries located beside the tracheal muscle were carefully isolated and transiently occluded. In this dissecting process, we used a surgical microscope (SZ61, Olympus, Shinjuku, Japan) to enhance surgical accuracy and avoid vagus nerve impairment. At the same time, to monitor the electroencephalograph (EEG), electrodes were placed in two bilateral burr holes. To induce the ischemic condition, we drained blood (8~10 cc) from the femoral artery and set systemic arterial blood pressure within the range of 40 ± 10 mmHg. The isolated bilateral common carotid arteries were occluded with a surgical clamp (Fine Science Tools, Foster city, CA, USA). When the systemic arterial blood pressure range was within 40 ± 10 mmHg and the electroencephalograph reached a sustained isoelectric point, we maintained these conditions for 7 min to induce the global cerebral ischemic condition. After occlusion, blood circulation to the brain was restored by unclamping the device and reperfusing the removed blood. During this period, the systemic arterial blood pressure and EEG signal were consistently monitored and vital signs of the experimental animal were periodically monitored and adjusted as required. Experimental animal groups were immediately administered NAC to the intraperitoneal space following restoration of blood perfusion and normal EEG activity; vehicle groups were given the same volume of 0.9% normal saline.

4.3. Experimental Procedures and NAC Administration

Planned experimental procedures were 3 days and 7 days. These time points mean the termination of GCI surgery. In the acute phase (3 days), GCI-induced neurodegenerative cascades were assessed using several immunostaining processes for histological evaluation. In the chronic phase (7 days), GCI-induced cognitive decline and neurological deficits were verified by behavior tests. Vehicle and NAC was administered daily during the acute and chronic phases. To investigate the effects of NAC post-administration to reduce GCI-induced hippocampal neurodegenerations, we used NAC (Sigma-Aldrich, St. Louis, MO, USA). NAC was dissolved with 0.9% normal saline and injected into the intraperitoneal space once per day for 3 days and 7 days at a dose of 150 mg/kg in the present study. Vehicle groups (sham and GCI) were administered the same volumes of 0.9% normal saline only.

4.4. Sample Preparation

Three and seven days after global cerebral ischemia, all experimental rats were deeply anesthetized with urethane (1.5 g/kg) dissolved in 0.9% normal saline to obtain brain samples. For removal of whole blood, deeply anesthetized experimental animals were intracardially perfused with 0.9% normal saline followed by 4% paraformaldehyde (PFA) dissolved in phosphate-buffered saline (PBS) for brain sample fixation. Terminated sample fixation occurred during perfusion, and the whole brain sample was obtained and was immersed in 4% PFA for one hour. After post-fixation, samples were moved into 30% sucrose for cryoprotection. Brain samples sank to the bottom of the tube, and the whole brain was coronally sectioned into thicknesses of 30 μm each using a cryostat microtome (CM1850, Leica, Wetzlar, Germany).

4.5. Assessment of Neuronal Death

Three or seven days after GCI, we estimated hippocampal neuronal death using Fluoro-Jade B (FJB, Histo-Chem, Jefferson, AR) staining method (as previously described in detail [12]). Approximately nine to ten coronal brain sections (based on 30-μm thickness, 270- to 300-μm intervals) were obtained from 3.00 to 4.68 mm caudal to bregma. These brain sections were mounted on gelatin-coated slides, and photographed using a fluorescence microscope (SZ61, Olympus, Shinjuku, Japan, FITC green fluorescence excitation-emission wavelength: 460–490 nm). To avoid experimenter bias, a blind observer counted the whole number of FJB-positive neurons in the hippocampal subiculum, cornu ammonis 1 (CA1), and CA2 regions. Collected data were used for statistical analysis.

4.6. Zinc Fluorescence Staining

To verify the alteration of intraneuronal free zinc concentrations under normal or ischemic conditions, fresh brain samples were obtained without transcardial perfusion process, immediately frozen, and stored in a freezer at a consistently maintained temperature of −80 °C. These samples were coronally sectioned with 10-μm thickness in cryostat and sectioned samples were stained with *N*-(6-methoxy-8-quinolyl)-para-toluenesulfonamide (TSQ; Molecular Probes, Eugene, OR, USA) [25]. In the present study, the concentration of TSQ solution contained 4.5 μM TSQ, 140 mM sodium barbital, and 140 mM sodium acetate. The TSQ-zinc binding fluorescence signal was verified using an Olympus fluorescence microscope (DAPI blue fluorescence excitation-emission wavelength: ~400 nm).

4.7. Histological Analysis

Before initiation of brain tissue immunostaining, blocking of endogenous peroxidase activity is important. The previously described activity was blocked by 1.2% hydrogen peroxide incubation for 20 min at room temperature. After incubation and washing in 0.01% phosphate buffered saline, sections were incubated with several antibodies to analyze our data. In this study, the antibodies used were as follows: 4HNE (diluted 1:500, Alpha Diagnostic Intl. Inc., San Antonio, TX, USA), GFAP (diluted 1:1000, Abcam, Cambridge, UK), Iba-1 (diluted 1:500, Abcam), TRPM2 (diluted 1:400, Abcam), and NeuN (diluted 1:500, Billerica, Millipore Co., Burlington, MA, USA).

4.8. Behavior Outcome Assessment

1. Adhesive removal test: to test whether NAC restores against GCI-induced sensorimotor deficit, all of the experimental groups were evaluated using the adhesive removal test method for 7 consecutive days after GCI. This behavior test was conducted with reference to previous descriptions [13,71]. To explain again in detail, the overall evaluation process was divided into the following steps. Before initiation, rats were acclimated to the transparent testing box (size: 45 × 35 × 20 cm) for 1 min. After acclimating, two pieces of adhesive tape (1 × 1 cm) were attached to the palm of each forepaw and the experimental animal's behavioral outcome for removal time in the testing box was observed (maximum time: 120 s). Removal time is defined as recognizing and detaching the adhesive tape on the bilateral forepaw. If the adhesive tape was detached by shaking its forepaw or bringing its forepaw to its mouth, time was recorded. This process was conducted for five trials with a minute between each trial. In the case that adhesive tapes could not be detached during the test period, removal time was considered the maximum.

2. Modified neurological severity score: to test whether NAC administration attenuated GCI-induced neurological deficits, we used the modified neurological severity score (mNSS) method [72]. The mNSS includes several evaluations of motor (muscle status, abnormal movement), sensory (visual, tactile, and proprioceptive), balance, and reflex functionality. The mNSS assessment criteria were graded from 0 (normal, no deficits) to 18 (maximum points). To explain in detail, this functional test is based on several criteria that include raising subject by the tail and observing flexion (3 points), walking on the floor (3 points), sensory test (2 points), maintaining balance on a beam (6 points), and absence of reflex/abnormal movements (4 points) [73]. This procedure was conducted consecutively for 7 days after GCI. A higher score means a more severe condition.

4.9. Statistical Analysis

All data analyses were conducted using a blind test to reduce researcher bias. All quantified results in the present study were displayed as the mean value ± standard error of mean (SEM), and statistically significant differences were considered at $p < 0.05$. Behavioral data were analyzed by analysis of variance (ANOVA) using Statistical Package for the Social Sciences (SPSS, Chicago, IL, USA). Comparisons between vehicle and NAC-treated groups were conducted with the Mann–Whitney U test, and other comparisons between 4 groups and the remaining data were analyzed by the Kruskal–Wallis test with post-hoc Bonferroni correction. Non-parametric tests were chosen because of the small sample size and lack of normally distributed data.

Author Contributions: D.K.H. researched the data, and wrote and edited the manuscript; A.R.K., S.H.L., J.H.J., B.S.K., D.H.K. and M.K.P. researched the data; K.-H.P., M.-S.L. reviewed and edited the manuscript. B.Y.C. and S.W.S. contributed to conceptualization and supervision of the study, and editing and writing of the manuscript. All authors read and approved the final manuscript.

Funding: This study was supported by funding from the National Research Foundation of Korea (NRF) (NRF-2019R1A6A3A13093671) to D.K.H. and (NRF-2019R1A2C4004912) to B.Y.C. This work was also supported by the Brain Research Program through the NRF, funded by the Ministry of Science, Information and Communication Technology and Future Planning (NRF-2017M3C7A1028937, 2018R1A4A1020922 and 2020R1A2C2008480) to S.W.S.

Conflicts of Interest: The authors declare no conflict of interest.

References

1. Lo, E.H.; Dalkara, T.; Moskowitz, M.A. Mechanisms, challenges and opportunities in stroke. *Nat. Rev. Neurosci.* **2003**, *4*, 399–415. [CrossRef]
2. Cho, C.H.; Byun, H.R.; Jover-Mengual, T.; Pontarelli, F.; Dejesus, C.; Cho, A.R.; Zukin, R.S.; Hwang, J.Y. Gadd45b Acts as Neuroprotective Effector in Global Ischemia-Induced Neuronal Death. *Int. Neurourol. J.* **2019**, *23*, S11–S21. [CrossRef] [PubMed]
3. Tasca, C.I.; Dal-Cim, T.; Cimarosti, H. In vitro oxygen-glucose deprivation to study ischemic cell death. *Methods Mol. Biol.* **2015**, *1254*, 197–210. [CrossRef] [PubMed]
4. Mehta, S.L.; Pandi, G.; Vemuganti, R. Circular RNA Expression Profiles Alter Significantly in Mouse Brain After Transient Focal Ischemia. *Stroke* **2017**, *48*, 2541–2548. [CrossRef]
5. Zhou, L.; Lin, Q.; Wang, P.; Yao, L.; Leong, K.; Tan, Z.; Huang, Z. Enhanced neuroprotective efficacy of bone marrow mesenchymal stem cells co-overexpressing BDNF and VEGF in a rat model of cardiac arrest-induced global cerebral ischemia. *Cell Death Dis.* **2017**, *8*, e2774. [CrossRef]
6. Hu, H.J.; Song, M. Disrupted Ionic Homeostasis in Ischemic Stroke and New Therapeutic Targets. *J. Stroke Cereb. Dis.* **2017**, *26*, 2706–2719. [CrossRef] [PubMed]
7. Harukuni, I.; Bhardwaj, A. Mechanisms of brain injury after global cerebral ischemia. *Neurol. Clin.* **2006**, *24*, 1–21. [CrossRef]
8. Patel, A.R.; Ritzel, R.; McCullough, L.D.; Liu, F. Microglia and ischemic stroke: A double-edged sword. *Int. J. Physiol. Pathophysiol. Pharmacol.* **2013**, *5*, 73–90.
9. Barreto, G.; White, R.E.; Ouyang, Y.; Xu, L.; Giffard, R.G. Astrocytes: Targets for neuroprotection in stroke. *Cent. Nerv. Syst. Agents. Med. Chem.* **2011**, *11*, 164–173. [CrossRef]
10. Koh, J.Y.; Suh, S.W.; Gwag, B.J.; He, Y.Y.; Hsu, C.Y.; Choi, D.W. The role of zinc in selective neuronal death after transient global cerebral ischemia. *Science* **1996**, *272*, 1013–1016. [CrossRef]
11. Hong, D.K.; Kho, A.R.; Choi, B.Y.; Lee, S.H.; Jeong, J.H.; Lee, S.H.; Park, K.H.; Park, J.B.; Suh, S.W. Combined Treatment With Dichloroacetic Acid and Pyruvate Reduces Hippocampal Neuronal Death After Transient Cerebral Ischemia. *Front. Neurol.* **2018**, *9*, 137. [CrossRef] [PubMed]
12. Hong, D.K.; Choi, B.Y.; Kho, A.R.; Lee, S.H.; Jeong, J.H.; Kang, B.S.; Kang, D.H.; Park, K.H.; Suh, S.W. Carvacrol Attenuates Hippocampal Neuronal Death after Global Cerebral Ischemia via Inhibition of Transient Receptor Potential Melastatin 7. *Cells* **2018**, *7*, 231. [CrossRef] [PubMed]
13. Kho, A.R.; Choi, B.Y.; Lee, S.H.; Hong, D.K.; Lee, S.H.; Jeong, J.H.; Park, K.H.; Song, H.K.; Choi, H.C.; Suh, S.W. Effects of Protocatechuic Acid (PCA) on Global Cerebral Ischemia-Induced Hippocampal Neuronal Death. *Int. J. Mol. Sci.* **2018**, *19*, 1420. [CrossRef]

14. Petrovic-Djergovic, D.; Goonewardena, S.N.; Pinsky, D.J. Inflammatory Disequilibrium in Stroke. *Circ. Res.* **2016**, *119*, 142–158. [CrossRef] [PubMed]

15. Lee, S.Y.; Kim, J.H. Mechanisms underlying presynaptic Ca2+ transient and vesicular glutamate release at a CNS nerve terminal during in vitro ischaemia. *J. Physiol.* **2015**, *593*, 2793–2806. [CrossRef]

16. Bonaventura, A.; Liberale, L.; Vecchie, A.; Casula, M.; Carbone, F.; Dallegri, F.; Montecucco, F. Update on Inflammatory Biomarkers and Treatments in Ischemic Stroke. *Int. J. Mol. Sci.* **2016**, *17*, 1967. [CrossRef]

17. Shuttleworth, C.W.; Weiss, J.H. Zinc: New clues to diverse roles in brain ischemia. *Trends Pharmacol. Sci.* **2011**, *32*, 480–486. [CrossRef]

18. Weiss, J.H.; Sensi, S.L.; Koh, J.Y. Zn(2+): A novel ionic mediator of neural injury in brain disease. *Trends Pharmacol. Sci.* **2000**, *21*, 395–401. [CrossRef]

19. Kawahara, M.; Tanaka, K.I.; Kato-Negishi, M. Zinc, Carnosine, and Neurodegenerative Diseases. *Nutrients* **2018**, *10*, 147. [CrossRef]

20. Lee, S.R. Critical Role of Zinc as Either an Antioxidant or a Prooxidant in Cellular Systems. *Oxid. Med. Cell Longev.* **2018**, *2018*, 9156285. [CrossRef]

21. Portbury, S.D.; Adlard, P.A. Zinc Signal in Brain Diseases. *Int. J. Mol. Sci.* **2017**, *18*, 2506. [CrossRef] [PubMed]

22. Choi, B.Y.; Hong, D.K.; Jeong, J.H.; Lee, B.E.; Koh, J.Y.; Suh, S.W. Zinc transporter 3 modulates cell proliferation and neuronal differentiation in the adult hippocampus. *Stem. Cells* **2020**, *38*, 994–1006. [CrossRef] [PubMed]

23. McAllister, B.B.; Dyck, R.H. Zinc transporter 3 (ZnT3) and vesicular zinc in central nervous system function. *Neurosci. Biobehav. Rev.* **2017**, *80*, 329–350. [CrossRef]

24. Kho, A.R.; Choi, B.Y.; Kim, J.H.; Lee, S.H.; Hong, D.K.; Lee, S.H.; Jeong, J.H.; Sohn, M.; Suh, S.W. Prevention of hypoglycemia-induced hippocampal neuronal death by N-acetyl-L-cysteine (NAC). *Amino Acids* **2017**, *49*, 367–378. [CrossRef] [PubMed]

25. Choi, B.Y.; Lee, S.H.; Choi, H.C.; Lee, S.K.; Yoon, H.S.; Park, J.B.; Chung, W.S.; Suh, S.W. Alcohol dependence treating agent, acamprosate, prevents traumatic brain injury-induced neuron death through vesicular zinc depletion. *Transl. Res.* **2019**, *207*, 1–18. [CrossRef]

26. Higashi, Y.; Aratake, T.; Shimizu, S.; Shimizu, T.; Nakamura, K.; Tsuda, M.; Yawata, T.; Ueba, T.; Saito, M. Influence of extracellular zinc on M1 microglial activation. *Sci. Rep.* **2017**, *7*, 43778. [CrossRef]

27. Qi, Z.; Liang, J.; Pan, R.; Dong, W.; Shen, J.; Yang, Y.; Zhao, Y.; Shi, W.; Luo, Y.; Ji, X.; et al. Zinc contributes to acute cerebral ischemia-induced blood-brain barrier disruption. *Neurobiol. Dis.* **2016**, *95*, 12–21. [CrossRef]

28. Suh, S.W.; Hamby, A.M.; Gum, E.T.; Shin, B.S.; Won, S.J.; Sheline, C.T.; Chan, P.H.; Swanson, R.A. Sequential release of nitric oxide, zinc, and superoxide in hypoglycemic neuronal death. *J. Cereb. Blood. Flow. Metab.* **2008**, *28*, 1697–1706. [CrossRef]

29. Park, S.E.; Song, J.H.; Hong, C.; Kim, D.E.; Sul, J.W.; Kim, T.Y.; Seo, B.R.; So, I.; Kim, S.Y.; Bae, D.J.; et al. Contribution of Zinc-Dependent Delayed Calcium Influx via TRPC5 in Oxidative Neuronal Death and its Prevention by Novel TRPC Antagonist. *Mol. Neurobiol.* **2019**, *56*, 2822–2835. [CrossRef]

30. Sen, C.K. Antioxidant and redox regulation of cellular signaling: Introduction. *Med. Sci. Sports Exerc.* **2001**, *33*, 368–370. [CrossRef]

31. Kerksick, C.; Willoughby, D. The antioxidant role of glutathione and N-acetyl-cysteine supplements and exercise-induced oxidative stress. *J. Int. Soc. Sports Nutr.* **2005**, *2*, 38–44. [CrossRef] [PubMed]

32. Thanan, R.; Oikawa, S.; Hiraku, Y.; Ohnishi, S.; Ma, N.; Pinlaor, S.; Yongvanit, P.; Kawanishi, S.; Murata, M. Oxidative stress and its significant roles in neurodegenerative diseases and cancer. *Int. J. Mol. Sci.* **2014**, *16*, 193–217. [CrossRef] [PubMed]

33. Liu, Z.; Zhou, T.; Ziegler, A.C.; Dimitrion, P.; Zuo, L. Oxidative Stress in Neurodegenerative Diseases: From Molecular Mechanisms to Clinical Applications. *Oxid. Med. Cell Longev.* **2017**, *2017*, 2525967. [CrossRef] [PubMed]

34. Choi, B.Y.; Kim, I.Y.; Kim, J.H.; Lee, B.E.; Lee, S.H.; Kho, A.R.; Jung, H.J.; Sohn, M.; Song, H.K.; Suh, S.W. Decreased cysteine uptake by EAAC1 gene deletion exacerbates neuronal oxidative stress and neuronal death after traumatic brain injury. *Amino Acids* **2016**, *48*, 1619–1629. [CrossRef]

35. Rushworth, G.F.; Megson, I.L. Existing and potential therapeutic uses for N-acetylcysteine: The need for conversion to intracellular glutathione for antioxidant benefits. *Pharmacol. Ther.* **2014**, *141*, 150–159. [CrossRef]

36. Tardiolo, G.; Bramanti, P.; Mazzon, E. Overview on the Effects of N-Acetylcysteine in Neurodegenerative Diseases. *Molecules* **2018**, *23*, 3305. [CrossRef]

37. Jiang, L.H.; Yang, W.; Zou, J.; Beech, D.J. TRPM2 channel properties, functions and therapeutic potentials. *Expert Opin. Ther. Targets* **2010**, *14*, 973–988. [CrossRef]

38. Sita, G.; Hrelia, P.; Graziosi, A.; Ravegnini, G.; Morroni, F. TRPM2 in the Brain: Role in Health and Disease. *Cells* **2018**, *7*, 82. [CrossRef]

39. Turlova, E.; Feng, Z.P.; Sun, H.S. The role of TRPM2 channels in neurons, glial cells and the blood-brain barrier in cerebral ischemia and hypoxia. *Acta Pharmacol. Sin.* **2018**, *39*, 713–721. [CrossRef]

40. McHugh, D.; Flemming, R.; Xu, S.Z.; Perraud, A.L.; Beech, D.J. Critical intracellular Ca2+ dependence of transient receptor potential melastatin 2 (TRPM2) cation channel activation. *J. Biol. Chem.* **2003**, *278*, 11002–11006. [CrossRef]

41. Belrose, J.C.; Jackson, M.F. TRPM2: A candidate therapeutic target for treating neurological diseases. *Acta Pharmacol. Sin.* **2018**, *39*, 722–732. [CrossRef] [PubMed]

42. Zhu, T.; Zhao, Y.; Hu, H.; Zheng, Q.; Luo, X.; Ling, Y.; Ying, Y.; Shen, Z.; Jiang, P.; Shu, Q. TRPM2 channel regulates cytokines production in astrocytes and aggravates brain disorder during lipopolysaccharide-induced endotoxin sepsis. *Int. Immunopharmacol.* **2019**, *75*, 105836. [CrossRef] [PubMed]

43. Haraguchi, K.; Kawamoto, A.; Isami, K.; Maeda, S.; Kusano, A.; Asakura, K.; Shirakawa, H.; Mori, Y.; Nakagawa, T.; Kaneko, S. TRPM2 contributes to inflammatory and neuropathic pain through the aggravation of pronociceptive inflammatory responses in mice. *J. Neurosci.* **2012**, *32*, 3931–3941. [CrossRef] [PubMed]

44. Nilius, B.; Flockerzi, V. Mammalian transient receptor potential (TRP) cation channels. Preface. *Handb. Exp. Pharmacol.* **2014**, *223*, v–vi. [PubMed]

45. Yu, P.; Wang, Q.; Zhang, L.H.; Lee, H.C.; Zhang, L.; Yue, J. A cell permeable NPE caged ADP-ribose for studying TRPM2. *PLoS ONE* **2012**, *7*, e51028. [CrossRef] [PubMed]

46. Miller, B.A.; Cheung, J.Y. TRPM2 protects against tissue damage following oxidative stress and ischaemia-reperfusion. *J. Physiol.* **2016**, *594*, 4181–4191. [CrossRef]

47. Jang, B.G.; Won, S.J.; Kim, J.H.; Choi, B.Y.; Lee, M.W.; Sohn, M.; Song, H.K.; Suh, S.W. EAAC1 gene deletion alters zinc homeostasis and enhances cortical neuronal injury after transient cerebral ischemia in mice. *J. Trace Elem. Med. Biol.* **2012**, *26*, 85–88. [CrossRef]

48. Suh, S.W.; Garnier, P.; Aoyama, K.; Chen, Y.; Swanson, R.A. Zinc release contributes to hypoglycemia-induced neuronal death. *Neurobiol. Dis.* **2004**, *16*, 538–545. [CrossRef]

49. Suh, S.W.; Frederickson, C.J.; Danscher, G. Neurotoxic zinc translocation into hippocampal neurons is inhibited by hypothermia and is aggravated by hyperthermia after traumatic brain injury in rats. *J. Cereb. Blood Flow Metab.* **2006**, *26*, 161–169. [CrossRef]

50. Ye, M.; Yang, W.; Ainscough, J.F.; Hu, X.P.; Li, X.; Sedo, A.; Zhang, X.H.; Zhang, X.; Chen, Z.; Li, X.M.; et al. TRPM2 channel deficiency prevents delayed cytosolic Zn2+ accumulation and CA1 pyramidal neuronal death after transient global ischemia. *Cell Death Dis.* **2014**, *5*, e1541. [CrossRef]

51. Medvedeva, Y.V.; Ji, S.G.; Yin, H.Z.; Weiss, J.H. Differential Vulnerability of CA1 versus CA3 Pyramidal Neurons After Ischemia: Possible Relationship to Sources of Zn2+ Accumulation and Its Entry into and Prolonged Effects on Mitochondria. *J. Neurosci.* **2017**, *37*, 726–737. [CrossRef] [PubMed]

52. Hara, Y.; Wakamori, M.; Ishii, M.; Maeno, E.; Nishida, M.; Yoshida, T.; Yamada, H.; Shimizu, S.; Mori, E.; Kudoh, J.; et al. LTRPC2 Ca2+-permeable channel activated by changes in redox status confers susceptibility to cell death. *Mol. Cell* **2002**, *9*, 163–173. [CrossRef]

53. Takahashi, N.; Kozai, D.; Kobayashi, R.; Ebert, M.; Mori, Y. Roles of TRPM2 in oxidative stress. *Cell Calcium.* **2011**, *50*, 279–287. [CrossRef] [PubMed]

54. Berk, M.; Malhi, G.S.; Gray, L.J.; Dean, O.M. The promise of N-acetylcysteine in neuropsychiatry. *Trends Pharmacol. Sci.* **2013**, *34*, 167–177. [CrossRef] [PubMed]

55. Mandal, S.; Das, G.; Askari, H. Interactions of N-acetyl-l-cysteine with metals (Ni 2+, Cu 2+ and Zn 2+): An experimental and theoretical study. *Struct. Chem.* **2014**, *25*, 43–51. [CrossRef]

56. Allen, N.J.; Barres, B.A. Neuroscience: Glia-more than just brain glue. *Nature* **2009**, *457*, 675–677. [CrossRef]

57. Phatnani, H.; Maniatis, T. Astrocytes in neurodegenerative disease. *Cold Spring Harb. Perspect Biol.* **2015**, *7*. [CrossRef]

58. Lakhan, S.E.; Kirchgessner, A.; Hofer, M. Inflammatory mechanisms in ischemic stroke: Therapeutic approaches. *J. Transl. Med.* **2009**, *7*, 97. [CrossRef]

59. Kauppinen, T.M.; Higashi, Y.; Suh, S.W.; Escartin, C.; Nagasawa, K.; Swanson, R.A. Zinc triggers microglial activation. *J. Neurosci.* **2008**, *28*, 5827–5835. [CrossRef]

60. Malko, P.; Syed Mortadza, S.A.; McWilliam, J.; Jiang, L.H. TRPM2 Channel in Microglia as a New Player in Neuroinflammation Associated With a Spectrum of Central Nervous System Pathologies. *Front. Pharmacol.* **2019**, *10*, 239. [CrossRef]

61. Li, F.; Abuarab, N.; Sivaprasadarao, A. Reciprocal regulation of actin cytoskeleton remodelling and cell migration by Ca2+ and Zn2+: Role of TRPM2 channels. *J. Cell Sci.* **2016**, *129*, 2016–2029. [CrossRef] [PubMed]

62. Sensi, S.L.; Ton-That, D.; Sullivan, P.G.; Jonas, E.A.; Gee, K.R.; Kaczmarek, L.K.; Weiss, J.H. Modulation of mitochondrial function by endogenous Zn2+ pools. *Proc. Natl. Acad Sci. USA* **2003**, *100*, 6157–6162. [CrossRef] [PubMed]

63. Kim, Y.H.; Kim, E.Y.; Gwag, B.J.; Sohn, S.; Koh, J.Y. Zinc-induced cortical neuronal death with features of apoptosis and necrosis: Mediation by free radicals. *Neuroscience* **1999**, *89*, 175–182. [CrossRef]

64. Suh, S.W.; Gum, E.T.; Hamby, A.M.; Chan, P.H.; Swanson, R.A. Hypoglycemic neuronal death is triggered by glucose reperfusion and activation of neuronal NADPH oxidase. *J. Clin. Investig.* **2007**, *117*, 910–918. [CrossRef]

65. Noh, K.M.; Kim, Y.H.; Koh, J.Y. Mediation by membrane protein kinase C of zinc-induced oxidative neuronal injury in mouse cortical cultures. *J. Neurochem.* **1999**, *72*, 1609–1616. [CrossRef]

66. Li, X.; Jiang, L.H. A critical role of the transient receptor potential melastatin 2 channel in a positive feedback mechanism for reactive oxygen species-induced delayed cell death. *J. Cell Physiol.* **2019**, *234*, 3647–3660. [CrossRef]

67. Di, A.; Gao, X.P.; Qian, F.; Kawamura, T.; Han, J.; Hecquet, C.; Ye, R.D.; Vogel, S.M.; Malik, A.B. The redox-sensitive cation channel TRPM2 modulates phagocyte ROS production and inflammation. *Nat. Immunol.* **2011**, *13*, 29–34. [CrossRef] [PubMed]

68. Mortadza, S.S.; Sim, J.A.; Stacey, M.; Jiang, L.H. Signalling mechanisms mediating Zn(2+)-induced TRPM2 channel activation and cell death in microglial cells. *Sci. Rep.* **2017**, *7*, 45032. [CrossRef]

69. Kilkenny, C.; Browne, W.J.; Cuthill, I.C.; Emerson, M.; Altman, D.G. Improving bioscience research reporting: The ARRIVE guidelines for reporting animal research. *PLoS Biol.* **2010**, *8*, e1000412. [CrossRef]

70. Sanderson, T.H.; Wider, J.M. 2-vessel occlusion/hypotension: A rat model of global brain ischemia. *J. Vis. Exp.* **2013**. [CrossRef]

71. Bouet, V.; Boulouard, M.; Toutain, J.; Divoux, D.; Bernaudin, M.; Schumann-Bard, P.; Freret, T. The adhesive removal test: A sensitive method to assess sensorimotor deficits in mice. *Nat. Protoc.* **2009**, *4*, 1560–1564. [CrossRef] [PubMed]

72. Xing, Z.; Xia, Z.; Peng, W.; Li, J.; Zhang, C.; Fu, C.; Tang, T.; Luo, J.; Zou, Y.; Fan, R.; et al. Xuefu Zhuyu decoction, a traditional Chinese medicine, provides neuroprotection in a rat model of traumatic brain injury via an anti-inflammatory pathway. *Sci. Rep.* **2016**, *6*, 20040. [CrossRef] [PubMed]

73. Chen, J.; Sanberg, P.R.; Li, Y.; Wang, L.; Lu, M.; Willing, A.E.; Sanchez-Ramos, J.; Chopp, M. Intravenous administration of human umbilical cord blood reduces behavioral deficits after stroke in rats. *Stroke* **2001**, *32*, 2682–2688. [CrossRef] [PubMed]

International Journal of
Molecular Sciences

Article

Pyridoxine Deficiency Exacerbates Neuronal Damage after Ischemia by Increasing Oxidative Stress and Reduces Proliferating Cells and Neuroblasts in the Gerbil Hippocampus

Hyo Young Jung [1], Woosuk Kim [2], Kyu Ri Hahn [1], Min Soo Kang [3], Tae Hyeong Kim [3], Hyun Jung Kwon [4], Sung Min Nam [5], Jin Young Chung [6], Jung Hoon Choi [3], Yeo Sung Yoon [1], Dae Won Kim [4], Dae Young Yoo [7,*] and In Koo Hwang [1,*]

[1] Department of Anatomy and Cell Biology, College of Veterinary Medicine, and Research Institute for Veterinary Science, Seoul National University, Seoul 08826, Korea; hyoyoung@snu.ac.kr (H.Y.J.); hkinging@snu.ac.kr (K.R.H.); ysyoon@snu.ac.kr (Y.S.Y.)
[2] Department of Biomedical Sciences, and Research Institute for Bioscience and Biotechnology, Hallym University, Chuncheon 24252, Korea; tank3430@hallym.ac.kr
[3] Department of Anatomy, College of Veterinary Medicine and Institute of Veterinary Science, Kangwon National University, Chuncheon 24341, Korea; imkangms@kangwon.ac.kr (M.S.K.); xogudsla9402@kangwon.ac.kr (T.H.K.); jhchoi@kangwon.ac.kr (J.H.C.)
[4] Department of Biochemistry and Molecular Biology, Research Institute of Oral Sciences, College of Dentistry, Gangneung-Wonju National University, Gangneung 25457, Korea; donuts25@gwnu.ac.kr (H.J.K.); kimdw@gwnu.ac.kr (D.W.K.)
[5] Department of Anatomy, College of Veterinary Medicine, Konkuk University, Seoul 05030, Korea; lovingvet@gmail.com
[6] Department of Veterinary Internal Medicine and Geriatrics, College of Veterinary Medicine, Kangwon National University, Chuncheon 24341, Korea; jychung77@gmail.com
[7] Department of Anatomy, College of Medicine, Soonchunhyang University, Cheonan 31151, Korea
* Correspondence: dyyoo@sch.ac.kr (D.Y.Y.); vetmed2@snu.ac.kr (I.K.H.)

Received: 30 June 2020; Accepted: 29 July 2020; Published: 4 August 2020

Abstract: We investigated the effects of pyridoxine deficiency on ischemic neuronal death in the hippocampus of gerbil ($n = 5$ per group). Serum pyridoxal 5′-phosphate levels were significantly decreased in Pyridoxine-deficient diet (PDD)-fed gerbils, while homocysteine levels were significantly increased in sham- and ischemia-operated gerbils. PDD-fed gerbil showed a reduction in neuronal nuclei (NeuN)-immunoreactive neurons in the medial part of the hippocampal CA1 region three days after. Reactive astrocytosis and microgliosis were found in PDD-fed gerbils, and transient ischemia caused the aggregation of activated microglia in the stratum pyramidale three days after ischemia. Lipid peroxidation was prominently increased in the hippocampus and was significantly higher in PDD-fed gerbils than in Control diet (CD)-fed gerbils after ischemia. In contrast, pyridoxine deficiency decreased the proliferating cells and neuroblasts in the dentate gyrus in sham- and ischemia-operated gerbils. Nuclear factor erythroid-2-related factor 2 (Nrf2) and brain-derived neurotrophic factor (BDNF) levels also significantly decreased in PDD-fed gerbils sham 24 h after ischemia. These results suggest that pyridoxine deficiency accelerates neuronal death by increasing serum homocysteine levels and lipid peroxidation, and by decreasing Nrf2 levels in the hippocampus. Additionally, it reduces the regenerated potentials in hippocampus by decreasing BDNF levels. Collectively, pyridoxine is an essential element in modulating cell death and hippocampal neurogenesis after ischemia.

Keywords: pyridoxine deficiency; ischemia; gerbil; homocysteine; cell death; glia; neurogenesis

1. Introduction

Vitamin B_6 vitamers consist of pyridine derivatives such as naïve and phosphorylated forms of pyridoxine, pyridoxal, and pyridoxamine. Intakes of B_6 vitamers are absorbed in the intestine and transformed into its active form, pyridoxal 5′-phosphate (PLP), in the liver. However, among B_6 vitamers, naïve forms can cross the blood–brain barrier in the brain [1]. PLP acts as a coenzyme or cofactor in more than 100 reactions associated with energy metabolism and neurotransmitter synthesis. In addition, vitamin B_6 has antioxidant properties that quench reactive oxygen [2] and reduce the formation of advanced glycation end-products [3,4].

Brain ischemia is one of the major life-threatening diseases worldwide, and it debases the quality of life in survivors. Mongolian gerbils (*Meriones unguiculatus*) are used in animal models for brain ischemia because they have incomplete posterior communicating arteries that cause brain ischemia only by occlusion of the common carotid artery for 5 min in the neck region [5]. However, more sophisticated models with complete interruption of cerebral blood flow have been implemented in rats [6–8] and mice [9], which are more widely used in research. Several studies demonstrate that PLP has neuroprotective effects against various neurological diseases including ischemia [10,11], vascular dementia [12], Parkinson's disease [13], and cortical damage [14]. In addition, pyridoxine treatment increases proliferating cells and neuroblasts in the dentate gyrus [15]. There is conflicting evidence of dietary vitamin B_6 on cardiovascular diseases in human studies [16–19]. Plasma vitamin B_6 levels are inversely related to risk and incidence of cardiovascular diseases in the United States, Japan, and Korea [16–18], while a Finnish study demonstrates that there is no relationship between vitamin B_6 intake and cardiovascular diseases [19].

In contrast, feeding with pyridoxine deficient diets causes cognitive impairment in normal mice [20] and in animal models for Alzheimer's disease [21]. Additionally, pyridoxine deficiency increases homocysteine levels because pyridoxine is used in homocysteine metabolism [22]. Increased homocysteine is considered as a risk factor for stroke [23], and it facilitates the production of reactive oxygen species by auto-oxidation [24]. However, there are no studies on the effects of pyridoxine deficiency on neuronal cell death after ischemia in the hippocampus. In this study, we examined the effects of pyridoxine deficiency on ischemia-induced cell death based on the oxidative stress in the hippocampus after 5 min of forebrain ischemia.

2. Results

2.1. Pyridoxine Deficiency Decreases PLP and Increases Homocysteine Levels in Serum

In the Control diet (CD)-fed sham group, PLP and homocysteine levels were 32.5 ± 12.1 μmol/L and 4.91 ± 2.05 μmol/L in the serum, respectively. In the Pyridoxine-deficient diet (PDD)-fed sham group, PLP levels were dramatically decreased to 0.298 ± 0.205 μmol/L, while homocysteine levels were significantly increased to 57.4 ± 16.9 μmol/L. Transient forebrain ischemia decreased PLP levels in CD- and PDD-fed groups, although the statistical significance was not detected between CD- and PDD-fed groups. PLP levels were significantly lower in the PDD-fed group compared to the CD-fed group three and four days after ischemia. In contrast, homocysteine levels were maintained with significant increases in PDD-the fed group compared to that in the CD-fed group three and four days after ischemia, although homocysteine levels were slightly higher after ischemia compared to that in the sham group. The two-way analysis of variance (ANOVA) test showed that there were no interactions between ischemia and PDD diets in PLP and homocysteine levels in the serum (Figure 1A,B).

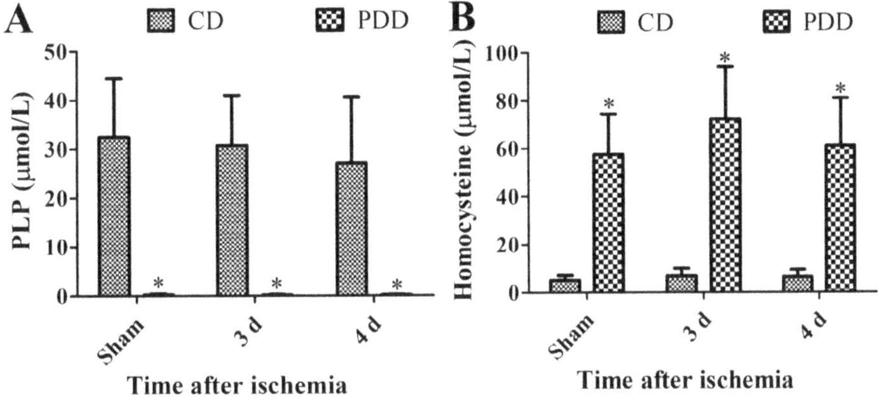

Figure 1. Pyridoxine deficiency decreases serum pyridoxal 5′-phosphate (PLP) levels (**A**) and increases serum homocysteine levels (**B**) in the control diet (CD)- and pyridoxine-deficient diet (PDD)-fed gerbils of sham- and ischemia-operated groups. Data were analyzed with a two-way ANOVA test followed by Bonferroni post-hoc tests (n = 7 per group; * p < 0.05, significant difference between CD- and PDD-fed group). All data are expressed as mean with standard deviation.

2.2. Pyridoxine Deficiency Causes Early Neuronal Death after Ischemia

In the CD-fed sham group, neuronal nuclei (NeuN)-positive neurons were abundantly observed in all hippocampal regions, including the CA1 region, and the same observation was also obtained in the PDD-fed sham group. In the PDD-fed sham group, the number of NeuN-positive neurons was slightly decreased to 94.8% of the CD-fed sham group. In the CD-fed ischemic group, numerous NeuN-positive neurons were found in the hippocampus three days after ischemia, while NeuN-positive neurons were decreased in the medial side of the CA1 region and not in the lateral region three days after ischemia. The number of NeuN positive neurons was significantly decreased in the PDD-fed group compared to that in the CD-fed sham group to 83.2% of the CD-fed sham group. Four days after ischemia, NeuN-positive neurons were prominently decreased in the hippocampal CA1 region of the CD- and PDD-fed ischemia group to 4.9% and 4.4% of the CD-fed sham group. However, there were no significant differences in the number of NeuN-positive neurons between groups. The two-way ANOVA test demonstrated that there were no interactions between ischemia and PDD diets in neuronal numbers in the hippocampal CA1 region (Figure 2A,B).

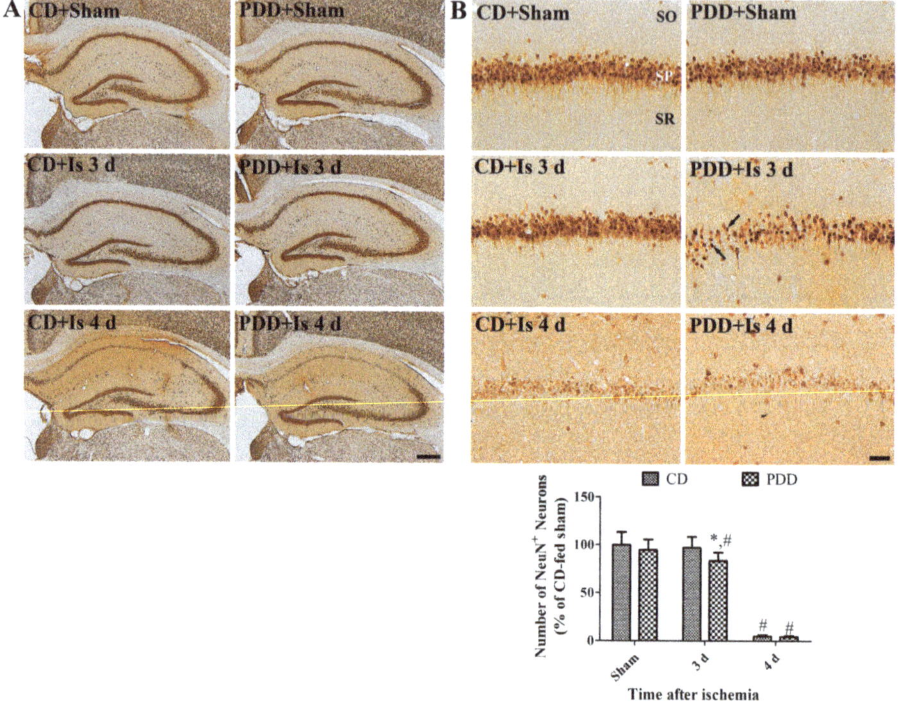

Figure 2. Immunohistochemical staining for neuronal nuclei (NeuN) in the whole hippocampus (**A**) and its magnified CA1 region (**B**) in the CD- and PDD-fed gerbils of sham- and ischemia-operated groups. Note that fewer NeuN-positive cells (arrows) are found in the medial side of the stratum pyramidale (SP) in the PDD-fed gerbils three days after ischemia. SO, stratum oriens; SR, stratum radiatum. Scale bar = 400 μm (**A**), 50 μm (**B**). The number of NeuN-immunoreactive nuclei in the CA1 region compared to the CD-fed sham group per section, in all the groups, is shown (n = 7 per group; * $p < 0.05$, significant difference between CD- and PDD-fed group; # $p < 0.05$, significant difference between sham- and ischemia-operated groups). All data are expressed as mean with standard deviation.

2.3. Pyridoxine Deficiency Facilitates the Activation of Astrocytes and Microglia after Ischemia

In the CD-fed sham group, glial fibrillary acidic protein (GFAP)-immunoreactive astrocytes and ionized calcium-binding adapter molecule 1 (Iba-1)-immunoreactive microglia had small cytoplasm with long and distinct processes. In the PDD-fed sham group, they had enlarged cytoplasm with thickened processes in the hippocampal CA1 region, and GFAP and Iba-1 immunoreactivities were significantly increased in this group, compared to that in the CD-fed sham group. Three days after ischemia/reperfusion, GFAP-immunoreactive astrocytes showed similar morphology in CD- and PDD-fed groups compared to the PDD-fed sham group, and GFAP immunoreactivity was similarly observed. In contrast, Iba-1-immunoreactive microglia showed similar morphology in CD-fed ischemic group three days after ischemia/reperfusion, but in the PDD-fed group, Iba-1-immunoreactive microglia were abundantly found in the medial side of the stratum pyramidale, where the neuronal death occurred. Additionally, Iba-1-immunoreactive microglia that were detected in the stratum pyramidale had round cytoplasm without processes, and Iba-1 immunoreactivity was significantly increased. Four days after ischemia/reperfusion, GFAP-immunoreactive astrocytes had punctuated cytoplasm with thickened processes in CD- and PDD-fed groups. Iba-1-immunoreactive microglia were abundantly observed in the stratum pyramidale of the CA1 region as well as the stratum oriens and radiatum in CD-

and PDD-fed groups. There were no significant differences in the GFAP and Iba-1 immunoreactivities between CD- and PDD-fed groups after ischemia. However, the two-way ANOVA test indicated that ischemia and PDD diets significantly increased the GFAP (df = 2, F = 5.341, p = 0.0093) and Iba-1 immunoreactivity (df = 2, F = 19.06, p < 0.0001) (Figure 3A,B).

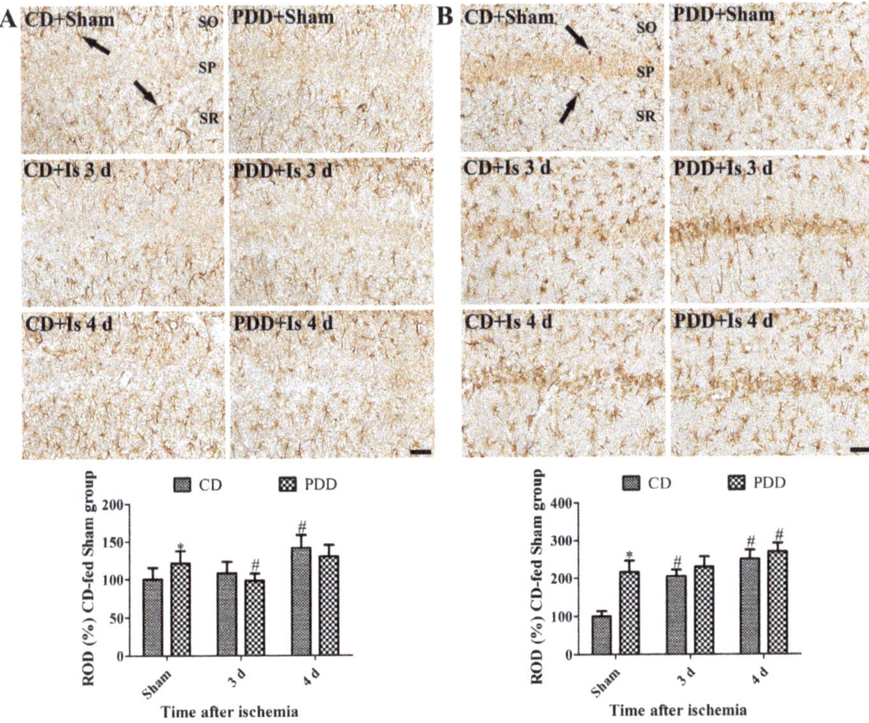

Figure 3. Immunohistochemical staining for GFAP (**A**) and Iba-1 (**B**) in the hippocampal CA1 region in the CD- and PDD-fed gerbils of sham- and ischemia-operated groups. Note that GFAP-immunoreactive astrocytes (arrows) and Iba-1-immunoreactive microglia (arrows) show morphological changes after PDD-diet and ischemic damage. SO, stratum oriens; SP, stratum pyramidale; SR, stratum radiatum. Scale bar = 50 μm. Relative optical density (ROD) corresponding to the percentage of GFAP and Iba-1 immunoreactivity value in the hippocampal CA1 region of the CD-fed sham group per section is shown (n = 7 per group; * p < 0.05, significant difference between CD- and PDD-fed groups; # p < 0.05, significant difference between sham- and ischemia-operated groups). All data are expressed as mean with standard deviation.

2.4. Pyridoxine Deficiency Decreases Ischemia-Induced Proliferating Cells and Neuroblasts

In all groups, Ki67-positive nuclei and cell bodies of doublecortin (DCX)-immunoreactive neuroblasts were mainly found in the subgranular zone of the dentate gyrus, and processes of neuroblasts branched out to the molecular layer of the dentate gyrus. However, there were significant changes in the number of proliferating Ki67-positive cells and the immunoreactivity of DCX-immunoreactive neuroblasts among groups. In the sham group, the DCX immunoreactivity was significantly decreased in the PDD-fed group compared to that in the CD-fed group, while the number of Ki67-positive cells was slightly decreased. Three and four days after ischemia/reperfusion, the number of Ki67-positive cells and the DCX immunoreactivity were significantly increased compared to the sham group, and there were fewer Ki67-positive cells and lower DCX immunoreactivity in the

PDD-fed group compared to that in the CD-fed group. However, the two-way ANOVA test indicated that ischemia and PDD diets significantly changes the number of proliferating Ki67-positive cells (df = 2, F = 8.816, p = 0.0008), and there were no interaction effects of ischemia and PDD diets in the density of DCX-immunoreactive neuroblasts (Figure 4A,B).

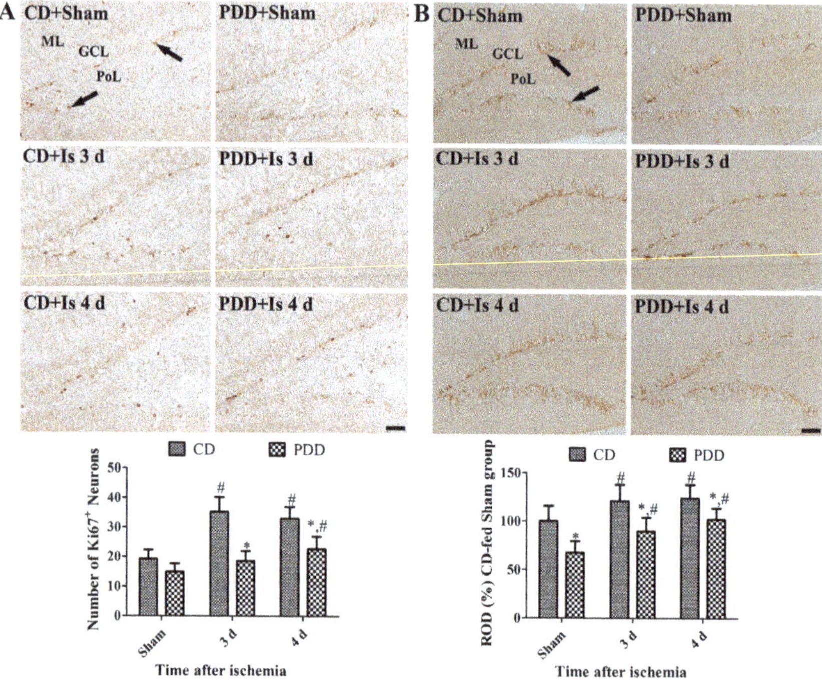

Figure 4. Immunohistochemical staining for Ki67 (**A**) and doublecortin (DCX) (**B**) in the dentate gyrus in the CD- and PDD-fed gerbils of sham- and ischemia-operated groups. Note that fewer proliferating Ki67-positive cells (arrows) and lower DCX immunoreactivity (arrows) are found in the PDD-fed gerbils than in the CD-fed ones. SO, stratum oriens; SP, stratum pyramidale; SR, stratum radiatum. Scale bar = 50 μm (**A**) and 100 μm (**B**). The number of Ki67-positive nuclei in the dentate gyrus and the ROD of DCX compared to the CD-fed sham group per section, in all the groups, are shown (n = 7 per group; * p < 0.05, significant difference between CD- and PDD-fed group; # p < 0.05, significant difference between sham- and ischemia-operated groups). All data are expressed as mean with standard deviation.

2.5. Pyridoxine Deficiency Increases Lipid Peroxidation and Decreases Nrf2 and BDNF Expression

In the PDD-fed sham group, malondialdehyde (MDA) levels in the hippocampus were higher (43.5%) compared to that in the CD-fed sham group. MDA levels in CD- and PDD-fed groups were dramatically increased three hours after ischemia to 368.0% and 395.0% of their respective sham gerbils and thereafter, MDA levels decreased with time after ischemia in both CD- and PDD-fed groups. Twenty-four hours after ischemia, MDA levels in CD- and PDD-fed groups showed 180.4% and 222.5% of their respective sham group. MDA levels were significantly higher in the PDD-fed group compared to that in the CD-fed group after ischemia, and this was not observed in the sham group (Figure 5A).

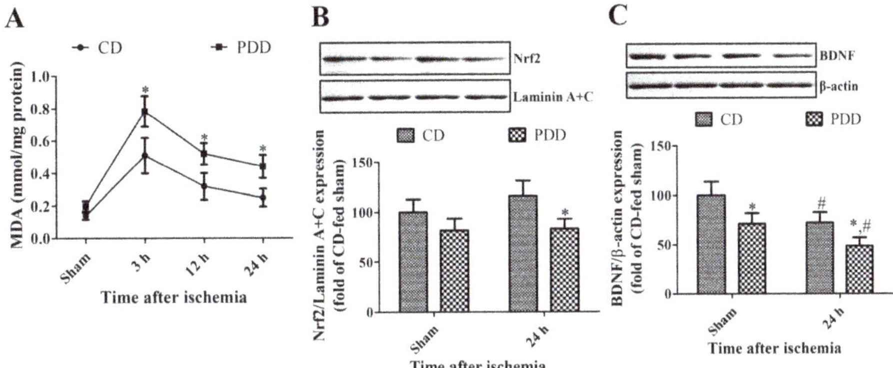

Figure 5. Measurement of malondialdehyde (MDA) levels (**A**) in the hippocampus and Western blot analysis of nuclear Nrf2 (**B**) and total mature brain-derived neurotrophic factor (BDNF) levels (**C**) in the CD- and PDD-fed gerbils of sham- and ischemia-operated groups (n = 6–7 per group; * p < 0.05, significant difference between CD- and PDD-fed groups; # p < 0.05, significant difference between sham- and ischemia-operated groups). All data are expressed as mean with standard deviation.

In the PDD-fed sham group, nuclear factor erythroid-2-related factor 2 (Nrf2) levels in nuclear fraction decreased compared to that in the CD-fed sham group, although statistical significance was not detected between groups. Twenty-four hours after ischemia, nuclear Nrf2 levels were slightly, but not significantly, increased in the CD-fed group, but they were maintained in the PDD-fed group. At this time point, nuclear Nrf2 showed significantly lower levels (71.6% of the CD-fed group) in the PDD-fed group (Figure 5B).

In the sham group, brain-derived neurotrophic factor (BDNF) levels in the PDD-fed gerbils were significantly decreased to 71.1% of CD-fed gerbils. Moreover, BDNF levels were significantly decreased in the hippocampus 24 h after ischemia/reperfusion in CD- and PDD-fed gerbils, and significantly lower levels (67.5% of CD-fed gerbils) of BDNF were found in the PDD-fed gerbils 24 h after ischemia (Figure 5C).

The two-way ANOVA test indicated that ischemia and PDD diets significantly changed MDA levels (df = 3, F = 3.807, p = 0.0193) in the hippocampus, while there were no interactive effects of ischemia and PDD diets on Nrf2 and BDNF expression in the hippocampus.

3. Discussion

Vitamin B$_6$ is one of the essential vitamins that maintain the health of the nervous system in mammals, which should be taken in sufficient quantity from the food because it is not synthesized in mammals [25,26]. Deficiency of vitamin B$_6$ causes cognitive function impairment, hippocampal neurogenesis reduction, cardiovascular disease, and Alzheimer's disease [20,27]. In the present study, we observed the effects of pyridoxine deficiency on cell death in the hippocampal CA1 region and on neuroblasts in the dentate gyrus after transient forebrain ischemia. We observed the changes in serum levels of PLP and homocysteine because vitamins B$_6$, B$_{12}$, and folic acid have been reported to be associated with the regulation of the methionine–homocysteine cycle [22]. PDD-fed gerbils showed a significant reduction in serum PLP levels in sham-operated animals as well as ischemic animals three and four days after ischemia. This result is consistent with our previous study that PDD-fed mice showed a significant reduction in serum and hippocampal PLP levels [20]. In contrast, pyridoxine deficiency significantly increased serum homocysteine levels in sham- and ischemia-operated animals and showed hyperhomocysteinemia, which is characterized by > 15 μmol/L homocysteine levels in the blood. Vitamin B$_6$, B$_{12}$, and folic acid are associated with homocysteine metabolism [18], and our previous study showed the presence of hyperhomocysteinemia in folic acid deficient gerbils, and this condition increased the DNA damage in the hippocampus after ischemia [28]. Moreover, many studies demonstrated that hyperhomocysteinemia caused neuronal damage, cognitive impairments, and psychiatric diseases [29–33].

To confirm whether pyridoxine deficiency facilitates neuronal death in the hippocampal CA1 region, we conducted immunohistochemical staining of NeuN, a marker for mature neurons, in the hippocampus three and four days after ischemia because neuronal death was detected in the CA1 region four days after ischemia. In the sham group, NeuN-positive neurons slightly decreased in the CA1 region, although there is no statistical significance between the CD- and PDD-fed gerbils. A study showed that hyperhomocysteinemia significantly increased neuronal death in the rat hippocampus [34]. In the present study, we observed that pyridoxine deficiency significantly reduced the number of NeuN-positive neurons in the medial side of the CA1 region, and this suggests that pyridoxine deficiency facilitates neuronal death in the CA1 region after ischemia. However, we did not observe any significant changes in the number of NeuN-positive nuclei four days after ischemia in CD- and PDD-fed gerbils because most of the neurons showed degeneration in the CA1 region. Pyridoxine deficiency also showed activation of astrocytes and microglia in the hippocampus of sham-operated gerbils, and microglial aggregation was found in the medial side of the stratum pyramidale three days after ischemia/reperfusion because of neuronal death in this region. These results suggest that pyridoxine deficiency facilitates glial activation in the hippocampal CA1 region of sham- and ischemia-operated gerbils three days after ischemia. Our study provides new insight into the relationship between pyridoxine deficiency and glial activation in the hippocampus. We could not elucidate the mechanisms of glial activation in the hippocampus of PDD-fed gerbils. One hypothesis is hyperhomocysteinemia induced by pyridoxine deficiency because treatment with homocysteine significantly increased reactive gliosis in astrocytes and microglia in rats [35], and excess homocysteine causes toxicity in the brain [36,37].

We examined the possible mechanisms of pyridoxine deficiency in the hippocampal neuronal death after ischemia based on oxidative stress due to the presence of antioxidant potentials in vitamin B_6, which can quench hydroxyl radicals similar with vitamin C. In contrast, transient forebrain ischemia increases oxidative stress, including DNA damage and lipid peroxidation in the hippocampus [24,38–41]. In the present study, lipid peroxidation measured by MDA levels was significantly increased 3 h after ischemia in the CD-fed gerbils and thereafter decreased with time after ischemia. This result is consistent with previous studies that MDA products increased as an accumulation of aldehyde in various neurological disorders such as Alzheimer's disease, ischemia, and Parkinson's disease [24,40–45]. Additionally, MDA levels in this study were significantly higher in PDD-fed gerbils than in CD-fed ones 3, 12, and 24 h after ischemia, and not in the sham-operated group. This result suggests that PDD increases reactive aldehyde in the hippocampus, which aggravates neuronal death in the hippocampal CA1 region after transient forebrain ischemia. Changes in Nrf2 expression were also observed in the hippocampus because it is believed to be a key transcription factor in decreasing reactive oxygen species [46]. There have been reports that Nrf2 decreased ischemic damage by reducing oxidative stress [47], while Nrf2 deficient mice showed vulnerability to oxidative stress [48]. In the present study, lower Nrf2 expression levels were found in the PDD-fed gerbils than in the CD-fed gerbils 24 h after ischemia, and this result suggests that the reduction of Nrf2 may be associated with early neuronal damage after transient forebrain ischemia.

In this study, we also observed the effects of pyridoxine deficiency on regenerative potentials based on proliferating cells and neuroblasts in the dentate gyrus because we observed the reduction of neuroblasts in the mice hippocampus [20]. In gerbil brain, the proliferating cells and neuroblasts were reduced in the dentate gyrus of sham- and ischemia-operated animals, and this result suggests that pyridoxine deficiency reduces the regeneration potentials in ischemic brain. Additionally, we observed BDNF levels in the hippocampus because it modulates the hippocampal neurogenesis [49,50]. Pyridoxine deficiency and/or brain ischemia decreased BDNF levels in the hippocampus, and BDNF levels were lowest in PDD-fed ischemic gerbils. This result supports the immunohistochemical finding for proliferating cells and neuroblasts in the dentate gyrus. Interestingly, Nrf2 is one of the essential elements to regulate the hippocampal neurogenesis [51] and homocysteine reduced cell differentiation in chicken embryonic brain [52]. However, neurogenesis in the dentate gyrus is unlikely to be related to regeneration; rather, migrating neuroblasts from SVZ may contribute to regeneration [53,54]. Moreover, the initial increase in neurogenesis in the dentate gyrus may be accompanied by its significant decrease at a later stage after ischemia [7] due to the limited pool of stem cells in this region [55].

In conclusion, pyridoxine deficiency facilitates neuronal death in the hippocampal CA1 region after transient forebrain ischemia by increasing homocysteine levels in the serum and lipid peroxidation in the brain, as well as reducing the Nrf2 levels in the hippocampus. Moreover, pyridoxine deficiency reduces the proliferating cells and neuroblasts, probably by reducing BDNF levels. This result suggests that pyridoxine is an essential element in fighting neuronal death and in increasing the regenerative potentials after ischemia.

4. Materials and Methods

4.1. Experimental Animals

Male Mongolian gerbils (5 weeks of age) were purchased from Japan SLC, Inc. (Shizuoka, Japan). The animals were housed and cared based on the Guide for the Care and Use of Laboratory Animals (8th edition, 2011). Experimental protocols were approved by the Institutional Animal Care and Use Committee (IACUC) of Seoul National University (SNU-190408-2) on 14 May 2019.

After a week of acclimation, gerbils were fed with pyridoxine-deficient diet (PDD, D10001, Research Diets) and its control diet (CD, D15501R, control, Research Diets, New Brunswick, NJ, USA) for 56 days as described in the previous study [20] since the half-life of the elimination of pyridoxine exceeds 15 to 20 days [56]. All animals used in this study were sacrificed on the 56th day of diet feeding.

4.2. Ischemic Surgery

Animals were anesthetized with 2.5% isoflurane (Baxter, Deerfield, IL, USA) and then carotid arteries were isolated from adjacent tissue and occluded using aneurysm clips for 5 min as previously reported [57]. Blood flow through carotid arteries was monitored in the central artery of the retinae using an ophthalmoscope (HEINE K180®; HEINE Optotechnik, Herrsching, Germany). Body temperature (37 ± 0.5 °C) was regulated by a thermometric blanket under the monitoring using a rectal temperature probe (TR-100; Fine Science Tools, Foster City, CA, USA) until recovered from anesthesia. Two animals with incomplete ischemic induction were excluded. The sham group received the same procedures except for carotid artery occlusion.

4.3. Immunohistochemistry

To show the morphological evidence of the changes in neurons, astrocytes, and microglia in the hippocampus, immunohistochemical staining was conducted for NeuN, GFAP, and Iba-1, respectively, as previously described [20,57]. In addition, proliferating cells and neuroblasts were visualized with the immunohistochemistry of Ki67 and DCX. Briefly, animals ($n = 5$ per group) were anesthetized with a mixture of 75 mg/kg alfaxalone and 10 mg/kg xylazine on the 56th day of diet feeding, and blood was obtained by cardiac puncture in the right ventricle. Thereafter, animals were perfused transcardially, and the brain was coronally sectioned with a 30 µm thickness between 2.0 and 2.7 mm caudal to the bregma based on gerbil stereotaxic coordinates [58]. Four sections located 150 µm apart were selected and incubated with each antibody; mouse anti-NeuN antibody (1:1000; Merck Millipore, Temecula, CA, USA), rabbit anti-GFAP antibody (1:1000; Merck Millipore), rabbit anti-Iba-1 (1:500; Wako, Osaka, Japan), rabbit anti-Ki67 (1:1000; Abcam, Cambridge, UK), or rabbit anti-DCX (1:2000; Abcam). Immunoreaction was visualized with 3,3′-diaminobenzidine tetrachloride (Sigma, St. Louis, MO, USA) in 0.1 M Tris-HCl buffer (pH 7.2). Sections were dehydrated and mounted on gelatin-coated slides in Canada balsam (Kanto Chemical, Tokyo, Japan).

4.4. High-Performance Liquid Chromatography (HPLC) Analysis

Blood samples were obtained by cardiac puncture in gerbils for immunohistochemical staining described in Section 2.3. PLP and homocysteine levels were measured in the serum as described before [20]. Briefly, serum was injected onto a C_{18} reverse-phase column (250 mm × 4.6 mm, 5 μm; Agilent Technologies, Santa Clara, CA, USA) in an HPLC system (Agilent 1100 series, Agilent Technologies, Santa Clara, CA, USA) equipped with an electrochemical detector.

4.5. Malondialdehyde Assay

To elucidate the effects of pyridoxine deficiency on lipid peroxidation induced by ischemia, transient forebrain ischemia was induced with occlusion of common carotid arteries, and the animals (n = 5 per group) were sacrificed at 3, 12, and 24 h after ischemia/reperfusion on the 56th day of diet feeding. Both hippocampi were quickly removed, and MDA levels were measured with a Bioxytech MDA-586 kit (Oxis Research, Portland, OR, USA) as described in the previous study [59]. MDA levels were normalized to the protein concentration and the assay was triplicated.

4.6. Western Blot Analysis

Nrf2 and BDNF levels were assessed by Western blotting as described in the previous study [16]. Briefly, animals (n = 5 per group) were sacrificed 24 h after ischemia/reperfusion under deep anesthesia with a mixture of 75 mg/kg alfaxalone and 10 mg/kg xylazine on the 56th day of diet feeding. The left hippocampus was homogenized briefly, and the nuclear and cytosolic fractions were separated using extraction kits following the manufacturer's instructions (Abcam). Homogenized proteins were loaded onto sodium dodecyl sulfate polyacrylamide gel electrophoresis. Thereafter, the gel was transferred onto a nitrocellulose membrane (Pall Crop, East Hills, NY, USA), and the membrane was incubated with rabbit anti-Nrf2 (1:1000, Abcam) and rabbit anti-BDNF (1:5000, Abcam). Protein bands were visualized using a chemiluminescence solution (GE Healthcare, Buckinghamshire, UK) and were normalized versus laminin A + C and β-actin levels, respectively, as demonstrated in the previous study [20].

4.7. Data Analysis

Four sections per antibody between 2.0 mm and 2.7 mm caudal to the bregma [27] were examined using an image analysis system and ImageJ software v. 1.5 (National Institutes of Health, Bethesda, MD, USA). Digital images of the whole dentate gyrus and midpoint of the CA1 region were captured with a BX51 light microscope (Olympus, Tokyo, Japan) equipped with a digital camera (DP72, Olympus). The number of NeuN- and Ki67-positive nuclei were counted in the hippocampal CA1 region and dentate gyrus, respectively, using an image analysis system (Optimas 6.5, CyberMetrics, Scottsdale, AZ, USA). Intensities for GFAP, Iba-1, and DCX were evaluated by ROD obtained after transforming the mean gray level using this formula: ROD = log(256/mean gray level). The ROD of background staining was determined in unlabeled portions of the sections using Photoshop CC software (Adobe Systems Inc., San Jose, CA, USA), and this value was subtracted to correct for nonspecific staining using ImageJ v. 1.50 software (National Institutes of Health, Bethesda, MD, USA). The data are expressed as the percentage of the vehicle-treated group values (set to 100%).

4.8. Statistical Analysis

The data obtained represent the mean with standard deviation. Differences among means were statistically analyzed using the two-way ANOVA test, followed by Bonferroni post-hoc tests. Statistical significance was considered at $p < 0.05$.

Author Contributions: H.Y.J., W.K., K.R.H., M.S.K., T.H.K., H.J.K., S.M.N., J.Y.C., Y.S.Y., D.W.K., D.Y.Y. and I.K.H. conceived the study. H.Y.J., D.Y.Y. and I.K.H. designed the study and wrote the manuscript. H.Y.J., W.K., K.R.H., M.S.K., T.H.K. and S.M.N. conducted the animal experiments. H.J.K., J.Y.C. and D.W.K. conducted biochemical experiments. S.M.N., J.Y.C., J.H.C., Y.S.Y. and D.W.K. participated in designing and discussing the study. All authors have read and agreed to the published version of the manuscript.

Funding: This work was supported by the Promising-Pioneering Researcher Program through Seoul National University (SNU) in 2015 and by the National Research Foundation of Korea (NRF) grant funded by the Korea government (MSIP) (No. 2019R1A2C1005440). In addition, this study was partially supported by the Research Institute for Veterinary Science of Seoul National University.

Conflicts of Interest: The authors declare that they have no competing interests.

References

1. Spector, R.; Johanson, C.E. Vitamin transport and homeostasis in mammalian brain: Focus on Vitamins B and E. *J. Neurochem.* **2007**, *103*, 425–438. [CrossRef] [PubMed]
2. Bilski, P.; Li, M.Y.; Ehrenshaft, M.; Daub, M.E.; Chignell, C.F. Vitamin B6 (pyridoxine) and its derivatives are efficient singlet oxygen quenchers and potential fungal antioxidants. *Photochem. Photobiol.* **2000**, *71*, 129–134. [CrossRef]
3. Hellmann, H.; Mooney, S. Vitamin B6: A molecule for human health? *Molecules* **2010**, *15*, 442–459. [CrossRef] [PubMed]
4. Booth, A.A.; Khalifah, R.G.; Hudson, B.G. Thiamine pyrophosphate and pyridoxamine inhibit the formation of antigenic advanced glycation end-products: Comparison with aminoguanidine. *Biochem. Biophys. Res. Commun.* **1996**, *220*, 113–119. [CrossRef]
5. Martínez, N.S.; Machado, J.M.; Pérez-Saad, H.; Coro-Antich, R.M.; Berlanga-Acosta, J.A.; Salgueiro, S.R.; Illera, G.G.; Alba, J.S.; del Barco, D.G. Global brain ischemia in Mongolian gerbils: Assessing the level of anastomosis in the cerebral circle of Willis. *Acta Neurobiol. Exp. (Wars.)* **2012**, *72*, 377–384.
6. Atochin, D.N.; Chernysheva, G.A.; Aliev, O.I.; Smolyakova, V.I.; Osipenko, A.N.; Logvinov, S.V.; Zhdankina, A.A.; Plotnikova, T.M.; Plotnikov, M.B. An improved three-vessel occlusion model of global cerebral ischemia in rats. *Brain Res. Bull.* **2017**, *132*, 213–221. [CrossRef]
7. Khodanovich, M.; Kisel, A.; Kudabaeva, M.; Chernysheva, G.; Smolyakova, V.; Krutenkova, E.; Wasserlauf, I.; Plotnikov, M.; Yarnykh, V. Effects of fluoxetine on hippocampal neurogenesis and neuroprotection in the model of global cerebral ischemia in rats. *Int. J. Mol. Sci.* **2018**, *19*, 162. [CrossRef]
8. Khodanovich, M.Y.; Kisel', A.A.; Chernysheva, G.A.; Smol'yakova, V.I.; Kudabaeva, M.S.; Krutenkova, E.P.; Tyumentseva, Y.A.; Plotnikov, M.B. p-Tyrosol enhances the production of new neurons in the hippocampal CA1 field after transient global cerebral ischemia in rats. *Bull. Exp. Biol. Med.* **2019**, *168*, 224–228. [CrossRef]
9. Taguchi, N.; Nakayama, S.; Tanaka, M. Fluoxetine has neuroprotective effects after cardiac arrest and cardiopulmonary resuscitation in mouse. *Resuscitation* **2012**, *83*, 652–656. [CrossRef]
10. Yamashima, T.; Zhao, L.; Wang, X.D.; Tsukada, T.; Tonchev, A.B. Neuroprotective effects of pyridoxal phosphate and pyridoxal against ischemia in monkeys. *Nutr. Neurosci.* **2001**, *4*, 389–397. [CrossRef]
11. Hwang, I.K.; Yoo, K.Y.; Kim, D.H.; Lee, B.H.; Kwon, Y.G.; Won, M.H. Time course of changes in pyridoxal 5′-phosphate (vitamin B6 active form) and its neuroprotection in experimental ischemic damage. *Exp. Neurol.* **2007**, *206*, 114–125. [CrossRef]
12. Li, P.; Zhu, M.L.; Pan, G.P.; Lu, J.X.; Zhao, F.R.; Jian, X.; Liu, L.Y.; Wan, G.R.; Chen, Y.; Ping, S.; et al. Vitamin B6 prevents isocarbophos-induced vascular dementia in rats through N-methyl-D-aspartate receptor signaling. *Clin. Exp. Hypertens.* **2018**, *40*, 192–201. [CrossRef]
13. Wei, Y.; Lu, M.; Mei, M.; Wang, H.; Han, Z.; Chen, M.; Yao, H.; Song, N.; Ding, X.; Ding, J.; et al. Pyridoxine induces glutathione synthesis via PKM2-mediated Nrf2 transactivation and confers neuroprotection. *Nat. Commun.* **2020**, *11*, 941. [CrossRef]
14. Kuypers, N.J.; Hoane, M.R. Pyridoxine administration improves behavioral and anatomical outcome after unilateral contusion injury in the rat. *J. Neurotrauma.* **2010**, *27*, 1275–1282. [CrossRef]
15. Yoo, D.Y.; Kim, W.; Kim, D.W.; Yoo, K.Y.; Chung, J.Y.; Youn, H.Y.; Yoon, Y.S.; Choi, S.Y.; Won, M.H.; Hwang, I.K. Pyridoxine enhances cell proliferation and neuroblast differentiation by upregulating the GABAergic system in the mouse dentate gyrus. *Neurochem. Res.* **2011**, *36*, 713–721. [CrossRef]
16. Page, J.H.; Ma, J.; Chiuve, S.E.; Stampfer, M.J.; Selhub, J.; Manson, J.E.; Rimm, E.B. Plasma vitamin B$_6$ and risk of myocardial infarction in women. *Circulation* **2009**, *120*, 649–655. [CrossRef] [PubMed]

17. Ishihara, J.; Iso, H.; Inoue, M.; Iwasaki, M.; Okada, K.; Kita, Y.; Kokubo, Y.; Okayama, A.; Tsugane, S.; JPHC Study Group. Intake of folate, vitamin B6 and vitamin B12 and the risk of CHD: The Japan Public Health Center-Based Prospective Study Cohort I. *J. Am. Coll. Nutr.* **2008**, *27*, 127–136. [CrossRef]

18. Jeon, J.; Park, K. Dietary vitamin B6 intake associated with a decreased risk of cardiovascular disease: A prospective cohort study. *Nutrients* **2019**, *11*, 1484. [CrossRef]

19. Marniemi, J.; Alanen, E.; Impivaara, O.; Seppänen, R.; Hakala, P.; Rajala, T.; Rönnemaa, T. Dietary and serum vitamins and minerals as predictors of myocardial infarction and stroke in elderly subjects. *Nutr. Metab. Cardiovasc. Dis.* **2005**, *15*, 188–197. [CrossRef]

20. Jung, H.Y.; Kim, W.; Hahn, K.R.; Kwon, H.J.; Nam, S.M.; Chung, J.Y.; Yoon, Y.S.; Kim, D.W.; Yoo, D.Y.; Hwang, I.K. Effects of pyridoxine deficiency on hippocampal function and its possible association with V-type proton ATPase subunit B2 and heat shock cognate protein 70. *Cells* **2020**, *9*, 1067. [CrossRef] [PubMed]

21. Hasegawa, T.; Mikoda, N.; Kitazawa, M.; LaFerla, F.M. Treatment of Alzheimer's disease with anti-homocysteic acid antibody in 3xTg-AD male mice. *PLoS ONE* **2010**, *5*, e8593. [CrossRef] [PubMed]

22. Brosnan, J.T.; Jacobs, R.L.; Stead, L.M.; Brosnan, M.E. Methylation demand: A key determinant of homocysteine metabolism. *Acta Biochim. Pol.* **2004**, *51*, 405–413. [CrossRef] [PubMed]

23. McCully, K.S. Hyperhomocysteinemia and arteriosclerosis: Historical perspectives. *Clin. Chem. Lab. Med.* **2005**, *43*, 980–986. [CrossRef] [PubMed]

24. Rahman, K. Studies on free radicals, antioxidants, and co-factors. *Clin. Interv. Aging* **2007**, *2*, 219–236. [PubMed]

25. Kennedy, D.O. B Vitamins and the Brain: Mechanisms, Dose and Efficacy– Review. *Nutrients* **2016**, *8*, 68. [CrossRef]

26. Calderón-Ospina, C.A.; Nava-Mesa, M.O. B Vitamins in the nervous system: Current knowledge of the biochemical modes of action and synergies of thiamine, pyridoxine, and cobalamin. *Cns Neurosci. Ther.* **2020**, *26*, 5–13. [CrossRef]

27. Spinneker, A.; Sola, R.; Lemmen, V.; Castillo, M.J.; Pietrzik, K.; González-Gross, M. Vitamin B6 status, deficiency and its consequences–an overview. *Nutr. Hosp.* **2007**, *22*, 7–24.

28. Hwang, I.K.; Yoo, K.Y.; Suh, H.W.; Kim, Y.S..; Kwon, D.Y.; Kwon, Y.G.; Yoo, J.H.; Won, M.H. Folic acid deficiency increases delayed neuronal death, DNA damage, platelet endothelial cell adhesion molecule-1 immunoreactivity, and gliosis in the hippocampus after transient cerebral ischemia. *J. Neurosci. Res.* **2008**, *86*, 2003–2015. [CrossRef]

29. Kim, H.; Lee, K.J. Serum homocysteine levels are correlated with behavioral and psychological symptoms of Alzheimer's disease. *Neuropsychiatr. Dis. Treat.* **2014**, *10*, 1887–1896.

30. Mattson, M.P.; Shea, T.B. Folate and homocysteine metabolism in neural plasticity and neurodegenerative disorders. *Trends Neurosci.* **2003**, *26*, 137–146. [CrossRef]

31. Stanger, O.; Fowler, B.; Piertzik, K.; Huemer, M.; Haschke-Becher, E.; Semmler, A.; Lorenzl, S.; Linnebank, M. Homocysteine, folate and vitamin B12 in neuropsychiatric diseases: Review and treatment recommendations. *Expert Rev. Neurother.* **2009**, *9*, 1393–1412. [CrossRef] [PubMed]

32. Zhang, C.E.; Wei, W.; Liu, Y.H.; Peng, J.H.; Tian, Q.; Liu, G.P.; Zhang, Y.; Wang, J.Z. Hyperhomocysteinemia increases beta-amyloid by enhancing expression of gamma-secretase and phosphorylation of amyloid precursor protein in rat brain. *Am. J. Pathol.* **2009**, *174*, 1481–1491. [CrossRef] [PubMed]

33. Agrawal, A.; Ilango, K.; Singh, P.K.; Karmakar, D.; Singh, G.P.; Kumari, R.; Dubey, G.P. Age dependent levels of plasma homocysteine and cognitive performance. Behav. *Brain Res.* **2015**, *283*, 139–144.

34. Lehotský, J.; Tothová, B.; Kovalská, M.; Dobrota, D.; Beňová, A.; Kalenská, D.; Kaplán, P. Role of Homocysteine in the Ischemic Stroke and Development of Ischemic Tolerance. *Front. Neurosci.* **2016**, *10*, 538.

35. Kumar, M.; Sandhir, R. Hydrogen sulfide suppresses homocysteine-induced glial activation and inflammatory response. *Nitric Oxide* **2019**, *90*, 15–28. [CrossRef]

36. Beard, R.S., Jr.; Reynolds, J.J.; Bearden, S.E. Hyperhomocysteinemia increases permeability of the blood–brain barrier by NMDA receptor-dependent regulation of adherens and tight junctions. *Blood* **2011**, *118*, 2007–2014. [CrossRef]

37. Kruman, I.I.; Culmsee, C.; Chan, S.L.; Kruman, Y.; Guo, Z.; Penix, L.; Mattson, M.P. Homocysteine elicits a DNA damage response in neurons that promotes apoptosis and hypersensitivity to excitotoxicity. *J. Neurosci.* **2000**, *20*, 6920–6926. [CrossRef]

38. Won, M.H.; Kang, T.C.; Jeon, G.S.; Lee, J.C.; Kim, D.Y.; Choi, E.M.; Lee, K.H.; Choi, C.D.; Chung, M.H.; Cho, S.S. Immunohistochemical detection of oxidative DNA damage induced by ischemia-reperfusion insults in gerbil hippocampus in vivo. *Brain Res.* **1999**, *836*, 70–78. [CrossRef]

39. Candelario-Jalil, E.; Mhadu, N.H.; Al-Dalain, S.M.; Martínez, G.; León, O.S. Time course of oxidative damage in different brain regions following transient cerebral ischemia in gerbils. *Neurosci. Res.* **2001**, *41*, 233–241. [CrossRef]

40. Wang, Q.; Tompkins, K.D.; Simonyi, A.; Korthuis, R.J.; Sun, A.Y.; Sun, G.Y. Apocynin protects against global cerebral ischemia-reperfusion-induced oxidative stress and injury in the gerbil hippocampus. *Brain Res.* **2006**, *1090*, 182–189. [CrossRef]

41. Yoo, D.Y.; Cho, S.B.; Jung, H.Y.; Kim, W.; Lee, K.Y.; Kim, J.W.; Moon, S.M.; Won, M.H.; Choi, J.H.; Yoon, Y.S.; et al. Protein disulfide-isomerase A3 significantly reduces ischemia-induced damage by reducing oxidative and endoplasmic reticulum stress. *Neurochem. Int.* **2019**, *122*, 19–30. [CrossRef]

42. Lovell, M.A.; Ehmann, W.D.; Mattson, M.P.; Markesbery, W.R. Elevated 4-hydroxynonenal in ventricular fluid in Alzheimer's disease. *Neurobiol. Aging* **1997**, *18*, 457–461. [CrossRef]

43. Montine, K.S.; Olson, S.J.; Amarnath, V.; Whetsell, W.O., Jr.; Graham, D.G.; Montine, T.J. Immunohistochemical detection of 4-hydroxy-2-nonenal adducts in Alzheimer's disease is associated with inheritance of APOE4. *Am. J. Pathol.* **1997**, *150*, 437–443.

44. Sayre, L.M.; Zelasko, D.A.; Harris, P.L.; Perry, G.; Salomon, R.G.; Smith, M.A. 4-Hydroxynonenal-derived advanced lipid peroxidation end products are increased in Alzheimer's disease. *J. Neurochem.* **1997**, *68*, 2092–2097. [CrossRef]

45. Ishrat, T.; Hoda, M.N.; Khan, M.B.; Yousuf, S.; Ahmad, M.; Khan, M.M.; Ahmad, A.; Islam, F. Amelioration of cognitive deficits and neurodegeneration by curcumin in rat model of sporadic dementia of Alzheimer's type (SDAT). *Eur. Neuropsychopharmacol.* **2009**, *19*, 636–647. [CrossRef]

46. Wakabayashi, N.; Slocum, S.L.; Skoko, J.J.; Shin, S.; Kensler, T.W. When NRF2 talks, who's listening? *Antioxid. Redox Signal.* **2010**, *13*, 1649–1663. [CrossRef]

47. Narayanan, S.V.; Dave, K.R.; Saul, I.; Perez-Pinzon, M.A. Resveratrol Preconditioning Protects Against Cerebral Ischemic Injury via Nuclear Erythroid 2-Related Factor 2. *Stroke* **2015**, *46*, 1626–1632. [CrossRef]

48. Loboda, A.; Damulewicz, M.; Pyza, E.; Jozkowicz, A.; Dulak, J. Role of Nrf2/HO-1 system in development, oxidative stress response and diseases: An evolutionarily conserved mechanism. *Cell. Mol. Life Sci.* **2016**, *73*, 3221–3247. [CrossRef]

49. Waterhouse, E.G.; An, J.J.; Orefice, L.L.; Baydyuk, M.; Liao, G.Y.; Zheng, K.; Lu, B.; Xu, B. BDNF promotes differentiation and maturation of adult-born neurons through GABAergic transmission. Version 2. *J. Neurosci.* **2012**, *32*, 14318–14330. [CrossRef]

50. von Bohlen Und Halbach, O.; von Bohlen Und Halbach, V. BDNF effects on dendritic spine morphology and hippocampal function. *Cell Tissue Res.* **2018**, *373*, 729–741. [CrossRef]

51. Kärkkäinen, V.; Pomeshchik, Y.; Savchenko, E.; Dhungana, H.; Kurronen, A.; Lehtonen, S.; Naumenko, N.; Tavi, P.; Levonen, A.L.; Yamamoto, M.; et al. Nrf2 regulates neurogenesis and protects neural progenitor cells against Aβ toxicity. *Stem Cells* **2014**, *32*, 1904–1916.

52. Cecchini, M.S.; Bourckhardt, G.F.; Jaramillo, M.L.; Ammar, D.; Müller, Y.M.R.; Nazari, E.M. Exposure to homocysteine leads to cell cycle damage and reactive gliosis in the developing brain. *Reprod. Toxicol.* **2019**, *87*, 60–69. [CrossRef]

53. Nemirovich-Danchenko, N.M.; Khodanovich, M.Y. New neurons in the post-ischemic and injured brain: Migrating or resident? *Front. Neurosci.* **2019**, *13*, 588. [CrossRef]

54. Nakatomi, H.; Kuriu, T.; Okabe, S.; Yamamoto, S.; Hatano, O.; Kawahara, N.; Tamura, A.; Kirino, T.; Nakafuku, M. Regeneration of hippocampal pyramidal neurons after ischemic brain injury by recruitment of endogenous neural progenitors. *Cell* **2002**, *110*, 429–441. [CrossRef]

55. Encinas, J.M.; Michurina, T.V.; Peunova, N.; Park, J.H.; Tordo, J.; Peterson, D.A.; Fishell, G.; Koulakov, A.; Enikolopov, G. Division-coupled astrocytic differentiation and age-related depletion of neural stem cells in the adult hippocampus. *Cell Stem Cell* **2011**, *8*, 566–579. [CrossRef]

56. Brown, M.J.; Beier, K. *StatPearls: Vitamin B6 Deficiency (Pyridoxine)*; StatPearls Publishing: Treasure Island, FL, USA, 2019.

57. Kim, W.; Hahn, K.R.; Jung, H.Y.; Kwon, H.J.; Nam, S.M.; Kim, T.H.; Kim, J.W.; Yoo, D.Y.; Kim, D.W.; Choi, J.H.; et al. Cuprizone Affects Hypothermia-Induced Neuroprotection and Enhanced Neuroblast Differentiation in the Gerbil Hippocampus after Ischemia. *Cells* **2020**, *9*, 1438. [CrossRef]
58. Radtke-Schuller, S.; Schuller, G.; Angenstein, F.; Grosser, O.S.; Goldschmidt, J.; Budinger, E. Brain atlas of the Mongolian gerbil (*Meriones unguiculatus*) in CT/MRI-aided stereotaxic coordinates. *Brain Struct. Funct.* **2016**, *221* (Suppl. 1), 1–272. [CrossRef]
59. Moon, S.M.; Choi, G.M.; Yoo, D.Y.; Jung, H.Y.; Yim, H.S.; Kim, D.W.; Hwang, I.K.; Cho, B.M.; Chang, I.B.; Cho, S.M.; et al. Differential Effects of Pioglitazone in the Hippocampal CA1 Region Following Transient Forebrain Ischemia in Low- and High-Fat Diet-Fed Gerbils. *Neurochem. Res.* **2015**, *40*, 1063–1073. [CrossRef]

 International Journal of
Molecular Sciences

Article

Preventive Triple Gene Therapy Reduces the Negative Consequences of Ischemia-Induced Brain Injury after Modelling Stroke in a Rat

Vage Markosyan [1], Zufar Safiullov [1], Andrei Izmailov [1], Filip Fadeev [1], Mikhail Sokolov [1], Maksim Kuznetsov [1], Dmitry Trofimov [1], Evgeny Kim [1], Grayr Kundakchyan [2], Airat Gibadullin [1], Ilnur Salafutdinov [2], Leniz Nurullin [3], Farid Bashirov [1] and Rustem Islamov [1,*]

[1] Department of Medical Biology and Genetics, Kazan State Medical University, 420012 Kazan, Russia; vage.markosyan@gmail.com (V.M.); redblackwhite@mail.ru (Z.S.); gostev.andrei@gmail.com (A.I.); philip.fadeyev@gmail.com (F.F.); supermihon@yandex.ru (M.S.); qmaxksmu@yandex.ru (M.K.); t.dima19961996@gmail.com (D.T.); profzh@yandex.ru (E.K.); halsoulo@gmail.com (A.G.); faridbashirov@yandex.ru (F.B.)
[2] Institute of Fundamental Medicine and Biology, Kazan [Volga Region] Federal University, 420008 Kazan, Russia; grayr23@gmail.com (G.K.); sal.ilnur@gmail.com (I.S.)
[3] Kazan Institute of Biochemistry and Biophysics, Federal Research Center of Kazan Scientific Center of Russian Academy of Sciences, 119991 Kazan, Russia; leniz2001@mail.ru
* Correspondence: rustem.islamov@gmail.com

Received: 26 August 2020; Accepted: 17 September 2020; Published: 18 September 2020

Abstract: Currently, the main fundamental and clinical interest for stroke therapy is focused on developing a neuroprotective treatment of a penumbra region within the therapeutic window. The development of treatments for ischemic stroke in at-risk patients is of particular interest. Preventive gene therapy may significantly reduce the negative consequences of ischemia-induced brain injury. In the present study, we suggest the approach of preventive gene therapy for stroke. Adenoviral vectors carrying genes encoding vascular endothelial growth factor (VEGF), glial cell-derived neurotrophic factor (GDNF) and neural cell adhesion molecule (NCAM) or gene engineered umbilical cord blood mononuclear cells (UCB-MC) overexpressing recombinant VEGF, GDNF, and NCAM were intrathecally injected before distal occlusion of the middle cerebral artery in rats. Post-ischemic brain recovery was investigated 21 days after stroke modelling. Morphometric and immunofluorescent analysis revealed a reduction of infarction volume accompanied with a lower number of apoptotic cells and decreased expression of Hsp70 in the peri-infarct region in gene-treated animals. The lower immunopositive areas for astrocytes and microglial cells markers, higher number of oligodendrocytes and increased expression of synaptic proteins suggest the inhibition of astrogliosis, supporting the corresponding myelination and functional recovery of neurons in animals receiving preventive gene therapy. In this study, for the first time, we provide evidence of the beneficial effects of preventive triple gene therapy by an adenoviral- or UCB-MC-mediated intrathecal simultaneous delivery combination of *vegf165*, *gdnf*, and *ncam1* on the preservation and recovery of the brain in rats with subsequent modelling of stroke.

Keywords: stroke; preventive gene therapy; adenoviral vector; VEGF; GDNF; NCAM; human umbilical cord blood mononuclear cells

1. Introduction

The current options for ischemic stroke treatment are extremely limited and are aimed at restoring blood flow in the ischemic area by intravenous infusion of recombinant tissue plasminogen activator and/or physical removal of the clots [1]. To date, the main fundamental and clinical interest is focused on developing a neuroprotective treatment of the penumbra region within the therapeutic window. The strategy of cell-, gene-, and gene-cell therapy for neuroprotection in stroke treatment has been proven by numerous experiments in animal models [2–4]. Besides brain-specific cell types, umbilical cord blood (UCB) is widely used for neuroprotection in the central nervous system (CNS) for different pathological conditions [5]. UCB cells are considered a valuable source of stem cells, growth and neurotrophic factors for cell therapy. The mononuclear fraction of UCB contains populations of different immature cells that are capable of differentiating into many cell types [6] and, thus, represent an alternative to embryonic stem cells for transplantation to patients with post-ischemic, post-traumatic and degenerative diseases [7,8]. To date, the following have been discovered in UCB: Hematopoietic stem cells (HSCs), endothelial progenitor cells, mesenchymal stem cells (MSCs), unrestricted somatic stem cells (USSCs), and side population cells (SP) [9–12].

Due to the immaturity of the immune system of a new-born, the use of UCB cells for cell therapy does not require matching of genes relating to HLA (Human Leucocyte Antigens) human tissue compatibility, as evidenced by the absence of an acute or chronic form of the disease "graft-versus-host" (graft versus host disease) [13,14]. In addition, with UCB cell transplantation, tumor transformation of cells in the recipient's body is practically prevented [15].

Another attractive reason for using UCB cells for cell therapy is their ability to produce various biologically active molecules, such as proteins which are antioxidant, angiogenic, neurotrophic, and growth factors [16–20]. Thus, transplantation of UCB cells can be aimed at replacing dead cells and at preventing the further death of surviving cells due to secreted biologically active molecules. Enhancement of the positive effects of UCB cells on tissue regeneration after their genetic modification is a relatively new and promising gene-cell approach in cell therapy to stimulate post-traumatic or post-ischemic brain injury [21,22]. Gene-modified UCB cells may provide addressed delivery of therapeutic genes and supply the expression of the recombinant molecules at the site of regeneration.

In our previous studies, we showed the positive effect of gene-modified umbilical cord blood mononuclear cells (UCB-MC), simultaneously producing three recombinant molecules—vascular endothelial growth factor (VEGF), glial cell-derived neurotrophic factor (GDNF) and neural cell adhesion molecule (NCAM)—in animal models of amyotrophic lateral sclerosis [23], spinal cord injury [24], and stroke [25]. The rationale of using a combination of two neurotrophic factors with cell adhesion molecules is based on the well-known neuroprotective effects of VEGF and GDNF [26,27], with the expression of NCAM increasing the homing and survivability of UCB-MC at the brain injury site [28] supporting local production of the therapeutic molecules. In the model of middle cerebral artery occlusion (MCAO) in rats, we demonstrated that intrathecal injection of genetically-engineered UCB-MC over-expressing VEGF, GDNF, and NCAM, four hours after MCAO results in a reduction of infarct volume, the positive reaction of neuroglial cells and an increase in synaptic protein expression. Thus, ex vivo gene modification may enhance the naïve neuroprotective properties of UCB-MC.

The development of treatment under the threat of stroke is of particular interest. Preventive therapy may highly reduce the consequences of a stroke-induced brain injury. In the present study, we suggest the approach of preventive cell-mediated gene therapy for stroke. The efficacy of gene-engineered UCB-MC overexpressing recombinant molecules-stimulants of neuroregeneration VEGF, GDNF, and NCAM, administered intrathecally 3 days before MCAO in rats, was investigated using morphometric and immunofluorescent methods.

2. Results

2.1. Molecular Analysis of Gene-Modified UCBC

The efficiency of UCB-MC transduction by Ad5-GFP adenoviral vector was confirmed after 72 h of UCB-MC+Ad5-GFP cultivation using fluorescent microscopy. In the cytoplasm of UCB-MC+Ad5-GFP, a specific green glow was detected (Figure 1A). By the flow cytometry method, it was established that the percentage of GFP-positive human UCB-MC at multiplicity of infection (MOI) equal to 10 reaches 29.5% (Figure 1B).

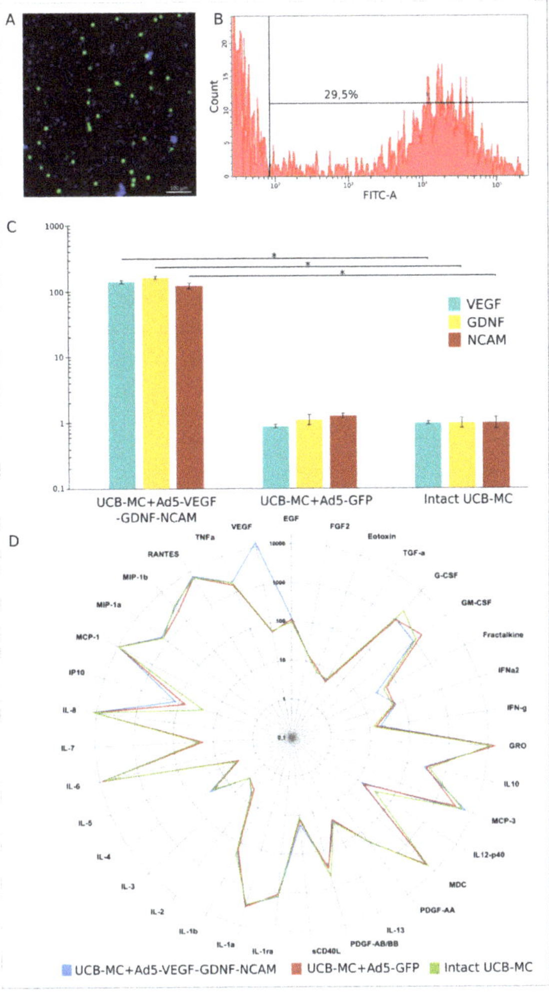

Figure 1. Molecular analysis of gene modified umbilical cord blood mononuclear cells (UCB-MC). (**A**,**B**)—Production of green fluorescent protein (GFP) in UCB-MC, 72 h after transduction with Ad5-GFP (MOI = 10). (**A**)—fluorescent microscopy shows a specific green glow in UCB-MC+Ad5-GFP. Cell nuclei are stained with Hoechst 33342 (blue glow). Scale bar = 200 μm. (**B**)—flow cytometry analysis demonstrates that 29.5% of UCB-MC effectively produce reporter GFP. (**C**)—Quantitative analysis of mRNA *vegf165*, *gdnf*, and *ncam1* levels in intact (naïve) UCB-MC and genetically-modified UCB-MC, 72 h after transduction with three adenoviral vectors simultaneously carrying the therapeutic genes (Ad5-VEGF, Ad5-GDNF, and Ad5-NCAM) or with Ad5-GFP (UCBC + Ad5-GFP), with a MOI = 10. Data from two independent experiments are represented as an average value of ± SE, *—$p < 0.05$. (**D**)—Radial comparative diagram of cytokine, chemokines and growth factors in supernatant obtained 72 h after incubation of gene modified UCB-MC (UCB-MC+Ad5- vascular endothelial growth factor (VEGF)- glial cell-derived neurotrophic factor (GDNF)- neural cell adhesion molecule (NCAM) and UCB-MC+Ad5-GFP) and intact UCB-MC.

Evaluation of mRNA levels of transgenes (*vegf165*, *gdnf*, and *ncam1*) in genetically-modified UCB-MC was performed by RT-PCR, 72 h after incubation of UCB-MC+Ad5-VEGF-GDNF-NCAM. The molecular analysis revealed that the content of mRNA *vegf165* increased 141.8 ± 8.24 times, *gdnf* mRNA 167.51 ± 6.85 times and *ncam1* mRNA 122.9 ± 13.5 times compared with intact (naïve) UCB-MC (Figure 1C).

Multiplex analysis of cytokines, chemokines and growth factors in supernatants harvested after the cultivation of naïve UCB-MC identified a wide range of pro- and anti-inflammatory cytokines, chemokines and growth factors (Figure 1D). However, the concentrations of Flt-3L, IL12-p70, IL-15, IL-17a, IL-9, and TNFb were lower than the detection level. Genetic modification of human UCB-MC by recombinant adenovirus Ad5-GFP does not affect the secretory profile of modified cells in relation to the investigated factors when compared to the naïve cells. UCB-MC simultaneously transduced with Ad5-VEGF, Ad5-GDNF, and Ad5-NCAM also preserves the qualitative and quantitative profile of expression of the investigated factors. However, as expected, in comparison with the naïve and UCB-MC+Ad5-GFP, we observed a 200-fold increase of VEGF level. The presented results correlate with PCR-RT data above and confirm the efficiency of UCB-MC transduction and its ability to synthesize and secrete recombinant molecules, as with the example of VEGF. In addition, the results support that adenoviral vector does not affect the production of the studied biologically active molecules by human UCB-MC.

2.2. Morphometric Analysis of Infarct Area

Three weeks after ischemic stroke modelling, cerebral infarction volume analysis revealed an infarct zone located in the parietal lobe (parietal cortex, area 1 (Par1), which corresponds to the site of MCAO (Figure 2A,B). Morphometric analysis of the brain cortex infarct cavities volume showed the differences between therapeutic and control groups (Figure 2C,D). The infarct cavities volume was significantly less in the Ad5-VEGF-GDNF-NCAM (0.177 [0.155; 0.197]) and UCB-MC+Ad5-VEGF-GDNF-NCAM (0.070 [0.014; 0.245]) groups when compared with the control saline (0.607 [0.568; 0.759]) and Ad5-GFP (0.817 [0.754; 0.865]) groups ($p < 0.05$). In the UCB-MC+Ad5-GFP (0.249 [0.119; 0.305]) group, the infarction volume did not differ from the gene-treated groups and was lower when compared with the saline group ($p < 0.05$) (Figure 2E).

2.3. Immunofluorescent Study of Brain

Comparative analysis of the molecular and cellular changes in the peri-infarct zone (Figure 2C) of experimental rats' brains revealed different patterns in the expression of cellular stress, apoptosis, and synaptic proteins and in the reorganization of neuroglial cells in relation to data obtained from intact animals.

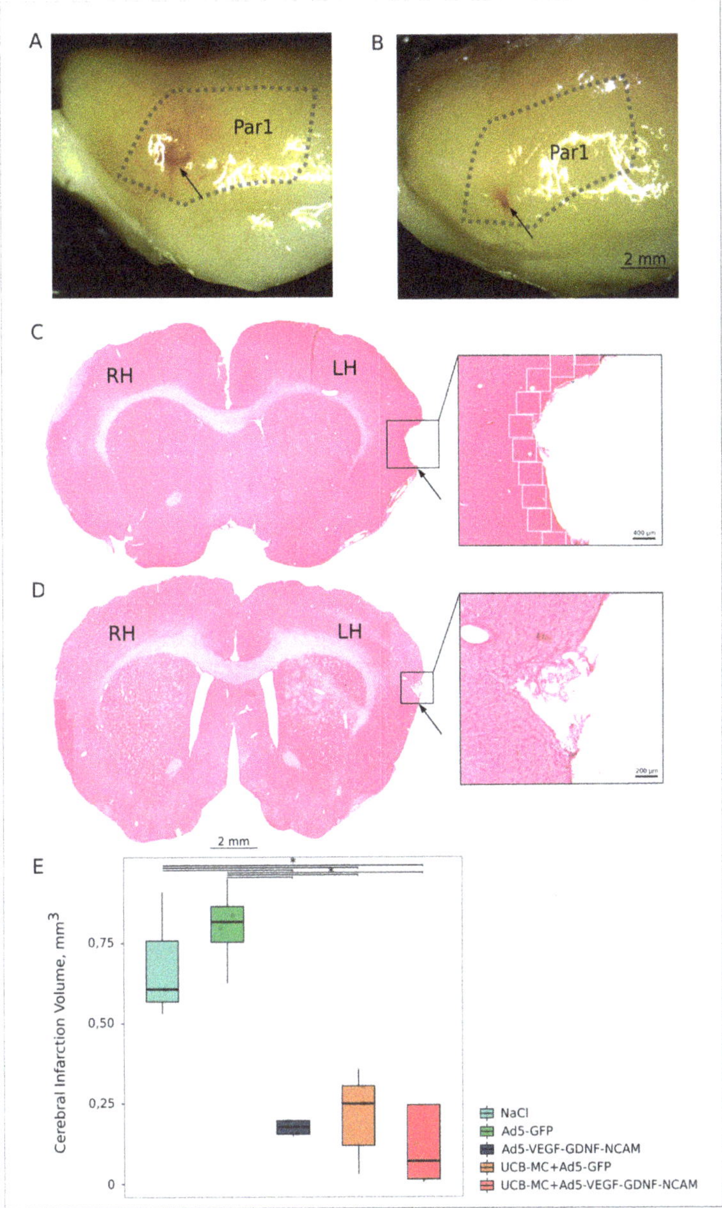

Figure 2. Ischemic stroke in rats 3 weeks after the distal middle cerebral artery occlusion. (**A**,**C**)—The brain from the control (NaCl) group. (**B**,**D**)—The brain from the therapeutic (UCB-MC+Ad5-VEGF-GDNF-NCAM) group. In (**A**) and (**B**) the stroke area in the parietal lobe of the brain is shown with the arrow. Par1—cortex parietals, area 1 is marked by a dotted line. In (**C**) and (**D**)—frontal sections of the brain in the stroke area stained with hematoxylin and eosin. RH—right hemisphere; LH—left hemisphere. Inserts with arrows show the stroke area with the maximum depth and the maximum radius of the infarct cavities at higher magnification. White boxes in the insert in (**C**) panel indicate areas with S = 0.05 mm^2 in the peri-infarct zone used for immunofluorescent analysis. (**E**)—comparative morphometric analysis of the infarct cavities volume in experimental groups; *—$p < 0.05$.

2.3.1. Cellular Stress and Apoptosis Proteins

The expression of heat shock protein 70 kDa (Hsp70) in neural and glial brain cells has constitutional and inductive options. Thus, Hsp70 expression is significantly upregulated in ischemic conditions [29]. In this study, analysis of the intensity of immunofluorescent brain cortex staining with antibodies against Hsp70 revealed a different response of brain cells in the area of ischemic damage in experimental animals when compared with intact animals (Figure 3A,G). The Hsp70 expression level in control (NaCl) (Figure 3B,G) and Ad5-GFP (Figure 3C,G) groups was 27.375 [24.940; 28.649] and 34.361 [31.315; 35.612], respectively, and was significantly higher than that of intact animals (18.422 [17.786; 19.510]). In the therapeutic groups with preventive gene therapy with Ad5-VEGF-GDNF-NCAM (20.419 [18.510; 22.950]) (Figure 3E,G) and UCB-MC+Ad5-VEGF-GDNF-NCAM (18.169 [17.462; 18.964]) (Figure 3F,G), Hsp70 immunoexpression was lower compared to the control groups ($p < 0.05$) and did not differ from intact group ($p < 0.05$). In the UCB-MC+Ad5-GFP group (Figure 3D,G), the expression level of Hsp70 (17.651 [17.648; 20.832]) was lower than the control groups and did not differ from the intact and therapeutic groups.

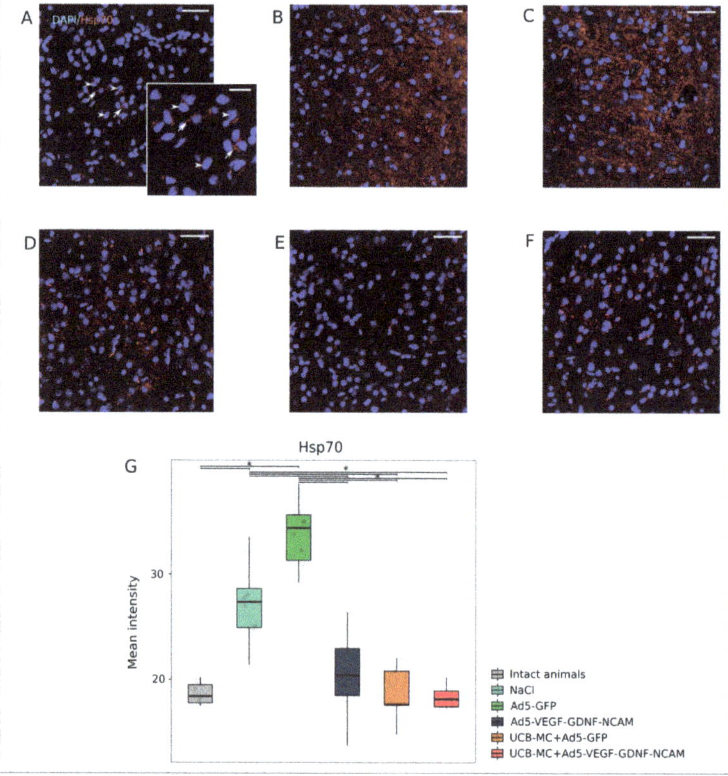

Figure 3. Immunoexpression of Hsp70 in the rat brain cortex, 21 days after the middle cerebral artery occlusion. Immunofluorescent staining with antibody to heat shock protein 70 kDa in intact (**A**), control NaCl (**B**) and Ad5-GFP (**C**), and therapeutic UCB-MC+Ad5-GFP (**D**), Ad5-VEGF-GDNF-NCAM (**E**) and UCB-MC+Ad5-VEGF-GDNF-NCAM (**F**) groups. Arrows indicate cytoplasmic localization of Hsp70 (red glow) in brain cells. Arrowheads point to cell nuclei visualized using DAPI (blue glow). Scale bar in (**A–F**) = 50 μm; scale bar in the insert = 20 μm. (**G**)—comparative analysis of fluorescence intensity level of Hsp70 in experimental groups; *—$p < 0.05$.

Activation of Caspase3 enzyme characterizes the irreversible phase of nuclear DNA degradation and subsequent cell death. In the present study, the activity of the pro-apoptotic protein Caspase3 was studied in the ischemic brain cortex in experimental rats (Figure 4A–F). It was found that the therapeutic groups Ad5-VEGF-GDNF-NCAM (12.00 [9.00; 13.00]) (Figure 4D,F) and UCB-MC+Ad5-VEGF-GDNF-NCAM (9.00 [7.50; 9.00]) (Figure 4E,F) had significantly less Caspase3-positive cells than the control NaCl (25.00 [22.00; 27.00]) group (Figure 4A,F). The number of apoptotic cells in the UCB-MC+Ad5-VEGF-GDNF-NCAM group was also lower than the Ad5-GFP control group (15.50 [11.75; 19.00]) (Figure 4B,F) ($p < 0.05$). In the UCBC+Ad5-GFP group (6.00 [3.25; 9.75]) (Figure 4C,F), the content of Caspase3-positive cells was lower than in the control groups ($p < 0.05$) and did not differ from other therapeutic groups.

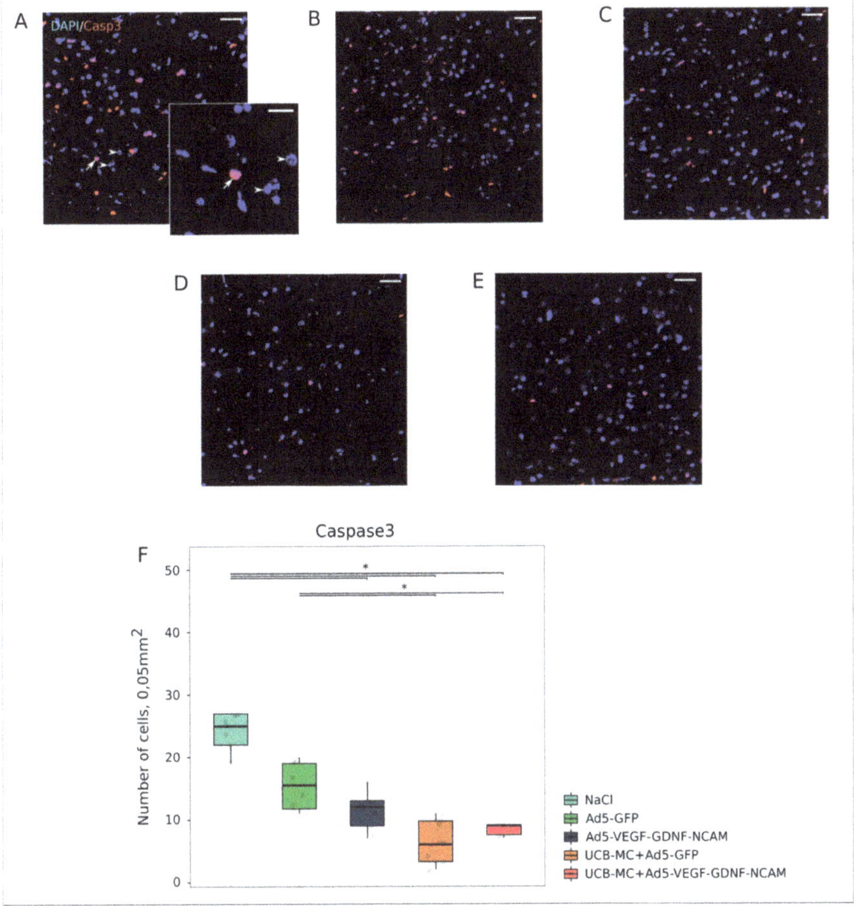

Figure 4. Count of Caspase3-positive cells in the rat brain cortex, 21 days after the middle cerebral artery occlusion. Immunofluorescent staining with antibody to Caspase3 in control NaCl (**A**) and Ad5-GFP (**B**), and therapeutic UCB-MC+Ad5-GFP (**C**), Ad5-VEGF-GDNF-NCAM (**D**) and UCB-MC+Ad5-VEGF-GDNF-NCAM (**E**) groups. Arrow indicates nuclear localization of Caspase3 (red glow) in brain cells. Arrowheads point to cell nuclei visualized using DAPI (blue glow). Scale bar in (**A–E**) = 50 µm; scale bar in the insert = 20 µm. (**F**)—comparative analysis of Caspase3-positive cells number in experimental groups; *—$p < 0.05$.

2.3.2. Neuroglia Cells

Analysis of GFAP-positive cells (astrocytes), Olig2-positive cells (oligodendrocytes), and Iba1-positive cells (microglia) revealed different reactions of neuroglia cells in the ischemic area of the brain in experimental animals.

Microglial cells are the dominant cell type involved in post-stroke neuroinflammation and the organization of the glial scar. In the intact group, the immunopositive area for microglia cells detected with Iba1 marker was lower (5.42 [4.96; 5.98]) (Figure 5A,G) than in the control NaCl (12.57 [10.31; 16.82]) (Figure 5B,G) and Ad5-GFP (14.43 [12.93; 16.01]) (Figure 5C,G) groups and in the therapeutic UCB-MC+Ad5-GFP (10.07 [7.67; 10.52]]) group (Figure 5D,G) ($p < 0.05$). The Iba1-positive area in the therapeutic Ad5-VEGF-GDNF-NCAM (5.60 [4.89; 6.69]) (Figure 5E,G) and UCBC+Ad5-VEGF-GDNF-NCAM (8.80 [6.37; 9.99]) (Figure 5F,G) groups was lower when compared with the control groups and did not differ from the intact group ($p < 0.05$).

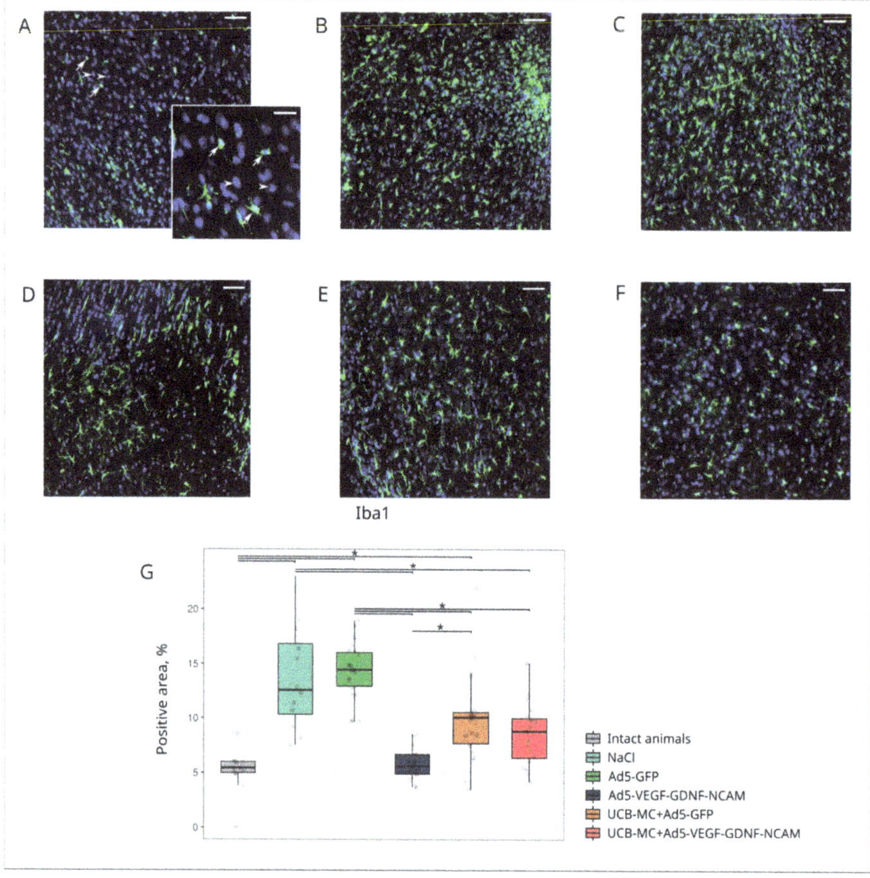

Figure 5. Immunoexpression of Iba1 in the rat brain cortex, 21 days after the middle cerebral artery occlusion. Immunofluorescent staining with antibody to microglia specific calcium-binding protein Iba1 in intact (**A**), control NaCl (**B**) and Ad5-GFP (**C**), and therapeutic UCB-MC+Ad5-GFP (**D**), Ad5-VEGF-GDNF-NCAM (**E**), and UCB-MC+Ad5-VEGF-GDNF-NCAM (**F**) groups. Arrows indicate cytoplasmic localization of Iba1 (green glow) in microglial cells. Arrowheads point to cell nuclei visualized using DAPI (blue glow). Scale bar in (**A–F**) = 50 μm; scale bar in the insert = 20 μm. (**G**)—comparative analysis of Iba1-positive areas in experimental groups; *—$p < 0.05$.

Astrocytes in the post-ischemic brain, in association with microglial cells, are involved in the formation of the glial scar, with physical and chemical properties inhibitory for brain recovery. Decreased astrogliosis often correlates with reduced volume of ischemic infarct [30]. The GFAP-positive areas in the brains of the intact group was lower (4.12 [3.36; 4.85]) (Figure 6A,G) than in the control NaCl (9.34 [8.08; 10.15]) (Figure 6B,G) and Ad5-GFP (8.92 [6.25; 12.42]]) (Figure 6C,G) groups ($p < 0.05$). In the therapeutic groups Ad5-VEGF-GDNF-NCAM (4.22 [3.36; 5.19]) (Figure 5E,G) and UCBC+Ad5-VEGF-GDNF-NCAM (3.43 [2.81; 4.19]) (Figure 5F,G), the GFAP-positive areas were lower than in the control (NaCl and Ad5-GFP) groups and did not differ from intact group ($p < 0.05$). In the UCBC+Ad5-GFP (Figure 5D,G) group, the GFAP-positive area did not differ in comparison with all experimental groups ($p < 0.05$).

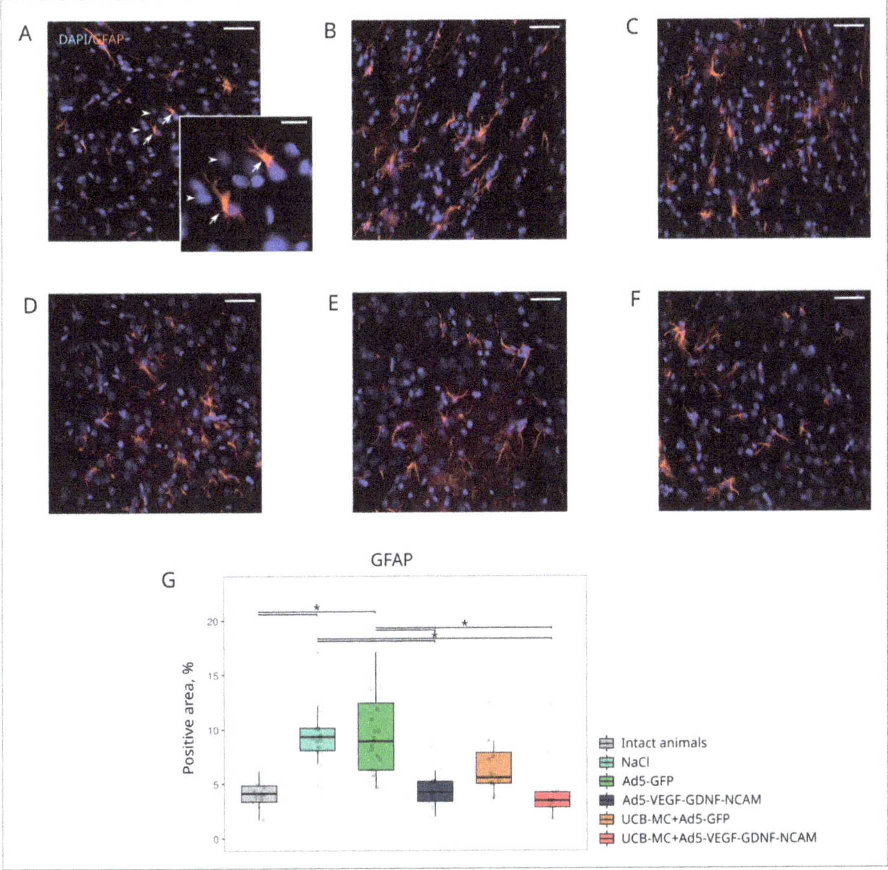

Figure 6. Immunoexpression of GFAP in the rat brain cortex, 21 days after the middle cerebral artery occlusion. Immunofluorescent staining with antibody to astrocyte specific cytoskeletal glial fibrillary acidic protein GFAP in intact (**A**), control NaCl (**B**) and Ad5-GFP (**C**), and therapeutic UCB-MC+Ad5-GFP (**D**), Ad5-VEGF-GDNF-NCAM (**E**) and UCB-MC+Ad5-VEGF-GDNF-NCAM (**F**) groups. Arrows indicate cytoplasmic localization of GFAP (red glow) in astrocytes. Arrowheads point to cell nuclei visualized using DAPI (blue glow). Scale bar in (**A–F**) = 50 µm; scale bar in the insert = 20 µm. (**G**)—comparative analysis of GFAP-positive areas in experimental groups; *—$p < 0.05$.

Postischemic brain damage is accompanied by the destruction of oligodendrocytes and the subsequent demyelination of neural processes [31]. The number of Olig2-positive cells was significantly reduced in the control NaCl (10.00 [9.00; 11.00]) (Figure 7B,G) and Ad5-GFP (10.00 [9.00; 11.00]) (Figure 7C,G) groups and in the UCBC+Ad5-GFP group (9.00 [9.00; 11.00]) (Figure 7D,G) relative to the intact group (14 [13; 19]) (Figure 7A,G) ($p < 0.05$). The number of Olig2-positive cells in the intact group did not differ from the therapeutic Ad5-VEGF-GDNF-NCAM (12.00 [9.00; 18.00]) (Figure 7E,G) and UCBC+Ad5-VEGF-GDNF-NCAM (14.00 [12.00; 16.00]) (Figure 7F,G) groups. It is important to note that there were more oligodendrocytes in the UCBC+Ad5-VEGF-GDNF-NCAM group than in the control groups and in the UCBC+Ad5-GFP group ($p < 0.05$), but the number did not differ from the Ad5-VEGF-GDNF-NCAM group.

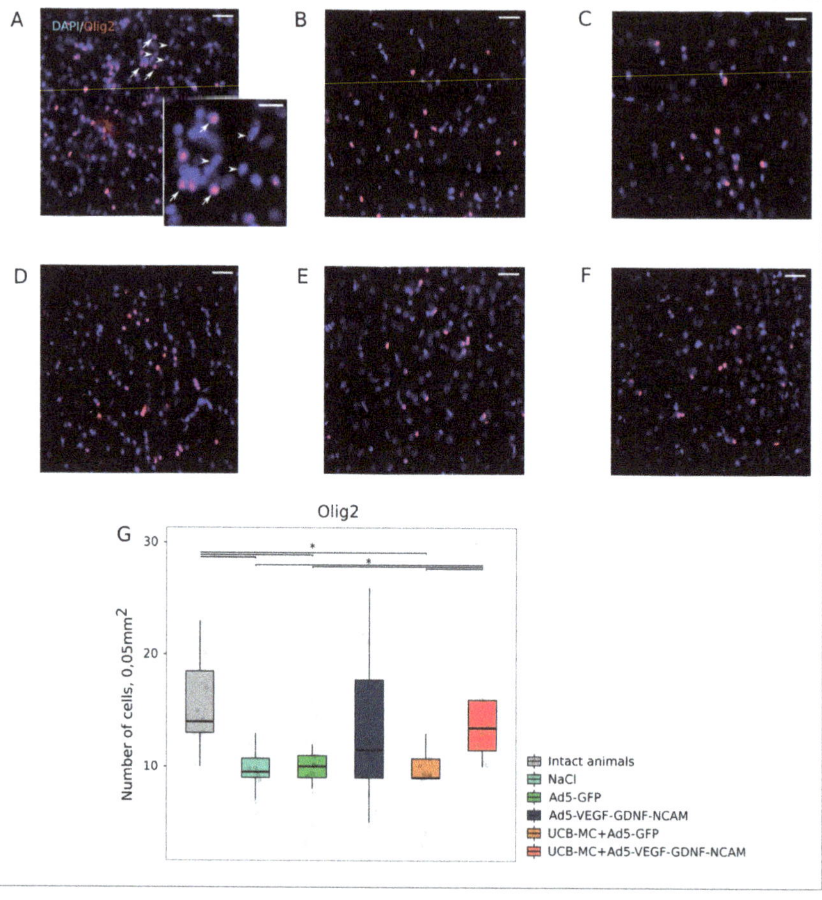

Figure 7. Count of Olig2-positive oligodendrocytes in the rat brain cortex, 21 days after the middle cerebral artery occlusion. Immunofluorescent staining with antibody to oligodendrocyte transcription factor Olig2 (red glow) in intact (**A**), control NaCl (**B**), and Ad5-GFP (**C**), and therapeutic UCB-MC+Ad5-GFP (**D**), Ad5-VEGF-GDNF-NCAM (**E**), and UCB-MC+Ad5-VEGF-GDNF-NCAM (**F**) groups. Arrows indicate nuclear localization of Olig2 (red glow) in oligodendrocytes. Arrowheads point to cell nuclei visualized using DAPI (blue glow). Scale bar in (**A**–**F**) = 50 μm; scale bar in the insert = 20 μm. (**G**)—comparative analysis of Olig2-positive cells number in experimental groups; *—$p < 0.05$.

2.3.3. Synaptic Proteins

Synaptothysin is associated with synaptic vesicles involved in nerve impulse transmission, and its content in the presynaptic area reflects the functional activity of neurons. The synaptophysin level of fluorescence was reduced in the control NaCl (55.00 [53.00; 61.00]) (Figure 8B,G) and Ad5-GFP (56.00 [48.00; 59.00]) (Figure 8C,G) groups when compared with the intact group (78.00 [61.00; 87.00]) (Figure 8A,G) ($p < 0.05$). Immunoexpression of synaptophysin in the therapeutic Ad5-VEGF-GDNF-NCAM (64.00 [61.00; 72.00]) (Figure 8E,G), UCBC+Ad5-VEGF-GDNF-NCAM (62.00 [61.00; 65.00]) (Figure 8F,G) and UCBC+Ad5-GFP groups (79.00 [76.00; 81.00]) (Figure 8D,G) was higher than in the control groups and did not differ from the intact group ($p < 0.05$).

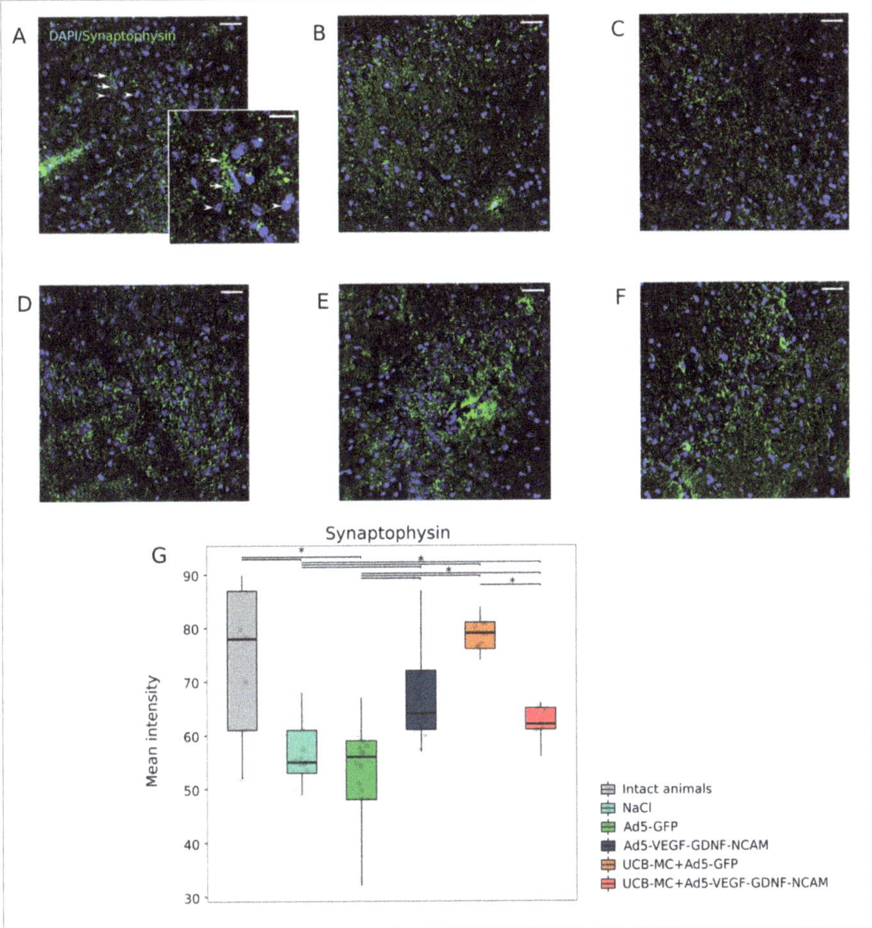

Figure 8. Immunoexpression of Synaptophysin in the rat brain cortex, 21 days after the middle cerebral artery occlusion. Immunofluorescent staining with antibody to synaptic vesicle membrane protein Synaptophysin in intact (**A**), control NaCl (**B**) and Ad5-GFP (**C**), and therapeutic UCB-MC+Ad5-GFP (**D**), Ad5-VEGF-GDNF-NCAM (**E**) and UCB-MC+Ad5-VEGF-GDNF-NCAM (**F**) groups. Arrows indicate Synaptophysin (green glow) in neurons. Arrowheads point to cell nuclei visualized using DAPI (blue glow). Scale bar in (**A—F**) = 50 μm; scale bar in the insert = 20 μm. (**G**)—comparative analysis of fluorescence intensity level of Synaptophysin in experimental groups; *—$p < 0.05$.

Postsynaptic density protein 95 kDa (PSD95) is associated with the functional activity of the postsynaptic membrane, providing its stability and plasticity in interactions with neurotransmitters. As for the intact group, the immune expression of PSD95 (65.0 [62.00; 69.00]) (Figure 9A,G) was reduced in all experimental groups except for the group UCBC+Ad5-VEGF-GDNF-NCAM (55.00 [50.00; 57.00]) ($p < 0.05$) (Figure 9F,G). However, the level of PSD95 fluorescence was significantly higher in the therapeutic Ad5-VEGF-GDNF-NCAM (62.50 [57.75; 67.00]) (Figure 9E,G), UCBC+Ad5-VEGF-GDNF-NCAM and UCBC+Ad5-GFP (50.00 [48.50; 57.50]) (Figure 9D,G) groups, in comparison with the control group NaCl (43.00 [38.50; 46.00]) (Figure 9B,G), which did not differ from the Ad5-GFP group (49.00 [46.00; 51.00]) (Figure 9C,G) ($p < 0.05$).

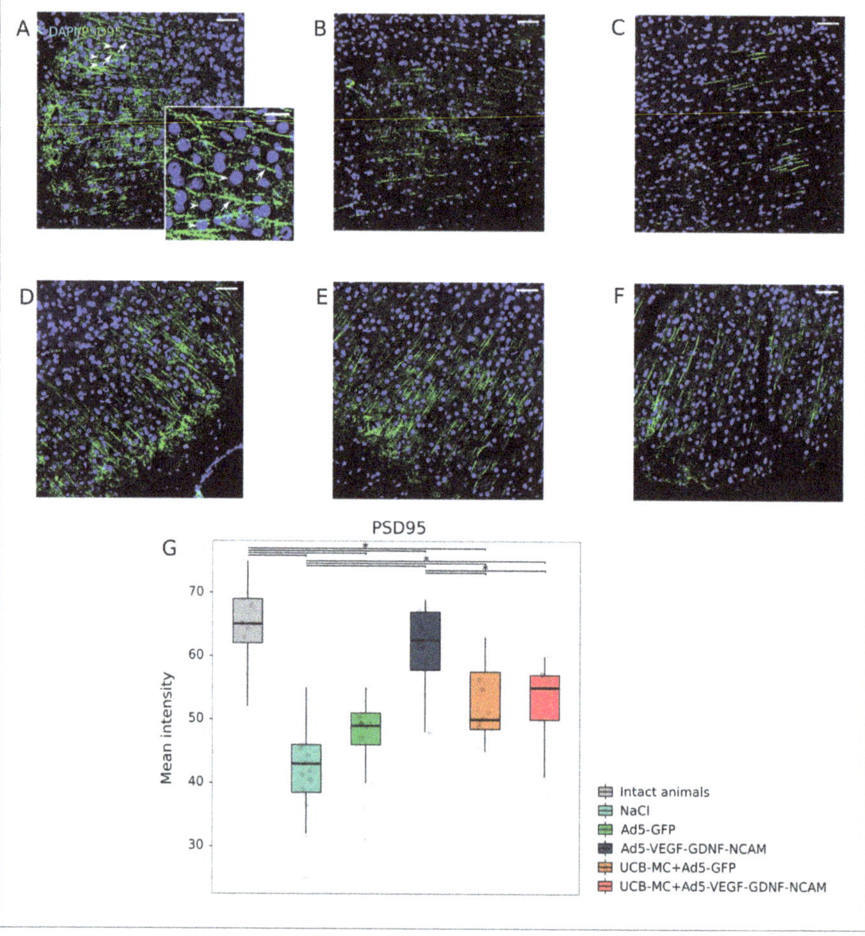

Figure 9. Immunoexpression of PSD95 in the rat brain cortex 21 days after the middle cerebral artery occlusion. Immunofluorescent staining with antibody to postsynaptic membrane protein PSD95 in intact (**A**), control NaCl (**B**) and Ad5-GFP (**C**), and therapeutic UCB-MC+Ad5-GFP (**D**), Ad5-VEGF-GDNF-NCAM (**E**) and UCB-MC+Ad5-VEGF-GDNF-NCAM (**F**) groups. Arrows indicate PSD95 (green glow) in neurons. Arrowheads point to cell nuclei visualized using DAPI (blue glow). Scale bar in (**A–F**) = 50 μm; scale bar in the insert = 20 μm. (**G**)—Comparative analysis of fluorescence intensity level of PSD95 in experimental groups; *—$p < 0.05$.

2.4. Expression of Recombinant Molecules in the Rat Brain after Intrathecal Adenoviral-Mediated Delivery of Transgenes

The expression of the reporter *gfp* and therapeutic *vegf165*, *gdnf* and *ncam1* genes was studied in the stroke area of rat brains, 25 days after an intrathecal injection of Ad5-GFP or Ad5-VEGF-GDNF-NCAM. Green fluorescent protein was detected in brain cells by laser microscopy (Figure 10A). Immunofluorescent analysis using specific antibodies to VEGF, GDNF, and NCAM also revealed target molecules in the brain cells (Figure 10B–D). In the peri-infarct zone, we observed an approximately equal distribution of the cells producing reporter GFP or recombinant VEGF, GDNF, and NCAM molecules.

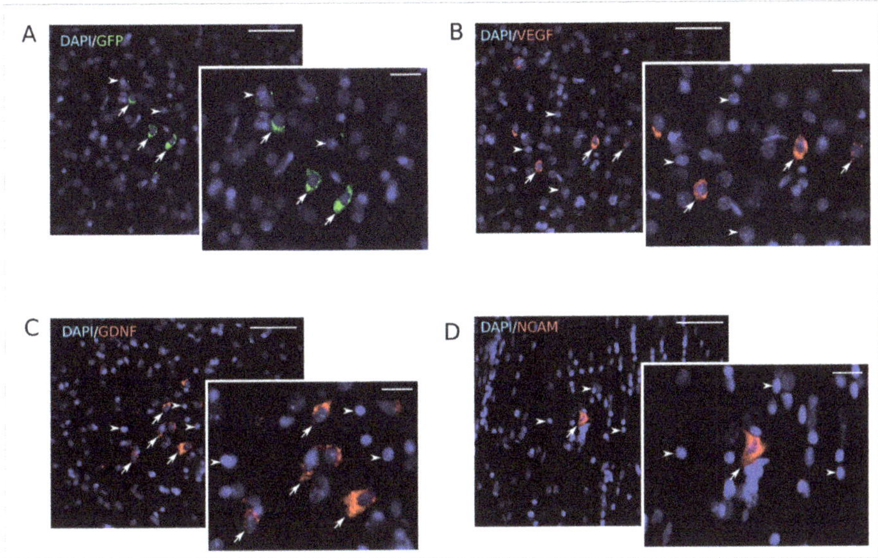

Figure 10. Expression of recombinant molecules in rat brain on 25th day after intrathecal injection of Ad5-GFP and Ad5-VEGF-GDNF-NCAM. (**A**)—Specific green glow of GFP in the transduced brain cells. (**B**)—Immune reaction with antibody to vascular endothelial growth factor (VEGF) in brain cells (red glow). (**C**)—Immune reaction with antibody to glial-derived neurotrophic factor (GDNF) (red glow). (**D**)—Immune reaction with antibody to neuronal cell adhesion molecule (NCAM) (red glow). Arrows point to the transduced brain cells. Arrowheads indicate cell nuclei visualized using DAPI (blue glow). Scale bar in (**A–D**) = 50 μm; scale bar in the inserts = 20 μm.

2.5. Expression of Recombinant Molecules in Rat Brain after Intrathecal UCB-MC-Mediated Delivery of Transgenes

The synthesis of the reporter GFP in UCB-MC in vivo was studied on day 24 after intrathecal injection of UCB-MC transduced with Ad-GFP. Antibody to human nuclear antigen (HNA) was used for the identification of UCB-MC in the rat brain cortex. HNA-positive cells producing GFP were revealed in the peri-infarct zone (Figure 11A). The production of therapeutic molecules (VEGF, GDNF, and NCAM) in transplanted, genetically-modified UCB-MC was studied using a double immunofluorescent staining method with antibodies against HNA and to one of the recombinant molecules. HNA-positive cells producing recombinant human molecules VEGF (Figure 11B), GDNF (Figure 11C), and NCAM (Figure 11D) were detected in the left hemisphere, 24 days after the intrathecal injection of UCB-MC+Ad5-VEGF-GDNF-NCAM.

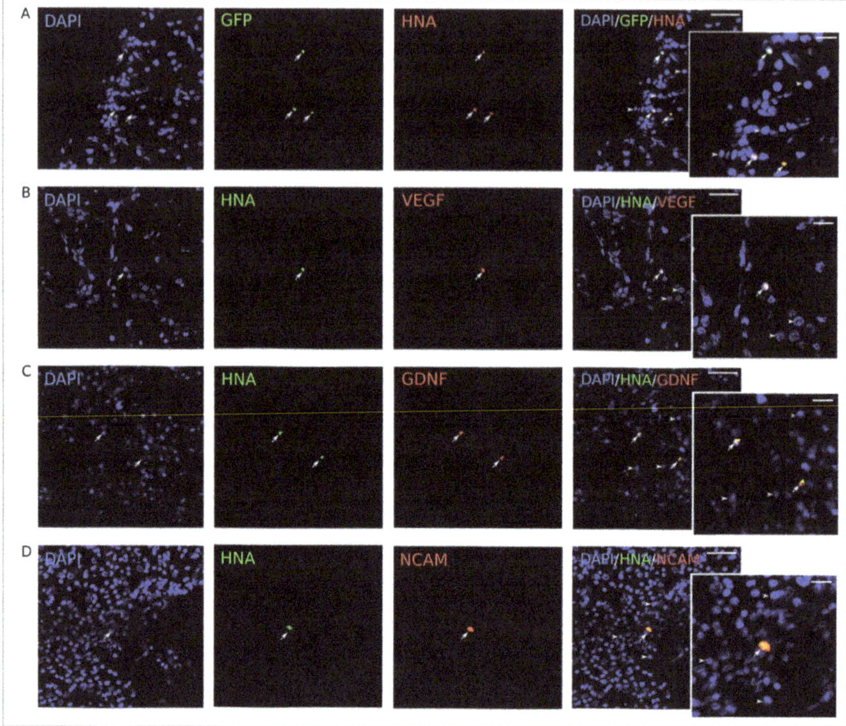

Figure 11. Expression of recombinant molecules in rat brain on the 24th day after intrathecal injection of UCB-MC+Ad5-GFP and UCB-MC+Ad-VEGF-GDNF-NCAM. UCB-MC+Ad-VEGF-GDNF-NCAM are visualized with antibody against the human nuclear antigen (HNA). Cell nuclei are counterstained with DAPI (blue glow). (**A**)—Specific green glow in UCB-MC+Ad5-GFP. (**B**)—Immune reaction with antibody against vascular endothelial growth factor (VEGF) (red glow). (**C**)—Immune response with antibody against glial-derived neurotrophic factor (GDNF) (red glow). (**D**)—Immune response with antibody against neuronal cell adhesion molecule (NCAM) (red glow). Arrows point to the UCB-MC. Arrowheads indicate cell nuclei visualized using DAPI (blue glow). Scale bar in (**A–D**) = 50 μm; scale bar in the inserts = 20 μm.

2.6. Multiplex Analysis of Cytokines, Chemokines and Growth Factors in Blood (Serum) and Cerebrospinal Fluid (Liquor) Samples of Experimental and Intact Animals

Granulocyte colony-stimulating factor (G-CSF), granulocyte and monocyte colony-stimulating factor (GM-CSF), interferon-γ (IFN-γ), interleukin-4 (IL-4), interleukin-10 (IL-10), interleukin-12 (IL-12), and VEGF are considered cytokines with neuroprotective action, while interleukin-1 (IL-1), interleukin-6 (IL-6), interleukin-17 (IL-17), and tumor necrosis factor-α (TNF-α) can cause secondary death of nerve cells. In the present study, preliminary analysis of endogenous cytokines, chemokines and growth factors in cerebrospinal fluid in animals from the control group NaCl showed no significant differences in comparison with intact rats, which probably indicates the cessation of inflammatory response 21 days after modelling a stroke. In addition, intrathecal injection of a combination of adenoviral vectors carrying cDNA encoding VEGF, GDNF, and NCAM, or genetically-modified UCB-MC+Ad5-VEGF-GDNF-NCAM on day 21 of the experiment does not reliably affect the cytokine levels in liquor samples when compared with intact rats.

3. Discussion

It is known that ischemic stroke leads to massive loss of neurons and glial cells immediately after acute arterial occlusion. Neuron death caused by necrosis and apoptosis is the result of blood flow impairment, oxidative stress, mitochondrial dysfunction, and excitotoxicity. Unfortunately, these irreversible changes rapidly spread beyond the epicenter of the ischemic damage, resulting in progressive neurodegeneration on the following days, and even over the following week [32]. However, during the first hours, an ischemic penumbra with blood supply below a normal level is formed around the ischemic core with irreversible changes. The viability of cells in the penumbra is limited to a 3–6-h period of the therapeutic window, during which it is possible to restrain the entry into apoptosis of neurons and glial cells and thus prevent the increase of infarction volume.

Despite the improved efficiency of modern therapy for ischemic brain stroke, recovery of patients is not satisfactory. Today, specific therapy for acute stroke is aimed at restoring blood circulation in the ischemic area and maintaining the metabolism of brain cells. However, the drugs used in practical medicine are not effective in controlling the death of nerve cells in the ischemic penumbra, which results in an extensive increase in brain infarction. It is known that there are natural limitations neuroregeneration in the CNS; therefore, there is a need to develop new effective approaches to maintain the viability of nerve cells in the ischemic penumbra in the 3–6-h therapeutic window.

It is known that neuroontogenesis (addressed migration of brain cells, directed axon growth and establishment of intercellular contacts) involves an information exchange between neurons, which is realized through neurotrophic factors produced in some neurons or in non-neural cells that affect other neurons. In postnatal ontogenesis, there is no formation of new cortical neurons. Therefore, dying neurons are not restored. This does not mean that there is no regeneration in brain cortex; however, as regeneration is carried out due to intracellular regeneration of damaged neurons, growth of their neuritis and restoration of synaptic connections between regenerating and intact neurons. During neuroregeneration, neurotrophic factors support neuronal survival, stimulate axonal growth, and establish lost synaptic contacts.

At present, the most promising of the actively developed strategies to prevent brain cell death in penumbra is the development of gene and cellular technologies. Gene therapy is mainly aimed at the delivery of therapeutic genes encoding neurotrophic factors. Among them, the most promising are genes encoding neurotrophic factors (BDNF, CNTF, GDNF, VEGF), anti-apoptotic proteins (Bcl-2, Bcl-XL), heat shock proteins (Hsp25, Hsp70), and anti-inflammatory molecules (IL-1RA). The neuroprotective effect of these factors has been experimentally proven, but there is no unequivocal, let alone exhaustive, answer to the question of which of these factors may be recommended as neuroprotective factors in practical medicine. It has been established that combinations of several neurotrophic factors may have a more pronounced effect on nerve cell survival [25,28].

Other equally important issues in the strategy for the gene therapy of stroke include the development of technology to deliver transgenes to the brain. Difficulties in the delivery of therapeutic genes to the cerebral infarction area are one of the main reasons for the unavailability of effective gene therapy in treating post-ischemic negative consequences in the brain. Direct (in vivo) gene therapy provides for the delivery of transgenes into the recipient brain plasmid or viral genetic vectors [33]. Cell-mediated (ex vivo) gene therapy is based on the delivery of therapeutic genes using cells that serve as carriers of transgenes, as well as producers of recombinant protein molecules.

For the treatment of stroke, various methods of gene therapy are known, the effectiveness of which has been proven in numerous experiments on animals [2,4,34]. Injections of viral vectors carrying transgenes into the ventricles or the infarction area are mainly used to deliver the therapeutic genes to the brain. Genetic vectors based on Sendai virus vectors containing *gdnf* or *ngf* genes were injected 30 min after stroke simulation [35]. The adeno-associated viral vector carrying *gdnf* was injected 2 days after the stroke modelling [36]. In other studies, viral vectors carrying therapeutic genes were delivered before modelling the stroke. The positive effect was shown after local delivery to the brain of *gdnf* or *cntf* genes—7 days [37], *bdnf*—2 weeks [38], *gdnf*—4 weeks [39], or *ngf* and *bdnf*—4 to 5 weeks [40] before stroke modelling.

The list of genes employed in gene therapy for stroke is quite long. Of these, we find *vegf*, *gdnf*, and *ncam* to be the most promising.

In addition to angiogenic action, VEGF exhibits the properties of a typical neurotrophic factor. It supports survival of sensitive [41] and motor neurons [42] and stimulates proliferation of astrocytes [43], neural stem cells [44], and Schwann's cells [41]. GDNF has a pronounced neuroprotective effect on dopaminergic brain neurons and cholinergic spinal cord motoneurons [45] and stimulates the growth of nerve processes [46]. NCAM (CD56) is expressed on the surface of neurons and glial cells. Intercellular interactions mediated by NCAM in neuro-ontogenesis and posttraumatic regeneration provide survival and migration of neurons, directed neurite growth and synaptogenesis.

The efficiency of cell therapy for stroke treatment in experiments using neural precursors derived from embryonic stem cells [47], induced pluripotent cells [48], MSCs isolated from red bone marrow [49] or UCB [50] suggests the use of these cells as carriers of therapeutic genes for delivery to the brain. Thus, MSCs were used for delivery of *bdnf* [51], *pigf* [52], and *vegf165* [53] to the brain. Of particular interest is the transduction of cell carriers by two or more expression vectors [28]. This approach allows simultaneous overexpression of several molecules/stimulants of neuroregeneration to be obtained. In our studies, in addition to the gene encoding neurotrophic factors, a gene encoding NCAM was delivered into UCB-MC, which, according to the obtained data, promoted the addressed migration of transplanted cells into the CNS after intravenous injection, increased their survival in the recipient's tissues and supported prolonged production of recombinant therapeutic molecules [28]. In clinical investigations, autologous cells isolated from the red bone marrow (mononuclear and MSCs) or peripheral blood (CD34+) are predominantly used in the cell therapy of patients after stroke [54,55].

The most promising cell carriers of therapeutic genes are UCB-MC [56–58]. The basis for their application is the suitability for both allografting and autotransplantation in humans, availability, and the ease of obtaining and storage. An important factor is the absence of legal, ethical, and religious prohibitions related to blood cell transplantation. In our previous study, we showed that intrathecal injection of adenoviral vectors carrying *vegf*, *gdnf*, and *ncam*, or genetically-modified UCB-MC+Ad5-VEGF-GDNF-NCAM, 4 h after stroke modelling in rats, had a positive effect on the morpho-functional recovery of the post-ischemic brain [25]. Adenoviral vectors and genetically-modified UCB-MC with cerebrospinal fluid reached the ischemic area and delivered the production of recombinant VEGF, GDNF, and NCAM, lasting up to 21 days in the experiment.

Other important issues in the strategy for the treatment of ischemic stroke include the development of approaches to enhance the viability of neurons with the threat of a stroke. Patients with transient ischemic attacks, arterial hypertension, atrial fibrillation, disorders of lipid metabolism with high cholesterol and diabetes are at high risk of ischemic stroke. Preventive therapy aimed at increasing the survivability of neurons in at-risk patients may prevent severe post-ischemic consequences in the brain, or improve the outcome of the disease. Currently, in medical practice, measures to prevent stroke are based on the use of anticoagulants and prosthetics of blood vessels. At the same time, the preventive methods that are able to considerably decrease the death of neurons in the "ischemic penumbra" during the 3–6-h "therapeutic window" are unknown. Enhancement of the viability of nerve cells at risk of stroke is also associated with the delivery of therapeutic genes that encode molecules to the brain, which inhibit neuronal death and stimulate neuroregeneration. In this study, for the first time, we propose the approach of preventive gene therapy to improve the viability of brain neurons under threat of ischemic stroke to contain neuronal death in the first hours of a stroke. The use of leucocytes for delivery of therapeutic genes (*vegf165*, *gdnf*, and *ncam1*) in the brain was based on their biological properties. Leucocytes are cells with high secretory and migration potentials, which suggest their exclusive role as cell carriers for addressed delivery and effective expression of transgenes. The results obtained in the study demonstrate that preventive intrathecal adenoviral- or UCB-MC-mediated delivery of *vegf165*, *gdnf*, and *ncam1* results in a reduction of apoptosis and, consequently, the infraction volume.

In addition to the decrease in expression of proteins of cellular stress and restraining neuronal death in the area of ischemic damage, we found evidence of the restoration of functional activity of neurons (increase in expression of synaptic proteins), maintenance of myelinization (increase in the number of oligodendrocytes) and an obstacle to astrogliosis development (decrease in the immunopositive areas for astrocytes and microglial cells markers). Importantly, transplantation of gene-modified UCB-MC is safer and more efficacious compared with direct gene therapy. These data are in line with our results using the same gene and gene-cell constructs for ischemic stroke treatment in rats and allow us to conclude that preventive gene therapy may be effective in overcoming the negative consequences of ischemic stroke in the rehabilitation period.

Recently, for personalized ex vivo gene therapy, we suggested the use of gene-modified leucoconcentrate (GML) prepared from patient's peripheral blood and chimeric adenoviral vectors (Ad5/35F) carrying one or a combination of therapeutic genes [59]. Taken together with the concept of GML-therapy and the data of this study, we propose the use of GML carrying *vegf165*, *gdnf*, and *ncam1* for personalized preventive gene therapy in the threat of stroke.

4. Materials and Methods

4.1. Preparation and Molecular Analysis of Gene-Modified UCB-MC

Human (*vegf165*, *gdnf*, *ncam1*) and reporter (*gfp*) genes were inserted into recombinant replication-defective adenovirus serotype 5 (Ad5) in the Gamaleya Research Institute of Epidemiology and Microbiology (Moscow, Russia), as described previously [25]. Viral vectors were grown in HEK-293 cell culture, purified by exclusion chromatography and the titres of Ad5-VEGF (2.6×10^9 PFU/mL), Ad5-GDNF (1.7×10^{10} PFU/mL), Ad5-NCAM (2.4×10^{10} PFU/mL) and Ad5-GFP (1.2×10^{10} PFU/mL) were determined in the HEK-293 cell culture using the plaque formation technique.

The fraction of mononuclear cells from UCB was isolated onto a density barrier by the standard technique of sedimentation, seeded in 10 cm culture dishes and simultaneously transduced with three adenoviral vectors at an equal ratio: Ad5-VEGF (1/3)+Ad5-GDNF (1/3)+Ad5-NCAM (1/3) or with Ad5-GFP with MOI of 10, as described previously [25]. To confirm the efficacy of UCB-MC transduction, the gene-modified cells were cultured for 96 h. Transduction with therapeutic genes (*vegf165*, *gdnf*, *ncam1*) was confirmed by a real-time polymerase chain reaction (RT-PCR) and with reporter green fluorescent protein (*gfp*) using fluorescent microscopy, as described previously [57].

For direct (in vivo) preventive gene therapy, a mixture of 2×10^7 virus particles in an equal ratio of Ad5-VEGF (1/3)+Ad5-GDNF (1/3)+Ad5-NCAM (1/3) in 20 µL of saline and Ad5-GFP in 20 µL of saline were prepared. For cell-mediated (ex vivo) preventive gene therapy, 2×10^6 UCB-MC+Ad5-VEGF+Ad5-GDNF+Ad5-NCAM in 20 µL of saline and 2×10^6 UCB-MC+Ad5-GFP in 20 µL of saline were prepared.

The efficiency of UCB-MC genetic modification by adenoviral vector, bearing green fluorescent protein reporter gene, was analyzed 72 h after cell transduction. The production of green fluorescent protein UCB-MC+Ad5-GFP was examined with the Axio Oberver Z1 inverted fluorescence microscope (Carl Zeiss, Germany). The number of transduced UCBC was estimated using a BD FACSAria III flow cytomography fluorimeter (BD Bioscience, New York, NY, USA) and BD FACS Diva7 software (BD Bioscience, New York, NY, USA).

mRNA levels of *vegf165*, *gdnf*, and *ncam1* transgenes in the UCB-MC, simultaneously transduced by a combination of three adenoviral vectors Ad5-VEGF, Ad5-GDNF and Ad5-NCAM, were analyzed with PCR-RT. Extraction of common RNA from UCB-MC+Ad5-VEGF-GDNF-NCAM and UCB-MC+Ad5-GFP, 72 h after cell incubation, was carried out using a set Yellow Solve (Silex), according to the manufacturer's instructions. Reference samples of common RNA were derived from non-transduced UCBC. The cDNA synthesis was performed using six nucleotide random primers and reverse transcriptase RevertAid Reverse Transcriptase (Thermo Fisher Scientific). Quantitative level analysis mRNA of the target cDNA was performed using a CFX 96 Real-Time PCR System thermocirculator (BioRad, Hercules, CA, USA), using the TaqMan system. Sequences of primers and probes of the reaction mixture are presented in Table 1. Received data was normalized by the 18S rRNA reference gene. For each target gene, the results were obtained in two independent experiments and presented as a mean value ± SE, $p < 0.05$.

Table 1. Experimental groups of animals.

Groups	Preparation for Animals Treatment	Number of Animals
Control	20 µL 0.9% NaCl intrathecally injected 4 days before ischemic stroke modelling	5
Ad5-GFP	Ad5 carrying *gfp* in 20 µL of saline intrathecally injected 4 days before ischemic stroke modelling	6
Ad5-VEGF+Ad5-GDNF+Ad5-NCAM	Mixture of the three Ad5 carrying *vegf165*, *gdnf* and *ncam1* in 20 µL of saline intrathecally injected 4 days before ischemic stroke modelling	6
UCBC+Ad5-GFP	2×10^6 UCBC transduced with Ad5 carrying *gfp* in 20 µL of saline intrathecally injected 3 days before ischemic stroke modelling	6
UCBC+Ad5-VEGF-GDNF-NCAM	2×10^6 UCBC simultaneously transduced with three Ad5 carrying *vegf165*, *gdnf* and *ncam1* in 20 µL of saline intrathecally injected 3 days before ischemic stroke modelling	6

Supernatant obtained 72 h after incubation of gene-modified UCB-MC (UCB-MC+Ad5-VEGF -GDNF-NCAM and UCB-MC+Ad5-GFP) and intact UCB-MC were used for multiplex profiling. In this work, commercially available panels, Human Cytokine/Chemokine Magnetic Bead Panel (HCYTMAG-60K-PX41), were applied, containing the following analytes: EGF, FGF2, Eotoxin, TGF-a, G-CSF, Flt-3L, GM-CSF, Fractalkine, IFNa2, IFN-g, GRO, IL10, MCP-3, IL12-p40, MDC, IL12-p70, PDGF-AA, IL-13, PDGF-AB/BB, IL-15, sCD40L, IL-17a, IL-1ra, IL-1a, IL-9, IL-1bIL-2, IL-3, IL-4, IL-5, IL-6, IL-7, IL-8, IP10, MCP-1, MIP-1a, MIP-1b, RANTES, TNFa, TNFb, and VEGF (Millipore). The quantitative analysis of target molecules was performed on a Bio-Plex200 System (BioRad, USA) according to the protocol recommended by the manufacturer and using Bio-Plex Manager 4.1 software (Bio-RadLaboratories). The results were statistically processed using the following methods, using linear models implemented in the limma package. Fold Change was used as an estimate for the effect size, obtained for pairwise comparison.

4.2. Animals and Treatments

The animal protocols were conducted in strict compliance with the guidelines established by the Kazan State Medical University Animal Care and Use Committee (approval No. 10, 2017). Experiments were performed on adult female Wistar rats (weight of 200–250 g) obtained from Pushchino Laboratory (Pushchino, Russia). Animals were housed one per cage according to approved procedures for the use of animals in laboratory experiments.

For preventive gene therapy, rats were deeply anesthetized intraperitoneally with Zolitil 100 (Virbac Laboratoires, France) 3 mg/kg and Xyla (Interchemie werken "De Adelaar" B.V., Netherlands) 4.8 mg/kg. A laminectomy was made over the L4–L5 vertebral level and gene-modified UCBC or adenoviral vectors were infused intrathecally. The rats were divided into five experimental groups according to the injected substances (Table 1).

Ischemic stroke in the rats was induced 3 days after intrathecal injection of UCB-MC+Ad5-GFP and UCB-MC+Ad5-VEGF-GDNF-NCAM or 4 days after intrathecal injection of 0.9% NaCl, Ad5-GFP and Ad5-VEGF-GDNF-NCAM. The ischemic stroke was induced by permanent MCAO, as described previously [25]. In brief, to reduce blood flow in the circle of Willis, the right common carotid artery was ligated with surgical silk. In the left side, a 4–5 mm hole was drilled in the temporal bone and MCAO was performed by thermocoagulation using an operating microscope.

To reduce suffering and distress, post-operative care included analgesic therapy with Ketamine (Dr. Reddy's Laboratories, Ltd., Hyderabad, Andhra Pradesh, India) intramuscularly (2.5 mg/kg) once-a-day and antibacterial therapy with Ceftriaxone (Sandoz, Austria) intramuscularly (50 mg/kg) once-a-day for 5 days. Based on our previous study [25] for evaluation of preventive gene therapy, experimental animals were sacrificed three weeks after MCAO.

4.3. Morphometric and Immunofluorescent Analysis of the Brain

For histological analysis, rats were deeply anesthetized by intraperitoneal injection of sodium pentobarbital (60 mg/kg) and intracardially perfused with 4% paraformaldehyde (PFA, Sigma) in phosphate-buffered saline (PBS, pH 7.4). The brains of the rats were isolated from the skull, post-fixed in 4% paraformaldehyde and cryoprotected in 30% sucrose. Frozen frontal sections of the brain were obtained through the epicenter of ischemic injury using a cryostat (Microm HM 560, Thermo Scientific, Waltham, MA, USA).

Morphometric analysis of the volume of the cerebral infarction included macroscopic evaluation of the infarct area of the whole brain by capturing digital images and microscopic study every 10 frontal brain sections (with 200 μm interval) through the ischemic injury area, after staining with hematoxylin and eosin. To calculate the infarction volume, maximal depth and maximal diameter of the infarct cavities were estimated using ImageJ software (NIH). The calculation was carried out as $V = 1/3 \times$ maximal depth $\times \pi \times$ (maximal diameter/2)^2.

Immunofluorescent staining was performed on 20 μm sections of frontal brain using primary antibodies (Ab) reacting with cell specific markers (Table 2). The viability of neural cells was evaluated using Ab to pro-apoptotic protein (Caspase3) and heat shock protein 70 kDa (Hsp70). Synaptic function of preserved neurons was assessed with Ab against synaptophysin and postsynaptic density protein 95 kDa (PSD95). Responses of glial cells were assessed with Ab to GFAP for astrocytes, Ab to Olig2 for oligodendrocytes and Ab to Iba1 for microglia. Anti-human nuclear antigen (HNA) antibodies were used for identification of human UCBC in the rat brain. Double immunofluorescent staining with Ab to HNA and Ab to human VEGF, GDNF, and NCAM were employed to analyze the expression of those recombinant molecules in UCB-MC. For immunofluorescent labelling, sections were incubated with secondary antibodies (Table 2). Appropriate secondary Alexa Fluor 647 and 488 conjugated Abs were used for immunofluorescent labelling of the target molecules. Cell nuclear counterstaining was performed with DAPI (10 μg/mL in PBS, Sigma), sections were embedded in glycerol (GalenoPharm, Saint Petersburg, Russia) and observed under a LEICA TCS SP5 MP microscope (Leica Microsystems, Wetzlar, Germany).

Digital images were captured using identical confocal settings in areas of 0.05 mm^2 and analyzed with ImageJ software (NIH) in 10 captured fields in the peri-infarct zone (Figure 2C) in each section, as described previously [25]. The number of immunopositive cells for apoptosis protein Caspase3 and oligodendroglial cells transcription factor Olig2 were counted in regard to nuclear counterstaining with DAPI. Neuroglial markers for astrocytes (GFAP) and microglia (Iba1) were evaluated as the immunopositive areas and presented in percentages. The level of synaptic proteins (synaptophysin and PSD95) and Hsp70 immunoexpression in brain sections was evaluated as mean pixel intensities.

Table 2. Antibodies used in immunofluorescent analysis.

Antibody against:	Host	Dilution	Source
Caspase3	Rabbit	1:200	Abcam
Glial cell-derived neurotrophic factor (GDNF)	Rabbit	1:100	Santa Cruz
Glial fibrillary acidic protein (GFAP)	Mouse	1:200	Santa Cruz
Ionized calcium binding adaptor molecule 1 (Iba1)	Rabbit	1:150	Biocare Medical
Human Nuclear Antigen (HNA)	Mouse	1:150	Millipore
Heat shock protein 70 kDa (Hsp70)	Rabbit	1:200	Abcam
Oligodendrocyte transcription factor 2 (Olig2)	Rabbit	1:100	Santa Cruz
Neural cell adhesion molecule (NCAM)	Rabbit	1:100	Santa Cruz
Postsynaptic density protein 95 kDa (PSD95)	Rabbit	1:200	Abcam
Synaptophysin	Rabbit	1:100	Abcam
Vascular endothelial growth factor (VEGF)	Goat	1:300	Sigma
Rabbit IgG conjugated with Alexa 647	Donkey	1:200	Invitrogen
Mouse IgG conjugated with Alexa 488	Donkey	1:200	Invitrogen

4.4. Multiplex Cytokine Analysis of the Cerebrospinal Fluid

Multiplex profiling was performed in cerebrospinal fluid (liquor) of experimental and intact animals using the commercially available Bio-Plex Pro™ Rat Cytokine 23-Plex Assay panel, containing the following analytes: G-SCF, GM-CSF, GRO/KC, IFN-g, IL-10, IL-12p70, IL-13, IL-18, IL-1a, IL-1b, IL-2, IL-4, IL-5, IL-6, IL-7, IL17A, M-CSF, MCP-1, MIP1a, MIP3a, RANTES, TNFa, and VEGF. Quantitative analysis of target molecules was performed on a Bio-Plex200 System analyzer (BioRad, USA), as recommended by the manufacturer, using Bio-Plex Manager 4.1 software (Bio-RadLaboratories). The statistical processing of obtained results was carried out using linear models implemented in the limma package.

4.5. Statistics

Statistical data analysis and visualizations were performed using R version 3.5.3 (R Foundation for Statistical Computing, Vienna, Austria). Sample distributions of quantitative values were visualized using box plots, and descriptive statistics are presented as: (Median [1st quartile; 3rd quartile]). The Kruskal–Wallis test was used to compare morphometric and immunofluorescent staining data between experimental groups when a null hypothesis was rejected; we used Dunn's-test as the post hoc method. Differences were considered statistically significant where $p < 0.05$.

5. Conclusions

Methods of preventive gene therapy of ischemic stroke in at-risk patients are currently unknown. In this study, for the first time, we provide evidence of the beneficial effects of preventive triple gene therapy by an adenoviral- or UCB-MC-mediated intrathecal simultaneous delivery combination of *vegf165*, *gdnf*, and *ncam1* on the preservation and recovery of the brain in rats with subsequent modelling of stroke. The obtained data represent a novel and potentially successful approach for ischemic stroke treatment in patients.

Author Contributions: Conceptualization, R.I.; Data curation, V.M. and M.K.; Formal analysis, M.K.; Investigation, V.M., F.F., M.S., D.T., E.K., G.K., A.G. and I.S.; Methodology, Z.S., A.I. and F.F.; Project administration, V.M., Z.S., A.I., F.F. and M.S.; Supervision, F.B. and R.I.; Visualization, A.I. and L.N.; Writing—original draft and review & editing, R.I. All authors have read and agreed to the published version of the manuscript.

Funding: This research was funded by the grant of Russian Science Foundation No. 19-75-10030.

Acknowledgments: Some experiments were carried out as part of the state assignment of the Federal Research Center "Kazan Scientific Center of the Russian Academy of Sciences".

Conflicts of Interest: The authors declare no conflict of interest.

References

1. Karlupia, N.; Manley, N.C.; Prasad, K.; Schäfer, R.; Steinberg, G.K. Intraarterial transplantation of human umbilical cord blood mononuclear cells is more efficacious and safer compared with umbilical cord mesenchymal stromal cells in a rodent stroke model. *Stem Cell Res. Ther.* **2014**, *5*, 45. [CrossRef] [PubMed]
2. Rhim, T.; Lee, M. Targeted delivery of growth factors in ischemic stroke animal models. *Expert Opin. Drug Deliv.* **2016**, *13*, 709–723. [CrossRef] [PubMed]
3. Yamashita, T.; Deguchi, K.; Nagotani, S.; Kamiya, T.; Abe, K. Gene and Stem Cell Therapy in Ischemic Stroke. *Cell Transplant.* **2009**, *18*, 999–1002. [CrossRef] [PubMed]
4. Ooboshi, H. Gene Therapy as a Novel Pharmaceutical Intervention for Stroke. *Curr. Pharm. Des.* **2011**, *17*, 424–433. [CrossRef] [PubMed]
5. Yang, W.Z.; Zhang, Y.; Wu, F.; Min, W.P.; Minev, B.; Zhang, M.; Luo, X.L.; Ramos, F.; Ichim, T.E.; Riordan, N.H.; et al. Safety evaluation of allogeneic umbilical cord blood mononuclear cell therapy for degenerative conditions. *J. Transl. Med.* **2010**, *8*, 75. [CrossRef]
6. Harris, D.T.; Rogers, I. Umbilical cord blood: A unique source of pluripotent stem cells for regenerative medicine. *Curr. Stem Cell Res. Ther.* **2007**, *2*, 301–309. [CrossRef]
7. Arien-Zakay, H.; Lazarovici, P.; Nagler, A. Tissue regeneration potential in human umbilical cord blood. *Best Pract. Res. Clin. Haematol.* **2010**, *23*, 291–303. [CrossRef]
8. Arien-Zakay, H.; Lecht, S.; Nagler, A.; Lazarovici, P. Human umbilical cord blood stem cells: Rational for use as a neuroprotectant in ischemic brain disease. *Int. J. Mol. Sci.* **2010**, *11*, 3513–3528. [CrossRef]
9. Erices, A.; Conget, P.; Minguell, J.J. Mesenchymal progenitor cells in human umbilical cord blood. *Br. J. Haematol.* **2000**, *109*, 235–242. [CrossRef]
10. Kögler, G.; Sensken, S.; Airey, J.A.; Trapp, T.; Müschen, M.; Feldhahn, N.; Liedtke, S.; Sorg, R.V.; Fischer, J.; Rosenbaum, C.; et al. A new human somatic stem cell from placental cord blood with intrinsic pluripotent differentiation potential. *J. Exp. Med.* **2004**, *200*, 123–135. [CrossRef]
11. Gluckman, E. Ten years of cord blood transplantation: From bench to bedside. *Br. J. Haematol.* **2009**, *147*, 192–199. [CrossRef] [PubMed]
12. Pimentel-Coelho, P.M.; Rosado-de-Castro, P.H.; Barbosa da Fonseca, L.M.; Mendez-Otero, R. Umbilical cord blood mononuclear cell transplantation for neonatal hypoxic–ischemic encephalopathy. *Pediatr. Res.* **2012**, *71*, 464–473. [CrossRef] [PubMed]
13. Goldstein, G.; Toren, A.; Nagler, A. Transplantation and Other Uses of Human Umbilical Cord Blood and Stem Cells. *Curr. Pharm. Des.* **2007**, *13*, 1363–1373. [CrossRef] [PubMed]
14. Rocha, V.; Labopin, M.; Sanz, G.; Arcese, W.; Schwerdtfeger, R.; Bosi, A.; Jacobsen, N.; Ruutu, T.; De Lima, M.; Finke, J.; et al. Transplants of umbilical-cord blood or bone marrow from unrelated donors in adults with acute leukemia. *N. Engl. J. Med.* **2004**, *351*, 2276–2285. [CrossRef]
15. Balassa, K.; Rocha, V. Anticancer cellular immunotherapies derived from umbilical cord blood. *Expert Opin. Biol. Ther.* **2018**, *18*, 121–134. [CrossRef]
16. Arien-Zakay, H.; Lecht, S.; Bercu, M.M.; Tabakman, R.; Kohen, R.; Galski, H.; Nagler, A.; Lazarovici, P. Neuroprotection by cord blood neural progenitors involves antioxidants, neurotrophic and angiogenic factors. *Exp. Neurol.* **2009**, *216*, 83–94. [CrossRef]
17. Bachstetter, A.D.; Pabon, M.M.; Cole, M.J.; Hudson, C.E.; Sanberg, P.R.; Willing, A.E.; Bickford, P.C.; Gemma, C. Peripheral injection of human umbilical cord blood stimulates neurogenesis in the aged rat brain. *BMC Neurosci.* **2008**, *9*, 22. [CrossRef]
18. Dasari, V.R.; Spomar, D.G.; Li, L.; Gujrati, M.; Rao, J.S.; Dinh, D.H. Umbilical cord blood stem cell mediated downregulation of Fas improves functional recovery of rats after spinal cord injury. *Neurochem. Res.* **2008**, *33*, 134–149. [CrossRef]
19. Schira, J.; Gasis, M.; Estrada, V.; Hendricks, M.; Schmitz, C.; Trapp, T.; Kruse, F.; Kögler, G.; Wernet, P.; Hartung, H.P.; et al. Significant clinical, neuropathological and behavioural recovery from acute spinal cord trauma by transplantation of a well-defined somatic stem cell from human umbilical cord blood. *Brain* **2012**, *135*, 431–446. [CrossRef]

20. Xiao, J.; Nan, Z.; Motooka, Y.; Low, W.C. Transplantation of a novel cell line population of umbilical cord blood stem cells ameliorates neurological deficits associated with ischemic brain injury. *Stem Cells Dev.* **2005**, *14*, 722–733. [CrossRef]

21. Ikeda, Y.; Fukuda, N.; Wada, M.; Matsumoto, T.; Satomi, A.; Yokoyama, S.-I.; Saito, S.; Matsumoto, K.; Kanmatsuse, K.; Mugishima, H. Development of angiogenic cell and gene therapy by transplantation of umbilical cord blood with vascular endothelial growth factor gene. *Hypertens. Res.* **2004**, *27*, 119–128. [CrossRef] [PubMed]

22. Chen, H.K.; Hung, H.F.; Shyu, K.G.; Wang, B.W.; Sheu, J.R.; Liang, Y.J.; Chang, C.C.; Kuan, P. Combined cord blood stem cells and gene therapy enhances angiogenesis and improves cardiac performance in mouse after acute myocardial infarction. *Eur. J. Clin. Investig.* **2005**, *35*, 677–686. [CrossRef]

23. Islamov, R.R.; Sokolov, M.E.; Bashirov, F.V.; Fadeev, F.O.; Shmarov, M.M.; Naroditskiy, B.S.; Povysheva, T.V.; Shaymardanova, G.F.; Yakupov, R.A.; Chelyshev, Y.A.; et al. A pilot study of cell-mediated gene therapy for spinal cord injury in mini pigs. *Neurosci. Lett.* **2017**, *644*, 67–75. [CrossRef] [PubMed]

24. Izmailov, A.A.; Povysheva, T.V.; Bashirov, F.V.; Sokolov, M.E.; Fadeev, F.O.; Garifulin, R.R.; Naroditsky, B.S.; Logunov, D.Y.; Salafutdinov, I.I.; Chelyshev, Y.A.; et al. Spinal Cord Molecular and Cellular Changes Induced by Adenoviral Vector- and Cell-Mediated Triple Gene Therapy after Severe Contusion. *Front. Pharmacol.* **2017**, *8*, 813. [CrossRef] [PubMed]

25. Sokolov, M.E.; Bashirov, F.V.; Markosyan, V.A.; Povysheva, T.V.; Fadeev, F.O.; Izmailov, A.A.; Kuztetsov, M.S.; Safiullov, Z.Z.; Shmarov, M.M.; Naroditskyi, B.S.; et al. Triple-Gene Therapy for Stroke: A Proof-of-Concept in Vivo Study in Rats. *Front. Pharmacol.* **2018**, *9*, 111. [CrossRef] [PubMed]

26. Mothe, A.J.; Tator, C.H. Review of transplantation of neural stem/progenitor cells for spinal cord injury. *Int. J. Dev. Neurosci.* **2013**, *31*, 701–713. [CrossRef]

27. Razavi, S.; Ghasemi, N.; Mardani, M.; Salehi, H. Remyelination improvement after neurotrophic factors secreting cells transplantation in rat spinal cord injury. *Iran. J. Basic Med. Sci.* **2017**, *20*, 392–398. [CrossRef] [PubMed]

28. Safiullov, Z.Z.; Garanina, E.E.; Izmailov, A.A.; Garifulin, R.R.; Fedotova, V.Y.; Salafutdinov, I.I.; Rizvanov, A.A.; Islamov, R.R. Homing and survivability of genetically modified mononuclear umbilical cord blood cells after transplantation into transgenic G93A mice with amyotrophic lateral sclerosis. *Genes Cells* **2015**, *X*, 1–4.

29. Franklin, T.B.; Krueger-Naug, A.M.; Clarke, D.B.; Arrigo, A.-P.; Currie, R.W. The role of heat shock proteins Hsp70 and Hsp27 in cellular protection of the central nervous system. *Int. J. Hyperth.* **2005**, *21*, 379–392. [CrossRef]

30. Barreto, G.E.; White, R.; Ouyang, Y.; Xu, L.G.; Giffard, R. Astrocytes: Targets for Neuroprotection in Stroke. *Cent. Nerv. Syst. Agents Med. Chem.* **2012**, *11*, 164–173. [CrossRef]

31. Dewar, D.; Underhill, S.M.; Goldberg, M.P. Oligodendrocytes and ischemic brain injury. *J. Cereb. Blood Flow Metab.* **2003**, *23*, 263–274. [CrossRef] [PubMed]

32. Du, C.; Hu, R.; Csernansky, C.A.; Hsu, C.Y.; Choi, D.W. Very delayed infarction after mild focal cerebral ischemia: A role for apoptosis? *J. Cereb. Blood Flow Metab.* **1996**, *16*, 195–201. [CrossRef] [PubMed]

33. Abdellatif, A.A.; Pelt, J.L.; Benton, R.L.; Howard, R.M.; Tsoulfas, P.; Ping, P.; Xu, X.-M.; Whittemore, S.R. Gene delivery to the spinal cord: Comparison between lentiviral, adenoviral, and retroviral vector delivery systems. *J. Neurosci. Res.* **2006**, *84*, 553–567. [CrossRef] [PubMed]

34. Craig, A.J.; Housley, G.D. Evaluation of Gene Therapy as an Intervention Strategy to Treat Brain Injury from Stroke. *Front. Mol. Neurosci.* **2016**, *9*, 34. [CrossRef] [PubMed]

35. Shirakura, M.; Inoue, M.; Fujikawa, S.; Washizawa, K.; Komaba, S.; Maeda, M.; Watabe, K.; Yoshikawa, Y.; Hasegawa, M. Postischemic administration of Sendai virus vector carrying neurotrophic factor genes prevents delayed neuronal death in gerbils. *Gene Ther.* **2004**, *11*, 784–790. [CrossRef] [PubMed]

36. Mätlik, K.; Abo-Ramadan, U.; Harvey, B.K.; Arumäe, U.; Airavaara, M. AAV-mediated targeting of gene expression to the peri-infarct region in rat cortical stroke model. *J. Neurosci. Methods* **2014**, *236*, 107–113. [CrossRef]

37. Hermann, D.M.; Kilic, E.; Kügler, S.; Isenmann, S.; Bähr, M. Adenovirus-mediated GDNF and CNTF pretreatment protects against striatal injury following transient middle cerebral artery occlusion in mice. *Neurobiol. Dis.* **2001**, *8*, 655–666. [CrossRef]

38. Zhang, J.; Yu, Z.; Yu, Z.; Yang, Z.; Zhao, H.; Liu, L.; Zhao, J. rAAV-mediated delivery of brain-derived neurotrophic factor promotes neurite outgrowth and protects neurodegeneration in focal ischemic model. *Int. J. Clin. Exp. Pathol.* **2011**, *4*, 496–504.

39. Arvidsson, A.; Kirik, D.; Lundberg, C.; Mandel, R.J.; Andsberg, G.; Kokaia, Z.; Lindvall, O. Elevated GDNF levels following viral vector-mediated gene transfer can increase neuronal death after stroke in rats. *Neurobiol. Dis.* **2003**, *14*, 542–556. [CrossRef]

40. Andsberg, G.; Kokaia, Z.; Klein, R.L.; Muzyczka, N.; Lindvall, O.; Mandel, R.J. Neuropathological and behavioral consequences of adeno-associated viral vector-mediated continuous intrastriatal neurotrophin delivery in a focal ischemia model in rats. *Neurobiol. Dis.* **2002**, *9*, 187–204. [CrossRef]

41. Sondell, M.; Lundborg, G.; Kanje, M. Vascular endothelial growth factor has neurotrophic activity and stimulates axonal outgrowth, enhancing cell survival and Schwann cell proliferation in the peripheral nervous system. *J. Neurosci.* **1999**, *19*, 5731–5740. [CrossRef] [PubMed]

42. Islamov, R.R.; Chintalgattu, V.; Pak, E.S.; Katwa, L.C.; Murashov, A.K. Induction of VEGF and its Flt-1 receptor after sciatic nerve crush injury. *Neuroreport* **2004**, *15*, 2117–2121. [CrossRef] [PubMed]

43. Rosenstein, J.M.; Mani, N.; Silverman, W.F.; Krum, J.M. Patterns of brain angiogenesis after vascular endothelial growth factor administration in vitro and in vivo. *Proc. Natl. Acad. Sci. USA* **1998**, *95*, 7086–7091. [CrossRef] [PubMed]

44. Jin, L.; Neff, T.; Blau, C.A. Marrow sensitization to 5-fluorouracil using the ligands for Flt-3 and c-Kit. *Exp. Hematol.* **1999**, *27*, 520–525. [CrossRef]

45. Cheng, H.; Wu, J.-P.; Tzeng, S.-F. Neuroprotection of glial cell line-derived neurotrophic factor in damaged spinal cords following contusive injury. *J. Neurosci. Res.* **2002**, *69*, 397–405. [CrossRef]

46. Iannotti, C.; Li, H.; Yan, P.; Lu, X.; Wirthlin, L.; Xu, X.M. Glial cell line-derived neurotrophic factor-enriched bridging transplants promote propriospinal axonal regeneration and enhance myelination after spinal cord injury. *Exp. Neurol.* **2003**, *183*, 379–393. [CrossRef]

47. Drury-Stewart, D.; Song, M.; Mohamad, O.; Guo, Y.; Gu, X.; Chen, D.; Wei, L. Highly efficient differentiation of neural precursors from human embryonic stem cells and benefits of transplantation after ischemic stroke in mice. *Stem Cell Res. Ther.* **2013**, *4*, 93. [CrossRef]

48. Yuan, T.; Liao, W.; Feng, N.-H.; Lou, Y.-L.; Niu, X.; Zhang, A.-J.; Wang, Y.; Deng, Z.-F. Human induced pluripotent stem cell-derived neural stem cells survive, migrate, differentiate, and improve neurologic function in a rat model of middle cerebral artery occlusion. *Stem Cell Res. Ther.* **2013**, *4*, 73. [CrossRef]

49. He, B.; Yao, Q.; Liang, Z.; Lin, J.; Xie, Y.; Li, S.; Wu, G.; Yang, Z.; Xu, P. The Dose of Intravenously Transplanted Bone Marrow Stromal Cells Determines the Therapeutic Effect on Vascular Remodeling in a Rat Model of Ischemic Stroke. *Cell Transplant.* **2016**, *25*, 2173–2185. [CrossRef]

50. Zhu, H.; Poon, W.; Liu, Y.; Leung, G.K.-K.; Wong, Y.; Feng, Y.; Ng, S.C.P.; Tsang, K.S.; Sun, D.T.F.; Yeung, D.K.; et al. Phase III Clinical Trial Assessing Safety and Efficacy of Umbilical Cord Blood Mononuclear Cell Transplant Therapy of Chronic Complete Spinal Cord Injury. *Cell Transplant.* **2016**, *25*, 1925–1943. [CrossRef]

51. Nomura, T.; Honmou, O.; Harada, K.; Houkin, K.; Hamada, H.; Kocsis, J.D. IV Infusion of brain-derived neurotrophic factor gene-modified human mesenchymal stem cells protects against injury in a cerebral ischemia model in adult rat. *Neuroscience* **2005**, *136*, 161–169. [CrossRef] [PubMed]

52. Liu, H.; Honmou, O.; Harada, K.; Nakamura, K.; Houkin, K.; Hamada, H.; Kocsis, J.D. Neuroprotection by PlGF gene-modified human mesenchymal stem cells after cerebral ischaemia. *Brain* **2006**, *129*, 2734–2745. [CrossRef] [PubMed]

53. Chen, Y.; Zhu, H.; Liao, J.; Yi, Y.; Wang, G.; Tong, L.; Ge, J. Regulation of naotai recipe on the expression of HIF-1α/VEGF signaling pathway in cerebral ischemia/reperfusion rats. *Zhongguo Zhong xi yi jie he za zhi Zhongguo Zhongxiyi jiehe zazhi = Chin. J. Integr. Tradit. West. Med.* **2014**, *34*, 1225–1230.

54. Wang, X.-L.; Zhao, Y.-S.; Hu, M.-Y.; Sun, Y.-Q.; Chen, Y.-X.; Bi, X.-H. Umbilical cord blood cells regulate endogenous neural stem cell proliferation via hedgehog signaling in hypoxic ischemic neonatal rats. *Brain Res.* **2013**, *1518*, 26–35. [CrossRef]

55. Chernykh, E.R.; Shevela, E.Y.; Starostina, N.M.; Morozov, S.A.; Davydova, M.N.; Menyaeva, E.V.; Ostanin, A.A. Safety and Therapeutic Potential of M2 Macrophages in Stroke Treatment. *Cell Transplant.* **2016**, *25*, 1461–1471. [CrossRef]

56. Hernández, J.; Torres-Espín, A.; Navarro, X. Adult stem cell transplants for spinal cord injury repair: Current state in preclinical research. *Curr. Stem Cell Res. Ther.* **2011**, *6*, 273–287. [CrossRef]

57. Islamov, R.R.; Rizvanov, A.A.; Mukhamedyarov, M.A.; Salafutdinov, I.I.; Garanina, E.E.; Fedotova, V.Y.; Solovyeva, V.V.; Mukhamedshina, Y.O.; Safiullov, Z.Z.; Izmailov, A.A.; et al. Symptomatic improvement, increased life-span and sustained cell homing in amyotrophic lateral sclerosis after transplantation of human umbilical cord blood cells genetically modified with adeno-viral vectors expressing a neuro-protective factor and a neur. *Curr. Gene Ther.* **2015**, *15*, 266–276. [CrossRef]
58. Park, D.-H.; Lee, J.-H.; Borlongan, C.V.; Sanberg, P.R.; Chung, Y.-G.; Cho, T.-H. Transplantation of Umbilical Cord Blood Stem Cells for Treating Spinal Cord Injury. *Stem Cell Rev. Rep.* **2011**, *7*, 181–194. [CrossRef]
59. Islamov, R.R.; Bashirov, F.V.; Sokolov, M.E.; Izmailov, A.A.; Fadeev, F.O.; Markosyan, V.A.; Davleeva, M.A.; Zubkova, O.V.; Smarov, M.M.; Logunov, D.Y.; et al. Gene-modified leucoconcentrate for personalized ex vivo gene therapy in a mini pig model of moderate spinal cord injury. *Neural Regen. Res.* **2020**, *16*, 357–361. [CrossRef]

International Journal of
Molecular Sciences

Article

Lipid Emulsion Improves Functional Recovery in an Animal Model of Stroke

Motomasa Tanioka [1,2], Wyun Kon Park [3], Joohyun Park [2,4], Jong Eun Lee [2,4] and
Bae Hwan Lee [1,2,*]

[1]   Department of Physiology, Yonsei University College of Medicine, Seoul 03722, Korea; hpark@yuhs.ac
[2]   Brain Korea 21 PLUS Project for Medical Science, Yonsei University College of Medicine, Seoul 03722, Korea;
      jhpark922@yuhs.ac (J.P.); jelee@yuhs.ac (J.E.L.)
[3]   Department of Anesthesiology and Pain Medicine, Anesthesia and Pain Research Institute,
      Yonsei University College of Medicine, Seoul 03722, Korea; wkp7ark@yuhs.ac
[4]   Department of Anatomy, Yonsei University College of Medicine, Seoul 03722, Korea
*   Correspondence: bhlee@yuhs.ac; Tel.: +82-2-2228-1711

Received: 1 September 2020; Accepted: 3 October 2020; Published: 6 October 2020

Abstract: Stroke is a life-threatening condition that leads to the death of many people around the world. Reperfusion injury after ischemic stroke is a recurrent problem associated with various surgical procedures that involve the removal of blockages in the brain arteries. Lipid emulsion was recently shown to attenuate ischemic reperfusion injury in the heart and to protect the brain from excitotoxicity. However, investigations on the protective mechanisms of lipid emulsion against ischemia in the brain are still lacking. This study aimed to determine the neuroprotective effects of lipid emulsion in an in vivo rat model of ischemic reperfusion injury through middle cerebral artery occlusion (MCAO). Under sodium pentobarbital anesthesia, rats were subjected to MCAO surgery and were administered with lipid emulsion through intra-arterial injection during reperfusion. The experimental animals were assessed for neurological deficit wherein the brains were extracted at 24 h after reperfusion for triphenyltetrazolium chloride staining, immunoblotting and qPCR. Neuroprotection was found to be dosage-dependent and the rats treated with 20% lipid emulsion had significantly decreased infarction volumes and lower Bederson scores. Phosphorylation of Akt and glycogen synthase kinase 3-β (GSK3-β) were increased in the 20% lipid-emulsion treated group. The Wnt-associated signals showed a marked increase with a concomitant decrease in signals of inflammatory markers in the group treated with 20% lipid emulsion. The protective effects of lipid emulsion and survival-related expression of genes such as Akt, GSK-3β, Wnt1 and β-catenin were reversed by the intra-peritoneal administration of XAV939 through the inhibition of the Wnt/β-catenin signaling pathway. These results suggest that lipid emulsion has neuroprotective effects against ischemic reperfusion injury in the brain through the modulation of the Wnt signaling pathway and may provide potential insights for the development of therapeutic targets.

Keywords: neuroprotection; stroke; ischemia; middle cerebral artery occlusion; reperfusion injury; lipid emulsion; excitotoxicity

---

1. Introduction

Stroke is an acute life-threatening condition in which poor blood perfusion in the brain causes cell death. Life style modification and pharmacological interventions to manage risk factors have been used to lower the incidence of strokes worldwide [1]. Despite such efforts, stroke is still associated with high mortality and morbidity across the globe [2]. Ischemic reperfusion injury in the brain is detrimental and elicits major dysfunctions within the body, such as impairments in movement, cognition and other vital functions [3]. Ischemic reperfusion injury in the brain encompasses abnormal production of oxygen radicals that may exacerbate initial ischemic injuries [4]. Such injury can occur in the current clinical therapies for stroke, such as thrombectomy. These surgeries are effective to ameliorate ischemic damage in the brain by the elimination of blockages [5]. However, it is currently arduous to prevent further oxidative damage caused by reperfusion. Again, secondary damage caused by reperfusion is intractable and could render leaving the reversible areas, such as the penumbra, vulnerable to oxidative damage [6]. Meanwhile, increases in inflammatory markers have been shown to reflect ischemic injuries in clinical settings. Cytokines, such as interleukin-1β (IL-1β), interleukin-6 (IL-6), interleukin-8 (IL-8), tumor necrosis factor-α (TNF-α), ficolin-1 and others, have been reported to increase in relations with the incident of stroke [7–10]. Therefore, decreasing inflammatory activity may be a goal for effective therapies. In previous studies, the administration of recombinant interleukin-1 receptor antagonist in rat models of middle cerebral artery occlusion (MCAO) successfully provided protection against ischemic injury [11,12]. However, the protective effects of interleukin-1 receptor antagonist have not been able to translate to neuroprotection in clinical trials but remain as a potential therapy through continued research [7].

Previous studies have implicated the importance of the canonical Wnt signaling pathway in the case of ischemic reperfusion injury [13,14]. The activation of glycogen synthase kinase-3β (GSK-3β) and the downregulation of protein kinase B (Akt) have been reported as important factors that contribute to the death of neurons [15,16]. The modulation of these proteins has been reported to affect survival of neurons in excitotoxic conditions [16,17]. The activity of GSK-3β leads to the degradation of downstream survival markers such as β-catenin through ubiquitination. Furthermore, canonical Wnt signals, which are lipid-modified glycoproteins, have been reported to inactivate GSK-3β through phosphorylation and thereby promote cell survival [18]. A recent study reported that motor exercise stimulated the canonical Wnt/β-catenin pathway for recovering from focal cerebral ischemic reperfusion injury in juvenile rats. Motor activity regulated canonical Wnt/β-catenin pathway and promoted neurogenesis and myelin repair [19]. Another study reported that electroacupuncture in multiple acupoints of a paralyzed limb stimulated the proliferation of neural progenitor cells through the activation of the Wnt/β-catenin signaling pathway and suppression of GSK-3β [20].

Lipid emulsion (LE) was approved for clinical use as a component of parenteral nutrition in 1962 [21]. LE is composed of 20% soybean oil, 1.2% egg yolk phospholipids, 2.25% glycerin and water for injection (Intralipid™ 20%, Fresenius Kabi, Uppsala, Sweden). The major fatty acids that constitute soybean oil are linoleic acid (44–62%), oleic acid (19–30%), palmitic acid (7–14%), linolenic acid (4–11%) and stearic acid (1.4–5.5%). In 1998, Weinberg et al. [22] shed light on the cardioprotective properties of LE against local anesthetic toxicity. Clinical reports have utilized the administration of LE for resuscitation from cardiac toxicity induced by lidocaine [23], ropivacaine [24] or bupivacaine [25]. The mechanism of protection of cardiac cells from excitotoxic conditions by LE depends on the phosphorylation of Akt [26] and GSK-3β [27] that contributes to cell survival. LE also has been reported to provide cardioprotection against ischemic reperfusion injury in the isolated rat heart [28]. While research into the utilization of LE against various cardiotoxic conditions is being actively conducted, investigations delineating the protective properties of LE in the central nervous system are still lacking. Neurons and cardiomyocytes share key similarities such as excitability and conductibility, which lead to their vulnerability to excitotoxic conditions. Relevant reports have claimed that distinct cardioprotective properties of LE might provide therapeutic targets or diagnostic tools for neuroprotection [29,30]. Our recent study has reported the protective effects of LE against kainic acid-induced excitotoxicity when administered directly into the brain, thereby revealing the potential neuroprotective aspects of LE [31]. However, the neuroprotective effects of LE against ischemic reperfusion injury in the brain have not been clearly elucidated.

Therefore, the present study investigated the neuroprotective effects of LE against ischemic reperfusion injury and elucidates the mechanism involved in the protection process. We examined the neuroprotective roles of LE in an in vivo rat model of the middle cerebral artery occlusion (MCAO) and reperfusion. We assessed the neurological deficits using the modified Bederson score [32] and extracted the brain to measure the severity of infarctions in experimental groups. We assessed the changes in protein and mRNA expression of distinct genes related to cell survival, Wnt/β-catenin signaling pathway and inflammation. We verified that XAV939, a Wnt/β-catenin signaling pathway inhibitor, reversed the protective effects. Based on our results, we propose that LE provides neuroprotection against ischemic reperfusion injury in the brain by regulating the Wnt/β-catenin signaling pathway.

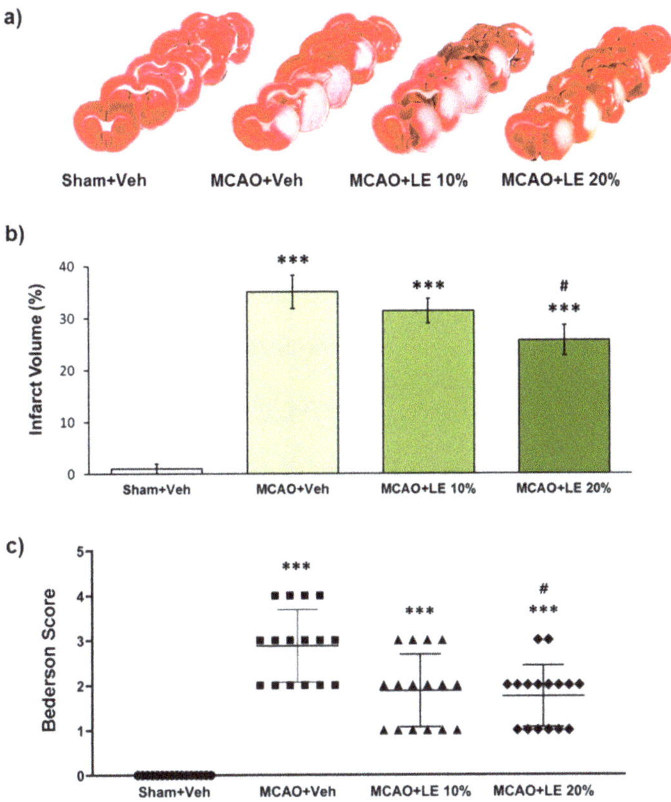

Figure 1. Neuroprotective effects of lipid emulsion (LE) or vehicle (Veh) after the middle cerebral artery occlusion (MCAO) and reperfusion injury. (**a**) Triphenyltetrazolium chloride (TTC)-stained brain slices for infarction measurement. Decrease in infarction volumes were observed in MCAO+LE 10% and MCAO+LE 20% groups. Sham group did not show ischemic reperfusion injury; (**b**) Measurement of infarction volume from TTC staining. The MCAO+Veh group increased significantly in infarction volume compared to Sham+Veh group. Both MCAO+LE 10% and MCAO+LE 20% decreased in infarction volume but only the MCAO+LE 20% group had significant difference compared to the MCAO+Veh group; (**c**) Bederson scores of the experimental groups. The MCAO+Veh group increased in Bederson score significantly compared to Sham+Veh group. Both MCAO+LE 10% and MCAO+LE 20% groups decreased in Bederson scores compared to MCAO+Veh group; however, only the MCAO+LE 20% group decreased significantly. Data are presented as mean ± standard error of the mean (SEM); $n = 16$ for each group; *** $p < 0.001$ vs. Sham+Veh, # $p < 0.05$ vs. MCAO+Veh. Statistical analysis for the measurement of infarction volume was performed using one-way analysis of variance (ANOVA) followed by Tukey's multiple comparison test. Statistical analysis for Bederson scores was performed using Kruskal-Wallis non-parametric test followed by Dunn's post hoc test.

2. Results

2.1. Dosage-Dependent Reduction in Infarction and Behavior by LE

The severity of infarction was measured by triphenyltetrazolium chloride (TTC) staining and neurological deficit assessment. The severity of infarction was visible through the unstained areas of the brain in Figure 1a. The Sham+Vehicle (Veh) group did not experience notable infarction. The MCAO+Veh group suffered an injury of approximately 35% of the left hemisphere. The MCAO+LE 10% group exhibited decrease in infarction volume to about 31%, which was not significant compared to MCAO+Veh group ($p > 0.05$, one-way ANOVA). There was a significant decrease in infarction to about 26% in the MCAO+LE 20% group compared to the MCAO+Veh group ($p < 0.05$, one-way ANOVA followed by Tukey's multiple comparison test) (Figure 1b). In behavioral test, all experimental groups suffered a certain degree of behavioral deficit except for the Sham+Veh group. The MCAO+Veh group had an average Bederson score of 3. The MCAO+LE 10% group scored lower in the behavior test and did not differ significantly compared to the MCAO+Veh group ($p > 0.05$, Kruskal-Wallis non-parametric test). The administration of LE 20% significantly decreased the average Bederson score to approximately 2 ($p < 0.05$, Kruskal-Wallis non-parametric test followed by Dunn's post hoc test). The majority of the MCAO+LE 20% group achieved a Bederson score of 2 or under, while the majority MCAO+Veh group recorded between 2 and 4. (Figure 1c).

2.2. Wnt/β-Catenin-Dependent Reduction in Infarction and Behavior by LE

Experimental groups were administered with intra-peritoneal (i.p.) injection of XAV939 to inhibit the Wnt/β-catenin signaling pathway induced by LE. The control group was administered with DMSO instead of XAV939. The severity of infarction was visible through the unstained areas of the brain as depicted in Figure 2a. The DMSO+Sham+Veh group did not experience notable infarction. Approximately 33% of the left hemisphere of the DMSO+MCAO+Veh group suffered an injury, while a significant decrease in infarction to about 24% was observed in the DMSO+MCAO+LE 20% group ($p < 0.05$, one-way ANOVA followed by Tukey's multiple comparison test). The XAV939+MCAO+LE 20% failed to protect the brain with approximately 29% infarction volume, which was not significantly different from the DMSO+MCAO+Veh group ($p > 0.05$, one-way ANOVA followed by Tukey's multiple comparison test) (Figure 2b). All experimental groups suffered a certain degree of infarction except for the DMSO+Sham+Veh group. The DMSO+Sham+Veh group did not experience notable neurological deficits. The DMSO+MCAO+Veh group recorded an average Bederson score of 3. The administration of LE 20% decreased the Bederson score to approximately 2 ($p < 0.05$, Kruskal-Wallis non-parametric test followed by Dunn's post hoc test). The XAV939+MCAO+Veh group achieved an average Bederson score of 3, which was similar to the DMSO+MCAO+Veh group ($p > 0.05$, Kruskal-Wallis non-parametric test) (Figure 2c). The i.p. injection of DMSO did not induce significant differences in the experimental group.

2.3. LE Dosage-Dependent Alleviation of Ischemic Reperfusion Injury through the Wnt/β-Catenin Signaling Pathway and Reduction of Inflammatory Protein Markers

The administration of LE affected the protein expression of survival and inflammation-related signals (Figure 3a). The expression of total Akt did not differ among the experimental groups (Figure 3b). The MCAO+Veh group showed significant decrease in phosphorylation of Akt levels (pAkt) compared to the Sham+Veh group ($p < 0.05$, one-way ANOVA followed by Tukey's multiple comparison test). The MCAO+LE 10% group also exhibited an increase in pAkt levels. However, such an increase was not statistically significant compared to the MCAO+Veh group ($p > 0.05$, one-way ANOVA). The significantly elevated level of pAkt in MCAO+LE 20% group compared to the MCAO+Veh group indicated the increased survival of neurons after ischemic reperfusion injury ($p < 0.05$, one-way ANOVA followed by Tukey's multiple comparison test) (Figure 3c). The ratio of pAkt and Akt was markedly decreased in the MCAO+Veh group compared to the Sham+Veh group ($p < 0.05$, one way

ANOVA followed by Tukey's multiple comparison test). pAkt/Akt increased in expression in a dosage dependent manner in MCAO+LE 10% and MCAO+LE 20% groups; but only the MCAO+LE 20% group was significantly different from the MCAO+Veh group ($p < 0.05$, one-way ANOVA followed by Tukey's multiple comparison test) (Figure 3d).

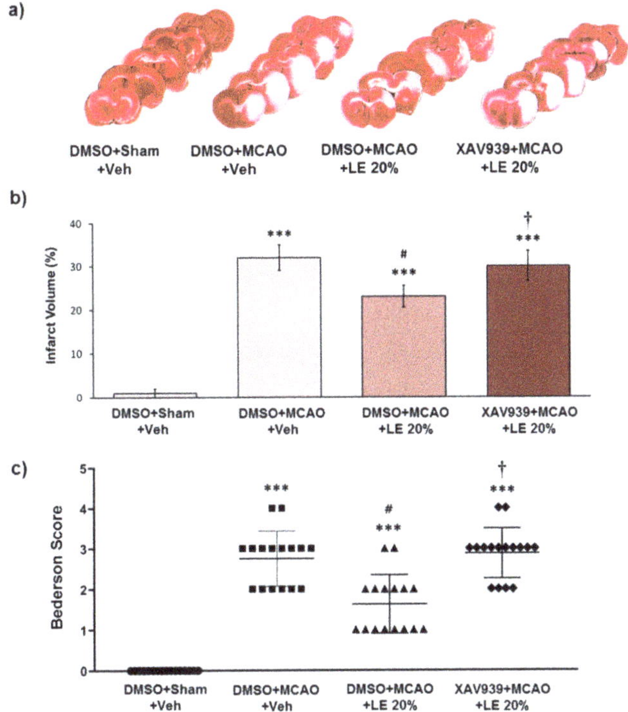

Figure 2. Neuroprotective effects of LE or vehicle on the MCAO and reperfusion injury after the administration of DMSO or XAV939. (**a**) TTC-stained brain slices for infarction measurement. Decrease in infarction volume was observed in the DMSO+MCAO+LE 20% group but not in the XAV939+MCAO+LE 20% group. Sham group (DMSO+Sham+Veh) did not experience ischemic reperfusion injury; (**b**) Measurement of infarction volume by TTC staining. The DMSO+MCAO+Veh group exhibited significant increase in infarction volume compared to the DMSO+Sham+Veh group. DMSO+MCAO+LE 20% group decreased significantly with respect to infarction volume when compared to the DMSO+MCAO+Veh. The XAV939+MCAO+LE 20% group had significantly increased infraction volume compared to the DMSO+MCAO+LE 20% group; (**c**) Bederson scores of experimental groups. The DMSO+MCAO+Veh group significantly increased in Bederson scores compared to the DMSO+Sham+Veh group. The DMSO+MCAO+LE 20% group showed significant decrease in Bederson scores compared to the DMSO+MCAO+Veh groups. The XAV939+MCAO+LE 20% group significantly increased in Bederscon scores compared to the DMSO+MCAO+LE 20% group. Data are presented as mean ± standard error of the mean (SEM); $n = 16$ for each group; *** $p < 0.001$ vs. DMSO+Sham+Veh, # $p < 0.05$ vs. DMSO+MCAO+Veh, † $p < 0.05$ vs. DMSO+MCAO+LE 20%. Statistical analysis for the measurement of infarction volume was performed using one-way ANOVA followed by Tukey's multiple comparison test. Statistical analysis for Bederson scores was performed using Kruskal-Wallis non-parametric test followed by Dunn's post hoc test.

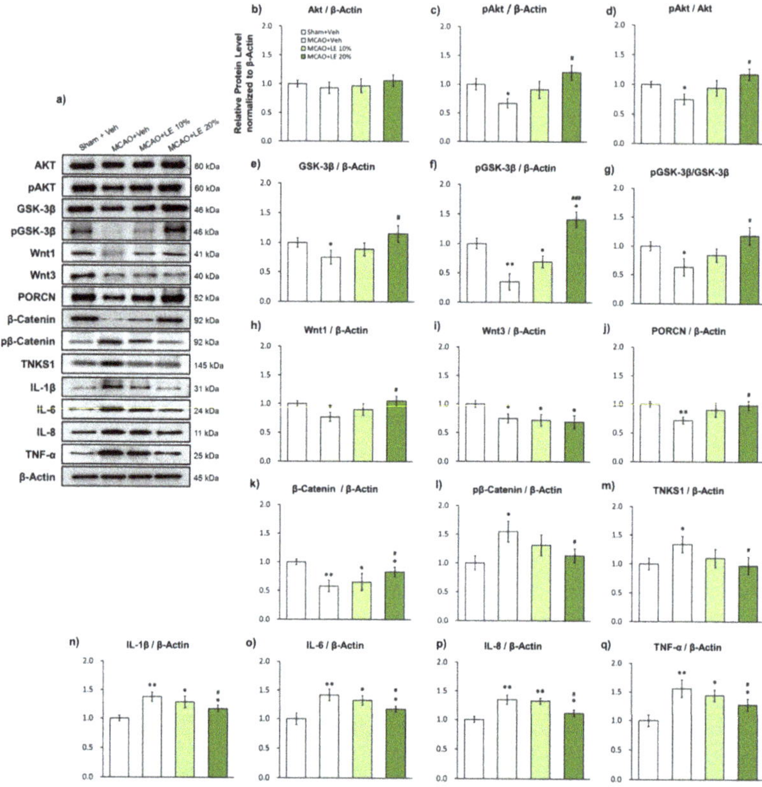

Figure 3. Effects of LE or vehicle on protein expressions after MCAO and reperfusion injury. (**a**) Representative Western blots indicating the expression of specific proteins in the penumbra region of the left hemisphere; (**b–d**) Expression and phosphorylation of Akt in the experimental groups. pAkt levels in MCAO+Veh group decreased significantly compared to the Sham+Veh group. pAkt was significantly increased in the MCAO+LE 20% group compared to the MCAO+Veh group; (**e–g**) Expression and phosphorylation levels of GSK-3β (pGSK-3β) in the experimental groups. pGSK-3β and GSK-3β levels in the MCAO+Veh group was significantly lower than the Sham+Veh group. The MCAO+LE 20% had significantly increased pGSK-3β and GSK-3β levels compared to the MCAO+Veh group; (**h–m**) Wnt signal-related protein expressions of experimental groups. Decreased levels of Wnt1, Wnt3, PORCN and β-catenin were observed in the MCAO+Veh group compared to the Sham+Veh group. The MCAO+LE 20% group had significantly increased protein levels of Wnt1, PORCN and β-catenin compared to MCAO+Veh group. Increased levels of pβ-catenin and tankyrase 1 were observed in the MCAO+Veh group compared to the Sham+Veh group. The MCAO+LE 20% group had significantly decreased levels of pβ-catenin and tankyrase 1 compared to the MCAO+Veh group; (**n–q**) Inflammatory protein expressions of experimental groups. Significantly increased expressions of inflammatory markers were observed in MCAO-injured groups compared to the Sham+Veh group. The MCAO+LE 20% group had significantly decreased inflammatory protein expression levels compared to the MCAO+Veh group. Data are presented as mean ± standard error of the mean (SEM); $n = 8$ for each group; * $p < 0.05$, ** $p < 0.01$ vs. Sham+Veh, # $p < 0.05$, ### $p < 0.001$ vs. MCAO+Veh, one-way ANOVA followed by Tukey's multiple comparison test.

Significant decrease in total GSK-3β expression was observed in the MCAO+Veh group compared to the Sham+Veh group ($p < 0.05$, one-way ANOVA followed by Tukey's multiple comparison test). The MCAO+LE 10% group showed substantial increase in GSK-3β levels but this increment was not statistically significant compared to the MCAO+Veh group ($p > 0.05$, one-way ANOVA). The total GSK-3β expression of the MCAO+LE 20% was also significantly increased compared to the MCAO+Veh group ($p < 0.05$, one-way ANOVA followed by Tukey's multiple comparison test) (Figure 3e). The phosphorylation of GSK-3β (pGSK-3β) was significantly decreased in the MCAO+Veh group compared to the Sham+Veh group ($p < 0.01$, one-way ANOVA), which might be indicating a diminished Wnt activity. The phosphorylation of GSK-3β was significantly increased in the MCAO+LE 10% ($p < 0.05$, one-way ANOVA followed by Tukey's multiple comparison test) and MCAO+LE 20% ($p < 0.001$, one-way ANOVA followed by Tukey's multiple comparison test) group when compared to the MCAO+Veh group. There was also a significant increase in pGSK-3β in the MCAO+LE 20% group compared to the Sham+Veh group ($p < 0.05$, one-way ANOVA followed by Tukey's multiple comparison test), thereby indicating an increased activity of the Wnt/β-catenin signaling pathway (Figure 3f). The pGSK-3β/GSK-3β activity exhibited marked decrease in the MCAO+Veh group compared to the Sham+Veh group ($p < 0.05$, one-way ANOVA followed by Tukey's multiple comparison test). The pGSK-3β/GSK-3β expression for MCAO+LE 10% did not differ significantly compared to the MCAO+Veh group ($p > 0.05$, one-way ANOVA). pGSK-3β/GSK-3β activity significantly increased in the MCAO+LE 20% group compared to the MCAO+Veh group ($p < 0.05$, one-way ANOVA followed by Tukey's multiple comparison test) (Figure 3g).

Wnt1, a canonical Wnt signal and upstream marker of GSK-3β, was significantly decreased in the MCAO+Veh group compared to the Sham+Veh group ($p < 0.05$, one-way ANOVA followed by Tukey's multiple comparison test) and significantly increased in the MCAO+LE 20% group compared to the MCAO+Veh group ($p < 0.05$, one-way ANOVA followed by Tukey's multiple comparison test) (Figure 3h). All experimental groups ($p < 0.05$, one-way ANOVA followed by Tukey's multiple comparison test), excluding the Sham+Veh group, displayed robust decrease in the neurogenesis marker, Wnt3. Although Wnt3 is one of canonical Wnt signals, it does not seem to affect the Wnt/β-catenin signaling pathway induced by LE (Figure 3i). Porcupine (PORCN), a key regulator of Wnt proteins [33], decreased significantly in the MCAO+Veh group compared to the Sham+Veh group ($p < 0.05$, one-way ANOVA followed by Tukey's multiple comparison test). PORCN expression increased in the MCAO+LE 20% group compared to the MCAO+Veh group, implying elevated activity in the modulation of Wnt proteins (Figure 3j). The downstream survival marker of GSK-3β, β-catenin, was decreased in all MCAO injury groups compared to the Sham+Veh group due to infarction. However, β-catenin was significantly preserved in the MCAO+LE 20% group compared to the MCAO+Veh group ($p < 0.05$, one-way ANOVA followed by Tukey's multiple comparison test) (Figure 3k). Phosphorylation of β-catenin (pβ-catenin) was increased significantly in the MCAO+Veh group compared to the Sham+Veh group ($p < 0.05$, one-way ANOVA followed by Tukey's multiple comparison test), indicating an elevated level of β-catenin degradation. The MCAO+LE 20% group had a significantly lower expression level of pβ-catenin compared to the MCAO+Veh group ($p < 0.05$, one-way ANOVA followed by Tukey's multiple comparison test), which supported the survival of cells (Figure 3l). The expression level of tankyrase 1 was significantly increased in the MCAO+Veh group compared to the Sham+Veh group ($p < 0.05$, one-way ANOVA followed by Tukey's multiple comparison test), indicating the accumulation of tankyrase 1 through increased axis inhibition protein (AXIN) stabilization for β-catenin degradation. The MCAO+LE 20% group had a significantly lower expression level of tankyrase 1 compared to the MCAO+Veh group ($p < 0.05$, one-way ANOVA followed by Tukey's multiple comparison test) (Figure 3m).

Inflammatory protein markers of ischemic reperfusion damage, IL-1β, IL-6, IL-8 and TNF-α [34,35], increased significantly for all MCAO-injured groups when compared to the Sham+Veh group. However, attenuated levels of inflammatory markers were observed in the MCAO+LE 20% group compared to the MCAO+Veh group ($p < 0.05$, one-way ANOVA followed by Tukey's multiple comparison test), indicating the reduction of ischemic reperfusion injury (Figure 3n–q).

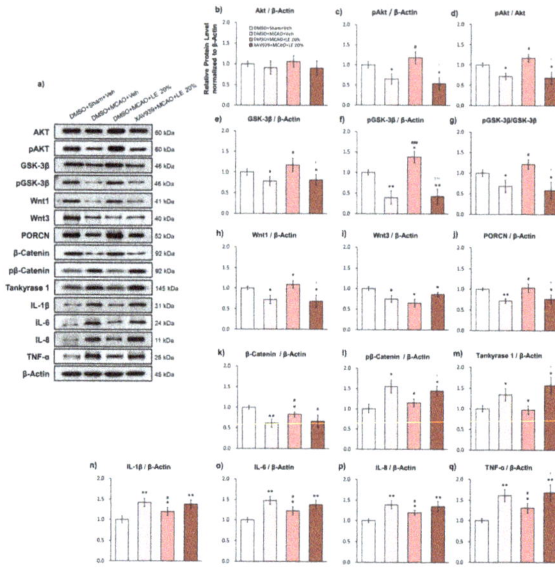

Figure 4. Effects of LE or vehicle on protein expression on the MCAO and reperfusion injury after the administration of DMSO or XAV939. (**a**) Representative Western blots of proteins in the penumbra region of the left hemisphere; (**b–d**) Levels of Akt and pAkt in the experimental groups. The DMSO+MCAO+Veh group and XAV939+MCAO+LE 20% group decreased significantly in pAkt level compared to DMSO+Sham+Veh group. pAkt was significantly increased in the DMSO+MCAO+LE 20% group compared to the DMSO+MCAO+Veh group and XAV939+MCAO+LE 20% group; (**e–g**) Levels of GSK-3β and pGSK-3β in the experimental groups. GSK-3β and pGSK-3β levels of DMSO+MCAO+Veh group and XAV939+MCAO+LE 20% group were significantly lower than the DMSO+Sham+Veh group. The DMSO+MCAO+LE 20% had significantly increased pGSK-3β and GSK-3β levels compared to the DMSO+MCAO+Veh and XAV939+MCAO+LE 20% groups; (**h–m**) Wnt signal-related protein expressions of experimental groups. Significantly decreased levels of Wnt1, Wnt3, PORCN and β-catenin were observed in the DMSO+MCAO+Veh group compared to the DMSO+Sham+Veh group. The DMSO+MCAO+LE 20% group had significantly increased protein levels of Wnt1 and PORCN compared to the DMSO+MCAO+Veh group. The XAV939+MCAO+LE 20% group had decreased expression levels of Wnt1 and PORCN compared to the DMSO+MCAO+LE 20% group. The β–catenin expression level in the DMSO+MCAO+LE 20% and XAV939+MCAO+LE 20% groups did not differ significantly. There was a significant increase in pβ–catenin and tankyrase 1 in the DMSO+MCAO+Veh compared to the DMSO+Sham+Veh group. Pβ–catenin and tankyrase1 was significantly decreased in the DMSO+MCAO+LE 20% compared to the DMSO+MCAO+Veh group. The XAV939+MCAO+LE 20% group had significantly increased pβ–catenin and tankyrase 1 expression levels compared to the DMSO+MCAO+LE 20% group; (**n–q**) Inflammatory protein expressions in the experimental groups. Significantly increased expressions of inflammatory markers were observed in the MCAO-injured groups compared to the DMSO+Sham+Veh group. The DMSO+MCAO+LE 20% group had significantly decreased inflammatory protein expression levels compared to the DMSO+MCAO+Veh group. Significant decrease in inflammatory protein expressions were observed in the XAV939+MCAO+LE 20% group compared to the DMSO+Sham+Veh group. TNF-α expression level was increased significantly in the XAV939+MCAO+LE 20% group compared to the DMSO+MCAO+LE 20% group. Data are presented as mean ± standard error of the mean (SEM); $n = 8$ for each group; * $p < 0.05$, ** $p < 0.01$ vs. DMSO+Sham+Veh, # $p < 0.05$, ### $p < 0.001$ vs. DMSO+MCAO+Veh, † $p < 0.05$, ††† $p < 0.001$ vs. DMSO+MCAO+LE 20%, one-way ANOVA followed by Tukey's multiple comparison test.

2.4. Wnt/β-Catenin-Dependent Alleviation of Ischemic Reperfusion Injury and Reversal of Protection-Related Proteins

The administration of XAV939 inhibited the activity of the Wnt/β-catenin signaling pathway, which was reflected in the protein expression (Figure 4a) of distinct genes. The expression of total Akt did not differ among the experimental groups (Figure 4b). The DMSO+MCAO+Veh group and XAV939+MCAO+LE 20% group significantly showed substantial decrease in pAkt levels compared to the DMSO+Sham+Veh group ($p < 0.05$, one-way ANOVA followed by Tukey's multiple comparison test). There was significant elevation in the level of pAkt in DMSO+MCAO+LE 20% group compared to the DMSO+MCAO+Veh group ($p < 0.05$, one-way ANOVA followed by Tukey's multiple comparison test) and XAV939+MCAO+LE 20% group ($p < 0.01$, one-way ANOVA followed by Tukey's multiple comparison test) (Figure 4c). The pAkt/Akt expression was decreased significantly in the DMSO+MCAO+Veh and XAV939+MCAO+LE 20% groups compared to the DMSO+Sham+Veh group ($p < 0.05$, one-way ANOVA followed by Tukey's multiple comparison test). The expression of pAkt/Akt in the DMSO+MCAO+LE 20% group was significantly increased compared to DMSO+MCAO+Veh and XAV939+MCAO+LE 20% groups ($p < 0.05$, one-way ANOVA followed by Tukey's multiple comparison test) (Figure 4d).

Significant decrease in total GSK-3β expression was observed in the DMSO+MCAO+Veh and XAV939+MCAO+LE 20% groups when compared to the DMSO+Sham+Veh group ($p < 0.05$, one-way ANOVA followed by Tukey's multiple comparison test). Total GSK-3β expression of the DMSO+MCAO+LE 20% group was also significantly increased compared to the DMSO+MCAO+Veh and XAV939+MCAO+LE 20% groups ($p < 0.05$, one-way ANOVA followed by Tukey's multiple comparison test) (Figure 4e). The level of pGSK-3β was significantly decreased in the DMSO+MCAO+Veh and XAV939+MCAO+LE 20% groups compared to the DMSO+Sham+Veh group ($p < 0.01$, one-way ANOVA followed by Tukey's multiple comparison test), which might be indicative of a compromised Wnt activity. The phosphorylation of GSK-3β was significantly increased in the DMSO+MCAO+LE 20% group compared to the DMSO+Sham+Veh group ($p < 0.05$, one-way ANOVA followed by Tukey's multiple comparison test). There was also a significant increase in the pGSK-3β levels in the DMSO+MCAO+LE 20% group compared to the DMSO+MCAO+Veh and XAV939+MCAO+LE 20% groups ($p < 0.001$, one-way ANOVA followed by Tukey's multiple comparison test) (Figure 4f). The pGSK-3β/GSK-3β activity was significantly decreased in the DMSO+MCAO+Veh and XAV939+MCAO+LE 20% groups when compared to the DMSO+Sham+Veh group ($p < 0.05$, one-way ANOVA followed by Tukey's multiple comparison test) and significantly increased in the DMSO+MCAO+LE 20% group compared to the DMSO+MCAO+Veh and XAV939+MCAO+LE 20% groups ($p < 0.05$, one-way ANOVA followed by Tukey's multiple comparison test). The administration of XAV939 effectively decreased the phosphorylation GSK-3β induced by LE (Figure 4g).

Wnt1 was significantly decreased in the DMSO+MCAO+Veh and XAV939+MCAO+LE 20% groups compared to the DMSO+Sham+Veh group ($p < 0.05$, one-way ANOVA followed by Tukey's multiple comparison test) and significantly increased in the DMSO+MCAO+LE 20% group compared to the DMSO+MCAO+Veh and XAV939+MCAO+LE 20% groups ($p < 0.05$, one-way ANOVA followed by Tukey's multiple comparison test) (Figure 4h). All experimental groups, excluding the DMSO+Sham+Veh group, showed steep decrease ($p < 0.05$, one-way ANOVA followed by Tukey's multiple comparison test) in Wnt3 (Figure 4i). PORCN decreased significantly in the DMSO+MCAO+Veh group ($p < 0.01$, one-way ANOVA followed by Tukey's multiple comparison test) and XAV939+MCAO+LE 20% group ($p < 0.05$, one-way ANOVA followed by Tukey's multiple comparison test) compared to the DMSO+Sham+Veh group. PORCN expression increased in expression in the DMSO+MCAO+LE 20% group compared to the DMSO+MCAO+Veh and XAV939+MCAO+LE 20% groups ($p < 0.05$, one-way ANOVA followed by Tukey's multiple comparison test) (Figure 4j). β-catenin was decreased in all MCAO injury groups compared to the DMSO+Sham+Veh group due to infarction. Significantly attenuated level of β-catenin was also observed in the XAV939+MCAO+LE 20% group, while DMSO+MCAO+LE 20% group were preserved significantly compared to DMSO+MCAO+Veh group ($p < 0.05$, one-way ANOVA followed by Tukey's multiple comparison test). Average β-catenin expression decreased in the XAV939+MCAO+LE 20% group compared to the DMSO+MCAO+LE 20% group but not to a significant level ($p > 0.05$ one-way ANOVA) (Figure 4k). pβ-Catenin was increased in the DMSO+MCAO+Veh group compared to the DMSO+Sham+Veh group ($p < 0.05$, one-way ANOVA followed by Tukey's multiple comparison test), indicating lowered cellular survival. The level of pβ-catenin was significantly decreased in the DMSO+MCAO+LE 20% group compared to the DMSO+MCAO+Veh group ($p < 0.05$, one-way ANOVA followed by Tukey's multiple comparison test). The pβ-catenin expression level was significantly increased in the XAV939+MCAO+LE 20% group compared to the DMSO+MCAO+LE 20% group ($p < 0.05$, one-way ANOVA followed by Tukey's multiple comparison test) (Figure 4l). The accumulation of tankyrase 1 was significant in the DMSO+MCAO+Veh compared to DMSO+Sham+Veh ($p < 0.05$, one-way ANOVA followed by Tukey's multiple comparison test). The tankyrase 1 expression of DMSO+MCAO+LE 20% group significantly decreased compared to the DMSO+MCAO+Veh group ($p < 0.05$, one-way ANOVA followed by Tukey's multiple comparison test). The XAV939+MCAO+LE 20% group had significantly higher levels of tankyrase 1 expression levels compared to the DMSO+MCAO+LE 20% group ($p < 0.05$, one-way ANOVA followed by Tukey's multiple comparison test). The administration of XAV939 successfully inhibited tankyrase 1 activity which increased degradation of β-catenin (Figure 4m).

Inflammatory protein markers of ischemic reperfusion damage, IL-1β, IL-6, IL-8 and TNF-α, also significantly increased for all MCAO-injured groups compared to the DMSO+Sham+Veh group. Attenuated levels of inflammatory markers were observed in the DMSO+MCAO+LE 20% group compared to the DMSO+MCAO+Veh group ($p < 0.05$, one-way ANOVA followed by Tukey's multiple comparison test). The expression levels of the XAV939+MCAO+LE 20% group were not significantly different from the DMSO+MCAO+Veh group ($p > 0.05$, one-way ANOVA) The TNF-α expression level of the XAV939+MCAO+LE 20% group increased significantly compared to the DMSO+MCAO+LE 20% group ($p < 0.05$, one-way ANOVA followed by Tukey's multiple comparison test) (Figure 4n–q).

2.5. LE Dosage-Dependent mRNA Expression Against Ischemic Reperfusion Injury

According to the results of qPCR, Wnt1 can be implicated as one of the main regulators for neuroprotection. in the MCAO+Veh group, Wnt1 expression level was approximately 0.4 folds compared to the Sham+Veh group ($p < 0.05$, one-way ANOVA followed by Tukey's multiple comparison test). The mRNA expression of Wnt1 signals of MCAO+LE 20% was upregulated by approximately 2.4 folds compared to the Sham+Veh group ($p < 0.01$, one-way ANOVA followed by Tukey's multiple comparison test) and 4 folds compared to the MCAO+Veh group ($p < 0.001$, one-way ANOVA followed by Tukey's multiple comparison test) (Figure 5a). Wnt3 expression was attenuated in all MCAO-injured groups compared to the Sham+Veh group. The MCAO+LE 10% ($p < 0.05$, one-way ANOVA followed by Tukey's multiple comparison test) and MCAO+LE 20% groups' ($p < 0.001$, one-way ANOVA followed by Tukey's multiple comparison test) Wnt3 expression levels were significantly decreased compared to the MCAO+Veh group (Figure 5b). Mki67, a cell proliferation marker, increased in all MCAO-injured groups compared to the Sham+Veh group. The Mki67 expression level increased significantly in the MCAO+LE 20% group compared to the MCAO+Veh group ($p < 0.05$, one-way ANOVA followed by Tukey's multiple comparison test), which may have been affected by the elevated expression level of Wnt1 (Figure 5c). The Wnt regulator, Porcn, decreased significantly in the MCAO+Veh group compared to the Sham+Veh group ($p < 0.05$, one-way ANOVA followed by Tukey's multiple comparison test). Significant increase of Porcn was observed in the MCAO+LE 20% group compared to the MCAO+Veh group ($p < 0.05$, one-way ANOVA followed by Tukey's multiple comparison test), which may be induced by the increase in Wnt1 activity (Figure 5d). Inflammatory markers, IL-1β, IL-6, IL-8 and TNF-α, significantly increased in MCAO-injured groups compared to the Sham+Veh group. Significantly lower inflammatory mRNA markers were observed in the MCAO+LE 20%

group compared to the MCAO+Veh group ($p < 0.05$, one-way ANOVA followed by Tukey's multiple comparison test) (Figure 5e–h).

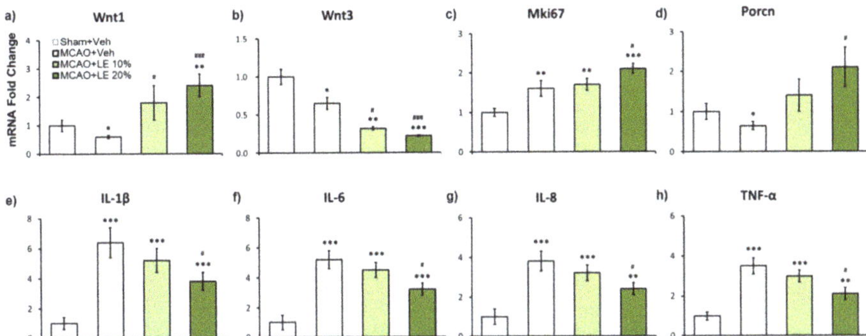

Figure 5. Effects of LE or vehicle on mRNA expression after the MCAO and reperfusion injury. (**a,b**) Wnt expressions in experimental groups. The Wnt1 mRNA expression of the MCAO+Veh group was significantly decreased compared to the Sham+Veh group. Significantly increased expression of Wnt1 was expressed in the MCAO+LE 10% and MCAO+LE 20% groups compared to the MCAO+Veh group. Wnt3 expressions were significantly lower in MCAO injury groups compared to the Sham+Veh group. Significantly decreased Wnt3 expressions were observed in the MCAO+LE 10% and MCAO+LE 20% groups compared to the MCAO+Veh group; (**c**) Mki67 expression was increased in all MCAO-injury groups compared to the Sham+Veh group. Significant increase in Mki67 expression was observed in the MCAO+LE 20% group compared to the MCAO+Veh group; (**d**) Porcn expression was significantly decreased in the MCAO+Veh group compared to the Sham+Veh group. Significant increase of Porcn was observed in the MCAO+LE 20% group compared to the MCAO+Veh group; (**e–h**) mRNA expression of inflammatory markers. MCAO-injury groups expressed significantly increased levels of inflammatory markers compared to the Sham+Veh group. Significantly decreased inflammatory expression levels were observed in the MCAO+LE 20% group compared to the MCAO+Veh group. Data are presented as mean $\pm$ standard error of the mean (SEM); $n = 8$ for each group; * $p < 0.05$, ** $p < 0.01$, *** $p < 0.001$ vs. Sham+Veh, # $p < 0.05$, ### $p < 0.001$ vs. MCAO+Veh, one-way ANOVA followed by Tukey's multiple comparison test. Wnt subfamily mRNA expressions are shown in Supplementary Materials S1.

2.6. Wnt-Dependent mRNA Expression of LE Against Ischemic Reperfusion Injury

XAV939 injection inhibited Wnt activity in mRNA levels to reverse neuroprotection. in the DMSO+MCAO+Veh group ($p < 0.05$, one-way ANOVA followed by Tukey's multiple comparison test) and XAV939+MCAO+LE 20% group, Wnt1 mRNA expression levels were approximately 0.5 folds and 0.6 folds compared to the DMSO+Sham+Veh group, respectively. Expression of Wnt1 signals in the DMSO+MCAO+LE 20% was upregulated by approximately 2.2 folds compared to the DMSO+Sham+Veh group ($p < 0.01$, one-way ANOVA followed by Tukey's multiple comparison test) and 4.2 folds compared to the DMSO+MCAO+Veh group ($p < 0.001$, one-way ANOVA followed by Tukey's multiple comparison test). There was a significant decrease in Wnt1 expression in the XAV939+MCAO+LE 20% group compared to the DMSO+MCAO+LE 20% group ($p < 0.01$, one-way ANOVA followed by Tukey's multiple comparison test) (Figure 6a). Wnt3 expression was attenuated in all MCAO-injured groups compared to the DMSO+Sham+Veh group. The DMSO+MCAO+LE 20% group's Wnt3 expression was significantly decreased compared to the DMSO+MCAO+Veh group ($p < 0.001$, one-way ANOVA followed by Tukey's multiple comparison test). The XAV939+MCAO+LE 20% group was also showed significant decreased in Wnt3 expression level compared to the DMSO+Sham+Veh group ($p < 0.01$, one-way ANOVA followed by Tukey's multiple comparison test) but displayed marked increase when compared to the DMSO+MCAO+LE

20% group ($p < 0.001$, one-way ANOVA followed by Tukey's multiple comparison test) (Figure 6b). Mki67 levels increased in all MCAO-injured groups compared to the DMSO+Sham+Veh group. in the DMSO+MCAO+LE 20% group, Mki67 expression level increased significantly compared to the DMSO+MCAO+Veh group ($p < 0.05$, one-way ANOVA followed by Tukey's multiple comparison test); while in the XAV939+MCAO+LE 20% group, it was significantly downregulated compared to the DMSO+MCAO+LE 20% group ($p < 0.05$, one-way ANOVA followed by Tukey's multiple comparison test) (Figure 6c). Porcn decreased significantly in the DMSO+MCAO+Veh group ($p < 0.05$, one-way ANOVA followed by Tukey's multiple comparison test) and XAV939+MCAO+LE 20% group compared to the DMSO+Sham+Veh group. Significant increase of Porcn was observed in the DMSO+MCAO+LE 20% group when compared to the DMSO+MCAO+Veh group ($p < 0.05$, one-way ANOVA followed by Tukey's multiple comparison test). The XAV939+MCAO+LE 20% group decreased in Porcn compared to the DMSO+Sham+Veh group but did not differ significantly. The XAV939+MCAO+LE 20% group expression levels of Porcn were significantly decreased compared to the DMSO+MCAO+LE 20% group ($p < 0.05$, one-way ANOVA followed by Tukey's multiple comparison test) (Figure 6d). Inflammatory markers, IL-1β, IL-6, IL-8 and TNF-α, significantly increased in MCAO-injured groups compared to the DMSO+Sham+Veh group. Significantly lower inflammatory mRNA markers were observed in the DMSO+MCAO+LE 20% group compared to the DMSO+MCAO+Veh group ($p < 0.05$, one-way ANOVA). XAV939 injection significantly reversed attenuated levels of inflammation by LE 20% (Figure 6e–h).

3. Discussion

Our results suggest that LE provides neuroprotection against ischemic reperfusion injury in a dosage-dependent manner. Intra-arterial injection of 20% LE during reperfusion after MCAO was able to reduce infarction significantly. The decreases in infarction volumes have been reflected in the improved performance in the neurological deficit assessment. Rats injected with LE 20% possessed better control over their paralyzed limb. Increased levels of Wnt1 signaling in both protein and mRNA levels in 20% LE-treated rats were observed, which in turn enhanced the resistance to reperfusion injury following ischemia. Especially, a notable increase in Wnt1 mRNA expression was observed, which may have led to increased Wnt1 protein expression. The elevated levels of Wnt1 are consistent with previous studies that have indirectly stimulated Wnt/β-catenin signaling pathway for cell survival after ischemic injury [20,36]. The expression levels of Wnt3 decreased in all MCAO groups which may be due to a decline in cell populations as a result of ischemic reperfusion injury [37]. Although Wnt3 is also known as a canonical Wnt signal, not all canonical Wnt signals seem to be involved in the protection process. As in our previous study [31], Wnt3 did not seem to be involved in protection process in response to the administration of LE. In addition, the phosphorylation of GSK-3β increased due to 20% LE injection, which inactivated the β-catenin destruction complex. β-catenin was preserved to promote cell survival, which resulted in enhanced resistance to the ischemic reperfusion damage. Elevation of PORCN expression may have been due to the increase in demand for Wnt1 to resist damage; however, a clear link between PORCN and LE cannot be formulated at this stage. A previous study on the ischemic reperfusion injury in the heart has reported that palmitic acid of LE affects the lipid-modification of Wnt ligands for protection [38]. Clear distinctions between cardiac cells and neurons exist but Wnt modifications seem to occur regardless of their metabolic factors. Inflammatory markers, IL-1β, IL-6, IL-8 and TNF-α, are known to increase after ischemic reperfusion injury [34,35]. Inflammation by oxidative stress was significantly decreased in the 20% LE-treated group, indicating an anti-inflammatory action of LE in the central nervous system. The preservation of β-catenin through the activation of the Wnt signaling pathway induces the transcription of T cell factor/lymphoid enhancer factor, which results in decreased inflammatory activity [39]. Therefore, the expressions of inflammatory cytokines, including IL-1β, IL-6 and TNF-α, might be reduced. The decrease in the expression of IL-8 is assumed to be reduced by an overall decline in inflammatory activity.

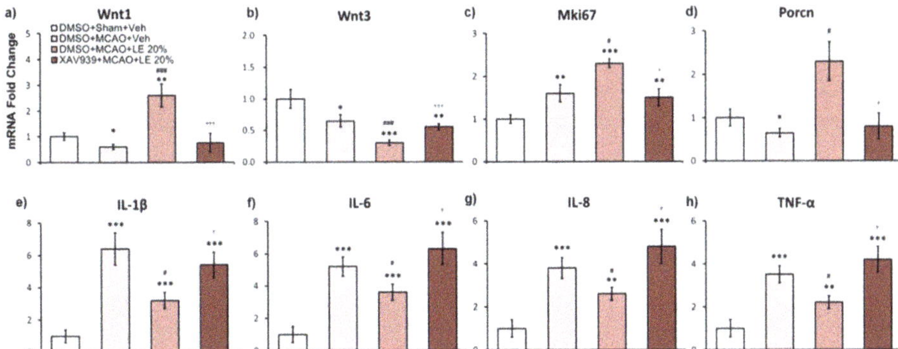

Figure 6. Effects of LE or vehicle on mRNA expression on the MCAO and reperfusion injury after the administration of DMSO or XAV939. (**a,b**) Wnt expressions in experimental groups. The Wnt1 mRNA expression of the DMSO+MCAO+Veh group was significantly decreased compared to the DMSO+Sham+Veh group. Significantly increased expression of Wnt1 was expressed Wnt1 decreased significantly in the XAV939+MCAO+LE 20% group compared to the DMSO+MCAO+LE 20% group. There was no significant difference in Wnt1 expression in the XAV939+MCAO+LE 20% group compared to the DMSO+MCAO+Veh group. Wnt3 expressions were significantly lower in MCAO-injury groups compared to the DMSO+Sham+Veh group. Significantly decreased Wnt3 expressions were observed in the DMSO+MCAO+LE 20% group compared to the DMSO+MCAO+Veh and XAV939+MCAO+LE 20% groups; (**c**) Mki67 expression was increased in all MCAO-injury groups compared to the DMSO+Sham+Veh group. Significant increase in Mki67 expression was observed in the DMSO+MCAO+LE 20% group compared to the DMSO+MCAO+Veh. The XAV939+MCAO+LE 20% group had a significant decrease in Mki67 expression compared to the DMSO+MCAO+LE 20% group; (**d**) Porcn expression was significantly decreased in the DMSO+MCAO+Veh group compared to the DMSO+Sham+Veh group. Significant increase of Porcn was observed in the DMSO+MCAO+LE 20% group compared to the DMSO+MCAO+Veh. The XAV939+MCAO+LE 20% group had a significant decrease in Porcn expression compared to the DMSO+MCAO+LE 20% group. There was no significant difference in Porcn expression in the XAV939+MCAO+LE 20% group compared to the DMSO+MCAO+Veh group; (**e–h**) The mRNA expression of inflammatory markers. MCAO injury groups expressed significantly increased levels of inflammatory markers compared to the DMSO+Sham+Veh group. Significantly decreased inflammatory expression levels were observed in the DMSO+MCAO+LE 20% group compared to DMSO+MCAO+Veh group. Significantly increased levels of inflammatory markers were observed in the XAV939+MCAO+LE 20% group compared to the DMSO+MCAO+LE 20% group and DMSO+Sham+Veh group but there was no significant difference compared to the DMSO+MCAO+Veh group. Data are presented as mean ± standard error of mean (SEM); $n = 8$ for each group; * $p < 0.05$, ** $p < 0.01$, *** $p < 0.001$ vs. DMSO+Sham+Veh, # $p < 0.05$, ### $p < 0.001$ vs. DMSO+MCAO+Veh, † $p < 0.05$, ††† $p < 0.001$ vs. DMSO+MCAO+LE 20%, one-way ANOVA followed by Tukey's multiple comparison test. Wnt subfamily mRNA expressions are shown in Supplementary Materials S1.

The Wnt/β-catenin signaling pathway has been considered a potential therapeutic target in preventing cell death for an extensive amount of time [18,40,41]. The down-regulation of Wnt ligands and increased antagonistic activity have been observed in neurodegenerative and excitotoxic disorders [42–44]. In order to verify the protective mechanism of LE, which increased Wnt1, we were required to interrupt the protective process. XAV939, a Wnt/β-catenin signaling pathway inhibitor, has been commonly utilized to inhibit Wnt activity in many studies [45,46]. in the present study, the i.p. injection of XAV939 successfully inhibited Wnt activity and reversed the protective effects induced by LE. The XAV939+MCAO+LE 20% group exhibited infarction volumes nearly as arge as our control, the DMSO+MCAO+Veh group. Akin to earlier studies, our results show that the Wnt/β-catenin signaling pathway might be a potential therapeutic target for neuroprotection. The administration of XAV939 inhibited the mRNA expression level of Wnt1, indicating that Wnt1 may be dependent on the activity of GSK-3β. XAV939 is known to inhibit tankyrase 1, which prevents the degradation of axis inhibition protein 2 (AXIN2), allowing the accumulation of the GSK-3β destruction complex [47]. Therefore, pGSK-3β was attenuated in the XAV939+MCAO+LE 20% group, which reversed the protective effect of LE by the degradation of the downstream survival marker, β-catenin (Figure 4k). We can infer that LE has effects on the upstream signals of GSK-3β and not on the downstream signals of GSK-3β (Figure 7). If LE affected the GSK-3β downstream marker, β-catenin, XAV939 might not have affected the protection effect. In addition, if LE affected GSK-3β directly, a less effective or partial reversal of protection may have occurred. We observed that the infarction damage was not significantly different between the DMSO+MCAO+Veh group and XAV939+MCAO+LE 20% group, implying an effective reversal from the effects of 20% LE injection.

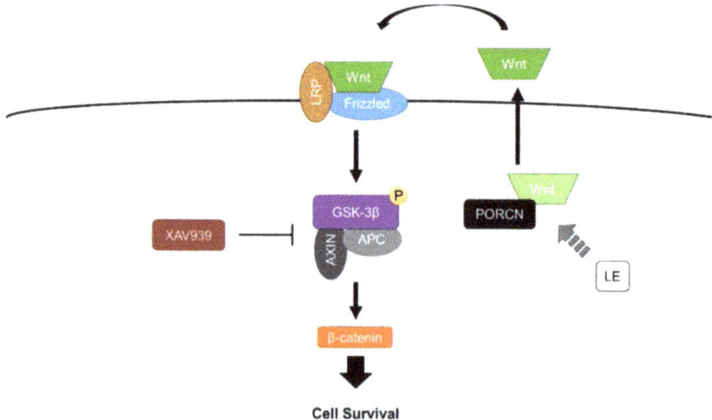

Figure 7. Schematic diagram of the protective mechanism of LE in ischemic reperfusion injury. LE increases the lipid modification of Wnt proteins by PORCN. Lipid-modified/active Wnt proteins are released to the extra-cellular space for the binding to Frizzled/LRP receptors. The activated receptors induce the phosphorylation of GSK-3β, which prevents the destruction complex from the degradation of β-catenin by proteasomes. Preserved levels of β-catenin lead to cell survival. The injection of XAV939 prevents the phosphorylation of GSK-3β, which allows β-catenin degradation for cell death. Abbreviations: PORCN—porcupine, Wnt—wingless integration, LRP—low-density lipoprotein receptor-related protein, GSK-3β—glycogen synthase kinase 3β, APC—adenomatous polyposis coli, AXIN—axis inhibition protein.

The protective roles of fatty acids have received much attention in studies regarding stroke. Different compositions of lipid emulsions, such as omega-3, have been reported to provide neuroprotection against ischemic stroke injuries [48]. Emulsions of n-3 fatty acids have been shown to effectively reduce infarction volume and inflammation in the ischemic brain in neonatal mice [49]. in the present study, we observed significant neuroprotective effects of LE against ischemic reperfusion injury but limitations exist. Although we were able to investigate protective properties of LE regarding the Wnt/β-catenin signaling pathway, we are yet to discover specific targets of LE. Further studies should consider other signals that are involved in the Wnt/β-catenin signaling pathway, such as frizzled-1, low-density lipoprotein receptor-related protein 5/6 or PORCN, that can be subjected for inhibition. Elucidation of specific signals may lead to more effective therapeutic targets in the brain. Through our study, we suggested that the neuroprotective mechanism of LE is dependent on the Wnt/β-catenin signaling pathway through the inhibition of GSK-3β phosphorylation. We assumed that lipid modification occurs through PORCN through the constituents of LE that interact with the protein.

As a result, the expression levels of Wnt1 increased, which also affects its downstream signals. The scope for future studies is considerable: LE may be utilized with other lipophilic anti-inflammatory drugs to further extend its neuroprotective properties. Furthermore, the investigation of different routes of delivery may also be considered with regard to invasiveness and safety.

In conclusion, the intra-arterial administration of 20% LE alleviated ischemic reperfusion injury induced by MCAO and reperfusion. Infarction volumes and Bederson scores were attenuated in a dosage-dependent manner. Protein and mRNA expression levels of the Wnt/β-catenin signaling pathway were elevated and inflammatory markers decreased significantly in the 20% LE-treated group. Especially, GSK-3β was phosphorylated significantly, which in turn preserved β-catenin to promote cell survival. The protective actions were reversed by the administration of XAV939 indicating that the protection mechanism of LE has been induced through the Wnt/β-catenin signaling pathway. These findings regarding the anti-inflammatory and neuroprotective properties of LE may provide a foundation for further research regarding excitotoxic neural injuries and functional recovery.

4. Materials and Methods

4.1. Animals

Adult Sprague-Dawley rats weighing 260-300 g (Koatec, Pyeongtaek, South Korea) were used for experiments in this study. Animals were housed in groups of three per cage under 12-h light/dark cycles, with free access to food and water. Animals were subjected to 7 days of acclimatization upon arrival at the Association for Assessment and Accreditation of Laboratory Animal Care (AAALAC)-accredited Yonsei University College of Medicine Animal Care Facilities. All experimental procedures were approved by the Institutional Animal Care and Use Committee of Yonsei University Health System (permit no.: 2019-0263, approval date: 22 January 2020) and performed according to the National Institutes of Health Guide for Care.

4.2. Middle Cerebral Artery Occlusion and Reperfusion

Rats were anesthetized by i.p. injection of 50 mg/kg sodium pentobarbital (Hanlim Pharmaceutical, Seoul, South Korea) and were placed on a heated mat to maintain body temperature of 37.0 ± 1.0 °C. The surgical and injection process is illustrated in Figure 8. The Koizumi method [50] of MCAO was used for inducing ischemic reperfusion injury. Cervical skin incision was made to expose the common carotid artery (CCA), external carotid artery (ECA) and internal carotid artery (ICA). The CCA was permanently ligated below the bifurcation to the ECA and ICA. The ECA was temporarily ligated while a small incision was made in the CCA to insert a silicone coated nylon filament (403723PK10Re, Doccol Corporation, Sharon, MA, USA) into the ICA until a mild resistance stopped the insertion. The length of the insertion was approximately 18-20 mm (Figure 8a). The rat was occluded for 90 min. and the filament was removed. The experimental group undergoing sham surgery did not have the filament inserted. Polyethylene tubing with an inner diameter of 0.28 mm and outer diameter of 0.61 mm (427401, Becton, Dickson and Company, Sparks, MD, USA) was inserted for injection of LE or vehicle (Figure 8b). Injections were made at 0.5 mL/min over 2 min. The polyethylene tubing was removed after the injection and the CCA above the incision was permanently ligated. The temporary ligation of the ECA was removed. A total of 374 rats were used for this study. The MCAO surgery was conducted on 310 rats and sham surgery was conducted on 64 rats. 84 rats that deceased during MCAO, reperfusion or intra-arterial injection were excluded from the experiment. None of Sham-operated rats deceased. 34 rats that underwent MCAO and reperfusion but did not show infarction due to surgical errors were also excluded from the study.

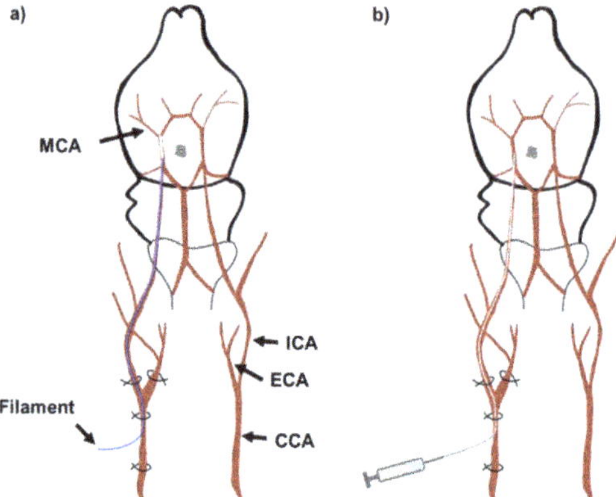

Figure 8. Illustration of middle cerebral artery occlusion and injection. (**a**) Insertion of nylon filament for MCAO. The CCA was ligated permanently while the ECA and ICA were temporarily ligated to stop blood flow. A small incision was made in the CCA for the filament insertion. The silicone-coated filament was inserted until it reached the MCA. The artery was occluded for 90 min. Then the filament was removed for reperfusion; (**b**) Insertion of polyethylene tubing for injection of LE or vehicle. The polyethylene tubing was inserted near to the MCA. After injection, the tubing was removed and the CCA near the bifurcation was permanently ligated. The temporary ligations on the ICA and ECA were removed. Abbreviations: CCA—common carotid artery, ECA—external carotid artery, ICA—internal carotid artery, MCA—middle cerebral artery.

4.3. Drug Treatment

LE (Intralipid™ 20%, Fresenius Kabi, Uppsala, Sweden) was dissolved in sterile 0.9% NaCl to a final concentration of 10% for the injection of the LE 10% group. Vehicle, 10% LE and 20% LE were intra-arterially administered during reperfusion after 90 min. of MCAO. Intra-peritoneal injection XAV939 (S1180, Selleck Chemicals, Houston, TX, USA) 40mg/kg or DMSO were injected for 2 days prior to the MCAO surgery and on the day of the surgery before MCAO.

4.4. Neurological Deficit Assessment

A neurological examination of experimental animals was performed in accordance with the modified Bederson method [51] 24 h after MCAO or sham surgery by a blinded researcher. The scoring criteria are as follows: score 0: no deficit; score 1: lost forelimb flexion; score 2: as for 1, plus decreased resistance to lateral push; score 3: unidirectional circling; score 4: longitudinal spinning or seizure activity; score 5: no movement.

4.5. Infarction Volume Assessment

The infarction volume was assessed by TTC staining of consecutive 2.0 mm coronal sections from bregma +4 mm to −6.0 mm. The sections were immersed into 2% TTC (T8877, Sigma Aldrich, St. Louis, MO, USA) solution for 30 min at 37.0 °C. The sections were transferred into 4% paraformaldehyde solution for 24 h. The sections were then photographed and analyzed using the ImageJ software (ImageJ, National Institute of Health, Bethesda, MD, USA). The infarction volume was calculated using the following formula [50]: (contralateral volume − non-infarct ipsilateral volume)/contralateral volume × 100%. The contralateral volume refers to the opposite brain hemisphere of the infarcted side.

4.6. Western Blotting

Frozen samples were homogenized for protein extraction in lysis buffer (ProPrep; Intron Biotechnology, Pyeongtaek, South Korea) with phosphatase inhibitors (Phosstop; Roche, Mannheim,

Germany). Supernatants were collected from homogenized samples that were centrifuged at 15,000 rpm for 15 min at 4 °C. Total protein concentrations were measured using a spectrophotometer (Nano Drop ND-1000, NanoDrop Technologies Inc., Wilmington, DE, USA) and equal amount of protein (30 mg per well) were resolved on 10% sodium dodecyl sulfate-polyacrylamide gel electrophoresis (SDS-PAGE) and transferred to a polyvinylidene difluoride membrane (Merck Millipore, Darmstadt, Germany) for 2 h. Phospho-proteins were detected by immunoblotting prior to their corresponding total protein levels, which were detected after the membrane had been stripped. Transferred proteins on membranes were fixed using 0.05% glutaraldehyde in Tris-buffered saline containing 0.05% Tween-20 (TBST) for 15 min at room temperature. The membrane was stained with Ponceau S (P7170, Sigma-Aldrich, St. Louis, MO, USA) for visualization of the transferred proteins and cut into strips according to the size of the target protein to minimize interactions between antibodies. Membranes were blocked with 5% bovine serum albumin (BSA) dissolved in TBST for 1 h at room temperature. The membranes were incubated overnight with primary antibodies diluted in 5% BSA in TBST at 4 °C. Rabbit was the host of all primary antibodies used for immunoblotting. The following antibodies were used—anti-Wnt1 (ab15251, 1:1000, Abcam, Cambridge, UK), anti-Wnt3 (ab32249, Abcam), anti-PORCN (ab105543, Abcam), anti-β-catenin (#9562, 1:3000, Cell Signaling Technology, Beverly, MA, USA), anti-Akt (#4691, 1:3000, Cell Signaling Technology), anti-Phospho-Akt (#9271, 1:1000, Cell Signaling Technology), anti-GSK3-β (#9315, 1:3000, Cell Signaling Technology), anti-Phospho-GSK3-β (#9336, 1:1000, Cell Signaling Technology), anti-pβ-catenin (#9561, 1:1000, Cell Signaling Technology), anti-tankyrase 1 (MBS8531631, 1:1000, MyBioSource, San Diego, CA, USA), anti-IL-1β (AB1832P, 1:1000, Sigma Aldrich), anti-IL-6 (ARC0062, 1:500, Invitrogen, Carlsbad, CA, USA), anti-IL-8 (MBS9385550, 1:500, MyBioSource), anti-TNF-α (AAR33, 1:1000, Bio-Rad, Hercules, CA, USA) and anti-β-actin (#4967, 1:10,000, Cell Signaling Technology). After overnight incubation of primary antibodies, the membranes were incubated with anti-rabbit horseradish peroxidase-conjugated secondary antibody (#7074, 1:10,000, Cell Signaling Technology). Visualization of immunoreactive proteins was performed with the application of chemiluminescent detection reagent (ECL™ Prime, GE Healthcare, Little Chalfont, UK) and images were taken by ImageQuant™ LAS 4000 (GE Healthcare). Protein immunoreactivity was measured using Multi-gauge software (Fuji Film Inc., Tokyo, Japan).

Table 1. Primer pairs for qPCR.

Gene Name	Forward Primer (5′-3′)	Reverse Primer (5′-3′)
Wnt1	GCAACCAAAGTCGCCAGAA	TATGTTCACGATGCCCCACCA
Wnt2b	GCTACCCAGACATCATGCG	ACACTCTCGGATCCATTCCC
Wnt3	AATTTGGTGGTCCCTGGC	GATAGAGCCGCAGAGCAGAG
Wnt4	GTTTCCAGTGGTCAGGATGC	AGGACTGTGAGAAGGCTACGC
Wnt5a	AAGGGAACGAATCCACGCC	ATACTGTCCTGCGACCTGCTTC
Wnt7a	CCAAGGTCTTCGTGGATGC	TGTAAGTTCATGAGGGTTCGG
Wnt7b	CGTGTTTCTCTGCTTTGGC	CACCACGGATGACAATGC
Wnt9a	GTACAGCAGCAAGTTTGTCAAGG	CACGAGGTTGTTGTGGAAGTCC
Wnt10a	CGGAACAAAGTCCCCTACG	AGGCGAAAGCACTCTCTCG
Wnt16	GCACTCTGTAACCAGGTCATGC	TGCAAGGTGGTGTCACAGG
Mki67	TTCAGTTCCGCCAATCCAAC	CCGTGCTGGTTCCTTTCCA
Porcn	CCTACCTCTTCCCCTACTTCA	CTTTGCGTTTCTTGTTGCGA
IL-1β	AATGCCTCGTGCTGTCTG	TCCATTGAGGTGGAGAGC
IL-6	ATGAAGTTTCTCTCCGCAAG	CAACAACATCAGTCCCAAG
IL-8	AGCCTTCCTGATTTCTGC	AGCACTCCTTGGCAAAAC
TNFα	CAGCCGATTTGCCATTTC	TCTTGATGGCAGAGAGGAG
β-Actin	GTCCACCCGCGAGTACAAC	TATCGTCATCCATGGCGAACTGG

4.7. qPCR

RNA was extracted by using the Hybrid-R kit (305-010; GeneAll Biotechnology, Seoul, Korea). The concentration of RNA was measured using a spectrophotometer (Nano Drop ND-1000, NanoDrop

Technologies Inc.). cDNA was prepared from 1 µg of total RNA using the PrimeScript 1st strand cDNA synthesis kit (Takara Bio, Shiga, Japan). PCR amplification was executed using the SYBR-Green reagent (Takara Bio) in the ABI 7500 real-time PCR system (Applied Biosystems, Foster City, CA, USA). PCR amplification was performed in 20 µL reaction volumes. Sequences for oligonucleotide primers were selected using the Gene Database of National Center for Biotechnology Information (NCBI) and Primer Express™ Software v3.0.1 (Thermo Fisher Scientific, Waltham, MA, USA). Primer pairs are listed in Table 1.

4.8. Statistical Analysis

Statistical evaluations were performed using one-way ANOVA or unpaired *t*-test, as indicated in figure legends. Post hoc analyses were performed using the Tukey's multiple comparisons test or as otherwise specified in the figure legends. All statistical analyses were performed using GraphPad Prism software (GraphPad Software Inc., San Diego, CA, USA). A *p*-value less than 0.05 was considered statistically significant for all analyses.

Supplementary Materials: Supplementary Materials can be found at http://www.mdpi.com/1422-0067/21/19/7373/s1.

Author Contributions: Conceptualization, M.T., W.K.P., J.P., J.E.L. and B.H.L.; Data curation, M.T. and J.P.; Formal analysis, M.T. and J.P.; Funding acquisition, B.H.L.; Investigation, M.T. and J.P.; Methodology, M.T., J.P. and J.E.L.; Project administration, M.T. and B.H.L.; Resources, M.T., W.K.P., J.P. and J.E.L.; Supervision, W.K.P., J.E.L. and B.H.L.; Validation, M.T., W.K.P., J.P., J.E.L. and B.H.L.; Visualization, M.T.; Writing—original draft, M.T.; Writing—review & editing, M.T., W.K.P. and B.H.L. All authors have read and agreed to the published version of the manuscript.

Funding: This study was supported by the Basic Research Program through the National Research Foundation (NRF) funded by the Ministry of Science and ICT (NRF-2020R1A2C3008481) and by the "Dongwha" Faculty Research Assistance Program of Yonsei University College of Medicine (6-2017-0164).

Conflicts of Interest: The authors declare no conflict of interest.

Abbreviations

AAALAC	Association and Accreditation of Laboratory Animal Care
ANOVA	analysis of variance
APC	adenomatous polyposis coli
AXIN	axis inhibition protein
CCA	common carotid artery
DMSO	dimethyl sulfoxide
ECA	external carotid artery
GSK3-β	glycogen synthase kinase-3β
ICA	internal carotid artery
IL	interleukin
LE	lipid emulsion
LRP	low-density lipoprotein receptor-related protein
MCA	middle cerebral artery
MCAO	middle cerebral artery occlusion
mRNA	messenger ribonucleic acid
PORCN	porcupine
qPCR	quantitative polymerase chain reaction
SEM	standard error of mean
TNF-α	tumor necrosis factor-α
Veh	vehicle
Wnt	wingless integration
XAV939	2-[4-(trifluoromethyl)phenyl]-7,8-dihydro-5H-thiino[4,3-d]pyrimidin-4-ol

References

1. Romero, J.R.; Morris, J.; Pikula, A. Stroke prevention: Modifying risk factors. *Ther. Adv. Cardiovasc. Dis.* **2008**, *2*, 287–303. [CrossRef] [PubMed]
2. Feigin, V.L.; Forouzanfar, M.H.; Krishnamurthi, R.; Mensah, G.A.; Connor, M.; Bennett, D.A.; Moran, A.E.; Sacco, R.L.; Anderson, L.; Truelsen, T.; et al. Global and regional burden of stroke during 1990–2010: Findings from the global burden of disease study 2010. *Lancet* **2014**, *383*, 245–254. [CrossRef]
3. Gottesman, R.F.; Hillis, A.E. Predictors and assessment of cognitive dysfunction resulting from ischaemic stroke. *Lancet Neurol.* **2010**, *9*, 895–905. [CrossRef]
4. Sun, M.-S.; Jin, H.; Sun, X.; Huang, S.; Zhang, F.-L.; Guo, Z.-N.; Yang, Y. Free radical damage in ischemia-reperfusion injury: An obstacle in acute ischemic stroke after revascularization therapy. *Oxid. Med. Cell Longev.* **2018**, *2018*, 3804979. [CrossRef] [PubMed]
5. Wardlaw, J.M.; Murray, V.; Berge, E.; del Zoppo, G.J. Thrombolysis for acute ischaemic stroke. *Cochrane Database Syst. Rev.* **2014**, *2014*, Cd000213. [CrossRef] [PubMed]
6. Pan, J.; Konstas, A.A.; Bateman, B.; Ortolano, G.A.; Pile-Spellman, J. Reperfusion injury following cerebral ischemia: Pathophysiology, MR imaging and potential therapies. *Neuroradiology* **2007**, *49*, 93–102. [CrossRef]
7. Sobowale, O.A.; Parry-Jones, A.R.; Smith, C.J.; Tyrrell, P.J.; Rothwell, N.J.; Allan, S.M. Interleukin-1 in stroke: From bench to bedside. *Stroke* **2016**, *47*, 2160–2167. [CrossRef]
8. Aref, H.M.A.; Fahmy, N.A.; Khalil, S.H.; Ahmed, M.F.; ElSadek, A.; Abdulghani, M.O. Role of interleukin-6 in ischemic stroke outcome. *Egypt J. Neurol. Psychiatr. Neurosurg.* **2020**, *56*, 12. [CrossRef]
9. Domac, F.M.; Misirli, H. The role of neutrophils and interleukin-8 in acute ischemic stroke. *Neurosciences (Riyadh)* **2008**, *13*, 136–141.
10. Zangari, R.; Zanier, E.R.; Torgano, G.; Bersano, A.; Beretta, S.; Beghi, E.; Casolla, B.; Checcarelli, N.; Lanfranconi, S.; Maino, A.; et al. Early ficolin-1 is a sensitive prognostic marker for functional outcome in ischemic stroke. *J. Neuroinflammation* **2016**, *13*, 16. [CrossRef]
11. Relton, J.K.; Rothwell, N.J. Interleukin-1 receptor antagonist inhibits ischaemic and excitotoxic neuronal damage in the rat. *Brain Res. Bull.* **1992**, *29*, 243–246. [CrossRef]
12. Yamasaki, Y.; Matsuura, N.; Shozuhara, H.; Onodera, H.; Itoyama, Y.; Kogure, K. Interleukin-1 as a pathogenetic mediator of ischemic brain damage in rats. *Stroke* **1995**, *26*, 676–680. [CrossRef] [PubMed]
13. Shruster, A.; Ben-Zur, T.; Melamed, E.; Offen, D. Wnt signaling enhances neurogenesis and improves neurological function after focal ischemic injury. *PLoS ONE* **2012**, *7*, e40843. [CrossRef] [PubMed]
14. Jean LeBlanc, N.; Menet, R.; Picard, K.; Parent, G.; Tremblay, M.-È.; ElAli, A. Canonical Wnt pathway maintains blood-brain barrier integrity upon ischemic stroke and its activation ameliorates tissue plasminogen activator therapy. *Mol. Neurobiol.* **2019**, *56*, 6521–6538. [CrossRef]
15. Endo, H.; Nito, C.; Kamada, H.; Yu, F.; Chan, P.H. Akt/GSK3beta survival signaling is involved in acute brain injury after subarachnoid hemorrhage in rats. *Stroke* **2006**, *37*, 2140–2146. [CrossRef]
16. Kim, U.J.; Lee, B.H.; Lee, K.H. Neuroprotective effects of a protein tyrosine phosphatase inhibitor against hippocampal excitotoxic injury. *Brain Res.* **2019**, *1719*, 133–139. [CrossRef]
17. Lee, K.H.; Won, R.; Kim, U.J.; Kim, G.M.; Chung, M.A.; Sohn, J.H.; Lee, B.H. Neuroprotective effects of FK506 against excitotoxicity in organotypic hippocampal slice culture. *Neurosci. Lett.* **2010**, *474*, 126–130. [CrossRef]
18. Wang, W.; Li, M.; Wang, Y.; Wang, Z.; Zhang, W.; Guan, F.; Chen, Q.; Wang, J. GSK-3β as a target for protection against transient cerebral ischemia. *Int. J. Med. Sci.* **2017**, *14*, 333–339. [CrossRef]
19. Cheng, J.; Shen, W.; Jin, L.; Pan, J.; Zhou, Y.; Pan, G.; Xie, Q.; Hu, Q.; Wu, S.; Zhang, H.; et al. Treadmill exercise promotes neurogenesis and myelin repair via upregulating Wnt/β-catenin signaling pathways in the juvenile brain following focal cerebral ischemia/reperfusion. *Int. J. Mol. Med.* **2020**, *45*, 1447–1463. [CrossRef]
20. Chen, B.; Tao, J.; Lin, Y.; Lin, R.; Liu, W.; Chen, L. Electro-acupuncture exerts beneficial effects against cerebral ischemia and promotes the proliferation of neural progenitor cells in the cortical peri-infarct area through the Wnt/β-catenin signaling pathway. *Int. J. Mol. Med.* **2015**, *36*, 1215–1222. [CrossRef]
21. Isaksson, B.; Hambraeus, L.; Vinnars, E.; Samuelson, G.; Larsson, J.; Asp, N.-G. In memory of Arvid Wretlind 1919–2002. *J. Food Nutr. Res.* **2002**, *46*, 117–118.

22. Weinberg, G.L.; VadeBoncouer, T.; Ramaraju, G.A.; Garcia-Amaro, M.F.; Cwik, M.J. Pretreatment or resuscitation with a lipid infusion shifts the dose-response to bupivacaine-induced asystole in rats. *Anesthesiology* **1998**, *88*, 1071–1075. [CrossRef] [PubMed]

23. Tierney, K.J.; Murano, T.; Natal, B. Lidocaine-induced cardiac arrest in the emergency department: Effectiveness of lipid therapy. *Int. J. Emerg. Med.* **2016**, *50*, 47–50. [CrossRef] [PubMed]

24. Litz, R.J.; Popp, M.; Stehr, S.N.; Koch, T. Successful resuscitation of a patient with ropivacaine-induced asystole after axillary plexus block using lipid infusion. *Anaesthesia* **2006**, *61*, 800–801. [CrossRef] [PubMed]

25. Rosenblatt, M.A.; Abel, M.; Fischer, G.W.; Itzkovich, C.J.; Eisenkraft, J.B. Successful use of a 20% lipid emulsion to resuscitate a patient after a presumed bupivacaine-related cardiac arrest. *Anesthesiology* **2006**, *105*, 217–218. [CrossRef]

26. Zaugg, M.; Lou, P.H.; Lucchinetti, E.; Gandhi, M.; Clanachan, A.S. Postconditioning with Intralipid emulsion protects against reperfusion injury in post-infarct remodeled rat hearts by activation of ROS-Akt/Erk signaling. *Transl. Res.* **2017**, *186*, 36–51. [CrossRef]

27. Rahman, S.; Li, J.; Bopassa, J.C.; Umar, S.; Iorga, A.; Partownavid, P.; Eghbali, M. Phosphorylation of GSK-3β mediates intralipid-induced cardioprotection against ischemia/reperfusion injury. *Anesthesiology* **2011**, *115*, 242–253. [CrossRef]

28. Li, J.; Ruffenach, G.; Kararigas, G.; Cunningham, C.M.; Motayagheni, N.; Barakai, N.; Umar, S.; Regitz-Zagrosek, V.; Eghbali, M. Intralipid protects the heart in late pregnancy against ischemia/reperfusion injury via Caveolin2/STAT3/GSK-3β pathway. *J. Mol. Cell Cardiol.* **2017**, *102*, 108–116. [CrossRef]

29. Anna, R.; Rolf, R.; Mark, C. Update of the organoprotective properties of xenon and argon: From bench to beside. *Intensive Care Med. Exp.* **2020**, *8*, 11. [CrossRef]

30. Auzmendi, J.; Puchulu, M.B.; Rodríguez, J.C.G.; Balaszczuk, A.M.; Lazarowski, A.; Merelli, A. EPO and EPO-receptor system as potential actionable mechanism for the protection of brain and heart in refractory epilepsy and SUDEP. *Curr. Pharm. Des.* **2020**, *26*, 1356–1364. [CrossRef]

31. Tanioka, M.; Park, W.K.; Shim, I.; Kim, K.; Choi, S.; Kim, U.J.; Lee, K.H.; Hong, S.K.; Lee, B.H. Neuroprotection from excitotoxic injury by local administration of lipid emulsion into the brain of rats. *Int. J. Mol. Sci.* **2020**, *21*, 2706. [CrossRef] [PubMed]

32. Fan, Y.; Luo, Q.; Wei, J.; Lin, R.; Lin, L.; Li, Y.; Chen, Z.; Lin, W.; Chen, Q. Mechanism of salvianolic acid B neuroprotection against ischemia/reperfusion induced cerebral injury. *Brain Res.* **2018**, *1679*, 125–133. [CrossRef] [PubMed]

33. Torres, V.I.; Godoy, J.A.; Inestrosa, N.C. Modulating Wnt signaling at the root: Porcupine and Wnt acylation. *Pharmacol. Therapeut.* **2019**, *198*, 34–45. [CrossRef] [PubMed]

34. Huang, L.; Ma, Q.; Li, Y.; Li, B.; Zhang, L. Inhibition of microRNA-210 suppresses pro-inflammatory response and reduces acute brain injury of ischemic stroke in mice. *Exp. Neurol.* **2018**, *300*, 41–50. [CrossRef]

35. Tarkowski, E.; Rosengren, L.; Blomstrand, C.; Wikkelsö, C.; Jensen, C.; Ekholm, S.; Tarkowski, A. Intrathecal release of pro- and anti-inflammatory cytokines during stroke. *Clin. Exp. Immunol.* **1997**, *110*, 492–499. [CrossRef]

36. Chong, Z.Z.; Shang, Y.C.; Hou, J.; Maiese, K. Wnt1 neuroprotection translates into improved neurological function during oxidant stress and cerebral ischemia through AKT1 and mitochondrial apoptotic pathways. *Oxid. Med. Cell Longev.* **2010**, *3*, 153–165. [CrossRef]

37. Morris, D.C.; Zhang, Z.G.; Wang, Y.; Zhang, R.L.; Gregg, S.; Liu, X.S.; Chopp, M. Wnt expression in the adult rat subventricular zone after stroke. *Neurosci. Lett.* **2007**, *418*, 170–174. [CrossRef]

38. Lou, P.-H.; Lucchinetti, E.; Zhang, L.; Affolter, A.; Schaub, M.C.; Gandhi, M.; Hersberger, M.; Warren, B.E.; Lemieux, H.; Sobhi, H.F.; et al. The mechanism of Intralipid®-mediated cardioprotection complex IV inhibition by the active metabolite, palmitoylcarnitine, generates reactive oxygen species and activates reperfusion injury salvage kinases. *PLoS ONE* **2014**, *9*, e87205. [CrossRef]

39. Suryawanshi, A.; Tadagavadi, R.K.; Swafford, D.; Manicassamy, S. Modulation of inflammatory responses by Wnt/β-Catenin signaling in dendritic cells: A novel immunotherapy target for autoimmunity and cancer. *Front. Immunol.* **2016**, *7*, 460. [CrossRef]

40. Yu, Z.; Cheng, C.; Liu, Y.; Liu, N.; Lo, E.H.; Wang, X. Neuroglobin promotes neurogenesis through Wnt signaling pathway. *Cell Death Dis.* **2018**, *9*, 945. [CrossRef]

41. Jia, L.; Piña-Crespo, J.; Li, Y. Restoring Wnt/β-catenin signaling is a promising therapeutic strategy for Alzheimer's disease. *Mol. Brain* **2019**, *12*, 104. [CrossRef] [PubMed]

42. Scott, E.L.; Brann, D.W. Estrogen regulation of Dkk1 and Wnt/β-Catenin signaling in neurodegenerative disease. *Brain Res.* **2013**, *1514*, 63–74. [CrossRef] [PubMed]

43. Rosi, M.C.; Luccarini, I.; Grossi, C.; Fiorentini, A.; Spillantini, M.G.; Prisco, A.; Scali, C.; Gianfriddo, M.; Caricasole, A.; Terstappen, G.C.; et al. Increased Dickkopf-1 expression in transgenic mouse models of neurodegenerative disease. *J. Neurochem.* **2010**, *112*, 1539–1551. [CrossRef] [PubMed]

44. Purro, S.A.; Galli, S.; Salinas, P.C. Dysfunction of Wnt signaling and synaptic disassembly in neurodegenerative diseases. *J. Mol. Cell Biol.* **2014**, *6*, 75–80. [CrossRef]

45. Ragab, N.; Viehweger, F.; Bauer, J.; Geyer, N.; Yang, M.; Seils, A.; Belharazem, D.; Brembeck, F.H.; Schildhaus, H.-U.; Marx, A.; et al. Canonical WNT/β-catenin signaling plays a subordinate role in rhabdomyosarcomas. *Front. Pediatr.* **2018**, *6*, 378. [CrossRef] [PubMed]

46. Wei, Z.Z.; Zhang, J.Y.; Taylor, T.M.; Gu, X.; Zhao, Y.; Wei, L. Neuroprotective and regenerative roles of intranasal Wnt-3a administration after focal ischemic stroke in mice. *J. Cereb. Blood Flow Metab.* **2018**, *38*, 404–421. [CrossRef] [PubMed]

47. Bao, R.; Christova, T.; Song, S.; Angers, S.; Yan, X.; Attisano, L. Inhibition of tankyrases induces Axin stabilization and blocks Wnt signalling in breast cancer cells. *PLoS ONE* **2012**, *7*, e48670. [CrossRef] [PubMed]

48. Berressem, D.; Koch, K.; Franke, N.; Klein, J.; Eckert, G.P. Intravenous treatment with a long-chain omega-3 lipid emulsion provides neuroprotection in a murine model of ischemic stroke—A pilot study. *PLoS ONE* **2016**, *11*, e0167329. [CrossRef]

49. Williams, J.J.; Mayurasakorn, K.; Vannucci, S.J.; Mastropietro, C.; Bazan, N.G.; Ten, V.S.; Deckelbaum, R.J. N-3 fatty acid rich triglyceride emulsions are neuroprotective after cerebral hypoxic-ischemic injury in neonatal mice. *PLoS ONE* **2013**, *8*, e56233. [CrossRef]

50. Morris, G.P.; Wright, A.L.; Tan, R.P.; Gladbach, A.; Ittner, L.M.; Vissel, B. A comparative study of variables influencing ischemic injury in the Longa and Koizumi methods of intraluminal filament middle cerebral artery occlusion in mice. *PLoS ONE* **2016**, *11*, e0148503. [CrossRef]

51. Lyu, C.; Zhang, Y.; Gu, M.; Huang, Y.; Liu, G.; Wang, C.; Li, M.; Chen, S.; Pan, S.; Gu, Y. IRAK-M deficiency exacerbates ischemic neurovascular injuries in experimental stroke mice. *Front. Cell Neurosci.* **2018**, *12*, 504. [CrossRef] [PubMed]

International Journal of
Molecular Sciences

Article

Signatures of the Consolidated Response of Astrocytes to Ischemic Factors In Vitro

Elena V. Mitroshina [1],*, Mikhail I. Krivonosov [2], Dmitriy E. Burmistrov [1], Maria O. Savyuk [1], Tatiana A. Mishchenko [1], Mikhail V. Ivanchenko [2] and Maria V. Vedunova [1],*

[1] Institute of Biology and Biomedicine, Lobachevsky State University of Nizhni Novgorod, 23 Prospekt Gagarina, 603950 Nizhny Novgorod, Russia; diman-burmistrov@yandex.ru (D.E.B.); mary.savyuk@bk.ru (M.O.S.); saharnova87@mail.ru (T.A.M.)
[2] Institute of Information, Technology, Mathematics and Mechanics, Lobachevsky State University of Nizhni Novgorod, 23 Prospekt Gagarina, 603950 Nizhny Novgorod, Russia; mike_live@mail.ru (M.I.K.); ivanchenko.mv@gmail.com (M.V.I.)
* Correspondence: helenmitroshina@gmail.com (E.V.M.); mvedunova@yandex.ru (M.V.V.); Tel.: +7-950-604-5137 (E.V.M.); +7-915-937-55-55 (M.V.V.)

Received: 29 September 2020; Accepted: 23 October 2020; Published: 26 October 2020

Abstract: Whether and under what conditions astrocytes can mount a collective network response has recently become one of the central questions in neurobiology. Here, we address this problem, investigating astrocytic reactions to different biochemical stimuli and ischemic-like conditions in vitro. Identifying an emergent astrocytic network is based on a novel mathematical approach that extracts calcium activity from time-lapse fluorescence imaging and estimates the connectivity of astrocytes. The developed algorithm represents the astrocytic network as an oriented graph in which the nodes correspond to separate astrocytes, and the edges indicate high dynamical correlations between astrocytic events. We demonstrate that ischemic-like conditions decrease network connectivity in primary cultures in vitro, although calcium events persist. Importantly, we found that stimulation under normal conditions with 10 μM ATP increases the number of long-range connections and the degree of corresponding correlations in calcium activity, apart from the frequency of calcium events. This result indicates that astrocytes can form a large functional network in response to certain stimuli. In the post-ischemic interval, the response to ATP stimulation is not manifested, which suggests a deep lesion in functional astrocytic networks. The blockade of Connexin 43 during ischemic modeling preserves the connectivity of astrocytes in the post-hypoxic period.

Keywords: astrocytes; astrocytic networks; ischemia; connexin 43; calcium activity

---

1. Introduction

For a long time, astrocytes have been thought to play an auxiliary role without generally affecting the higher functions of the central nervous system. However, recent experimental evidence has allowed for a more profound view on the functions of astrocytes and their role in adaptive processes in the CNS. Currently, there is no doubt that astrocytes are not simply passive transmitters of energy substrates and structural supports for neurons but also active participants in a large number of metabolic reactions [1]. Several studies have demonstrated the role of astrocytes in the pathological processes at sites of mechanical trauma and in the development of neurodegenerative diseases, such as Alzheimer's disease and Parkinson's disease [2–4].

While astrocytes are not electrically excitable cells, they are capable of producing and transmitting Ca2+ signals that can propagate between them, resulting in "calcium waves" [5]. It is known that astrocytes interact with neurons, mutually regulating the functional activity of each other [5]. At the same time, it remains unclear whether the astrocytes are able to respond to external stimuli and develop a consolidated response to stress independently on neurons. Investigating the collective dynamics of astrocytic calcium activity, particularly the coordinated activity, which would answer the long-standing problem of the existence of functional astrocytic networks, is highly anticipated.

In particular, a fundamental question is whether astrocytes are able to coordinate their activity over long distances. In other words, can a large-scale dynamic astrocytic network emerge in the absence of neurons? Another key question is whether a functional astrocytic network can be observed in the normal state or if its activity becomes manifested only after stimulation. Generally, the mechanisms that could possibly affect the strength of connections between cells in response to stress factors, thus reshaping the network, remain unknown. In this respect, elucidating the role of astrocytes in the neuroprotective function in the framework of functional astrocytic network reorganization would provide a novel perspective in understanding the processes of CNS adaptation.

Astrocytic interaction relies on gap junctions [6], which enable direct intercellular communication and transport of small molecules for maintaining homeostasis in the brain, glutamate, ATP, and Ca2+ [7]. Gap junctions are formed by two connexons (or hemichannels), and each of them consists of six connexin proteins. Overall, the brain expresses 11 connexins, among which connexin 43 (Cx43) is mostly attributed to astrocytes [6], affecting communication between them.

Understanding the influence of ischemia on the cellular network activity is also of paramount importance. While the resulting disturbance of neuronal activity has been well studied, much less is known regarding the alterations in functional consolidated astrocytic calcium activity. Arguably, elucidating the latter will pave the way to improving the therapy for ischemic stroke.

It is known that astrocytic gap junctions are normally open, while hemichannels formed by connexin 43 demonstrate a low probability of opening [8]. However, after an ischemic interval, the hemichannels become activated, which leads to the uptake of Na+ and Ca2+ into cells, accompanied by the release of ATP and other metabolites. This can cause changes in the calcium activity, an overload of Ca2+ up to toxic values, and osmotic imbalance up to cell death [9,10]. Additionally, ATP released from hemichannels can provoke neuroinflammation due to microglia and astrocyte activation [11]. Instructively, recent evidence demonstrates that hemichannel blockers, such as connexin 43 blocker GAP19, manifest neuroprotective effects in cerebral ischemia (reperfusion) [8].

Gap19 inhibits the opening of hemichannels without affecting gap junctions in astrocytes [12]. According to observations, selective inhibition of hemichannels Cx43 by Gap19 decreases the severity of a stroke and prevents the death of astrocytes in vivo and in vitro [13–15]. However, the effect of connexins 43 blockade on the consolidation of post-ischemic astrocytic calcium activity is currently unknown.

Primary astrocyte cultures could be considered as a model for studying the role of the consolidated astrocytic response because in such cultures, astrocyte–astrocyte interactions form without neuronal regulation, and the reactions of cells to various biochemical and physiological stimuli in isolation from neural activation can be observed. Accordingly, the study aimed to investigate the networks and related parameters of calcium activity in primary astrocytic cultures in a model of ischemic damage, as well as to estimate the neuroprotective effects of connexin 43 Gap19 blocker. For this purpose, we used the calcium event analysis tool for time-lapse fluorescence image recordings of astrocytic cultures [16] and developed an approach for reconstructing functional astrocytic networks based on inferred dynamic correlations. We demonstrate that model stress conditions and biochemical stimuli may produce a dramatic impact on the structure and statistics of these networks, suggesting the underlying changes in consolidated astrocytic activity and indicating the potential of the method for elucidating the effect of the other imposed conditions.

2. Results

2.1. ATP and Gap19 Effects on Functional Calcium Signaling in Astrocytes

The starting point of the study was estimating the level of consolidated calcium activity of primary astrocytic cultures under normal conditions. This relied on the novel method of computational analysis of time-lapse calcium imaging as described in the materials and methods.

Astrocytes can easily exchange low-molecular-weight compounds (peptides up to 2 kDa, nucleotides, sugars) between adjacent cells through gap junctions that unite neighboring cells into a single functional conglomerate. However, such signals cannot propagate over long distances, as the speed of such passive transport is small and depends on the gradient of concentrations of low-molecular-weight substances in the cytoplasm of adjacent cells [17]. In this regard, an informative evaluation of functionally consolidated signals is given by remote in space and correlated in time calcium events in astrocytes. The correlation graph method showed a low-grade formation of functional networks between distant astrocytes in the "control" group of primary astrocyte cultures under normal conditions, characterized by a large number of disconnected subgraphs (Figure 1).

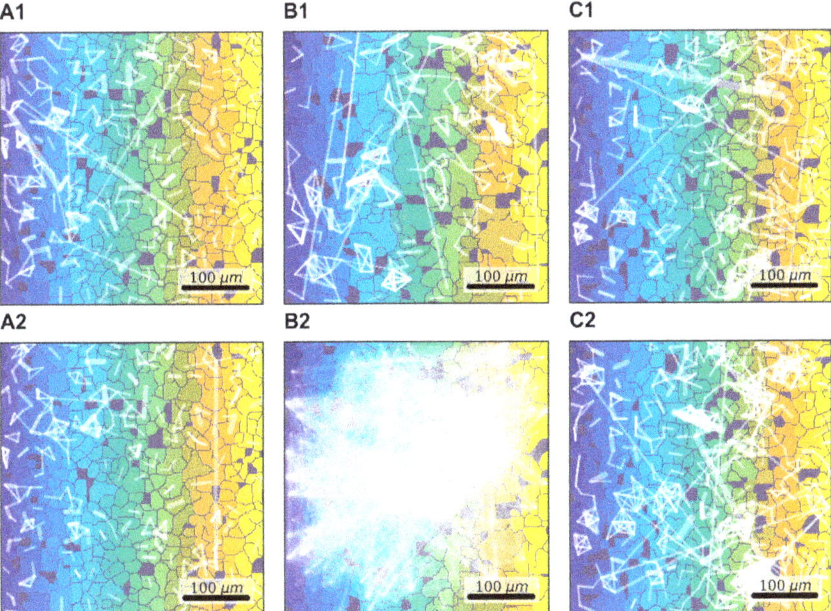

Figure 1. Representative examples of changes in correlation networks after ATP and connexin 43 blocker Gap19 application. (**A**) Control (PBS); (**B**) ATP (10 μM); (**C**) Gap19 (100 μM). (**1**)—before application; (**2**)—after application.

To examine the role of one of the most common types of connexins (connexins 43) in the consolidated astrocyte response, we used the selective blocker Gap19 at a concentration of 100 μM/mL. The application of Gap19 for 40 min did not affect the number of direct long-distance connections per astrocyte under normal conditions, which was 4.94 [3.94; 6.68] ("control" 3.95 [3.10; 4.47]). To validate the ability of astrocytes to become activated and produce a coordinated response to biochemical stimuli, we used ATP (10 μM), which led to a significant 3.5-fold increase in the average number of direct long-distant connections of an astrocyte (13.63 [3.61; 33.23]) over 10 min after ATP application (Figure 2).

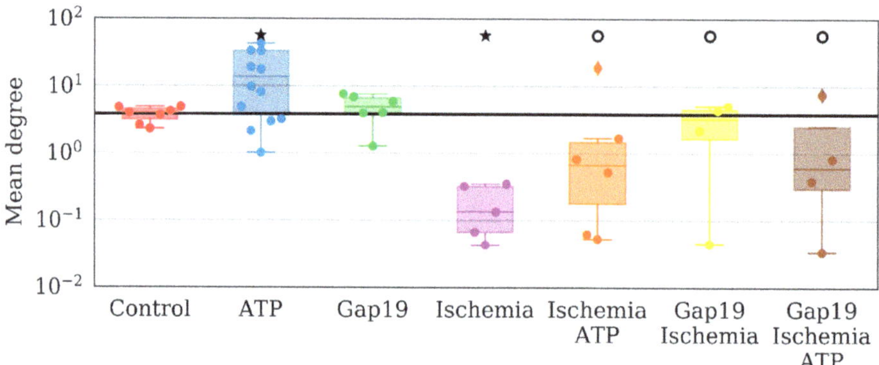

Figure 2. Average number of direct long-distance connections of a single astrocyte. ★—versus baseline, ○—versus "Ischemia", $p < 0.05$, the Kolmogorov–Smirnov test.

Analysis of the main parameters of functional calcium activity in primary astrocyte cultures revealed that ATP addition did not cause significant changes in the percentage of cells that exhibited calcium activity. The percentage of oscillated cells in the "ATP" group was 61.47 [43.7; 70.9]% ("control" 69.5 [61.37; 79.33]%, Figure 3A). At the same time, significant changes in the frequency (before ATP addition 1.959 [1.605; 2.177] osc./min, after ATP addition 2.874 [1.884; 3.635] osc./min, Figure 3C) and duration (before ATP addition 13.449 [11.542; 15.135] s, after ATP addition 10.421 [9.611; 11.492] s) of calcium oscillations are shown (Figure 3B).

Studies on the characteristics of astrocyte connectivity showed that ATP application led to a significant increase in the mean values of the maximum correlation between pairs of average astrocyte calcium levels over time by 20% relative to the baseline ($p < 0.05$ the Kolmogorov–Smirnov test) (Figure 4). In the "ATP" group, an absolute change from 0.096 [0.094; 0.099] to 0.126 [0.105; 0.156] was identified. The application of Gap19 did not significantly change this parameter.

An analysis of network characteristics revealed that ATP application leads to a significant increase in the number of long-distant astrocytic connections that can be considered an emergence of complex interactions between astrocytes. The total number of connections per series increases from 732 [548; 796] to 2133 [515; 5134] (the Kolmogorov–Smirnov (KS) test). Such interactions cannot be considered a result of a simple increase in the frequency of calcium oscillations. At the same time, the number of correlated distant astrocytes was also increased.

The most important parameters that characterize network activity are the distribution of calcium signals between astrocytes in a culture, including the time of signal propagation and frequency characteristics of calcium oscillations. The application of ATP increased the number of high-frequency events and suppressed low-frequency events (Figure 5). To characterize the frequency of calcium events, we computed the power spectral density of the cell calcium intensity time series, and took the frequencies corresponding to the 10th and 90th signal energy percentiles as the characteristic lower and upper frequency bounds. These bounds were then averaged over the whole image field.

These findings indicate changes in the activity profile of the entire system in monoastrocytic culture. The low-frequency events are most likely associated with long-term metabolic changes and are fairly stable under normal conditions.

The study of the effects of connexin 43 blocker (Gap19) on the main parameters of network activity showed that despite the great physiological importance of this type of astrocytic connexin, significant changes were detected only in the speed of signal propagation between astrocytes under normal conditions (Figure 6). The percentage of cells that exhibited Ca2+ activity did not change significantly and was 83.1 [78.7; 88.4]% (Figure 3A). There were also no significant alterations in the duration (15.130 [12.634; 16.740] s) or frequency (1.792 [1.524; 1.975] osc./min) of Ca2+ events (Figure 3B,C).

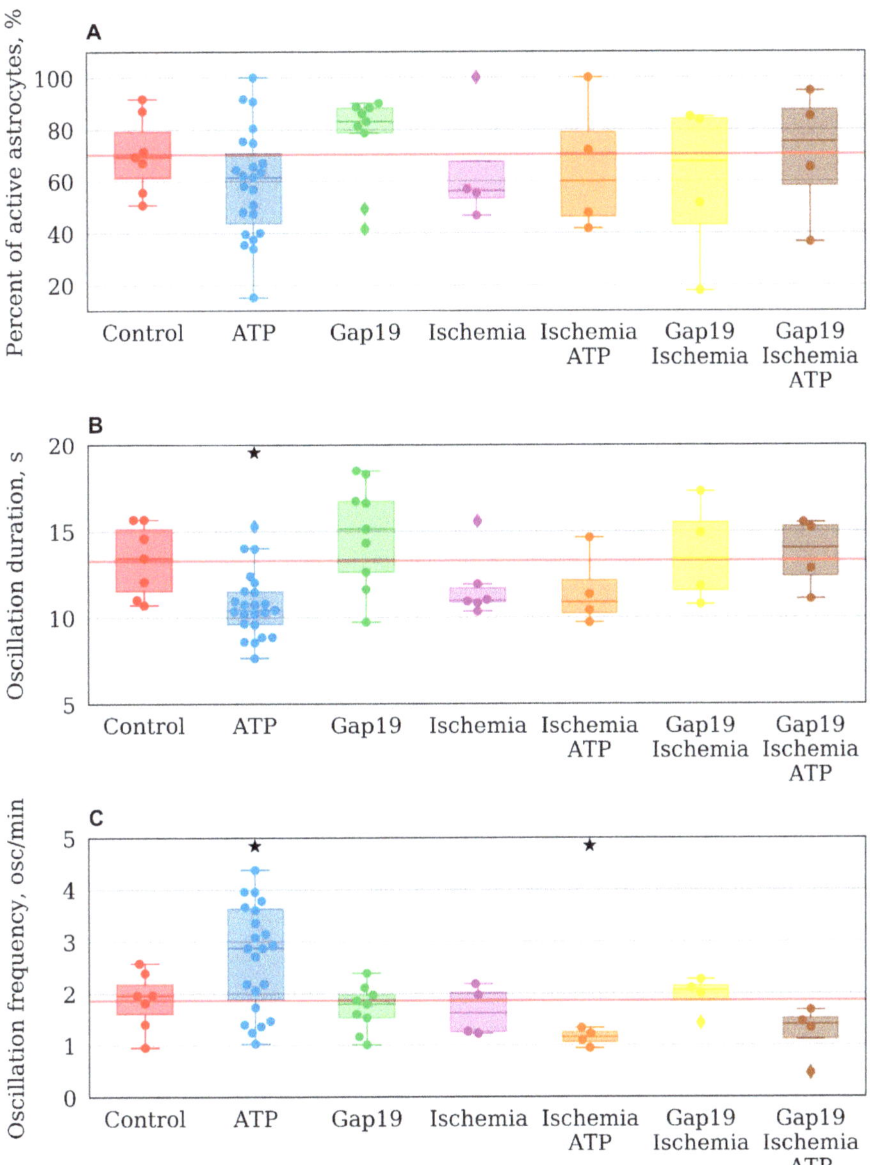

Figure 3. Main parameters of functional calcium activity in primary astrocyte cultures in the context of ATP and Gap19 influence under normal conditions and 7 days after ischemia-like condition modeling. (**A**) Proportion of astrocytes exhibiting calcium activity; (**B**) Duration of Ca2+ oscillations, s.; (**C**) Number of Ca2+ oscillations per min. ★—versus baseline (before application), $p < 0.05$, Wilcoxon rank-sum test.

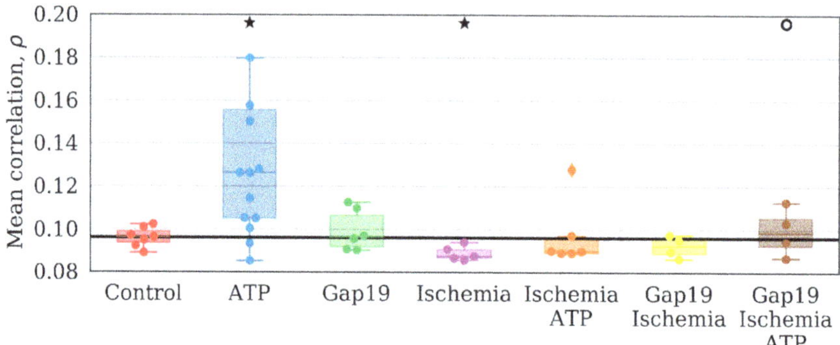

Figure 4. Mean value of the correlation maximum on the shift between pairs of average astrocyte calcium levels over time. ★—versus "Control", ○—versus "Ischemia", $p < 0.05$, the Kolmogorov–Smirnov test.

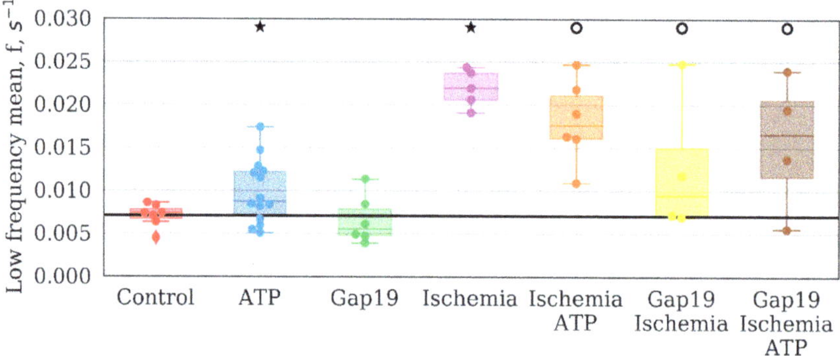

Figure 5. Average low-frequency component of calcium intensity in primary astrocyte cultures after ATP and Gap19 addition. ★ ATP—versus baseline, $p < 0.05$, t-criteria ($p = 0.004$); ★—versus "Control", $p < 0.05$, the Kolmogorov–Smirnov test ($p = 0.005$); ○—versus "Ischemia", $p < 0.05$, t-criteria ($p = 0.004$).

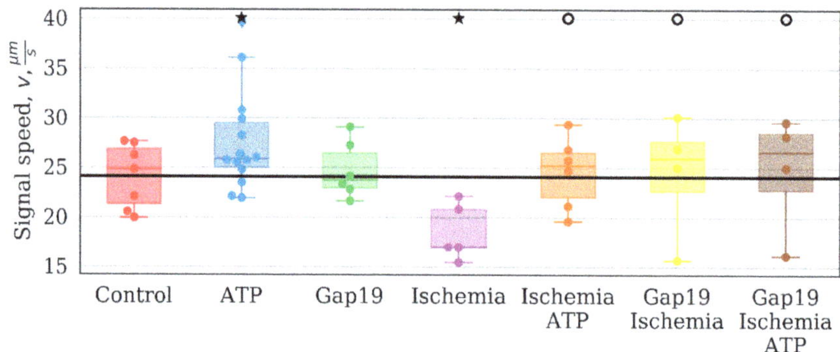

Figure 6. The rate of signal delay between astrocytes. ★ Gap19—versus baseline, $p < 0.05$, t-criteria ($p = 0.036$); versus "Control", $p < 0.05$, the Kolmogorov–Smirnov test ($p = 0.018$). ○—versus "Ischemia", $p < 0.05$, the Kolmogorov–Smirnov test ($p = 0.006$).

An analysis of network connectivity parameters did not reveal significant changes in the network characteristics of the astrocytic network under the selective blocker Gap19 application. Therefore, the role of connexin 43 in the formation of correlated calcium dynamics in primary astrocyte cultures under normal conditions is not significant.

2.2. Correlated Astrocytic Response to Ischemic Influence

The next stage of our study aimed to analyze changes in the adaptive capacity of the astrocytic network in response to modeling ischemia-like conditions. Hypoxia is one of the most common experimental models of brain cell damage [18]. Several studies have shown a significant decrease in the viability and functional activity of brain cells in models of 10-min acute hypoxia in vitro [18–20]. However, astrocytes are considered more resistant to hypoxic damage than neurons. In this regard, to create stress conditions that affect astrocytic network functional activity, we adapted a model of oxygen-glucose deprivation ("ischemia") lasting for 30 min (see the materials and methods). The results showed a significant decrease in the viability of primary astrocyte cultures in the "ischemia" group (see Supplementary Materials 1, Supplementary Materials Table S1), whereas blockage of Cx43 during ischemia modeling maintained the viability of astrocytes at the intact culture level ("control" 98.25 [96.99; 99.197]%, "ischemia" 95.59 [94.55; 97.35]%, "ischemia+Gap19" 99.21 [97.61; 100.0]%).

The number of cells that manifested calcium activity matched the control group and was 56.178 [53.258; 67.692]%. Furthermore, no statistically significant differences in the frequency and duration of Ca2+ oscillations were detected (1.612 [1.25; 2.01] osc./min and 10.967 [10.854;11.708] s, respectively) (Figure 3B,C).

Changes in the pattern of response to ATP stimulation in astrocytic cultures after ischemia-like condition modeling are of particular interest (see Supplementary Materials 1, Supplementary Materials Table S1). Under normal conditions, an increase in the frequency of Ca2+ oscillations after ATP application was observed. In contrast, the frequency of Ca2+ events in response to ATP application was decreased significantly in the "ischemia+ATP" group (1.156 [1.052;1.249] osc./min) ($p < 0.05$, Wilcoxon rank-sum test) (Figure 3C). The percentage of working cells and the duration of Ca2+ oscillations did not change. Therefore, ischemia leads to the loss of the functional response of astrocytes to the stimulating effect of ATP.

Analyzing astrocytic connectivity in the culture after ischemia modeling is of particular interest. While the primary astrocytic cultures in the post-ischemic period exhibited calcium events, their degree of synchrony and correlation was substantially different compared with that of the control cultures. In particular, the modeled ischemia-like conditions manifested as a significant decrease of all the parameters that characterized the connectivity of astrocytes. The number of direct long-distance connections was reduced by a factor of 30 (from 3.95 [3.1; 4.47] to 0.13 [0.07; 0.32]) (Figure 2). In the "ischemia" group, the average value of the maximum correlation of the shift between the average calcium level of neighboring astrocytes over time was 0.110 [0.107; 0.123], which is 2 times lower than that in the "control" group (0.242 [0.226; 0.263]; $p < 0.05$, the KS test). The number of long-distance connections in the "ischemia" group was 0.087 [0.087; 0.091] ("control" 0.096 [0.094; 0.099], $p < 0.05$, the KS test). Importantly, ischemia leads to changes in the astrocytic response to ATP stimulation in the post-ischemic period. Characteristic examples of changes in the dynamics of astrocytic calcium activity after ATP application in normal conditions and in model ischemia-like conditions are presented in Figure 7. Despite the statistically significant increase in distant connections after ATP application, similar to the control group (Figure 2), the average correlation level remained unchanged (Figure 4), which indicates the impairment of the functional astrocytic network.

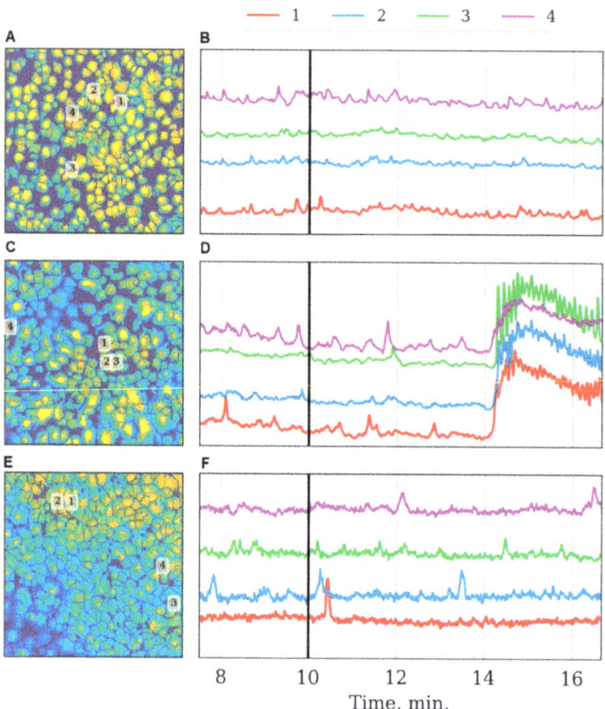

Figure 7. Characteristic examples of the dynamics of Oregon Green calcium sensor fluorescence in primary monoastrocytic cultures in vitro. (**A**,**B**)—normal conditions, PBS application; (**C**,**D**)—normal conditions, ATP applications; (**E**,**F**)—ischemic model, ATP application. (**A**,**C**,**E**)—examples of culture images segmented according to astrocyte positions. (**B**,**D**,**F**)—dynamics of calcium sensor fluorescence in selected astrocytes, note the correspondence between the astrocyte numbering and line colors. Vertical black lines indicate the moments of PBS or ATP application.

The frequency characteristics, particularly the average low-frequency component of calcium intensity, of the astrocytic network were also changed significantly after ischemia-like condition modeling. The average frequency of base-level oscillations was increased by a factor of 3 relative to the control values on day 7 after ischemia influence and was 0.0219 [0.0206; 0.0237] ($p < 0.05$, the KS test). Moreover, the signal delay rate between astrocytes was reduced significantly ("control": 24.81 [21.32; 26.85], "ischemia": 16.98 [16.96; 20.84], $p < 0.05$, the KS test).

Investigation of the role of connexin 43 in the regulation of astrocytic resistivity to ischemic conditions revealed that the Cx43 blockade during ischemia modeling improves the viability of cells in the post-ischemic period. On the seventh day from the start, the number of viable cells in the "ischemia+Gap19" group was not distinguishable from that in the control group (Supplementary Materials 1).

Assessment of the functional calcium activity in the "ischemia+Gap19" group did not reveal significant changes in the main parameters of functional Ca2+ activity compared with the "ischemia" and "control" group values on day 7 after induction of the ischemia model (Figure 3).

In contrast, the application of Gap19 during ischemia modeling maintained all parameters of network connectivity. The average number of long-distance connections of an astrocyte, the average value of the correlation between pairs of astrocytes, average low-frequency component of calcium intensity, and the signal delay rate were preserved at the level of the control values. Furthermore, partial normalization of the response to ATP application was observed, namely, the increasing average correlation of astrocytic activity, although the number of distant connections of a single astrocyte and the frequency of calcium oscillations were not significantly different. Thus, the blockade of hemichannels during ischemic modeling preserves the connectivity of astrocytes in the post-hypoxic period.

3. Discussion

Earlier studies showed that astrocytes interact with each other by transferring low-molecular-weight compounds through gap junctions [17]. This process not only enables the long-distance transfer of energy substrates but is also important for the activation of several ATP-dependent reactions in two adjacent astrocytes. Calcium waves can propagate between astrocytes and form complex spatial-temporal patterns of Ca2+ activity [5]. However, whether astrocytes can regulate metabolic activity and autonomously provide a systematic response to stress conditions independently of neurons remains unclear.

The most important issue in investigating the collective actrocytic response is defining calcium activity characteristics that would reflect the network behavior. Our proposed approach to dynamic astrocytic network construction allows us to identify functional connections between astrocytes and determine the presence and changes of network activity. The developed algorithm represents the astrocytic network as an oriented graph, the vertices of which correspond to separate astrocytes, while the edges designate above-threshold correlations between astrocytic events. When the correlation between a pair of astrocytes is maximal for a significant time shift between calcium signals, it is attributed to the directional coupling. In turn, such causality allows us to define and estimate the rate of calcium wave propagation in primary astrocyte culture.

Selecting a threshold for the correlation value is crucial for identifying and tracking astrocytic networks. The choice is not apparent since even statistically independent random time series can yield nonzero correlations over finite time windows. Moreover, the characteristic level of spurious correlations would not be universal and depends on particular signals. To circumvent the problem, we defined the 'baseline' correlation level as the correlation value for the most distant astrocytes in the observation field in the control case, assuming that little or no actual interaction exists between them. The threshold chosen in this way is equally applicable to astrocyte cultures subject to external stimuli. In this case, changes in the network statistics provide insight into the modification of the astrocytic interaction network.

The developed algorithms are integrated into the custom-written software package, which performs complete analysis and data processing from the obtained microscope images to the dynamic network construction and can be a convenient tool for analyzing image series of cell metabolic activity. The approach currently applied to study the behavior of astrocytic cultures can further be used to analyze network activity in neuronal cultures and mixed neuron-glial cultures. More generally, the developed algorithms allow the processing of series of images obtained from arbitrary cell cultures with some intracellular chemical dynamics.

Our studies revealed that astrocytes exhibit comparatively low network activity under normal conditions in vitro. The number of long-distance connections and the level of correlations increase significantly in response to an increase in intercellular ATP concentration, which cannot be explained by simple ATP-mediated activation of calcium homeostasis, as the calculated correlations are insensitive to the changes in the base level of calcium activity and the frequency of calcium oscillations. This demonstrates that astrocytes can become organized into a functional dynamic network in response to external stimulation.

Ischemia modeling has a significant impact on network activity, manifested by the decrease in all values that characterize network connectivity. The results also highlight important changes in astrocyte responses to ATP stimulation after ischemic influence. We hypothesize that the depression of the ATP-stimulated network emergence in this case is due to adaptive astrocytic reactions. Several studies addressed the changes in the functional activity of neuronal networks in hypoxic and ischemic lesions. Earlier, we demonstrated that hypoxia leads to the dramatic suppression of network interactions between neural cells, up to complete disruption, that is manifested in the alterations of both the calcium and bioelectric activity, as recorded by multi-electrode matrices [19,20]. At the same time, little is known regarding the network activity of mono-astrocytic cultures. Here, we are the first to show that while the calcium oscillations persist in the post-ischemic period, the correlations between cells are almost completely lost. Importantly, 10 µM ATP stimulation in normal conditions increases the frequency and improves correlations between calcium oscillations in distant astrocytes, such that a large-scale connected functional network emerges. In the post-ischemic period, such a response to ATP is absent, which suggests deep functional impairment of the astrocytic network.

There are experimental data showing changes in gliotransmitter regulation after hypoxic influence. It has been established that ATP release in the brain stem, presumably carried out by astrocytes, helps maintain cellular respiration and counteract hypoxia. It has been demonstrated that acute systemic hypoxia (a 5-min reduction in oxygen concentration in the breathable air to 10%) causes the release of a key gliotransmitter (ATP) in the brain stem areas responsible for the formation of breathing rhythm. Blockade of ATP receptors in the same brain stem area alleviates hypoxia-induced respiratory depression [21].

Interestingly, blockade of the main type of astrocytic connexins Cx43, which should provide a metabolically consolidated response under stress, preserves network parameters in primary astrocyte cultures for at least 7 days after the modeled ischemia-like conditions. A substantial increase in cell viability in primary astrocyte cultures in ischemia modeling and blockade of gap junctions was also observed. Accumulated experimental data indicate that Cx43 expression in astrocytes increases under hypoxic influence, and Cx43 plays an important role in cell death and neuronal damage caused by cerebral ischemia [22,23]. It is assumed that ischemia/reperfusion leads to an increase in the concentration of extracellular Ca2+ ions, the release of inflammatory factors, and, as a result, the activation of astrocytic hemichannels [24]. Hemichannels consisting of Cx43 can also be activated by kinase p38 and proinflammatory cytokines released by activated microglia [25]. This leads to an uncontrolled release of ATP, glutamate, and calcium, leading to calcium overload and tissue excitotoxicity [8]. Our data on the preservation of cell viability and network parameters of astrocytic cultures in the remote posthypoxic period are consistent with a study demonstrating that the use of the Cx43 blocker Gap26 improves neurological function in animals and decreases the incidence of heart attack in an ischemia model in vivo [26]. Therefore, Cx43 could be a promising therapeutic target in the development of methods for brain ischemia protection. This is an exciting area for upcoming research.

The current study revealed the emergent property of astrocytes to functionally unite in a large-scale network, and the Cx43 inhibition in ischemic conditions may be considered a favorable factor. In this regard, it is essential to further investigate astrocytic interactions as a factor in brain adaptation to ischemic damage, especially changes in these adaptive properties under physiological astrogliosis.

4. Materials and Methods

4.1. Ethics Statement

All experimental procedures were approved by the Bioethics Committee of Lobachevsky University and carried out in accordance with Act 708n (23 082010) of the Russian Federation National Ministry of Public Health, which states the rules of laboratory practice for the care and use of laboratory animals, and Council Directive 2010/63 EU of the European Parliament (22 September 2010) on the protection of animals used for scientific purposes. Newborn C57BL/6 (P1-P3) mice were killed by cervical vertebra dislocation.

4.2. Isolation of Primary Astrocyte Cultures

Primary astrocyte cultures were obtained from the cerebral cortex of newborn C57BL/6 mice (1–3 days after birth). The use of brain tissue in the early postnatal period minimizes the risk of the presence of nondifferentiated astrocytes in cultures.

Preparation and long-term culture of primary astrocyte cultures were performed in accordance with a protocol based on studies by [27] and [28] with several modifications. Surgically isolated cerebral cortices were cleared from the meninges and then mechanically dissected. To disrupt interconnections between cells, the tissue was additionally incubated with 0.25% trypsin solution (Thermo Fisher Scientific, Waltham, MA, USA) for 20 min in a CO_2 incubator (BINDER GmbH, Tuttlingen, Germany). Next, the cell suspension was washed three times in phosphate-buffered saline (PBS) and once in Dulbecco's Modified Eagle Medium (DMEM) containing 4.5 g/L glucose and supplemented with 0.5% L-glutamine, 1% B27 (Thermo Fisher, Waltham, MA, USA), 0.1% sodium pyruvate (Thermo Fisher, Waltham, MA, USA), and 10% fetal bovine serum (PanEco, Moscow, Russia). The suspension of dissociated cells was centrifuged at 800 rpm for 3 min. Then, the pellet was resuspended in culture medium, and, the obtained suspension was placed on coverslips pretreated with polyethylenimine solution (1 µg/mL) (Merck KGaA, Darmstadt, Germany), which provides effective cell attachment to the substrate. The initial cell density was 4500 cells/mm2. Primary astrocyte culture viability was maintained under constant conditions of 35.5 °C, 5% CO_2, and a humidified atmosphere in a cell culture incubator for more than 30 days. Half of the medium was replaced every third day.

4.3. Immunocytochemical Analysis

To verify the cellular content, primary astrocyte cultures were subjected to immunocytochemical staining on day 21 of culture development in vitro (DIV) (Supplementary Materials 1, Supplementary Materials Figure S1). The cultures were fixed with 4% paraformaldehyde for 15 min at room temperature, followed by incubation with a solution of 0.2% Triton X-100/PBS for effective cell permeabilization. For immunofluorescence reactions, the cultures were then incubated for 2 h in the presence of a polyclonal chicken anti-GFAP (glial fibrillary acidic protein, marker of differentiated astrocytes) primary antibody (Abcam, Cambridge, UK, 1:1000 dilution) and polyclonal goat anti βIII-tubulin (marker of differentiated neurons) primary antibody (Abcam, Cambridge, UK, 1:750 dilution). Next, the cultures were subjected to a 45-min incubation in the following secondary antibody mixture: mouse anti-chicken Alexa 647 (Thermo Fisher Scientific, Waltham, MA, USA, 1:100 dilution) and rabbit anti-mouse Alexa Fluor 555 (Thermo Fisher Scientific, Waltham, MA, USA, 1:100 dilution). The stained material was observed using a Zeiss 510 NLO fluorescence confocal microscope (Carl Zeiss, Oberkochen, Germany).

4.4. Calcium Imaging

The functional calcium activity of astrocytes was studied using an LSM 510 laser scanning microscope (Carl Zeiss, Oberkochen, Germany) with a W Plan-Apochromat 20×/1.0 objective. The calcium imaging technique allowed visualization of the functional architecture of cells in culture. We used the fluorescent calcium-sensitive dye Oregon Green 488 BAPTA-1 AM (OGB-1) (0.4 µM, Thermo Fisher Scientific, Waltham, MA, USA) dissolved in dimethylsulfoxide (DMSO) (Merck KGaA, Darmstadt, Germany) with 4% Pluronic F-127 (Thermo Fisher Scientific, Waltham, MA, USA). OGB-1 was added to the culture medium and incubated for 40 min in a CO2 incubator. The fluorescence of OGB1 was excited at 488 nm by argon laser radiation, and emission was recorded in the range of 500 to 530 nm. The dynamics of intracellular calcium concentration were measured by analysis of a time series of 512 × 512 pixel images capturing 420 µm × 420 µm fields of view that was recorded at 2 Hz. The following parameters of the functional calcium activity were assessed: duration of the calcium oscillations (time from the beginning to the end of an oscillation (s)), frequency of calcium oscillations (average number of oscillations per min), and percentage of working cells (ratio of the number of cells in which at least one oscillation was recorded among the total number of cells (%)) [29,30].

4.5. Ischemia-Like Condition Model

Ischemia-like conditions were modeled on day 21 of primary astrocyte culture development in vitro by replacing the culture medium with a medium with low oxygen (0.37 mL/L) free of glucose, lactate, and pyruvate for 30 min. The experiment was performed in a sealed chamber in which air was also replaced by argon gas.

4.6. Cell Viability Analysis

To identify the nuclei of dead cells and the total number of cell nuclei in the primary astrocyte cultures, propidium iodide (Merck KGaA, Darmstadt, Germany) and bis-benzimide (Merck KGaA, Darmstadt, Germany) were used. Solutions of 5 µg/mL propidium iodide and 1 µg/mL bis-benzimide were added separately to the culture medium 30 min before viability registration. The stained cultures were observed using a ZEISS Observer A1 inverted fluorescence microscope (Carl Zeiss, Oberkochen, Germany). The proportion of dead cells was calculated as the ratio of nuclei stained with propidium iodide to the total number of nuclei.

4.7. Biochemical Screening

To analyze the features of the functional activity of astrocytes, the following agents were added to the culture medium on day 28 of culture development in vitro: ATP (10 µM, Merck KGaA, Darmstadt, Germany) as an activator of calcium activity and Gap19 (100 µM, Merck KGaA, Darmstadt, Germany) to estimate the role of connexin blockade in the implementation of functional calcium reactions. The same volume of PBS was used as a control. The baseline activity was recorded for 10 min.

4.8. Calcium Event Analysis

Calcium event detection in astrocytes was performed by a previously developed algorithm [16]. Input data for the algorithm consist of two image series with the same dimensionality that are obtained from the same microscope: the astrocyte activity and images of a cell-free medium. The software pipeline includes the following steps:

1. To correctly measure relative changes of a signal, we subtract a constant offset, given by the time average of the cell-free image series.
2. In the second step, the two filters are applied sequentially: VBM3D signal filtering and spatial-temporal 3D box filtering.
3. Based on the cell-free image series, we evaluate the baseline of the calcium activity for each pixel. Next, the relative signal changes are estimated by referring to the baseline.
4. The astrocytic dynamic activity is defined as the signal that exceeds a spatial adaptive threshold above the baseline level. The adaptive threshold is computed by the statistical model of noise reduction adjusted according to the cell-free images (see details in [16]).
5. In the last step of the algorithm, spatiotemporal clustering of activity patterns into calcium events is performed by DBSCAN over time and by a window-based approach over space.

Initially, the entire image plane is considered, and events are identified as spatiotemporally connected areas with significant activity. Next, the watershed method is applied to the long-exposure calcium intensity image to discriminate the areas corresponding to individual astrocytes. Finally, calcium events are segmented according to astrocyte regions. Further analysis utilizes the obtained individual astrocyte calcium events and filtered intensity signals.

4.9. Construction of a Dynamic Astrocytic Network

The approach for reconstruction of an astrocytic network is based on calculating the Pearson correlation coefficient between filtered signals of each cell pair:

$$\rho_{ij} = \frac{\sum_{k=1}^{n} \check{x}_k^i \, \check{x}_k^j}{\sqrt{\sum_{k=1}^{n} \left(\check{x}_k^i\right)^2 \sum_{k=1}^{n} \left(\check{x}_k^j\right)^2}}, \tag{1}$$

$$\check{x}_k^i = x_k^i - \langle x_s^i \rangle_{k-w,k} \tag{2}$$

x_k^i—calcium signal of i-th cell at time k,

$\check{x}_k^i$—calcium signal minus moving average with window size w,

$\langle x_s^i \rangle_{k-w,k}$—average of signal in range $[k-w, k]$.

The astrocytic network is represented as an undirected graph, where vertices correspond to cells, and edges are drawn between the cells for which the correlation coefficient exceeds a certain threshold. Cellular signals are characterized by two main quantities: the level of intracellular calcium and the size of a calcium event inside the cell. Both are used to construct the network. Furthermore, the propagation of calcium signals between cells leads to detectable time delays in calcium elevations (Supplementary Material Movie 1) and in certain cases allows for assignment of a directed edge. Different time delays were probed to choose the one for which the correlation between a pair of cells would be maximal (Supplementary Material 2).

A threshold was set to reject spurious correlations between cell calcium signals that would be caused by coincidence rather than actual interaction. The choice of the threshold was based on the following. Since the direct interaction between astrocytes is local, the mutual influence should decrease with distance. Therefore, the baseline level of correlations can be estimated from values that are found for remote astrocytes. Given the typical size of an astrocyte up to 40 µm, we referred to the correlation level between cells separated by at least 100 µm as the baseline [31].

The implementation of this approach is illustrated in Figure 8A,B, which shows a typical relationship between the level of correlation and the distance in pairs of astrocytes in the control experiment. Adjacent cells are marked in red. Three indicative groups of points are highlighted in the figure. For the nearby and directly interacting astrocytes (distance between central points < 40 µm), the correlation coefficient can reach 0.9. For distant astrocytes (distance > 300 µm), the correlation does not exceed 0.3. We used this value as a threshold to distinguish a significant correlation between pairs of astrocytes. The third group of points is represented by pairs of astrocytes located at distances ranging from 40 to 300 µm with cross-correlation values exceeding 0.3. These properties are interpreted as the result of an indirect dynamic interaction between astrocytes and almost do not occur in the control group.

The correlation astrocytic network is constructed as follows: the vertices of the graph are mapped to astrocytes, and the presence of a significant level of correlation between pairs of astrocytes (correlation greater than 0.3) is indicated by an edge connecting the corresponding vertices. A characteristic example of the obtained network for the control group is presented in Figure 8C. Such networks typically display a sufficiently large number of small local groups.

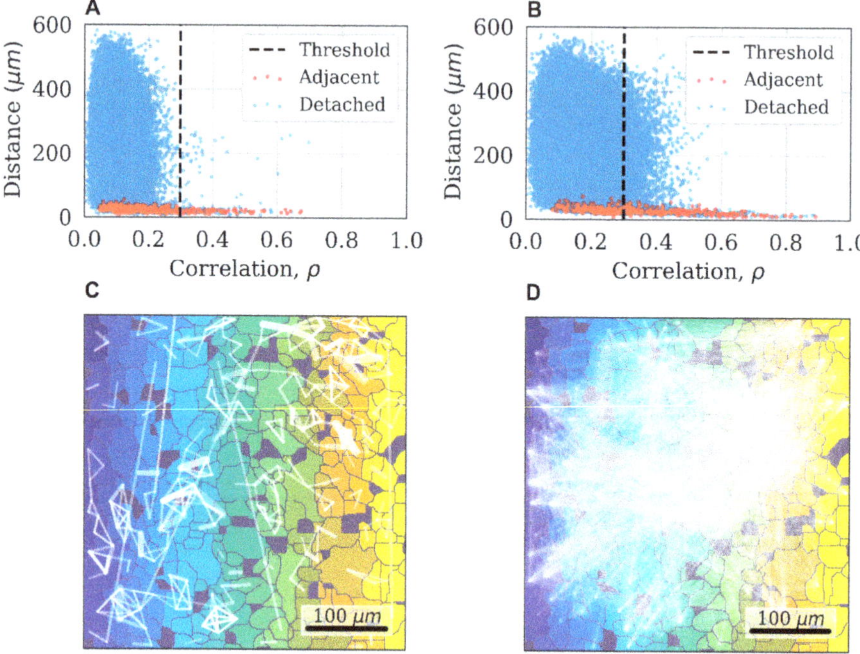

Figure 8. The distance–correlation relationship between pairs of neighboring (red) and distant (blue) astrocytes in the control state (**A**) and after applying ATP; (**B**) An example of a correlation network of astrocytes with a threshold $\rho > 0.3$ for the control state (**C**) and after applying ATP (**D**).

4.10. Dynamic Astrocytic Network Analysis

The impact of external factors on astrocytic culture can lead to changes in the correlation and dynamic properties of the system. As a result of exposure to ATP, the point cloud expands and shows higher correlation values at long distances (Figure 8A,B). This reflects an increase in the connectivity of the dynamical astrocytic network. While the original graph would contain several disconnected subgraphs, the resulting correlation astrocytic network manifests the so-called giant component, a connected subgraph of the size of the order of the entire network, due to the emerging large number of long-range functional connections (Figure 8D).

Network analysis focuses on the following key features: the number of functional connections between astrocyte pairs, the average number of connections between astrocytes, the average propagation speed of delays between signals, the average correlation level of network cells, and the frequency of cell signals, as described in detail in Supplementary Material 2.

4.11. Statistical Analysis

The astrocytic responses to various biochemical stimuli were investigated using statistical analysis. Astrocyte cultures were divided into different groups. The control group was designated as representing a normal state of astrocytic activity.

The influence of external factors on the astrocyte state was determined by comparing feature samples of the control and case groups by the one-sided two-sample Kolmogorov–Smirnov test. This is a nonparametric test that quantifies the distance between the empirical distribution functions of two samples [32]. The computed features of some groups violate the normality assumption for parametric tests. The normality of distributions is tested by applying the Kolmogorov–Smirnov test (KS test)

($p < 0.05$ for some groups, but not for all). To overcome this, a two-sample KS test was applied. Determination of the distribution shift direction was provided by a one-sided KS test. To assess whether group mean ranks differ, the Wilcoxon rank-sum test was applied to the data. A t-test was performed to compare group values with the initial baseline.

Statistical significance was determined using stats module of SciPy library [33]. The one-sided two-sample KS test is performed by the ks_2samp function. The Wilcoxon rank-sum test is performed by the Mann-Whitney function. Differences between groups were considered statistically significant if $p < 0.05$.

Descriptive statistics of each feature per group are represented as "M [Q1; Q3]", where M—median, Q1—first quartile (quantile 0.25), and Q3—third quartile (quantile 0.75) of the group samples.

5. Conclusions

We developed a novel mathematical approach to reconstruct functional astrocytic networks based on an analysis of dynamic correlations between calcium events. This approach revealed dramatic changes in the structure and properties of astrocytic networks in normal and ischemia-like conditions.

Supplementary Materials: The following are available online at http://www.mdpi.com/1422-0067/21/21/7952/s1, Figure S1: Morphology of primary astrocyte cultures on day 21 of cultivation in vitro, Table S1: Cell viability analysis of primary astrocyte cultures on day 7 after modeled stress factors, Video S1: the propagation of calcium signals between astrocytes.

Author Contributions: Conceptualization, E.V.M., M.V.I. and M.V.V.; Data curation, M.V.I.; Formal analysis, M.I.K. and D.E.B.; Investigation, E.V.M., M.I.K., D.E.B., M.O.S. and T.A.M.; Supervision, M.V.I. and M.V.V.; Visualization, M.I.K.; Writing—original draft, E.V.M., M.V.I. and M.V.V. All authors have read and agreed to the published version of the manuscript.

Funding: We acknowledge support by the grant of the Ministry of Education and Science of the Russian Federation Agreement No. 075-15-2020-808. This research was carried out using The Core Facilities «Molecular Biology and Neurophysiology».

Conflicts of Interest: The authors declare no conflict of interest.

References

1. Pannasch, U.; Rouach, N. Emerging role for astroglial networks in information processing: From synapse to behavior. *Trends Neurosci.* **2013**, *36*, 405–417. [CrossRef] [PubMed]
2. Siracusa, R.; Fusco, R.; Cuzzocrea, S. Astrocytes: Role and Functions in Brain Pathologies. *Front. Pharmacol.* **2019**, *10*, 1114. [CrossRef] [PubMed]
3. Booth, H.D.; Hirst, W.D.; Wade-Martins, R. The Role of Astrocyte Dysfunction in Parkinson's Disease Pathogenesis. *Trends Neurosci.* **2017**, *40*, 358–370. [CrossRef] [PubMed]
4. Rodríguez-Arellano, J.; Parpura, V.; Zorec, R.; Verkhratsky, A. Astrocytes in physiological aging and Alzheimer's disease. *Neuroscience* **2016**, *323*, 170–182. [CrossRef] [PubMed]
5. Semyanov, A.V. Spatiotemporal pattern of calcium activity in astrocytic network. *Cell Calcium* **2019**, *78*, 15–25. [CrossRef]
6. Verkhratsky, A.; Nedergaard, M. Physiology of Astroglia. *Physiol. Rev.* **2018**, *98*, 239–389. [CrossRef] [PubMed]
7. Naus, C.C.; Giaume, C. Bridging the gap to therapeutic strategies based on connexin/pannexin biology. *J. Transl. Med.* **2016**, *14*, 330. [CrossRef]
8. Kim, Y.; Davidson, J.; Green, C.R.; Nicholson, L.F.; O'Carroll, S.J.; Zhang, J. Connexins and Pannexins in cerebral ischemia. *Biochim. Biophys. Acta Biomembr.* **2018**, *1860*, 224–236. [CrossRef]
9. Davidson, J.O.; Geen, C.R.; Bennet, L.; Nicholson, L.F.; Danesh-Meyer, F.; O'Carroll, S.H.; Gunn, A.J. A key role for connexin hemichannels in spreading ischemic brain injury. *Curr. Drug Targets.* **2013**, *14*, 36–46. [CrossRef]
10. Retamal, M.A.; Alcayaga, J.; Verdugo, C.A.; Bultynck, G.; Leybaert, L.; Sáez, P.J.; Fernandez, R.; Leon, L.E.; Sáez, J.C. Opening of pannexin- and connexin-based channels increases the excitability of nodose ganglion sensory neurons. *Front. Cell. Neurosci.* **2014**, *8*. [CrossRef]

11. Sáez, P.J.; Shoji, K.F.; Retamal, M.A.; Harcha, P.A.; Ramírez, G.; Jiang, J.X.; von Bernhardi, R.; Sáez, J.C. ATP is Required and Advances Cytokine-Induced Gap Junction Formation in Microglia In Vitro. *Mediat. Inflamm.* **2013**, *2013*, 1–16. [CrossRef]

12. Abudara, V.; Bechberger, J.; Freitas-Andrade, M.; de Bock, M.; Wang, N.; Bultynck, G.; Naus, C.C.; Leybaert, L.; Giaume, C. The connexin43 mimetic peptide Gap19 inhibits hemichannels without altering gap junctional communication in astrocytes. *Front. Cell. Neurosci.* **2014**, *8*, 306. [CrossRef] [PubMed]

13. Chen, Y.; Wang, L.; Zhang, L.; Chen, B.; Yang, L.; Li, X.; Li, Y.; Yu, H. Inhibition of Connexin 43 Hemichannels Alleviates Cerebral Ischemia/Reperfusion Injury via the TLR4 Signaling Pathway. *Front. Cell. Neurosci.* **2018**, *12*, 372. [CrossRef] [PubMed]

14. Freitas-Andrade, M.; Wang, N.; Bechberger, J.F.; de Bock, M.; Lampe, P.D.; Leybaert, L.; Naus, C.C. Targeting MAPK phosphorylation of Connexin43 provides neuroprotection in stroke. *J. Exp. Med.* **2019**, *216*, 916–935. [CrossRef] [PubMed]

15. Yin, X.; Feng, L.; Ma, D.; Yin, P.; Wang, X.; Hou, S.; Hao, Y.; Zhang, J.; Xin, M.; Feng, J. Roles of astrocytic connexin-43, hemichannels, and gap junctions in oxygen-glucose deprivation/reperfusion injury induced neuroinflammation and the possible regulatory mechanisms of salvianolic acid B and carbenoxolone. *J. Neuroinflammation* **2018**, *15*, 1–24. [CrossRef]

16. Kustikova, V.; Krivonosov, M.; Epimashkin, A.; Denisov, P.; Zaikin, A.; Ivanchenko, M.; Meyerov, I.; Semyanov, A. CalciumCV: Computer Vision Software for Calcium Signaling in Astrocytes. In *Analysis of Images, Social Networks and Texts*; Springer: Cham, Switzerland, 2018; pp. 168–179.

17. Eugenin, E.A.; Basilio, D.; Sáez, J.C.; Orellana, J.A.; Raine, C.S.; Bukauskas, F.; Bennett, M.V.L.; Berman, J.W. The Role of Gap Junction Channels during Physiologic and Pathologic Conditions of the Human Central Nervous System. *J. Neuroimmune Pharmacol.* **2012**, *7*, 499–518. [CrossRef]

18. le Feber, J. In Vitro Models of Brain Disorders. In *In Vitro Neuronal Networks*; Advances in Neurobiology; Chiappalone, M., Pasquale, V., Frega, M., Eds.; Springer: Cham, Switzerland, 2019; Volume 22, pp. 19–49.

19. Mitroshina, E.V.; Mishchenko, T.A.; Usenko, A.V.; Epifanova, E.A.; Yarkov, R.S.; Gavrish, M.S.; Babaev, A.A.; Vedunova, M.V. AAV-Syn-BDNF-EGFP Virus Construct Exerts Neuroprotective Action on the Hippocampal Neural Network during Hypoxia In Vitro. *Int. J. Mol. Sci.* **2018**, *19*, 2295. [CrossRef]

20. Savyuk, M.; Krivonosov, M.; Mishchenko, T.; Gazaryan, I.; Ivanchenko, M.; Khristichenko, A.; Poloznikov, A.; Hushpulian, D.; Nikulin, S.; Tonevitsky, E.; et al. Neuroprotective Effect of HIF Prolyl Hydroxylase Inhibition in an In Vitro Hypoxia Model. *Antioxidants* **2020**, *9*, 662. [CrossRef]

21. Marina, N.; Kasymov, V.; Ackland, G.L.; Kasparov, S.; Gourine, A.V.; Roach, R.C.; Hackett, P.H.; Wagner, P.D. Astrocytes and Brain Hypoxia. In *Hypoxia*; Advances in Experimental Medicine and Biology; Roach, R., Hackett, P., Wagner, P., Eds.; Springer: Boston, MA, USA, 2016; pp. 201–207.

22. Ma, D.; Feng, L.; Cheng, Y.; Xin, M.; You, J.; Yin, X.; Hao, Y.; Cui, L.; Feng, J. Astrocytic gap junction inhibition by carbenoxolone enhances the protective effects of ischemic preconditioning following cerebral ischemia. *J. Neuroinflamm.* **2018**, *15*, 1–12. [CrossRef]

23. Davidson, J.; Green, C.; Bennet, L.; Gunn, A. Battle of the hemichannels—Connexins and Pannexins in ischemic brain injury. *Int. J. Dev. Neurosci.* **2014**, *45*, 66–74. [CrossRef]

24. Xing, L.; Yang, T.; Cui, S.; Chen, G. Connexin Hemichannels in Astrocytes: Role in CNS Disorders. *Front. Mol. Neurosci.* **2019**, *12*, 23. [CrossRef] [PubMed]

25. Giaume, C.; Leybaert, L.; Naus, C.C.; Sáez, J.C. Connexin and pannexin hemichannels in brain glial cells: Properties, pharmacology, and roles. *Front. Pharmacol.* **2013**, *4*, 88. [CrossRef] [PubMed]

26. Li, X.; Zhao, H.; Tan, X.; Kostrzewa, R.M.; Du, G.; Chen, Y.; Zhu, J.; Miao, Z.; Yu, H.; Kong, J.; et al. Inhibition of connexin43 improves functional recovery after ischemic brain injury in neonatal rats. *Glia* **2015**, *63*, 1553–1567. [CrossRef] [PubMed]

27. Evren, V.; Apaydin, M.; Khalilnezhad, A.; Erbaş, O.; Taskiran, D. Protective effect of edaravone against manganese-induced toxicity in cultured rat astrocytes. *Environ. Toxicol. Pharmacol.* **2015**, *40*, 563–567. [CrossRef]

28. Ogier, C.; Bernard, A.; Chollet, A.M.; Le Diguardher, T.; Hanessian, S.; Charton, G.; Khrestchatisky, M.; Rivera, S. Matrix metalloproteinase-2 (MMP-2) regulates astrocyte motility in connection with the actin cytoskeleton and integrins. *Glia* **2006**, *54*, 272–284. [CrossRef]

29. Zakharov, Y.N.; Mitroshina, E.V.; Shirokova, O.; Emukhina, I. Calcium Transient Imaging as Tool for Neuronal and Glial Network Interaction Study. In *Models, Algorithms, and Technologies for Network Analysis*;

Goldengorin, B., Kalyagin, V., Pardalos, P., Eds.; Springer Proceedings in Mathematics & Statistics; Springer: New York, NY, USA, 2012; Volume 32, pp. 225–232.

30. Vedunova, M.; Sakharnova, T.; Mitroshina, E.; Perminova, M.; Pimashkin, A.; Zakharov, Y.; Dityatev, A.; Mukhina, I. Seizure-like activity in hyaluronidase-treated dissociated hippocampal cultures. *Front. Cell. Neurosci.* **2013**, *7*, 149. [CrossRef]

31. Haseleu, J.; Anlauf, E.; Blaess, S.; Endl, E.; Derouiche, A. Studying subcellular detail in fixed astrocytes: Dissociation of morphologically intact glial cells (DIMIGs). *Front. Cell. Neurosci.* **2013**, *7*, 7. [CrossRef]

32. Pratt, J.W.; Gibbons, J.D. Kolmogorov-Smirnov Two-Sample Tests. In *Concepts of Nonparametric Theory*; Springer Series in Statistics; Springer: New York, NY, USA, 1981; pp. 318–344.

33. Virtanen, P.; Gommers, R.; Oliphant, T.E.; Haberland, M.; Reddy, T.; Cournapeau, D.; Burovski, E.; Peterson, P.; Weckesser, W.; Bright, J.; et al. SciPy 1.0: Fundamental algorithms for scientific computing in Python. *Nat. Methods* **2020**, *17*, 261–272. [CrossRef]

Publisher's Note: MDPI stays neutral with regard to jurisdictional claims in published maps and institutional affiliations.

International Journal of
Molecular Sciences

Article

Failed Neuroprotection of Combined Inhibition of L-Type and ASIC1a Calcium Channels with Nimodipine and Amiloride

Jonas Ort [1,*], Benedikt Kremer [1], Linda Grüßer [2], Romy Blaumeiser-Debarry [2], Hans Clusmann [1], Mark Coburn [3], Anke Höllig [1,†] and Ute Lindauer [1,†]

[1] Department of Neurosurgery, Medical Faculty, RWTH Aachen University, 52074 Aachen, Germany;
bkremer@ukaachen.de (B.K.); hclusmann@ukaachen.de (H.C.); ahoellig@ukaachen.de (A.H.);
ulindauer@ukaachen.de (U.L.)
[2] Department of Anaesthesiology, Medical Faculty, RWTH Aachen University, 52074 Aachen, Germany;
lgruesser@ukaachen.de (L.G.); rblaumeiser-debarry@ukaachen.de (R.B.-D.)
[3] Department of Anaesthesiology and Intensive Care Medicine, University Hospital Bonn,
53127 Bonn, Germany; mark.coburn@ukbonn.de
* Correspondence: jort@ukaachen.de; Tel.: +49-241-800
† These authors have equal contributions.

Received: 7 November 2020; Accepted: 20 November 2020; Published: 24 November 2020

Abstract: Effective pharmacological neuroprotection is one of the most desired aims in modern medicine. We postulated that a combination of two clinically used drugs—nimodipine (L-Type voltage-gated calcium channel blocker) and amiloride (acid-sensing ion channel inhibitor)—might act synergistically in an experimental model of ischaemia, targeting the intracellular rise in calcium as a pathway in neuronal cell death. We used organotypic hippocampal slices of mice pups and a well-established regimen of oxygen-glucose deprivation (OGD) to assess a possible neuroprotective effect. Neither nimodipine (at 10 or 20 µM) alone or in combination with amiloride (at 100 µM) showed any amelioration. Dissolved at 2.0 Vol.% dimethyl-sulfoxide (DMSO), the combination of both components even increased cell damage ($p = 0.0001$), an effect not observed with amiloride alone. We conclude that neither amiloride nor nimodipine do offer neuroprotection in an in vitro ischaemia model. On a technical note, the use of DMSO should be carefully evaluated in neuroprotective experiments, since it possibly alters cell damage.

Keywords: neuroprotection; neural injury; nimodipine; subarachnoid haemorrhage; acid-sensing ion channels; amiloride; oxygen-glucose deprivation

---

1. Introduction

Despite years of research pharmacological neuroprotection remains challenging. Neuronal cell death represents the smallest common denominator occurring within the pathophysiological cascade of acute cerebrovascular or traumatic brain diseases. For example, subarachnoid haemorrhage occurs with an incidence of 9/100,000 per year [1], while traumatic brain injury is one of the leading causes for death and disability in the young with reported incidences of 262/100,000 per year [2]. Considering the dreads of brain damage for patients and their dependants, therapeutic means of effective neuroprotection is possibly one of the most aimed-for means in modern medicine.

Although the pathomechanisms of neuronal cell death are yet to be fully understood, one of the first and most frequently investigated events is the elevation of intracellular calcium concentration. Elevated intracellular calcium represents a central part within the early phenomenon of excitotoxicity as well as all the pathways resulting in or from these raised ion concentrations [3–7]. Various drugs blocking the excitotoxicity-induced calcium influx have failed to show effective neuroprotection in the first place or failed in overcoming the translational roadblock from in vivo or animal in vitro experiments to actual clinical application (e.g., trials of glutamate-receptor antagonists) [8,9]. Beside calcium influx via glutamate receptor overactivation, calcium influx may also occur via voltage-gated calcium channels (VGCC) located within neuronal membranes. Additionally, VGCCs are also expressed by vascular smooth muscle cells responsible for intracellular calcium rise and vasoconstriction when activated.

The L-type calcium channel inhibitor nimodipine is well established as an oral agent for the management of delayed cerebral ischaemia (DCI) for patients after subarachnoid haemorrhage (SAH) [10–15]. Although nimodipine induces vascular smooth muscle relaxation [16] and hence is widely believed to reduce vasospasms, research has shown that the beneficial effects in SAH patients with DCI may not be primarily caused by this effect on larger cerebral vessels [12,17], but possibly by a direct neuroprotective effect. Experimental studies have been rather controversial with some arguing towards direct neuroprotection [18–21] and some against [22–25]. In brief, the inhibition of voltage-gated L-type calcium channels is believed to provide protection against raised intracellular calcium concentration once the cell depolarises in ischaemic conditions. However, it remains unclear whether this effect also counts for the observed beneficial effect of VGCC inhibitors in SAH.

Other possible drug targets regarding cerebral ischaemia are sodium- and (to a lesser extent) calcium-permeable acid-sensing ion channels (ASICs), with the ASIC1a subtype channel as the most prominent amongst them [26]. ASIC1a channels can be responsible for neuronal damage in acidosis and provide a neither voltage-gated nor glutamate-dependent mechanism for calcium influx. Amiloride can block this non-specifically [27–29]. Many experiments have already shown that neuroprotection can be achieved by blocking ASIC1a and thus targeting these channels might be promising [9,30,31]. Some studies point out connections between the affinity of ASIC for H^+ depending and the extracellular concentration of calcium [32,33], as it occurs in brain ischemia [34].

In our experiment, we investigated the possible neuroprotective effect of two clinically used drugs, which both block calcium channels: The L-type calcium channel inhibitor nimodipine and the acid-sensing ion channel (ASIC1a) blocker amiloride.

We hypothesised a possible synergistic neuroprotective effect of the two compounds being used in combination since ASIC channels are known to be influenced by external and internal calcium levels [32,33]. To assess our postulation, we used a well-established in vitro model of oxygen–glucose deprivation (OGD) in organotypic hippocampus slices of mice with propidium iodide (PI) staining for cell death assessment. Nimodipine was investigated for neuroprotective effects alone and in combination with amiloride to observe a possible synergistic interaction.

2. Results

2.1. OGD Damage

Comparing our control group ($n = 102$) with our OGD group ($n = 96$) a significant difference was seen ($p < 0.0001$, Mann–Whitney test), thus demonstrating significant and robust cell damage by our OGD regimen (Figure 1).

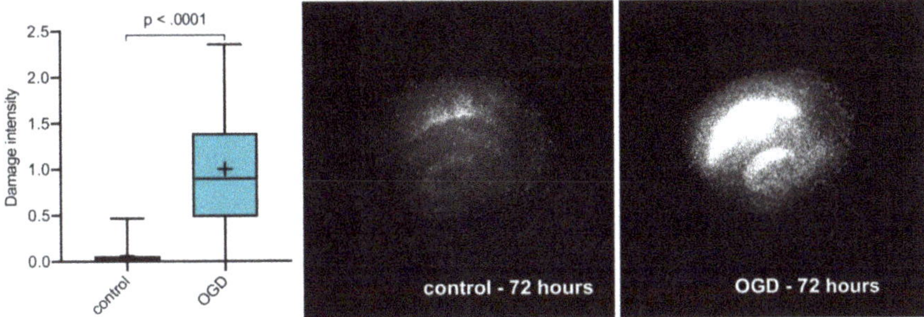

Figure 1. Damage intensity of the control group ($n = 101$) vs. oxygen–glucose deprivation (OGD) group ($n = 96$). Mann–Whitney U test was used to determine whether our OGD model caused adequate damage. Slice images shown are example slices for control and OGD groups, both at 72 h with typical means for the grey-scale value. Slice images are red-channel-filtered and contrast-enhanced.

2.2. Effect of DMSO as Vehicle on OGD-Induced Damage

To rule out possible effects of our vehicle, slices were incubated receiving dimethyl-sulfoxide (DMSO) without any drug in addition either as controls (Figure 2, left) or undergoing OGD (Figure 2, right) at levels of 0.1, 1.0, or 2.0 Vol.% DMSO. For pure vehicle control slices without OGD, there was a tendency towards a concentration dependent effect; however, with significant cell impairment observed only with the highest concentration of 2.0 Vol.% DMSO ($p = 0.0001$, Kruskal–Wallis with post hoc Dunn's test). Following OGD, there was a slight albeit far not significant tendendy towards a protective effect of DMSO (Figure 2, right: DMSO concentration of 1.0 Vol.%, $p = 0.2237$) compared with the OGD slices without vehicle or with other DMSO concentrations.

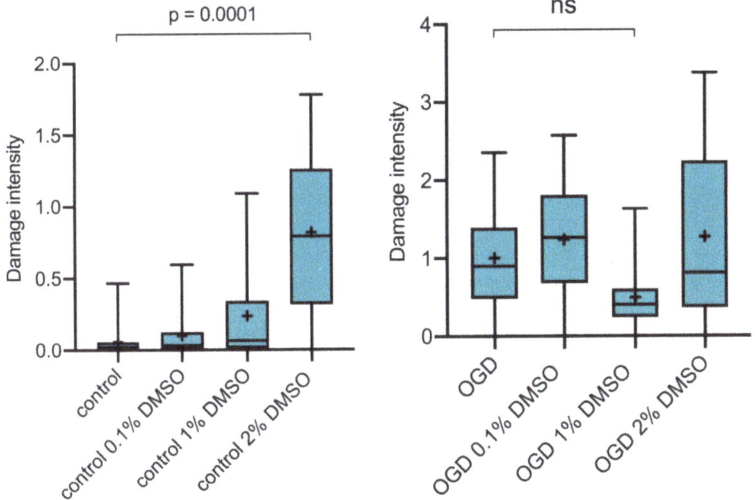

Figure 2. (**Left**): Slices receiving 2.0 Vol.% dimethyl-sulfoxide (DMSO) alone ($n = 20$) without undergoing OGD showed significantly ($p = 0.0001$) more cell damage. This effect can be observed for 0.1 ($n = 25$) and 1.0 Vol.% ($n = 12$) as well but at a much lower level without statistical significance. (**Right**): No significant effects of DMSO in slices undergoing OGD were observed for 0.1 ($n = 41$), 1.0 ($n = 19$) or 2.0 ($n = 45$) Vol.% DMSO compared to the OGD group without DMSO ($n = 96$). The figure depicts a tendency towards less cell damage at 1.0 Vol.% DMSO.

2.3. Effect of Nimodipine at 10 or 20 µM Dissolved in Varying DMSO Concentrations

L-type calcium channel blockage with nimodipine at either 10 or 20 µM dissolved in 0.1 or 1.0 Vol.% DMSO, respectively, did not show significant neuroprotection (Figure 3).

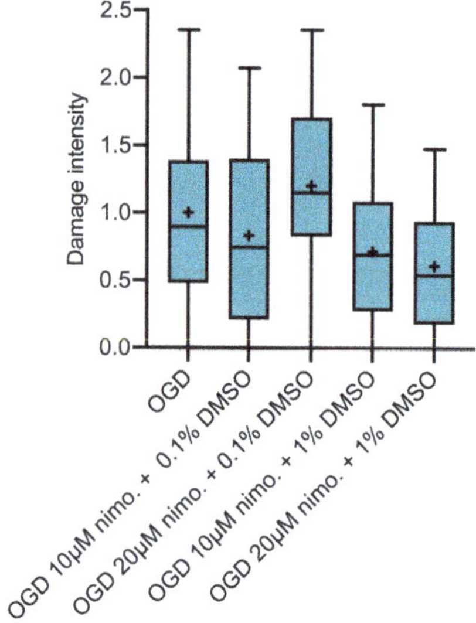

Figure 3. Slices undergoing OGD received nimodipine at either 10 or 20 µM dissolved in 0.1 or 1.0 Vol.% DMSO. No significant differences in cell damage were observed at 72 h after OGD (OGD n = 96; dissolved in 0.1 Vol.% DMSO: 10µM nimodipine n = 77, 20 µM nimodipine n = 64; dissolved in 1.0 Vol.% DMSO: 10 µM nimodipine n = 30, 20 µM nimodipine n = 25).

2.4. Effect of Nimodipine at 10 or 20 µM in Combination with 100 µM Amiloride Dissolved in Varying DMSO Concentrations

To observe a neuroprotective effect of combined blockage of L-type calcium channels and ASIC1a, a combination of 10 or 20 µM nimodipine and 100 µM amiloride dissolved in 1.1 Vol.% DMSO was applied. No significant effect for the combination of nimodipine and amiloride or amiloride alone at 1.0 Vol.% DMSO could be observed (Figure 4). Further, no significant difference was detected in slices receiving nimodipine at 10 or 20 µM in 1.0 Vol.% DMSO or with the vehicle alone. These slices showed a tendency towards less severe cell damage when compared to the OGD group and slices receiving regimens with 0.1 Vol.% DMSO (also Figure 4).

Lastly, treatment combination of 100 µM amiloride and either 10 or 20 µM nimodipine was investigated while using a final concentration of 2.0 Vol.% DMSO. No combination of nimodipine and amiloride or amiloride alone showed any significant neuroprotective effect. However, slices receiving the combination of 100 µM amiloride and 10µM (p = 0.0001, Kruskal–Wallis with post hoc Dunn's test, Cohen's d = 1.196243) or 20 µM (p = 0.0001, Kruskal–Wallis with post hoc Dunn's test, Cohen's d = 1.219638) nimodipine at 2.0 Vol.% DMSO displayed significantly higher levels of cell damage at 72 h after OGD compared to the OGD group without any treatment (Figure 5).

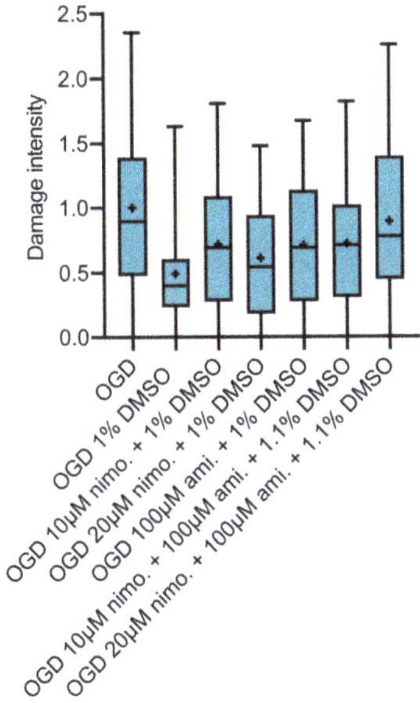

Figure 4. Slices receiving 10 (*n* = 30) or 20 µM (*n* = 25) nimodipine in 1.0 Vol.% DMSO compared to slices receiving 100 µM amiloride alone in 1 Vol.% DMSO (*n* = 50) or in combination with 10 (*n* = 56) or 20 µM (*n* = 59) nimodipine in 1.1 Vol.% DMSO. No significant difference in cell damage intensity was observed at 72 h after OGD.

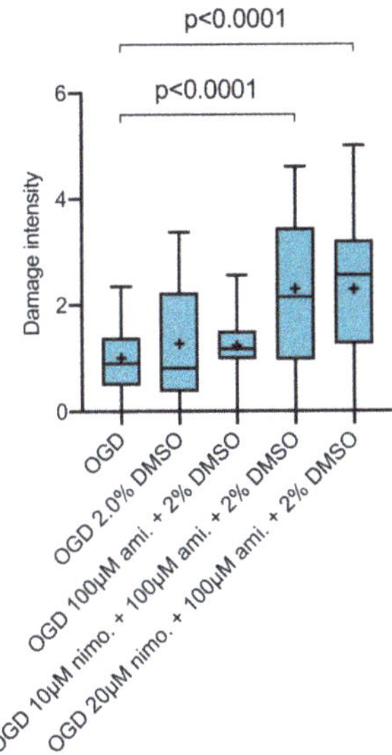

Figure 5. Slices receiving a combination of either 10 (*n* = 47) or 20 µM (*n* = 46) nimodipine with 100 µM amiloride in 2.0 Vol.% DMSO showed significant impairment of cell viability at 72 h (Kruskal–Wallis test with post hoc Dunn's test, *p* < 0.0001) compared to OGD slices (*n* = 96). This effect was not observed for slices receiving 100 µM amiloride in 2.0 Vol.% DMSO (*n* = 43) or 2.0 Vol.% DMSO alone (*n* = 45).

3. Discussion

In our study, we investigated the clinically commonly used drugs nimodipine and amiloride to test for neuroprotective effects in an in vitro model of brain ischemia using OGD. No protective effect could be observed for nimodipine or amiloride alone or in combination. Our model was already used in the past to prove neuroprotective effects of other drugs, such as the noble gas argon, which showed a protection of >80% in the best suited protocol [35]. DMSO, which was used as vehicle to dissolve both drugs, induced significant impairment of cell integrity at 2.0 Vol.% in our control slices (that did not undergo OGD) compared with the control group without DMSO. Paradoxically, this effect was not observed at the same concentration in slices undergoing OGD. There was even a tendency towards ameliorated levels of cell damage in OGD slices at 1.0 Vol.% DMSO (Figure 2, right). Interestingly, the combination of nimodipine and amiloride significantly increased cell damage when dissolved at 2.0 Vol.% DMSO. This effect was not apparent for 2.0 Vol.% DMSO alone or with the same concentration as vehicle for 100 µM amiloride. Nimodipine alone in 2.0 Vol.% DMSO was not investigated.

3.1. The Role of Calcium for Cell Death and Failure to Provide Neuroprotection

The exact mechanisms of neuronal cell death in ischemia are yet to be fully understood. Dirnagl et al. postulated in 1999 that there are several main mechanisms: excitotoxicity (i.e., damage through an overshooting release of neurotransmitters, mainly glutamate, followed by intracellular calcium overload), oxidative stress, cortical spreading depolarisations, inflammation, and apoptosis [36]. Calcium influx has long been labelled as "final common pathway" [3] of toxic cell death; however, this term might be misleading, since calcium itself activates a cascade of intracellular reactions and thus the possible mechanisms are manifold [37], e.g., a two-step model was suggested consisting of neuronal swelling in a first step and a delayed entrance of calcium in a second [23].

The main hypothesis of our study was that neuroprotection can be achieved by preventing the intracellular rise in calcium concentration by a combined blockade of two postulated calcium entry pathways, while blocking the L-Type calcium channel with nimodipine and the ASIC1a channel using amiloride.

Nimodipine has been reported to improve functional outcome in ischaemic and haemorrhagic stroke in in vivo animal experiments as well as in clinical trials [13,18,20,38,39] and is still the only therapeutic option in the treatment of DCI after SAH [40]. However, the effect of L-type channel inhibition might primarily be targeted to smooth muscle cell function to prevent large as well as small vessel contractions. A possible additional effect on neurons (and astrocytes) might be overrated. To separate direct neuroprotective from vascular effects, models without the dependency of blood circulation (such as cell culture or slice culture models) have been used to investigate this postulated effect of nimodipine and other VGCC inhibitors. Although there are data showing that nimodipine decreases membrane depolarization and thus calcium influx in OGD [21], Kass et al. for instance pointed out that it is rather the loss of calcium extrusion mechanisms (i.e., the dysfunction of the Ca^{2+}-ATPase and the Na^+-Ca^{2+}-Exchanger due to the loss of ATP in hypoxic condition [41]) and not an increased calcium influx that is responsible for cell damage [42]. In addition, the question remains how relevant the portion of the L-type channel for calcium overload within ischaemic damage is. Some data suggest Q-type and N-type channels are far more critical with regard to neuronal injury and that the L-type channel accounts for less than 10% of total damage [24]. Most of the intracellular calcium overload appears to be the result of excitotoxic activation of N-methyl-D-aspartate (NMDA) receptors that become permeable for the ion [41,43]. Moreover, only targeting calcium as single ion might be too simplistic. Goldberg and colleagues demonstrated in experiments that if only calcium is removed, the neuronal damage is enhanced. Only if calcium, sodium, and potassium are removed from the culture medium, protection is achieved. The group concluded that there are two phases to neuronal ischaemic damage pathways: the first one caused by acute swelling, the second one depending on the rise in intracellular calcium concentrations [23]. Thus, there are calcium-independent effects. Regarding these publications, it seems reasonable to conclude that nimodipine does play a role in the calcium metabolism of an ischaemia-exposed neuron; however, the L-type calcium channel alone is likely not central enough in the complex system of calcium-mediated cell death to act as a promising drug target. Our data show no evidence of a protective effect against OGD in our hippocampal slice culture model. This points towards a minor role for L-type calcium channels within the pathophysiological cascade of early ischaemic cell damage.

3.2. Amiloride and Combined L-Type Calcium Channel and ASICs Inhibition

ASICs have been identified to be accountable for increased levels of intracellular calcium in ischaemic conditions. Under acidic conditions, as it occurs during ischaemia, ASIC1a channels become permeable for calcium and provide a mechanism for calcium influx that is not dependent on depolarisation or excitotoxicity [29]. In our study, the ASIC channel inhibitor amiloride did not show any significant neuroprotective effect. This contrasts with other studies showing a protective effect of ASIC inhibition on brain injury [27]. Positive effects of ASIC inhibition were most prominent in a combined acidosis and OGD model. That may explain the insufficient protection of amiloride in our pure OGD model in which the OGD induced acidosis may be less pronounced. Besides, a combination of L-type calcium channel and ASIC inhibition by nimodipine plus amiloride did also not show a protective effect in our model. These findings point against a significant damaging role of calcium entry in parenchymal cells via VGCCs and ASICS during OGD in models not depending on an intact blood supply. In our study, the amount of damage was investigated 72 h after the insult. It can therefore not be ruled out that an early, only transient, and thus not sustained effect of calcium entry blockade via VGCCs and ASICs may have occurred.

Interestingly, in slices treated with regimes using the possibly harmful concentration of 2.0 Vol.% DMSO, we observed a significant further increase in cell damage only for slices where nimodipine and amiloride were applied in combination. DMSO at this high concentration already caused damage in our control slices without OGD. It is also known from the literature that DMSO higher than 5% may have harmful effects on biological tissue and cells [44,45]. Safe concentrations are described up to 3% in hippocampal neurons [44]. In combination with OGD, 2 Vol.% DMSO did not induce an increase in damage beyond the OGD induced damage. In addition, slices receiving DMSO at this high concentration with additional 100μM amiloride did not significantly differ from the OGD control group. We did not investigate nimodipine alone in a concentration of 2.0 Vol.% DMSO. Thus, we cannot draw conclusions whether this effect is caused by the combination of amiloride and nimodipine in 2.0 Vol.% DMSO, or whether nimodipine combined with high concentration of DMSO alone would have the same effect. The high DMSO concentration of 2 Vol.% may have affected the system in addition to OGD, inducing a setting of higher damage-susceptibility, where calcium channel inhibition (equal whether via nimodipine alone or in combination with amiloride) is even harmful instead of neuroprotective. If we assume a combined mechanism, impairment of regulatory systems responsible for ASIC function by the simultaneous L-type calcium channel inhibition with nimodipine might be the answer. This could either be caused by a reduction in intracellular calcium or by the missing reduction in extracellular calcium. Paukert et al., for example, showed that extracellular calcium influences ASIC function and can be competitively inhibited by calcium ions. Furthermore, they suggest that higher extracellular calcium shifts ASIC sensitivity to more acidic pH levels [33]. Admittedly, we would rather expect less ASIC activity and thus less damage if we follow this argumentation. A further explanation may be based on amiloride's unspecific effects. Amiloride does also interfere with other channels, e.g., Na^+/H^+- or Na^+/Ca^{2+}-exchangers and even T-type calcium channels [46]. As aforementioned, calcium extrusion mechanisms appear to be crucial for neuronal integrity. It could be imagined that unspecific blockage mechanisms in combination with L-type inhibition prevent calcium homeostasis mechanisms in the damaging environment of possible toxic effects of DMSO and OGD. Lastly, the concentration of 2.0 Vol.% DMSO is a comparably high choice for the solvent. We recommend avoiding this concentration if feasible.

3.3. Nimodipine—Most Important Effect on the Vasculature

Despite years of research, nimodipine is the only drug available that shows an improved functional outcome and positive effects on mortality after SAH. However, the exact underlying mechanism is still unclear [40]. As pointed out in this publication, a direct neuroprotective effect appears to be unlikely. After the CONSCIOUS-1 study [47], the scientific community changed their perspective on vasospasms of large vessels and DCI. The latter is now considered a complex condition with possible pathways in micro-thrombi [48,49], micro-vasculature spasms (mostly described in experiments and possibly a cause for the secondary development of micro-thrombi) [50–52], neuroinflammation [53,54], and cortical spreading depolarization [55]. Indeed, there is evidence from the literature that nimodipine interacts with several of the abovementioned mechanisms rather than directly acting on neurons, finally resulting in neuroprotection and better functional outcome. That explains a lack of neuroprotection in our model using organotypic slice cultures.

3.4. DMSO as Solvent

DMSO is a commonly used solvent in slice experiments similar to the one presented here. In healthy slices, DMSO applied for 72 h displayed significant damage to the slices at the highest concentration of 2 Vol.% tested. The lower concentrations only induced a concentration-dependent tendency towards slightly enhanced but not relevant damage. In contrast to this effect on slices without OGD, while applying DMSO following OGD, 1.0 Vol.% DMSO alone appears to induce a slight, although not statistically significant, reduction in cell damage. The highest concentration of 2 Vol.% DMSO did not show a protective effect but also did not significantly add to the OGD induced damage. This suggests identical damage pathways for DMSO at this high concentration and OGD. DMSO seems to have a U-shaped concentration-dependent effect on slices after OGD. Neuroprotective effects through DMSO have been described in different experimental designs and models [44,56,57], and thus, DMSO was even suggested as a treatment option for ischaemic brain conditions [58]. Lu and Mattson, e.g., report that DMSO inhibits the glutamate-induced (excitotoxic) calcium influx in hippocampal rat neurons at DMSO levels from 0.5 to 2.0% [44]. Suppression of excitotoxicity may thus also be a possible explanation for our observation of slightly protective effects at 1.0 Vol.%, since neuronal connections stay intact in organotypic hippocampal slices [59]. Another suggested explanation is DMSO's property as a scavenger for free oxygen radicals [58]. However, in our control slices, we observed a dose-depending increase in cell damage with significantly more damage for DMSO at 2.0 Vol.% compared to slices without DMSO or with 0.1 Vol.%. Contrary to the inhibiting effect on excitotoxicity, Galvao et al. have reported apoptosis-inducing effects in retinal cells using DMSO concentrations as low as 1% [60]. Zhang et al. confirmed these observations in experiments with neurons and astrocytes, reporting neuronal alterations at 0.5% DMSO [61]. In addition, even the smallest DMSO concentrations are described to have effects on cell metabolism, and the solvent possibly accumulates in brain slices [62]. From these diverse literature findings, a clear concentration-dependent effect—regardless of protective or harmful—cannot be identified. We consider it an important technical note to this paper that for brain slice experiments addressing mechanisms of neuroprotection, DMSO should be avoided if feasible. If DMSO is utilised, it should only be done so with the utmost caution and proper utilisation of vehicle controls.

3.5. Quality of Slices—Need for Defined Exclusion Criteria

In our group, we recognised that in many publications using organotypic hippocampus slices the inclusion criteria for slices are unclear or not exactly defined. Organotypic slice cultures are highly sensitive to external influences and can therefore be inhomogeneous with regard to cell damage [63]. We claim that a standardised inclusion pipeline would benefit further research conducted using this method. With the here-suggested pipeline (see Materials and Methods), we are confident that we enhanced the quality of data used for statistical analysis in our experiments.

3.6. Limitations

There are several limitations that may be considered given our conclusions:

Firstly, the described interactions of DMSO in the concentrations used may have interfered with the neuroprotective effects of both tested substances. However, this may only account for the highest concentration of 2.0 Vol.% of DMSO and not for the lower concentrations, not showing significant effects on the outcome in control as well as OGD. Secondly, we did not use a specific imaging technique to depict calcium concentrations. By applying well-established concentrations of nimodipine and amiloride, we are confident that an effective channel blockade was achieved, preventing calcium influx via these ion channels. Thirdly, we planned smaller groups for vehicle controls and control groups from the beginning to minimise the number of required animals. Thus, some control groups have limited sample size compared to treatment groups, and further power would be eligible regarding our observation of a possible neuroprotective effect of DMSO at a concentration of 1.0 Vol.%. Lastly, we did not investigate the effect of nimodipine in 2.0 Vol.% DMSO. Hence, it is not possible to draw specific conclusions from our observation of enhanced damage of combined treatment with nimodipine and amiloride at 2.0 Vol.% DMSO.

4. Materials and Methods

4.1. Mediums

Preparation medium (Gey's balanced salt solution (Sigma-Aldrich, Munich, Germany), 5 mg/mL D-(+)Glucose (Roth, Karlsruhe, Germany)) [63] was used for initial slice manufacturing. The growth medium used for slice culturing consisted of 50% Eagle minimal essential medium with Earle's salts (Sigma-Aldrich), 25% Hank's balanced salt solution (Sigma-Aldrich), 25% heat inactive horse serum (Sigma-Aldrich) with additional 5 mg/mL D-(+)Glucose (Roth, Karlsruhe, Germany), 1 Vol.% antibiotic/antimycotic solution [penicillin G GIBCO™, 10,000 units/mL, streptomycin sulphate 10 mg/mL, amphotericin B 25 μg/mL] (Thermo Fisher Scientific, Waltham, MA, USA), 5 μL/mL medium L-Glutamine solution (Sigma-Aldrich) and 10 μL/mL HEPES buffer solution (Sigma-Aldrich) [63]. For experiments, the experimental medium (75% Eagle minimal essential medium with Earle's salts (Sigma-Aldrich), 25% Hank's balanced salt solution (Sigma-Aldrich) with additional, 5 mg/mL D-(+)Glucose (Roth, Karlsruhe, Germany), 1 Vol.% antibiotic/antimycotic solution [penicillin G GIBCO™, 10,000 units/mL, streptomycin sulphate 10 mg/mL, amphotericin B 25 μg/mL] (Thermo Fisher Scientific, Waltham, MA, USA), 5 μL/mL L-Glutamine solution (Sigma-Aldrich) and 10 μL/mL HEPES buffer solution (Sigma-Aldrich)) or OGD medium (75% Eagle minimal essential medium with Earle's salts (Sigma-Aldrich), 25% Hank's balanced salt solution (Sigma-Aldrich) with additional 1 Vol.% antibiotic/antimycotic solution [penicillin G GIBCO™, 10,000 units/mL, streptomycin sulphate 10 mg/mL, amphotericin B 25 μg/mL] (Thermo Fisher Scientific, Waltham, MA, USA), 5 μL/mL L-Glutamine solution (Sigma-Aldrich) and 10 μL/mL HEPES buffer solution (Sigma-Aldrich)) were used, respectively. In essence, OGD medium is simply experimental medium without D-(+)Glucose.

4.2. Slice Preparation and Cultivation

The experiments in this article were strictly conducted according to institutional and governmental guidelines (TierSchG) with institutional permission by the animal protection representative of the Institute of Animal Research at the RWTH Aachen University Hospital and the local institutional committee (LANUV North Rhine-Westphalia, TV-11141A4). After decapitation of 4–7-day-old mice pups (C57BL/6N from Charles Rivers Laboratories, Sulzfeld, Germany and from Janvier Labs, La Rochelle, France, *n* = 138), their brains were extracted and instantaneously immersed into ice-cold preparation medium. The hippocampus slices were prepared using an already established method [58,63]. In brief, the brains were sagittally divided in half, and the frontal pole, as well as the cerebellum, was resected. Using a McIlwain Tissue Chopper (The Mickle Laboratory Engineering Co. ltd. [Now: Cavey Laboratory Engineering Co. ltd], Gomshall, UK), the brains were sliced into

400 µM thick slices from which the hippocampus was then carefully dissected. The hippocampus slices were then transferred onto MilliCell tissue culture inserts (MilliCell-CM, Millipore Corporation, Billerica, MA, USA) placed in 1 mL growth medium. Slices were cultivated at 37 °C and 5% CO2 for 14 days with the growth medium exchanged one day after the preparation and every following third day. On average, 9.1 slices per pup were prepared. Slices of one animal were allocated in two wells, which were then randomly allocated to the experimental groups to avoid allocation of slices of the same animal to only one experimental group. Overview in Figure 6a.

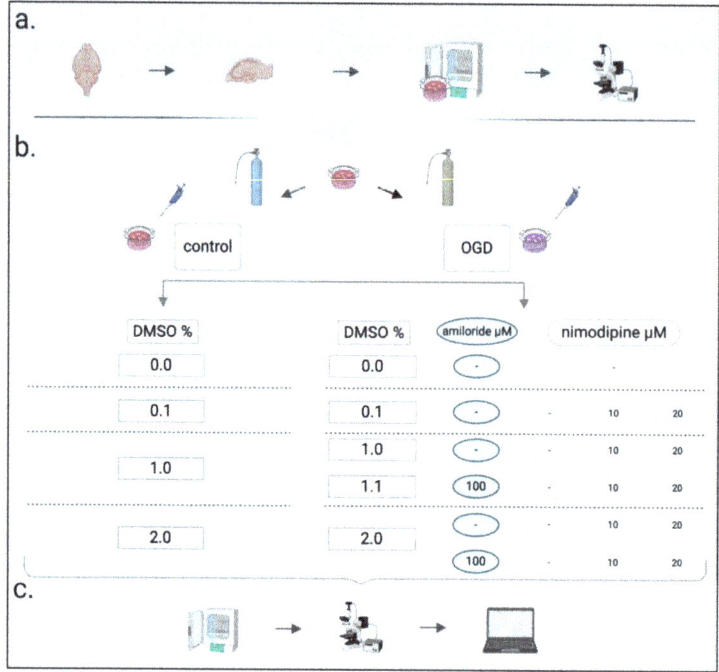

Figure 6. Experimental design. (**a**) Slice preparation, culturing for 14 days and baseline propidium iodide (PI) imaging. (**b**) OGD experiments with overview of experimental groups. (**c**) Incubation for 72 h after OGD, PI imaging for cell death assessment at 72 h, analysis of experimental groups. Created with BioRender.com.

4.3. Imaging

To obtain baseline images, immediately before the OGD experiment growth medium was exchanged for the experimental medium with additional 3 µL/mL propidium iodide (PI) and then incubated again for a minimum of 30 min at 37 °C and 5% CO2. Propidium iodide stains DNA of cells with impaired cell membrane and was used to assess the number of cells damaged [64] using a fluorescence microscope (Zeiss Axioplan, Carl Zeiss MicroImaging GmbH, Jena, Germany) (exposure time was calculated for every imaging session and typically ranged between 15.500 and 16.500 milliseconds) and MetaVue software (MetaVue, Molecular Devices, Sunnyvale, CA, USA). Imaging was performed at baseline for all slices and 72 h after experiments. Imaging is depicted in Figure 6a,c.

4.4. Oxygen–Glucose Deprivation (OGD)

For OGD (Figure 6b), first OGD medium was aerated with 95% N_2, 5% CO_2 for 30 min using a Spectron flowmeter FLM-32 (Spectron Gas Control Systems GmbH, Frankfurt, Germany) at a rate of 15% at 0.2 bar to desaturate the OGD medium from oxygen and warmed up afterwards. The medium was quickly exchanged with OGD medium for the OGD groups and normal experimental medium for control groups, respectively, before immediately being transferred into air-tight experimental chambers (750 mL volume). Chambers containing OGD slices were then flushed with 95% N_2, 5% CO_2 at a rate of 100% at 0.5 bar for 6 min (resulting in a flow of 2.73 L/min) to guarantee a sufficiently hypoxic environment [35]. The chambers were then sealed, and OGD was sustained for 60 min to receive a reasonable amount of cell damage [23]. After OGD, all slices were changed back to experimental medium with additional 3 µL/mL propidium iodide (PI) and randomly allocated to neuroprotective protocols. Thereafter, the slices were again incubated at 37 °C and 5% CO_2 for 72 h.

4.5. Neuroprotective Protocols

Neuroprotective drugs were dissolved using dimethyl-sulfoxide (DMSO) (Sigma-Aldrich, St. Louis, MO, USA) [65]. Vehicle controls were performed for OGD and control groups for 0.1 Vol.%, 1.0 Vol.% and 2.0 Vol.% DMSO. For OGD groups, nimodipine was applied at 10 or 20 µM dissolved in either 0.1 Vol.% or 1.0 Vol.% DMSO and in combination with amiloride at 100 µM [27] again at 10 or 20 µM with 1.1 Vol.% or 2.0 Vol.% DMSO, respectively. Finally, we applied amiloride at 100 µM alone with either 1.0 Vol.% or 2.0 Vol.% DMSO. All agents remained within the medium for 72 h after OGD or time control, ending the experiments with the final imaging. An overview of all protective protocols can be seen in Figure 6b and in Appendix A.

4.6. Cell Death Assessment

PI resulted in red staining of damaged cells. To assess the amount of cell death within each slice, we used python to split channels for each picture into red, green, and blue and then created a corresponding picture in grey values for the red channel only. Again, using python, histograms were created for each picture depicting the corresponding grey-scale values from 0 to 255. A threshold for pixels below a grey-scale value of 100 was used to filter background fluorescence [35,63,66]. Finally, all pixel values were summed to resulting in one value per pixel representing the total damage for each slice.

4.7. Pre-Statistical Image Processing

To further process the slices for total damage analysis, the following exclusion criteria were defined. Every slice was excluded when [1.] more than one slice was shown on the corresponding images at 72 h ($n = 6$ excluded), when [2.] the dentate gyrus was not reliably identifiable ($n = 57$ excluded), when [3.] the CA1-region was not reliably identifiable ($n = 28$ excluded), when [4.] we identified unexplained dark spots in the picture ($n = 31$ excluded), when [5.] PI clots were observed on the image ($n = 18$ excluded), when [6.] slices were not plane but exhibited "wrinkling" that likely occurred in the cultivation process ($n = 78$ excluded), or when [7.] slices presented an inhomogeneous margin, likely due to inadequate preparation technique ($n = 67$ excluded). In total, 227 slices were excluded, and 1032 slices were used for further analysis. It should be noted that several slices showed more than one of the abovementioned features. We provide examples and possible explanations for our observations in the Appendix B.

In addition, to only include slices showing no preparation and cultivation induced damage already before OGD, we specified a maximal pre-damage threshold at the mean plus one standard deviation of all 1032 slices so far identified as useful. Seventy-four slices at 0 h presented a level of pre-experimental damage that was above that threshold. Finally, we defined a further threshold for the slices at 72 h after OGD to obviate extreme outliers in each of the experimental groups. For each of our 20 experimental groups, we established an individual threshold of the mean plus two times the standard deviation as a maximal damage plausibly caused by our OGD method (this excluded 33 slices at 72 h). By applying these criteria, comparability between slices was enhanced, as pre-damaged slices were restrained from entering the experiment and unrealistic outliers were excluded. The experimenter applied all exclusion criteria blinded to the slice allocation to the experimental groups. In total, 334 slices were eliminated, and 925 slices were included in our final analysis.

4.8. Statistical Analysis

Statistical Analysis was performed using GraphPad Prism version 8.2.0 for Windows (GraphPad Software, San Diego, CA, USA). Our data were normalised with the mean of the untreated OGD group as reference. Using a Kolmogorov–Smirnov test, we concluded that the assumption of a normal distribution was not met by our data. A two-tailed Mann–Whitney test between our control and our OGD group was used to verify our OGD model caused adequate damage. We used the Kruskal–Wallis test comparing the mean of rank of each group with every other group between the 72 h data of all OGD groups and for our control groups, respectively. Dunn's test was used to correct for multiple comparisons. For all statistical analysis, a p-value <0.05 was considered significant. To calculate Cohen's D, a short Python script was used.

5. Conclusions

The cellular mechanisms of neuronal ischaemic damage are incredibly complex and effective neuroprotection is—as desired as it might be for all professionals working in that field—still a long way down the road. The idea of only targeting one of the mechanisms, while ischemia activates a symphony of potentially harmful pathways, is probably too simplistic. Nevertheless, the role of calcium is crucial, and the interplay of different calcium channels and their respective effect on cell injury needs further research efforts. Nimodipine remains a hot topic in SAH research mainly due to its vascular effect. Based on encouraging recent findings of a possible calcium-independent effect of nimodipine on microglia [67], we suggest to additionally add a perspective of possible neuro-regeneration to the list of experimental questions, as well as further observing effects on the microvasculature.

We again want to stress that in vitro experiments using DMSO as a solvent should be evaluated critically, since the popular drug vehicle interacts with neuronal damage mechanisms.

Author Contributions: Conceptualization, J.O., U.L., M.C., and A.H.; methodology, J.O., B.K., L.G., U.L., and M.C.; software, J.O.; validation, J.O., U.L., and A.H.; formal analysis, J.O.; investigation, J.O.; resources, H.C. and M.C.; data curation, J.O.; writing—original draft preparation, J.O.; writing—review and editing, all.; visualization, J.O.; supervision, U.L., R.B.-D., and A.H.; project administration, U.L.; funding acquisition, H.C., U.L., and A.H. All authors have read and agreed to the published version of the manuscript.

Funding: This research received no external funding.

Acknowledgments: The authors want to thank Joachim Weis and the Institute of Neuropathology, RWTH Aachen University for their support especially regarding the use of laboratory infrastructure. In addition, we thank Tamara Fechter and Lisa Liebenstund for their support in laboratory work and maintenance. Lastly, the first author wants to thank Patrick Emonts for his tutoring in python and the help in the initial scripts used for the data analysis.

Conflicts of Interest: The authors declare no conflict of interest.

Abbreviations

ASIC	Acid-sensing ion channel
DCI	Delayed cerebral ischaemia
DMSO	Dimethyl-sulfoxide
OGD	Oxygen–glucose deprivation
PI	Propidium iodide
SAH	Subarachnoid haemorrhage

Appendix A

Table A1. Overview of all experimental regimes.

OGD/Control	Vol.% DMSO	Nimodipine Concentration in µM	Amiloride Concentration in µM	n Slices	Median	95% CI (Lower–Upper)
Control	-	-	-	101	0.02308	0.01774–0.03227
Control	0.1	-	-	25	0.03409	0.01500–0.05189
Control	1.0	-	-	12	0.06632	0.01006–0.3592
Control	1.0	-	100	63	0.1470	0.09480–0.1877
Control	2.0	-	-	20	0.7912	0.3633–1.253
Control	2.0	-	100	6	0.02141	0.000–0.1145
OGD	-	-	-	96	0.8978	0.8099–1.118
OGD	0.1	-	-	41	1.261	0.8406–1.697
OGD	0.1	10	-	77	0.7449	0.4920–1.027
OGD	0.1	20	-	64	1.152	0.9478–1.471
OGD	1.0	-	-	19	0.4013	0.2374–0.6065
OGD	1.0	10	-	30	0.6925	0.3331–1.012
OGD	1.0	20	-	25	0.5422	0.2462–0.8696
OGD	1.0	-	100	50	0.6901	0.4543–0.8942
OGD	1.1	10	100	56	0.7106	0.4539–0.8876
OGD	1.1	20	100	59	0.7777	0.5228–1.001
OGD	2.0	-	-	45	0.8082	0.5730–1.515
OGD	2.0	-	100	43	1.172	1.067–1.353
OGD	2.0	10	100	47	2.147	1.775–3.085
OGD	2.0	20	100	46	2.563	1.526–2.939

Appendix B

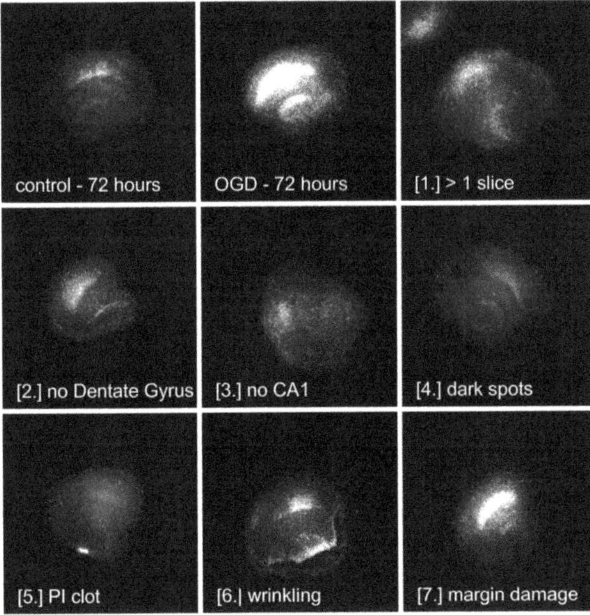

Figure A1. Example slices for our exclusion criteria. Upper left and middle picture show normal slices at 72 h after OGD or control, respectively. [1.] Image files with more than one slice were excluded. If the dentate gyrus [2.] or the CA1 region [3.] were not or only unreliably identifiable, slices were excluded. [4.] Dark spots and [7.] inhomogeneous margins are likely due to imprecise preparation technique, while [6.] wrinkling is most likely caused by problems in the culturing process. Lastly, [5.] on a few slices, PI clots were present. Again, slice images are red-channel-filtered and contrast-enhanced. Contrast enhancement does not change grey scale values and is purely performed to ensure a more comfortable visual evaluation of slice images.

The python scripts can be found at https://github.com/jonasort/OGD-analysis-scripts-in-Python. All questions may be asked to the corresponding author.

References

1. De Rooij, N.K.; Linn, F.H.; van der Plas, J.A.; Algra, A.; Rinkel, G.J. Incidence of subarachnoid haemorrhage: A systematic review with emphasis on region, age, gender and time trends. *J. Neurol. Neurosurg. Psychiatry* **2007**, *78*, 1365–1372. [CrossRef] [PubMed]
2. Peeters, W.; van den Brande, R.; Polinder, S.; Brazinova, A.; Steyerberg, E.W.; Lingsma, H.F.; Maas, A.I. Epidemiology of traumatic brain injury in Europe. *Acta Neurochir.* **2015**, *157*, 1683–1696. [CrossRef] [PubMed]
3. Schanne, F.A.; Kane, A.B.; Young, E.E.; Farber, J.L. Calcium dependence of toxic cell death: A final common pathway. *Science* **1979**, *206*, 700–702. [CrossRef] [PubMed]
4. Siesjo, B.K. Calcium and cell death. *Magnesium* **1989**, *8*, 223–237. [PubMed]
5. Siesjo, B.K.; Bengtsson, F.; Grampp, W.; Theander, S. Calcium, excitotoxins, and neuronal death in the brain. *Ann. N. Y. Acad. Sci.* **1989**, *568*, 234–251. [CrossRef]
6. Dubinsky, J.M. Examination of the role of calcium in neuronal death. *Ann. N. Y. Acad. Sci.* **1993**, *679*, 34–42. [CrossRef]
7. Dubinsky, J.M. Intracellular calcium levels during the period of delayed excitotoxicity. *J. Neurosci.* **1993**, *13*, 623–631. [CrossRef]
8. Wahlgren, N.G.; Ahmed, N. Neuroprotection in cerebral ischaemia: Facts and fancies–the need for new approaches. *Cerebrovasc. Dis.* **2004**, 153–166. [CrossRef]
9. O'Bryant, Z.; Vann, K.T.; Xiong, Z.G. Translational strategies for neuroprotection in ischemic stroke–focusing on acid-sensing ion channel 1a. *Transl. Stroke Res.* **2014**, *5*, 59–68. [CrossRef]
10. D'Souza, S. Aneurysmal Subarachnoid Hemorrhage. *J. Neurosurg. Anesthesiol.* **2015**, *27*, 222–240. [CrossRef]
11. Pala, A.; Schick, J.; Klein, M.; Mayer, B.; Schmitz, B.; Wirtz, C.R.; Konig, R.; Kapapa, T. The influence of nimodipine and vasopressors on outcome in patients with delayed cerebral ischemia after spontaneous subarachnoid hemorrhage. *J. Neurosurg.* **2019**. [CrossRef]
12. Lawton, M.T.; Vates, G.E. Subarachnoid Hemorrhage. *N. Eng. J. Med.* **2017**, *377*, 257–266. [CrossRef] [PubMed]
13. Dorhout Mees, S.M.; Rinkel, G.J.; Feigin, V.L.; Algra, A.; van den Bergh, W.M.; Vermeulen, M.; van Gijn, J. Calcium antagonists for aneurysmal subarachnoid haemorrhage. *Cochrane Database Syst. Rev.* **2007**. [CrossRef] [PubMed]
14. Pickard, J.D.; Murray, G.D.; Illingworth, R.; Shaw, M.D.; Teasdale, G.M.; Foy, P.M.; Humphrey, P.R.; Lang, D.A.; Nelson, R.; Richards, P.; et al. Effect of oral nimodipine on cerebral infarction and outcome after subarachnoid haemorrhage: British aneurysm nimodipine trial. *BMJ* **1989**, *298*, 636–642. [CrossRef] [PubMed]
15. Ohman, J.; Heiskanen, O. Effect of nimodipine on the outcome of patients after aneurysmal subarachnoid hemorrhage and surgery. *J. Neurosurg.* **1988**, *69*, 683–686. [CrossRef] [PubMed]
16. Towart, R.; Kazda, S. The cellular mechanism of action of nimodipine (BAY e 9736), a new calcium antagonist. *Br. J. Pharmacol.* **1979**, *67*, 409P–410P. [CrossRef] [PubMed]
17. Petruk, K.C.; West, M.; Mohr, G.; Weir, B.K.; Benoit, B.G.; Gentili, F.; Disney, L.B.; Khan, M.I.; Grace, M.; Holness, R.O.; et al. Nimodipine treatment in poor-grade aneurysm patients. Results of a multicenter double-blind placebo-controlled trial. *J. Neurosurg.* **1988**, *68*, 505–517. [CrossRef]
18. Uematsu, D.; Greenberg, J.H.; Hickey, W.F.; Reivich, M. Nimodipine attenuates both increase in cytosolic free calcium and histologic damage following focal cerebral ischemia and reperfusion in cats. *Stroke* **1989**, *20*, 1531–1537. [CrossRef]

19. Greiner, C.; Schmidinger, A.; Hulsmann, S.; Moskopp, D.; Wolfer, J.; Kohling, R.; Speckmann, E.J.; Wassmann, H. Acute protective effect of nimodipine and dimethyl sulfoxide against hypoxic and ischemic damage in brain slices. *Brain Res.* **2000**, *887*, 316–322. [CrossRef]

20. Krieglstein, J.; Lippert, K.; Poch, G. Apparent independent action of nimodipine and glutamate antagonists to protect cultured neurons against glutamate-induced damage. *Neuropharmacology* **1996**, *35*, 1737–1742. [CrossRef]

21. Pisani, A.; Calabresi, P.; Tozzi, A.; D'Angelo, V.; Bernardi, G. L-type Ca2+ channel blockers attenuate electrical changes and Ca2+ rise induced by oxygen/glucose deprivation in cortical neurons. *Stroke* **1998**, *29*, 196–201; discussion 202. [CrossRef] [PubMed]

22. Kass, I.S.; Cottrell, J.E.; Chambers, G. Magnesium and cobalt, not nimodipine, protect neurons against anoxic damage in the rat hippocampal slice. *Anesthesiology* **1988**, *69*, 710–715. [CrossRef] [PubMed]

23. Goldberg, M.P.; Choi, D.W. Combined oxygen and glucose deprivation in cortical cell culture: Calcium-dependent and calcium-independent mechanisms of neuronal injury. *J. Neurosci* **1993**, *13*, 3510–3524. [CrossRef] [PubMed]

24. Small, D.L.; Monette, R.; Buchan, A.M.; Morley, P. Identification of calcium channels involved in neuronal injury in rat hippocampal slices subjected to oxygen and glucose deprivation. *Brain Res.* **1997**, *753*, 209–218. [CrossRef]

25. Martinez-Sanchez, M.; Striggow, F.; Schroder, U.H.; Kahlert, S.; Reymann, K.G.; Reiser, G. Na(+) and Ca(2+) homeostasis pathways, cell death and protection after oxygen-glucose-deprivation in organotypic hippocampal slice cultures. *Neuroscience* **2004**, *128*, 729–740. [CrossRef]

26. Waldmann, R.; Champigny, G.; Bassilana, F.; Heurteaux, C.; Lazdunski, M. A proton-gated cation channel involved in acid-sensing. *Nature* **1997**, *386*, 173–177. [CrossRef]

27. Xiong, Z.G.; Zhu, X.M.; Chu, X.P.; Minami, M.; Hey, J.; Wei, W.L.; MacDonald, J.F.; Wemmie, J.A.; Price, M.P.; Welsh, M.J.; et al. Neuroprotection in ischemia: Blocking calcium-permeable acid-sensing ion channels. *Cell* **2004**, *118*, 687–698. [CrossRef]

28. Xiong, Z.G.; Chu, X.P.; Simon, R.P. Ca2+ -permeable acid-sensing ion channels and ischemic brain injury. *J. Membr. Biol* **2006**, *209*, 59–68. [CrossRef]

29. Yermolaieva, O.; Leonard, A.S.; Schnizler, M.K.; Abboud, F.M.; Welsh, M.J. Extracellular acidosis increases neuronal cell calcium by activating acid-sensing ion channel 1a. *Proc. Natl. Acad. Sci. USA* **2004**, *101*, 6752–6757. [CrossRef]

30. Pignataro, G.; Simon, R.P.; Xiong, Z.G. Prolonged activation of ASIC1a and the time window for neuroprotection in cerebral ischaemia. *Brain* **2007**, *130*, 151–158. [CrossRef]

31. Osmakov, D.I.; Andreev, Y.A.; Kozlov, S.A. Acid-sensing ion channels and their modulators. *Biochemistry* **2014**, *79*, 1528–1545. [CrossRef] [PubMed]

32. Paukert, M.; Babini, E.; Pusch, M.; Grunder, S. Identification of the Ca2+ blocking site of acid-sensing ion channel (ASIC) 1: Implications for channel gating. *J. Gen. Physiol.* **2004**, *124*, 383–394. [CrossRef] [PubMed]

33. Babini, E.; Paukert, M.; Geisler, H.S.; Grunder, S. Alternative splicing and interaction with di- and polyvalent cations control the dynamic range of acid-sensing ion channel 1 (ASIC1). *J. Biol. Chem.* **2002**, *277*, 41597–41603. [CrossRef] [PubMed]

34. Harris, R.J.; Symon, L.; Branston, N.M.; Bayhan, M. Changes in extracellular calcium activity in cerebral ischaemia. *J. Cereb. Blood Flow Metab.* **1981**, *1*, 203–209. [CrossRef] [PubMed]

35. Loetscher, P.D.; Rossaint, J.; Rossaint, R.; Weis, J.; Fries, M.; Fahlenkamp, A.; Ryang, Y.M.; Grottke, O.; Coburn, M. Argon: Neuroprotection in in vitro models of cerebral ischemia and traumatic brain injury. *Crit. Care* **2009**, *13*, R206. [CrossRef] [PubMed]

36. Dirnagl, U.; Iadecola, C.; Moskowitz, M.A. Pathobiology of ischaemic stroke: An integrated view. *Trends Neurosci.* **1999**, *22*, 391–397. [CrossRef]

37. Ogawa, S.; Kitao, Y.; Hori, O. Ischemia-induced neuronal cell death and stress response. *Antioxid. Redox. Signal.* **2007**, *9*, 573–587. [CrossRef]

38. Gelmers, H.J.; Gorter, K.; de Weerdt, C.J.; Wiezer, H.J. A controlled trial of nimodipine in acute ischemic stroke. *N. Engl. J. Med.* **1988**, *318*, 203–207. [CrossRef]

39. Lee, K.S.; Lee, K.C.; Choi, J.U. Clinical trial of a calcium channel blocker in patients with aneurysmal subarachnoid hemorrhage–prevention of delayed ischemic deficits. *Yonsei Med. J.* **1987**, *28*, 126–130. [CrossRef]

40. Rowland, M.J.; Hadjipavlou, G.; Kelly, M.; Westbrook, J.; Pattinson, K.T. Delayed cerebral ischaemia after subarachnoid haemorrhage: Looking beyond vasospasm. *Br. J. Anaesth.* **2012**, *109*, 315–329. [CrossRef]

41. Won, S.J.; Kim, D.Y.; Gwag, B.J. Cellular and molecular pathways of ischemic neuronal death. *J. Biochem. Mol. Biol.* **2002**, *35*, 67–86. [CrossRef] [PubMed]

42. Kass, I.S.; Lipton, P. Calcium and long-term transmission damage following anoxia in dentate gyrus and CA1 regions of the rat hippocampal slice. *J. Physiol.* **1986**, *378*, 313–334. [CrossRef] [PubMed]

43. Benveniste, H.; Jorgensen, M.B.; Diemer, N.H.; Hansen, A.J. Calcium accumulation by glutamate receptor activation is involved in hippocampal cell damage after ischemia. *Acta Neurol. Scand.* **1988**, *78*, 529–536. [CrossRef] [PubMed]

44. Lu, C.; Mattson, M.P. Dimethyl sulfoxide suppresses NMDA- and AMPA-induced ion currents and calcium influx and protects against excitotoxic death in hippocampal neurons. *Exp. Neurol.* **2001**, *170*, 180–185. [CrossRef] [PubMed]

45. Yuan, C.; Gao, J.Y.; Guo, J.C.; Bai, L.; Marshall, C.; Cai, Z.Y.; Wang, L.M.; Xiao, M. Dimethyl Sulfoxide Damages Mitochondrial Integrity and Membrane Potential in Cultured Astrocytes. *PLoS ONE* **2014**, *9*. [CrossRef] [PubMed]

46. Xiong, Z.G.; Pignataro, G.; Li, M.; Chang, S.Y.; Simon, R.P. Acid-sensing ion channels (ASICs) as pharmacological targets for neurodegenerative diseases. *Curr. Opin. Pharmacol.* **2008**, *8*, 25–32. [CrossRef] [PubMed]

47. Macdonald, R.L.; Kassell, N.F.; Mayer, S.; Ruefenacht, D.; Schmiedek, P.; Weidauer, S.; Frey, A.; Roux, S.; Pasqualin, A.; CONSCIOUS-1 Investigators. Clazosentan to overcome neurological ischemia and infarction occurring after subarachnoid hemorrhage (CONSCIOUS-1): Randomized, double-blind, placebo-controlled phase 2 dose-finding trial. *Stroke* **2008**, *39*, 3015–3021. [CrossRef]

48. Vergouwen, M.D.; Vermeulen, M.; Coert, B.A.; Stroes, E.S.; Roos, Y.B. Microthrombosis after aneurysmal subarachnoid hemorrhage: An additional explanation for delayed cerebral ischemia. *J. Cereb. Blood Flow Metab.* **2008**, *28*, 1761–1770. [CrossRef]

49. Suzuki, S.; Kimura, M.; Souma, M.; Ohkima, H.; Shimizu, T.; Iwabuchi, T. Cerebral microthrombosis in symptomatic cerebral vasospasm–a quantitative histological study in autopsy cases. *Neurol. Med. Chir.* **1990**, *30*, 309–316. [CrossRef]

50. Friedrich, B.; Muller, F.; Feiler, S.; Scholler, K.; Plesnila, N. Experimental subarachnoid hemorrhage causes early and long-lasting microarterial constriction and microthrombosis: An in-vivo microscopy study. *J. Cereb. Blood Flow Metab.* **2012**, *32*, 447–455. [CrossRef]

51. Wellman, G.C.; Koide, M. Impact of subarachnoid hemorrhage on parenchymal arteriolar function. *Acta Neurochir. Suppl.* **2013**, *115*, 173–177. [CrossRef] [PubMed]

52. Budohoski, K.P.; Guilfoyle, M.; Helmy, A.; Huuskonen, T.; Czosnyka, M.; Kirollos, R.; Menon, D.K.; Pickard, J.D.; Kirkpatrick, P.J. The pathophysiology and treatment of delayed cerebral ischaemia following subarachnoid haemorrhage. *J. Neurol. Neurosurg. Psychiatry* **2014**, *85*, 1343–1353. [CrossRef] [PubMed]

53. Lucke-Wold, B.P.; Logsdon, A.F.; Manoranjan, B.; Turner, R.C.; McConnell, E.; Vates, G.E.; Huber, J.D.; Rosen, C.L.; Simard, J.M. Aneurysmal Subarachnoid Hemorrhage and Neuroinflammation: A Comprehensive Review. *Int. J. Mol. Sci.* **2016**, *17*, 497. [CrossRef] [PubMed]

54. de Oliveira Manoel, A.L.; Macdonald, R.L. Neuroinflammation as a Target for Intervention in Subarachnoid Hemorrhage. *Front. Neurol.* **2018**, *9*, 292. [CrossRef]

55. Dreier, J.P.; Major, S.; Manning, A.; Woitzik, J.; Drenckhahn, C.; Steinbrink, J.; Tolias, C.; Oliveira-Ferreira, A.I.; Fabricius, M.; Hartings, J.A.; et al. Cortical spreading ischaemia is a novel process involved in ischaemic damage in patients with aneurysmal subarachnoid haemorrhage. *Brain* **2009**, *132*, 1866–1881. [CrossRef]

56. Di Giorgio, A.M.; Hou, Y.; Zhao, X.; Zhang, B.; Lyeth, B.G.; Russell, M.J. Dimethyl sulfoxide provides neuroprotection in a traumatic brain injury model. *Restor. Neurol. Neurosci.* **2008**, *26*, 501–507.

57. Shimizu, S.; Simon, R.P.; Graham, S.H. Dimethylsulfoxide (DMSO) treatment reduces infarction volume after permanent focal cerebral ischemia in rats. *Neurosci. Lett.* **1997**, *239*, 125–127. [CrossRef]

58. Phillis, J.W.; Estevez, A.Y.; O'Regan, M.H. Protective effects of the free radical scavengers, dimethyl sulfoxide and ethanol, in cerebral ischemia in gerbils. *Neurosci. Lett.* **1998**, *244*, 109–111. [CrossRef]

59. Stoppini, L.; Buchs, P.-A.; Muller, D. A simple method for organotypic cultures of nervous tissue. *J. Neurosci. Methods* **1991**, *37*, 173–182. [CrossRef]

60. Galvao, J.; Davis, B.; Tilley, M.; Normando, E.; Duchen, M.R.; Cordeiro, M.F. Unexpected low-dose toxicity of the universal solvent DMSO. *FASEB J.* **2014**, *28*, 1317–1330. [CrossRef]

61. Zhang, C.; Deng, Y.; Dai, H.; Zhou, W.; Tian, J.; Bing, G.; Zhao, L. Effects of dimethyl sulfoxide on the morphology and viability of primary cultured neurons and astrocytes. *Brain Res. Bull.* **2017**, *128*, 34–39. [CrossRef] [PubMed]

62. Nasrallah, F.A.; Garner, B.; Ball, G.E.; Rae, C. Modulation of brain metabolism by very low concentrations of the commonly used drug delivery vehicle dimethyl sulfoxide (DMSO). *J. Neurosci. Res.* **2008**, *86*, 208–214. [CrossRef] [PubMed]

63. Grüßer, L.; Blaumeiser-Debarry, R.; Krings, M.; Kremer, B.; Höllig, A.; Rossaint, R.; Coburn, M. Argon attenuates the emergence of secondary injury after traumatic brain injury within a 2-hour incubation period compared to desflurane: An in vitro study. *Med. Gas. Res.* **2017**, *7*, 93–100. [CrossRef] [PubMed]

64. Macklis, J.D.; Madison, R.D. Progressive incorporation of propidium iodide in cultured mouse neurons correlates with declining electrophysiological status: A fluorescence scale of membrane integrity. *J. Neurosci. Methods* **1990**, *31*, 43–46. [CrossRef]

65. Brayton, C.F. Dimethyl sulfoxide (DMSO): A review. *Cornell Vet.* **1986**, *76*, 61–90. [PubMed]

66. Krings, M.; Höllig, A.; Liu, J.; Grüßer, L.; Rossaint, R.; Coburn, M. Desflurane impairs outcome of organotypic hippocampal slices in an in vitro model of traumatic brain injury. *Med. Gas. Res.* **2016**, *6*, 3–9. [CrossRef]

67. Schampel, A.; Volovitch, O.; Koeniger, T.; Scholz, C.J.; Jorg, S.; Linker, R.A.; Wischmeyer, E.; Wunsch, M.; Hell, J.W.; Ergun, S.; et al. Nimodipine fosters remyelination in a mouse model of multiple sclerosis and induces microglia-specific apoptosis. *Proc. Natl. Acad. Sci. USA* **2017**, *114*, E3295–E3304. [CrossRef]

Publisher's Note: MDPI stays neutral with regard to jurisdictional claims in published maps and institutional affiliations.

International Journal of
Molecular Sciences

Article

Liver Growth Factor "LGF" as a Therapeutic Agent for Alzheimer's Disease

Rafael Gonzalo-Gobernado [1,2,†], Juan Perucho [1,†], Manuela Vallejo-Muñoz [1,†], Maria José Casarejos [1], Diana Reimers [1], Adriano Jiménez-Escrig [1,3], Ana Gómez [1], Gonzalo M. Ulzurrun de Asanza [1] and Eulalia Bazán [1,*]

[1]   Servicio de Neurobiología, Instituto Ramón y Cajal de Investigación Sanitaria (IRYCIS), 28034 Madrid, Spain; rd.gonzalo@cnb.csic.es (R.G.-G.); juan.perucho@outlook.com (J.P.); manuela.vmqm@gmail.com (M.V.-M.); m.jose.casarejos@hrc.es (M.J.C.); diana.reimers@hrc.es (D.R.); adriano.jimenez@hrc.es (A.J.E.); go.s6506@hotmail.com (A.G.); gonzalo.munoz@hrc.es (G.M.U.d.A.)

[2]   National Centre for Biotechnology (CNB), CSIC, 28049 Madrid, Spain

[3]   Servicio de Neurología, Hospital Ramón y Cajal, 28034 Madrid, Spain

*   Correspondence: eulalia.bazan@hrc.es; Tel.: +34-913-368-168

†   These authors contributed equally to this work.

Received: 23 October 2020; Accepted: 29 November 2020; Published: 2 December 2020

Abstract: Alzheimer's disease (AD) is a progressive degenerative disorder and the most common cause of dementia in aging populations. Although the pathological hallmarks of AD are well defined, currently no effective therapy exists. Liver growth factor (LGF) is a hepatic albumin–bilirubin complex with activity as a tissue regenerating factor in several neurodegenerative disorders such as Parkinson's disease and Friedreich's ataxia. Our aim here was to analyze the potential therapeutic effect of LGF on the APPswe mouse model of AD. Twenty-month-old mice received intraperitoneal (i.p.) injections of 1.6 µg LGF or saline, twice a week during three weeks. Mice were sacrificed one week later, and the hippocampus and dorsal cortex were prepared for immunohistochemical and biochemical studies. LGF treatment reduced amyloid-β (Aβ) content, phospho-Tau/Tau ratio and the number of Aβ plaques with diameter larger than 25 µm. LGF administration also modulated protein ubiquitination and HSP70 protein levels, reduced glial reactivity and inflammation, and the expression of the pro-apoptotic protein Bax. Because the administration of this factor also restored cognitive damage in APPswe mice, we propose LGF as a novel therapeutic tool that may be useful for the treatment of AD.

Keywords: liver growth factor; Alzheimer's disease; inflammation; neuroprotection; microglia; Tg2576 transgenic mice; amyloid-beta

---

1. Introduction

Alzheimer's disease (AD), the most common cause of dementia, was discovered by Alois Alzheimer in 1901. The prevalence of AD is about 1% in individuals aged 60–64, but shows an exponential increase with age, so that in people aged 85 years or older the prevalence is between 24–33% in the Western World. The hallmarks of the disease are extracellular deposits of plaques composed of amyloid-β (Aβ) and intracellular neurofibrillary tangles, composed of hyperphosphorylated tau (phospho-tau), a normal axonal protein that binds to microtubules [1]. The relationship among Aβ deposits, tangle formation and neurodegeneration, and cell death is not clear at present. The areas more vulnerable for the neurodegeneration processes are the cortex and the hippocampus, affecting mechanisms of spatial, semantic and episodic memories [2]. Other processes implicated in the pathological pathways in AD are cholinergic dysfunction, neuroinflammation, calcium channels, oxidation, iron chelation, abnormalities in the mitochondrial DNA and lipid metabolism, among others [3].

Therapeutic strategies in AD trying to ameliorate or eliminate its main disturbances, include the elimination of Aβ deposits, vaccination and immunization against Aβ, inhibitors of β or γ-secretases (enzymes excising Aβ from APP), antifibrillisation agents, statins (inhibitors of cholesterol synthesis), neuroprotectors, antioxidants and anti-inflammatory drugs, among others (reviewed in [4]). However, although the pathological hallmarks of AD are well defined, no effective therapy exists currently.

Liver growth factor (LGF) is a hepatic mitogen purified in 1986 by Dr. Diaz-Gil's group [5]. This 64 kDa factor is an albumin–bilirubin complex that stimulates the proliferation of different cell types, and promotes regeneration of damaged tissues including the brain (reviewed in [6]). Thus, administration of LGF was able to stimulate axonal growth in the striatum, to increase the number of dopaminergic neurons in the damaged substantia nigra, and to improve rotational behavior stimulated by apomorphine in experimental Parkinson's disease [7,8]. LGF also promoted the proliferation and migration of neural progenitors from the forebrain subventricular zone [9], and improved the viability, differentiation and integration of stem cell grafts into the host tissue [10]. Besides, our previous studies have reported the remarkable anti-inflammatory and antioxidant activities of LGF in extracerebral tissues [11–13], and in several experimental models of neurodegeneration where its main cellular target in the brain appears to be microglia [7,9,14,15]. Both effects are considered to be closely related with Alzheimer's disease pathology [16,17].

To analyze the potential therapeutic effectiveness of LGF in AD, we have used Tg2576 transgenic mice (from now APPswe mice) that over-express the Swedish APP mutation (K670N/M671L) on a C57Bl/6 9 SJL background [18]. From 12 months onwards, these mice show Aβ plaque depositions in the hippocampus and cerebral cortex, reactive gliosis, inflammation and cognitive impairment that are neuropathological features of AD [19]. Chronic LGF administration to a 20–21-month-old APPswe mouse, significantly reduces Aβ and phospho-Tau protein levels and modulates protein ubiquitination and the expression of the heat shock protein 70 (Hsp70) that is involved in Aβ clearance. LGF also reduces microglia activation, astrogliosis and the expression of the apoptosis-associated speck-like protein containing a CARD (ASC), which is a component of the inflammatory response and cognitive impairment in AD pathology. Moreover, LGF up-regulates the expression of the Nuclear factor erythroid 2-related factor 2 (Nrf2), which plays an important role against oxidative stress. Because these beneficial effects correlate with a better cognitive outcome in APPswe mice, we may propose LGF as a potential novel therapeutic tool that may be useful for the treatment of AD.

2. Results

2.1. LGF Improves Behavioral Working Memory in APPswe Mice

The alternance index in the Y-maze, a parameter considered as an indicator of mnemonic function, was significantly reduced in the APPswe mice with respect to WT (Figure 1A). This important marker of working memory was recovered in APP-LGF treated mice (Figure 1A). APP mice also exhibited a reduced number of entries in comparison with WT mice, indicating a low general locomotor activity (18.8 ± 0.9 ($n = 10$) and 11 ± 1.8* ($n = 11$) in WT and APP mice, respectively * $p \leq 0.05$ vs. WT). Instead, no significant differences were observed in the number of entries after LGF treatment in comparison with APP mice (APP-LGF: 13 ± 1.7 ($n = 12$)).

The marble-burying test is a useful model of neophobia, anxiety, and obsessive-compulsive behavior. As shown in Figure 1B, the marble burying test index was significantly reduced in APPswe aged mice, and LGF recovered this index of potential hippocampal affection.

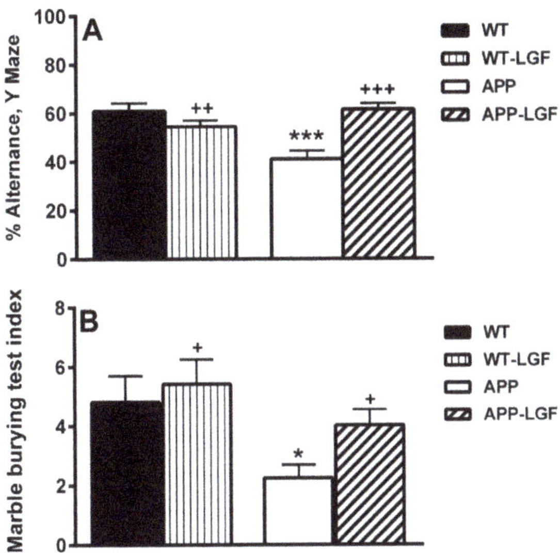

Figure 1. Liver growth factor (LGF) improves cognitive behavior in APPswe mice. Panel (**A**) shows how the alternance index in the Y-maze is significantly reduced in APP mice, and how LGF treatment recovers this important marker of working memory. As shown in panel (**B**), the marble burying test index is significantly reduced in APP mice, and LGF recovers this potential index of hippocampal and cortical affection. Results represent the mean ± SEM of 11 to 16 independent mice. The statistical analysis was performed by one-way ANOVA followed by Newman–Keuls test. * $p \leq 0.05$, *** $p \leq 0.001$ vs. wild type (WT). + $p \leq 0.01$, ++ $p \leq 0.01$, +++ $p \leq 0.001$ vs. APP mice.

2.2. LGF Modulates β-Amyloid Protein Expression in Hippocampus and Cerebral Cortex of APPswe Mice

β-amyloid protein accumulation in the Central Nervous System (CNS) is the most important feature of the experimental model of AD used in this study. As shown in Figure 2A, APPswe mice expressed the APP 695 human protein, and several Aβ peptides of different molecular weight lower than 60 kDa which expression was up-regulated in the hippocampus and cerebral cortex, in comparison with WT mice (Figure 2B). The immunohistochemical analysis also revealed the presence of Aβ-positive plaques in the hippocampus (40 ± 7 ($n = 5$) Aβ-positive plaques/mm^2) and cerebral cortex (38 ± 6 ($n = 5$) Aβ-positive plaques/mm^2) of APPswe mice. These plaques had different sizes with a $78 \pm 9\%$ of them showing a diameter lower than 25 μm in both brain structures (Figure 2C). In the APP-LGF experimental group, the total number of plaques was not significantly different to that observed in APPswe mice (54.6 ± 3.2 ($n = 4$) and 44 ± 3.5 ($n = 4$) Aβ-positive plaques/mm^{2}in the hippocampus and cerebral cortex, respectively). However, LGF treatment significantly reduced Aβ protein levels (Figure 2B), and the number of plaques with a diameter higher than 25 μm compared to APPswe mice treated with the vehicle (Figure 2D). Besides, LGF slightly augmented, but not significantly, augmented the percentage of plaques with a diameter lower than 25 μm by 1.5 ± 0.1-fold and 1.3 ± 0.14-fold in the hippocampus and cerebral cortex, respectively.

2.3. Effects of LGF in Tau Phosphorylation and Protein Ubiquitination

A neuropathological feature of AD is the accumulation of Tau and ubiquitin in the neurofibrillary tangles (NFT). At 20 months of age, phospho-Tau/Tau ratio was increased by 2.5-fold in the hippocampus of APPswe mice, and LGF administration partially reduced this parameter to similar values of those observed in WT mice (Figure 3A,D). Phospho-Tau/Tau ratio was similar in the cerebral cortex of

APPswe and WT mice, but LGF treatment significantly reduced this ratio below WT levels (Figure 3A). A similar effect was observed in the cerebral cortex of WT-LGF treated mice (Figure 3A).

Protein ubiquitination is essential in an important number of processes including protein degradation by the proteosome. Accumulation of polyubiquitinated proteins was observed in the hippocampus and cerebral cortex of APPswe mice, and LGF administration reduced this parameter in both structures (Figure 3B,D).

HSP-70 is a chaperone that allows the correct folding of mis-folded proteins that could lead to their aggregation. As shown in Figure 3C, the hippocampus of APP-LGF treated mice showed significantly higher levels of HSP-70 than WT or APPswe mice. HSP-70 protein expression was down-regulated in the cerebral cortex of APPswe mice, and LGF treatment restored its levels to control values (Figure 3C,D).

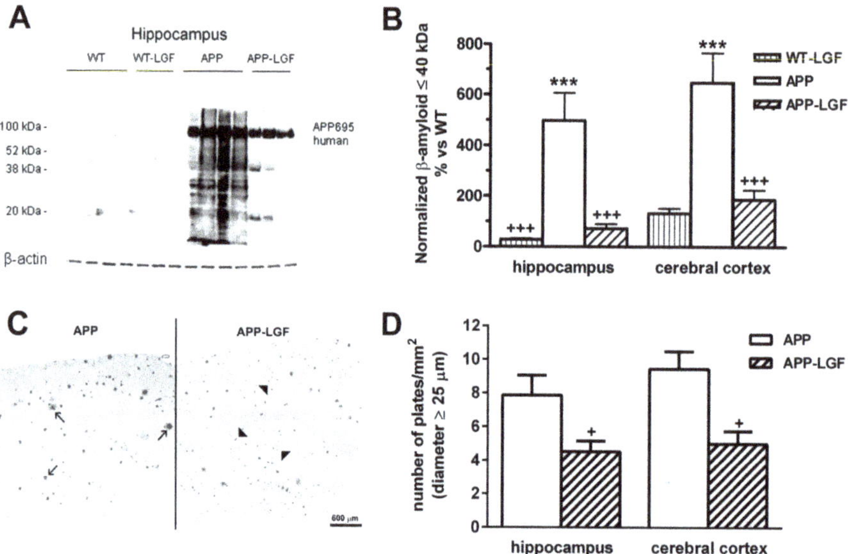

Figure 2. LGF reduces amyloid-β accumulation and the size of amyloid-β positive plaques in the hippocampus and cerebral cortex of APPswe mice. Panels (**A,B**) show a Western blot representative image of Aβ aggregates in the hippocampus (**A**) and the quantification of Aβ-aggregates ≤40 kDa in the hippocampus and cerebral cortex (**B**). Note that Aβ protein levels are up-regulatedin the hippocampus and cerebral cortex of APPswe mice and that LGF treatment significantly reduces its expression in both structures. Panel (**C**) shows a representative image of Aβ-positive plaques of large (**C**, black arrows) and small size (**C**, black arrowheads) in one hemisphere of APP mice brain (scale bar: 600 μm). Note that in APP-LGF mice, large plaques are reduced, while small plaques increase. Panel (**D**) shows how LGF treatment significantly reduces the number of plaques with a diameter higher than 25 μm in the hippocampus and cerebral cortex. Results represent the mean ± SEM of 7 to 8 (**B**) and 4 to 5 (**D**) independent mice in each experimental group. The statistical analysis was performed by one-way ANOVA followed by Newman–Keuls test. *** $p \leq 0.001$ vs. WT. + $p \leq 0.05$, +++ $p \leq 0.001$ vs. APP mice.

2.4. LGF Modulates Microglia Activation and Reduces Inflammation in APPswe Mice

Recent reports suggest that microglia play an important role in the evolution of AD. The immunohistochemical analysis of the hippocampus and cerebral cortex of APPswe mice showed that the number of microglial cells that expressed the Ionized Calcium Binding Protein-1 (Iba1) was significantly higher than in WT control mice (Figure 4A). Most of the Iba1-positive cells associated with the plaques were located around them (Figure 4B), and a few showed Iba1-positive processes penetrating into the plaques (Figure 4C). Iba1 protein levels were also up-regulated in both structures (Figure 4D,H), suggesting the presence of activated microglia in these mice. Increased proliferation has also been associated with microglia activation. As shown in Figure 4E, the expression of PCNA was significantly increased in the hippocampus and cerebral cortex of APPswe mice. Similar to this finding, BrdU incorporation was higher in these mice than in the WT group in both structures (Figure 4F), and more than $61 \pm 4\%$ ($n = 8$) of the proliferating cells were Iba1-positive microglia (Figure 4G).

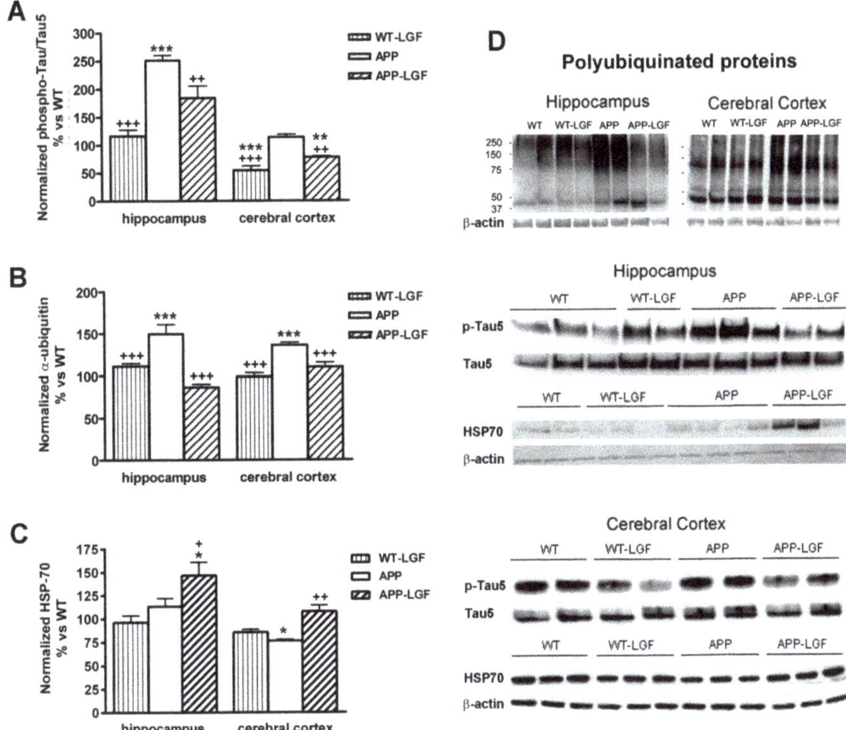

Figure 3. Effects of LGF in tau pathology, ubiquitinated protein levels and HSP70 heat shock protein expression. Panel (**A**), shows the relative optical density of phospho-tau/tau ratio analyzed by Western blot and expressed as % vs. WT. Note how LGF treatment reduces the expression of phosphorylated tau in the hippocampus and cerebral cortex in APP mice. Panels (**B,C**) show the quantification of accumulated ubiquitinated proteins (**B**) and HSP-70 (**C**) in the hippocampus and cerebral cortex of APPswe mice. Note how LGF reduces the accumulation of ubiquitinated proteins (**B**) and increases HSP70 expression (**C**). Panel (**D**) shows representative Western blot images of phospho-tau, total tau, Hsp-70 and β-actin as charge control. Results represent the mean ± SEM of 4 to 8 independent mice. The statistical analysis was performed by one-way ANOVA followed by Newman–Keuls test. * $p \leq 0.05$, ** $p \leq 0.01$, *** $p \leq 0.001$ vs. WT. + $p \leq 0.05$, ++ $p \leq 0.01$, +++ $p \leq 0.001$ vs. APPswe mice.

The APP-LGF group of animals showed a significant reduction in Iba1 (Figure 4D,H) and PCNA protein expression (Figure 4E,H), and a slight but reduced number of BrdU-positive cells in comparison with APPswe mice (Figure 4F). Besides, LGF treatment increased from 6 ± 2.9 ($n = 3$) to 19 ± 11 ($n = 4$) the percentage of Iba1-positive cells, which prolongations penetrated profoundly in the plaques in the hippocampus.

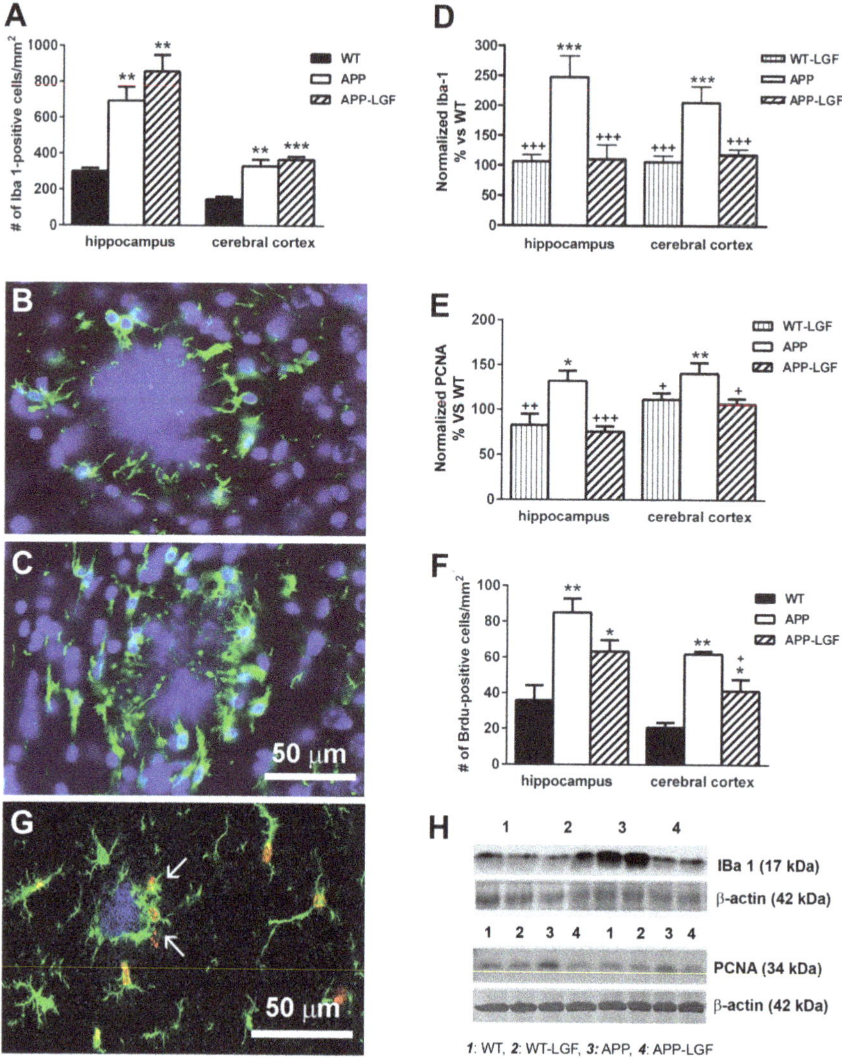

Figure 4. Liver growth factor modulates microglia cell activity in the hippocampus and cerebral cortex of APPswe mice. Panel (**A**) shows the quantification of Iba1-positive microglia cells in APPswe mice. Panels (**B,C**) show different responses of Iba1-positive cells: in (**B**) microglial prolongations do not penetrate inside the deposit, while in (**C**) they actively phagocytize the plaque (scale bar: 50 µm). Panels (**D,E**) show quantitative Western blot analysis of Iba1 (**D**) and PCNA (**E**) protein expression. Note how LGF administration significantly reduces the over-expression of Iba1 (**D**) and PCNA (**E**) in the hippocampus and cerebral cortex. LGF also reduces BrdU cell incorporation (**F**) as an index of cellular proliferation. Note how BrdU (**G**, red) is associated with Iba1-positive cells (**G**, green, white arrows). Panel (**H**) show representative Western blots of Iba1 and PCNA and their respective β-actin as charge control. The results represent the mean ± SEM of 4 (**A,F**) and 8 to 10 (**D,E**). independent mice. The statistical analysis was performed by one-way ANOVA followed by Newman–Keuls test. * $p \leq 0.05$, ** $p \leq 0.01$, *** $p \leq 0.001$ vs. WT. + $p \leq 0.05$, ++ $p \leq 0.01$, +++ $p \leq 0.001$ vs. APPswe mice.

GFAP is a protein expressed by astrocytes for which up-regulation has been associated with inflammatory states in chronic processes as AD and aging. As shown in Figure 5A, GFAP was over-expressed in the hippocampus and cerebral cortex of APPswe mice, and LGF treatment significantly reduced its levels in both structures. On the other hand, LGF decreased ASC protein levels that were up-regulated in the hippocampus and the cerebral cortex of APPswe aged mice (Figure 5B,D). ASC is involved in the production of Interleukin-1beta (Il-1beta) but, neither APP nor APP-LGF treated mice showed any significant change in the levels of this pro-inflammatory cytokine in the hippocampus (113 ± 6 ($n = 6$) and 101 ± 7 ($n = 6$) % of IL-1beta expression vs. WT in APP and APP-LGF treated mice, respectively) and the cerebral cortex (102 ± 11 ($n = 5$) and 92 ± 12 ($n = 6$) % of IL-1beta expression vs. WT in APP and APP-LGF treated mice, respectively). Similar results were observed when we analyzed the potential effects of LGF in Tumor Necrosis Factor-alpha (TNF-alpha) protein expression in the hippocampus (92 ± 6 ($n = 6$) and 88 ± 10 ($n = 8$) % of TNF-alpha expression vs. WT in APP and APP-LGF treated mice, respectively) and cerebral cortex (95 ± 6 ($n = 5$) and 107 ± 9 ($n = 5$) % of TNF-alpha expression vs. WT in APP and APP-LGF treated mice, respectively). This latter cytokine mediates LGF-induced neuroregeneration/neuroprotection [6,20].

LGF also modulated the expression of the transcription factor Nrf2, and the class B scavenger receptor CD36, two proteins that regulate oxidative stress and inflammatory responses in AD. As shown in Figure 5D, Nrf2 protein levels were reduced in the hippocampus and cerebral cortex of APPswe mice, and LGF treatment up-regulated Nrf2 protein expression to reach control values. LGF treatment reduced the levels of CD36 that is over-expressed in the hippocampus of APPswe mice (Figure 5C,E). Besides, LGF up-regulated CD36 protein expression in the hippocampus of WT mice (Figure 5C,E), but it lacked activity in the cerebral cortex, a brain structure where neither APP nor APP-LGF mice showed any alteration in CD36 protein levels (Figure 5C).

2.5. LGF Modulates the Expression of Proteins Involved in Cell Survival

Our previous studies showed how in an experimental model of Parkinson's disease, LGF potentiates cell survival through the regulation of Bcl2 and Bax protein expression [9]. In the hippocampus of APPswe mice the Bcl2/Bax ratio was significantly reduced, and LGF treatment increased this ratio to control values (Figure 6A). This effect was due to a reduction in the expression of the pro-apoptotic protein Bax which levels were significantly up-regulated in the hippocampus of APPswe mice (Figure 6B).

Akt is another protein involved in cell survival modulated by LGF in neurodegeneration [9]. As shown in Figure 6C, phospho-Akt/Akt ratio was up-regulated in the hippocampus of APPswe mice where phospho-Akt levels were significantly increased in comparison with WT mice (Figure 6D). The APP-LGF group of mice also showed an increased phospho-Akt/Akt ratio and higher levels of phospho-Akt than control mice (Figure 6C,D). Neither Bcl2/Bax nor phospho-Akt/Akt ratio were affected in the cerebral cortex at any of the experimental conditions used in this study (Figure 6A,B).

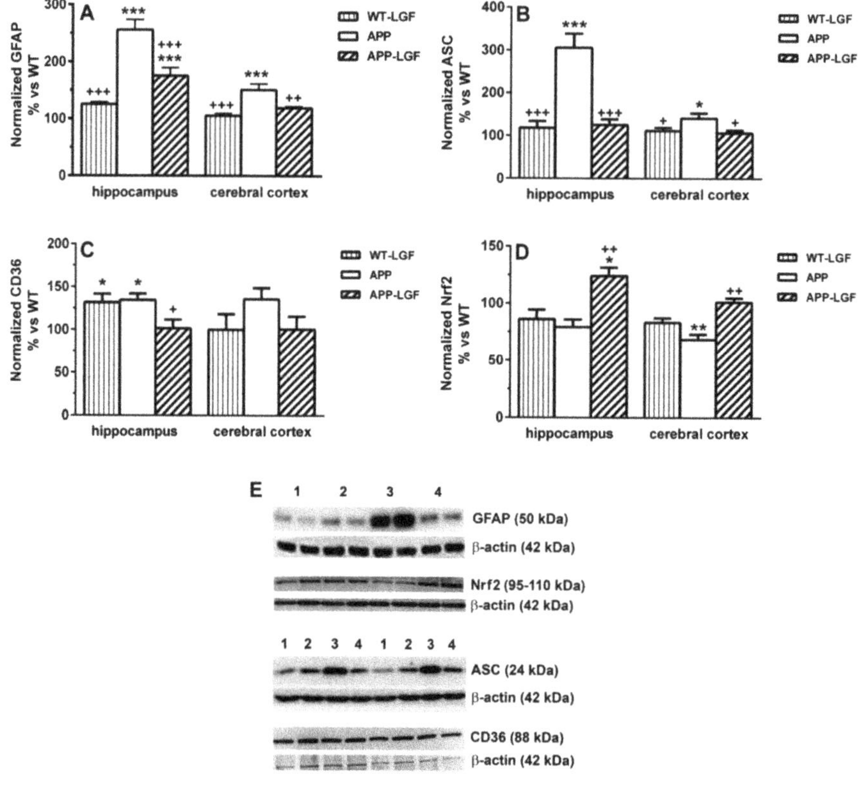

Figure 5. LGF reduces the inflammatory response in APPswe mice. Panel (**A**) shows how LGF significantly reduces GFAP protein expression in the cerebral cortex and hippocampus of APPswe. LGF administration also decreases the expression of ASC (**B**), which is a protein involved in the production of pro-inflammatory cytokines. Panel (**C**) shows quantitative Western blot analysis of the receptor scavenger CD36 that is over-expressed in the hippocampus of APP mice. Panel (**D**) represents the quantitative analysis of the transcription factor Nrf2 as an oxidative stress regulator. Representative Western blots of GFAP, Nrf2, ASC, CD36 and their respective β-actin as charge are shown in panel (**E**). Results represent the mean ± SEM of 4 to 10 independent mice. The statistical analysis was performed by one-way ANOVA followed by Newman–Keuls test. * $p \leq 0.05$, ** $p \leq 0.01$, *** $p \leq 0.001$ vs. WT. + $p \leq 0.05$, ++ $p \leq 0.01$, +++ $p \leq 0.001$ vs. APPswe mice.

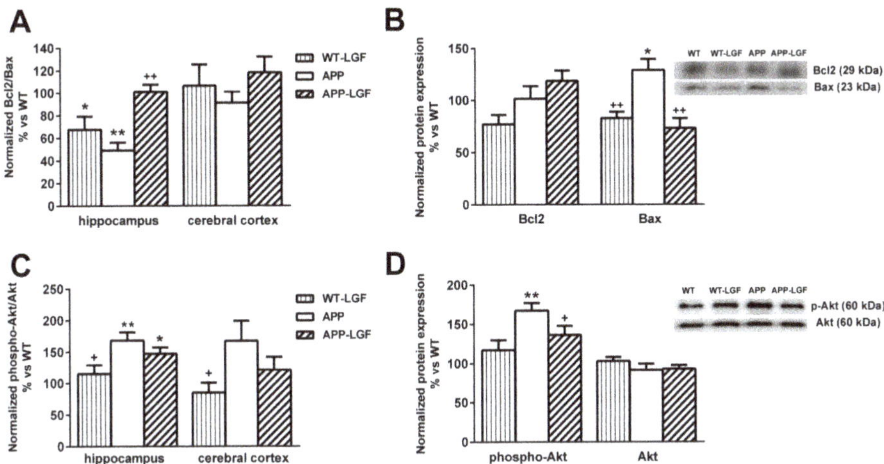

Figure 6. Effect of LGF on the expression of proteins involved in cell survival. Protein expression was analyzed by Western blot using specific antibodies against Bcl2, Bax, p-Akt and Akt. Panel (**A**), shows that LGF increases Bcl-2/Bax ratio in the hippocampus. This effect is mainly due to a reduction in Bax protein levels (**B**). Panels (**C,D**) show that phospho-Akt/Akt ratio (**C**) and phospho-Akt protein levels (**D**) are significantly increased in the hippocampus of APP and APP-LGF treated mice. Results represent the mean ± SEM of 4 to 8 independent mice. The statistical analysis was performed by one-way ANOVA followed by Newman–Keuls test. * $p \leq 0.05$, ** $p \leq 0.01$, vs. WT.+ $p \leq 0.05$, ++ $p \leq 0.01$ vs. APPswe mice.

3. Discussion

In the experimental model of AD used in this study (APPswe mouse), LGF treatment reduced Aβ and phospho-Tau protein levels as well as the number of Aβ plaques with a diameter higher than 25 µm. LGF administration also modulated protein ubiquitination and HSP70 protein levels, reduced glial reactivity and inflammation, and up-regulated the expression of proteins involved in cell survival. These LGF beneficial effects correlate with a better cognitive outcome in APPswe mice.

Amyloid-β aggregation and deposition of Aβ plaques are considered as causative agents in AD, so many therapies have been directed to modify Aβ levels (reviewed in [21]). Under our experimental conditions, LGF modulated Aβ by reducing its levels to WT values in the hippocampus and cerebral cortex. This reduction was observed in the expression of those Aβ peptides with a molecular weight lower than 60 kDa. This fraction includes the Aβ1-42 peptide that has been extensively studied due to its abundance in patients predisposed to AD [22] and mouse models of AD [19]. Here, we also show that LGF treatment reduced the accumulation of polyubiquitinated proteins, and up-regulated HSP70 protein expression in both structures. The high levels of ubiquitinated proteins observed in APPswe mice indicate that proteins are not correctly proteolyzed, probably due to the generation of free radicals and/or an inflammatory environment promoted by the accumulation of Aβ HSP70 is a chaperone which overexpression could suppress the progression of AD by promoting Aβ clearance (reviewed in [23]). HSP70-mediated Aβ clearance from the brain has been associated with the up-regulation of insulin-degrading enzyme (IDE) and transforming growth factor β1 (TGF-β1) [24]. This study did not evaluate IDE and TGF-β1 expression but LGF administration significantly reduced plasmatic levels of insulin in an experimental model of atherosclerosis in mice [25], and modulated the synthesis of TGF-β1 in bile duct-ligated rats [11].

Another major histopathological characteristic in AD is neurofibrillary tangles composed of aggregates of hyperphosphorylated forms of the protein associated with Tau microtubules [26,27]. As shown here, LGF treatment reduced phospho-Tau/Tau ratio in the hippocampus of APPswe mice.

Interestingly, Tau homeostasis is controlled through the action of molecular chaperones, such as HSP70 [28,29] that, as noted above, was up-regulated in this structure. In the case of the cerebral cortex, phosphoTau/Tau ratio in APP mice was not significantly different from untreated WT mice but, in this brain structure, LGF treatment reduced phospho-Tau/Tau ratio in both, WT and APP mice. At present we do not have a possible explanation for this event. We could suggest that this is a potential specific effect of LGF not associated to the accumulation of Aβ plaques and inflammation observed in APP mice. However, according to our results it does not seem the case because similar results were not observed in the hippocampus of WT mice where, as already mentioned, LGF did not reduced p-Tau/Tau ratio. Besides HSP70, which was also up-regulated in the cerebral cortex of APP-LGF mice, LGF could modulate Tau phosphorylation in this brain structure by different mechanisms not analyzed in this study. Further studies must be conducted in order to elucidate these discrepancies.

Activated microglia and reactive astrocytes are two pathological markers observed in AD [30–32]. Both cell types are considered cell targets of LGF in the CNS [6,7,9,20]. According to the present results, LGF significantly reduced proliferation (as measured by BrdU incorporation and PCNA protein expression) and Iba1 protein levels, which are two features of activated microglia. These results are in agreement with previous studies that show how chronic administration of LGF reduced microglia cell reactivity in an experimental model of Parkinson's disease in rats [9] and in ataxic rats [15]. Our immunohistochemical studies also suggest that the administration of LGF could prime microglia to acquire a phagocytic phenotype. Thus, although the number of Iba1-positive cells was not reduced in APP-LGF treated mice, this treatment increased by 3-fold the percentage of Iba1-positive cells that invaded Aβ plaques and reduced the number of Aβ large plaques while increasing the number of smaller plaques, which would explain the slight, but not significant increase observed in this parameter. How LGF modulates microglia activity in APPswe mice is unknown but, as mentioned above, LGF could stimulate TGFβ1expression that is a cytokine that stimulates Aβ clearance through activation of phagocytic microglia [33]. On the other hand, activation of phagocytic microglia has been associated with increased levels of interleukin 6 and TNF-alpha [34]. This latter cytokine seems to play an important role in the mitogenic cascade of LGF in the liver, and in LGF-induced neuroregeneration (reviewed in [6]) and neuroprotection [20].

Reactive astrogliosis is a well-known hallmark of AD that is identified, among others, by an increased expression of GFAP. We also found that LGF reduced GFAP over-expression in the hippocampus and cerebral cortex of APPswe mice, suggesting that astrocytes are potential targets of LGF in this experimental model of AD. Microglia cell reactivity to Aβ and phagocytic activity are also under the regulation of astrocytes [35] so, their stimulation with LGF could lead to the release of factors with antioxidant and/or anti-inflammatory effects that could modulate both events. In this regard, we should mention that LGF modulates the expression of TNF-alpha in cultures of mesencephalic glia [20], and significantly reduced by 36 ± 11% the expression of ASC in these cultures ($p \leq 0.05$ vs. untreated cultures).

The anti-oxidative and anti-inflammatory activities of LGF have also been observed in vivo. Thus, chronic LGF decreased oxidized glutathione concentration in ataxic mice [14], and reduced TNF-alpha protein levels in 6-OHDA-lesioned rats [9]. As presented here, LGF modulates the expression of several proteins that regulate oxidative stress and inflammatory responses in AD. ASC is an adaptor protein that participates in the production of pro-inflammatory cytokines, and contributes to the inflammatory response and cognitive impairment in AD pathology [36–39]. Chronic administration of LGF down-regulated ASC protein levels in the hippocampus and cerebral cortex of APPswe mice. In these mice, ASC is mainly expressed by microglia, which is associated with Aβ plaques (Reimers et al. manuscript in preparation). Because ASC released by microglia forms an aggregate complex with Aβ that promotes inflammation [37,39], we may argue that ASC reduction is associated with a potential anti-inflammatory effect of LGF in this experimental model of AD. Although we were unable to detect any significant change in TNF-alpha and Il-1beta protein expression at the experimental time analyzed in this study, we cannot exclude the possibility that LGF modulates the expression of these cytokines at earlier times of treatment, as observed for TNF-alpha in 6-OHDA-lesioned rats [9]. In addition, small changes in the levels of both proteins may not be detected by Western blot analysis, so that it may be necessary to use other methods with higher sensitivity as ELISA. Other studies have reported that reduced levels of ASC stimulate phagocytosis in glial cells in a mouse model of AD [40], so LGF-induced down-regulation of ASC could also mediate the increase in phagocytic activity proposed in our study.

CD36 is a class B scavenger receptor involved in phagocytosis which expression in microglia plays a key role in the Aβ-mediated activation of this cell type (reviewed in [41]). Moreover, CD36 over-expression in microglia seems to mediate the inflammatory processes associated to Aβ pathology in transgenic mice [42], and the production of ROS in response to Aβ in AD human brains [43,44]. Similarly, the increased levels of CD36 observed in the hippocampus of APPswe mice could contribute to creating an inflammatory environment in this structure. According to our results, LGF treatment could modulate inflammation and/or ROS production in the APPswe hippocampus by decreasing CD36 protein expression and, in consequence, its interaction with Aβ in this structure. In fact, the inhibition of Aβ interaction with CD36 has been proposed as a potential therapeutic strategy in AD [45]. An interesting observation of our study is that LGF up-regulated CD36 levels in the hippocampus of WT mice. These high levels of CD36 could promote the phagocytic activity of microglia to improve the clearance of debris that is accumulated in the normal brain during the processes of aging [46,47]. Although similar results were observed in the cerebral cortex of APPswe and APP-LGF treated mice, no statistically significant differences were detected. Perhaps CD36 levels are not sufficient to be detected through Western blot in this structure, or we need to study a higher number of mice, or mice at different ages. De facto, we were unable to detect CD36 up-regulation in the hippocampus of thirteen months old APPswe mice (100 ± 7 (n = 4) and 104 ± 6 (n = 4) % of CD36 expression in WT and APPswe mice, respectively).

Nrf2 is a transcription factor that regulates oxidative stress [48,49] and anti-inflammatory responses [50,51]. Nrf2 mRNA and protein deficiency have been observed in AD patients [52,53] and AD models in mice [54,55]. As shown here, LGF up-regulated Nrf2 protein levels that were reduced in the cerebral cortex and the hippocampus of APPswe mice. Nrf2 up-regulation ameliorated cognitive impairment in APP Knock-in AD mice [56], and in APPswe/PSEN1dE9 hypoxic mice, the intranasal administration of Nrf2 modulated CD36 protein expression, reduced Aβ deposition, and improved spatial memory [57]. In both studies, these beneficial effects were associated to a reduction in oxidative stress and neuroinflammation promoted by the increased levels of Nrf2. Since in our mice LGF significantly ameliorates cognitive impairment, we may consider that up-regulation of Nrf2 could underlie this behavioral improvement through the activation of anti-oxidant and anti-inflammatory responses, as has been reported in other experimental pathologies[58].

Our previous results indicate a potential neuroprotective role of LGF in the brainstem and cerebellum of ataxic rats [15], and in the substantia nigra and striatum of 6-hydroxydopamine-lesioned rats [9]. In both experimental models of neurodegeneration, LGF raised the ratio Bcl2/Bax due to the up-regulation of the anti-apoptotic protein Bcl2, which is involved in the survival of neural cells [59]. In the experimental model of AD used in this study, LGF also restored Bcl2/Bax ratio to control values, but this effect was due to the down-regulation of the pro-apoptotic protein Bax that was over-expressed in the hippocampus of APPswe mice. Up-regulation of Bax has been observed in neurons that exhibit neurofibrillary tangle pathology [60], and APP transgenic mice [61]. Besides, a neuroblastoma cell line co-transfected with Swedish mutant APP showed a higher Bax/Bcl2 ratio [62]. LGF-induced down-regulation of Bax could contribute to decreasing the excessive neuronal apoptosis reported in AD [63]. The increase in Bax levels observed in APPswe could result in neuronal death. However, APPswe mice did not show any reduction in the expression of the specific nuclear neuronal marker NeuN in the hippocampus (100 ± 8 (n = 6) and 99 ± 10 (n = 5) % of NeuN expression in WT and APP mice, respectively), and cerebral cortex (100 ± 7 (n = 6) and 99 ± 12 (n = 5) % of NeuN expression in WT and APP mice, respectively). These results agree with the lack of neuronal death previously reported in this experimental model of AD [64,65]. Akt, is a serine/threonine kinase which phosphorylated form plays a critical role in the regulation of neuronal survival [66,67] that has been involved in the pathogenesis of AD [68]. Our results showed an increased phospho-Akt/Akt ratio in the APPswe hippocampus due to the higher levels of phospho-Akt observed in this brain structure. Phospho-Akt may protect neuronal cell death through different mechanisms including the phosphorylation and inhibition of cytosolic pro-apoptotic proteins (reviewed in [69]). Because the phosphorylation of Bax by active Akt affects its translocation to the mitochondria [70], we may consider the up-regulation of phospho-Akt in the hippocampus as an intrinsic mechanism to protect neurons of Bax-mediated apoptotic cell death in the APPswe mice.

From our study, we cannot define if the observed effects of LGF are due to a direct or indirect action of the factor on the brain of APPswe mice. A potential direct action would imply that LGF could cross the blood–brain barrier by binding to specific receptors found in the endothelium known as RAGE receptors (receptors for advanced glycation end-products) [71]. In fact, LGF show some structural similarities with the advanced glycation end products (AGE) which structure is mainly based on albumins bound to glucose with the ability to change albumin conformation [72]. RAGE receptors are also expressed by astrocytes and microglia [73], and their activation elicits TNF-alpha release from these cells [73].

Present results show how LGF treatment also improves memory performance impairment in APPswe mice. The APPswe mice in comparison with the WT animals, had a reduced ambulatory behavior in the Y-maze, as shown by the total number of entrances and this pattern was unchanged by LGF treatment, so we may consider that motor activity and anxiety were affected in this experimental model of AD. However, the alternance index in the Y-maze, which is an indicator of mnemonic function, was significantly reduced in these mice. LGF treatment recovered this important parameter of working memory to similar values to those observed in WT mice. Other studies have reported a reduction in the exploratory behavior from 10 months old APPswe mice that are coincident with the presence of Aβ plaque depositions in the hippocampus and cerebral cortex (reviewed in [74]). Thus, LGF-induced reduction in Aβ plaques size and levels could contribute to the cognitive improvement observed in our study. In addition, LGF could ameliorate the exploratory behavior in these mice through the reduction of glial reactivity, and its anti-inflammatory and anti-oxidative responses referred to above.

The marble-burying test index was lower in the APPswe mice than in WT mice, and LGF rescued this behavioral parameter. Other studies have shown reduced scrabble behavior and buried marbles in mice with cortical and hippocampal lesions [75] and in Tg-APP/PS1 mice [76]. In the latter study, the behavioral deficits were attributed to the accumulation of Aβ in the brain, so we may consider that the beneficial effects observed in APP-LGF treated mice could be in part due to the reduction in Aβ levels promoted by the factor.

4. Materials and Methods

4.1. Ethics Statement

The Ethics Committee of the Hospital Ramón y Cajal, Madrid (animal facilities ES280790002001) approved on 11 May 2015 all the protocols related to the use of laboratory animals. All procedures associated with animal experiments were in accordance with Spanish legislation (RD 53/2013) and the European Union Council Directive (2010/63/EU).

4.2. LGF Purification

LGF was obtained and purified from serumof rats that were subjected to a bile duct ligation for 5 weeks, as previously described by Díaz-Gil et al. [77]. The quantitation of the purity of LGF was performed by HPLC [78] and serum samples that showed the highest concentrations of the growth factor were chosen to continue with the process. The method included three chromatographic procedures using DEAE-cellulose, Sephadex G-150, and hydroxylapatite. The absence of contaminants and other growth factors (purity) in the LGF samples was also evaluated according to the standard criteria described by Díaz-Gil et al. [5,77,79]. All LGF preparations showed a single band in SDS-PAGE electrophoresis and were lyophilized and stored at 4 °C until use. Finally, LGF aliquots were dissolved in saline prior to intraperitoneal administration.

4.3. Transgenic Animals

The Tg2576 line over-expressing human APP695 was used as a model of AD in this study. This line contains the human mutated APPswe that includes the double mutation M671L and K670N under the control of the prion protein promoter [18]. This mouse line of AD was generously donated by Dr. Carro [80]. These mice develop age-dependent AD-type neuropathology [18]. Tg2576 mice show elevated levels of the major soluble form of brain amyloid (Aβ1-40) at 1 year old. An intercross between APP Tg2576 and Wild-type (WT) C57BL6 was performed in order to obtain WT littermate controls.

Transgenic (Tg) animals were generated by breeding the mice according to the following diagram: male APPswe × female WT → APPswe (50/57.6) and WT (50/42.4)—(the numbers in brackets show the expected/found genotype frequencies of the offspring expressed in %).

Eighteen WT mice were divided into 2 experimental groups, WT control (WT, *n* = 11) and LGF treated group (WT-LGF, *n* = 7). APPswe mice were also divided in APPswe control (APP, *n* = 12) and LGF treated APPswe (APP-LGF, *n* = 16). Four mice per group were used for histological assays, and 4 to 10 mice were used for biochemical studies. All mice included in each experimental group were used for behavioral studies. The animals were studied at 21 months of age.

4.4. Genotype by PCR

The genotype of each animal was confirmed as described by Carro et al. [80]. Genomic DNA was extracted from the mouse tail using the High Pure PCR template preparation kit (Roche, Barcelona, Spain) according to the manufacturer's instructions. Briefly, 150 ng of DNA was amplified by polymerase chain reaction (PCR) using the specific primers described in [80] to detect mutant human APP sequence.

4.5. LGF and BrdU Administration

Twenty-month-old mice received 2 weekly 100 μL i.p. injections of saline used as a vehicle (WT and APP groups) or LGF (1.6 μg/100 μL) (WT-LGF and APP-LGF groups) for 3 weeks. This optimal dose of LGF has been used in different model systems using an identical or similar schedule [13,14]. Mice were sacrificed 1 week after the last treatment with vehicle or LGF (Figure 7).

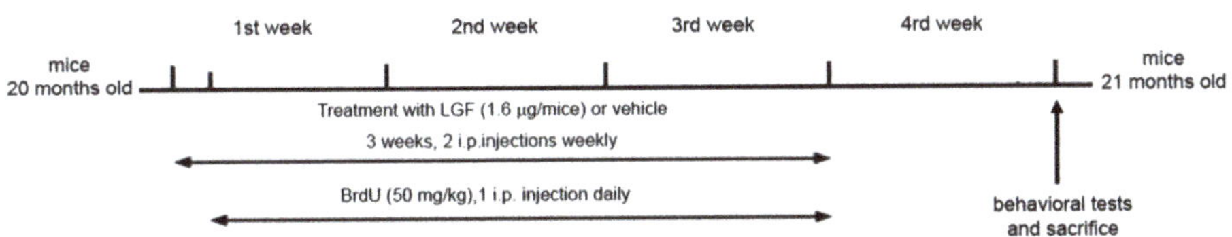

Figure 7. Representative scheme showing the methodological procedure followed for the administration of LGF or vehicle to 20 months old WT and APPswe mice.

To determine whether the i.p. administration of LGF increased cell proliferation, a group of WT and APP animals were injected daily for 3 weeks with the mitotic marker 5-bromodeoxyuridine (BrdU) (50 mg/kg i.p.), starting 24 h after the beginning of LGF/saline injections.

4.6. Behavioral Studies

Exploratory behavior and body weight were measured before treatment initiation to allocating the animals to the experimental groups (WT, WT-LGF, APP, APP-LGF). After 3 weeks of treatment with vehicle or LGF, mice were subjected to behavioral analysis. Two behavioral tests were carried out in this study: Y-maze test and marble burying test.

4.6.1. Y-Maze

Spatial working memory was measured using the Y-maze behavioral paradigm. The Y-maze test is based on the natural exploratory preference of mice to alternate arms when exploring a new environment. The spontaneous alternation behavior of mice in the Y-maze was analyzed to assess short term memory impairment in mice, following the methodology previously reported by Perucho et al. [81,82]. Testing takes place in a Y-shaped maze with three white, opaque plastic arms spaced at an angle of 120 degrees. Briefly, after acclimatizing the mice to the behavioral room (30 min), each mouse was placed at the same end of one arm of the Y-maze and allowed to move freely through the maze for 5 min. The number of arms entered, as well as the number of spontaneous alternations, defined as a sequence of entries in each of 3 consecutive arms without repetition, were recorded to determinate the alternation rate. The percentage of alternation score was calculated as follows:

$$\text{Spontaneous alternation}\% = \left(\frac{\text{number of spontaneous alternations}}{\text{total number of arm entries} - 2}\right) \times 100 \tag{1}$$

4.6.2. Marble Burying Test

Marble burying was assessed as an additional measure of exploratory digging. This test is a useful paradigm to study repetitive compulsive-like behaviors based on the observation of rodents burying objects in their bedding, a phenomenon dependent on hippocampal function [75]. Briefly, mice were individually housed in a cage filled with 5 cm of wood chip bedding for a 30 min testing period of acclimation, and continued throughout the test session. Tests were performed between 18:00 and 20:00 h, under standard laboratory conditions of temperature, $22 \pm 2\ ^\circ\text{C}$, 12 h light:dark cycle, relative humidity 50–60% and with food and water restriction. The mice were placed individually for 30 min in plastic cages, with 9 marbles placed on top of the bedding material, in a 3×3 grid with 4 cm center-to-center spacing. A new cage, clean marbles and fresh bedding were used for each mouse. At the end of that time, each mouse was returned to its home cage and the number of marbles uncovered; completely buried (not visible); 2/3 covered (only top visible) and half covered, were counted. This classification was used to obtain a "buried index", as an indication of activity and interaction with their environment [83].

4.7. Antibodies and Immunochemicals for Immunohistochemistry

The primary antibodies used in this study were: rabbit polyclonal anti-Iba1 (1:100; Wako Chemicals USA, Inc., Los Angeles, CA, USA), mouse monoclonal anti-5-bromodeoxyuridine (BrdU, 1:25; DakoCytomation, Glostrup, Denmark), and mouse monoclonal anti-Aβ (6E10, 1:100; Covance, Emeryville, CA, USA). The secondary antibodies used were: Alexa Fluor-568 goat anti-mouse IgG, and Alexa Fluor-488 goat anti-rabbit IgG (1:400; both from Molecular Probes; Eugene, OR, USA), fluorescein-conjugated goat anti-mouse IgG (1:25; Jackson ImmunoResearch Laboratories Inc., West Grove, PA, USA).

4.8. Tissue Processing, Immunohistochemistry and Morphometric Analyses

Four weeks after the beginning of LGF or saline treatment transgenic mice were perfused intracardially under deep anesthesia with 10 mL of heparinized isotonic saline, followed by 40 mL of 4% paraformaldehyde. Brains were postfixed in the same solution for 24 h at 4 °C, cryoprotected and frozen, before sectioning into 20 μm-thick coronal sections in a cryostat. For quantitative measurements and immunohistochemical analysis of the dorsal hippocampus and dorsal cortex, coronal sections were obtained at the antero-posterior level of −2.5 to −3 mm from Bregma [84].

Tissue sections were mounted on coated FLEX IHC microscope slides (DakoCytomation, Glostrup, Denmark), treated with sodium acetate 10 mM, pH 6.0, at 95 °C for 4 min, and blocked with 5% normal goat serum (NGS) and 0.1% Triton-X 100 in phosphate buffer saline (PBS), pH 7.4, for 30 min. Primary antibodies diluted in 0.5% NGS in PBS, were applied for 24 h at 4 °C, and were visualized using fluorescent secondary anti-rabbit and anti-mouse antibodies diluted in 0.5% NGS in PBS. The slides were coverslipped in a medium containing p-phenylenediamine and bisbenzimide (Hoechst 33342; Sigma, St. Louis, MO, USA) for detection of nuclei. For double immunolabeling with anti-Iba1 and anti-BrdU antibodies, the former immunostaining was performed prior to the immunodetection of BrdU incorporated into nuclei. The latter detection needed pretreatment of sections with 2N HCl at 37 °C for 30 min, previous to the NGS blocking step. In the case of Aβ immunodetection in plaques, sections were pretreated with 70% formic acid for 20 min at RT, followed by the blocking step.

For quantification of proliferating BrdU-positive cells, microglial Iba1-positive cells, and Aβ 6E10-positive plaques, panoramic views were obtained from one representative coronal section for each immunostaining technique and for each brain. These panoramic views, where the complete transversal surface of dorsal hippocampus and dorsal cortex of each cerebral hemisphere were visualized (3 mm^2 for each structure), were obtained by using the stitching tool of the software NIS ELEMENTS C, version BR 3.2 (Nikon Instruments Inc., Tokio, Japan), coupled to a Nikon ECLIPSE Ti-e microscope (Nikon Instruments Inc., Tokio, Japan) with a motorized stage, and the 10× objective. Four WT, 4 APP and 4 APPL-GF were analyzed for these evaluations. The data entered in the corresponding graphs were: (a) the mean ± SEM obtained from the data of both cerebral hemispheres (number of plaques/mm^2); (b) the mean of 2 different regions in the cortex and of 4 different regions in the hippocampus of one cerebral hemisphere (density of Iba1-positive cells); and (c) the sum of data from both hemispheres (density of BrdU-positive cells).

4.9. Brain Regions and Tissue Preparation for Biochemical Analysis

After decapitation the brain was extracted and the hippocampus and cortex were free-hand dissected, then the samples were frozen on dry ice for biochemical studies (Western blot). For the protein extraction, dried tissue samples were weighed and placed in six volumes (w/v) of phosphate-buffered saline (PBS) with protease inhibitor cocktail 1× (Calbiochem, Temecula, CA, USA) and 20 mM N-ethylmaleimide to inactivate deubiquitinating enzymes and subjected to two 30-s rounds of sonication. The lysates were immediately boiled for 5 min and centrifuged at 12,000× g at 4 °C for 30 min. The supernatant-PBS-inhibitors was defined as the soluble fraction and was used for protein analysis by Western blot. To obtain the total Aβ (soluble plus insoluble) fraction, the initial homogenate (200 μL) was extracted with 5 M guanidine in 6 mM Tris–HCl, pH 8.0 by rotating the sample at room temperature overnight. The sample obtained after guanidine extraction represented the total fraction. Soluble and total fractions were used for protein analysis by Western blot.

4.10. Protein Analysis

Aliquots of 20–30 μg of protein were added to sample loading buffer 2× (50 mM Tris pH 6.8, 10% glycerol, 2% SDS, 5% β-mercaptoethanol and 0.1% bromophenol blue). The electrophoresis was performed using SDS-polyacrylamide gels (10–15%) and then the samples were electro-blotted using nitrocellulose membranes (0.45 μm). The blots were blocked using TTBS solution (5% dry skimmed milk, 137 mM NaCl plus 0.1% Tween-20 and 20 mM Tris-HCl pH 7.6) at room temperature for 2 h. Once the blocking of non-specific binding was performed, the membranes were incubated overnight at 4 °C with specific primary antibodies diluted in blocking solution. Then, the membranes were washed twice for 10 min using blocking solution followed by another two washes with TTBS for 5 min. β-actin was used as a loading control and housekeeping protein. The membranes were developed by enhanced chemiluminescence detection using a commercial kit (Bio-Rad Laboratories Inc., Hercules, CA, USA) and quantified by computer-assisted video densitometry using the Bio-Rad *Quantity One* software (Bio-Rad Laboratories Inc, Hercules, CA, USA). The data of the proteins of interest analyzed in the study were normalized with respect to β-actin levels.

4.10.1. Primary Antibodies

The primary antibodies used in this study were: Mouse monoclonal amyloid-β antibody 6E10 (BioLegend Inc., San Diego, CA, USA), diluted 1:1000. Mouse monoclonal anti-HSP70 (1:1000) and rabbit polyclonal antibody against ASC (1:500) were from Santa Cruz (Temecula, CA, USA). Mouse monoclonal anti tau-5 (for measurement of total tau protein) antibody (Chemicon, Madrid, Spain) diluted 1:5000 and rabbit polyclonal to phospho-tau (phospho serine199 + serine202) (1:1000) were from Abcam (Cambridge, UK). Mouse monoclonal anti-glial fibrillar acid protein (GFAP) antibody diluted 1:5000 was from Chemicon (Madrid, Spain) and rabbit polyclonal antibody for anti-Iba1, 1:1000 was for WAKO (Japan). Rabbit anti-proliferating cell nuclear antigen (PCNA, 1:75; SantaCruz Biotechnology, Santa Cruz, CA, USA), and mouse monoclonal anti-ubiquitin antibody (1/500) was from Chemicon (Chemicon International Inc., Temecula, CA, USA). Mouse anti Bcl2 (1:250) and rabbit anti-Bax (1:250) were from Santa Cruz (Santa Cruz Biotechnology). Rabbit anti-phospho-Akt (Ser473P) (1:2000) and rabbit anti-Akt (1:2000) were from Cell Signaling Technology, Beverly, MA, USA. Rabbit anti Nrf2 (1:1000) and rabbit anti CD36 (1:500) were from Thermo Fisher Scientific, Rosemont, IL, USA. To correct and quantify the protein charge mouse monoclonal anti-β-actin antibody diluted 1:10,000 was used from Sigma.

4.10.2. Secondary Antibodies

Goat anti-mouse-IGg-HRP and anti-rabbit-IGg-HRP secondary antibodies diluted 1:1000 were purchased from Bio-Rad (Bio-Rad Laboratories Inc., Hercules, CA, USA).

4.11. Statistical Analysis

Results are expressed as mean ± SEM of (n) independent animals. Statistical analysis was performed with the GraphPad Prism software, version 6.01 (La Jolla, CA, USA). Before analysis, the Shapiro–Wilk test was used to test normality. One-way ANOVA followed by the Newman–Keuls multiple comparison test were performed. Differences were considered significant when $p \leq 0.05$.

5. Conclusions

Liver growth factor is an albumin–bilirubin complex that exhibits remarkable neuroprotective, anti-inflammatory, and anti-oxidant activities in several models of neurodegenerative diseases. In the experimental model of AD used in this study (APPswe mouse), chronic LGF treatment reduced Aβ and phospho-Tau protein levels which are the main neuropathological features associated to the disease. LGF also reduced microglia and astroglia cell reactivity and modulated the expression of ASC CD36 and Nrf2 which are proteins involved in the regulation of inflammatory and oxidative processes. Because these beneficial effects correlated with an improvement in the cognitive deficits observed in the APPswe mice, we may consider LGF as a potential new therapeutic factor for AD.

6. Patents

US 8,642,551 B2, 4 February 2014

CE, num. 09732019.6-1456, Ref. EP-883, April 2015

US 14/140.014, 26 May 2015

Author Contributions: Experimental Design, R.G.-G., J.P., D.R., M.J.C. and E.B.; performed the experiments, R.G.-G., J.P., M.V.-M., D.R., A.G., M.J.C., G.M.U.d.A. and E.B.; data Analysis: R.G.-G., M.J.C., D.R., A.J.E. and E.B. paper writing, R.G.-G., D.R., M.J.C. and E.B. All authors have read and agreed to the published version of the manuscript.

Funding: R.G.-G. was the recipient of a Research Supporting Staff Grant Contract (Spanish Health Research Fund (FIS)).

Acknowledgments: We are grateful to Miriam Novillo Pérez for technical help. We also thank the technical help provided by the Confocal Microscopy Unit of the Ramón y Cajal Institute for Health Research (IRYCIS). Our group also wants to thank Juan José Díaz Gil, who passed away on 4 November 2016, for the discovery of the Liver Growth Factor and for his exceptional dedication to pursuing knowledge about its effectiveness against neurodegenerative diseases. He will always be our great mentor and dear friend.

Conflicts of Interest: The authors declare no conflict of interest.

References

1. Bird, T.D. Alzheimer Disease Overview. In *GeneReviews®*; Adam, M.P., Ardinger, H.H., Pagon, R.A., Wallace, S.E., Bean, L.J.H., Stephens, K., Amemiya, A., Eds.; University of Washington, Seattle: Seattle, WA, USA; pp. 1993–2020.

2. De-Paula, V.J.; Radanovic, M.; Diniz, B.S.; Forlenza, O.V. Alzheimer's disease. *Subcell. Biochem.* **2012**, *65*, 329–352. [CrossRef] [PubMed]

3. Serrano-Pozo, A.; Frosch, M.P.; Masliah, E.; Hyman, B.T. Neuropathological alterations in Alzheimer disease. *Cold Spring Harb. Perspect. Med.* **2011**, *1*, a006189. [CrossRef] [PubMed]

4. Gandy, S.; DeKosky, S.T. Toward the treatment and prevention of Alzheimer's disease: Rational strategies and recent progress. *Annu. Rev. Med.* **2013**, *64*, 367–383. [CrossRef] [PubMed]

5. Díaz-Gil, J.J.; Escartin, P.; Garcia-Canero, R.; Trilla, C.; Veloso, J.J.; Sanchez, G.; Moreno-Caparros, A.; Enrique de Salamanca, C.; Lozano, R.; Gavilanes, J.G.; et al. Purification of a liver DNA-synthesis promoter from plasma of partially hepatectomized rats. *Biochem. J.* **1986**, *235*, 49–55. [CrossRef]

6. Gonzalo-Gobernado, R.; Calatrava-Ferreras, L.; Perucho, J.; Reimers, D.; Casarejos, M.J.; Herranz, A.S.; Jimenez-Escrig, A.; Diaz-Gil, J.J.; Bazan, E. Liver growth factor as a tissue regenerating factor in neurodegenerative diseases. *Recent Pat. CNS Drug Discov.* **2014**, *9*, 173–180. [CrossRef] [PubMed]

7. Gonzalo-Gobernado, R.; Reimers, D.; Herranz, A.S.; Diaz-Gil, J.J.; Osuna, C.; Asensio, M.J.; Baena, S.; Rodriguez-Serrano, M.; Bazan, E. Mobilization of neural stem cells and generation of new neurons in 6-OHDA-lesioned rats by intracerebroventricular infusion of liver growth factor. *J. Histochem. Cytochem.* **2009**, *57*, 491–502. [CrossRef]

8. Reimers, D.; Herranz, A.S.; Diaz-Gil, J.J.; Lobo, M.V.; Paino, C.L.; Alonso, R.; Asensio, M.J.; Gonzalo-Gobernado, R.; Bazan, E. Intrastriatal infusion of liver growth factor stimulates dopamine terminal sprouting and partially restores motor function in 6-hydroxydopamine-lesioned rats. *J. Histochem. Cytochem.* **2006**, *54*, 457–465. [CrossRef]

9. Gonzalo-Gobernado, R.; Calatrava-Ferreras, L.; Reimers, D.; Herranz, A.S.; Rodriguez-Serrano, M.; Miranda, C.; Jimenez-Escrig, A.; Diaz-Gil, J.J.; Bazan, E. Neuroprotective activity of peripherally administered liver growth factor in a rat model of Parkinson's disease. *PLoS ONE* **2013**, *8*, e67771. [CrossRef]

10. Reimers, D.; Osuna, C.; Gonzalo-Gobernado, R.; Herranz, A.S.; Diaz-Gil, J.J.; Jimenez-Escrig, A.; Asensio, M.J.; Miranda, C.; Rodriguez-Serrano, M.; Bazan, E. Liver growth factor promotes the survival of grafted neural stem cells in a rat model of Parkinson's disease. *Curr. Stem. Cell Res. Ther.* **2012**, *7*, 15–25. [CrossRef]

11. Díaz-Gil, J.J.; Garcia-Monzon, C.; Rua, C.; Martin-Sanz, P.; Cereceda, R.M.; Miquilena-Colina, M.E.; Machin, C.; Fernandez-Martinez, A.; Garcia-Canero, R. The anti-fibrotic effect of liver growth factor is associated with decreased intrahepatic levels of matrix metalloproteinases 2 and 9 and transforming growth factor beta 1 in bile duct-ligated rats. *Histol. Histopathol.* **2008**, *23*, 583–591. [CrossRef]

12. Díaz-Gil, J.J.; Munoz, J.; Albillos, A.; Rua, C.; Machin, C.; Garcia-Canero, R.; Cereceda, R.M.; Guijarro, M.C.; Trilla, C.; Escartin, P. Improvement in liver fibrosis, functionality and hemodynamics in CCl4-cirrhotic rats after injection of the Liver Growth Factor. *J. Hepatol.* **1999**, *30*, 1065–1072. [CrossRef]

13. Martinez-Galan, L.; del Puerto-Nevado, L.; Perez-Rial, S.; Diaz-Gil, J.J.; Gonzalez-Mangado, N.; Peces-Barba, G. Liver growth factor improves pulmonary fibrosis secondary to cadmium administration in mice. *Arch. Bronconeumol.* **2010**, *46*, 20–26. [CrossRef]

14. Calatrava-Ferreras, L.; Gonzalo-Gobernado, R.; Reimers, D.; Herranz, A.S.; Casarejos, M.J.; Jimenez-Escrig, A.; Regadera, J.; Velasco-Martin, J.; Vallejo-Munoz, M.; Diaz-Gil, J.J.; et al. Liver Growth Factor (LGF) Upregulates Frataxin Protein Expression and Reduces Oxidative Stress in Friedreich's Ataxia Transgenic Mice. *Int. J. Mol. Sci.* **2016**, *17*, 2066. [CrossRef] [PubMed]

15. Calatrava-Ferreras, L.; Gonzalo-Gobernado, R.; Reimers, D.; Herranz, A.S.; Jimenez-Escrig, A.; Diaz-Gil, J.J.; Casarejos, M.J.; Montero-Vega, M.T.; Bazan, E. Neuroprotective Role of Liver Growth Factor "LGF" in an Experimental Model of Cerebellar Ataxia. *Int. J. Mol. Sci.* **2014**, *15*, 19056–19073. [CrossRef]

16. Heneka, M.T.; O'Banion, M.K.; Terwel, D.; Kummer, M.P. Neuroinflammatory processes in Alzheimer's disease. *J. Neural. Transm. (Vienna)* **2010**, *117*, 919–947. [CrossRef]

17. Huang, W.J.; Zhang, X.; Chen, W.W. Role of oxidative stress in Alzheimer's disease. *Biomed. Rep.* **2016**, *4*, 519–522. [CrossRef]

18. Hsiao, K.; Chapman, P.; Nilsen, S.; Eckman, C.; Harigaya, Y.; Younkin, S.; Yang, F.; Cole, G. Correlative memory deficits, Abeta elevation, and amyloid plaques in transgenic mice. *Science* **1996**, *274*, 99–102. [CrossRef]

19. Lee, Y.W.; Kim, D.H.; Jeon, S.J.; Park, S.J.; Kim, J.M.; Jung, J.M.; Lee, H.E.; Bae, S.G.; Oh, H.K.; Son, K.H.; et al. Neuroprotective effects of salvianolic acid B on an Abeta25-35 peptide-induced mouse model of Alzheimer's disease. *Eur. J. Pharmacol.* **2013**, *704*, 70–77. [CrossRef]

20. Gonzalo-Gobernado, R.; Reimers, D.; Casarejos, M.J.; Calatrava Ferreras, L.; Vallejo-Munoz, M.; Jimenez-Escrig, A.; Diaz-Gil, J.J.; Ulzurrun de Asanza, G.M.; Bazan, E. Liver Growth Factor Induces Glia-Associated Neuroprotection in an in Vitro Model of Parkinson's Disease. *Brain Sci.* **2020**, *10*, 315. [CrossRef]

21. Arbor, S.C.; LaFontaine, M.; Cumbay, M. Amyloid-beta Alzheimer targets—Protein processing, lipid rafts, and amyloid-beta pores. *Yale J. Biol. Med.* **2016**, *89*, 5–21. [PubMed]

22. Butterfield, D.A.; Swomley, A.M.; Sultana, R. Amyloid beta-peptide (1-42)-induced oxidative stress in Alzheimer disease: Importance in disease pathogenesis and progression. *Antioxid. Redox Signal.* **2013**, *19*, 823–835. [CrossRef] [PubMed]

23. Lu, R.C.; Tan, M.S.; Wang, H.; Xie, A.M.; Yu, J.T.; Tan, L. Heat shock protein 70 in Alzheimer's disease. *Biomed. Res. Int.* **2014**, *2014*, 435203. [CrossRef] [PubMed]

24. Hoshino, T.; Murao, N.; Namba, T.; Takehara, M.; Adachi, H.; Katsuno, M.; Sobue, G.; Matsushima, T.; Suzuki, T.; Mizushima, T. Suppression of Alzheimer's disease-related phenotypes by expression of heat shock protein 70 in mice. *J. Neurosci.* **2011**, *31*, 5225–5234. [CrossRef] [PubMed]

25. Surra, J.C.; Guillen, N.; Barranquero, C.; Arbones-Mainar, J.M.; Navarro, M.A.; Gascon, S.; Arnal, C.; Godino, J.; Guzman, M.A.; Diaz-Gil, J.J.; et al. Sex-dependent effect of liver growth factor on atherosclerotic lesions and fatty liver disease in apolipoprotein E knockout mice. *Histol. Histopathol.* **2010**, *25*, 609–618. [PubMed]

26. Avila, J. Alzheimer disease: Caspases first. *Nat. Rev. Neurol.* **2010**, *6*, 587–588. [CrossRef]

27. Hernandez, F.; Avila, J. Tauopathies. *Cell Mol. Life Sci.* **2007**, *64*, 2219–2233. [CrossRef] [PubMed]

28. Dou, F.; Netzer, W.J.; Tanemura, K.; Li, F.; Hartl, F.U.; Takashima, A.; Gouras, G.K.; Greengard, P.; Xu, H. Chaperones increase association of tau protein with microtubules. *Proc. Natl. Acad. Sci. USA* **2003**, *100*, 721–726. [CrossRef]

29. Miyata, Y.; Koren, J.; Kiray, J.; Dickey, C.A.; Gestwicki, J.E. Molecular chaperones and regulation of tau quality control: Strategies for drug discovery in tauopathies. *Future Med. Chem.* **2011**, *3*, 1523–1537. [CrossRef]

30. Dossi, E.; Vasile, F.; Rouach, N. Human astrocytes in the diseased brain. *Brain Res. Bull.* **2018**, *136*, 139–156. [CrossRef]

31. Sarlus, H.; Heneka, M.T. Microglia in Alzheimer's disease. *J. Clin. Investig.* **2017**, *127*, 3240–3249. [CrossRef]

32. Taipa, R.; Ferreira, V.; Brochado, P.; Robinson, A.; Reis, I.; Marques, F.; Mann, D.M.; Melo-Pires, M.; Sousa, N. Inflammatory pathology markers (activated microglia and reactive astrocytes) in early and late onset Alzheimer disease: A post mortem study. *Neuropathol. Appl. Neurobiol.* **2018**, *44*, 298–313. [CrossRef] [PubMed]

33. von Bernhardi, R.; Eugenin-von Bernhardi, L.; Eugenin, J. Microglial cell dysregulation in brain aging and neurodegeneration. *Front. Aging Neurosci.* **2015**, *7*, 124. [CrossRef] [PubMed]

34. Kakimura, J.; Kitamura, Y.; Takata, K.; Umeki, M.; Suzuki, S.; Shibagaki, K.; Taniguchi, T.; Nomura, Y.; Gebicke-Haerter, P.J.; Smith, M.A.; et al. Microglial activation and amyloid-beta clearance induced by exogenous heat-shock proteins. *FASEB J.* **2002**, *16*, 601–603. [CrossRef] [PubMed]

35. Von Bernhardi, R.; Ramirez, G. Microglia-astrocyte interaction in Alzheimer's disease: Friends or foes for the nervous system? *Biol. Res.* **2001**, *34*, 123–128. [CrossRef] [PubMed]

36. Franklin, B.S.; Bossaller, L.; de Nardo, D.; Ratter, J.M.; Stutz, A.; Engels, G.; Brenker, C.; Nordhoff, M.; Mirandola, S.R.; Al-Amoudi, A.; et al. The adaptor ASC has extracellular and 'prionoid' activities that propagate inflammation. *Nat. Immunol.* **2014**, *15*, 727–737. [CrossRef] [PubMed]

37. Friker, L.L.; Scheiblich, H.; Hochheiser, I.V.; Brinkschulte, R.; Riedel, D.; Latz, E.; Geyer, M.; Heneka, M.T. beta-Amyloid Clustering around ASC Fibrils Boosts Its Toxicity in Microglia. *Cell Rep.* **2020**, *30*, 3743–3754. [CrossRef]

38. Scott, X.O.; Stephens, M.E.; Desir, M.C.; Dietrich, W.D.; Keane, R.W.; de Rivero Vaccari, J.P. The Inflammasome Adaptor Protein ASC in Mild Cognitive Impairment and Alzheimer's Disease. *Int. J. Mol. Sci.* **2020**, *21*, 4674. [CrossRef]

39. Venegas, C.; Kumar, S.; Franklin, B.S.; Dierkes, T.; Brinkschulte, R.; Tejera, D.; Vieira-Saecker, A.; Schwartz, S.; Santarelli, F.; Kummer, M.P.; et al. Microglia-derived ASC specks cross-seed amyloid-beta in Alzheimer's disease. *Nature* **2017**, *552*, 355–361. [CrossRef]

40. Couturier, J.; Stancu, I.C.; Schakman, O.; Pierrot, N.; Huaux, F.; Kienlen-Campard, P.; Dewachter, I.; Octave, J.N. Activation of phagocytic activity in astrocytes by reduced expression of the inflammasome component ASC and its implication in a mouse model of Alzheimer disease. *J. Neuroinflamm.* **2016**, *13*, 20. [CrossRef]

41. Doens, D.; Fernandez, P.L. Microglia receptors and their implications in the response to amyloid beta for Alzheimer's disease pathogenesis. *J. Neuroinflamm.* **2014**, *11*, 48. [CrossRef]

42. Martin, E.; Boucher, C.; Fontaine, B.; Delarasse, C. Distinct inflammatory phenotypes of microglia and monocyte-derived macrophages in Alzheimer's disease models: Effects of aging and amyloid pathology. *Aging Cell* **2017**, *16*, 27–38. [CrossRef] [PubMed]

43. Coraci, I.S.; Husemann, J.; Berman, J.W.; Hulette, C.; Dufour, J.H.; Campanella, G.K.; Luster, A.D.; Silverstein, S.C.; El-Khoury, J.B. CD36, a class B scavenger receptor, is expressed on microglia in Alzheimer's disease brains and can mediate production of reactive oxygen species in response to beta-amyloid fibrils. *Am. J. Pathol.* **2002**, *160*, 101–112. [CrossRef]

44. Ricciarelli, R.; D'Abramo, C.; Zingg, J.M.; Giliberto, L.; Markesbery, W.; Azzi, A.; Marinari, U.M.; Pronzato, M.A.; Tabaton, M. CD36 overexpression in human brain correlates with beta-amyloid deposition but not with Alzheimer's disease. *Free Radic. Biol. Med.* **2004**, *36*, 1018–1024. [CrossRef] [PubMed]

45. Wilkinson, K.; Boyd, J.D.; Glicksman, M.; Moore, K.J.; El Khoury, J. A high content drug screen identifies ursolic acid as an inhibitor of amyloid beta protein interactions with its receptor CD36. *J. Biol. Chem.* **2011**, *286*, 34914–34922. [CrossRef] [PubMed]

46. Conde, J.R.; Streit, W.J. Microglia in the aging brain. *J. Neuropathol. Exp. Neurol.* **2006**, *65*, 199–203. [CrossRef] [PubMed]

47. Mosher, K.I.; Wyss-Coray, T. Microglial dysfunction in brain aging and Alzheimer's disease. *Biochem. Pharmacol.* **2014**, *88*, 594–604. [CrossRef]

48. Itoh, K.; Chiba, T.; Takahashi, S.; Ishii, T.; Igarashi, K.; Katoh, Y.; Oyake, T.; Hayashi, N.; Satoh, K.; Hatayama, I.; et al. An Nrf2/small Maf heterodimer mediates the induction of phase II detoxifying enzyme genes through antioxidant response elements. *Biochem. Biophys. Res. Commun.* **1997**, *236*, 313–322. [CrossRef]

49. Suzuki, T.; Muramatsu, A.; Saito, R.; Iso, T.; Shibata, T.; Kuwata, K.; Kawaguchi, S.I.; Iwawaki, T.; Adachi, S.; Suda, H.; et al. Molecular Mechanism of Cellular Oxidative Stress Sensing by Keap1. *Cell Rep.* **2019**, *28*, 746–758. [CrossRef]

50. Itoh, K.; Mochizuki, M.; Ishii, Y.; Ishii, T.; Shibata, T.; Kawamoto, Y.; Kelly, V.; Sekizawa, K.; Uchida, K.; Yamamoto, M. Transcription factor Nrf2 regulates inflammation by mediating the effect of 15-deoxy-Delta(12,14)-prostaglandin j(2). *Mol. Cell Biol.* **2004**, *24*, 36–45. [CrossRef]

51. Kobayashi, E.H.; Suzuki, T.; Funayama, R.; Nagashima, T.; Hayashi, M.; Sekine, H.; Tanaka, N.; Moriguchi, T.; Motohashi, H.; Nakayama, K.; et al. Nrf2 suppresses macrophage inflammatory response by blocking proinflammatory cytokine transcription. *Nat. Commun.* **2016**, *7*, 11624. [CrossRef]

52. Kerr, F.; Sofola-Adesakin, O.; Ivanov, D.K.; Gatliff, J.; Gomez Perez-Nievas, B.; Bertrand, H.C.; Martinez, P.; Callard, R.; Snoeren, I.; Cocheme, H.M.; et al. Direct Keap1-Nrf2 disruption as a potential therapeutic target for Alzheimer's disease. *PLoS Genet* **2017**, *13*, e1006593. [CrossRef] [PubMed]

53. Youssef, P.; Chami, B.; Lim, J.; Middleton, T.; Sutherland, G.T.; Witting, P.K. Evidence supporting oxidative stress in a moderately affected area of the brain in Alzheimer's disease. *Sci. Rep.* **2018**, *8*, 11553. [CrossRef] [PubMed]

54. Joshi, G.; Gan, K.A.; Johnson, D.A.; Johnson, J.A. Increased Alzheimer's disease-like pathology in the APP/PS1DeltaE9 mouse model lacking Nrf2 through modulation of autophagy. *Neurobiol. Aging.* **2015**, *36*, 664–679. [CrossRef] [PubMed]

55. Rojo, A.I.; Pajares, M.; Garcia-Yague, A.J.; Buendia, I.; van Leuven, F.; Yamamoto, M.; Lopez, M.G.; Cuadrado, A. Deficiency in the transcription factor NRF2 worsens inflammatory parameters in a mouse model with combined tauopathy and amyloidopathy. *Redox. Biol.* **2018**, *18*, 173–180. [CrossRef]

56. Uruno, A.; Matsumaru, D.; Ryoke, R.; Saito, R.; Kadoguchi, S.; Saigusa, D.; Saito, T.; Saido, T.C.; Kawashima, R.; Yamamoto, M. Nrf2 Suppresses Oxidative Stress and Inflammation in App Knock-In Alzheimer's Disease Model Mice. *Mol. Cell Biol.* **2020**, *40*. [CrossRef]

57. Wang, C.Y.; Wang, Z.Y.; Xie, J.W.; Cai, J.H.; Wang, T.; Xu, Y.; Wang, X.; An, L. CD36 upregulation mediated by intranasal LV-NRF2 treatment mitigates hypoxia-induced progression of Alzheimer's-like pathogenesis. *Antioxid. Redox Signal.* **2014**, *21*, 2208–2230. [CrossRef] [PubMed]

58. Giron-Martinez, A.; Perez-Rial, S.; Terron-Exposito, R.; Diaz-Gil, J.J.; Gonzalez-Mangado, N.; Peces-Barba, G. Proliferative Activity of Liver Growth Factor is Associated with an Improvement of Cigarette Smoke-Induced Emphysema in Mice. *PLoS ONE* **2014**, *9*, e112995. [CrossRef]

59. Frebel, K.; Wiese, S. Signalling molecules essential for neuronal survival and differentiation. *Biochem. Soc. Trans.* **2006**, *34*, 1287–1290. [CrossRef]

60. Eckert, A.; Marques, C.A.; Keil, U.; Schussel, K.; Muller, W.E. Increased apoptotic cell death in sporadic and genetic Alzheimer's disease. *Ann. N. Y. Acad. Sci.* **2003**, *1010*, 604–609. [CrossRef]

61. Feng, Z.; Qin, C.; Chang, Y.; Zhang, J.T. Early melatonin supplementation alleviates oxidative stress in a transgenic mouse model of Alzheimer's disease. *Free Radic. Biol. Med.* **2006**, *40*, 101–109. [CrossRef]

62. Yan, Y.; Gong, K.; Ma, T.; Zhang, L.; Zhao, N.; Zhang, X.; Tang, P.; Gong, Y. Protective effect of edaravone against Alzheimer's disease-relevant insults in neuroblastoma N2a cells. *Neurosci. Lett.* **2012**, *531*, 160–165. [CrossRef] [PubMed]

63. Obulesu, M.; Lakshmi, M.J. Apoptosis in Alzheimer's disease: An understanding of the physiology, pathology and therapeutic avenues. *Neurochem. Res.* **2014**, *39*, 2301–2312. [CrossRef] [PubMed]

64. King, D.L.; Arendash, G.W. Behavioral characterization of the Tg2576 transgenic model of Alzheimer's disease through 19 months. *Physiol. Behav.* **2002**, *75*, 627–642. [CrossRef]

65. Takeuchi, A.; Irizarry, M.C.; Duff, K.; Saido, T.C.; Hsiao Ashe, K.; Hasegawa, M.; Mann, D.M.; Hyman, B.T.; Iwatsubo, T. Age-related amyloid beta deposition in transgenic mice overexpressing both Alzheimer mutant presenilin 1 and amyloid beta precursor protein Swedish mutant is not associated with global neuronal loss. *Am. J. Pathol.* **2000**, *157*, 331–339. [CrossRef]

66. Noshita, N.; Lewen, A.; Sugawara, T.; Chan, P.H. Evidence of phosphorylation of Akt and neuronal survival after transient focal cerebral ischemia in mice. *J. Cereb. Blood Flow Metab.* **2001**, *21*, 1442–1450. [CrossRef] [PubMed]

67. Zhao, H.; Sapolsky, R.M.; Steinberg, G.K. Phosphoinositide-3-kinase/akt survival signal pathways are implicated in neuronal survival after stroke. *Mol. Neurobiol.* **2006**, *34*, 249–270. [CrossRef]

68. Matsuda, S.; Nakagawa, Y.; Tsuji, A.; Kitagishi, Y.; Nakanishi, A.; Murai, T. Implications of PI3K/AKT/PTEN Signaling on Superoxide Dismutases Expression and in the Pathogenesis of Alzheimer's Disease. *Diseases* **2018**, *6*, 28. [CrossRef]

69. Chong, Z.Z.; Li, F.; Maiese, K. Activating Akt and the brain's resources to drive cellular survival and prevent inflammatory injury. *Histol. Histopathol.* **2005**, *20*, 299–315. [CrossRef]

70. Gardai, S.J.; Hildeman, D.A.; Frankel, S.K.; Whitlock, B.B.; Frasch, S.C.; Borregaard, N.; Marrack, P.; Bratton, D.L.; Henson, P.M. Phosphorylation of Bax Ser184 by Akt regulates its activity and apoptosis in neutrophils. *J. Biol. Chem.* **2004**, *279*, 21085–21095. [CrossRef]

71. Sagare, A.P.; Winkler, E.A.; Bell, R.D.; Deane, R.; Zlokovic, B.V. From the liver to the blood-brain barrier: An interconnected system regulating brain amyloid-beta levels. *J. Neurosci. Res.* **2011**, *89*, 967–968. [CrossRef]

72. Thornalley, P.J. Cell activation by glycated proteins. AGE receptors, receptor recognition factors and functional classification of AGEs. *Cell Mol. Biol. (Noisy-Le-Grand)* **1998**, *44*, 1013–1023. [PubMed]

73. Park, I.H.; Yeon, S.I.; Youn, J.H.; Choi, J.E.; Sasaki, N.; Choi, I.H.; Shin, J.S. Expression of a novel secreted splice variant of the receptor for advanced glycation end products (RAGE) in human brain astrocytes and peripheral blood mononuclear cells. *Mol. Immunol.* **2004**, *40*, 1203–1211. [CrossRef] [PubMed]

74. Ameen-Ali, K.E.; Wharton, S.B.; Simpson, J.E.; Heath, P.R.; Sharp, P.; Berwick, J. Review: Neuropathology and behavioural features of transgenic murine models of Alzheimer's disease. *Neuropathol. Appl. Neurobiol.* **2017**, *43*, 553–570. [CrossRef] [PubMed]

75. Deacon, R.M. Digging and marble burying in mice: Simple methods for in vivo identification of biological impacts. *Nat. Protoc.* **2006**, *1*, 122–124. [CrossRef] [PubMed]

76. Kim, T.K.; Han, H.E.; Kim, H.; Lee, J.E.; Choi, D.; Park, W.J.; Han, P.L. Expression of the plant viral protease NIa in the brain of a mouse model of Alzheimer's disease mitigates Abeta pathology and improves cognitive function. *Exp. Mol. Med.* **2012**, *44*, 740–748. [CrossRef] [PubMed]

77. Díaz-Gil, J.J.; Rua, C.; Machin, C.; Cereceda, R.M.; Garcia-Canero, R.; de Foronda, M.; Perez de Diego, J.; Trilla, C.; Escartin, P. Hepatic growth induced by injection of the liver growth factor into normal rats. *Growth Regul.* **1994**, *4*, 113–122. [PubMed]

78. Singh, J.; Bowers, L.D. Quantitative fractionation of serum bilirubin species by reversed-phase high-performance liquid chromatography. *J. Chromatogr.* **1986**, *380*, 321–330. [CrossRef]

79. Díaz-Gil, J.J.; Sanchez, G.; Trilla, C.; Escartin, P. Identification of biliprotein as a liver growth factor. *Hepatology* **1988**, *8*, 484–486. [CrossRef]

80. Carro, E.; Trejo, J.L.; Gomez-Isla, T.; LeRoith, D.; Torres-Aleman, I. Serum insulin-like growth factor I regulates brain amyloid-beta levels. *Nat. Med.* **2002**, *8*, 1390–1397. [CrossRef]

81. Perucho, J.; Casarejos, M.J.; Gomez, A.; Solano, R.M.; de Yebenes, J.G.; Mena, M.A. Trehalose protects from aggravation of amyloid pathology induced by isoflurane anesthesia in APP(swe) mutant mice. *Curr. Alzheimer Res.* **2012**, *9*, 334–343. [CrossRef]

82. Perucho, J.; Casarejos, M.J.; Rubio, I.; Rodriguez-Navarro, J.A.; Gomez, A.; Ampuero, I.; Rodal, I.; Solano, R.M.; Carro, E.; Garcia de Yebenes, J.; et al. The effects of parkin suppression on the behaviour, amyloid processing, and cell survival in APP mutant transgenic mice. *Exp. Neurol.* **2010**, *221*, 54–67. [CrossRef] [PubMed]

83. Casarejos, M.J.; Perucho, J.; Gomez, A.; Munoz, M.P.; Fernandez-Estevez, M.; Sagredo, O.; Fernandez Ruiz, J.; Guzman, M.; de Yebenes, J.G.; Mena, M.A. Natural cannabinoids improve dopamine neurotransmission and tau and amyloid pathology in a mouse model of tauopathy. *J. Alzheimers Dis.* **2013**, *35*, 525–539. [CrossRef] [PubMed]

84. Franklin, K.; Paxinos, G. *The Mouse Brain in Stereotaxic Coordinates*, 3rd ed.; Academic Press: San Diego, CA, USA, 2007.

Publisher's Note: MDPI stays neutral with regard to jurisdictional claims in published maps and institutional affiliations.

International Journal of
Molecular Sciences

Article

Neuroprotective Effect of Bean Phosphatidylserine on TMT-Induced Memory Deficits in a Rat Model

Minsook Ye [1], Bong Hee Han [2], Jin Su Kim [3], Kyungsoo Kim [1,*] and Insop Shim [2,*]

[1] Department of Biomedicine & Health Sciences, College of Medicine, The Catholic University of Korea, Seoul 06591, Korea; jh486ms22@naver.com
[2] Department of Physiology, College of Medicine, Kyung Hee University, Seoul 02435, Korea; hanbh10@hanmail.net
[3] Division of RI-Convergence Research, Korea Institute of Radiological and Medical Sciences, Seoul 01812, Korea; kjs@kirams.re.kr
*   Correspondence: kskim@catholic.ac.kr (K.K.); ishim@khu.ac.kr (I.S.); Tel.: +82-2-2245-6284 (K.K.); +82-2-961-0698 (I.S.); Fax: +82-2-537-8979 (K.K.); +82-2-963-2175 (I.S.)

Received: 24 April 2020; Accepted: 9 July 2020; Published: 11 July 2020

Abstract: Background: Trimethyltin (TMT) is a potent neurotoxin affecting various regions of the central nervous system, including the neocortex, the cerebellum, and the hippocampus. Phosphatidylserine (PS) is a membrane phospholipid, which is vital to brain cells. We analyzed the neuroprotective effects of soybean-derived phosphatidylserine (Bean-PS) on cognitive function, changes in the central cholinergic systems, and neural activity in TMT-induced memory deficits in a rat model. Methods: The rats were randomly divided into an untreated normal group, a TMT group (injected with TMT + vehicle), and a group injected with TMT + Bean-PS. The rats were treated with 10% hexane (TMT group) or TMT + Bean-PS (50 mg·kg^{-1}, oral administration (p.o.)) daily for 21 days, following a single injection of TMT (8.0 mg/kg, intraperitoneally (i.p.)). The cognitive function of Bean-PS was assessed using the Morris water maze (MWM) test and a passive avoidance task (PAT). The expression of acetylcholine transferase (ChAT) and acetylcholinesterase (AchE) in the hippocampus was assessed via immunohistochemistry. A positron emission tomography (PET) scan was used to measure the glucose uptake in the rat brain. Results: Treatment with Bean-PS enhanced memory function in the Morris water maze (MWM) test. Consistent with the behavioral results, treatment with Bean-PS diminished the damage to cholinergic cells in the hippocampus, in contrast to those of the TMT group. The TMT+Bean-PS group showed elevated glucose uptake in the frontal lobe of the rat brain. Conclusion: These results demonstrate that Bean-PS protects against TMT-induced learning and memory impairment. As such, Bean-PS represents a potential treatment for neurodegenerative disorders, such as Alzheimer's disease.

Keywords: neuroprotection; neurodegenerative disorder; choline acetyltransferase (ChAT); trimethyltin (TMT); bean phosphatidylserine (Bean-PS)

---

1. Introduction

Trimethyltin (TMT) intoxication is regarded as an appropriate model of chronic neuronal degeneration associated with cognitive impairment, and is therefore useful in the study of Alzheimer's disease (AD) [1,2]. The organotin trimethyltin chloride (TMT) is a neurotoxin that induces neuronal degeneration in the central nervous system (CNS) [3]. In particular, TMT injection leads to substantial damage of the hippocampus, which is implicated in memory [4]. Necrosis of hippocampal pyramidal cells and granule cells, produced by TMT, has been associated with the disruption of normal behavioral patterns [5], hippocampal physiological activity [6,7], and neurochemical markers of endogenous hippocampal neurotransmitters [7,8]. Rats exposed to TMT show behavioral, biochemical, and histological deficits [4]. For example, granule cells in the dentate gyrus, and cornu ammonis 1 (CA1) and cornu ammonis 3 (CA3) pyramidal cells, were significantly impregnated with TMT [9–11]. TMT intoxication attenuates hippocampal-dependent behavior in the Morris water maze [12] and the passive avoidance test [13]. TMT injection causes massive neuronal death, accompanied by enhanced hippocampal neurogenesis in the rat brain [14].

Several studies have shown that the cognitive dysfunction was associated with damaged cholinergic neurons in the brains of animal models of AD. Damage to the cholinergic system in the brain is closely associated with memory deficits. These anatomical and behavioral findings in TMT-intoxicated rats have been used to develop an attractive model of degenerative diseases such as AD [15]. AD has also been correlated with the loss of cholinergic neurons and decreased levels of acetylcholine (ACh) and choline acetyltransferase (ChAT). Lesions in these pathways lead to decreased ACh release, resulting in learning and memory dysfunction [16]. Current therapeutic measures are designed to increase levels of ACh in the brains of AD patients via suppression of acetylcholinesterase. Drugs, including the cholinesterase inhibitors, donepezil, galantamine, and rivastigmine, slow the breakdown of synaptic ACh, prolong its ability to stimulate post-synaptic receptors, and amplify the natural pattern of ACh release in the brain [17,18]. The drugs currently approved for the treatment of AD act by countering the acetylcholine deficits, leading to symptomatic relief, improved cognitive function, and enhanced acetylcholine levels in the brain.

Phosphatidylserine (PS) is the major anionic phospholipid found in the inner leaflet of eukaryotic cell membranes. PS-supplemented rodents showed enhanced memory, learning capacity, and other cognitive parameters [19]. In human studies, the efficacy of bovine-brain-derived PS (BC-PS) has been reported in patients with dementia [20]. Treatment with BC-PS improved memory function, especially delayed recall in the elderly with memory complaints. Although PS extracts from bovine cortex are known to be effective in improving memory function in humans and animals, alternative sources of PS are increasingly in demand [21]. It has been demonstrated that soybean-derived PS, one of the most promising alternatives, improved memory function in humans as much as BC-PS [22]. In addition, the treatment of rodents with BC-PS improved scopolamine-induced amnesia in the passive avoidance test, and cognitive disorders in senile subjects [23].

However, few studies have reported the effects of soybean-derived phosphatidylserine (Bean-PS) on cognitive improvement and its underlying mechanisms. The purpose of this study is to investigate whether Bean-PS prevents the neurodegeneration of the hippocampus, and the impairment of learning and memory, induced by TMT. The study demonstrated that Bean-PS improved cognitive function, and activated cholinergic systems in the hippocampus in addition to neural activation, in rats with TMT-induced cognitive deficits.

2. Results

2.1. Morris Water Maze Test

Figure 1A represents the escape latency intervals recorded during successive training trials. In the acquisition trials, the TMT group showed deteriorated cognitive deficits compared with the normal group, reflected by the increased escape latency (*** $p < 0.001$, on Days 1 and 4, ** $p < 0.01$, on Days 2 and 3). The TMT + Bean-PS group demonstrated amelioration of spatial memory and learning ability relative to the TMT group starting from Day 2 ($p < 0.05$) and Day 4 ($p < 0.05$). In the retention test, the times spent on the platform varied significantly among the groups. The TMT group spent less time on the platform than the normal group ($F_{2,\,18} = 6.42$, $p < 0.05$). The TMT + Bean-PS group did not affect the time spent in the platform area as seen in Figure 1B.

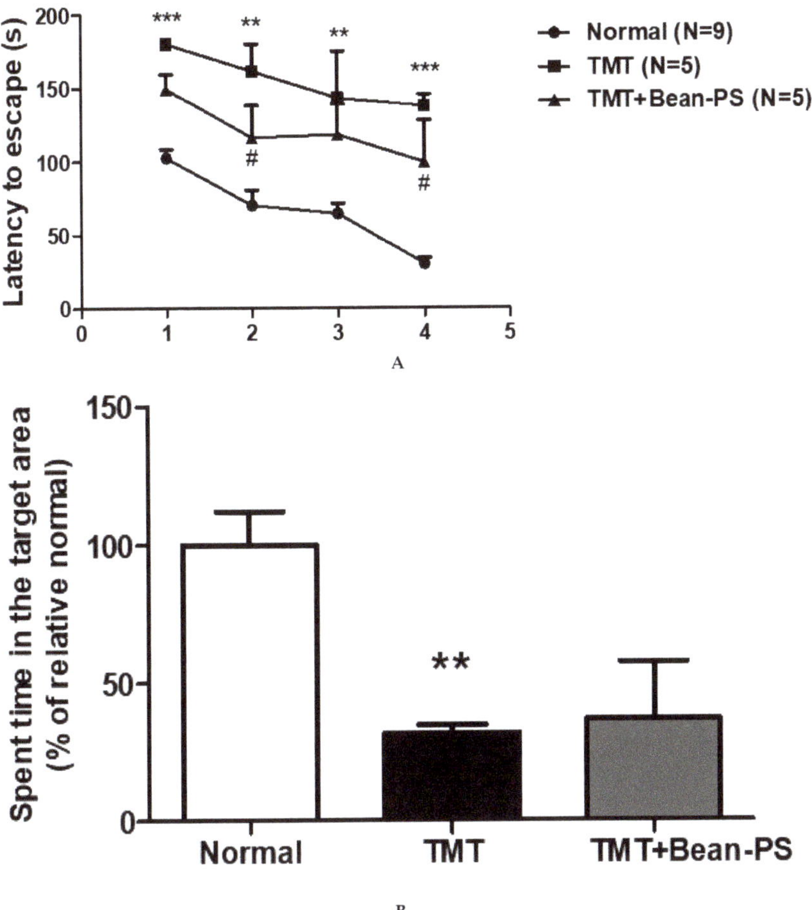

Figure 1. Effects of soy-bean-derived phosphatidylserine (Bean-PS) on spatial learning and memory dysfunction in trimethyltin (TMT)-induced rats. (**A**) The latency in escaping onto the hidden platform during the acquisition test. In the acquisition test, the task entailed three trials each day over 4 days. The values are presented as mean ± S.E.M. ** $p < 0.01$, *** $p < 0.001$ vs. normal group, # $p < 0.05$ vs. TMT group, respectively. (**B**)Retention was tested on Day 5. Results are expressed as means ± S.E.M. ** $p < 0.01$ vs. normal group. Normal group ($N = 9$); TMT group ($N = 5$); TMT-Bean-PS ($N = 5$).

2.2. Passive Avoidance Test

The passive avoidance test was conducted to determine the ability of working memory and learning. As shown in Figure 2, there was no significant difference among the three groups.

2.3. ChAT and AchE Immunoreactivity

The results of the evaluations of the ChAT-positive cells per section from the different hippocampal formations are shown in Figure 3. The ChAT activity in the hippocampus of the normal group was significantly higher than that of the TMT group. In particular, there were significant differences in the hippocampal CA1 ($F_{2, 18} = 8.53$, $p < 0.01$) and CA3 ($F_{2, 18} = 22.94$, $p < 0.001$).

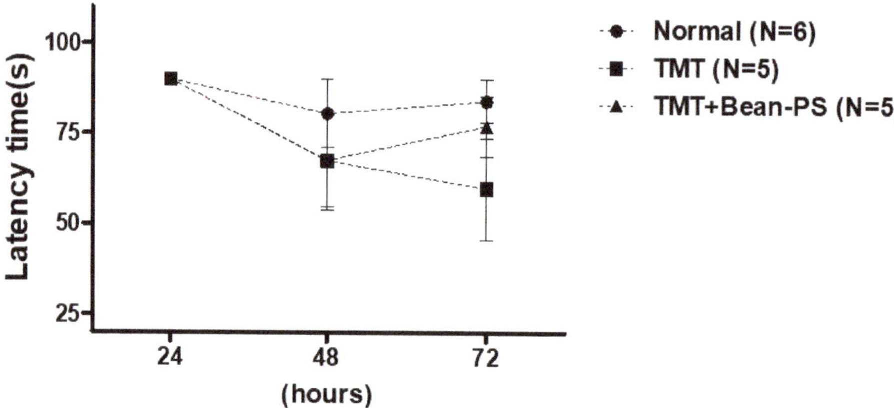

Figure 2. Effect of Bean-PS on escape latency into the dark phase of the retention test during the passive avoidance task. Each value is expressed ± S.E.M. Normal group ($N = 6$); TMT group ($N = 5$); TMT-Bean-PS ($N = 5$).

The numbers of ChAT-positive cells in the TMT+Bean-PS group were higher than those in the TMT group, particularly in CA1 ($p < 0.05$). Immunoreactivity of ChAT in the TMT + Bean-PS group was significantly increased in CA1 ($p < 0.05$), as seen in Figure 3A,B. The TMT + Bean-PS group in the CA3 region showed no difference compared with that of the TMT group as seen in Figure 3A,B. The results of the evaluations of the acetylcholinesterase (AchE) immunoreactive cells per section from the different hippocampi are shown in Figure 4A,B. The AChE activity in the hippocampus of the TMT group was significantly lower than that of the normal group ($p < 0.001$). In particular, there were significant differences in both CA1 ($F_{2,\ 15} = 12.07$, $p < 0.01$) and CA3 ($F_{2,\ 15} = 10.85$, $p < 0.01$). However, the AChE reactivity in the TMT+Bean-PS group was higher than that in the TMT group, particularly in CA1 ($p < 0.01$) and CA3 ($p < 0.01$).

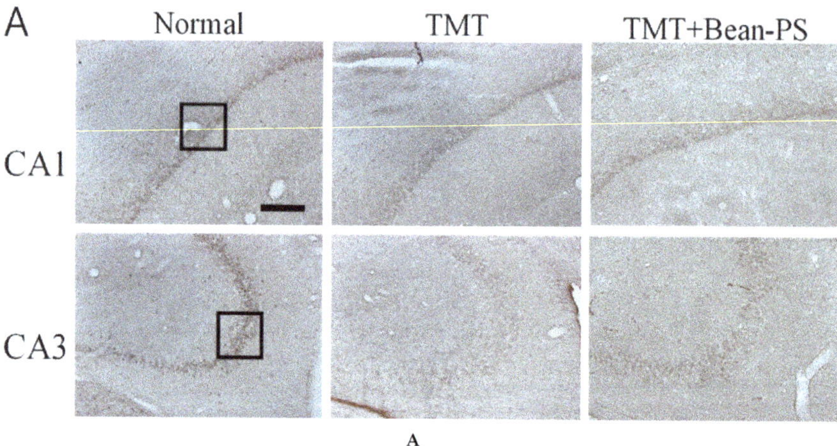

Figure 3. *Cont.*

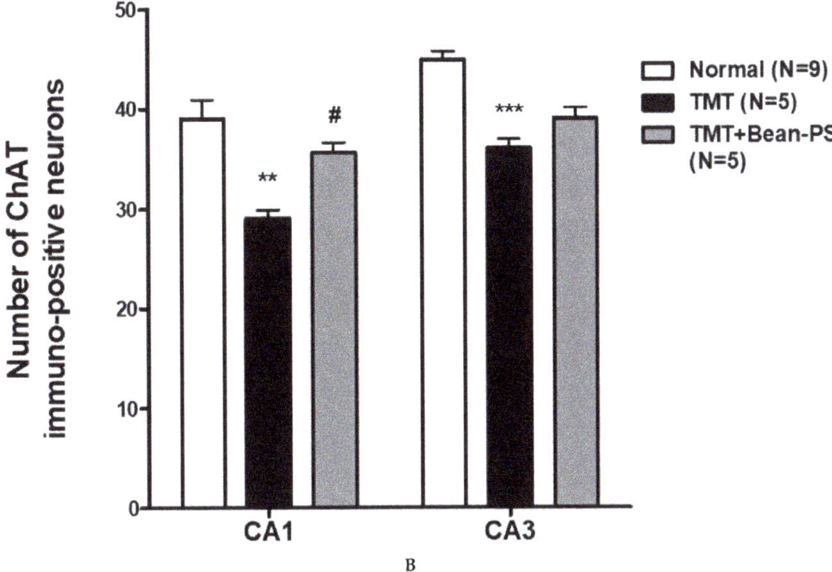

Figure 3. Effect of Bean-PS on the number of choline acetyltransferase (ChAT)-positive neurons in the hippocampus. Representative photographs and the number of positive neurons are indicated in (**A**) and (**B**). Black square represents region of CA1 and CA3 in the hippocampus and the scale bar represents 200 μm. Results are expressed as means ± S.E.M. *** $p < 0.001$, ** $p < 0.01$ vs. normal group, # $p < 0.05$ vs. TMT group. Normal group ($N = 9$); TMT group ($N = 5$); TMT-Bean-PS ($N = 5$).

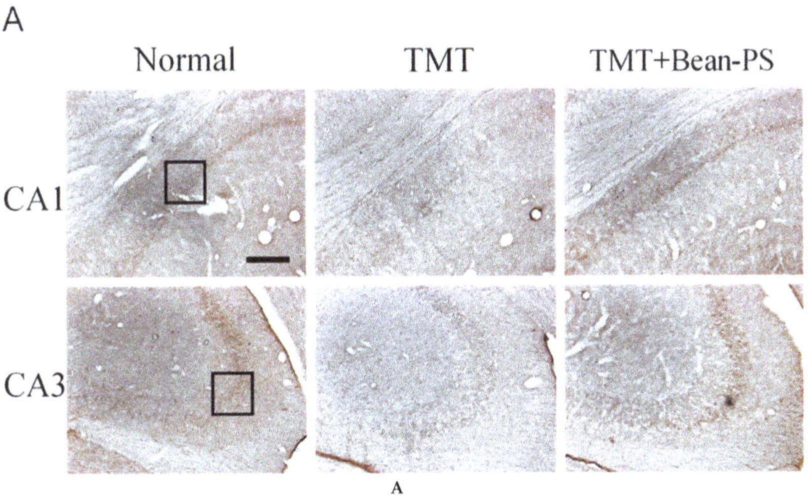

Figure 4. *Cont.*

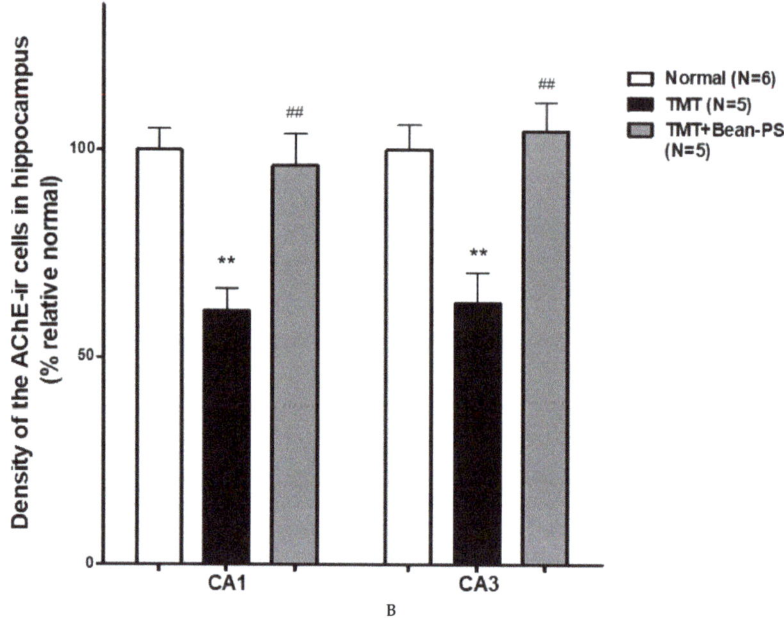

Figure 4. Effect of Bean-PS on the density of acetylcholinersterase (AchE)-immunostained nuclei in the hippocampus. Representative photographs and the density of AchE-immunostained nuclei are indicated in (**A**) and (**B**). Black square represents region of CA1 and CA3 in the hippocampus and the scale bar represents 200 μm. The values are presented as means ± S.E.M. ** $p < 0.01$ vs. normal group, ## $p < 0.01$ vs. TMT group, respectively. Normal group ($N = 6$); TMT group ($N = 5$); TMT-Bean-PS ($N = 5$).

2.4. Brain Glucose Metabolism

Figure 5 and Table 1 show the results of ^{18}F-fluorodeoxyglucose (FDG) uptake measurements from the different brain regions. The cerebral glucose activity of the TMT group was markedly reduced in the hippocampus and frontal lobe compared with that of the normal group in statistical parametric mapping (SPM) analysis of FDG-positron emission tomography (PET) (Figure 5A and Table 1A, $p < 0.05$). In addition, the activity of the TMT + Bean-PS group was significantly increased in the frontal lobe and hippocampus compared with the TMT group (Figure 5B and Table 1B, $p < 0.05$).

Table 1. The changes of Z values in the hippocampus and frontal lobe. The results of voxel-wise comparison between the TMT + Bean-PS and TMT group datasets are shown (Table 1A). In the statistical parametric mapping (SPM) analysis, the cerebral glucose uptake of TMT+Bean-PS datasets was significantly increased in the hippocampus and frontal lobe compared with that of the TMT group (Table 1B).

A		
Brain Area	**Coordinates (x, y, z)**	**Z Value**
Frontal lobe	(8, 1, 10)	1.95
Hippocampus (Right)	(9, −1, 11)	1.69
Hippocampus (Left)	(−9, −1, 10)	1.66
B		
Brain Area	**Coordinates (x, y, z)**	**Z Value**
Frontal lobe	(2, 4, 8)	3.24
Hippocampus (Right)	(7, −3, 3)	4.57
Hippocampus (Left)	(−5, −5, 1)	3.3

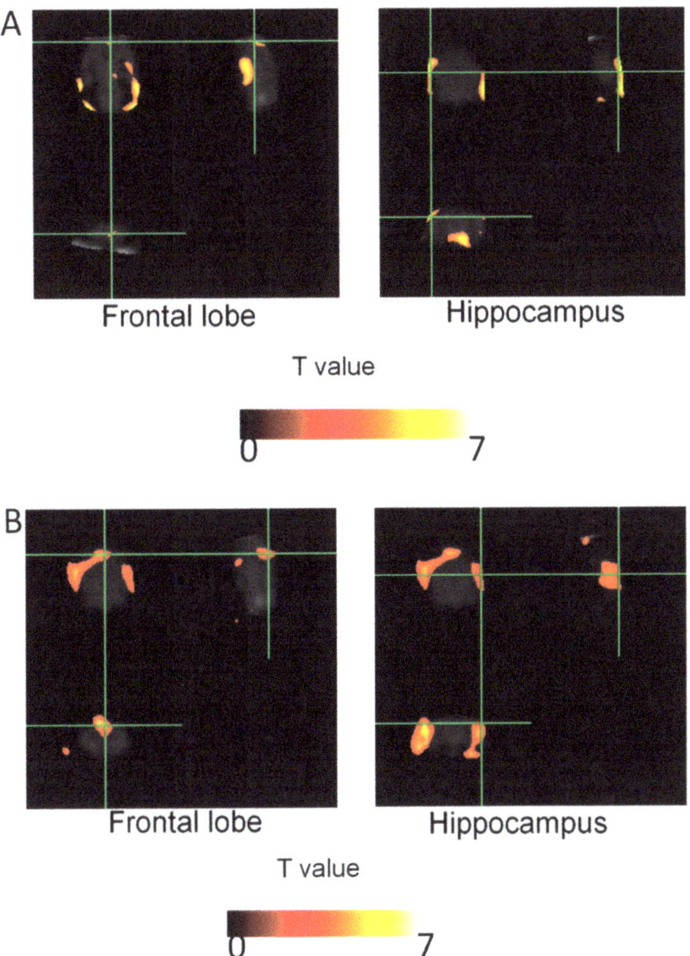

Figure 5. Brain regions showing regional glucose metabolism. (**A**) Brain regions where the regional ^{18}F-fluodeoxyglucose (FDG) uptake in the TMT group ($N = 3$) was markedly lower than in the normal group ($N = 3$) (hippocampus and frontal lobe). (**B**) Brain region where reginal glucose uptake in the TMT + Bean-PS group ($N = 3$) was significantly higher than in the TMT group (frontal lobe and hippocampus). Green line: cross hair

3. Discussion

AD is characterized by deterioration in memory, thinking, and the ability to carry out activities. An estimated 50 million people worldwide manifest dementia, and almost 10 million new cases are reported every year [24]. There is no established therapeutic agent currently available to treat dementia. Numerous new treatments are being investigated, in various stages of clinical trials. This study suggested that injection of TMT caused critical deficits in performance during tests of cognitive function, as well as causing corresponding signs of neurodegeneration, including decreased cholinergic neurons in the hippocampus. Our results showed that administration of Bean-PS reduced the TMT-induced learning and memory deficits in the Morris water maze, and suppressed TMT-induced reduction in ChAT and AchE in the hippocampus.

TMT exposure triggers severe behavioral and cognitive defects in both humans [25] and experimental animals [26]. TMT injection induced damage to the hippocampal pyramidal neurons, and also in the associated areas in rats [26–33]. Furthermore, previous studies have reported that TMT injection increases the risk of neuronal cell death, via possible excitotoxicity, intracellular calcium overload, and mitochondrial damage [34]. Several behavioral studies have suggested that the performance of TMT-induced rats is poor in memory and learning tasks [35,36]. The Morris water maze test has been used to test permanent spatial learning ability and reference memory, and utilized to evaluate cognitive-enhancing agents for the treatment of neurocognitive disorders [37,38].

Our study indicated that spatial memory impairment was ameliorated in the TMT+Bean-PS groups during the training days in contrast with that of the TMT group. In addition, the data demonstrated that Bean-PS preserved the TMT-induced reduction in spatial retention. These results suggested that Bean-PS improved learning and memory deficits in TMT-intoxicated rats used as experimental models of neurodegeneration for the study of Alzheimer-like diseases [15,39].

The animal model used in this study clearly demonstrates the functional significance of hippocampal neurodegeneration induced by TMT. The cholinergic system is involved in information processing associated with hippocampal learning and memory [40]. The hippocampus carries information derived from the related regions of the brain that are involved in learning and memory [41], and any damage to the cholinergic system may result in altered behavioral responses [42]. In particular, the loss of cholinergic function has been associated with a decline in cognition during aging and in AD [43]. Thus, the expression of AchE and ChAT in the hippocampus and its relation to TMT-induced cognitive impairment in rats was examined. Bean-PS also continuously improved the activity of cholinergic neurons in the hippocampus, which eventually restored the cholinergic pathway [44].

Based on the cholinergic hypothesis, patients with senile dementia show a selective and irreversible lack of cholinergic function in the brain [45]. Therefore, in patients with AD, treatment with cholinesterase inhibitors and ChAT activators may compensate for decreased ACh levels.

It is quite probable that the observed improvement in spatial learning deficits following the treatment with Bean-PS of rats was related to the enhanced release of ACh. According to the results from the Morris water maze (MWM), exposure to Bean-PS ameliorated the TMT-induced deficits in learning and memory. In addition, treatment with Bean-PS decreased cell loss, increased central cholinergic function, and prevented degeneration of cholinergic neurons mediating cognitive processes [46].

Our previous study reported that treatment with Bean-PS dissolved in medium-chain triglyceride (MCT) improved cognitive function and enhanced the neural activity in rats with TMT-induced learning and memory deficits. The active biological ingredient in the resulting extract is affected by the solvents [47]. Therefore, we investigated how the activity of Bean-PS in 10% hexane solvent affected TMT-induced memory deficits in rats.

The present study analyzed the effect of 10% hexane solvent on Bean-PS-attenuated TMT-induced cognitive defects in the Morris water maze, and found a protective effect in contrast to TMT-induced reduction in ChAT and AchE in the hippocampal areas.

Exposure to Bean-PS resulted in upregulation of glucose uptake in the hippocampus and frontal lobe. Administration of Bean-PS may have robust therapeutic potential as a treatment for neurodegenerative disorders, such as AD.

4. Material and Methods

4.1. Animals and Experimental Design

Male Sprague-Dawley rats weighing 250–580 g each were purchased from Samtaco Animal Corp. All rats were housed individually in a room at 23 °C (room temperature) under an alternating 12 h light/dark cycle. The rats were fed a commercial diet and provided with tap water ad libitum throughout the study. Food and water were made accessible ad libitum. This experiment was conducted in accordance with the National Institutes of Health Guide for the Care and Use of Laboratory Animals, revised in 1996, and was approved by the Institutional Animal Care and Use Committee of Kyung Hee University (KHUAP(SE)-18-073, 05/30/2018). In this study, the rats were randomly assigned to three groups: untreated, naïve (Normal, $n = 9$); TMT injected with vehicle (TMT, $n = 5$); and TMT injected along with 50 mg/kg^{-1} Bean-PS (Bean-PS, $n = 5$). The rats were injected intraperitoneally (i.p.) with TMT (8.0 mg/kg, body weight) dissolved in 0.9% saline and then returned to their home cages. The TMT+Bean-PS mixture (50 mg/kg, oral administration (p.o.)) was dissolved in 10% hexane and orally administered for two weeks after TMT-induced neurodegeneration.

4.2. Drug Treatment

PS was produced from soy lecithin by enzymatic transphophatidylation, and comprised a mixture of 90% phosphatidylserine (PS), 2% phosphatidylcholine (PC), and 6% phosphatidic acid (PA). The Bean-PS contained palmitic (17.7%), palmitoleic (1.3%), stearic (1.3%), oleic (14.4%), limoleic (61.2%), linolenic (1.4%), eicosapentaenoic acid (EPA), and other fatty acids (9.4%). PS and 10% hexane, used as a solvent for PS, were provided by Doosan Co. Glonet BG (Youngin, Korea) and the PS was stored in a freezer ($-60\,^{\circ}$C).

4.3. Morris Water Maze (MWM)

From the 17th day after the treatment with Bean-PS, the Morris water maze test was performed for 5 days. The black, plastic pool used in the Morris water maze test measured 200 cm in diameter and was filled to a depth of 35 cm with clear water maintained at $23 \pm 2\,^{\circ}$C. It was inserted in a submerged platform using external cues around the pool and within the behavioral room. For behavioral analysis, a personal computer was used to assort a charge coupled device (CCD). A probe test was run after the last trial on Day 4. In the experiment, the rats were exposed to an acquisition trial on each of 4 consecutive days. On Day 5, they were also trained in retention tests. The rats were allowed to search for the platform for 180 s during the acquisition test. On Day 5 of the retention test, the rats underwent a 1 min probe trial, in which the platform was eliminated from the pool. Probe trials were run at 1 min intervals. The animals in the performance test of each water maze trial were evaluated with a personal computer during the behavioral analysis (S-mart program, Barcelona, Spain).

4.4. Passive Avoidance Task (PAT)

A passive avoidance task was performed after 21 days of the treatment with Bean-PS. A test chamber was composed of two platforms, one light (white partition, 20 cm × 20 cm × 30 cm) and the other dark (black partition, 20 cm × 20 cm × 30 cm). A guillotine door opening (6 cm × 6 cm) was made on the floor in the center of the divider between the two partitions. Stainless steel bars, of diameter approximately 5 mm, were spaced at about 1 cm apart. Electric shocks (5 V, 0.5 mA, 10 s) were transferred to the grid on the floor of the dark compartment. All rats were again located on the platform, and the escape latency was recorded in the retention test.

4.5. Immunohistochemical Staining

After completion of the ^{18}F fluorodeoxyglucose (FDG), rats were anesthetized with sodium pentobarbital (100 mg·kg^{-1}, intraperitoneally). Rats were perfused with saline solution, followed by 4% paraformaldehyde (PFA). The brain was anatomized from the skull, post-fixed in PFA overnight, and stored in a 30% sucrose in PBS until it subsided. The brain was embedded and serially sectioned on a cryostat (Leica Microsystem Co., Ltd., Wetzlar, Germany) at 30 μm thickness in the coronal plane and the sections were collected in phosphate buffered saline (PBS). The primary antibodies against the following specific antigens were used: ChAT and AchE. The primary antibody was diluted from the concentrate with the blocking solution (0.2% phosphate buffered saline with tween 20 (PBST), 2% blocking serum in PBST). The primary antibody was infused into the brain sections for 72 h at 4 °C. After several rinses in PBST, the sections were incubated with secondary biotinylated antibodies against rabbit immunoglobulin G (IgG) or sheet IgG (Vector Laboratories, Burlingame, CA, USA) for

2 h. After washing with PBST, the sections were incubated with an appropriate biotinylated secondary antibody and processed with an avidin–biotin complex kit (Vectastain ABC kit; Vector Laboratories, Burlingame, CA, USA). The staining was conducted using 0.05% 3,3'-Diaminobenzidine (DAB) in the presence of 0.003% H_2O_2 in 0.1 M PB. After rinsing with 0.1 M PB, the stained tissue sections were mounted on the slide. The images were captured using a DP2-BSW imaging system (Olympus, CA, USA) and processed using Adobe Photoshop Cells. Those testing positive for ChAT and AChE were counted on a grid that was placed on CA1 and CA3 in the hippocampus area. The number of cells was counted at a magnification of 100×, using a rectangular microscopic grid measuring 200×200 μm^2. The cells were counted in 3 sections per rat within the hippocampus.

4.6. Image Processing and Analysis

After completion of the behavioral test, animals underwent prior fasting over 12–15 h and bedding was changed during the fasting period to avoid ingestion of bedding. The animals were treated with 500 mCi/100 g 18F-FDG via tail intravenous insjsmsjection, and the animals inhaled 2% isoflurane in 100% oxygen (Forane solution; ChoongWae Pharma, Korea) until the positron emission tomography (PET) scan. A transverse resolution of <1.8 mm was used at the center [48]. The emission data were acquired at an energy window setting (350–650 keV, 30 min). The acquired data in the emission list mode were arranged into three-dimensional (3D) sinograms and reconstructed using 3D Reprojection (3DRP) methods. To distinguish the cerebral glucose metabolism, between TMT + Bean-PS and TMT group datasets, a voxel-based statistical analysis was performed using SPM. In summary, the area of the brain was masked using the rectangular method. The PET data were spatially reconstructed onto a rat brain template, smoothed using a 3 mm isotropic Gaussian kernel, and counted. A Statistical Parametric Mapping 8 program was used in the voxel-wise t-test between the TMT + Bean-PS and TMT group datasets ($p < 0.05$, K > 50).

4.7. Statistical Analysis

Results were assessed using one-way or two-way analysis of variance (ANOVA), and repeated measures of ANOVA, followed by Tukey's post hoc test, using SPSS 15.0 for Windows, for statistical analysis, for multi-group comparisons, and Student's *t* test for single comparisons, using Prism 5.01 software (Graphpad Software Inc., San Diego, CA, USA). In all statistical assessments, $p < 0.05$ was considered to be statistically significant. Results are expressed as the mean ± standard error of the mean.

5. Conclusions

Treatment with Bean-PS attenuated the TMT-induced memory impairment in the MWM test. Furthermore, the Bean-PS improved cognitive function, and generated a neuroprotective effect by increasing the expression of cholinergic neurons and glucose metabolism. Thus, Bean-PS is a potential agent that can be used to prevent and protect against neurodegenerative diseases such as dementia and AD. It may represent a good therapeutic agent for of the management of AD.

Author Contributions: Conceptualization, M.Y., K.K., and I.S.; Data curation, M.Y., J.S.K., and B.H.H.; Formal analysis, M.Y., J.S.K., and B.H.H.; Investigation, M.Y., K.K., and I.S.; Methodology, M.Y., J.S.K., and I.S.; Project administration, K.K., and I.S.; Software, M.Y. and J.S.K.; Validation, M.Y. and J.S.K.; Writing—original draft, M.Y.; Writing—review & editing, M.Y., K.K., and I.S. All authors have read and agreed to the published version of the manuscript.

Funding: This work was supported by the Korean Institute of Planning and Evaluation for Technology in Food, Agriculture, Forestry and Fisheries (IPET 11080-03-3-HD020), funded by the Ministry of Agriculture, Food and Rural Affairs (MAFRA).

Conflicts of Interest: The authors have no conflicts of interest to declare. The funders had no role in the design of the study; in the collection, analyses, or interpretation of data; in the writing of the manuscript; or in the decision to publish the results.

References

1. Jung, E.-Y.; Lee, M.-S.; Ahn, C.J.; Cho, S.-H.; Bae, H.; Shim, I. The Neuroprotective Effect of Gugijihwang-Tang on Trimethyltin-Induced Memory Dysfunction in the Rat. *Evid. Based Complement. Altern. Med.* **2013**, *2013*, 1–6. [CrossRef]
2. Kang, J.Y.; Park, S.K.; Guo, T.J.; Ha, J.S.; Lee, D.S.; Kim, J.M.; Lee, U.; Kim, D.; Heo, H.J. Reversal of Trimethyltin-Induced Learning and Memory Deficits by 3,5-Dicaffeoylquinic Acid. *Oxidative Med. Cell. Longev.* **2016**, *2016*, 1–13. [CrossRef] [PubMed]
3. Chen, J.; Huang, C.; Zheng, L.; Simonich, M.; Bai, C.; Tanguay, R.L.; Dong, Q. Trimethyltin chloride (TMT) neurobehavioral toxicity in embryonic zebrafish. *Neurotoxicol. Teratol.* **2011**, *33*, 721–726. [CrossRef]
4. Liu, Z.; Jiang, C.; Li, R.; Zhao, S.; Li, W.; Tang, X. The neuroprotective effect of lithium chloride on cognitive impairment through glycogen synthase kinase-3beta inhibition in intracerebral hemorrhage rats. *Eur. J. Pharmacol.* **2018**, *840*, 50–59. [CrossRef]
5. Corvino, V.; Marchese, E.; Giannetti, S.; Lattanzi, W.; Bonvissuto, D.; Biamonte, F.; Mongiovì, A.M.; Michetti, F.; Geloso, M.C. The neuroprotective and neurogenic effects of neuropeptide Y administration in an animal model of hippocampal neurodegeneration and temporal lobe epilepsy induced by trimethyltin. *J. Neurochem.* **2012**, *122*, 415–426. [CrossRef]
6. Dyer, R.S.; Boyes, W.K. Trimethyltin reduces recurrent inhibition in rats. *Neurobehav. Toxicol. Teratol.* **1984**, *6*, 367–371.
7. Naalsund, L.U.; Allen, C.N.; Fonnum, F. Changes in neurobiological parameters in the hippocampus after exposure to trimethyltin. *NeuroToxicology* **1985**, *6*, 145–158.
8. Valdes, J.J.; Mactutus, C.F.; Santos-Anderson, R.M.; Dawson, R.; Annau, Z. Selective neurochemical and histological lesions in rat hippocampus following chronic trimethyltin exposure. *Neurobehav. Toxicol. Teratol.* **1983**, *5*, 357–361.
9. Chang, L.W.; Dyer, R.S. Early effects of trimethyltin on the dentate gyrus basket cells: A morphological study. *J. Toxicol. Environ. Heal. Part A* **1985**, *16*, 641–653. [CrossRef]
10. Oderfeld-Nowak, B.; Zaremba, M. GM1 ganglioside potentiates trimethyltin-induced expression of interleukin-1 beta and the nerve growth factor in reactive astrocytes in the rat hippocampus: An immunocytochemical study. *Neurochem. Res.* **1998**, *23*, 443–453. [CrossRef]
11. Koczyk, D.; Oderfeld-Nowak, B. Long-term microglial and astroglial activation in the hippocampus of trimethyltin-intoxicated rat: Stimulation of NGF and TrkA immunoreactivities in astroglia but not in microglia. *Int. J. Dev. Neurosci.* **2000**, *18*, 591–606. [CrossRef]
12. Halladay, A.; Wilson, D.; Wagner, G.; Reuhl, K. Trimethyltin-induced alterations in behavior are linked to changes in PSA-NCAM expression. *NeuroToxicol.* **2006**, *27*, 137–146. [CrossRef] [PubMed]
13. Kim, C.R.; Choi, S.J.; Kwon, Y.K.; Kim, J.K.; Kim, Y.-J.; Park, G.G.; Shin, N.-H. Cinnamomum loureirii Extract Inhibits Acetylcholinesterase Activity and Ameliorates Trimethyltin-Induced Cognitive Dysfunction in Mice. *Boil. Pharm. Bull.* **2016**, *39*, 1130–1136. [CrossRef]
14. Da Silva, I.F.; Lima, L.C.F.; Graceli, J.B.; Rodrigues, L.C.D.M. Organotins in Neuronal Damage, Brain Function, and Behavior: A Short Review. *Front. Endocrinol.* **2018**, *8*, 366. [CrossRef] [PubMed]
15. Woodruff, M.L.; Baisden, R.H.; Cannon, R.L.; Kalbfleisch, J.; Freeman, J.N. Effects of trimethyltin on acquisition and reversal of a light-dark discrimination by rats. *Physiol. Behav.* **1994**, *55*, 1055–1061. [CrossRef]
16. Kobayashi, H.; Saito, F.; Yuyama, A. Effects of organotins on the cholinergic system in the chicken brain in vitro. *Toxicol. Vitr.* **1992**, *6*, 337–343. [CrossRef]
17. Francis, P.T.; Palmer, A.M.; Snape, M.; Wilcock, G.K. The cholinergic hypothesis of Alzheimer's disease: A review of progress. *J. Neurol. Neurosurg. Psychiatry* **1999**, *66*, 137–147. [CrossRef]
18. Van Marum, R. Current and future therapy in Alzheimer's disease. *Fundam. Clin. Pharmacol.* **2008**, *22*, 265–274. [CrossRef]
19. Lee, B.; Sur, B.-J.; Han, J.-J.; Shim, I.; Her, S.; Lee, H.-J.; Hahm, D.-H. Krill phosphatidylserine improves learning and memory in Morris water maze in aged rats. *Prog. Neuro-Psychopharm. Boil. Psychiatry* **2010**, *34*, 1085–1093. [CrossRef]
20. Amaducci, L.; Crook, T.H.; Lippi, A.; Bracco, L.; Baldereschi, M.; Latorraca, S.; Piersanti, P.; Tesco, G.; Sorbi, S. Use of Phosphatidylserine in Alzheimer's Disease. *Ann. N. Y. Acad. Sci.* **1991**, *640*, 245–249. [CrossRef]

21. Friedland, R.P.; Petersen, R.B.; Rubenstein, R. Bovine Spongiform Encephalopathy and Aquaculture. *J. Alzheimer's Dis.* **2008**, *17*, 277–279. [CrossRef]

22. Kato-Kataoka, A.; Sakai, M.; Ebina, R.; Nonaka, C.; Asano, T.; Miyamori, T. Soybean-Derived Phosphatidylserine Improves Memory Function of the Elderly Japanese Subjects with Memory Complaints. *J. Clin. Biochem. Nutr.* **2010**, *47*, 246–255. [CrossRef]

23. Jorissen, B.; Brouns, F.; Van Boxtel, M.; Ponds, R.; Verhey, F.; Jolles, J.; Riedel, W. The influence of soy-derived phosphatidylserine on cognition in age-associated memory impairment. *Nutr. Neurosci.* **2001**, *4*, 121–134. [CrossRef]

24. Wimo, A.; Winblad, B. Economical Aspects of Dementia. *Hendbook Clin. Neurol.* **2008**, *89*, 137–146. [CrossRef]

25. Fortemps, E.; Amand, G.; Bomboir, A.; Lauwerys, R.; Laterre, E.C. Trimethyltin poisoning report of two cases. *Int. Arch. Occup. Environ. Health* **1978**, *41*, 1–6. [CrossRef]

26. Dyer, R.S. Physiological methods for assessment of Trimethyltin exposure. *Neurobehav. Toxicol. Teratol.* **1982**, *4*, 659–664.

27. Brown, A.W.; Aldridge, W.N.; Street, B.W.; Verschoyle, R.D. The behavioral and neuropathologic sequelae of intoxication by trimethyltin compounds in the rat. *Am. J. Pathol.* **1979**, *97*, 59–82.

28. Chang, L.W.; Tiemeyer, T.M.; Wenger, G.R.; McMillan, D.E. Neuropathology of trimethyltin intoxication. III. Changes in the brain stem neurons. *Environ. Res.* **1983**, *30*, 399–411. [CrossRef]

29. Balaban, C.; Callaghan, J.; Billingsle, M. Trimethyltin-induced neuronal damage in the rat brain: Comparative studies using silver degeneration stains, immunocytochemistry and immunoassay for neuronotypic and gliotypic proteins. *Neuroscience* **1988**, *26*, 337–361. [CrossRef]

30. Chang, L.W.; Dyer, R.S. A time-course study of trimethyltin induced neuropathology in rats. *Neurobehav. Toxicol. Teratol.* **1983**, *5*, 443–459.

31. Chang, L.W.; Dyer, R.S. Trimethyltin induced pathology in sensory neurons. *Neurobehav. Toxicol. Teratol.* **1983**, *5*, 673–696.

32. Chang, L.W.; Wenger, G.R.; McMillan, D.E.; Dyer, R.S. Species and strain comparison of acute neurotoxic effects of trimethyltin in mice and rats. *Neurobehav. Toxicol. Teratol.* **1983**, *5*, 337–350.

33. Chang, L.W.; Dyer, R.S. Septotemporal gradients of trimethyltin-induced hippocampal lesions. *Neurobehav. Toxicol. Teratol.* **1985**, *7*, 43–49. [PubMed]

34. Noraberg, J.; Gramsbergen, J.B.P.; Fonnum, F.; Zimmer, J. Trimethyltin (TMT) neurotoxicity in organotypic rat hippocampal slice cultures. *Brain Res.* **1998**, *783*, 305–315. [CrossRef]

35. Swartzwelder, H.S.; Hepler, J.; Holahan, W.; King, S.E.; Leverenz, H.A.; A Miller, P.; Myers, R.D. Imparied maze performance in the rat caused by trimethyltin treatment: Problem-solving deficits and perseveration. *Neurobehav. Toxicol. Teratol.* **1982**, *4*, 169–176.

36. Andersson, H.; Luthman, J.; Lindqvist, E.; Olson, L. Time-course of trimethyltin effects on the monoaminergic systems of the rat brain. *NeuroToxicology* **1995**, *16*, 201–210. [PubMed]

37. Koczyk, D.; Skup, M.; Zaremba, M.; Oderfeld-Nowak, B. Trimethyltin-induced plastic neuronal changes in rat hippocampus are accompanied by astrocytic trophic activity. *Acta Neurobiol. Exp.* **1996**, *56*, 237–241.

38. Zhao, T.F.; Xu, C.X.; Li, Z.W.; Xie, F.; Zhao, Y.T.; Wang, S.Q.; Luo, C.H.; Lu, R.S.; Ni, G.L.; Ku, Z.Q.; et al. (Effect of Tremella fuciformis Berk on acute radiation sickness in dogs (author's transl)). *Zhongguo Yi Xue Ke Xue Yuan Xue Bao* **1982**, *4*, 20–23.

39. Walsh, T.J.; Gallagher, M.; Bostock, E.; Dyer, R.S. Trimethyltin impairs retention of a passive avoidance task. *Neurobehav. Toxicol. Teratol.* **1982**, *4*, 163–167.

40. Squire, L.R.; Davis, H.P. The Pharmacology of Memory: A Neurobiological Perspective. *Annu. Rev. Pharmacol. Toxicol.* **1981**, *21*, 323–356. [CrossRef]

41. Lanier, L.P.; Isaacson, R.L. Activity changes related to the location of lesions in the hippocampus. *Behav. Boil.* **1975**, *13*, 59–69. [CrossRef]

42. Agrawal, A.K.; Roy, A.; Seth, K.; Raghubir, R.; Seth, P. Restorative potential of cholinergic rich transplants in cholchicine induced lesioned rats: A comparative study of single and multiple micro-transplantation approach. *Int. J. Dev. Neurosci.* **2003**, *21*, 191–198. [CrossRef]

43. Muir, J.L. Acetylcholine, Aging, and Alzheimer's Disease. *Pharmacol. Biochem. Behav.* **1997**, *56*, 687–696. [CrossRef]

44. Shinjo, H.; Ueki, A.; Miwa, C.; Morita, Y. Effect of entorhinal cortex lesion on hippocampal cholinergic system in rat in operant learning task as studied by in vivo brain microdialysis. *J. Neurol. Sci.* **1998**, *157*, 13–18. [CrossRef]

45. Giacobini, E. Long-term stabilizing effect of cholinesterase inhibitors in the therapy of Alzheimer' disease. *J. Neural. Transm. Suppl.* **2002**, *62*, 181–187. [CrossRef]

46. Lorenzini, C.A.; Baldi, E.; Bucherelli, C.; Sacchetti, B.; Tassoni, G. Role of dorsal hippocampus in acquisition, consolidation and retrieval of rat's passive avoidance response: A tetrodotoxin functional inactivation study. *Brain Res.* **1996**, *730*, 32–39. [CrossRef]

47. Ajanal, M.; Gundkalle, M.B.; Nayak, S.U. Estimation of total alkaloid in Chitrakadivati by UV-Spectrophotometer. *Anc. Sci. Life* **2012**, *31*, 198–201. [CrossRef]

48. Bao, Q.; Newport, D.; Chen, M.; Stout, D.B.; Chatziioannou, A.F. Performance evaluation of the inveon dedicated PET preclinical tomograph based on the NEMA NU-4 standards. *J. Nucl. Med.* **2009**, *50*, 401–408. [CrossRef]

International Journal of
Molecular Sciences

Article

Exercise-Induced Elevated BDNF Level Does Not Prevent Cognitive Impairment Due to Acute Exposure to Moderate Hypoxia in Well-Trained Athletes

Zofia Piotrowicz [1,*], Małgorzata Chalimoniuk [2], Kamila Płoszczyca [3], Miłosz Czuba [3,4] and Józef Langfort [1]

[1] Institute of Sport Sciences, The Jerzy Kukuczka Academy of Physical Education, 40-065 Katowice, Poland; langfort@imdik.pan.pl

[2] Department of Tourism and Health in Biała Podlaska, The Józef Piłsudski University of Physical Education, 00-968 Warsaw, Poland; malgorzata.chalimoniuk@awf-bp.edu.pl

[3] Department of Kinesiology, Institute of Sport, 01-982 Warsaw, Poland; kamila.ploszczyca@insp.waw.pl (K.P.); milosz.czuba@insp.waw.pl (M.C.)

[4] Faculty of Health Sciences, Jan Dlugosz University, 42-200 Czestochowa, Poland

* Correspondence: zofia.a.piotrowicz@gmail.com

Received: 16 July 2020; Accepted: 31 July 2020; Published: 4 August 2020

Abstract: Exposure to acute hypoxia causes a detrimental effect on the brain which is also manifested by a decrease in the ability to perform psychomotor tasks. Conversely, brain-derived neurotrophic factor (BDNF), whose levels are elevated in response to exercise, is a well-known factor in improving cognitive function. Therefore, the aim of our study was to investigate whether the exercise under hypoxic conditions affects psychomotor performance. For this purpose, 11 healthy young athletes performed a graded cycloergometer exercise test to volitional exhaustion under normoxia and acute mild hypoxia ($FiO_2 = 14.7\%$). Before, immediately after exercise and after a period of recovery, choice reaction time (CRT) and number of correct reactions (NCR) in relation to changes in serum BDNF were examined. Additionally, other selected factors which may modify BDNF production, i.e., cortisol (C), nitrite, catecholamines (adrenalin-A, noradrenaline-NA, dopamine-DA, serotonin-5-HT) and endothelin-1 (ET-1), were also measured. Exercise in hypoxic conditions extended CRT by 13.8% ($p < 0.01$) and decreased NCR (by 11.5%) compared to rest ($p < 0.05$). During maximal workload, NCR was lower by 9% in hypoxia compared to normoxia ($p < 0.05$). BDNF increased immediately after exercise in normoxia (by 29.3%; $p < 0.01$), as well as in hypoxia (by 50.0%; $p < 0.001$). There were no differences in BDNF between normoxia and hypoxia. Considering the fact that similar levels of BDNF were seen in both conditions but cognitive performance was suppressed in hypoxia, acute elevation of BDNF did not compensate for hypoxia-induced cognition impairment. Moreover, neither potentially negative effects of C nor positive effects of A, DA and NO on the brain were observed in our study.

Keywords: brain-derived neurotrophic factor; moderate hypoxia; physical exercise; psychomotor function; reaction time; cortisol; catecholamines; nitrite; endotheline-1; lactate

---

1. Introduction

Hypoxia is a condition in which some organ(s) or a whole organism is deprived of adequate oxygen supply. Except in very short or static exercises [1–3], hypoxia negatively affects exercise performance [4,5]. In particular, the maximal aerobic workload that can be sustained during exercise involving large muscle groups (e.g., cycling, running) is considerably lower in hypoxia compared with normoxia. The origin of human performance limitation in hypoxia is attributed to a decrease in maximal oxygen uptake (VO_{2max}). Dempsey and Wagner [6] observed that each 1% decrement in

SaO$_2$% below the 95% level approximates to a 1–2% decrement in VO$_{2max}$. Diminished VO$_{2max}$ in hypoxia is accompanied by a lowered O$_2$ partial pressure in arterial blood (PaO$_2$), which reduces O$_2$ delivery to tissues and negatively affects muscle metabolism and contraction [7,8], leading to so-called peripheral fatigue. There is also evidence that maximal cardiac output and maximal heart rate (HR$_{max}$) during maximal exercise in hypoxia are decreased [9] and the decrease in HR$_{max}$ is linearly related to the decrease in SaO$_2$% [10]. This effect can be reversed by oxygen administration during hypoxia exposure in both acute [11] and chronic hypoxia [9,11–13].

The aforementioned factors do not fully explain the hypoxia-induced reduction in exercise performance. As biochemical, electromyographic and mechanical signs of muscle fatigue are reduced in severe hypoxia compared with normoxia, peripheral (muscle) fatigue may not be the main factor responsible for impaired exercise performance [12,14]. It is well recognized that metabolites produced in working muscles can directly modulate central nervous system (CNS) functions by changes in sensory nerve impulses. Moreover, chemical messengers originating from working muscle are released into the circulation [15] and can affect brain function after their translocation to the CNS. An alternative hypothesis that may, at least partially, explain reduced exercise performance in hypoxic condition attributes it to so-called central fatigue [16]. This assumption is supported by studies showing impairment of cognitive performance [17] by reduced O$_2$ delivery. Several studies [18,19] have reported that moderate levels of hypoxia degraded the ability to perform psychomotor tasks, and the main cause of cognitive impairment is the low PaO$_2$ regardless of the type of hypoxia (normobaric vs. hypobaric) [20].

The negative impact of hypoxia on cognitive functions is manifested by memory deterioration, reduced learning ability, decreased concentration, and psychomotor performance [21]. One of the best indicators of the speed and efficiency of mental processes is choice reaction time (CRT) and the number of correct reactions (NCR), especially if these variables are used to assess the cognitive function within the same group of participants [22,23]. However, in previous studies, the effect of hypoxia on CRT was ambiguous and most likely depended on the time of exposure [24,25] and level of hypoxia: moderate vs. severe [24,26,27]. Some studies show that acute exposure to severe hypoxia led to an increase in CRT [24,27]. However, prolonged exposure to moderate hypoxia did not disturb CRT [24,27]. Animal studies reveal that hypoxia causes neuronal injuries in the hippocampus and cortex, leading to functional and behavioral deficits [28–30]. Likewise, data obtained from neuroimaging proved that intermittent hypoxia may result in a decrease in the volume of the hippocampus in humans [31].

Some previous studies performed in normoxia indicated that exercise of low or moderate intensity improves psychomotor performance [32–37], while other studies showed a significant decrease in psychomotor performance during heavy exhaustive exercise [34,37,38]. It is not known if these aforementioned effects can be modulated by a hypoxia-induced deleterious influence on the CNS. Most recent data suggest that an essential role in these phenomena is played by brain-derived neurotrophic factor (BDNF) (for review, see [39,40]).

BDNF plays a key role in the physiology of the developing and mature CNS, showing a high affinity for the TrkB receptor. Consequently, it is responsible for neurogenesis, differentiation, survival, and remodeling of neurons, and it also positively affects synaptogenesis, synaptic plasticity, and long-term potentiation [41–43]. The upregulation of BDNF may influence brain functions including learning and memory [44].

Several lines of evidence suggest a link between BDNF and physical activity. Both acute and chronic aerobic activity were effective for increasing peripheral BDNF concentrations [45]. An elevated level of BDNF was also seen in active sportsman compared to sedentary individuals [46,47]. Another factor that can be considered a stimulator of BDNF production within the brain is nitric oxide (NO). A role for NO in increased BDNF production in response to exercise has been recently evidenced [48,49]. Exercises of extreme intensity or duration are known to greatly elevate blood cortisol (C) [50,51], while high circulating corticosterone has been shown to suppress brain production of BDNF in rats [52].

Of importance, BDNF induces expression of the monocarboxylate transporter that enables the use of lactate as an alternative energy source [53].

There are also very limited data about the efficacy of hypoxia exposure on psychomotor performance where subjects performed exercise with low and high intensity. Some data suggest that exposure to hypoxia reduces cognitive functions [17]. Also, mechanisms of hypoxia's effect on the CNS are still poorly understood. One possible candidate which might take part in this phenomenon is BDNF. It has been shown that cognitive impairment is noticeable in neurodegenerative diseases, which is associated with a lower serum BDNF level as compared to healthy individuals [54]. Moreover, the level of this decrease depends on the degree of cognitive impairment [54]. On the other hand, BDNF is thought to be responsible for improving cognitive function as a result of exercise effort [55]. Moreover, it has been proven that both resting and post-exercise peripheral BDNF levels correspond to its brain production [45,56] and its level is elevated in response to exercise effort [46].

Therefore, the aim of this study was to examine the impact of a single bout of exercise to volitional exhaustion during acute exposure of well-trained endurance athletes to moderate hypoxia on psychomotor performance. For this purpose, we measured a peripheral level of BDNF and CRT and NCR as indices of psychomotor performance during graded cycloergometer exercise test. Furthermore, we examined a level of selected circulating biochemical factors, such as C, NO pathway-related metabolites (nitrite, endothelin-1(ET-1), catecholamines), because they are known to affect BDNF expression/production [48,49,52,57–60] and their expression can be influenced by both exercise and hypoxia [34,48–51,61–68].

2. Results

2.1. Maximal Workload and Respiratory Variables

The paired sample t-test showed that maximal workload (WR_{max}) decreased significantly ($p < 0.001$) by 16.3% in hypoxia (3000 m) compared to the initial measurements in normoxia (Figure 1). The same trend of changes was observed in VO_{2max} values. The values of VO_{2max} decreased significantly ($p < 0.001$) in hypoxia compared to normoxia respectively by 14.5% (Figure 1). Additionally, there were statistically significant changes in delta values of blood lactate concentration (ΔLA) after the incremental test between normoxia and hypoxia 3000 m. The Wilcoxon test showed that ΔLA increased significantly ($p < 0.01$) by 6.05% despite the significant reduction in WR_{max} in hypoxia 3000 m compared to the measurements in normoxia (Figure 2).

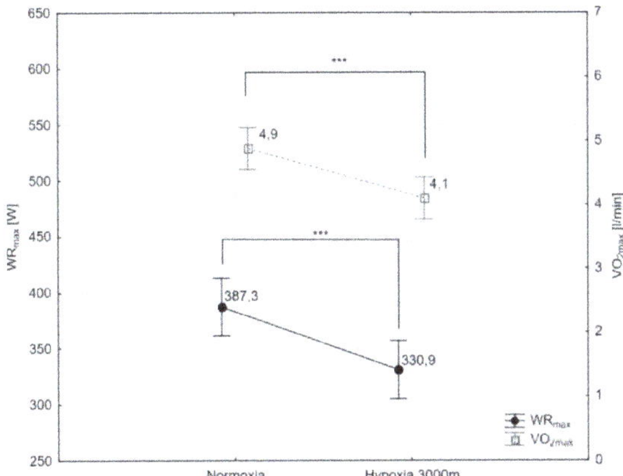

Figure 1. Maximal workload (WR_{max}), and maximal oxygen uptake (VO_{2max}) during incremental test performed in different conditions. *** $p < 0.001$.

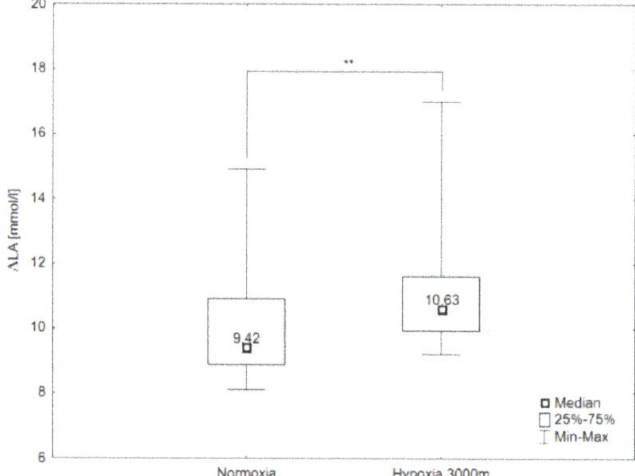

Figure 2. Delta values of blood lactate concentration (ΔLA) during incremental test performed in different conditions. ** $p < 0.01$.

2.2. Choice Reaction Time and Number of Correct Reactions

No significant interaction (condition x time of measure) effect was found on CRT but only a significant main effect of time of measurement (at rest, max and after 3 min of recovery) on CRT values (F = 21.88; $p < 0.001$) was observed. Additionally, there was a significant interaction (condition × time of measure) effect in the NCR (F = 3.44; $p < 0.05$) during the incremental test.

The post-hoc Tukey's test showed that CRT decreased significantly ($p < 0.05$) by 9.7% after 3 min of recovery after the incremental test ($CRT_{after\ 3'\ of\ recovery}$) compared to initial values observed at rest (CRT_{rest}) in normoxia. Additionally, the values of $CRT_{after\ 3'\ of\ recovery}$ decreased significantly in normoxia ($p < 0.001$) and hypoxia ($p < 0.01$) compared to CRT during maximal workload of the incremental test (CRT_{max}) respectively by 17.7 and 12.2% However, CRTmax increased significantly ($p < 0.01$) by 13.8% compared to CRT_{rest} in hypoxic conditions (Figure 3).

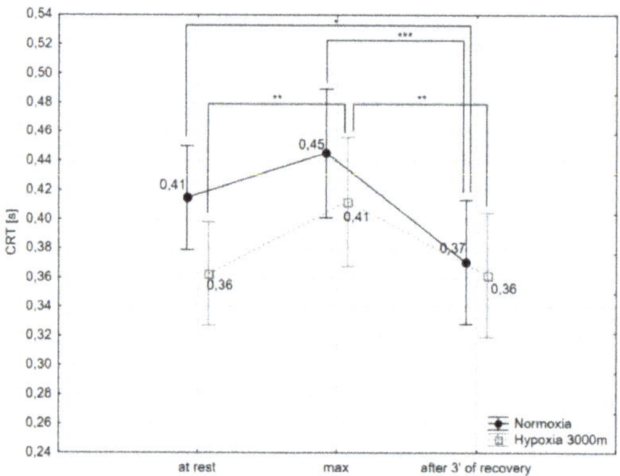

Figure 3. Choice reaction time (CRT) at rest, during maximal effort (max) and after 3 min of the recovery period in normoxia and hypoxia (3000 m). * $p < 0.05$; ** $p < 0.01$; *** $p < 0.001$.

The post-hoc Tukey's test showed that number of correct reactions (NCR) during maximal workload of the incremental test (NCR_{max}) decreased significantly ($p < 0.05$) by 9% in hypoxic conditions compared to normoxia. NCR_{max} values in hypoxia were significantly lower compared to NCR at rest (11.5%; $p < 0.001$) and NCR after 3 min of recovery after the incremental test (9.7%; $p < 0.01$) (Figure 4).

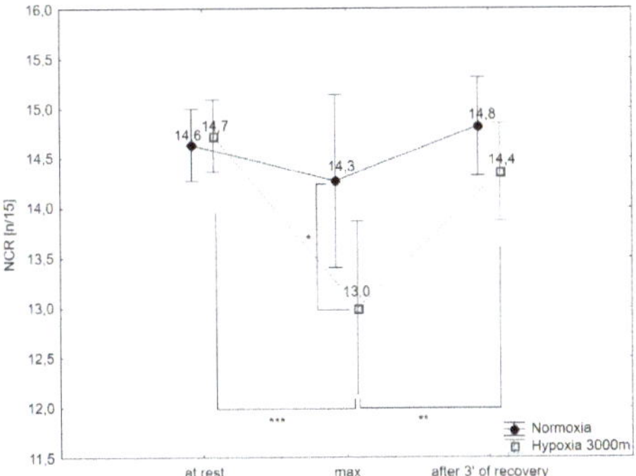

Figure 4. Number of correct reactions (NCR) at rest, during maximal effort (max) and after 3 min of the recovery period in normoxia and hypoxia (3000m). * $p < 0.05$; ** $p < 0.01$; *** $p < 0.001$.

2.3. Brain-Derived Neurotrophic Factor and Selected Biochemical Variables

There was a significant interaction (condition x time of measure) effect in the BDNF (F = 3.66; $p < 0.05$) serum concentrations.

The post-hoc Tukey's test showed that BDNF concentration increased significantly immediately after the incremental test ($BDNF_{max}$) in normoxia (by 29.3%; $p < 0.01$), as well as in hypoxia (by 50.0% $p < 0.001$). Additionally, BDNF concentration decreased significantly after a 1 h recovery period ($BDNF_{rest}$) by 20.7% ($p < 0.01$) in normoxia and by 41.3% ($p < 0.001$) in hypoxia (Figure 5). There were no statistically significant differences in BDNF concentration between normoxia and hypoxia trials.

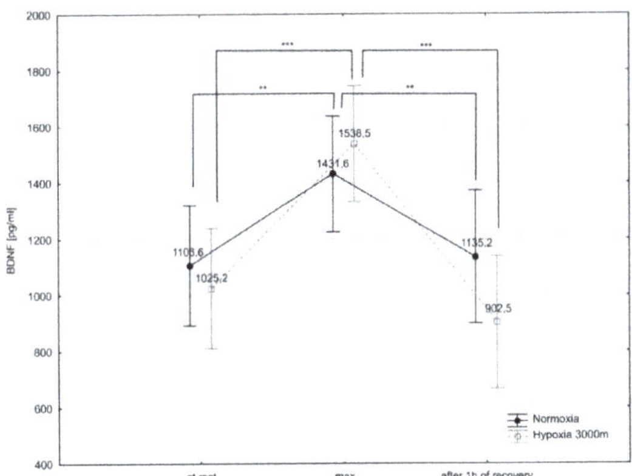

Figure 5. Brain-derived neurotrophic factor (BDNF) serum concentration at rest, during maximal effort (max) and after 1 h recovery of the recovery period in normoxia and hypoxia (3000 m). ** $p < 0.01$; *** $p < 0.001$.

The Friedman test showed a statistically significant effect of the time of measurement on selected biochemical variables such as NO_2^-, C, adrenalin (A) and dopamine (DA) in both conditions (normoxia and hypoxia) (Table 1).

The post-hoc Friedman test showed that NO_2^- concentration measured immediately after the incremental test was significantly higher ($p < 0.05$) compared to the NO_2^- concentration observed at rest and 1 h after the incremental test in normoxia as well as hypoxia. However, the Wilcoxon test showed that were no significant differences between normoxia and hypoxia in NO_2^- concentrations (Table 1).

Table 1. Mean values of selected biochemical variables registered in the different conditions (normoxia, hypoxia 3000 m) at rest, after incremental test (max) and after 1 h recovery period (after 1 h).

| Variables | Normoxia (N) | | | | | | Hypoxia 3000 m (H3) | | | | | | Significance of Differences |
| | at Rest (1) | | Max (2) | | after 1 h (3) | | at Rest (4) | | Max (5) | | after 1 h (6) | | ($*$ $p < 0.05$; $**$ $p < 0.01$) |
	$x \pm SD$	Me	$x \pm SD$	Me	$x \pm SD$	Me	$x \pm SD$	Me	$x \pm SD$	Me	$x \pm SD$	Me	
ET-1 (pg/mL)	2.5 ± 0.8	2.5	2.4 ± 1.7	2.3	2.0 ± 1.1	1.8	2.3 ± 0.8	2.5	2.3 ± 1.0	2.61	2.7 ± 1.2	2.5	
NO_2^- (pg/mL)	28.7 ± 12.1	23.2	47.6 ± 21.2	38.8	30.2 ± 12.7	25.6	33.8 ± 16.3	39.1	52.3 ± 16.3	49.8	29.5 ± 9.3	26.3	**N:** $X^2 = 13.81$; $p = 0.001$ 1–2 $*$; 2–3 $*$ **H3:** $X^2 = 16.9$; $p = 0.002$ 4–5 $*$; 5–6 $*$
C (pg/mL)	8.7 ± 4.1	7.1	11.5 ± 3.8	11.1	12.4 ± 1041	9.8	7.2 ± 2.7	6.7	8.2 ± 2.48	8.8	9.4 ± 5.9	7.2	**N vs. H3:** 2–5 $*$ ($U = 29.0$; $p = 0.041$)
NA (pg/L)	338.9 ± 213.4	337.2	555.6 ± 25.6	575.6	406.8 ± 28.4	329.3	334.6 ± 153.2	238.7	1109.2 ± 1045.4	348.5	575.2 ± 359	435.2	
A (pg/L)	53.9 ± 29.4	48.8	479 ± 358.1	369	112.5 ± 102.3	117.1	206.9 ± 70.7	228.5	1436.1 ± 622.9	1384.7	827.4 ± 263.7	742.7	**N:** $X^2 = 16.9$; $p = 0.002$ 1–2 $*$; 2–3 $*$ **H3:** $X^2 = 18.2$; $p = 0.001$ 4–5 $*$; 4–6 $*$ **N vs. H3:** 1–4 $**$ ($U = 1.0$; $p = 0.001$) 2–5 $**$ ($U = 9.0$; $p = 0.001$) 3–6 $**$ ($U = 0.0$; $p = 0.001$)
DA (pg/L)	7.6 ± 10.9	4.3	8.9 ± 5.8	7.8	5.6 ± 1.4	5.7	7.3 ± 2.5	6.7	12.5 ± 2.7	13.2	12.1 ± 5.5	13.7	**N:** $X^2 = 10.1$; $p = 0.006$ 2–3 $*$ **H3:** $X^2 = 10.4$; $p = 0.005$ 4–5 $*$; 4–6 $*$ **N vs. H3:** 2–5 $**$ ($U = 17.0$; $p = 0.008$) 3–6 $**$ ($U = 16.0$; $p = 0.006$)
5-HT (pg/L)	155.1 ± 104.2	135.4	212.6 ± 148.1	150.3	192.1 ± 90.1	190.7	136.5 ± 67.1	156.7	205.9 ± 124.3	150.4	156.4 ± 82.1	148.3	

x—arithmetic means; SD—standard deviations; Me—median; ET-1—endothelin-1; C—cortisol; NA—noradrenaline; A—adrenalin; DA—dopamine; 5-HT—serotonin; $*$ $p < 0.05$; $**$ $p < 0.001$.

There was no significant effect of the time of measurement (at rest, max, 1 h after test) on C concentrations in both conditions. However, there was a significant effect of conditions (normoxia vs. hypoxia) on C concentration. The Wilcoxon test showed that C concentration measured immediately after the incremental test (C_{max}) was significantly higher ($p < 0.05$) by 20.7% in normoxic conditions compared to hypoxia (Table 1).

Moreover, the post-hoc Friedman test showed that A concentration measured immediately after the incremental test (A_{max}) was significantly higher ($p < 0.05$) compared to the A concentration observed at rest (A_{rest}; by 656.1%) and 1 h after the incremental test ($A_{1h\ after}$; by 215,1%) in normoxia. However, in hypoxia, A_{max} and $A_{1h\ after}$ were significantly higher ($p = 0.001$) compared to A_{rest} respectively by 505.9% and 86.4%. The Wilcoxon test showed that A_{rest}, A_{max} and $A_{1h\ after}$ in hypoxia were significantly higher ($p = 0.001$) compared to these values in normoxia, respectively by 368.2%, 275% and 534.2% (Table 1).

The post-hoc Friedman test showed that DA concentration measured 1 h after the incremental test ($DA_{1h\ after}$) significantly decreased ($p < 0.05$) by 26.9% compared to the DA concentration measured immediately after the incremental test (DA_{max}) in normoxia. However, in hypoxia, DA_{max} and $DA_{1h\ after}$ significantly increased ($p < 0.05$) compared to the DA concentration at rest, respectively by 97% and 104.5%. Additionally, the Wilcoxon test showed that DA_{max} and $DA_{1h\ after}$ in hypoxia were significantly higher ($p < 0.01$) compared to these values in normoxia, respectively by 69.2% and 140.3% (Table 1).

3. Discussion

Physical training is planned to improve physical fitness and performance, which confers numerous positive effects on the whole body function [69]. A pivotal role in regulation of these changes is assigned to the brain, particularly the prefrontal cortex, which takes part in regulation of many executive functions to prepare humans for situations demanding high levels of working memory, attention and cognitive flexibility [70]. On the other hand, altitude training, which nowadays has become a standard training protocol in many sports to increase exercise capacity [71–73], causes cognitive frailty [74]. The cognitive decline is most intensely manifested after acute hypoxic exposure and is more profound in athletes than non-trained individuals [75]. Therefore, the primary purpose of this study was to examine the impact of a single bout of exercise to volitional exhaustion during acute exposure well endurance-trained athletes to moderate hypoxia on psychomotor performance. We also measured serum BDNF, some selected hormones (C and catecholamines), the NO pathway metabolites (nitrite), as well as ET-1, as a possible candidate that may be involved in modulation of this phenomenon. We used measurements of CRT and NCR as indicators of psychomotor skills because these variables were demonstrated to be a dependable measure of cognition in different experimental approaches [22].

Results from this investigation revealed that exercise to volitional exhaustion extended CRT in both experimental conditions but only the impact of hypoxic conditions on this variable was statistically significant. A similar alteration, but with statistically significant changes in both cases, was seen in NCR (Figure 4). Additionally, NCR was significantly increased in response to exercise to volitional exhaustion performed in hypoxia as compared with normoxia. These results are in agreement with previous data reported by others which showed cognitive impairment of trained subjects at high altitude [75] and additionally suggested that NCR which specified response accuracy was a more sensitive tool for estimating cognitive state than CRT. Interestingly, the aforementioned exercise-induced effect was transient in both experimental conditions and studied variables returned to basal values within a few minutes after cessation of exercise.

Increasingly more evidence supports the action of BDNF as an underlying factor that elicits exercise/training-induced beneficial changes CNS [76–78]. Moreover, there are also suggestions that the negative impact on cognitive function in hypoxia can be at least partially explained by the simultaneous decrease of BDNF [79,80]. Studies with animal models have shown that this neurotrophin is produced, among other tissues, in brain by motor neurons [81] and intermittent hypoxia increases BDNF levels in neurons of the primary motor cortex [82]. Also, central BDNF cannot be measured in living humans. It has been suggested that the brain is the main source of the increased BDNF in circulation [83,84]. If so, and considering the fact that in our study similar circulating levels of BDNF were seen in both experimental groups but cognitive performance was suppressed in hypoxia, one could conclude that acute elevation of BDNF did not compensate for hypoxia-induced cognition impairment. Previous evidence suggests that BDNF plays a key role in memory and learning [85] and is a vital regulator of neuronal function and plasticity [86,87]. However, these aforementioned effects appeared almost selectively in response to a repetitive or chronic stimulus or in studies in vitro. Results of the current research are in line with reports of some studies using cognitive tasks such as executive function or attention which were unrelated to changes in BDNF after acute exercise [88–90]. One reason which at least theoretically might be considered for the acute action of BDNF is its involvement in synaptic transmission [91]. The results from animal studies indicate that during hypoxia neurons can within minutes alter synaptic transmission [92,93]. Support for a link between exposure to hypoxia and run-down of synaptic transmission has been widely documented in in vitro studies [94,95] and in rats subjected to severe (6100 m) hypoxia [96]. The mechanism by which severe hypoxia induced cognitive impairment was accompanied by a decrease in Acetylcholine (Ach) level and increase in Acetylcholinesterase (AChE) in the cortex [96].

Apart from BDNF, C has been identified as a possible endocrinological mediator of exercise which may modulate brain function [97,98]. Both hypoxia and exercise stimulate the gland cortex to release C. While it is well recognized that elevation of C has been observed in response to acute exercise of higher than moderate intensities, data from hypoxic studies are less consistent. In response to this environmental stimulus, increases or lack of changes have been reported. In our study, when maximal effort was performed in moderate hypoxia the C level was significantly lower compared to normoxia. As this hormone impairs the prefrontal cortex [99,100], the region of the brain that controls more of our cognitive function, one can argue that its deteriorative impact on psychomotor performance, if any, was negligible. However, there is a study showing similar exercised increased plasma C in normoxia and acute hypobaric hypoxia (3000 m altitude) [101]. This discrepancy may be the result of different research designs between aforementioned and our studies (normobaric vs. hypobaric hypoxia as well as cyclists vs. cross-country skiers and ice hockey players).

Previous studies demonstrated that physically active persons have shorter CRT than sedentary ones [102] and the regular U-shaped curve was obtained in athletes when CRT values were plotted against A and noradrenaline (NA) during graded incremental exercise to volitional exhaustion [34,62]. The latter results indicate that CRT exceeds the resting values at exercise loads close to maximal, and catecholamines may play a role in this phenomenon. Although most studies agree that catecholamine levels increase at rest and exercise at high altitude, brief or moderate hypoxia does not always elevate their levels [103,104], and this is especially true for NA [105]. The present finding reflects this phenomenon in the case of NA and serotonin (5-HT) while indicating elevated A as a potential player in cognitive control in our experimental paradigm. In favor of such an assumption is the significantly elevated A level during maximal effort in hypoxia with simultaneous statistically significant extension of CRT. Previous studies on patients with psychological trauma indicate that A and DA are involved in the activation of the prefrontal cortex [106] and similarly stress and aggressive behavior were shown to increase turnover of both A and DA in this area in rodents [107]. Acute hypoxia is considered as a systemic stress factor and was also seen to reduce cognitive function in rats, which was associated with DA signaling in the prefrontal cortex [63]. This finding is additionally supported by increased errors in the cognitive test, which were associated with reduced DA signaling in the prefrontal cortex [108]. However, the lack of significant differences in DA between our investigated groups probably excludes DA participation in the modulation of cognitive function in our subjects. It is worth noting that there is also a study [66] showing that during a short episode of anoxia, an increase in A level can have a protective effect against its disruptive effects. In the present study, this effect did not occur in response to exercise either in normoxia or in moderate hypoxia despite an elevated level of

A in both cases. Collectively, the aforementioned results lead to the conclusion that A probably reveals biphasic action on cognitive function in hypoxic conditions, i.e., a positive effect during short-term and negative after prolonged actions.

It is well established that hypoxia releases a diffusible vasoconstrictor and vasodilator substance and this process can affect blood flow to the brain. Some evidence supports the view that an increase of blood flow to the brain may induce cognitive improvement by delivering elevated oxygenated hemoglobin [109]. In accordance with the above knowledge, we have measured in this research ET-1 and NO metabolites as the most potent representatives of endothelial released vasoconstrictors [110] and vasodilators [111], respectively. In humans, no changes in ET-1 response to maximal exercise in acute hypobaric hypoxia (3000) were noted as compared to normoxia [101]. In our study, both exercise and exercise in hypoxia also provoked no changes in ET-1, confirming occurrence of the above-mentioned effect in normobaric hypoxia, and suggesting that ET-1 had no important influence on blood low regulation in the acute response to moderate hypoxia.

Nitric oxide is mainly generated in the body by endothelial cells, but it is also produced in the CNS, where it is closely involved in neurotransmission and modulation of neuron metabolism [112]. A previous study conducted with diabetic patients indicated that increased bioavailability of NO was a factor that might enhance cognitive function [113]. However, the latest data questioned this possibility when psychomotor performance was tested in hypoxic conditions [114]. Our research conducted in a normobaric hypoxic chamber revealed an inconsiderable rise in serum NO_2^- levels both in basal and exercise conditions as compared to normoxia. Since simultaneously cognitive performance was blunted it implied that NO production under these circumstances was likely too low to affect cognitive functions or NO was not a crucial player in this phenomenon. These assumptions are partially in line with recent findings which do not support a beneficial effect of NO_3^- supplementation on cognitive function in sedentary males at moderate and very high simulated altitude [115]. However, it is clear that if during a profound reduction of absolute work under hypoxia the NO_2^- level was higher than in normoxia then an additional exercise-independent system was responsible for NO_2^- formation. This pathway is activated in parallel during exercise under hypoxic conditions, yet it needs to be identified.

Physical exercise involves markedly increased activity of many brain structures [116]. The metabolites produced in the muscles, which can diffuse into the CNS and can be utilized as a fuel to sustain increased energy requirements, may participate in this process [117]. Such possibility underscores the importance of the muscle produced lactate (LA) during exercise which can be transferred to neurons via monocarboxylate carriers and used in addition to LA delivered via astrocytes-neurons lactate shuttle [118] as energy fuel during neuronal activation as well. On the other hand, LA was recognized as a signaling molecule in the brain [119]. Among others, it can bind a receptor of the G protein coupled receptor family (GPRs) [120] and thereby cause a decrease of cyclic adenosine monophosphate (cAMP) level. This raises the possibility of interaction between LA and A in metabolism regulation on a subcellular level. If true, this phenomenon should have been more strongly connected with hypoxic conditions because A level during maximal effort in hypoxia was significantly elevated while lowering LA (as compared to controls). This was accompanied with simultaneous statistically significant extension of CRT.

4. Materials and Methods

4.1. Participants

Eleven cyclists (20 ± 1.4 years of age) were recruited for the study as volunteers. All participants had current valid medical examinations and showed no contraindications that would exclude them from the study. They declared that for at least one month before testing they did not take either medications or dietary supplements. Written informed consent was obtained prior to study commencement.

The basic anthropometric data of the volunteers (body height—BH, body mass—BM, fat content—FAT) are presented in Table 2. The experimental procedures involved, and the related risks were explained to all the participants verbally, informed written consent was taken from each participant and they could withdraw at any time of the study. The research project was conducted according to the Helsinki Declaration and was approved (no. 5/2013, approval date: 26.06.2013) by the Ethics Committee for Scientific Research at the Jerzy Kukuczka Academy of Physical Education in Katowice, Poland.

Table 2. Mean values of body height (BH), body mass (BM) and fat content (FAT) of study participants ($n = 11$).

BH (cm)	BM (kg)	FAT (%)
180.5 ± 6.5	70.3 ± 6.8	9.4 ± 3.1

The subjects participating in the study were tested on two randomized occasions separated by 5 days duration in normoxic and hypoxic conditions. Participants were allocated to conditions using a computer-generated randomized list [121].

Hypoxic conditions were created using a normobaric hypoxia chamber (LOSA HYP/HYOP-2/3NU system, LOWOXYGEN SYSTEMS, Berlin, Germany) that is in use in the Laboratory of Hypoxia of the Jerzy Kukuczka Academy of Physical Education and the selected hypoxia was an equivalent of 3000 m altitude ($FiO_2 = 14.7\%$).

On each occasion, the participants were subject to two graded ergocycle tests each performed under normobaric normoxic and normobaric hypoxic conditions (3000 m asl). Before each test, body mass and body composition of each participant was determined using a model Inbody 720 (Biospace Co., Tokio, Japan) body composition analyzer using electrical impedance measurements.

4.2. Ergocycle Graded Exercise Test

The ergocycle tests were performed on a model Excalibur Sport (Lode BV, Groningen Netherlands) cycloergometer, beginning at a work load of 40 W, which was increased by 40 W every 3 min until volitional exhaustion. If a subject terminated the test before completing a given workload, then the maximum workload was calculated from the formula WRmax = WRk + (t/T × WRp) [122], where WRk—previous workload, t—exercise duration with the work-load until premature failure, T—duration of each workload, WRp—the amount of workload by which exercise intensity increased during the test.

During the tests, heart rate (HR), minute ventilation (VE), breathing frequency (BF), oxygen uptake (VO_2) and carbon dioxide content in expired air (VCO_2) were recorded in the subjects with the MetaMax 3B gas analyzer (Cortex, Leipzig, Germany). Fingertip capillary blood samples for the assessment of LA concentration (Biosen C-line Clinic, EKF-diagnostic GmbH, Barleben, Germany) were drawn at rest and at the end of each step of the test, as well as during the 3rd, 6th, 9th, and 12th minute of recovery. Additionally, capillary rest and post-exercise blood samples were used to determine acid-base equilibrium and oxygen saturation of hemoglobin (Rapid Lab 248, Siemens/Bayer Diagnostics, Erlangen, Germany).

4.3. Psychomotor Performance Determination

The choice reaction time and NCR were selected as indices of psychomotor performance as described previously [26]. Briefly, the CRT console was mounted on the wall in front of the ergometer at eye level, 1.5 m away from the subject. The test included 15 positive (red light or a sound) and 15 negative (green and yellow lights) stimuli applied in a randomized order in 1 to 4 s intervals. The subjects were asked to press and then to release, as quickly as possible, the button of the switch devised kept in the right hand in response to the red light, the button in the left hand in response to the sound and not react to the negative stimuli. The total time for each CRT was 107 s. The stimuli and the subjects' responses were recorded using the reaction time measuring device (MRK 432, ZEAM, Zabrze, Poland). The reaction time was determined to the nearest 0.01 s. The results are presented at the mean reaction time of 15 responses to positive stimuli. The subjects were familiarized with the procedure a week before the study by practicing the task both at rest and during cycling.

4.4. Venous Blood

The participants were cannulated into the antecubital vein on the day of ergocycle testing 15 min prior to the breakfast. Venous blood samples (2 samples per time point) were collected 10 min later, then immediately after cessation of each ergocycle test, and 1 h after each ergocycle test. One sample of each pair was taken using ethylenediaminetetraacetic (EDTA) tubes (for morphology analysis); the other one was drawn using no anti-coagulant tubes and processed for serum for the other biochemical assays (BDNF, ET-1, C, catecholamines). After 30 min, blood samples were centrifuged at $1500 \times g$ for 15 min. The sera obtained were stored at $-80\ ^\circ$C until analyzed.

4.5. Determination of Brain-Derived Neurotrophic Factor, Cortisol and Endothelin-1 Concentrations

Serum BDNF, C, and ET-1 concentrations were determined using a commercially available Quantikine ELISA kit (R&D Systems, Minneapolis, MN, USA) according to the procedure supplied by the manufacturer. This method allows measurement of BDNF, C, and ET-1 in the range of 0.372–4000 pg/mL, 0.030–100 ng/mL, and 0.031–50 ng/mL respectively. The intra-assay coefficient of variance was <4.0%, <8%, <4%, respectively. To quantify the level, a standard curve was performed using a standard solution.

4.6. Determination of Catecholamines by HPLC Method

Adrenalin, NA, DA and 5-HT were assayed in the serum using high performance liquid chromatography (HPLC, Gynkotek, Copenhagen, Denmark) with electrochemical detection using Coulochem III model 520 (ESSA, Copenhagen, Denmark).

Serum samples were mixed with 0.1 M perchloric acid containing 22.5 ng/mL ascorbic acid (ASC, Sigma-Aldrich, St. Louis, MO, USA). After centrifugation at 15,000 g, 10 min, at $4\ ^\circ$C, supernatant was filtered through a nylon syringe filter (Millipore, 0.22 µm, Merck KGaA, Darmstadt, Germany). Samples of 20 µL filtrate were injected into a high performance liquid chromatography system (Gynkotek, Copenhagen, Denmark) equipped with a Hypersil Gold (15 cm × 4.6 mm) column (Thermo Electron Corporation, Kleinostheim, Germany). The samples were eluted by a mobile phase made of 107 mM of $Na_2HPO_4 \times 2H_2O$, 107 mM citric acid, 0.3 mM octane-1 sulfonic acid sodium salt (OSA), 0.2 µM of EDTA, pH, 4.6, 1.5% methanol and 1.5% acetonitrile at a flow rate of 0.8 mL/min. The column temperature was set at $25\ ^\circ$C. Peaks were detected by electrochemical detection (Coulochem III, ESSA, Copenhagen, Denmark) at potentials of E1 = −50 mV and E2 = +400 mV. Data were collected and analyzed using Chromeleon software run on a PC (Gynkotek, Copenhagen, Denmark). DA and 5-HT contents in the sample were calculated by extrapolating the peak area from a standard cure.

4.7. Determination of Nitrite Concentration

Serum nitrite concentration were determined using a commercially colorimetric kit (R&D Systems, Minneapolis, MN, USA) according to the procedure supplied by the manufacturer.

4.8. Statistical Analysis

The results of the study were analyzed using Statistica 13.0 software (StatSoft, Cracow, Poland). The results are presented as arithmetic means (x) and standard deviations (SD). The statistical significance was set at $p < 0.05$. Prior to all statistical analyses, the normality of the distribution of variables was verified using the Shapiro-Wilk test. The paired sample t-test was used to determine the significance of differences in VO_{2max}, WR_{max} between the two trials in different conditions. Furthermore, due to the lack of normality of distribution, the Wilcoxon test was used to determine the significance of differences in delta values in lactate concentrations (Delta LA) between the trials. The intergroup differences between the research trials (condition × time of measurement) were determined using the two-way ANOVA for repeated measures. When significant differences were found, the post hoc Tukey's test was used. If the normality assumption was violated, the Friedman test was applied, whereas when significant differences were found, we used the post hoc Friedman test. The comparisons of repeated measurements (normoxia vs. hypoxia) were assessed by the Wilcoxon signed-rank test.

5. Conclusions

In conclusion, the results of this study showed that maximal physical exercise, regardless of whether it is performed under normoxia or mild normobaric hypoxia (equivalent of 3000 m altitude) caused a similar increase in BDNF concentration in the blood of well-trained athletes. Despite the fact that BDNF has been known to possess a protective effect on the brain, an elevated BDNF level did not protect our participants from cognitive impairment due to acute exposure to hypoxia, because indices of this variable, i.e., CRT and NCR, were worse in hypoxic conditions. All examined potential circulating factors that are known to affect BDNF expression/production (C, nitrite, ET-1, catecholamines, LA) most likely did not affect psychomotor functions in our experimental paradigm.

Author Contributions: Conceptualization Z.P., M.C. (Miłosz Czuba), and J.L.; methodology, Z.P., M.C. (Miłosz Czuba), and J.L.; formal analysis, K.P. and M.C. (Miłosz Czuba); investigation, Z.P., M.C. (Małgorzata Chalimoniuk) and M.C. (Miłosz Czuba); data curation, M.C. (Małgorzata Chalimoniuk) and K.P.; writing—original draft preparation, Z.P., M.C. (Małgorzata Chalimoniuk), K.P., M.C. (Miłosz Czuba), J.L.; writing—review and editing, Z.P., M.C. (Miłosz Czuba) and J.L.; visualization, Z.P. and K.P.; supervision, M.C. (Małgorzata Chalimoniuk) and J.L.; funding acquisition, J.L. All authors have read and agreed to the published version of the manuscript.

Funding: The research was supported by the grant No. 2013/09/B/NZ7/00726 from the National Science Centre of Poland.

Conflicts of Interest: The authors declare no conflict of interest. The funders had no role in the design of the study; in the collection, analyses, or interpretation of data; in the writing of the manuscript, or in the decision to publish the results.

Abbreviations

BDNF	brain-derived neurotrophic factor
ET-1	endothelin-1
5-HT	serotonin
CRT	choice reaction time
NCR	number of correct reactions
DA	dopamine
LA	lactate
NA	noradrenaline
C	cortisol
A	adrenalin

References

1. Bendahan, D.; Badier, M.; Jammes, Y.; Confort-Gouny, S.; Salvan, A.M.; Guilot, C.; Conzzone, P. Metabolic and myoelectrical effects of acute hypoxaemia during isometric contraction of forearm muscles in humans: A combined 31P-magnetic resonance spectroscopy-surface electromyogram (MRS-SEMG) study. *Clin. Sci.* **1998**, *94*, 279–286. [CrossRef] [PubMed]

2. Calbet, J.A.; De Paz, J.A.; Garatachea, N.; Cabeza de Vaca, S.; Chavarren, J. Anaerobic energy provision does not limit Wingate exercise performance in endurance-trained cyclists. *J. Appl. Physiol.* **2003**, *94*, 668–676. [CrossRef] [PubMed]

3. Perrey, S. Decrease in cerebral oxygenation influences central motor output in humans. *Acta Physiol.* **2009**, *196*, 279–281. [CrossRef] [PubMed]

4. Amann, M.; Eldridge, M.W.; Lovering, A.T.; Stickland, M.K.; Pegelow, D.F.; Dempsey, J.A. Arterial oxygenation influences central motor output and exercise performance via effects on peripheral locomotor muscle fatigue in humans. *J. Physiol.* **2006**, *575*, 937–952. [CrossRef]

5. Peltonen, J.E.; Rusko, H.K.; Rantamaki, J.; Sweins, K.; Nittymaki, S.; Vitasalo, J.T. Effects of oxygen fraction in inspired air on force production and electromyogram activity during ergometer rowing. *Eur. J. Appl. Physiol.* **1997**, *76*, 495–503. [CrossRef]

6. Dempsey, J.A.; Wagner, P.D. Exercise-induced arterial hypoxemia. *J. Appl. Physiol.* **1999**, *87*, 1997–2006. [CrossRef]

7. Adams, R.P.; Welch, H.G. Oxygen uptake, acid-base status, and performance with varied inspired oxygen fractions. *J. Appl. Physiol.* **1980**, *49*, 863–868. [CrossRef]

8. Hogan, M.C.; Richardson, R.S.; Haseler, L.J. Human muscle performance and PCr hydrolysis with varied inspired oxygen fraction: A 31P-MRS study. *J. Appl. Physiol.* **1999**, *86*, 1367–1373. [CrossRef]

9. Peltonen, J.E.; Tikkanen, H.O.; Ritola, J.J.; Ahotupa, M.; Rusko, H.K. Oxygen uptake response during maximal cycling in hyperoxia, normoxia and hypoxia. *Aviat. Space Environ. Med.* **2001**, *72*, 904–911.

10. Benoit, H.; Busso, T.; Castells, J.; Geyssant, A.; Denis, C. Decrease in peak heart rate with acute hypoxia in relation to sea level VO(2max). *Eur. J. Appl. Physiol.* **2003**, *90*, 514–519. [CrossRef]

11. Lundby, C.; Moeller, P.; Kanstrup, I.L.; Olsen, N.V. Heart rate response to hypoxic exercise: Role of dopamine D2-receptors and effect of oxygen supplementation. *Clin. Sci.* **2001**, *101*, 377–383. [CrossRef] [PubMed]

12. Kayser, B.; Narici, M.; Binzoni, T.; Grassi, B.; Cerretelli, P. Fatigue and exhaustion in chronic hypobaric hypoxia: Influence of exercising muscle mass. *Am. J. Physiol.* **1994**, *76*, 634–640. [CrossRef] [PubMed]

13. Boushel, R.; Calbet, J.A.; Rådegran, G.; Sondergaard, H.; Wagner, P.D.; Saltin, B. Parasympathetic neural activity accounts for the lowering of exercise heart rate at high altitude. *Circulation* **2001**, *104*, 1785–1791. [CrossRef] [PubMed]

14. Amann, M.; Romer, L.M.; Subudhi, A.W.; Pegelow, D.F.; Dempsey, J.A. Severity of arterial hypoxaemia affects the relative contributions of pripheral muscle fatigue to exercise performance in healthy humans. *J. Physiol.* **2007**, *581*, 389–403. [CrossRef]

15. Pedersen, B.K.; Febbraio, M.A. Muscle as an endocrine organ: Focus on muscle-derived interleukin-6. *Phys. Rev.* **2008**, *88*, 1379–1406. [CrossRef]

16. Davis, J.M.; Bailey, S.P. Possible mechanisms of central nervous system fatigue during exercise. *Med. Sci. Sports Exerc.* **1997**, *29*, 45–57. [CrossRef]

17. Lieberman, P.; Protopapas, A.; Reed, E.; Youngs, J.W.; Kankim, B.G. Cognitive defects at altitude. *Nature* **1994**, *372*, 325. [CrossRef]

18. Blogg, S.L.; Gennser, M. Cerebral blood flow velocity and psychomotor performance during acute hypoxia. *Aviat. Space Environ. Med.* **2006**, *77*, 107–113.

19. Temme, L.A.; Still, D.L.; Acromite, M.T. Hypoxia and flight performance of military instructor pilots in a flight simulator. *Aviat. Space Environ. Med.* **2010**, *81*, 654–659. [CrossRef]

20. McMorris, T.; Hale, B.H.; Barwood, M.; Costello, J.; Corbett, J. Effect of acute hypoxia on cognition: A systematic review and meta-regression analysis. *Neurosci. A Biobehav. Rev.* **2017**, *74*, 225–232. [CrossRef]

21. Lal, C.; Strange, C.; Bachman, D. Neurocognitive impairment in obstructive sleep apnea. *Chest* **2012**, *141*, 1601–1610. [CrossRef] [PubMed]

22. Jensen, A.R. *Clocking the Mind: Mental Chronometry and Individual Differences*, 1st ed.; Elsevier Science: London, UK, 2006; pp. 43–54.

23. Salthouse, T.A.; Hedden, T. Interpreting reaction time measures in between-group comparisons. *J. Clin. Exp. Neuropsychol.* **2002**, *24*, 858–872. [CrossRef] [PubMed]

24. Caldwell, H.G.; Coombs, G.B.; Tymko, M.M.; Nowak-Fluck, D.; Ainslie, P.N. Severity-dependent influence of isocapnic hypoxia on reaction time is independent of neurovascular coupling. *Physiol. Behav.* **2018**, *1*, 262–269. [CrossRef] [PubMed]

25. De Aquino-Lemos, V.; Santos, R.V.T.; Antunes, H.K.M.; Lira, F.S.; Bittar, I.G.L.; Caris, L.V.; Tufik, S.; de Mello, M.T. Acute physical exercise under hypoxia improves sleep, mood and reaction time. *Physiol. Behav.* **2016**, *154*, 90–99. [CrossRef]

26. Ochi, G.; Kanazawa, J.; Hyodo, K.; Suwabe, K.; Shimizu, T.; Fukuie, T.; Byun, K.; Soya, H. Hypoxia-induced lowered executive function depends on arterial oxygen desaturation. *J. Physiol. Sci.* **2018**, *68*, 847–853. [CrossRef]

27. Pramsohler, S.; Wimmer, S.; Kopp, M.; Gatterer, H.; Faulhaber, M.; Burtscher, M.; Netzer, N.C. Normobaric hypoxia overnight impairs cognitive reaction time. *BMC Neurosi.* **2017**, *18*, 43. [CrossRef]

28. Hartman, R.E.; Lee, J.M.; Zipfel, G.J.; Wozniak, D.F. Characterizing learning deficits and hippocampal neuron loss following transient global cerebral ischemia in rats. *Brain Res.* **2005**, *1043*, 48–56. [CrossRef]

29. Maiti, P.; Singh, S.B.; Muthuraju, S.; Veleri, S.; Ilavazhagen, G. Hypobaric hypoxia damages the hippocampal pyramidal neurons in the rat brain. *Brain Res.* **2007**, *1175*, 1–9. [CrossRef]

30. Hota, S.K.; Barhwal, K.; Singh, S.B.; Ilavazhagan, G. Chronic hypobaric hypoxia induced apoptosis in CA1 region of hippocampus: A possible role of NMDAR mediated p75NTR upregulation. *Exp. Neurol.* **2008**, *212*, 5–13. [CrossRef]

31. Gale, S.D.; Hopkins, R.O. Effects of hypoxia on the brain: Neuroimaging and neuropsychological findings following carbon monoxide poisoning and obstructive sleep apnea. *J. Int. Neuropsychol. Soc.* **2004**, *10*, 60–71. [CrossRef]

32. Levitt, S.; Gutin, B. Multiple-choice reaction time and movement time during physical exertion. *Res. Q.* **1971**, *42*, 406–410. [CrossRef]

33. Sjöberg, H. Relations between heart rate, reaction speed and subjective effort at different work loads on a bicycle ergometer. *J. Hum. Stress* **1975**, *1*, 21–27. [CrossRef] [PubMed]

34. Chmura, J.; Nazar, K.; Kaciuba-Uscilko, H. Choice reaction time during graded exercise in relation to blood lactate and plasma catecholamine thresholds. *Int. J. Sports Med.* **1994**, *15*, 172–176. [CrossRef] [PubMed]

35. Chmura, J.; Krysztofiak, H.; Ziemba, A.W.; Nazar, K.; Kaciuba-Uscilko, H. Psychomotor performance during prolonged exercise above and below lactate threshold. *Eur. J. Appl. Physiol.* **1998**, *77*, 77–80. [CrossRef] [PubMed]

36. Travlos, A.K.; Marisi, D.Q. Information processing and concentration as a function of fitness level and exercise-induced activation to exhaustion. *Percept. Mot. Skills* **1995**, *80*, 15–26. [CrossRef]

37. Kruk, B.; Chmura, J.; Krzeminski, K.; Ziemba, A.W.; Nazar, K.; Pekkarinen, H.; Kaciuba-Uscilko, H. Influence of caffeine, cold and exercise on multiple choice reaction time. *Psychopharmacology* **2001**, *157*, 197–201. [CrossRef]

38. Bender, V.L.; McGlynn, L. The effect of various levels of strenuous to exhausting exercise on reaction time. *Eur. J. Appl. Physiol.* **1976**, *35*, 95–110. [CrossRef]

39. Rothman, S.M.; Griffioen, K.J.; Wan, R.; Mattson, M.P. Brain-derived neurotrophic factor as a regulator of systemic and brain energy metabolism and cardiovascular health. *Ann. N. Y. Acad. Sci.* **2012**, *1264*, 49–63. [CrossRef]

40. Huang, T.; Larsen, K.T.; Ried-Larsen, M.; Møller, N.C.; Andersen, L.B. The effects of physical activity and exercise on brain-derived neurotrophic factor in healthy humans: A review. *Scand. J. Med. Sci. Sports* **2014**, *24*, 1–10. [CrossRef]

41. Hohn, A.; Leibrock, J.; Bailey, K.; Barde, Y.A. Identyfication and characterization of a novel member of the nerve growth factor/brain-derived neurotrophic factor family. *Nature* **1990**, *344*, 339–341. [CrossRef]

42. Binder, D.K.; Scharfman, H.E. Brain-derived neurotrophic factor. *Growth Factors* **2004**, *22*, 123–131. [CrossRef] [PubMed]

43. Małczyńska, P.; Piotrowicz, Z.; Drabarek, D.; Langfort, J.; Chalimoniuk, M. Rola mózgowego czynnika neurotroficznego (BDNF) w procesach neurodegeneracji oraz w mechanizmach neuroregeneracji wywołanej wzmożoną aktywnością fizyczną. *Postępy Biochem.* **2019**, *65*, 2–8. [CrossRef] [PubMed]

44. Tyler, W.J.; Alonso, M.; Bramham, C.R.; Pozzo-Miller, L.D. From acquisition to consolidation: On the role of brain-derived neurotrophic factor signaling in hippocampal-dependent learning. *Learn. Mem.* **2002**, *9*, 224–237. [CrossRef]

45. Knaepen, K.; Goekint, M.; Heyman, E.M.; Meeusen, R. Neuroplasticity-exercise-induced response of peripheral brain-derived neurotrophic factor: A systematic review of experimental studies in human subjects. *Sports Med.* **2010**, *40*, 765–801. [CrossRef] [PubMed]

46. Zoladz, J.A.; Pilc, A.; Majerczak, J.; Grandys, M.; Zapart-Bukowska, J.; Duda, K. Endurance training increases plasma brain-derived neurotrophic factor concentration in young healthy men. *J. Physiol. Pharmacol.* **2008**, *59*, 119–132. [PubMed]

47. Correia, P.R.; Scorza, F.A.; Gomes da Silva, S.; Pansani, A.; Toscano-Silva, M.; de Almeida, A.C.; Arida, R.M. Increased basal plasma brain-derived neurotrophic factor levels in sprint runners. *Neurosci. Bull.* **2011**, *27*, 325–329. [CrossRef] [PubMed]

48. Cheng, A.; Wang, S.; Cai, J.; Rao, M.S.; Mattson, M.P. Nitric oxide acts in a positive feedback loop with BDNF to regulate neural progenitor cell proliferation and differentiation in the mammalian brain. *Dev. Biol.* **2003**, *258*, 319–333. [CrossRef]

49. Chen, M.J.; Ivy, A.S.; Russo-Neustadt, A.A. Nitric oxide synthesis is required for exercise-induced increases in hippocampal BDNF and phosphatidylinositol 3′ kinase expression. *Brain Res. Bull.* **2006**, *68*, 257–268. [CrossRef]

50. Zajac, A.; Poprzecki, S.; Zebrowska, A.; Chalimoniuk, M.; Langfort, J. Arginine and ornithine supplementation increases growth hormone and insulin-like growth factor-1 serum levels after heavy-resistance exercise in strength-trained athletes. *J. Strength Cond. Res.* **2010**, *24*, 1082–1090. [CrossRef]

51. Dessypris, A.; Wägar, G.; Fyhrquist, F.; Mäkinen, T.; Welin, M.G.; Lamberg, B.A. Marathon run: Effects on blood cortisol-ACTH, iodothyronines-TSH and vasopressin. *Acta Endocrinol. (Copenh.)* **1980**, *95*, 151–157. [CrossRef]

52. Yau, S.Y.; Lau, B.W.; Zhang, E.D.; Lee, J.C.; Li, A.; Lee, T.M.; Ching, Y.P.; Xu, A.M.; So, K.F. Effects of voluntary running on plasma levels of neurotrophins, hippocampal cell proliferation and learning and memory in stressed rats. *Neuroscience* **2012**, *222*, 289–301. [CrossRef] [PubMed]

53. Robinet, C.; Pellerin, L. Brain-derived neurotrophic factor enhances the expression of the monocarboxylate transporter 2 through translational activation in mouse cultured cortical neurons. *J. Cereb. Blood Flow Metab.* **2010**, *30*, 286–298. [CrossRef] [PubMed]

54. Ng, T.K.S.; Ho, C.S.H.; Tam, W.W.S.; Kua, E.H.; Ho, R.C. Decreased Serum Brain-Derived Neurotrophic Factor (BDNF) Levels in Patients with Alzheimer's Disease (AD): A Systematic Review and Meta-Analysis. *Int. J. Mol. Sci.* **2019**, *20*, 257. [CrossRef] [PubMed]

55. Liu, P.Z.; Nusslock, R. Exercise-Mediated Neurogenesis in the Hippocampus via BDNF. *Front. Neurosci.* **2018**, *12*, 52. [CrossRef]

56. Pan, W.; Banks, W.A.; Fasold, M.B.; Bluth, J.; Kastin, A.J. Transport of brain-derived neurotrophic factor across the blood-brain barrier. *Neuropharmacology* **1998**, *37*, 1553–1561. [CrossRef]

57. Ward, R.; Abdul, Y.; Ergul, A. Endothelin-1 inhibition improves the mBDNF/proBDNF ratio in endothelial cells and HT22 neurons under high glucose/palmitate growth conditions. *Physiol. Res.* **2018**, *67* (Suppl. 1), S237–S246. [CrossRef]

58. Hyman, C.; Hofer, M.; Barde, Y.A.; Juhasz, M.; Yancopoulos, G.D.; Squinto, S.P.; Lindsay, R.M. BDNF is a neurotrophic factor for dopaminergic neurons of the substantia nigra. *Nature* **1991**, *350*, 230–232. [CrossRef]

59. Popova, N.K.; Naumenko, V.S. Neuronal and behavioral plasticity: The role of serotonin and BDNF systems tandem. *Expert Opin. Ther. Targets* **2019**, *23*, 227–239. [CrossRef]

60. Venezia, A.C.; Quinian, E.; Roth, S.M. A single bout of exercise increase hippocampal Bdnf: Influence of chronic exercise and noradrenaline. *Genes Brain Behav.* **2017**, *16*, 800–811.

61. Cooke, M.; Cruttenden, R.; Mellor, A.; Lumb, A.; Pattman, S.; Burnett, A.; Boot, C.; Burnip, L.; Boos, C.; O'Hara, J.; et al. A pilot investigation into the effects of acute normobaric hypoxia, high altitude exposure and exercise on serum angiotensin-converting enzyme, aldosterone and cortisol. *J. Renin Angiotensin Aldosterone Syst.* **2018**, *19*, 1470320318782782. [CrossRef]

62. Schneider, D.A.; McGuiggin, M.E.; Kamimorim, G.H. A comparison of the blood lactate and plasma catecholamine thresholds in untrained male subjects. *Int. J. Sports Med.* **1992**, *13*, 52. [CrossRef] [PubMed]

63. Miguel, P.M.; Deniz, B.F.; Deckmann, I.; Confortim, H.D.; Diaz, R.; Diaz, R.; Laureano, D.P.; Silveira, P.P.; Pereira, L.O. Prefrontal Cortex Dysfunction in Hypoxic-Ischaemic Encephalopathy Contributes to Executive Function Impairments in Rats: Potential Contribution for Attention-Deficit/Hyperactivity Disorder. *World J. Biol. Psychiatry* **2018**, *19*, 547–560. [CrossRef] [PubMed]

64. Newsholme, E.A.; Acworth, I.N.; Blomstrand, E. Amino acids, brain neurotransmitters and a functional link between muscle brain that is important in sustained exercise. In *Advances in Biochemistry*; Benzi, G., Ed.; John Libby Eurotext: London, UK, 1987; pp. 127–138.

65. Wang, G.J.; Volkow, N.D.; Fowler, J.S.; Franceschi, D.; Logan, J.; Pappas, N.R.; Wong, C.T.; Netusilm, N. PET studies of the effects of aerobic exercise on human striatal dopamine release. *J. Nucl. Med.* **2000**, *41*, 1352–1356. [PubMed]

66. El-Khodor, B.F.; Boksa, P. Long-term reciprocal changes in dopamine levels in prefrontal cortex versus nucleus accumbens in rats born by Caesarean section compared to vaginal birth. *Exp. Neurol.* **1997**, *145*, 118–129. [CrossRef]

67. Piotrowicz, Ż.; Chalimoniuk, M.; Ploszczyca, K.; Czuba, M.; Langfort, J. Acute normobaric hypoxia does not affect the simultaneous exercise-induced increase in circulating BDNF and GDNF in young healthy men: A feasibility study. *PLoS ONE* **2019**, *14*, e0224207. [CrossRef]

68. Wang, H.; Niu, F.; Fan, W.; Shi, J.; Zhang, J.; Li, B. Modulating effects of predconditioning exercise in the expression of ET-1 and BNP via HIF-1a in ischemically injured brain. *Metab. Brain Dis.* **2019**, *34*, 1299–1311. [CrossRef]

69. Cotman, C.W.; Engesser-Cesar, C. Exercise Enhances and Protects Brain Function. *Exerc. Sport Sci. Rev.* **2002**, *30*, 75–79. [CrossRef]

70. Ludyga, S.; Gerber, M.; Brand, S.; Holsboer-Trachsler, E.; Puhse, U. Acute Effects of Moderate Aerobic Exercise on Specific Aspects of Executive Function in Different Age and Fitness Groups: A Meta-Analysis. *Psychophysiology* **2016**, *53*, 1611–1626. [CrossRef]

71. Czuba, M.; Wilk, R.; Karpiński, J.; Chalimoniuk, M.; Zajac, A.; Langfort, J. Intermittent hypoxic training improves anaerobic performance in competitive swimmers when implemented into a direct competition mesocycle. *PLoS ONE* **2017**, *12*, e0180380. [CrossRef]

72. Czuba, M.; Bril, G.; Płoszczyca, K.; Piotrowicz, Z.; Chalimoniuk, M.; Roczniok, R.; Zembron Lacny, A.; Gerasimuk, D.; Langfort, J. Intermittent hypoxic training at lactate threshold intensity improves aiming performance in well-trained biathletes with little change of cardiovascular variables. *Biomed. Res. Int.* **2019**, *2019*, 1287506. [CrossRef]

73. Millet, G.P.; Roels, B.; Schmitt, L.; Woorons, X.; Richalet, J.P. Combining hypoxic methods for peak performance. *Sports Med.* **2010**, *40*, 1–25. [CrossRef] [PubMed]

74. Dobashi, S.; Horiuchi, M.; Endo, J.; Kiuchi, M.; Koyama, K. Cognitive Function and Cerebral Oxygenation during Prolonged Exercise under Hypoxia in Healthy Young Males. *High Alt. Med. Biol.* **2016**, *17*, 214–221. [CrossRef] [PubMed]

75. Li, P.; Zhang, G.; You, H.Y.; Zheng, R.; Gao, Y.Q. Training-dependent cognitive advantage is suppressed at high altitude. *Physiol. Behav.* **2012**, *106*, 439–445. [CrossRef] [PubMed]

76. Chang, Y.K.; Labban, J.D.; Gapin, J.I.; Etnier, J.L. The effects of acute exercise on cognitive performance: A meta-analysis. *Brain Res.* **2012**, *1453*, 87–101. [CrossRef]

77. Archer, T.; Josefsson, T.; Lindwall, M. Effects of Physical Exercise on Depressive Symptoms and Biomarkers in Depression. *CNS Neurol. Disord. Drug Targets* **2014**, *13*, 1640–1653. [CrossRef]

78. Campos, C.; Rocha, N.B.; Lattari, E.; Nardi, A.E.; Machado, S. Exercise Induced Neuroplasticity to Enhance Therapeutic Outcomes of Cognitive Remediation in Schizophrenia: Analyzing the Role of Brai Nderived Neurotrophic Factor. *CNS Neurol. Disord. Drug Targets* **2017**, *16*, 638–651. [CrossRef]

79. Das, A.K.; Dhar, P.; Sharma, V.K.; Barhwal, K.; Hota, S.K.; Norboo, T.; Singh, S.B. High Altitude With Monotonous Environment Has Significant Impact on Mood and Cognitive Performance of Acclimatized Lowlanders: Possible Role of Altered Serum BDNF and Plasma Homocysteine Level. *J. Affect. Disord.* **2018**, *237*, 94–103. [CrossRef]

80. Wang, W.H.; He, G.P.; Xiau, P.; Gu, C.; Chen, H.Y. Relationship Between Brain-Derived Neurotrophic Factor and Cognitive Function of Obstructive Sleep Apnea/Hypopnea Syndrome Patients. *Asian Pac. J. Trop. Med.* **2012**, *5*, 906–910. [CrossRef]

81. Halievski, K.; Nath, S.; Katsuno, M.; Adachi, H.; Sobue, G.; Breedlove, S.; Lieberman, A.; Jordan, C. Disease Affects Bdnf Expression in Synaptic and Extrasynaptic Regions of Skeletal Muscle of Three SBMA Mouse Models. *Int. J. Mol. Sci.* **2019**, *20*, 1314. [CrossRef]

82. Satriotomo, I.; Nichols, N.I.; Dale, E.A.; Emery, A.T.; Dahlberg, J.M.; Mitcell, G.S. Repetitive Acute Intermittent Hypoxia Increases Growth/Neurotrophic Factor Expression in Non-Respiratory Motor Neurons. *Neuroscience* **2016**, *332*, 449–488. [CrossRef]

83. Karege, F.; Schwald, M.; Cisse, M. Postnatal developmental profile of brain-derived neurotrophic factor in rat brain and platelets. *Neurosci. Lett.* **2002**, *328*, 261–264. [CrossRef]

84. Rasmussen, P.; Brassard, P.; Adser, H.; Pedersen, M.V.; Leick, L.; Hart, E.; Secher, N.H.; Pedersen, B.K.; Pilegaard, H. Evidence for a release of brain-derived neurotrophic factor from the brain during exercise. *Exp. Physiol.* **2009**, *94*, 1062–1069. [CrossRef] [PubMed]

85. Erickson, K.I.; Voss, M.W.; Prakas, R.S.; Basak, C.; Szabo, A.; Chaddock, L.; Kim, J.S.; Heo, S.; Alves, H.; White, S.M. Exercise Training Increases Size of Hippocampus and Improves Memory. *Proc. Natl. Acad. Sci. USA* **2011**, *108*, 3017–3322. [CrossRef] [PubMed]

86. Lipski, R.H.; Marini, A.M. Neurotrophic factor in neuronal survival and behavior-related plasticity. *Ann. N. Y. Acad. Sci.* **2007**, *1122*, 130–143. [CrossRef] [PubMed]

87. Huang, E.J.; Reichardt, L.E. Neurotrophins: Roles in neuronal development and function. *Annu. Rev. Neurosci.* **2009**, *24*, 677–736. [CrossRef]

88. Ferris, L.T.; Williams, J.S.; Shen, C.L. The Effect of Acute Exercise on Serum Brain-Derived Neurotrophic Factor Levels and Cognitive Function. *Med. Sci. Sports Exerc.* **2007**, *39*, 728–734. [CrossRef]

89. Tsai, C.L.; Chen, F.C.; Pan, C.Y.; Wang, C.H.; Huang, T.H.; Chen, T.C. Impact of Acute Aerobic Exercise and Cardiorespiratory Fitness on Visuospatial Attention Performance and Serum BDNF Levels. *Psychoneuroedocrinology* **2014**, *41*, 121–131. [CrossRef]

90. Slusher, A.L.; Patterson, V.T.; Schwartz, C.S.; Acevedo, E.O. Impact of high intensity exercise on execrative function and brain derived neurotrophic factor in healthy college aged males. *Phsiol. Behav.* **2018**, *191*, 116–122. [CrossRef]

91. Sasi, M.; Vignoli, B.; Cannossa, M.; Blum, R. Neurobiology of Local and Intercellular BDNF Signaling. *Plugers Arch. Eur. J. Physiol.* **2017**, *469*, 593–610. [CrossRef]

92. Fowler, C.; Gervitz, L.; Hamilton, E.; Walker, J.A. Systemic Hypoxia and the Depression of Synaptic Transmission in Rat Hippocampus after Carotid Artery Occlusion. *J. Physiol.* **2003**, *550*, 961–972. [CrossRef]

93. Lanigan, S.; Corcoran, A.E.; Wall, A.; Mukandala, G.; O'Connor, J.J. Acute Hypoxic Exposure and Prolyl-Hydroxylase Inhibition Improves Synaptic Transmission Recovery Time from a Subsequent Hypoxic Insult in Rat Hippocampus. *Brain Res.* **2018**, *1701*, 212–218. [CrossRef] [PubMed]

94. Kline, D.D.; Ramirez-Navarro, A.; Kunze, D.L. Adaptive Depression in Synaptic Transmission in the Nucleus of the Solitary Tract after In Vivo Chronic Intermittent Hypoxia: Evidence for Homeostatic Plasticity. *J. Neurosci.* **2007**, *27*, 4663–4673. [CrossRef] [PubMed]

95. Jonas, E.; Hickman, J.A.; Hardwick, J.M.; Kaczmarek, L.K. Exposure to Hypoxia Rapidly Induces Mitochondrial Channel Activity within a Living Synapse. *J. Biol. Chem.* **2005**, *280*, 4491–4497. [CrossRef] [PubMed]

96. Muthuraju, S.; Maiti, P.; Solanki, P.; Sharma, A.K.; Amitabh; Singh, S.B.; Prasad, D.; Ilavazhagan, G. Acetylcholinesterase Inhibitors Enhance Cognitive Functions in Rats Following Hypobaric Hypoxia. *Behav. Brain Res.* **2009**, *203*, 1–14. [CrossRef]

97. Echouffo-Tcheugui, J.B.; Conner, S.C.; Himali, J.J.; Maillard, P.; DeCarli, C.S.; Beiser, A.S.; Vasan, R.S.; Seshadri, S. Circulating Cortisol and Cognitive and Structural Brain Measures: The Framingham Heart Study. *Neurology* **2018**, *91*, e1961–e1970. [CrossRef]

98. Sroykham, W.; Wongsawat, Y. Effects of Brain Activity, Morning Salivary Cortisol, and Emotion Regulation on Cognitive Impairment in Elderly People. *Medicine* **2019**, *98*, e16114. [CrossRef]

99. Stomby, A.; Boraxbekk, C.J.; Lundquist, A.; Nordin, A.; Nilsson, L.G.; Adolfsson, R.; Nyberg, L.; Olsson, T. Higher Diurnal Salivary Cortisol Levels Are Related to Smaller Prefrontal Cortex Surface Area in Elderly Men and Women. *Eur. J. Endocrinol.* **2016**, *175*, 117–126. [CrossRef]

100. Dominguez, G.; Henkous, N.; Prevot, T.; David, V.; Guillou, J.L.; Belzung, C.; Mons, N.; Béracochéa, D. Sustained Corticosterone Rise in the Prefrontal Cortex Is a Key Factor for Chronic Stress-Induced Working Memory Deficits in Mice. *Neurobiol. Stress* **2019**, *10*, 100161. [CrossRef]

101. Vuolteenaho, O.; Koistinen, P.; Martikkala, V.; Takala, T. Leppaluoto. Effect of physical exercise in hypobaric conditions o atrial natriuretic peptide secretion. *Am. J. Physiol.* **1992**, *263*, 647–652.

102. Rikki, R.E.; Edwards, M.E. Effects of a three-year exercise program on motor function and cognitive processing speed I older woe. *Res. Qart. Exerc. Sport* **1991**, *62*, 61–67.

103. Bouissou, P.; Peronnet, F.; Brisson, G.; Helie, R.; Ledoux, M. Metabolic and endocrine responses to graded exercise under acute hypoxia. *Eur. J. Appl. Physiol.* **1986**, *55*, 290–294. [CrossRef] [PubMed]

104. Bouissou, P.; Peronnet, F.; Brisson, G.; Helie, R.; Ledoux, M. Fluid-electrolyte shift and renin-aldosterone responses to exercise under hypoxia. *Horm. Met. Res.* **1987**, *19*, 331–334. [CrossRef] [PubMed]

105. Rostrup, M. Catecholamines, hypoxia and high altitude. *Acta Phsiol. Scand.* **1988**, *11*, 389–399. [CrossRef]

106. Weber, D.A.; Reynolds, C.R. Clinical Perspectives on Neurobiological Effects of Psychological Trauma. *Neuropsychol. Rev.* **2004**, *14*, 115–129. [CrossRef] [PubMed]

107. Reader, T.A. Distribution of Catecholamines and Serotonin in the Rat Cerebral Cortex: Absolute Levels and Relative Proportions. *J. Neural Transm.* **1981**, *50*, 13–27. [CrossRef] [PubMed]

108. Miguel, P.M.; Pereira, L.O.; Barth, B.; Filho, E.J.M.; Pokhvisneva, I.; Nguyen, T.T.T.; Garg, E.; Razzolini, B.R.; Koh, D.X.P. Prefrontal Cortex Dopamine Transporter Gene Network Moderates the Effect of Perinatal Hypoxic-Ischemic Conditions on Cognitive Flexibility and Brain Gray Matter Density in Children. *Biol. Psychiatry* **2019**, *86*, 621–630. [CrossRef]

109. Ogh, S.; Tsukamoto, H.; Hirasawa, A.; Hasegawa, H.; Hirose, N.; Hashimoto, T. The Effect of Changes in Cerebral Blood Flow on Cognitive Function during Exercise. *Physiol. Rep.* **2014**, *2*, e12163. [CrossRef]

110. Adachi, M.; Yang, Y.; Furiuchi, Y.; Miiyamoto, C. Cloning and characterization of cDNA encoding human A-type endothelin receptor. *Biochem. Biophys. Res. Commun.* **1991**, *180*, 1265–1272. [CrossRef]

111. Palmer, R.M.; Ferrige, A.G.; Moncada, S. Nitric Oxide Release Accounts for the Biological Activity of Endothelium-Derived Relaxing Factor. *Nature* **1987**, *327*, 524–526. [CrossRef]

112. Garthwaite, J. Nitric oxide as a multimodal transmitter in the brain: Discovery and current status. *Br. J. Pharmacol.* **2019**, *176*, 197–211. [CrossRef]

113. Gilchrist, M.; Winyard, P.G.; Fulford, J.; Anning, C.; Shore, A.C.; Benjamin, N. Dietary nitrate supplementation improves reaction time in type 2 diabetes: Development and application of a novel nitrate-depleted beetroot juice placebo. *Nitric Oxide* **2014**, *19*, 333–337. [CrossRef] [PubMed]

114. Lefferts, W.K.; Hughes, W.E.; White, C.N.; Brutsaert, T.D.; Heffernan, K.S. Effect of acute nitrate supplementation on neurovascular coupling and cognitive performance in hypoxia. *Appl. Physiol. Nutr. Metab.* **2016**, *41*, 133–141. [CrossRef] [PubMed]

115. Shannon, O.M.; Duckworth, L.; Barlow, M.J.; Deighton, K.; Matu, J.; Williams, E.L.; Woods, D.; Xie, L.; Stephan, B.C.M.; Siervo, M.; et al. Effects of Dietary Nitrate Supplementation on Physiological Responses, Cognitive Function, and Exercise Performance at Moderate and Very-High Simulated Altitude. *Front. Physiol.* **2017**, *8*, 401. [CrossRef] [PubMed]

116. Basso, J.C.; Suzuki, W.A. The Effects of Acute Exercise on Mood, Cognition, Neurophysiology, and Neurochemical Pathways: A Review. *Brain Plast.* **2017**, *2*, 127–152. [CrossRef] [PubMed]

117. Adeva-Andany, M.; Lopez-Ojen, M.; Funcasta-Calderon, R.; Ameneiros-Rodríguez, E.; Donapetry-García, C.; Vila-Altesor, M.; Rodríguez-Seijas, J. Comprehensive review on Lactate Metabolism in Human Health. *Mitochondrion* **2014**, *17*, 76–100. [CrossRef]

118. Pellerin, L.; Pellegri, G.; Bittar, P.G.; Charnay, Y.; Bouras, C.; Martinm, J.L.; Stella, N.; Magistretti, P.J. Evidence supporting the existence of an activity-dependent astrocyte neuron lactate shuttle. *Dev. Neurosci.* **1998**, *20*, 291–299. [CrossRef]

119. Carrard, A.; Elsayed, M.; Margineanu, M.; Boury-Jamot, B.; Fragnière, L.; Meylan, E.M.; Petit, J.M.; Fiumelli, H.; Magistretti, P.J.; Martin, J.L. Peripheral administration of lactate produces antidepressant-like effects. *Mol. Psychiatry* **2018**, *23*, 392–399. [CrossRef]

120. Liu, C.; Wu, J.; Zhu, J.; Kuei, C.; Yu, J.; Shelton, J.; Sutton, S.W.; Li, X.; Yun, S.J.; Mirzadegan, T. Lactate inhibits lipolysis in fat cells through activation of an orphan G-protein-coupled receptor, GPR81. *J. Biol. Chem.* **2009**, *284*, 2811–2822. [CrossRef]

121. Urbaniak, G.C.; Plous, S. Research Randomizer (Version 4.0) [Computer Software]. 2013. Available online: http://www.randomizer.org/ (accessed on 25 March 2017).

122. Kuipers, H.; Verstappen, F.T.J.; Keizer, H.A.; Guerten, P.; van Kranenburg, G. Variability of aerobic performance in the laboratory and its physiological correlates. *Int. J. Sports Med.* **1985**, *6*, 197–201. [CrossRef]

International Journal of
Molecular Sciences

Article

Peroxisome Proliferator-Activated Receptor γ Coactivator 1α Activates Vascular Endothelial Growth Factor That Protects Against Neuronal Cell Death Following Status Epilepticus through PI3K/AKT and MEK/ERK Signaling

Jyun-Bin Huang [1,2,†], Shih-Pin Hsu [3,†], Hsiu-Yung Pan [1,2], Shang-Der Chen [2,4,5],
Shu-Fang Chen [2,4], Tsu-Kung Lin [2,4,6], Xuan-Ping Liu [5], Jie-Hau Li [5], Nai-Ching Chen [2,4],
Chia-Wei Liou [2,4,6], Chung-Yao Hsu [7], Hung-Yi Chuang [8] and Yao-Chung Chuang [2,4,5,7,9,*]

[1]   Department of Emergency Medicine, Kaohsiung Chang Gung Memorial Hospital, Kaohsiung 83301, Taiwan;
     u9001135@gmail.com (J.-B.H.); fornever@cgmh.org.tw (H.-Y.P.)
[2]   College of Medicine, Chang Gung University, Taoyuan 33302, Taiwan; chensd@adm.cgmh.org.tw (S.-D.C.);
     fangoe1@yahoo.com.tw (S.-F.C.); tklin@adm.cgmh.org.tw (T.-K.L.); naiging@yahoo.com.tw (N.-C.C.);
     cwliou@ms22.hinet.net (C.-W.L.)
[3]   Department of Neurology, E-Da Hospital/School of Medicine, I-Shou University, Kaohsiung 824, Taiwan;
     a.pin.hsu@gmail.com
[4]   Department of Neurology, Kaohsiung Chang Gung Memorial Hospital, Kaohsiung 83301, Taiwan
[5]   Institute for Translation Research in Biomedicine, Kaohsiung Chang Gung Memorial Hospital,
     Kaohsiung 83301, Taiwan; pphome67@yahoo.com.tw (X.-P.L.); jiehau1060301@gmail.com (J.-H.L.)
[6]   Mitochondrial Research Unit, Kaohsiung Chang Gung Memorial Hospital and Chang Gung University
     College of Medicine, Kaohsiung 83301, Taiwan
[7]   Department of Neurology, School of Medicine, College of Medicine, Kaohsiung Medical University Hospital,
     Kaohsiung Medical University, Kaohsiung 80708, Taiwan; cyhsu61@gmail.com
[8]   Department of Occupational and Environmental Medicine, Kaohsiung Medical University Hospital and
     School of Public Health, Kaohsiung Medical University, Kaohsiung 80708, Taiwan; hychuang@gmail.com
[9]   Department of Biological Science, National Sun Yat-sen University, Kaohsiung 80424, Taiwan
*   Correspondence: ycchuang@adm.cgmh.org.tw
†   These authors contributed equally to this work.

Received: 7 September 2020; Accepted: 28 September 2020; Published: 30 September 2020

Abstract: Status epilepticus may cause molecular and cellular events, leading to hippocampal neuronal cell death. Peroxisome proliferator-activated receptor γ coactivator 1-α (PGC-1α) is an important regulator of vascular endothelial growth factor (VEGF) and VEGF receptor 2 (VEGFR2), also known as fetal liver kinase receptor 1 (Flk-1). Resveratrol is an activator of PGC-1α. It has been suggested to provide neuroprotective effects in epilepsy, stroke, and neurodegenerative diseases. In the present study, we used microinjection of kainic acid into the left hippocampal CA3 region in Sprague Dawley rats to induce bilateral prolonged seizure activity. Upregulating the PGC-1α pathway will increase VEGF/VEGFR2 (Flk-1) signaling and further activate some survival signaling that includes the mitogen activated protein kinase kinase (MEK)/mitogen activated protein kinase (ERK) and phosphatidylinositol 3-kinase (PI3K)/protein kinase B (AKT) signaling pathways and offer neuroprotection as a consequence of apoptosis in the hippocampal neurons following status epilepticus. Otherwise, downregulation of PGC-1α by siRNA against *pgc-1α* will inhibit VEGF/VEGFR2 (Flk-1) signaling and suppress pro-survival PI3K/AKT and MEK/ERK pathways that are also accompanied by hippocampal CA3 neuronal cell apoptosis. These results may indicate that the PGC-1α induced VEGF/VEGFR2 pathway may trigger the neuronal survival signaling, and the PI3K/AKT and MEK/ERK signaling pathways. Thus, the axis of PGC-1α/VEGF/VEGFR2 (Flk-1) and the triggering of downstream PI3K/AKT and MEK/ERK signaling could be considered an endogenous neuroprotective effect against apoptosis in the hippocampus following status epilepticus.

Keywords: neuroprotection; PGC-1α; vascular endothelial growth factor; vascular endothelial growth factor receptor 2; PI3K/AKT; MEK/ERK; status epilepticus; hippocampus

---

1. Introduction

Epilepsy is one of the most common serious brain conditions characterized by the recurrence of unprovoked seizures, affecting more than 70 million people worldwide [1]. Status epilepticus is a common neurological and medical emergency, which is associated with high morbidity, mortality, and health-care burden [1]. Sustained seizure activities during status epilepticus usually result in significant neuronal damage in the cerebral cortex, particularly in the hippocampus [2,3]. Evidence has shown both in human and animal studies that prolonged seizures may lead to a large number of changes of molecular and cellular cascades, including axonal sprouting, gliosis, network reorganization, activation of neuroinflammation, acquired channelopathies, oxidative stress, mitochondrial dysfunctions, angiogenesis, neurogenesis, and activation of some late cell death pathways, which contribute to neurodegeneration and brain damage [4–7].

Vascular endothelial growth factors (VEGFs) and their receptors (VEGFRs) have important roles in the formation, function, and maintenance of blood vessels and are essential regulators of angiogenesis and vascular permeability [8]. Recent evidence have shown that VEGFs also have crucial roles in other organ systems, including the central nervous system (CNS), kidney, lung, and liver, where they directly influence organ function and development [9]. Among the VEGF family, VEGF-A is currently considered to play an important role in neuroprotective effects on hypoxic motor neurons and amyotrophic lateral sclerosis [8]. The physiological functions of VEGF are majorly mediated by the receptors of VEGF (VEGFRs). VEGF-A can bind to and activate two tyrosine kinase receptors, including VEGFR-1 and VEGFR-2 [8]. VEGFR-2 is also referred to as the fetal liver kinase receptor 1 (Flk-1) [9], which mediates most of the endothelial growth and survival signals [8]. In addition, VGEFR2 has been reported on glia and neurons where they might be upregulated during neuronal perturbations [10–13]. In the nervous system, the most important VEGF receptor pathway is VEGF receptor 2 (VEGFR2) [9–11]. Under perturbations of neuronal cells, availability of VEGFA to bind to VGEFR2 (Flk-1) will increase and promote VGEFR2 (Flk-1) downstream signaling [9,12,14]. Recent evidence suggests that VEGF has therapeutic potential as a neuroprotective factor in many neurological diseases [12,15], such as stroke [16–18], epilepsy [14,19], and neurodegenerative diseases [12,20–22].

Both in the animal studies following sustained seizure activities [14,19,23], and in resected tissues from patients with focal cortical dysplasia and refractory epilepsy [24], VEGF is up-regulated in neural cells and glia in the hippocampus and pyramidal neurons of the cortex. In animal models, overexpression of VEGF mRNA and protein and blood-brain barrier (BBB) impairment in the hippocampus occurred early after electroconvulsive shock-induced seizures or pilocarpine-induced seizures [23–25]. The neuroprotective role of VEGF in ischemic stroke has been well studied. In epileptic seizures, such as status epilepticus, the literature on the neuroprotective role of the VEGF-A/VEGFR2 pathway is limited. Peroxisome proliferator-activated receptor γ (PPARγ) coactivator 1-α (PGC1-α) belongs to a small family of transcriptional coactivators identified as a cofactor for the nuclear hormone receptor PPARγ that possesses a common function in mitochondrial physiology and mitochondrial biogenesis [5,26]. Additionally, it is also involved in other metabolic processes, such as redox homeostasis, uncoupled respiration, and gluconeogenesis [26–28]. Hypoxia-induced upregulation of VEGF mRNA is associated with increases in the transcription factor hypoxia inducible factor (HIF)1-α, which is also upregulated after cerebral ischemia [11,18].

The trigger for increased neuronal VEGF after seizures and brain insults is unclear. Beyond well-studied VEGF regulator HIF-1α [11,18], activation of PGC-1α induces the expression of VEGF signaling, leading to the formation of new blood vessels and protecting from cell damage, in an HIF-1α independent pathway [29–31]. Therefore, PGC-1α in the regulation of VEGF/VEGFR2 signaling pathway in the neuronal cells might be a crucial mechanism in neuroprotection following status epilepticus.

Recent research showed that the upregulation of VEGF/VEGFR2 (Flk-1) occurred after sustained seizures and VEGF signaling offered neuroprotective effects against neuronal cell death in the hippocampus following status epilepticus [14,19,23]. When upregulation of VEGF in neurons and glial cells occurs after persistent epileptic seizures, it counteracts seizure-induced neurodegeneration via overexpression of VEGFR2 (Flk-1) [14]. A previous study [14] suggested that increased VEGF signaling pathway via overexpression of VEGFR2 may affect seizure activity even without altering angiogenesis. Therefore, VEGFR2 could be considered a novel target for developing therapy strategies against epileptic activities [14]. However, whether the VEGF/VEGFR2 (Flk-1) signaling pathway contributes neurogenesis and is counteractive to epileptogenesis remains unclear [14]. Growing evidence suggests that under stressful stimuli, such as prolonged epileptic seizures, the VEGF/VEGFR2 (Flk-1) signaling pathway may participate in mediating angiogenesis, neuronal migration, hippocampal cell proliferation, and anti-apoptosis, which provide neuroprotective effects through the downstream activation of survival signaling, including phosphatidylinositol 3-kinase (PI3K)/protein kinase B (AKT) and the mitogen activated protein kinase kinase (MEK)/mitogen activated protein kinase (ERK) [13,14,32,33].

In the present study, we propose that PGC-1α may be activated during experimental status epilepticus and regulate the VEGF/VEGFR2 signaling pathway, and further protects against apoptotic neuronal cell death in the hippocampus following status epilepticus through survival signaling and PI3K/AKT and MEK/ERK-dependent pathways.

2. Results

2.1. Temporal Changes of PGC-1α Expression in the Hippocampal CA3 following Status Epilepticus

Temporal changes of PGC-1α expression in the right hippocampal CA3 region were examined following status epilepticus. After unilateral microinjection of kainic acid (KA; 0.5 nmol) into the left CA3 subfield, *pgc-1α* mRNA expression exhibited a significantly incremental change in the right hippocampal CA3 area 1 h after the elicitation of prolonged seizure activities, followed by a progressive decrement that returned to base line at 24 h (Figure 1a). For detection of the PGC-1α protein level, western blot analysis also presented a significant augment of PGC-1α protein in the right hippocampal CA3 subfield 1–24 h after the status epilepticus induced by KA (Figure 1b).

2.2. Temporal Changes of VEGF and VEGFR2 Expression in the Hippocampal CA3 Region following Status Epilepticus

The expression of VEGF and VEGFR2 following KA-induced experimental status epilepticus were examined. Real-time PCR results showed the expression of *vegf* mRNA extracted from the right hippocampal CA3 region was significantly augmented at 3 h, 6 h, and 24 h after administration of KA (Figure 2a). Correspondingly, western blotting analysis also showed a significant increase of VEGF protein levels at 3, 6, and 24 h after experimental status epilepticus that peaked at 6 h (Figure 2b). Additionally, in the parallel experiments, the expression of *vegfr2* mRNA also exhibited a significant increase from 3 to 24 h after KA-induced status epilepticus (Figure 2c). By western blotting analysis, the VEGFR2 protein levels were revealed to be significantly increased in the right hippocampal CA3 tissue 3–24 h after the induction of status epilepticus (Figure 2d).

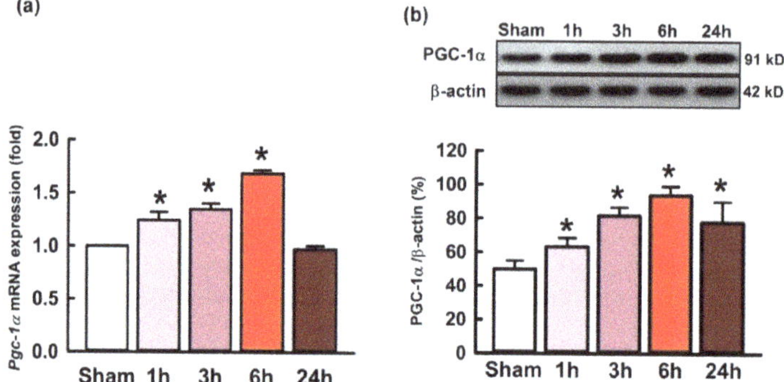

Figure 1. (**a**) Changes of the *pgc-1α* mRNA expression and (**b**) the protein levels of PGC-1α after kainic acid (KA; 0.5 nmol) was microinjected into the hippocampal CA3 subfield. Tissues were harvested from the right subfield CA3 of the hippocampus at 1, 3, 6, or 24 h after administration of KA or phosphate buffered saline (PBS) into the left hippocampal CA3. Values are presented as mean ± SEM of four animals per experimental group. * $p < 0.05$ versus the control group in the Scheffé multiple-range test.

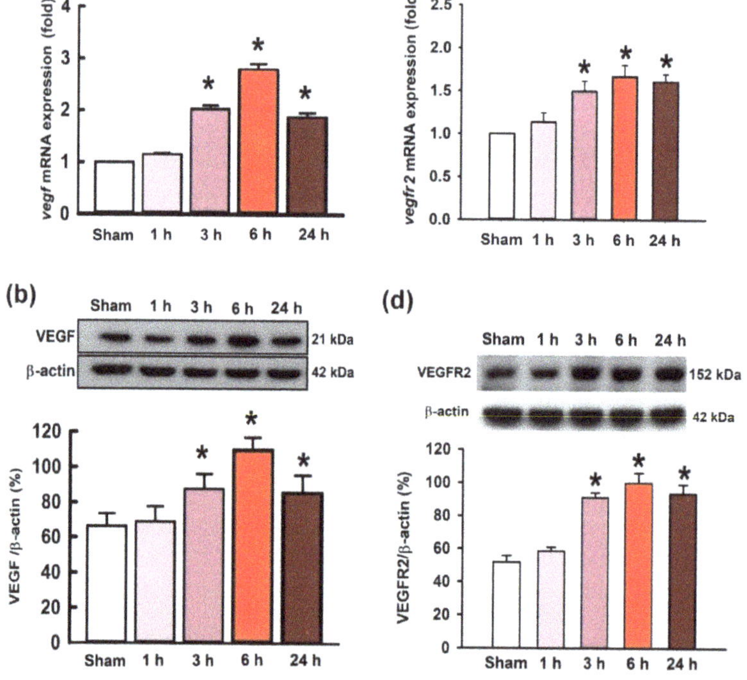

Figure 2. Upregulated changes of the expression of vascular endothelial growth factor (*vegf*) (**a**) and VEGF receptor 2 (*vegfr2*) (**c**) mRNA levels, and VEGF (**b**) and VEGF receptor 2 (VEGFR2) protein levels (**d**) after the administration of KA (0.5 nmol) into the hippocampal CA3 region. Samples were collected from the right hippocampal CA3 subfield at 1, 3, 6, or 24 h after the administration of KA or PBS into the left hippocampal CA3 subfield. Values are presented as mean ± SEM of quadruplicate analyses from four animals per experimental group. * $p < 0.05$ versus the control group in the Scheffé multiple-range test.

2.3. Effect of Resveratrol and Gene Knock-Down by Small Interfering RNA (siRNA) Against pgc-1α on VEGF Expression in the Hippocampus following Experimental Status Epilepticus

In a previous report [26], we demonstrated that resveratrol is a PGC-1α activator, and it can activate PGC-1α and the following signaling pathways and then promote mitochondrial biogenesis. To determine the causality of PGC-1α in modulation of VEGF on this experimental paradigm, we tested the effects of resveratrol, as a positive control, and siRNA against *pgc-1α*, as a negative control, on VEGF expression in the hippocampus following status epilepticus. Microinjection of resveratrol (100 μmol) into the hippocampal CA3 region significantly increased the expression of *vegf* mRNA (Figure 3a) and the level of VEGF protein (Figure 3b) in the CA3 region 6 h after the elicitation of sustained hippocampal seizure activities. On the other hand, the specificity of the gene knock-down strategy by siRNA against *pgc-1α* was tested for VEGF expression in the hippocampus. For confirming the specific effect of *pgc-1α* siRNA on VEGF expression, after a pre-treated microinjection with siRNA against *pgc-1α* (2 μg) into bilateral hippocampal CA3 region, significant decrease of the *vegf* mRNA level was shown in our real-time PCR data (Figure 3c), and western blot analysis also confirmed a drastically decline in the VEGF protein level in the hippocampal CA3 area 6 h after KA-induced status epilepticus (Figure 3d) compared with the pre-treatment of the control siRNA group.

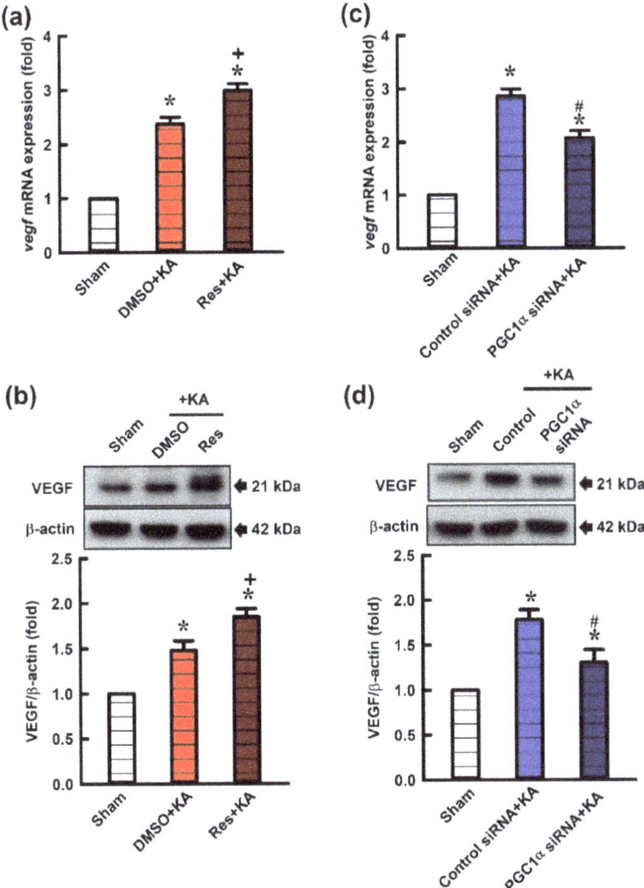

Figure 3. (**a**) Changes of *vegf* mRNA expression and (**b**) VEGF protein levels relative to β-actin from the CA3 of the hippocampus 6 h after microinjection of 0.5 nmol kainic acid (KA) or pretreatment with

microinjection of resveratrol (Res; 100 μmol) or 3% dimethyl sulfoxide (DMSO) into the hippocampal CA3 subfield. (**c**) Changes of *vegf* mRNA expression in the CA3 of the hippocampus 6 h after the microinjection of KA (0.5 nmol) into the left hippocampal CA3 subfield, and 24 h before pre-treatment with application into the bilateral CA3 subfield of control siRNA or siRNA for *pgc-1α* (2 μg). (**d**) Changes in VEGF relative to β-actin from the CA3 subfield of the hippocampus 6 h after the same experiments. Values are mean ± SEM of quadruplicate analyses from 6–8 animals per experimental group. * $p < 0.05$ versus the control group, + $p < 0.05$ versus DMSO+KA group, and # $p < 0.05$ versus control siRNA+KA group in the Scheffé multiple-range test.

To confirm that the changes in expression of VEGF by resveratrol treatment and *pgc-1α* gene knock-down observed in our above biochemical analyses, we applied double immunofluorescence staining to detect the intracellular expression of VEGF in hippocampal CA3 neurons (Figure 4). Within the unified fields of a laser scanning confocal microscope, signals for VEGF were weak in the neurons, which were strongly immunoreactive to the neuron-specific nuclear protein (NeuN) neuronal marker, of hippocampal CA3 in the control animals (Figure 4a–c). In contrast to the control, there was an increase in VEGF immunoreactivity in the neurons from the same field, the CA3 area, 6 h after KA-induced status epilepticus (Figure 4d–f). Pretreatment of resveratrol (100 μmol) augmented the PGC-1α immunoreactivity in the hippocampal CA3 neurons (Figure 4j–l).

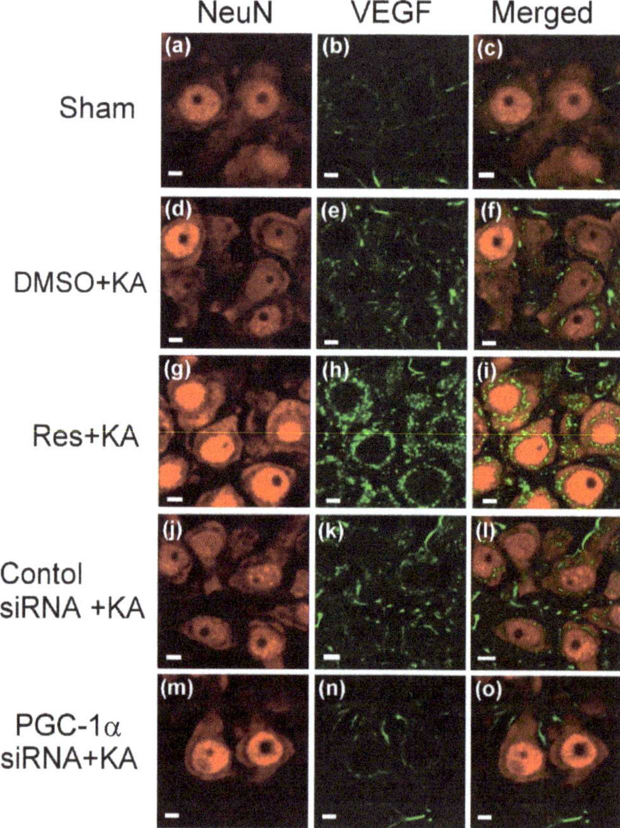

Figure 4. Confocal microscopic images of the right CA3b subregion of the hippocampus showing cells that were immunoreactive to neuron-specific nuclear protein (NeuN) (red fluorescence) or additionally

stained for VEGF (green fluorescence) in control rats (Figure 4a–c), or 6 h after administration of kainic acid (KA; 0.5 nmol) (Figure 4d–f) or pre-treatment with resveratrol (Res; 100 µmol) 30 min before KA-induced status epilepticus (Figure 4g–i). Induction of status epilepticus by KA in animals that received pre-treatment with application into the bilateral CA3 subfield 24 h after animals received control siRNA (Figure 4j–l) or siRNA against *pgc-1α* (2 µg) (Figure 4m–o). Scale bar, 10 µm.

In addition, in the parallel experiments, many VEGF-positive neurons were detected in the hippocampal CA3 area 6 h after experimental status epilepticus in animals pre-treated with pgc-1α control siRNA. However, reduced VEGF-positive cells were observed in the hippocampal CA3 subfield 6 h after status epilepticus in rats with pre-treated microinjection of control siRNA (Figure 4j–l) or siRNA against pgc-1α (2 µg) (Figure 4m–o) into the bilateral hippocampal CA3 field.

2.4. Effect of Resveratrol and Gene Knock-Down by Small Interfering RNA (siRNA) Against pgc-1α on Expression of VEGF Receptor 2 (VEGFR2) in the Hippocampus Following Experimental Status Epilepticus

To determine the causal effect of PGC-1α in regulation of the expression of VEGFR2 in this experimental paradigm, we tested resveratrol, a PGC-1α activator, and siRNA against *pgc-1α* on VEGF expression on VEGFR2 expression in the hippocampus following status epilepticus. Microinjection of resveratrol (100 µmol) into the hippocampal CA3 subfield underwent a significant increment of expression of *vegf* mRNA (Figure 5a) and the level of VEGFR2 protein (Figure 5b) in the CA3 region 6 h after the elicitation of sustained hippocampal seizure activities. Confirming the specificity of *pgc-1α* siRNA on the protein levels of VEGFR2, after a pre-treated microinjection of *pgc-1α* siRNA (2 µg) into the bilateral CA3 region, a significant elevation of the expression of *vegfr2* mRNA (Figure 5c) and diminishment of VEGF protein in the CA3 subfield 6 h after KA-induced status epilepticus (Figure 5d) compared with the rats with pre-treatment of control siRNA was observed.

2.5. Activation of VEGF/VEGFR2 Regulates the PI3K/Akt Survival Signaling Pathway in the Hippocampal Neurons Following Experimental Status Epilepticus

The activation of VEGF/VEGFR2 can subsequently activate PI3K/AKT signaling by altering the activation state of numerous downstream proteins relevant to cell proliferation and inhibiting apoptosis [34,35]. To determinate whether the PGC-1α regulates the survival signaling pathway, PI3K/Akt, through VEGF/VEGFR2 activation in the hippocampus following KA-induced status epilepticus, we delineated the regulatory effects of the PGC-1α activator and siRNA against *pgc-1α* on PI3K/Akt signaling. In the experiment, the phosphorylation state of PI3K (p-PI3K) and AKT (p-AKT), and total levels of PIK3 and AKT were measured by western blotting analysis. The ratio of p-PI3K/total PI3K and p-AKT/total AKT were raised 6 h after microinjection of 3% dimethyl sulfoxide (DMSO) and KA into the left hippocampal CA3 region compared with controls (Figure 6a,b).

In addition, the ratio of p-PI3K/total PI3K and p-AKT/total AKT were significantly enhanced 6 h after microinjection of resveratrol (100 µmol) followed by KA administration into the left hippocampal CA3 region (Figure 6a,b). On the other hand, the ratio of p-PI3K/total PI3K and p-AKT/total AKT were raised 6 h after administration of control siRNA with KA into left hippocampal CA3 region compared with the control group (Figure 6c,d). However, the ratio of p-PI3K/total PI3K and p-AKT/total AKT were inhibited in the hippocampal CA3 neuronal cells 24 h before pre-treated microinjection of siRNA against *pgc-1α* (2 µg) followed with KA microinjection into the hippocampus. Akt can be activated by p-PI3K and active PI3K (Figure 6c,d).

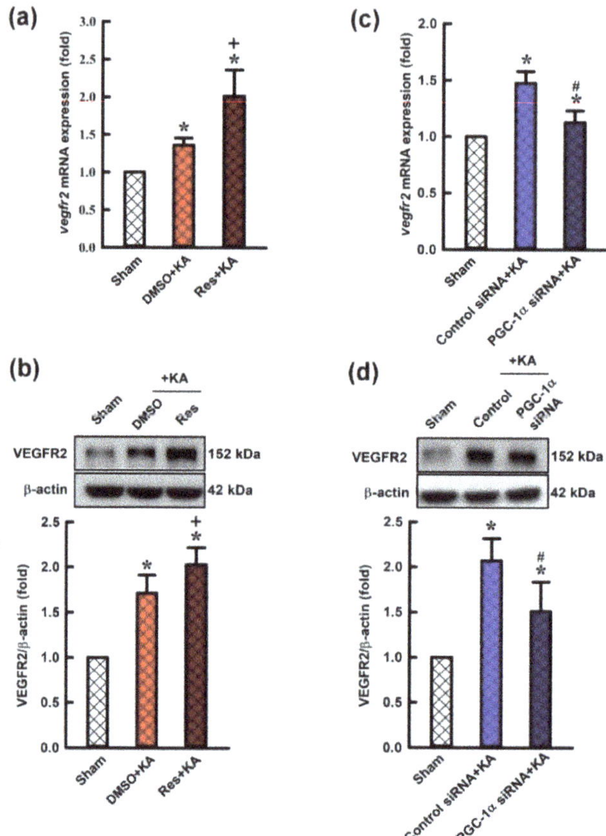

Figure 5. (**a**) Changes of *vegfr2* mRNA expression and (**b**) changes of VEGFR2 protein in the hippocampal CA3 tissues 6 h after administration with kainic acid (KA; 0.5 nmol) with pre-treatment with resveratrol (Res; 100 μmol), or 3% DMSO into bilateral CA3 subfield. (**c**) The *vegfr2* mRNA expression in the hippocampus 6 h after the administration of KA into the left CA3 area 24 h after pre-treatment of control siRNA or *pgc-1α* siRNA (2 μg) into the bilateral CA3 subfield. (**d**) Changes in VEGFR2 protein expression in the CA3 tissue 6 h after the same treatments. Values are presented as mean ± SEM from four animals per experimental group. * $p < 0.05$ versus control groups, + $p < 0.05$ versus DMSO+KA group, and # $p < 0.05$ versus control siRNA+KA group in the Scheffé multiple-range test.

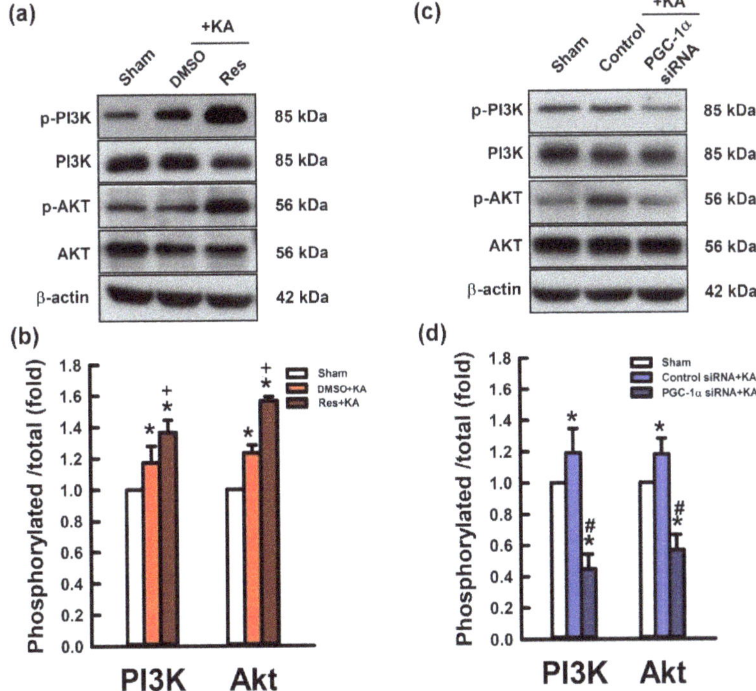

Figure 6. (a) Changes of phosphorylated phosphatidylinositol 3-kinase (p-PI3K), total PI3K, phosphorylated AKT (p-AKT), and total AKT expression and (b) p-PI3K protein levels relative to total PI3K and p-AKT relative to total AKT in the CA3 region of the hippocampus 6 h after microinjection of kainic acid (KA; 0.5 nmol) with pretreatment of microinjection of 3% DMSO or resveratrol (Res; 100 μmol) into the hippocampal CA3 subfield. (c) Ratio of activated phosphorylation of PI3K and AKT levels in the CA3 subfield of the hippocampus 6 h after the application of KA (0.5 nmol) into the left side of the hippocampal CA3 region 24 h before treatment with control siRNA or siRNA for *pgc-1α* (2 μg) application into the bilateral CA3 subfield. (d) Phosphorylation levels of PI3K and AKT in the CA3 region of the hippocampus after 6 h of the same experiments. Values presented as mean ± SEM from six animals per experimental group. * *p* < 0.05 versus the control group, + *p* < 0.05 versus DMSO+KA group, and # *p* < 0.05 versus control siRNA+KA group in the Scheffé multiple-range test.

2.6. *Activation VEGF/VEGFR2 Regulates the MEK/ERK Signaling Pathway in the Hippocampal Neurons Following Experimental Status Epilepticus*

VEGF/VEGFR2 activates the MEK/ERK signaling pathway as evidenced by an increase in ERK and MEK phosphorylation levels [36,37]. To determine the regulatory role of VEGE/VEGFR2 in the MEK/ERK pathway, we examined the protein levels of phosphorylated ERK (p-ERK) and MEK (p-MEK), and total ERK and MEK. Rats received pre-treatment with resveratrol or siRNA against *pgc-1α* and these proteins were analyzed by western blotting 6 h after KA-induced experimental status epilepticus. The ratio of p-ERK/total ERK and p-MEK/total MEK were increased 6h after microinjection of DMSO and KA into the left hippocampal CA3 region compared with control animals (Figure 7a,b). Additionally, the ratios of p ERK/total ERK and p-MEK/total MEK were significantly increased 6 h after microinjection of resveratrol (100 μmol) followed by KA administration into the left hippocampal CA3 (Figure 7a,b). Otherwise, the ratio of p-PI3K/total PI3K and p-AKT/total AKT were increased 6 h after administration of control siRNA with KA into the left hippocampal CA3 region compared with control rats (Figure 7c,d). However, the ratio of p-PI3K/total PI3K and p-AKT/total AKT were reduced in the hippocampal CA3 area 24 h before pre-treated microinjection of siRNA against *pgc-1α* (2 μg) followed by KA microinjection into the hippocampus (Figure 7c,d).

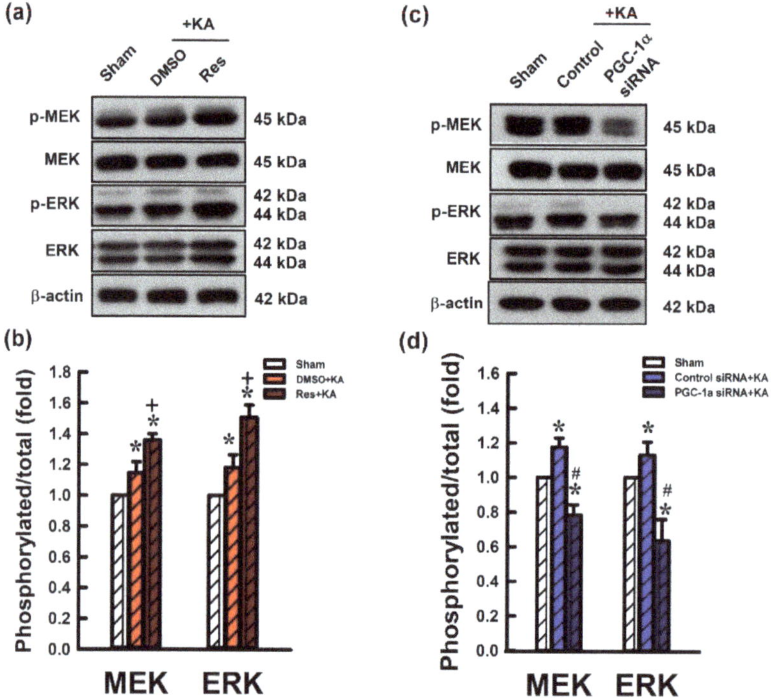

Figure 7. (**a**) Changes of phosphorylated mitogen activated protein kinase (MEK) (p-MEK), total MEK, phosphorylated MEK (p-ERK), and total ERK expression, and (**b**) ratio of phosphorylation level for MEK protein levels and total MEK and phosphorylation of ERK in the CA3 tissue of the hippocampus 6 h after application of kainic acid (KA; 0.5 nmol) with pre-treated microinjection of 3% DMSO or resveratrol (Res; 100 μmol) into the hippocampal CA3 subfield. (**c**) Changes of p-MEK, total MEK, p-ERK, and total ERK expression in the hippocampus CA3 region 6 h after the microinjection of KA (0.5 nmol) into the left CA3 region with pre-treatment of application into the bilateral CA3 subfield with control siRNA or siRNA for *pgc-1α* (2 μg) 24 h in advance. (**d**) Ratio of phosphorylation level for MEK protein levels and total MEK and phosphorylation of ERK in the CA3 region of the hippocampus 6 h after the same experiments. Values are mean ± SEM of quadruplicate analyses from six animals per experimental group. * $p < 0.05$ versus control group, + $p < 0.05$ versus DMSO+KA group, and # $p < 0.05$ versus control siRNA+KA group in the Scheffé multiple-range test.

2.7. Effect of VEGF/VEGFR2 on Apoptosis and Neuronal Survival in the Hippocampal CA3 Subfield Following Experimental Status Epilepticus

To further confirm the role of PGC-1α/VEGF/VEGFR2 signaling in neuroprotection from hippocampal damage due to prolonged seizure, we investigated the regulatory role of resveratrol on KA-induced hippocampal neuronal cell death. Significantly, pretreatment with resveratrol (100 μmol) attenuated the extent of cleaved caspases-3 expression in the hippocampal CA3 subfield 7 days after KA-induced status epilepticus (Figure 8a). Otherwise, the level of cleaved caspases-3 was significantly increased in the hippocampal CA3 area 24 h before pre-treated microinjection of siRNA against *pgc-1α* (2 μg) followed by KA microinjection into the hippocampus (Figure 8b). Mitigation of neuronal damage by resveratrol (100 μmol) treatment was demonstrated in the hippocampus in both DNA fragmentation qualitative (Figure 9a) and quantitative (Figure 9b) analyses, indexes for cell death, after inducing status epilepticus for 7 days. On the other hand, the level of qualitative (Figure 9c) and quantitative (Figure 9d) analysis of DNA fragmentation were aggravated in the hippocampal CA3 area 24 h before pre-treated microinjection of siRNA against *pgc-1α* (2 µg), followed by KA microinjection into the hippocampus.

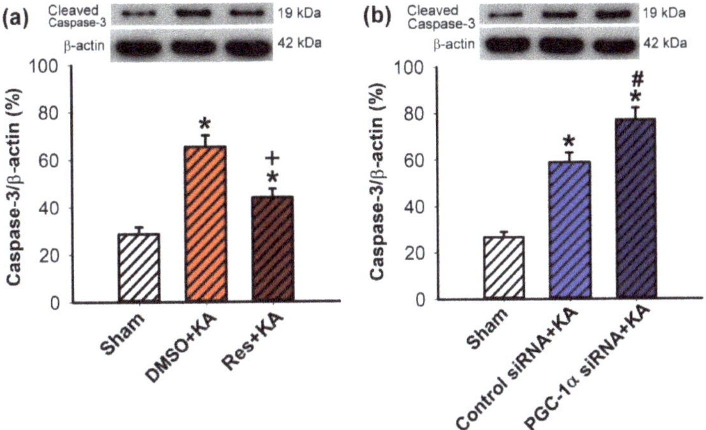

Figure 8. (**a**) Changes of cleaved caspase-3 levels 7 days after microinjection of 0.5 nmol of kainic acid (KA) with 24-h pretreatment with resveratrol (Res; 100 µmol) or 3% DMSO into the hippocampal CA3 subfield. (**b**) Increase of cleaved caspase-3 (19 kDa) 7 days after rats were microinjected in the bilateral CA3 subfields with control siRNA or siRNA for *pgc-1α* (2 µg) 24 h before the microinjection of KA into the left hippocampal CA3. Values are mean ± SEM of quadruplicate analyses from four animals per experimental group. * $p < 0.05$ versus control group, + $p < 0.05$ versus DMSO+KA group, and # $p < 0.05$ versus control siRNA+KA group in the Scheffé multiple-range test.

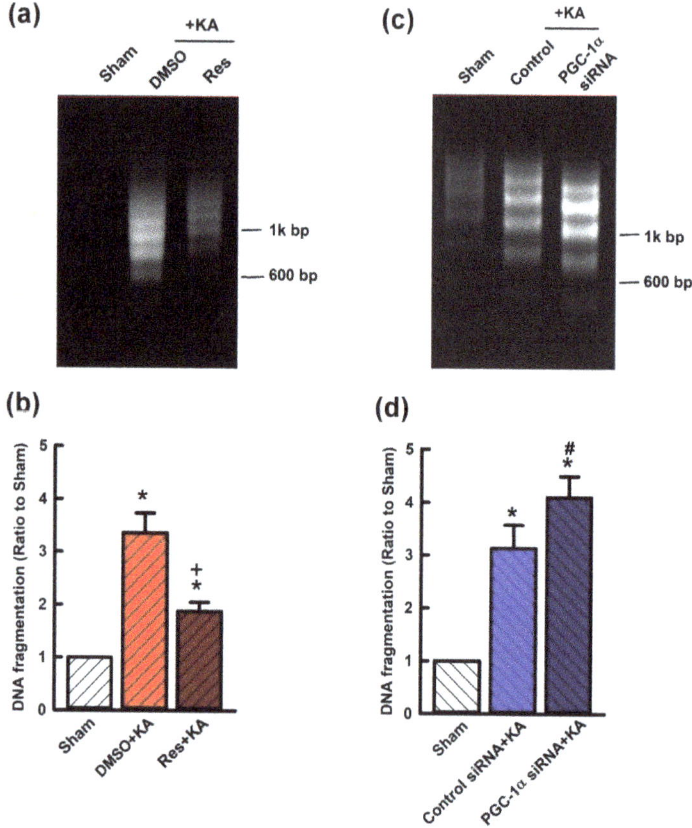

Figure 9. Qualitative (**a**),(**c**) or quantitative (**b**),(**d**) analysis of DNA fragmentation detected in samples harvested from the hippocampus 7 days after status epilepticus with or without Res and *pgc-1α* siRNA. Values are mean ± SEM from six animals per experimental group. (**a**),(**b**) Seven days after microinjection of 0.5 nmol kainic acid (KA) with 24-h pretreatment of microinjection of resveratrol (Res; 100 μmol) into the hippocampal CA3 subfield decreased fragmentation of DNA. Changes of qualitative (**c**) or quantitative (**d**) DNA fragmentation 7 days after microinjected control siRNA or siRNA for *pgc-1α* (2 μg) into bilateral CA3 subfields with 24 h before the microinjection of KA into the left hippocampal CA3. Values are mean ± SEM of quadruplicate analyses from four animals per experimental group. * $p < 0.05$ versus control group, + $p < 0.05$ versus DMSO+KA group, and # $p < 0.05$ versus control siRNA+KA group in the Scheffé multiple-range test.

3. Discussion

The concept of neuroprotection under brain insults, such as prolonged seizures (status epilepticus), should consist of not only the conservation of structural neuronal cell death, but also preservation of neuronal networks and functions [38,39]. In research on prevention of the deleterious effects of status epilepticus, the primary focus is on the development of chronic epilepsy and cognitive decline [38,39]. Thus, a more precise definition of neuroprotection in status epilepticus needs to include protection not just against neuronal death but also against neuronal and network dysfunction at the cellular and molecular levels [39,40]. However, the studies of neuroprotective mechanisms elicited by recurrent seizure activities, especially under status epilepticus, are limited. Considering the detrimental reaction in brain cells under status epilepticus, acute response proteins to counteract these detrimental effects elicited by sustained seizures can be a mechanism of endogenous neuroprotection against seizure-induced neuronal cell death [39,41,42]. Following status epilepticus, the mechanism of endogenous neuronal survival signals is evolutionarily conserved and may actuate extensive signaling pathways to offer the neuroprotective effect which therefore may be strong candidates for therapeutic strategies [39,42]. In our previous studies, both in human epilepsy [43] and animal studies following status epilepticus [5,26,41,44–46], several endogenous neuroprotective mechanisms to lessen neuronal damage were proposed, including PPARγ [41], mitochondrial uncoupling protein 2 (UCP2) [41], heat shock protein 70 [43,44], brain-derived neurotrophic factor [45], mitochondrial dynamin-related protein 1 [46], PGC1-α [5,26], and sirtuin 1 [5,26].

The present study demonstrated that activation of PGC-1α activity regulated VEGF/VEGFR2 (Flk-1) signaling and showed the neuroprotective effect against neuronal cell death in the hippocampus following experimental status epilepticus. Particularly, we noted that PGC-1α expression was enhanced in the rat hippocampus following status epilepticus and upregulated VEGF and VEGFR2 (Flk-1) expression. Administration of the activator of PGC-1α, resveratrol, was accompanied by increased VEGF and VEGFR2 expression, and promotion of PI3K/AKT and MEK/ERK pathways and decreased neuronal cell death in the hippocampus following status epilepticus. Downregulation of PGC-1α by pretreatment of siRNA against *pgc-1α* reduced VEGF and VEGFR2 expression and inhibited PI3K/AKT and MEK/ERK pathways, also accompanied by heightened caspase-3 activity, and augmented neuronal damage in the hippocampal CA3 subfield. We also noted that the mRNA levels for *vegf* and *vegfr2* were reduced after pretreatment of siRNA against *pgc-1α* compared with the levels of control siRNA groups. This suggests an additional mechanism for seizure-induced activation following status epilepticus. Therefore, our results indicated that seizure activities triggered the upregulation of the VEGF/VEGFR2 (Flk-1) pathway, activation of PI3K/AKT and MEK/ERK signaling, and worked against the hippocampal neuronal apoptotic cell death following status epilepticus. Therefore, our results showed VEGF/VEGFR2 and related downstream PI3K/AKT and MEK/ERK signaling can exert endogenous neuroprotection in status epilepticus.

Recently, our research revealed that PGC-1α is an important transcriptional regulator that acts through regulating the expression of reactive oxygen species (ROS), mitochondrial UCP2, superoxide dismutase 2, and mitochondrial biogenesis, which plays a beneficial part in neuroprotection in the CNS following global ischemia and status epilepticus [5,26,28,41,47]. In addition, recent studies have demonstrated that the PGC-1α signaling pathway exerts potential neuroprotective properties in many neurological diseases [26,48,49], such as Alzheimer's disease [50], Parkinson's disease [51], acute stroke [28,52], epilepsy [53], and status epilepticus [26,53–55]. Whether activation of PGC-1α following sustained seizures promotes the endogenous activation of VEGF/VEGFR2 signaling is unclear. However, recent evidence showed activation of PGC-1α induces the expression of VEGF/VEGFR2 signaling, leading to protection from apoptotic cell damage in the neuronal cells and providing neuroprotective effects [15,29–31]. Thus, the activation of PGC-1α may regulate the VEGF/VEGFR2 signaling pathway in the neuronal cells and might be a crucial mechanism in neuroprotection following status epilepticus.

VEGFs are strong endothelial cell mitogens and major regulators of neurogenesis and angiogenesis [8,56]. In addition to the well-established effects of VEGF, recent research has demonstrated pivotal roles for VEGF/VEGFR2 in a broad range of neurotrophic and neuroprotective effects in the CNS that may relate to neurogenesis and angiogenesis [11–13]. The neurotrophic and neuroprotective effects of VEGF are predominantly mediated by VEGFR2, also called fetal Flk-1 [14,57–59], or kinase insert-domain containing receptor (KDR) [60]. VEGF/VEGFR2 (Flk-1) signaling may have neuroprotective effects in many neurological diseases, such as hemorrhagic or ischemic stroke [16–18], traumatic brain injury [61], amyotrophic lateral sclerosis [20], Huntington's disease [62], Alzheimer's disease [12,21], and Parkinson's disease [12,22]. Recent evidence showed that during status epilepticus, VEGF is upregulated and it protects against seizure-induced neuronal cell death in the hippocampus [14,19,23]. In vitro, VEGF administration suppresses ictal and interictal epileptiform activity via the VEGFR2 (Flk-1) receptor. Thus, upregulated VEGF signaling through

VEGFR2 (Flk-1) overexpression may regulate epileptogenesis and ictogenesis in mice and counteract the focal epileptic seizure [14].

VEGF can directly stimulate the proliferation of neuronal progenitors; however, the possible mechanism of seizure-induced activation of endogenous PGC-1α that promotes the following VEGF signaling that contributes to neuroprotection in the hippocampus following status epilepticus is not clear. Thus, we investigated two downstream pathways of the VEGF/VEGFR2 (Flk-1) signaling pathways that included PI3K/AKT and MEK/ERK signaling [13,63–65]. Activation of VEGF has been suggested to regulate PI3K/AKT and MEK/ERK cascades. Furthermore, VEGF is known to be involved in the trophic and neuroprotective effects of growth factors via binding to its tyrosine kinase receptors, particularly of which VEGFR2 (Flk-1) is proposed to mediate most of the neuron-specific effects of VEGF via mediating PI3K/AKT and MEK/ERK signaling [10–13,33].

Mitogen-activated protein kinases (MAPK) are considered to be a family of Ser/Thr protein kinases and conserved in eukaryotes and involved in many cellular programs. The MEK/ERK pathway is one of the MAPK cascades involved in transducing cell survival signals via growth factor receptors [66,67]. Interestingly, growing evidence revealed that improving mitochondrial function with activation of PI3K/AKT and MEK/ERK pathways might tend to inhibit phosphorylation of c-Jun-NH2-terminal kinase (JNKs) and p38, the second and third major signaling cassettes in the MAPK pathway, which majorly respond to inflammatory and cellular stress to promote inflammation and cell death, in neuronal cells under many neurological conditions, both in vitro and in vivo studies [66–68]. Overall, MEK/ERK pathways have been reported to play a crucial role in protecting neuronal cells from death under hypoxia, global ischemia, epilepsy, status epilepticus, and Parkinson's disease [33,63,68,69]. Therefore, following status epilepticus, endogenous activation of PGC-1α may regulate the VEGF/VEGFR2 (Flk-1) signaling pathway that triggers PI3K/AKT and MEK/ERK cascades and also promotes the cAMP-CREB signaling axis [45,69], which regulates the anti-apoptotic B-cell lymphoma-2 (Bcl-2) family, the expression of autophagy [63] and lysosomal genes, and further contributes to neuron survival mechanisms in hippocampal neuronal cells [70–72].

Several studies showed that the PI3K/AKT signaling pathway can modulate cellular activities, such as neuronal proliferation, differentiation, cell survival, and synaptic plasticity, and this signaling cascade activation promotes an important cytoprotective mechanism that promotes neural survival associated with neurodegenerative disorders and epileptic seizures [13,73,74]. Binding of VEGF may further trigger VEGFR2 (Flk-1) to activate PI3K and then phosphorylate phosphatidylinositol 4,5-bisphosphate or PtdIns $(4,5)P_2$ (PIP$_2$) to the second messenger, phosphatidylinositol (3,4,5)-trisphosphate (PtdIns(3,4,5)P_3 (PIP$_3$), on the plasma membrane. PIP$_3$ directly or indirectly binds to AKT to induce structural changes and facilitate the phosphorylation of AKT amino acid residues under neurodegenerative diseases [13,73,74]. Therefore, under different acute or chronic neurological conditions, activated AKT can promote neuronal cell survival in several ways, including subsequently regulating numerous downstream molecules, such as cyclin D, glycogen synthase kinase-3β, and mechanistic targets of rapamycin complex 1, thereby regulating cell functions and improving cell survival advantage [75]. Moreover, phosphorylated AKT regulates cell survival and anti-apoptosis by targeting the pro- or anti-apoptotic mediators, such as Bcl-2, Bax, and caspases [76–78].

In addition to effects of the PI3K/AKT signaling pathway on cellular proliferation and survival, the activation of the MEK/ERK pathway encourages cell survival and prevents cell apoptosis [63,65]. Therefore, the role in neuroprotection of the MEK/ERK pathway is emphasized in many acute or chronic neurological diseases [13,63,79]. Growing evidence has shown that the MEK/ERK pathway plays a central role as a potential therapeutic target in neurological disorder diseases including epilepsy and status epilepticus [13,80,81]. The MEK/ERK pathway is also a vital message transmission pathway in embryonic and adult neurogenesis and plays a role in the regulation of brain physiological function [80]. When the cell suffers external stress, such as prolonged seizures, VEGF may stimulate cellular responses by binding to a special receptor, VEGFR2 (Flk-1), and this activates receptors in the form of the dimer which start signaling output and then triggers the activation of a series of signal transduction molecules, such as MEK/ERK signaling.

Evidence has shown that mitochondrial related autophagic pathways, such as the Jun N-terminal kinases-associated B-cell lymphoma-2 (JNK/Bcl-2) pathway, MEK/ERK signaling, and the sirtuin 1/forkhead box protein O1 pathway, may provide neuroprotective effects from neurotoxicity against dopamine neurons [82]. Activation of the MEK/ERK pathway by phosphorylation of threonine and tyrosine residues can quickly initiate ERK1/2-dependent biological processes in neurological diseases, including epilepsy and status epilepticus [39]. Recently, we noted exogenous resveratrol may provide neuroprotective effects through upregulation of PGC-1α, modulation of mitochondrial dynamics, and MEK/ERK regulated autophagy in rotenone-induced oxidative stress SH-SY5Y cell lines [63]. These results suggest activation PGC-1α regulated the VEGF/VEGFR2 pathway and further promoted MEK/ERK signaling which provided neuroprotective effects against epilepsy, status epilepticus, and neurodegenerative diseases.

4. Materials and Methods

4.1. Animals

The procedures of experimental status epilepticus in animals were carried out in compliance with the guidelines of Institutional Animal Care and Use Committee (the identification code: 2012122105; approval date was 01 August 2013) and were certificated by the experimental animal ethics committee at Chang Gung Memorial Hospital, Kaohsiung, Taiwan. A sample of 168 pathogen-free adult male Sprague Dawley rats (weight ranged from 280–350 g) were obtained from BioLASCO Taiwan Co. Ltd. (Taipei, Taiwan) and enrolled in the study. The animals were housed in the Center for Laboratory Animals in an environmentally controlled room (12 h/12 h light/dark cycle; 24 ± 1 °C). Tap water and rat chow in the laboratory were available ad libitum. All efforts were made to reduce the number of animals used and to minimize animal suffering during the experiment.

4.2. Experimental Status Epilepticus

The animal model of experimental status epilepticus that we used was well established previously [5,26,44]. Briefly, a stereotaxic headholder (Kopf, Tujunga, CA, USA) was used to fix the head of animals after they inhaled 3% of isoflurane for anesthesia, and the body temperature of animals was maintained at 37 °C by heating pads. KA (0.5 nmol; Tocris Cookson, Bristol, UK) dissolved in PBS (0.1 M, pH 7.4) was stereotaxically microinjected into the CA3 left hippocampal subfield (3.3–3.8 mm below the cortical surface, 2.4–2.7 mm from the midline, and 3.2–3.5 mm posterior to bregma) [5,26,44]. Microinjection of KA into the left hippocampal CA3 region caused progressive and accompanying augmented seizure-like hippocampal electroencephalographic (hEEG) activity that was routinely detected from the right hippocampal CA3 [83–85]. We therefore observed the seizure activity (right side hEEG) for 60 min, and then the seizure activities were terminated by diazepam (30 mg/kg, i.p.) [5,26]. To avoid post-surgical infection, sodium penicillin (10,000 IU; YF Chemical Corp., New Taipei City, Taiwan) was given intramuscularly. The animals were returned to the recovery room in individual cages. Animals with anesthesia and surgical preparations without additional experimental manipulations served as controls. The experimental schemes as the following Figure 10.

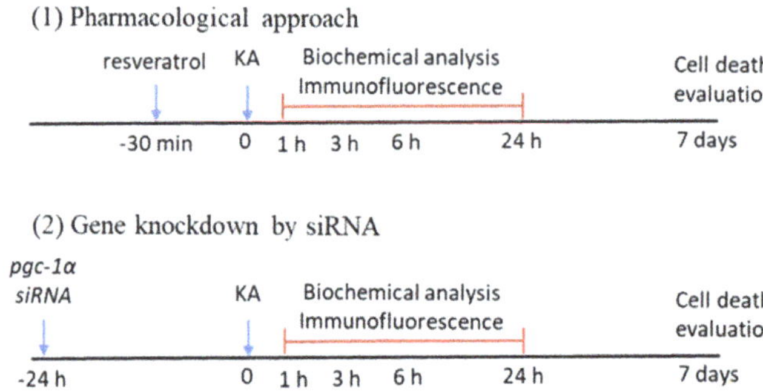

Figure 10. Two experimental schemes which included pharmacological pretreatment of resveratrol and gene knockdown by siRNA against *pgc-1α* following status epilepticus.

4.3. Pharmacological Pretreatments

The test agent included resveratrol (R5010, Sigma-Aldrich, St. Louis, MO, USA) [26,86], a PGC-1α activator, that was microinjected bilaterally and sequentially into the bilateral hippocampal CA3 areas [26]. The dose of resveratrol used was 100 µmol dissolved in 3% dimethyl sulfoxide (DMSO) (D5879, Sigma-Aldrich) as in our previous reports [26,41], at a volume of 150 nL on each side. Rats receiving microinjections of DMSO (3%) were used for the volume and vehicle controls. Each rat received only a single pharmacological pretreatment to prevent the confounding effects of drug–drug interactions 30 min before administration of KA (0.5 nmol) or PBS into the left hippocampal CA3 subfield.

4.4. Gene Knockdown by Microinjection of Small Interfering RNA (siRNA) Against pgc-1α into the Hippocampus

Gene knockdown was conducted with small interfering RNA (siRNA) against the *pgc-1α* gene. Multiple siRNA sequences were synthesized form by GE Healthcare (GE Healthcare, Chicago, IL, USA). All siRNAs were dissolved in an isotonic RNAi buffer (100 mM potassium acetate; 30 mM Hepes-KOH; 2 mM magnesium acetate; 26 mM NaCl, pH 7.4, at 37 °C). The siRNA against *pgc-1α* as following blow:

pgc-1α	Sequence
Sequence 1	5′-CGGUGGAUGAAGACGGAUU-3′
Sequence 2	5′-CAAUGAAUGCAGCGGUCUU-3′
Sequence 3	5′-GAACAAGACUAUUGAGCGA -3′
Sequence 4	5′-AUUCAAACUCAGACGAUUU-3′

Pooled siRNAs were mixed with Lipofectin reagent (1:1, 18292-011, Invitrogen, Carlsbad, CA, US) and microinjected bilaterally into the CA3 region of the hippocampus 24 h before the microinjected administration of KA. The pooled non-targeting siRNAs containing four scramble sequences were used as control siRNA.

4.5. Collection of Tissue Samples from the Hippocampus

At time-intervals (1, 3, 6, or 24; or 7 days) after microinjection of KA or PBS into the hippocampus, animals were anesthetized with 3% isoflurane and were perfused intracardially [83,85]. The brain was immediately removed and placed on gauze moistened with 0.9% ice-cold saline. To avoid the confounding effect of KA toxicity, we routinely collected samples from the CA3 region of the right hippocampus (this side is the hEEG recording side, with no KA injection). Our method permitted us to verify that our results from the analyzed samples were directly from sustained seizures (status epilepticus) and not indirectly due to KA toxicity [5,26,44]. Hippocampal samples were stored at −80 °C until use in biochemical analyses.

4.6. RNA Isolation and Reverse Transcription Real-Time Polymerase Chain Reaction

To quantitatively analyze the expression of *pgc-1α*, *vegf*, and *vegfr2 (Flk-1)* mRNA in the tissue of the hippocampal CA3, the brain was rapidly removed after PBS perfusion and total RNA from the hippocampus was extracted with a RNeasy Mini Kit (Qiagen, Duesseldorf, Germany) according to the manufacturer's protocol as in our previous reports [5,26,44]. Reverse transcriptase (RT) reaction was applied with the ImProm II^{TM} Reverse Transcription System (Promega, Madison, WI, USA) to acquire first-strand cDNA synthesis. Real-time polymerase chain reaction (PCR) for amplification of cDNA was performed using a LightCycler® 480 SYBR Green I Master (Roche Diagnostics, Mannheim, Germany). The qPCR of each sample was carried out in duplicate for all cDNA and for glyceraldehyde-3-phosphate dehydrogenase (GAPDH), the housekeeping gene used as a control [5,26,44]. The primer pairs for amplification of *pgc-1α*, *vegf*, and *vegfr2 (Flk-1)*, and *Gapdh* cDNA used in this study were as follows:

Gene	Forward Primer	Reverse Primer
pgc-1α	5′-GTTTCATTACCTACCGTTACAC-3′	5′-ATCGTCTGAGTTTGAATCTAGG-3′
vegf	5′-GCAGATGTGAATGCAGACCA-3′	5′-TTTCCCTTTCCTCGAACTGA-3′
vegfr2	5′-AAGCAAATGCTCAGCAGGAT-3′	5′-GAGGTAGGCAGGGAGAGTCC-3′
Gapdh	5′-AACGGCACAGTCAAGGCTGA-3′	5′- ACGCCAGTAGACTCCACGACAT -3′

The products of PCR were subsequently subjected to agarose gel electrophoresis for further confirmation of amplification specificity [5,26,44]. Fluorescence signals from the amplified products were quantitatively assessed using the LightCycler software program (version 3.5). The second derivative maximum mode was chosen with baseline adjustment set in the arithmetic mode. The relative change in target mRNA expression was determined by the fold-change analysis [5,26,44], in which fold change = $2^-[\Delta\Delta Ct]$, where $\Delta\Delta Ct = (Ct_{,\text{target mRNA}} - Ct_{,\text{GAPDH}})$.

4.7. Western Blotting Analysis

Western blot analysis was carried out on proteins extracted from total lysate or cytosolic fractions from hippocampal samples. The primary antiserum used included rabbit monoclonal or polyclonal antibodies against PGC-1α (sc-13067, Santa Cruz Biotechnology, Heidelberg, Germany), VEGF (5365-100, BioVision Huissen, Netherlands), VEGFR2 (9698S, Cell signaling, Danvers, MA, USA), PI3 Kinase (4257S, Cell signaling), Phospho-PI3 Kinase (4228S, Cell signaling), AKT (9272, Cell signaling), Phospho-AKT (4060S, Cell signaling), ERK1/2 (9102S, Cell signaling), Phospho-ERK (4370S, Cell signaling), Cleaved caspase-3 (9664, Cell signaling), and β-actin (ab8227, abcam, Cambridge, UK). The primary antibodies were diluted in 5% skimmed milk in TBST. The conditions for the membrane wash, detection of immunoreactive signals, and quantification of signal intensities on the blots were performed as previously described [5,26,44,46]. This was followed by incubation with horseradish peroxidase-conjugated secondary goat anti-rabbit IgG (111-035-045, Jackson ImmunoResearch, West Grove, PA, USA) to detect the first antibodies for PGC-1α, VEGF, VEGFR2, PI3 Kinase, Phospho-PI3 Kinase, AKT, Phospho-Akt, ERK1/2, Phospho-ERK Cleaved caspase-3, and β-actin. Specific antibody–antigen complexes were detected by an enhanced chemiluminescence western HRP substrate (Merck Millipore, Billerica, MA, USA). ImageJ software (National Institutes of Health, Bethesda, MD, USA) was applied to quantify the amount of proteins and expressed as the ratio relative to β-actin protein.

4.8. Double Immunofluorescence Staining and Laser Confocal Microscopy

Double immunofluorescence staining [5,26,44,46,84] was carried out using a goat polyclonal antiserum against VEGF (Santa Cruz Biotechnology) and a mouse monoclonal antiserum against a specific marker for neurons, neuron-specific nuclear protein (NeuN) (Chemicon, Temecula, CA, USA). A goat anti-mouse IgG conjugated with Alexa Fluor 568 (Molecular Probes, Eugene, OR, USA) and a goat anti-rabbit IgG conjugated with AlexaFluor 488 (Molecular Probes) served as the secondary antisera. The sections of right hippocampal CA3 tissue were inspected under an epifluorescence microscope (Olympus AX-51; Olympus, Kyoto, Japan). Under the epifluorescence microscope, the immunoreactivity for VEGF exhibited green fluorescence, and NeuN manifested red fluorescence.

4.9. Qualitative and Quantitative Analysis of DNA Fragmentation

Hippocampal CA3 tissues were subject to measurement of the apoptotic DNA fragmentation following status epilepticus [5,26,44,85]. After total DNA was extracted from the hippocampal CA3 tissues, nucleosomal DNA ladders were amplified by a PCR kit for DNA ladder assay (Cat.#: APO-DNA1; Maxim Biotech, San Francisco, CA, USA) to heighten the sensitivity according to the manufacturer's protocol [5,26,44,85]. For quantification of apoptotic DNA fragmentation, a cell death detection ELISA kit (Cat.#11774425001; Roche Molecular Biochemicals) was performed to detect the cytosolic level of histone-associated DNA fragments according to the manufacturer's protocol [5,26,44,85]. Proteins from hippocampal tissues served as the source of antigen, jointly with primary anti-histone antibody and secondary anti-DNA antibody coupled with horseradish peroxidase. The cytosolic nucleosomes were quantitatively measured using 2,2′-azino-di-[3-ethylbenzthiazoline] sulfonate as a substrate. Absorbance was determined at 405 nm and referenced at 490 nm using a Multiskan Spectrum reader (Thermo Scientific, Miami, OK, USA).

4.10. Statistical Analysis

The continuous variables were expressed as mean ± standard error of the mean (SEM). The statistical method was used to compare all the continuous variables with one-way analysis of variance (one-way ANOVA). Once the ANOVA was significant, for post hoc assessment, we used Scheffé multiple range test to assess the difference between groups, especially the experimental group and the control group. A $p < 0.05$ was considered statistically significant.

5. Conclusions

In the present study, our results suggested activation of VEGF/VEGFR2 (Flk-1) by PGC-1α subsequently activates PI3K/AKT and MEK/ERK signaling by alternating the activation state of numerous of downstream proteins relevant to cell proliferation and inhibiting apoptosis in the hippocampus following status epilepticus. Otherwise, downregulation of PGC-1α by siRNA against to *pgc-1α* will inhibit VEGF/VEGFR2 (Flk-1) signaling and suppress pro-survival PI3K/AKT and MEK/ERK pathways that also are accompanied by hippocampal CA3 neuronal cell apoptosis. These results may indicate that the PGC-1α induced VEGF/VEGFR2 pathway may trigger neuronal survival signaling and PI3K/AKT and MEK/ERK signaling pathways. Thus, the axis of PGC-1α/VEGF/VEGFR2 (Flk-1) and downstream PI3K/AKT and MEK/ERK signaling could be considered as an endogenous neuroprotective effect against apoptosis in the hippocampus following status epilepticus. Taking together, our previous research and the results from this study, new strategies for treatment, disease modification, and neuroprotection in epilepsy, particularly in status epilepticus, should be developed.

Author Contributions: Conceptualization, Y.-C.C., H.-Y.P., S.-P.H., and S.-D.C.; methodology, T.-K.L., X.-P.L., C.-W.L., and J.-H.L.; investigation, J.-B.H., S.-F.C., and Y.-C.C.; data curation, X.-P.L., J.-H.L., T.-K.L., C.-W.L., and N.-C.C.; writing—original draft preparation, J.-B.H., S.-P.H., and Y.-C.C.; writing—review and editing, Y.-C.C. and S.-D.C.; visualization, J.-B.H., S.-F.C., C.-Y.H., and H.-Y.P.; supervision, Y.-C.C.; project administration, Y.-C.C.; funding acquisition, Y.-C.C. and H.-Y.P.; statistical analysis, H.-Y.C. All authors have read and agreed to the published version of the manuscript.

Funding: This work was supported by research grants MOST102-2314-B-182A-090-MY3 from the Ministry of Science and Technology, Taiwan to Yao-Chung Chuang, and CMRPG8I0041 from the Chang Gung Medical Foundation to Hsiu-Yung Pan.

Conflicts of Interest: The authors declare no conflict of interest.

Abbreviations

PPARγ	Peroxisome proliferator-activated receptor γ
PGC1-α	Peroxisome proliferator-activated receptor γ coactivator 1α
VEGF	Vascular endothelial growth factor
VEGFR	Vascular endothelial growth factor
VEGFR2	Vascular endothelial growth factor 2
Flk1	Fetal liver kinase receptor 1
PI3K	Phosphatidylinositol 3-kinase
AKT	Protein kinase B
MEK	Mitogen activated protein kinase
ERK	Extracellular signal-regulated kinase
CNS	Central nervous system
KA	Kainic acid
siRNA	Small Interfering RNA
hEEG	Hippocampal electroencephalogram
NeuN	Neuron-specific nuclear protein
UCP2	Uncoupling protein 2

References

1. Thijs, R.D.; Surges, R.; O'Brien, T.J.; Sander, J.W. Epilepsy in adults. *Lancet* **2019**, *393*, 689–701. [CrossRef]
2. Trinka, E.; Brigo, F.; Shorvon, S. Recent advances in status epilepticus. *Curr. Opin. Neurol.* **2016**, *29*, 189–198. [CrossRef] [PubMed]
3. Betjemann, J.P.; Lowenstein, D.H. Status epilepticus in adults. *Lancet Neurol.* **2015**, *14*, 615–624. [CrossRef]
4. Chen, S.D.; Chang, A.Y.; Chuang, Y.C. The potential role of mitochondrial dysfunction in seizure-associated cell death in the hippocampus and epileptogenesis. *J. Bioenerg. Biomembr.* **2010**, *42*, 461–465. [CrossRef]
5. Chuang, Y.C.; Chen, S.D.; Jou, S.B.; Lin, T.K.; Chen, S.F.; Chen, N.C.; Hsu, C.Y. Sirtuin 1 regulates mitochondrial biogenesis and provides an endogenous neuroprotective mechanism against seizure-induced neuronal cell death in the hippocampus following status epilepticus. *Int. J. Mol. Sci.* **2019**, *20*, 3588. [CrossRef]
6. Haut, S.R.; Veliskova, J.; Moshe, S.L. Susceptibility of immature and adult brains to seizure effects. *Lancet Neurol.* **2004**, *3*, 608–617. [CrossRef]
7. Henshall, D.C.; Simon, R.P. Epilepsy and apoptosis pathways. *J. Cereb. Blood Flow Metab.* **2005**, *25*, 1557–1572. [CrossRef]
8. Takahashi, H.; Shibuya, M. The vascular endothelial growth factor (VEGF)/VEGF receptor system and its role under physiological and pathological conditions. *Clin. Sci.* **2005**, *109*, 227–241. [CrossRef]
9. Simons, M.; Gordon, E.; Claesson-Welsh, L. Mechanisms and regulation of endothelial VEGF receptor signalling. *Nat. Rev. Mol. Cell Biol.* **2016**, *17*, 611–625. [CrossRef]
10. Rosenstein, J.M.; Krum, J.M.; Ruhrberg, C. VEGF in the nervous system. *Organogenesis* **2010**, *6*, 107–114. [CrossRef]
11. Eichmann, A.; Simons, M. VEGF signaling inside vascular endothelial cells and beyond. *Curr. Opin. Cell Biol.* **2012**, *24*, 188–193. [CrossRef] [PubMed]
12. Shim, J.W.; Madsen, J.R. VEGF Signaling in Neurological Disorders. *Int. J. Mol. Sci.* **2018**, *19*, 275. [CrossRef] [PubMed]
13. Fournier, N.M.; Lee, B.; Banasr, M.; Elsayed, M.; Duman, R.S. Vascular endothelial growth factor regulates adult hippocampal cell proliferation through MEK/ERK-and PI3K/Akt-dependent signaling. *Neuropharmacology* **2012**, *63*, 642–652. [CrossRef] [PubMed]
14. Nikitidou, L.; Kanter-Schlifke, I.; Dhondt, J.; Carmeliet, P.; Lambrechts, D.; Kokaia, M. VEGF receptor-2 (Flk-1) overexpression in mice counteracts focal epileptic seizures. *PLoS ONE* **2012**, *7*, e40535. [CrossRef]
15. Zachary, I. Neuroprotective role of vascular endothelial growth factor: Signalling mechanisms, biological function and therapeutic potential. *Neurosignals* **2005**, *14*, 207–221. [CrossRef]
16. Geiseler, S.J.; Morland, C. The Janus face of VEGF in stroke. *Int. J. Mol. Sci.* **2018**, *19*, 1362. [CrossRef]

17. Li, C.; Zhang, B.; Zhu, Y.; Li, Y.; Liu, P.; Gao, B.; Tian, S.; Du, L.; Bai, Y. Post-stroke Constraint-induced movement therapy increases functional recovery, angiogenesis, and neurogenesis with enhanced expression of HIF-1alpha and VEGF. *Curr. Neurovasc. Res.* **2017**, *14*, 368–377. [CrossRef]

18. Ma, Y.; Qu, Y.; Fei, Z. Vascular endothelial growth factor in cerebral ischemia. *J. Neurosci. Res.* **2011**, *89*, 969–978. [CrossRef]

19. Nicoletti, J.N.; Lenzer, J.; Salerni, E.A.; Shah, S.K.; Elkady, A.; Khalid, S.; Quinteros, D.; Rotella, F.; Betancourth, D.; Croll, S.D. Vascular endothelial growth factor attenuates status epilepticus-induced behavioral impairments in rats. *Epilepsy Behav.* **2010**, *19*, 272–277. [CrossRef]

20. Kulshreshtha, D.; Vijayalakshmi, K.; Alladi, P.A.; Sathyaprabha, T.N.; Nalini, A.; Raju, T.R. Vascular endothelial growth factor attenuates neurodegenerative changes in the NSC-34 motor neuron cell line induced by cerebrospinal fluid of sporadic amyotrophic lateral sclerosis patients. *Neurodegener. Dis.* **2011**, *8*, 322–330. [CrossRef]

21. Harris, R.; Miners, J.S.; Allen, S.; Love, S. VEGFR1 and VEGFR2 in Alzheimer's Disease. *J. Alzheimers Dis.* **2018**, *61*, 741–752. [CrossRef] [PubMed]

22. Zou, J.; Chen, Z.; Wei, X.; Chen, Z.; Fu, Y.; Yang, X.; Chen, D.; Wang, R.; Jenner, P.; Lu, J.H.; et al. Cystatin C as a potential therapeutic mediator against Parkinson's disease via VEGF-induced angiogenesis and enhanced neuronal autophagy in neurovascular units. *Cell Death Dis.* **2017**, *8*, e2854. [CrossRef] [PubMed]

23. Nicoletti, J.N.; Shah, S.K.; McCloskey, D.P.; Goodman, J.H.; Elkady, A.; Atassi, H.; Hylton, D.; Rudge, J.S.; Scharfman, H.E.; Croll, S.D. Vascular endothelial growth factor is up-regulated after status epilepticus and protects against seizure-induced neuronal loss in hippocampus. *Neuroscience* **2008**, *151*, 232–241. [CrossRef]

24. Rigau, V.; Morin, M.; Rousset, M.C.; de Bock, F.; Lebrun, A.; Coubes, P.; Picot, M.C.; Baldy-Moulinier, M.; Bockaert, J.; Crespel, A.; et al. Angiogenesis is associated with blood-brain barrier permeability in temporal lobe epilepsy. *Brain* **2007**, *130*, 1942–1956. [CrossRef] [PubMed]

25. Newton, S.S.; Collier, E.F.; Hunsberger, J.; Adams, D.; Terwilliger, R.; Selvanayagam, E.; Duman, R.S. Gene profile of electroconvulsive seizures: Induction of neurotrophic and angiogenic factors. *J. Neurosci.* **2003**, *23*, 10841–10851. [CrossRef]

26. Chuang, Y.C.; Chen, S.D.; Hsu, C.Y.; Chen, S.F.; Chen, N.C.; Jou, S.B. Resveratrol promotes mitochondrial biogenesis and protects against seizure-induced neuronal cell damage in the hippocampus following status epilepticus by activation of the PGC-1alpha signaling pathway. *Int. J. Mol. Sci.* **2019**, *20*, 998. [CrossRef]

27. Austin, S.; St-Pierre, J. PGC1alpha and mitochondrial metabolism—emerging concepts and relevance in ageing and neurodegenerative disorders. *J. Cell Sci.* **2012**, *125*, 4963–4971. [CrossRef]

28. Chen, S.D.; Yang, D.I.; Lin, T.K.; Shaw, F.Z.; Liou, C.W.; Chuang, Y.C. Roles of oxidative stress, apoptosis, PGC-1alpha and mitochondrial biogenesis in cerebral ischemia. *Int. J. Mol. Sci.* **2011**, *12*, 7199–7215. [CrossRef]

29. Zhang, K.; Lu, J.; Mori, T.; Smith-Powell, L.; Synold, T.W.; Chen, S.; Wen, W. Baicalin increases VEGF expression and angiogenesis by activating the ERR{alpha}/PGC-1{alpha} pathway. *Cardiovasc. Res.* **2011**, *89*, 426–435. [CrossRef]

30. Thom, R.; Rowe, G.C.; Jang, C.; Safdar, A.; Arany, Z. Hypoxic induction of vascular endothelial growth factor (VEGF) and angiogenesis in muscle by truncated peroxisome proliferator-activated receptor gamma coactivator (PGC)-1alpha. *J. Biol. Chem.* **2014**, *289*, 8810–8817. [CrossRef]

31. Garcia-Quintans, N.; Prieto, I.; Sanchez-Ramos, C.; Luque, A.; Arza, E.; Olmos, Y.; Monsalve, M. Regulation of endothelial dynamics by PGC-1alpha relies on ROS control of VEGF-A signaling. *Free Radic. Biol. Med.* **2016**, *93*, 41–51. [CrossRef] [PubMed]

32. Kilic, U.; Kilic, E.; Jarve, A.; Guo, Z.; Spudich, A.; Bieber, K.; Barzena, U.; Bassetti, C.L.; Marti, H.H.; Hermann, D.M. Human vascular endothelial growth factor protects axotomized retinal ganglion cells in vivo by activating ERK-1/2 and Akt pathways. *J. Neurosci.* **2006**, *26*, 12439–12446. [CrossRef] [PubMed]

33. Ogaki, A.; Ikegaya, Y.; Koyama, R. Vascular abnormalities and the role of vascular endothelial growth factor in the epileptic brain. *Front. Pharmacol.* **2020**, *11*, 20. [CrossRef]

34. Abid, M.R.; Guo, S.; Minami, T.; Spokes, K.C.; Ueki, K.; Skurk, C.; Walsh, K.; Aird, W.C. Vascular endothelial growth factor activates PI3K/Akt/forkhead signaling in endothelial cells. *Arterioscler. Thromb. Vasc. Biol.* **2004**, *24*, 294–300. [CrossRef] [PubMed]

35. Song, F.; Hu, B.; Cheng, J.W.; Sun, Y.F.; Zhou, K.Q.; Wang, P.X.; Guo, W.; Zhou, J.; Fan, J.; Chen, Z.; et al. Anlotinib suppresses tumor progression via blocking the VEGFR2/PI3K/AKT cascade in intrahepatic cholangiocarcinoma. *Cell Death Dis.* **2020**, *11*, 573. [CrossRef]

36. Song, M.; Finley, S.D. Mechanistic insight into activation of MAPK signaling by pro-angiogenic factors. *BMC Syst. Biol.* **2018**, *12*, 145. [CrossRef]

37. Irusta, G.; Abramovich, D.; Parborell, F.; Tesone, M. Direct survival role of vascular endothelial growth factor (VEGF) on rat ovarian follicular cells. *Mol. Cell. Endocrinol.* **2010**, *325*, 93–100. [CrossRef]

38. Walker, M.C.; White, H.S.; Sander, J.W. Disease modification in partial epilepsy. *Brain* **2002**, *125*, 1937–1950. [CrossRef]

39. Walker, M. Neuroprotection in epilepsy. *Epilepsia* **2007**, *48* (Suppl. S8), 66–68. [CrossRef]

40. Sutula, T.P.; Hagen, J.; Pitkanen, A. Do epileptic seizures damage the brain? *Curr. Opin. Neurol.* **2003**, *16*, 189–195. [CrossRef]

41. Chuang, Y.C.; Lin, T.K.; Huang, H.Y.; Chang, W.N.; Liou, C.W.; Chen, S.D.; Chang, A.Y.; Chan, S.H. Peroxisome proliferator-activated receptors gamma/mitochondrial uncoupling protein 2 signaling protects against seizure-induced neuronal cell death in the hippocampus following experimental status epilepticus. *J. Neuroinflamm.* **2012**, *9*, 184. [CrossRef] [PubMed]

42. Simon, R.; Henshall, D.; Stoehr, S.; Meller, R. Endogenous mechanisms of neuroprotection. *Epilepsia* **2007**, *48* (Suppl. S8), 72–73. [CrossRef] [PubMed]

43. Chang, C.C.; Lui, C.C.; Lee, C.C.; Chen, S.D.; Chang, W.N.; Lu, C.H.; Chen, N.C.; Chang, A.Y.; Chan, S.H.; Chuang, Y.C. Clinical significance of serological biomarkers and neuropsychological performances in patients with temporal lobe epilepsy. *BMC Neurol.* **2012**, *12*, 15. [CrossRef] [PubMed]

44. Chang, C.C.; Chen, S.D.; Lin, T.K.; Chang, W.N.; Liou, C.W.; Chang, A.Y.; Chan, S.H.; Chuang, Y.C. Heat shock protein 70 protects against seizure-induced neuronal cell death in the hippocampus following experimental status epilepticus via inhibition of nuclear factor-kappaB activation-induced nitric oxide synthase II expression. *Neurobiol. Dis.* **2014**, *62*, 241–249. [CrossRef]

45. Yang, J.L.; Lin, Y.T.; Chuang, P.C.; Bohr, V.A.; Mattson, M.P. BDNF and exercise enhance neuronal DNA repair by stimulating CREB-mediated production of apurinic/apyrimidinic endonuclease 1. *Neuromol. Med.* **2014**, *16*, 161–174. [CrossRef]

46. Chen, S.D.; Zhen, Y.Y.; Lin, J.W.; Lin, T.K.; Huang, C.W.; Liou, C.W.; Chan, S.H.; Chuang, Y.C. Dynamin-related protein 1 promotes mitochondrial fission and contributes to the hippocampal neuronal cell death following experimental status epilepticus. *CNS Neurosci. Ther.* **2016**, *22*, 988–999. [CrossRef]

47. Chen, S.D.; Lin, T.K.; Yang, D.I.; Lee, S.Y.; Shaw, F.Z.; Liou, C.W.; Chuang, Y.C. Protective effects of peroxisome proliferator-activated receptors gamma coactivator-1alpha against neuronal cell death in the hippocampal CA1 subfield after transient global ischemia. *J. Neurosci. Res.* **2010**, *88*, 605–613. [CrossRef]

48. Zhang, Q.; Lei, Y.H.; Zhou, J.P.; Hou, Y.Y.; Wan, Z.; Wang, H.L.; Meng, H. Role of PGC-1alpha in Mitochondrial Quality Control in Neurodegenerative Diseases. *Neurochem. Res.* **2019**, *44*, 2031–2043. [CrossRef]

49. Li, P.A.; Hou, X.; Hao, S. Mitochondrial biogenesis in neurodegeneration. *J. Neurosci. Res.* **2017**, *95*, 2025–2029. [CrossRef]

50. Sweeney, G.; Song, J. The association between PGC-1alpha and Alzheimer's disease. *Anat. Cell Biol.* **2016**, *49*, 1–6. [CrossRef]

51. Tsunemi, T.; La Spada, A.R. PGC-1alpha at the intersection of bioenergetics regulation and neuron function: From Huntington's disease to Parkinson's disease and beyond. *Progr. Neurobiol.* **2012**, *97*, 142–151. [CrossRef] [PubMed]

52. Ham, P.B., III; Raju, R. Mitochondrial function in hypoxic ischemic injury and influence of aging. *Progr Neurobiol.* **2017**, *157*, 92–116. [CrossRef] [PubMed]

53. Valero, T. Mitochondrial biogenesis: Pharmacological approaches. *Curr. Pharm. Des.* **2014**, *20*, 5507–5509. [CrossRef]

54. Folbergrova, J.; Jesina, P.; Kubova, H.; Otahal, J. Effect of Resveratrol on oxidative stress and mitochondrial dysfunction in immature brain during epileptogenesis. *Mol. Neurobiol.* **2018**, *55*, 7512–7522. [CrossRef] [PubMed]

55. Wu, Z.; Xu, Q.; Zhang, L.; Kong, D.; Ma, R.; Wang, L. Protective effect of resveratrol against kainate-induced temporal lobe epilepsy in rats. *Neurochem. Res.* **2009**, *34*, 1393–1400. [CrossRef]

56. Yancopoulos, G.D.; Davis, S.; Gale, N.W.; Rudge, J.S.; Wiegand, S.J.; Holash, J. Vascular-specific growth factors and blood vessel formation. *Nature* **2000**, *407*, 242–248. [CrossRef]

57. Mesquita-Britto, M.H.R.; Mendonca, M.C.P.; Soares, E.S.; Sakane, K.K.; da Cruz-Hofling, M.A. Inhibition of VEGF-Flk-1 binding induced profound biochemical alteration in the hippocampus of a rat model of BBB breakdown by spider venom. A preliminary assessment using FT-IR spectroscopy. *Neurochem. Int.* **2018**, *120*, 64–74. [CrossRef]

58. Shimotake, J.; Derugin, N.; Wendland, M.; Vexler, Z.S.; Ferriero, D.M. Vascular endothelial growth factor receptor-2 inhibition promotes cell death and limits endothelial cell proliferation in a neonatal rodent model of stroke. *Stroke* **2010**, *41*, 343–349. [CrossRef]

59. Sondell, M.; Lundborg, G.; Kanje, M. Vascular endothelial growth factor has neurotrophic activity and stimulates axonal outgrowth, enhancing cell survival and Schwann cell proliferation in the peripheral nervous system. *J. Neurosci.* **1999**, *19*, 5731–5740. [CrossRef]

60. Ruiz de Almodovar, C.; Lambrechts, D.; Mazzone, M.; Carmeliet, P. Role and therapeutic potential of VEGF in the nervous system. *Physiol. Rev.* **2009**, *89*, 607–648. [CrossRef]

61. Ju, S.; Xu, C.; Wang, G.; Zhang, L. VEGF-C induces alternative activation of microglia to promote recovery from traumatic brain injury. *J. Alzheimers Dis.* **2019**, *68*, 1687–1697. [CrossRef] [PubMed]

62. Hsiao, H.Y.; Chen, Y.C.; Huang, C.H.; Chen, C.C.; Hsu, Y.H.; Chen, H.M.; Chiu, F.L.; Kuo, H.C.; Chang, C.; Chern, Y. Aberrant astrocytes impair vascular reactivity in Huntington disease. *Ann. Neurol.* **2015**, *78*, 178–192. [CrossRef] [PubMed]

63. Lin, K.L.; Lin, K.J.; Wang, P.W.; Chuang, J.H.; Lin, H.Y.; Chen, S.D.; Chuang, Y.C.; Huang, S.T.; Tiao, M.M.; Chen, J.B.; et al. Resveratrol provides neuroprotective effects through modulation of mitochondrial dynamics and ERK1/2 regulated autophagy. *Free Radic. Res.* **2018**, *52*, 1371–1386. [CrossRef] [PubMed]

64. Matsumoto, T.; Claesson-Welsh, L. VEGF receptor signal transduction. *Sci. STKE* **2001**, *2001*, re21. [CrossRef]

65. Hao, T.; Rockwell, P. Signaling through the vascular endothelial growth factor receptor VEGFR-2 protects hippocampal neurons from mitochondrial dysfunction and oxidative stress. *Free Radic. Biol. Med.* **2013**, *63*, 421–431. [CrossRef]

66. Plotnikov, A.; Zehorai, E.; Procaccia, S.; Seger, R. The MAPK cascades: Signaling components, nuclear roles and mechanisms of nuclear translocation. *Biochim. Biophys. Acta* **2011**, *1813*, 1619–1633. [CrossRef]

67. Javadov, S.; Jang, S.; Agostini, B. Crosstalk between mitogen-activated protein kinases and mitochondria in cardiac diseases: Therapeutic perspectives. *Pharmacol. Ther.* **2014**, *144*, 202–225. [CrossRef]

68. Zhu, H.; Zhang, Y.; Shi, Z.; Lu, D.; Li, T.; Ding, Y.; Ruan, Y.; Xu, A. The neuroprotection of liraglutide against ischaemia-induced apoptosis through the activation of the PI3K/AKT and MAPK pathways. *Sci. Rep.* **2016**, *6*, 26859. [CrossRef]

69. Yang, J.L.; Chen, W.Y.; Chen, S.D. The emerging role of GLP-1 receptors in DNA repair: Implications in neurological disorders. *Int. J. Mol. Sci.* **2017**, *18*, 1861. [CrossRef]

70. Martinez-Lopez, N.; Athonvarangkul, D.; Mishall, P.; Sahu, S.; Singh, R. Autophagy proteins regulate ERK phosphorylation. *Nat. Commun.* **2013**, *4*, 2799. [CrossRef]

71. Sun, J.; Ren, D.D.; Wan, J.Y.; Chen, C.; Chen, D.; Yang, H.; Feng, C.L.; Gao, J. Desensitizing mitochondrial permeability transition by ERK-cyclophilin D axis contributes to the neuroprotective effect of gallic acid against cerebral ischemia/reperfusion injury. *Front. Pharmacol.* **2017**, *8*, 184. [CrossRef] [PubMed]

72. Caglayan, B.; Caglayan, A.B.; Beker, M.C.; Yalcin, E.; Beker, M.; Kelestemur, T.; Sertel, E.; Ozturk, G.; Kilic, U.; Sahin, F.; et al. Evidence that activation of P2×7R does not exacerbate neuronal death after optic nerve transection and focal cerebral ischemia in mice. *Exp. Neurol.* **2017**, *296*, 23–31. [CrossRef] [PubMed]

73. Sanchez-Alegria, K.; Flores-Leon, M.; Avila-Munoz, E.; Rodriguez-Corona, N.; Arias, C. pi3k signaling in neurons: A central node for the control of multiple functions. *Int. J. Mol. Sci.* **2018**, *19*, 3725. [CrossRef]

74. Jha, S.K.; Jha, N.K.; Kar, R.; Ambasta, R.K.; Kumar, P. p38 MAPK and PI3K/AKT signalling cascades in Parkinson's disease. *Int. J. Mol. Cell. Med.* **2015**, *4*, 67–86. [PubMed]

75. Matsuda, S.; Ichimura, M.; Ogino, M.; Nakano, N.; Minami, A.; Murai, T.; Kitagishi, Y. Effective PI3K modulators for improved therapy against malignant tumors and for neuroprotection of brain damage after tumor therapy (Review). *Int. J. Oncol.* **2016**, *49*, 1785–1790. [CrossRef] [PubMed]

76. Yamaguchi, H.; Wang, H.G. The protein kinase PKB/Akt regulates cell survival and apoptosis by inhibiting Bax conformational change. *Oncogene* **2001**, *20*, 7779–7786. [CrossRef]

77. Korhonen, L.; Belluardo, N.; Mudo, G.; Lindholm, D. Increase in Bcl-2 phosphorylation and reduced levels of BH3-only Bcl-2 family proteins in kainic acid-mediated neuronal death in the rat brain. *Eur. J. Neurosci.* **2003**, *18*, 1121–1134. [CrossRef]

78. Brunet, A.; Bonni, A.; Zigmond, M.J.; Lin, M.Z.; Juo, P.; Hu, L.S.; Anderson, M.J.; Arden, K.C.; Blenis, J.; Greenberg, M.E. Akt promotes cell survival by phosphorylating and inhibiting a Forkhead transcription factor. *Cell* **1999**, *96*, 857–868. [CrossRef]

79. Cruz, C.D.; Cruz, F. The ERK 1 and 2 pathway in the nervous system: From basic aspects to possible clinical applications in pain and visceral dysfunction. *Curr. Neuropharmacol.* **2007**, *5*, 244–252. [CrossRef]

80. Sun, J.; Nan, G. The extracellular signal-regulated kinase 1/2 pathway in neurological diseases: A potential therapeutic target (Review). *Int. J. Mol. Med.* **2017**, *39*, 1338–1346. [CrossRef]

81. Chico, L.K.; Van Eldik, L.J.; Watterson, D.M. Targeting protein kinases in central nervous system disorders. *Nat. Rev. Drug Discov.* **2009**, *8*, 892–909. [CrossRef] [PubMed]

82. Zhang, C.; Yuan, X.; Hu, Z.; Liu, S.; Li, H.; Wu, M.; Yuan, J.; Zhao, Z.; Su, J.; Wang, X.; et al. Valproic acid protects primary dopamine neurons from MPP(+)-induced neurotoxicity: Involvement of GSK3beta phosphorylation by Akt and ERK through the mitochondrial intrinsic Apoptotic pathway. *Biomed. Res. Int.* **2017**, *2017*, 8124501. [CrossRef] [PubMed]

83. Chuang, Y.C.; Chang, A.Y.; Lin, J.W.; Hsu, S.P.; Chan, S.H. Mitochondrial dysfunction and ultrastructural damage in the hippocampus during kainic acid-induced status epilepticus in the rat. *Epilepsia* **2004**, *45*, 1202–1209. [CrossRef]

84. Chuang, Y.C.; Chen, S.D.; Lin, T.K.; Liou, C.W.; Chang, W.N.; Chan, S.H.; Chang, A.Y. Upregulation of nitric oxide synthase II contributes to apoptotic cell death in the hippocampal CA3 subfield via a cytochrome c/caspase-3 signaling cascade following induction of experimental temporal lobe status epilepticus in the rat. *Neuropharmacology* **2007**, *52*, 1263–1273. [CrossRef] [PubMed]

85. Chuang, Y.C.; Chen, S.D.; Liou, C.W.; Lin, T.K.; Chang, W.N.; Chan, S.H.; Chang, A.Y. Contribution of nitric oxide, superoxide anion, and peroxynitrite to activation of mitochondrial apoptotic signaling in hippocampal CA3 subfield following experimental temporal lobe status epilepticus. *Epilepsia* **2009**, *50*, 731–746. [CrossRef]

86. Lagouge, M.; Argmann, C.; Gerhart-Hines, Z.; Meziane, H.; Lerin, C.; Daussin, F.; Messadeq, N.; Milne, J.; Lambert, P.; Elliott, P.; et al. Resveratrol improves mitochondrial function and protects against metabolic disease by activating SIRT1 and PGC-1alpha. *Cell* **2006**, *127*, 1109–1122. [CrossRef]

International Journal of
Molecular Sciences

Article

Chronic Treatment of Ascorbic Acid Leads to Age-Dependent Neuroprotection against Oxidative Injury in Hippocampal Slice Cultures

Kyung Hee Lee [1], Un Jeng Kim [2], Myeounghoon Cha [2] and Bae Hwan Lee [2,3,*]

[1] Department of Dental Hygiene, Division of Health Science, Dongseo University, Busan 47011, Korea; kyhee@dongseo.ac.kr
[2] Department of Physiology, Yonsei University College of Medicine, Seoul 03722, Korea; mignon@yuhs.ac (U.J.K.); mhcha@yuhs.ac (M.C.)
[3] Brain Korea 21 PLUS Project for Medical Science, Yonsei University College of Medicine, Seoul 03722, Korea
* Correspondence: bhlee@yuhs.ac; Tel.: +82-2-2228-1711

Abstract: Increased oxidative damage in the brain, which increases with age, is the cause of abnormal brain function and various diseases. Ascorbic acid (AA) is known as an endogenous antioxidant that provides neuronal protection against oxidative damage. However, with aging, its extracellular concentrations and uptake decrease in the brain. Few studies have dealt with age-related functional changes in the brain to sustained ascorbate supplementation. This study aimed to investigate the susceptibility of hippocampal neurons to oxidative injury following acute and chronic AA administration. Oxidative stress was induced by kainic acid (KA, 5 μM) for 18 h in hippocampal slice cultures. After KA exposure, less neuronal cell death was observed in the 3 w cultured slice compared to the 9 w cultured slice. In the chronic AA treatment (6 w), the 9 w-daily group showed reduced neuronal cell death and increased superoxide dismutase (SOD) and Nrf2 expressions compared to the 9 w. In addition, the 9 w group showed delayed latencies and reduced signal activity compared to the 3 w, while the 9 w-daily group showed shorter latencies and increased signal activity than the 9 w. These results suggest that the maintenance of the antioxidant system by chronic AA treatment during aging could preserve redox capacity to protect hippocampal neurons from age-related oxidative stress.

Keywords: ascorbic acid; antioxidant; aging; organotypic hippocampal slice culture; neuroprotection

Citation: Lee, K.H.; Kim, U.J.; Cha, M.; Lee, B.H. Chronic Treatment of Ascorbic Acid Leads to Age-Dependent Neuroprotection against Oxidative Injury in Hippocampal Slice Cultures. *Int. J. Mol. Sci.* **2020**, *22*, 1608. https://doi.org/10.3390/ijms22041608

Academic Editor: Volkmar Lessmann
Received: 29 December 2020
Accepted: 2 February 2021
Published: 5 February 2021

Publisher's Note: MDPI stays neutral with regard to jurisdictional claims in published maps and institutional affiliations.

1. Introduction

Oxidative stress is caused by reactive oxygen species (ROS) and leads to structural and functional cellular changes in neurons, including apoptosis and necrosis [1]. Neurons are especially vulnerable to oxidative stress due to their high lipid content and high metabolic rates. Under physiological conditions, ROS are removed by a cellular antioxidant defense system that includes redox homeostasis and cellular signal transduction. Even though ROS are considered as damaging agents, ROS generation is initially necessary for cell function, as ROS act as intracellular messengers and serve as a platform for the transmission of physiological cellular redox signals in mitochondria and the cellular environment [2,3]. However, over-produced ROS lead to an accumulation of excess oxidant radicals that could damage the mitochondria and neuronal cells [4]. Increased ROS in the aging process progressively reduces the maintenance of tissue homeostasis and increases the likelihood of degenerative diseases. Studies on aging over the past several decades have shown that free radicals and oxidative stress increase with age and cause age-related increases in oxidative damage to many human cellular molecules [5]. These previous studies have implicated ROS as a source of oxidative damage to DNA, proteins, and lipids [6–9]. Mitochondria-generated ROS, as byproducts of mitochondrial respiration, are primary targets for oxidative damage and play an important role in aging [10,11]. Vitamins C and E are the main dietary antioxidants that protect erythrocytes from damage caused by ROS [12].

Ascorbic acid (AA), also known as vitamin C, is more highly concentrated in the brain than in other organs and plays an important role in neuronal differentiation and myelin formation [13]. For normal homeostasis of the nervous system, AA is considered to be the most important nutrient due to its crucial role in the brain's antioxidant defense system responsible for scavenging reactive oxygen and nitrogen species produced during cellular metabolism [14,15]. High concentrations of AA have been observed predominantly in neuron-rich areas of the hippocampus and neocortex in the human brain [16], and the ascorbate content in neurons compared to glia appears to be significantly different [17]. Therefore, AA is very important for brain function and maintenance. Furthermore, many degenerative central nervous system diseases, including Alzheimer's disease, multiple sclerosis, Parkinson's disease, and Huntington's disease [18], as well as psychiatric disorders, such as depression, anxiety disorders, and schizophrenia [13,14], are highly associated with AA deficiency. Brain levels of α-tocopherol, ascorbate, and glutathione have all been reported to decrease with aging, and the ascorbate level decrease was accompanied by decreased ascorbate synthesis and altered ascorbate transport characteristics [19]. These decreased ascorbate levels could either contribute to and/or result from the aging process. Interestingly, Michels et al. [20] demonstrated that declines in hepatic ascorbate in rats were associated with an age-related change in ascorbate uptake. Nevertheless, few studies have dealt with age-related functional changes in the brain to sustained ascorbate supplementation or the optical activity of neurons capable of restoring functional activity using optical imaging.

To investigate the results according to age and AA treatment effects, the present study evaluated the neuroprotective properties of ascorbate treatment for various age-related oxidation states in hippocampal slices, as well as the difference between acute and chronic administration of ascorbate during aging. In addition, optical images were used to evaluate the functional role of ascorbate-rescued neurons after oxidative damage caused by kinetic acid (KA).

2. Results

2.1. Different Neuroprotection Effects of AA

At 3 w or 9 w after hippocampal slice culture, 5 µM KA was applied for 18 h to induce oxidative injury and was included in fresh medium (vehicle group). AA with medium was replaced for 24 h after KA treatment (AA treatment). For the 9 w experiment groups, one was treated daily with AA from week 3 to week 9 (9 w-daily), and the other was treated only with medium from week 3 to week 9 (9 w). The experimental groups and the overall design are presented in Figure 1.

Figure 2 shows the different neuroprotective effects of AA in 3 w, 9 w, and 9 w-daily organotypic hippocampal slice cultures (OHSCs) before (Pre, 2A upper line) and 24 h after oxidative injury. KA treatment (5 µM) led to progressive cell death in the CA3 area of the hippocampus after 24 h (24 h, 2A bottom line). Hippocampal neurons in the 3 w KA + AA group showed reduced propidium iodide (PI) uptake compared to the 3 w KA + vehicle group. Similar to the results of 3 w group, the 9 w KA + AA and 9 w-daily KA + AA groups also showed decreased PI uptake compared to the 9 w KA + vehicle and 9 w-daily KA + vehicle groups. These results indicate the neuroprotective effects of AA after KA exposure. Statistical analysis of neuroprotective effects of AA is shown in Figure 2B. The 3 w KA + vehicle and KA + AA PI uptake values were significantly increased compared to the 3 w normal PI uptake (normal: 3.01 ± 1.58, KA + vehicle: 90.03 ± 3.02, KA + AA: 33.33 ± 4.75). In addition, 9 w of KA + vehicle and KA + AA and 9 w-daily KA + vehicle and KA + AA value were significantly increased (9 w normal: 3.80 ± 2.01, KA + vehicle: 91.83 ± 2.33, KA + AA: 74.46 ± 4.50; 9 w-daily normal: 2.73 ± 1.38, KA + vehicle: 76.54 ± 3.68, KA + AA: 59.37 ± 3.78). Moreover, 9 w KA + AA group showed significantly increased cell death compared to 3 w KA + AA group. These results indicate that the aging group (9 w) was more vulnerable compared to the young group (3 w) to cell death and less sensitive to the protective effect of AA treatment following an oxidative injury. In addition, 9 w-daily KA + vehicle and 9 w-daily KA + AA groups showed significantly reduced PI uptake compared to 9 w KA + vehicle and 9 w KA + AA groups. These results indicate that chronic AA treatment protects hippocampal neurons from KA-induced oxidative injury and that prolonged antioxidant treatment during the aging process has a neuroprotective effect.

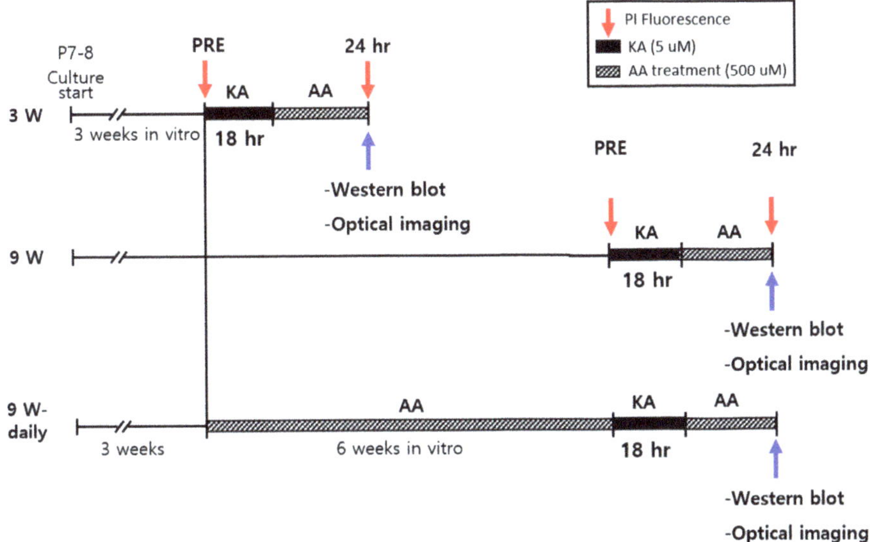

Figure 1. Experimental design and paradigm. Organotypic hippocampal slice cultures (OHSCs) were cultured for 3 w and 9 w. Slices were treated with propidium iodide (PI) 2 h before 5 µM kinetic acid (KA) treatment for all cultures (PRE). All received KA treatment except for the normal controls in each experimental group. After 18 h exposure to KA, the medium containing KA was changed to a fresh culture medium containing 500 µM ascorbic acid (AA). PI uptake, Western blot, and electrophysiological studies were conducted 24 h after AA treatment. Experimental groups were divided by culture time and treatment type, such as 3 w, 9 w, and 9 w-daily. In the 9 w-daily group, 500 µM AA was added to the culture medium for 6 w before studies began.

2.2. Activation of Antioxidant Signals by AA Treatment in Aging

In our aging model of OHSCs with time in culture, we observed age-related changes in the hippocampus. Synapsin-1 and PSD 95, pre- and post-synaptic components implicated in dendritic spine formation and neurotransmission, were observed to confirm the aging statuses of each culture condition. PSD 95 expression was significantly reduced in the 9 w group compared to 3 w group, while the 9 w-daily group showed significantly higher PSD 95 expression than the 9 w group. For synapsin-1, the 9 w group showed markedly lower expression levels than the 3 w and 9 w-daily groups. The expression levels of synapsin-1 and PSD 95 in the 9 w group was markedly reduced compared to the 3 w group, which indicated the age-related phenomena within the hippocampus in our aging model of OHSCs. As for the chronic AA treatment group, the expression levels of both synapsin-1 and PSD 95 proteins showed significant differences between the 9 w and 9 w-daily groups (Figure 3A). These results indicated that chronic AA treatment conserved the decreases in synaptic ability associated with neurodegeneration in aging.

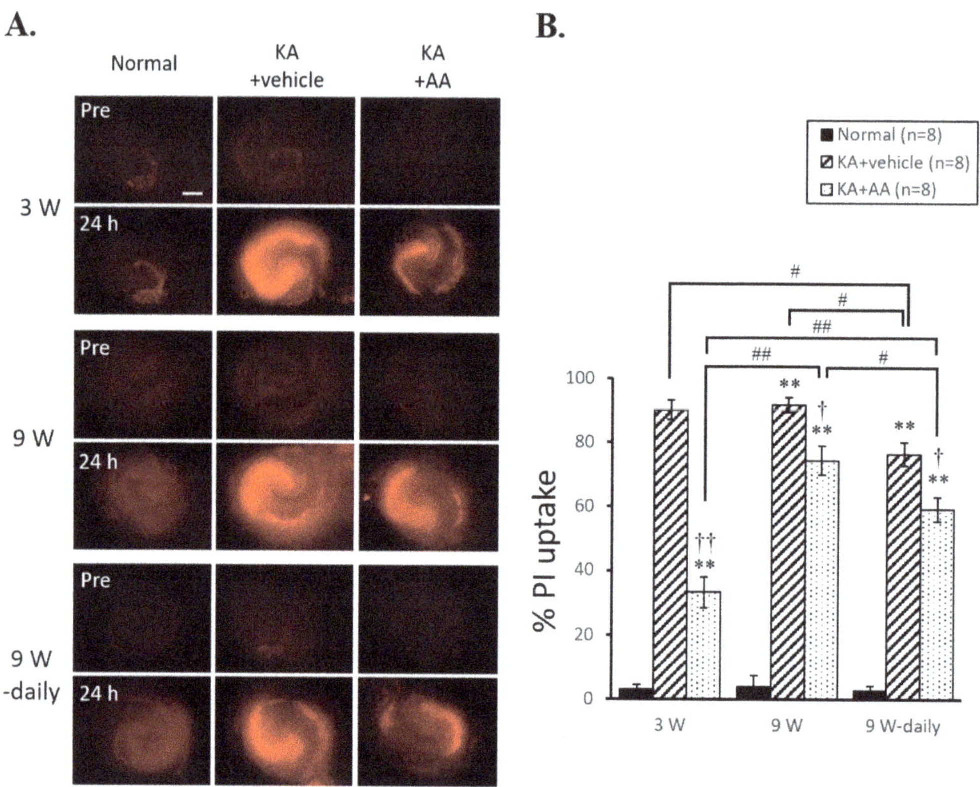

Figure 2. Neuroprotective effects of AA against KA-induced oxidative injury in hippocampal neurons. (**A**) Representative PI fluorescence images in OHSCs before (Pre) and 24 h after AA-treatment following oxidative injury. Representative images of PI uptake, as a marker of cell death, showed the effects of AA treatment at different time points. AA-only treatment (500 μM) showed no toxic effect in KA-untreated normal OHSCs (9 w-daily Normal). (**B**) KA treatment induced progressive cell death in the hippocampus compared to the normal group at each time point. After 24 h of AA treatment, PI signals in the CA3 area of the hippocampus were significantly reduced in the treated groups compared to no-treatment (vehicle) groups at each time point. After chronic AA treatment (9 w-daily), there was less cell death following the KA insult compared to the 9 w vehicle group, and the 9 w-daily AA treatment group also showed a significant decrease in the level of PI uptake compared to the 9 w AA treatment group. ** $p < 0.01$ compared to normal; † $p < 0.05$, †† $p < 0.01$ compared to vehicle; # $p < 0.05$, ## $p < 0.01$ comparing different time points: two-way ANOVA followed by Tukey's *post hoc* comparison. Scale bar: 500 μm.

To observe the antioxidant role of prolonged AA treatment on cell survival, we examined the expression of superoxide dismutase (SOD) and Nrf2, which are known to be factors in ROS-related cell survival signaling. At 3 w, the expression levels of SOD protein significantly increased in the KA + AA group compared to KA + vehicle group. In the aging model, the 9 w KA + AA group showed significantly higher SOD levels than the KA + vehicle group. Compared to the 3 w KA + AA group, the levels of SOD expression in the 9 w KA + AA group were significantly reduced. Moreover, SOD protein significantly increased in the 9 w-daily KA + AA group compared to the KA + vehicle group. The 9 w-daily KA + vehicle group also showed significantly increased SOD levels compared to the 9 w KA + vehicle group (Figure 3B). Nrf2 expression reflects the susceptibility of the brain to the damaging effects of ROS: the level of Nrf2 expression diminished with age, and consequently, its neuroprotective effect decreased. The expression levels of Nrf2 protein decreased in the vehicle group compared to the normal group at 3 w. With aging, the 9 w group showed no difference in Nrf2 expression between KA + AA and KA + vehicle groups. Compared to 3 w, Nrf2 expression was significantly decreased at 9 w for both KA + AA and KA + vehicle groups. In the 9 w-daily group, Nrf2 expression tended to be higher compared to the 9 w treatment groups (Figure 3C).

2.3. Effect of AA on Neuronal Activity

We performed optical imaging to observe cellular activities to visualize the active area based on the overall data and provide information on neuronal activation using a voltage-sensitive dye (VSD). Typical signal transmission and the spatial distribution of the cellular response after electrical stimulation in different age-related groups are shown in Figure 4A. As shown in Figure 4A, KA-untreated hippocampal neurons (normal groups at 3 w, 9 w, and 9 w-daily) demonstrated typical spatiotemporal changes using Schaffer collateral/commissural stimulation. However, limited activation accompanied by synaptic propagation was observed in the KA + vehicle group of OHSCs. In contrast, more activation (indicating more propagation) was observed in the KA + AA group compared to the KA + vehicle group in each age-related group.

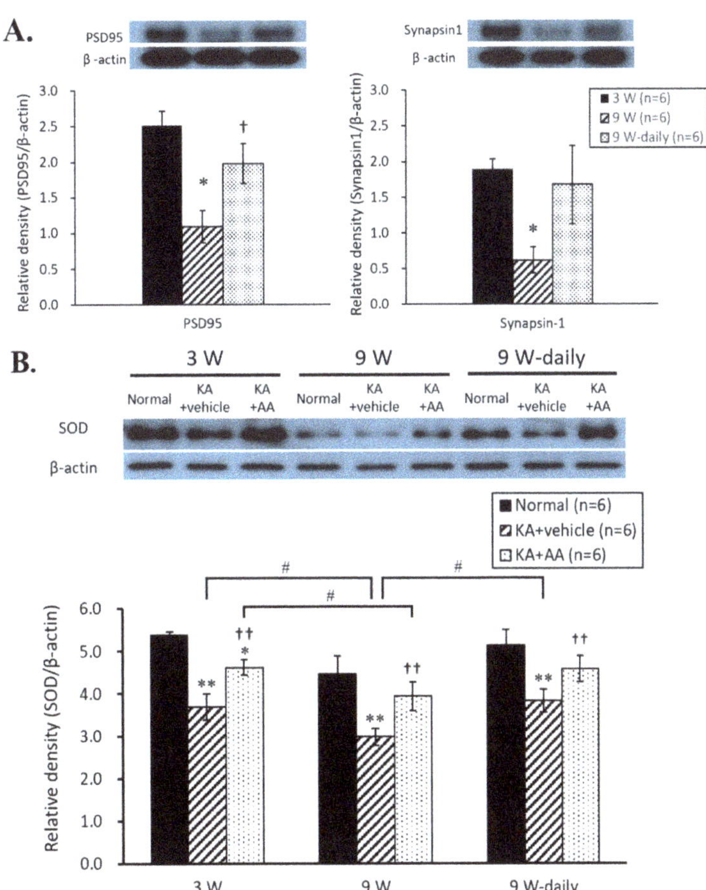

Figure 3. *Cont.*

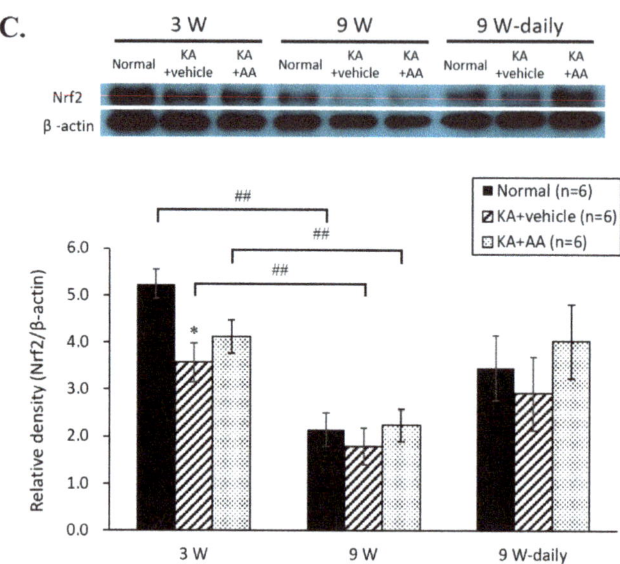

Figure 3. (A) Changes in synaptic protein expression by culture time of OHSCs. PSD 95 and synapsin-1 expression levels significantly reduced at 9 w compared to 3 w. In the chronic AA treatment group, the expression levels of PSD 95 and synapsin-1 proteins showed a significant difference between 9 w and 9 w-daily groups. (B) Western blot analysis of SOD and Nrf2 in OHSCs following oxidative injury. The expression of SOD in the vehicle group was significantly reduced compared to the normal group. The expression of SOD significantly increased 24 h after 500 µM AA treatment in the 3 w and 9 w groups, compared to the vehicle group. The chronic AA treatment (9 w-daily) group showed higher SOD expression than the 9 w group. (C) The expression level of Nrf2 increased in the AA treatment group compared to the vehicle group at 3 w. The 9 w group showed no difference in Nrf2 expression between the AA treatment and vehicle groups. However, Nrf2 expression was significantly decreased in both AA treatment and vehicle groups at 9 w compared to 3 w. In 9 w-daily groups, Nrf2 expression tended to increase compared to 9 w. The horizontal axis indicates each experimental group (3 w, 9 w, and 9 w-daily AA treatment), and the vertical axis represents the normalized level of each protein expression (ratio of each antibody expression/ β-actin expression). * $p < 0.05$, ** $p < 0.01$ compared to normal, † $p < 0.05$, †† $p < 0.01$ compared to vehicle, # $p < 0.05$, ## $p < 0.01$ comparing different time points: two-way ANOVA followed by Tukey's post hoc comparison.

N1 latency, a sign of synaptic activation, increased in both KA + vehicle and KA + AA groups compared to both 3 w and 9 w normal groups by optical signals. The statistical analysis of the neuronal activity of AA is shown in Figure 4B. At 3 w, the AA treatment group showed a significantly reduced latency compared to the KA + vehicle group (normal: 8.84 ± 0.91, KA + vehicle: 30.75 ± 3.66, KA + AA: 16.45 ± 0.97). In 9 w, there was no difference between KA + vehicle and KA + AA groups in N1 latency (9 w normal: 9.05 ± 0.64, KA + vehicle: 26.93 ± 4.27, KA + AA: 23.23 ± 1.56). Meanwhile, N1 latencies in the 9 w-daily groups were significantly shorter than those in the 9 w KA + vehicle and KA + AA groups (9 w-daily normal: 7.66 ± 0.46, KA + vehicle: 12.75 ± 2.52, KA + AA: 8.94 ± 0.81). These results showed that the chronic treatment AA (9 w-daily) in the aged group affected the neuronal activity as an indication of survival neuronal function (Figure 4B). In the activation area in Figure 4C, the KA + AA group showed significantly activated areas compared to the KA + vehicle group at 3 w (normal: 39.70 ± 3.81, KA + vehicle: 15.40 ± 3.58, KA + AA: 27.52 ± 3.00). In 9 w OHSCs, few activated areas were observed around the stimulating electrode in both the KA + vehicle and KA + AA groups (9 w normal: 38.41 ± 4.64, KA + vehicle: 9.39 ± 1.31, KA + AA: 14.38 ± 2.61). There was no difference between the 9 w-daily groups (9 w-daily normal: 37.81 ± 7.44, KA + vehicle: 26.37 ± 5.44, KA + AA: 30.90 ± 6.02). However, chronic AA treatment in the 9 w-daily KA + AA group elicited significantly more activated areas than the 9 w KA + AA group. In addition, the 9 w-daily KA + vehicle group showed significantly larger activated areas compared to the 9 w KA + vehicle group (Figure 4C). Therefore, synaptic propagation was better preserved and maintained after chronic AA treatment in the 9 w-daily group, and this group showed less vulnerability to oxidative injury compared to other 9 w groups.

A.

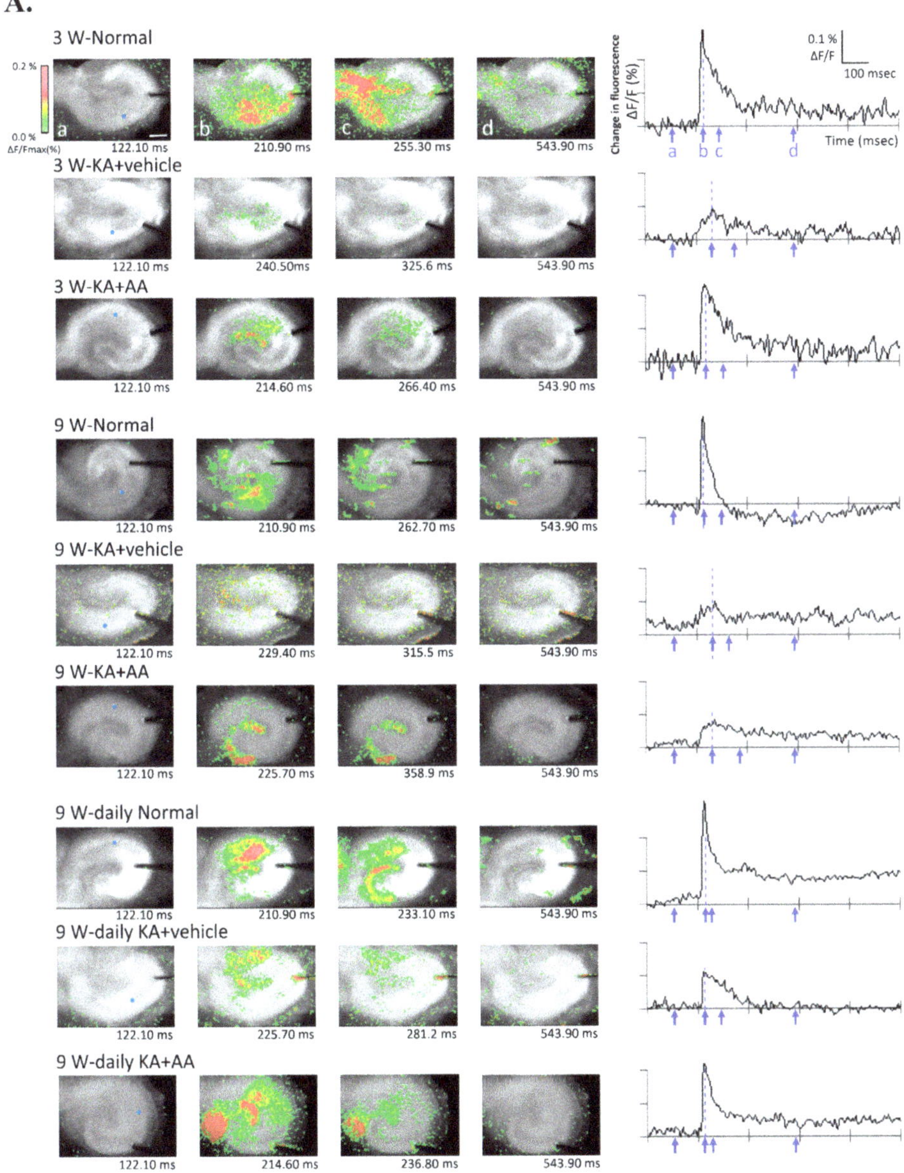

Figure 4. *Cont.*

B.

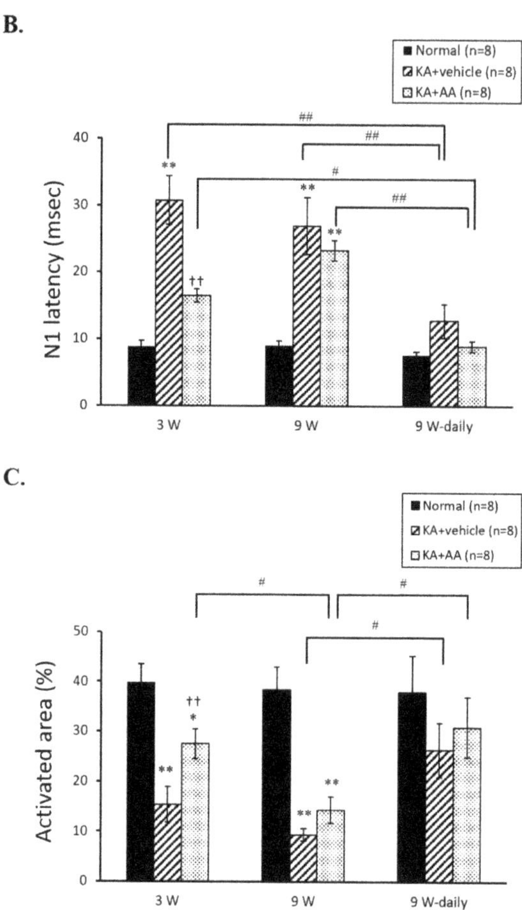

C.

Figure 4. Optical recordings were conducted with a voltage-sensitive dye (VSD) in hippocampal slices to assess AA treatment effects. (**A**) Representative pseudo-color optical images show the evoked excitatory neuronal signals and typical spatiotemporal changes that were observed in OHSCs after electrical stimulation. Each frame of the pseudo-color image (left side) is presented by an arrow below the wave form (right side) at each time point (a, before stimulation; b, peak of N1; c, half of the N1; d, post stimulation). The X-axis shows the time (msec), and the y-axis shows the percent change in fluorescent intensity (% $\Delta F/F$). The blue grid line shows the peak of the negative (N1) point, which was used to quantify the N1 latency. A colored scale (color calibration) shows the changes in optical signals (upper-left side). Signal transmission indicating spatiotemporal changes showed an increase in neuronal activity through long-lasting depolarization. The KA + vehicle groups at each time point showed a few signs of activation, only in the focal stimulation area. Spatiotemporal changes, reflecting the distribution of neuronal activity, increased after AA treatment. (**B**) Comparisons of optical signals in the different experimental groups by latency, an indication of basal synaptic transmission. Latencies were significantly delayed in the vehicle groups compared to normal groups at each time point. The latency of the optical signals was decreased in AA-treated hippocampal slices compared to the vehicle group at 3 w. For chronic AA treatment, there was a significant difference between the 9 w and 9 w-daily groups. (**C**) Quantification of pseudo-color activated areas that indicate neuronal activities through neural propagation. The activation area was decreased in the vehicle groups compared to the normal groups at each time point. In particular, the chronic AA-treated 9 w-daily group showed a significantly greater increase in activation area compared to the acute 9 w group. * $p < 0.05$, ** $p < 0.01$ compared to normal, tt $p < 0.01$ compared to vehicle, # $p < 0.05$, ## $p < 0.01$ comparing different time points: two-way ANOVA followed by Tukey's post hoc comparison. Scale bar: 500 μm.

3. Discussion

Providing cells with AA as an exogenous antioxidant retards their uptake of endogenous antioxidants. In agreement with previous research [21], AA treatment elicited significant reductions in neuronal cell death, compared to vehicle, through a reduction in ROS and oxidative stress by the activation of antioxidant pathways. However, the negative effect of oxidative decay caused by ROS is considered to be an irreversible progression in the biology of aging [22]. Research has shown that the number of neurons and astrocytes remain constant over 21 days in vitro (DIV, 3 w) among OHSCs by cell density analysis and that the density of synaptic contacts within 21 DIV OHSCs is two times higher than those in 7 DIV OHSCs [23]. Moreover, DIV 21 cultures have been found to show similarities to features of the acute slice at P21 [24]. Synapses formed in OHSCs arise from neurons that survive explanation, and synaptic differences in OHSCs according to culture time have been observed, with significant reductions in PSD 95 and evoked synaptic activity at 67-76 DIV [25]. In line with the study by Mielke et al. [25], we also observed significant decreases in PSD 95 and synapsin-1 expressions, which were assessed to reflect age-related phenomena within the hippocampus in our aging model of OHSCs.

Age-dependent changes in brain homeostasis and function occur gradually, particularly by the combination of ROS and the aging brain's impaired ability to repair increased oxidative stress damage [26]. Although we observed decreased neuronal cell death with acute AA treatment after 3 w and 9 w, compared to vehicle, PI uptake quantitation, as an indicator of cell death, was significantly different (based on age) between the 3 w and 9 w AA treatment groups. The 9 w group showed more vulnerability to oxidative injury and cell death compared to the 3 w group. Aging of the brain was also associated with an increase in seizure susceptibility, as well as with seizure-induced neuronal damage, after KA-induced oxidative injury [27,28]. While Siqueira et al. [19] reported that 2 weeks of chronic supplementation with ascorbate was unable to protect the hippocampus from age-related oxidative damage (H_2O_2), we observed that the 9 w-daily vehicle and 9 w-daily AA treatment groups showed significantly greater reductions in cell death than did the 9 w AA treatment group. As a consequence of aging, decreased ascorbate synthesis or altered ascorbate transport characteristics [19] may explain this reduced neuroprotective effect of acute AA treatment in the 9 w group compared to chronic AA treatment in the 9 w-daily group. We hypothesized that chronic AA treatment may lead to persistent regulation of reactions against increased reactive oxygen and nitrogen species with various target molecules during aging, a period during which the antioxidant system appears to be less functional in the brain. Mild prooxidant activity can enhance antioxidant defense systems through Nrf2 signaling [29–31]. In this regard, chronic treatment of AA, which is not only able to scavenge free radicals but also maintain an optimal ROS flow, may upregulate cytoprotective enzymatic antioxidants, thus improving neuronal survival. Chronic treatment of AA might act as a parahormetic phytochemical, which can be attributed to both antioxidant and mild prooxidant activities that affect the intracellular antioxidant defense system. As part of the antioxidant defense system, any decrease in SOD and Nrf2 levels would result in increased ROS [32]. In our Western blot analysis, the expressions of SOD and Nrf2 significantly decreased in the 9 w groups compared to the 3 w groups. Imbalances in ROS production due to impaired expression of SOD and decreased levels of antioxidant molecules occurred with aging, as well as in neurodegenerative diseases [33]. Similar to the PI results, chronic AA treatment (9 w-daily AA treatment group) significantly increased SOD levels. The neurons in the 9 w group suffered more severe attacks by free radicals compared to the neurons with chronic AA treatment, and AA supplementation might have scavenged free radicals from plasma directly. The present study indicates that the aging brain is highly susceptible to hippocampal cell loss by KA-induced oxidative stress, and that chronic AA treatment can aid the cellular antioxidant system against oxidative injury.

To assess neuronal survival and functional property, optical imaging of VSD was used to examine synaptic changes and strength in OHSCs. Optical imaging data represent the visualization of electrical signals from the population activity of postsynaptic neurons in the activated area as a distribution map of stimulus-induced activities [34]. Here, the differences in activities of surviving neurons were examined between 3 w and 9 w slice cultures that underwent AA treatment after oxidative injury. The 9 w group showed increased latencies of optical signals and less activated areas compared to the 3 w group, indicating reduced functionality of surviving neurons. This result may be related to decreased uptake of radiolabeled ascorbate in aged rats [19], as well as to a decline in sodium-dependent ascorbate transport during the aging process [20]. Siqueria et al. [19] reported a 40% decrease in radiolabeled ^{14}C-ascorbate uptake in 11-month-old rats compared to 4-month-old rats, and Michels et al. [20] described an age-related reduction in the plasma transport of dietary ascorbate. Neurons have a 10-fold-higher level of ascorbate than the glia, making them more sensitive to ascorbate deficiency and any reduction in total antioxidant capacity in an older hippocampus [19]. The VSD activation area is dependent on the spatial distribution of cellular responses to electrical stimulation. In this study, we demonstrated enhanced synaptic transmission resulting from chronic AA treatment in older 9 w OHSCs after oxidative injury. To our knowledge, there are no reports that chronic AA supplementation protects hippocampal cells in the brain from the age-related increases in susceptibility to oxidative damage. Therefore, with the optical recording results demonstrating better survival of hippocampal neurons and their connections, chronic AA supplementation may not only be neuroprotective, but also help restore the endogenous antioxidant system with advancing age.

4. Materials and Methods

4.1. Organotypic Hippocampal Slices Cultures (OHSCs)

Animal experiments were approved by the Institutional Animal Care and Use Committee of Yonsei University Health System (permit no.: 2018-0095, approval date: 8 May 2018). Postnatal Sprague-Dawley rats (6–7 d) were used according to the method of Stoppini et al. [35]. We used 45 rat pups, and 40 hippocampal slices were used in each group (one batch contained five hippocampal slices, n = 8 batches). Hippocampi were dissected and placed in Gey's salt solution (Sigma, Saint Louis, MO, USA) with glucose (6.5 mg/mL). Slices (350 μm thick) were cut parallel to the transverse axis of the hippocampus using a chopper (McIlwain tissue chopper; Mickle Laboratory Engineering Ltd., Surrey, UK). Millicell culture inserts (Millipore, Billerica, MA, USA) containing five slices were cultured in 6-well plates with medium (50% opti-MEM, 25% HBSS, 25% horse serum, 6.5 mg/mL glucose, pH adjusted to 7.2) for 3 w or 9 w.

4.2. Propidium Iodide (PI) Staining

To induce oxidative injury, 5 μM KA (Sigma) was applied for 18 h and was included with fresh medium in 3 w or 9 w after hippocampal slice culture (KA + vehicle). AA with medium was replaced for 24 h with 5 ug/mL of propidium iodide (PI) (5 μg/mL, Sigma), which was added to the culture medium (KA + AA). In chronic AA treatment, OHSCs were maintained in culture medium with 500 μM AA for 6 w before KA exposure (9 w-daily). As previously described [36], neuronal death was assessed by quantifying the fluorescence intensity of PI. This was based on the principle that PI is impermeable to normal plasma membranes and cells, and when cells are damaged, they migrate to the nucleus, where they form a complex with DNA, making the nucleus fluoresce. PI uptake images were captured with a fluorescence microscope digital camera (BX-51, Olympus, Tokyo, Japan) and quantified with the MetaMorph Imaging System (Universal Image Co, Downingtown, PA, USA). PI uptake was measured before (Pre) and 24 h after the application of drugs, and the value of the measured fluorescence area was expressed [(24 h − Pre/full kill − Pre) × 100] as the % PI uptake. N-methyl-D-aspartate (NMDA) (100 μM, Sigma) was applied to induce fulminant death of pyramidal neurons (full kill) at the termination of each experiment [37,38]. AA treatment alone (500 μM) showed no toxic effect in KA-untreated normal OHSCs (Normal group) at 3 w or 9 w.

4.3. Western Blot Analysis

Homogenized slice lysates were prepared with lysis buffer (PRO-PREP™, Intron Biotechnology, Burlington, MA, USA). Proteins were resolved by SDS-PAGE (100 µg/lane) using 10% (*w/v*) polyacrylamide gels and transferred to PVDF membranes (Millipore). Membranes were incubated with antibodies against anti-PSD 95 (1:5000, Abcam, Cambridge, MA, USA); anti-synapsin I (1:3000, Abcam); superoxide dismutase (SOD) (1:7500, Abcam); and Nrf2 (1:5000, Abcam) for 2 h at room temperature. As a loading control, membranes were also probed with anti-β-actin antibody (1:10,000, Abcam). The reaction was developed with an enhanced chemiluminescence Western blot analysis system (ECL, GE Healthcare, Marlborough, MA, USA). Signal intensities were analyzed using a gel-scanning integrated optical density software program (Multi-gauge 3.0, Fuji film, Tokyo, Japan).

4.4. Optical Recording

Hippocampal slices (*n* = 8) were stained with a voltage-sensitive dye (VSD) solution (Di-2-ANEPEQ, JPW114; Invitrogen, Carlsbad, CA, USA) [39]. Briefly, 0.2 mM Di-2-ANEPEQ staining solution was applied to the OHSCs in the plexiglass mesh-attached ring. The OHSCs were maintained in a humidified chamber with a mixture of O_2 and CO_2 gases for 25 min. The OHSCs were rinsed with artificial cerebrospinal fluid (ACSF: composition in mM: NaCl, 124; $NaHCO_3$, 26; glucose, 10; KCl, 3; $CaCl_2$, 2; $MgCl_2$, 1; HEPES, 10; pH 7.4) by dipping them through the plexiglass mesh-attached ring and kept for at least 1 h to be recovered before being used for the experiment. A stained hippocampal slice was attached to the chamber glass, which was pre-coated with 0.01% polyethylenimine for approximately 1 h and rinsed with distilled water. Stained OHSCs were stabilized in the recording chamber, which was continuously perfused with ACSF at 31 °C bubbled with mixed 95% O_2 and 5% CO_2 gases. The Schaffer collateral (SC)/commissural pathway was stimulated with a bipolar electrode (CBBPE75, FHC Inc, Bowdoin, ME, USA). Synaptic activity was observed using an optical imaging system (MiCAM02, Brain Vision Inc., Tsukuba, Japan) with a high-speed CCD camera. Optical images were recorded with 512 frames (1890 ms total) for the test stimulation and acquired at a sampling rate of 3.7 ms per frame. Trials were conducted every 20 s (0.05 Hz). Fluorescence intensity [ΔF/F: the change in the intensity of fluorescence (ΔF) in each pixel relative to the initial intensity of fluorescence (F)] was used to assess fractional changes in the amount of VSD fluorescence. Activated areas were determined by averaging images (spatial filter: 3 × 3 pixels, cubic filter: 3 × 3 pixels). Acquisition and analysis software (BV-Analyzer, Brain Vision Inc.) was used to display and analyze the optical images.

4.5. Statistical Analysis

Statistical analyses were performed using GraphPad Prism (GraphPad Software, San Diego, CA, USA). Data are expressed as means ± standard errors of the mean (SEM). The data were evaluated by two-way analysis of variance (ANOVA) followed by Tukey's multiple-range post hoc comparisons. Differences were considered significant when *p*-values were less than 0.05.

5. Conclusions

In this study, we demonstrated that chronic AA treatment has a neuroprotective influence on KA-induced oxidative stress in an aging hippocampus model. Based on PI staining, chronic AA treatment also reduced age-related neurotoxicity (9 w-daily AA treatment), and this was similar to its influence on oxidative injury in younger hippocampi (3 w AA treatment). Western blot analysis demonstrated that chronic AA treatment activated SOD, a well-known detoxifying enzyme, compared to acute AA treatment in the 9 w group. Therefore, we have shown that chronic AA treatment increases antioxidant activity and decreases neuronal cell death in this model. Using optical signals to observe the functional recovery of surviving neurons, AA treatment decreased latencies of the optical signal, indicating an increase in effective neuronal transmission. These results suggest that surviving neurons, protected by chronic AA treatment, show enhanced functional recovery in this aging model.

Author Contributions: All authors contributed substantially to the preparation of this study. K.H.L. designed the study and drafted the manuscript. U.J.K. performed experiments and corrected and analyzed the data. M.C. assisted with the drafting of the manuscript and analyzed optical data. B.H.L. oversaw the entire project and prepared the drafting of the manuscript. K.H.L. and U.J.K. were equal contributors to this study. All authors have read and agreed to the published version of the manuscript.

Funding: This work was supported by the Basic Science Research Program through the National Research Foundation of Korea (NRF) funded by the Ministry of Education (MOE) (NRF-2016R1D1A3B20-08194, NRF-2020R1A2C3008481).

Institutional Review Board Statement: All animal study was approved by Institutional Animal Care and Use Committee of Yonsei University Health System (permit no.: 2018-0095, approval date: 8 May 2018).

Informed Consent Statement: Not Applicable.

Data Availability Statement: Not Applicable.

Conflicts of Interest: The authors declare no conflict of interest.

References

1. Coimbra-Costa, D.; Alva, N.; Duran, M.; Carbonell, T.; Rama, R. Oxidative stress and apoptosis after acute respiratory hypoxia and reoxygenation in rat brain. *Redox Biol.* **2017**, *12*, 216–225. [CrossRef] [PubMed]
2. Santos, A.L.; Sinha, S.; Lindner, A.B. The Good, the Bad, and the Ugly of ROS: New Insights on Aging and Aging-Related Diseases from Eukaryotic and Prokaryotic Model Organisms. *Oxid. Med. Cell. Longev.* **2018**, *2018*, 1941285. [CrossRef] [PubMed]
3. Abate, G.; Vezzoli, M.; Sandri, M.; Rungratanawanich, W.; Memo, M.; Uberti, D. Mitochondria and cellular redox state on the route from ageing to Alzheimer's disease. *Mech. Ageing Dev.* **2020**, *192*, 111385. [CrossRef]
4. Starkov, A.A. The Role of Mitochondria in Reactive Oxygen Species Metabolism and Signaling. *Ann. N. Y. Acad. Sci.* **2008**, *1147*, 37–52. [CrossRef]
5. Martin, G.M.; Austad, S.N.; Johnson, T.E. Genetic analysis of ageing: Role of oxidative damage and environmental stresses. *Nat. Genet.* **1996**, *13*, 25–34. [CrossRef]
6. Cui, H.; Kong, Y.; Zhang, H. Oxidative Stress, Mitochondrial Dysfunction, and Aging. *J. Signal Transduct.* **2012**, *2012*, 646354. [CrossRef]
7. Cohen, G.; Heikkila, R.E. The generation of hydrogen peroxide, superoxide radical, and hydroxyl radical by 6-hydroxydopamine, dialuric acid, and related cytotoxic agents. *J. Biol. Chem.* **1974**, *249*, 2447–2452. [CrossRef]
8. Goncalves, R.L.; Rothschild, D.E.; Quinlan, C.L.; Scott, G.K.; Benz, C.C.; Brand, M.D. Sources of superoxide/H2O2 during mitochondrial proline oxidation. *Redox Biol.* **2014**, *2*, 901–909. [CrossRef]
9. Lee, K.H.; Cha, M.; Lee, B.H. Neuroprotective Effect of Antioxidants in the Brain. *Int. J. Mol. Sci.* **2020**, *21*, 7152. [CrossRef]
10. Kim, S.H.; Kim, H. Inhibitory Effect of Astaxanthin on Oxidative Stress-Induced Mitochondrial Dysfunction—A Mini-Review. *Nutrients* **2018**, *10*, 1137. [CrossRef]
11. Grimm, A.; Eckert, A. Brain aging and neurodegeneration: From a mitochondrial point of view. *J. Neurochem.* **2017**, *143*, 418–431. [CrossRef] [PubMed]
12. Xiong, Y.; Xiong, Y.; Zhou, S.; Sun, Y.; Zhao, Y.; Ren, X.; Zhang, Y.; Zhang, N. Vitamin C and E Supplements Enhance the Antioxidant Capacity of Erythrocytes Obtained from Aged Rats. *Rejuvenation Res.* **2016**, *20*, 85–92. [CrossRef] [PubMed]
13. Han, Q.-Q.; Shen, T.-T.; Wang, F.; Wu, P.-F.; Chen, J.-G. Preventive and Therapeutic Potential of Vitamin C in Mental Disorders. *Curr. Med. Sci* **2018**, *38*, 1–10. [CrossRef]
14. Kocot, J.; Luchowska-Kocot, D.; Kielczykowska, M.; Musik, I.; Kurzepa, J. Does Vitamin C Influence Neurodegenerative Diseases and Psychiatric Disorders? *Nutrients* **2017**, *9*, 659. [CrossRef]
15. May, J.M.; Qu, Z.-C. Ascorbic acid prevents oxidant-induced increases in endothelial permeability. *Biofactors* **2011**, *37*, 46–50. [CrossRef]
16. Mefford, I.N.; Oke, A.F.; Adams, R.N. Regional distribution of ascorbate in human brain. *Brain Res.* **1981**, *212*, 223–226. [CrossRef]
17. Rice, M.E.; Russo-Menna, I. Differential compartmentalization of brain ascorbate and glutathione between neurons and glia. *Neuroscience* **1997**, *82*, 1213–1223. [CrossRef]
18. Moretti, M.; Fraga, D.B.; Rodrigues, A.L.S. Preventive and therapeutic potential of ascorbic acid in neurodegenerative diseases. *CNS Neurosci. Ther.* **2017**, *23*, 921–929. [CrossRef]

19. Siqueira, I.R.; Elsner, V.R.; Leite, M.C.; Vanzella, C.; Moysés, F.d.S.; Spindler, C.; Godinho, G.; Battú, C.; Wofchuk, S.; Souza, D.O.; et al. Ascorbate uptake is decreased in the hippocampus of ageing rats. *Neurochem. Int.* **2011**, *58*, 527–532. [CrossRef]
20. Michels, A.J.; Joisher, N.; Hagen, T.M. Age-related decline of sodium-dependent ascorbic acid transport in isolated rat hepatocytes. *Arch. Biochem. Biophys.* **2003**, *410*, 112–120. [CrossRef]
21. Ballaz, S.J.; Rebec, G.V. Neurobiology of vitamin C: Expanding the focus from antioxidant to endogenous neuromodulator. *Pharm. Res.* **2019**, *146*, 104321. [CrossRef]
22. Monacelli, F.; Acquarone, E.; Giannotti, C.; Borghi, R.; Nencioni, A. Vitamin C, Aging and Alzheimer's Disease. *Nutrients* **2017**, *9*, 670. [CrossRef]
23. Buchs, P.A.; Stoppini, L.; Muller, D. Structural modifications associated with synaptic development in area CA1 of rat hippocampal organotypic cultures. *Brain Res. Dev. Brain Res.* **1993**, *71*, 81–91. [CrossRef]
24. De Simoni, A.; Griesinger, C.B.; Edwards, F.A. Development of rat CA1 neurones in acute versus organotypic slices: Role of experience in synaptic morphology and activity. *J. Physiol.* **2003**, *550*, 135–147. [CrossRef]
25. Mielke, J.G.; Comas, T.; Woulfe, J.; Monette, R.; Chakravarthy, B.; Mealing, G.A.R. Cytoskeletal, synaptic, and nuclear protein changes associated with rat interface organotypic hippocampal slice culture development. *Brain Res. Dev. Brain Res.* **2005**, *160*, 275–286. [CrossRef]
26. López-Alarcón, C.; Denicola, A. Evaluating the antioxidant capacity of natural products: A review on chemical and cellular-based assays. *Anal. Chim. Acta* **2013**, *763*, 1–10. [CrossRef]
27. Friedman, L.K.; Goldstein, B.; Rafiuddin, A.; Roblejo, P.; Friedman, S. Lack of resveratrol neuroprotection in developing rats treated with kainic acid. *Neuroscience* **2013**, *230*, 39–49. [CrossRef]
28. Kuruba, R.; Hattiangady, B.; Parihar, V.K.; Shuai, B.; Shetty, A.K. Differential Susceptibility of Interneurons Expressing Neuropeptide Y or Parvalbumin in the Aged Hippocampus to Acute Seizure Activity. *PLoS ONE* **2011**, *6*, e24493. [CrossRef]
29. Birringer, M. Hormetics: Dietary triggers of an adaptive stress response. *Pharm. Res.* **2011**, *28*, 2680–2694. [CrossRef]
30. Calabrese, V.; Cornelius, C.; Dinkova-Kostova, A.T.; Iavicoli, I.; Di Paola, R.; Koverech, A.; Cuzzocrea, S.; Rizzarelli, E.; Calabrese, E.J. Cellular stress responses, hormetic phytochemicals and vitagenes in aging and longevity. *Biochim. Biophys. Acta* **2012**, *1822*, 753–783. [CrossRef]
31. Forman, H.J.; Davies, K.J.; Ursini, F. How do nutritional antioxidants really work: Nucleophilic tone and para-hormesis versus free radical scavenging in vivo. *Free Radic. Biol. Med.* **2014**, *66*, 24–35. [CrossRef]
32. Puttachary, S.; Sharma, S.; Stark, S.; Thippeswamy, T. Seizure-Induced Oxidative Stress in Temporal Lobe Epilepsy. *Biomed. Res. Int.* **2015**, *2015*, 745613. [CrossRef] [PubMed]
33. Covarrubias-Pinto, A.; Acuna, A.I.; Beltran, F.A.; Torres-Diaz, L.; Castro, M.A. Old Things New View: Ascorbic Acid Protects the Brain in Neurodegenerative Disorders. *Int. J. Mol. Sci.* **2015**, *16*, 28194–28217. [CrossRef] [PubMed]
34. Weisenburger, S.; Vaziri, A. A Guide to Emerging Technologies for Large-Scale and Whole-Brain Optical Imaging of Neuronal Activity. *Annu. Rev. Neurosci.* **2018**, *41*, 431–452. [CrossRef]
35. Stoppini, L.; Buchs, P.A.; Muller, D. A simple method for organotypic cultures of nervous tissue. *J. Neurosci. Methods* **1991**, *37*, 173–182. [CrossRef]
36. Bruce, A.J.; Malfroy, B.; Baudry, M. beta-Amyloid toxicity in organotypic hippocampal cultures: Protection by EUK-8, a synthetic catalytic free radical scavenger. *Proc. Natl. Acad. Sci. USA* **1996**, *93*, 2312–2316. [CrossRef]
37. Borsello, T.; Croquelois, K.; Hornung, J.P.; Clarke, P.G.H. N-methyl-D-aspartate-triggered neuronal death in organotypic hippocampal cultures is endocytic, autophagic and mediated by the c-Jun N-terminal kinase pathway. *Eur. J. Neurosci.* **2003**, *18*, 473–485. [CrossRef] [PubMed]
38. Lee, K.H.; Park, J.H.; Won, R.; Lee, H.; Nam, T.S.; Lee, B.H. Inhibition of hexokinase leads to neuroprotection against excitotoxicity in organotypic hippocampal slice culture. *J. Neurosci. Res.* **2011**, *89*, 96–107. [CrossRef]
39. Tominaga, T.; Tominaga, Y.; Yamada, H.; Matsumoto, G.; Ichikawa, M. Quantification of optical signals with electrophysiological signals in neural activities of Di-4-ANEPPS stained rat hippocampal slices. *J. Neurosci. Methods* **2000**, *102*, 11–23. [CrossRef]

 International Journal of
Molecular Sciences

Article

GPR4 Knockout Improves the Neurotoxin-Induced, Caspase-Dependent Mitochondrial Apoptosis of the Dopaminergic Neuronal Cell

Md Ezazul Haque [1], Mahbuba Akther [1], Shofiul Azam [1], Dong-Kug Choi [1,2] and In-Su Kim [2,*]

[1]   Department of Applied Life Science, Graduate School, Konkuk University, Chungju 27478, Korea; mdezazulhaque@yahoo.com (M.E.H.); smritymahbuba@gmail.com (M.A.); shofiul_azam@hotmail.com (S.A.); choidk@kku.ac.kr (D.-K.C.)

[2]   Department of Biotechnology, Research Institute of Inflammatory Disease (RID), College of Biomedical and Health Science, Konkuk University, Chungju 27478, Korea

*   Correspondence: kis5497@kku.ac.kr; Tel.: +82-43-840-3905

Received: 11 September 2020; Accepted: 8 October 2020; Published: 12 October 2020

Abstract: In Parkinson's disease, mitochondrial oxidative stress-mediated apoptosis is a major cause of dopaminergic neuronal loss in the substantia nigra (SN). G protein-coupled receptor 4 (GPR4), previously recognised as an orphan G protein coupled-receptor (GPCR), has recently been claimed as a member of the group of proton-activated GPCRs. Its activity in neuronal apoptosis, however, remains undefined. In this study, we investigated the role of GPR4 in the 1-methyl-4-phenylpyridinium ion (MPP^+) and hydrogen peroxide (H_2O_2)-treated apoptotic cell death of stably GPR4-overexpressing and stably GPR4-knockout human neuroblastoma SH-SY5Y cells. In GPR4-OE cells, MPP^+ and H_2O_2 were found to significantly increase the expression levels of both mRNA and proteins of the pro-apoptotic Bcl-2-associated X protein (Bax) genes, while they decreased the anti-apoptotic B-cell lymphoma 2 (Bcl-2) genes. In addition, MPP^+ treatment activated Caspase-3, leading to the cleavage of poly (ADP-ribose) polymerase (PARP) and decreasing the mitochondrial membrane potential $(\Delta\Psi m)$ in GPR4-OE cells. In contrast, H_2O_2 treatment significantly increased the intracellular calcium ions (Ca^{2+}) and reactive oxygen species (ROS) in GPR4-OE cells. Further, chemical inhibition by NE52-QQ57, a selective antagonist of GPR4, and knockout of GPR4 by clustered regularly interspaced short palindromic repeats (CRISPR)/Cas9 decreased the Bax/Bcl-2 ratio and ROS generation, and stabilised the $\Delta\Psi m$, thus protecting the SH-SY5Y cells from MPP^+- or H_2O_2-induced apoptotic cell death. Moreover, the knockout of GPR4 decreased the proteolytic degradation of phosphatidylinositol biphosphate (PIP_2) and subsequent release of the endoplasmic reticulum (ER)-stored Ca^{2+} in the cytosol. Our results suggest that the pharmacological inhibition or genetic deletion of GPR4 improves the neurotoxin-induced caspase-dependent mitochondrial apoptotic pathway, possibly through the modulation of PIP_2 degradation-mediated calcium signalling. Therefore, GPR4 presents a potential therapeutic target for neurodegenerative disorders such as Parkinson's disease.

Keywords: apoptosis; neurodegeneration; GPR4 receptor; MPP^+; Parkinson's disease; CRISPR/cas9

---

1. Introduction

Parkinson's disease (PD) is a neurodegenerative disorder characterised by dopamine deficiency. An important pathological basis of PD is the loss of dopaminergic neurons due to apoptotic cell death in the substantia nigra (SN) of the brain [1]. An array of evidence suggests that reactive oxygen species (ROS)-induced oxidative stress is a major cause of the dopaminergic neuronal loss in the SN [2]. Mitochondria are key players in apoptosis during this neurodegeneration. In a cell undergoing apoptosis, mitochondria increase the production of oxyradicals and open the pores of its membranes, leading to the depolarisation of its transmembrane potential ($\Delta\psi$m) and the release of cytochrome C [3,4]. The B-cell lymphoma 2 (Bcl-2) family of proteins, which includes anti-apoptotic Bcl-2 and pro-apoptotic Bcl-2-associated X protein (Bax) members, plays a critical role in the initiation of the apoptotic pathway. When anti-apoptotic Bcl-2 localises in the mitochondrial inner membrane from the outer membrane, the intermembrane space protein, cytochrome C, is released. This binds to apoptotic peptidase activating factor 1 (APAF1) and forms a heptameric structure, known as the apoptosome. Apoptosomes recruit caspase 9 and activate a series of events, eventually activating Caspase-3 as a result [5]. Surprisingly, poly (ADP-ribose) polymerase (PARP), a nuclear protein involved in DNA repair that is specifically cleaved to a signature 89-kDa fragment, has been implicated as an early marker for apoptotic cell death in neurons [6]. The characteristic hallmarks of the mitochondrial oxidative damage-induced apoptosis pathway can be considered to be the opening of the mitochondrial permeability transition pore (mPTP), the collapse of the mitochondrial membrane potential ($\Delta\Psi$m), the release of cytochrome C, the activation of Caspase-3, and proteolytic degradation of PARP [7,8]. Surprisingly, the endoplasmic reticulum (ER) acts as a reservoir of calcium ions (Ca^{2+}). The ER can release Ca^{2+} through either its ryanodine (RYR) or inositol trisphosphate (IP3) receptors. Ca^{2+}, once released from the ER, is taken up by the mitochondrial Ca^{2+} uniporter (MCU) located in the mitochondrial inner membrane [9]. An increase in mitochondrial Ca^{2+} is associated with an increase of the mPTP opening and ROS generation, a decrease in $\Delta\Psi$m, and release of cytochrome C, as well as excitotoxicity and apoptosis [10–12]. Recent findings suggest that an increase in the Ca^{2+} released from the ER can increase the mitochondrial oxidant stress of the substantia nigra pars compacta (SNc) dopaminergic (DA) neurons [10,13]. Alteration of the intracellular calcium homeostasis PD cybrids has also been reported [14].

1-methyl-4-phenylpyridinium ion (MPP^+) and hydrogen peroxide (H_2O_2) are the most widely employed neurotoxins, due to their ability to mimic a PD-like syndrome with apoptotic cell death through mitochondrial oxidative damage, in both cellular and animal models of PD [15–17]. MPP^+, the active metabolite of 1-methyl-4-phenyl-1,2,3,6-tetrahydropyridine (MPTP), generates 5–7 times more ROS by selectively inhibiting mitochondrial complex I, thus initiating pro-apoptotic Bax/Bcl-2-dependent apoptotic cell death [15]. Besides, MPP^+- and H_2O_2-mediated oxidative stress have also been reported to cause mitochondrial oxidative stress-mediated apoptotic cell death [18]. Moreover, MPP^+- and H_2O_2-mediated cell death are associated with the alteration of intracellular Ca^{2+} homeostasis. An increase in intracellular Ca^{2+} due to ER stress also potentiates a decrease in $\Delta\Psi$m, the release of cytochrome C, and the induction of apoptosis [12,19–21]. Therefore, the inhibition of pro-apoptotic signalling, or the decrease of intracellular ROS and Ca^{2+}, may be beneficial for the protection of dopaminergic neuronal loss in PD.

G protein-coupled receptor 4 (GPR4), a proton-sensing receptor, is highly sensitive to the alteration of extracellular proton concentration [22]. It belongs to a small G protein-coupled proton-sensing receptor family that includes ovarian cancer G protein-coupled receptor 1 (OGR1), also referred to as GPR68, G2A, also termed GPR132, and T-cell death-associated gene 8 (TDAG8), also known as GPR65. These receptors signal through either to phosphoinositide mediated increase in intracellular Ca^{2+} or through modulating adenylate cyclase activity [23,24]. Information regarding the distribution and biology of GPR4 in the brain of an individual with PD is limited. An abundant level of GPR4 expression was observed in the retro-trapezoidal nucleus locus coeruleus, the cerebrovascular endothelium, the neurons of the dorsal raphe, and the lateral septum of a GPR4-knock-in mouse model [25]. Recently, the role of GPR4 in inflammation during ER stress pathway-mediated apoptotic cell death has been reported. In human umbilical vein endothelial cells (HUVEC) and other disease models, such as that of myocardial ischemic mice, knocking out GPR4 has been found to reduce cardiomyocyte apoptosis and improve cardiac function [26–28]. However, no study has yet to elucidate the role of GPR4 on apoptotic cell death in neurodegenerative disorders. Therefore, this work aims to forge an understanding of the role of GPR4 in neurotoxin-induced, mitochondrial oxidative stress-mediated apoptosis in a PD model. In particular, through our study of the overexpression and genetic deletion of GPR4, we investigate the role of GPR4 in the mitochondrial apoptosis pathway.

2. Results

2.1. Expression of GPR4 Is Upregulated in Neurotoxin-Stimulated Apoptosis in SH-SY5Y Cells

To investigate the concentrations of MPP^+ and H_2O_2 that precipitated a cell death of nearly 50% in the SH-SY5Y cells, 24 h serum-starved SH-SY5Y cells were treated with MPP^+ (0.25, 0.5, and 1 mM) or H_2O_2 (50, 75, and 125 μM) for 24 h. As is shown in Figure 1A, when treated with the various concentrations of MPP^+ (1 mM; 56.511 ± 1.55%) and H_2O_2 (125 μM; 53.12 ± 2.34%), half of the cell population in the MTT assay died. Furthermore, the mRNA and protein expressions of GPR4 in SH-SY5Y cells in both MPP^+- (1 mM) and H_2O_2- (125 μM) treated serum-free media gradually increased in a time-dependent manner (3–24 h; Figure 1B).

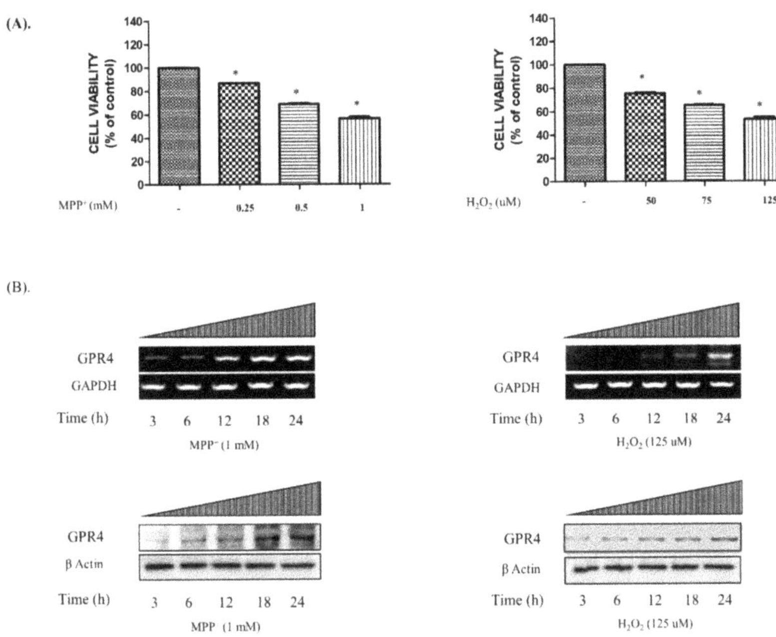

Figure 1. The cellular viability and the mRNA and protein expressions of GPR4 in MPP^+- and H_2O_2-treated SH-SY5Y cells. 24 h serum-starved SH-SY5Y cells were treated with different concentrations of MPP^+ (0.25, 0.5, and 1 mM) and H_2O_2 (50, 75, and 125 μM) for 24 h in serum-free culture media. (**A**) The cellular viability of the SH-SY5Y cell after treatment with different concentrations of MPP^+ and H_2O_2 in serum-free media for 24 h. (**B**) Reverse transcription-polymerase chain reaction (RT-PCR) and immunoblotting demonstrate the mRNA and protein expressions of GPR4 in SH-SY5Y cells at different time points (3, 6, 12, 18, and 24 h) after stimulation with MPP^+ (1 mM) and H_2O_2 (125 μM) in serum-free media. Glyceraldehyde-3-phosphate dehydrogenase (GAPDH) and β-actin were utilised as the internal controls. Mean ± standard error of the mean (SEM; $n = 3$) was employed to express the data. Tukey's multiple comparison test was performed using a one-way analysis of variance (ANOVA). Each * $p < 0.05$ refers to the other sample concentrations compared with the control cells.

2.2. Knockout of GPR4 Protects SH-SY5Y Cells from Neurotoxin-Stimulated Apoptosis in SH-SY5Y Cells

To assess the effect of GPR4 overexpression and knockout on MPP^+-induced apoptotic cell death, 24 h serum-starved SH-SY5Y cells were treated with MPP^+ (1 mM) for 24 h in serum-free media

(Figure 2). Following the MPP$^+$ (1 mM) treatment for 24 h in serum-free media, the number of SH-SY5Y viable cells decreased. Furthermore, the cells became rounded, displayed an increased neurite retraction, and were found to be loosely attached to the plate. Under bright-field optics, the GPR4-OE cells treated with MPP$^+$ (1 mM) exhibited less cell viability, with increased rounded cells, increased neurite retraction, and loose attachment to the surface. In contrast, the GPR4-KO cells treated with MPP$^+$ (1 mM) were more viable, strongly attached, neuronal shaped, and demonstrated less neuronal retraction than both the control and the GPR4-OE cells (Figure 2A).

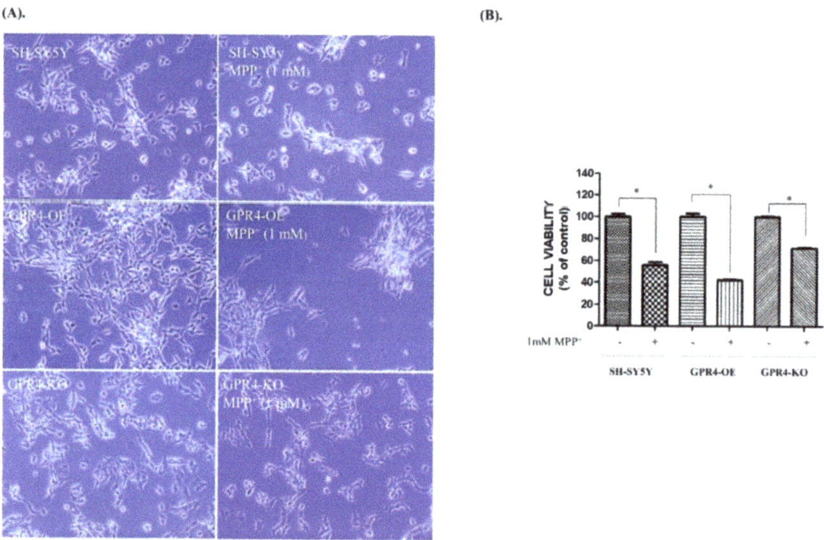

Figure 2. The cellular viability and morphology of MPP$^+$-treated SH-SY5Y cells that were stably GPR4-overexpressing (GPR4-OE) or GPR4-knockout (GPR4-KO). 24 h serum-starved SH-SY5Y cells were treated with MPP$^+$ (1 mM) for 24 h in serum-free culture media. (**A**) The morphology of SH-SY5Y GPR4-OE and GPR4-KO cells was observed through bright-field microscopy. (**B**) Cell viability was evaluated using an MTT assay. Mean ± SEM (n = 3) was employed to express the data. Tukey's multiple comparison test was performed using a one-way ANOVA. Each * $p < 0.05$ refers to the other sample concentrations compared with the control cells.

Cell viability was assessed with an MTT assay. The control SH-SY5Y cells presented a 55.67 ± 5.22% cell survival rate, whereas only 42.00 ± 2.01% of the GPR4-OE cells treated with MPP$^+$ (1 mM) survived. In contrast, the MPP$^+$-treated GPR4-KO cells had a significantly higher cell survival rate (71.63 ± 3.54%), at 15% higher than for the MPP$^+$-treated control SH-SY5Y cells and almost 30% higher than for the MPP$^+$-treated GPR4-OE cells (Figure 2B).

2.3. Knockout of GPR4 Decreases the Bax/Bcl-2 mRNA Ratio during Neurotoxin-Induced Apoptosis in SH-SY5Y Cells

To determine the role of GPR4 in both MPP$^+$- (1 mM) and H$_2$O$_2$- (125 µM) stimulated apoptotic cell death, we investigated the expression levels of the Bcl-2 family proteins (Bax and Bcl-2). Many studies suggest that the Bcl-2 family plays a critical role in the mitochondrial apoptotic pathway. Bax enhances the release of cytochrome C from the space of the mitochondrial intermembrane to the cytosol, resulting in apoptosis. In contrast, Bcl-2 prevents apoptosis through its prevention of cytochrome C release, thereby maintaining mitochondrial cellular integrity [29,30]. In this study, an RT-PCR was employed to assess the mRNA expression levels of GPR4, Bax, and Bcl-2 in 24 h serum-starved SH-SY5Y cells treated with either MPP$^+$ (1 mM) or H$_2$O$_2$ (125 µM; Figure 3A).

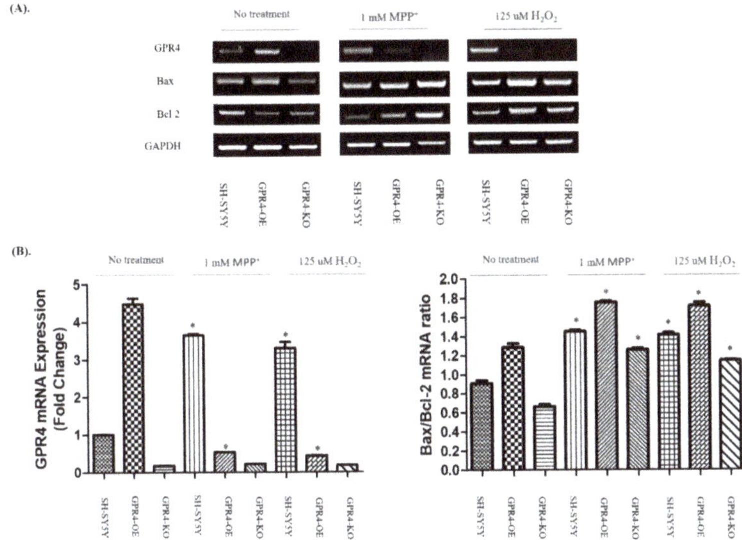

Figure 3. The measurement of GPR4 mRNA expression and the Bax/Bcl-2 mRNA ratio in MPP$^+$- and H$_2$O$_2$-treated SH-SY5Y cells that were stably GPR4-OE or GPR4-KO. (**A**) An RT-PCR illustrating the mRNA expression of pro-apoptotic Bax, anti-apoptotic Bcl-2, and GAPDH in SH-SY5Y, GPR4-OE, and GPR4-KO cells after stimulation with MPP$^+$ (1 mM) and H$_2$O$_2$ (125 μM) in serum-free media for 24 h. (**B**) A semi-quantification of the GPR4 mRNA and Bax/Bcl-2 mRNA expressions relative to GAPDH. This semi-quantification of the respective mRNA expression levels was performed on ImageJ software; GAPDH was utilised as an internal control. Mean ± SEM (n = 3) was employed to express the data. Tukey's multiple comparison test was performed using a one-way ANOVA. Each * $p < 0.05$ refers to the sample concentration compared with the same group of non-treated cells.

A semiquantitative analysis (Figure 3B) of the RT-PCR bands highlighted a more than 4-fold increase in the expression of GPR4 in the GPR4-OE cells without any treatment, compared with the non-treated SH-SY5Y cells. In comparison with the non-treated SH-SY5Y cells, neurotoxins increased the expression of GPR4 in the GPR4-OE cells by 3–4-fold (MPP$^+$, 3.65 ± 0.03; H$_2$O$_2$, 3.31 ± 0.17), whereas no significant difference in the GPR4 expression of the GPR4-OE cells (MPP$^+$, 0.53 ± 0.003; H$_2$O$_2$, 0.04 ± 0.003) was observed.

Interestingly, the ratio of Bax/Bcl-2 mRNA expression for non-treated GPR4-OE cells was slightly higher (1.40 ± 0.08) than that for the control SH-SY5Y cells, whereas the Bax/Bcl-2 mRNA ratio was slightly lower (0.73 ± 0.904) in the non-treated GPR4-KO group than that for the control SH-SY5Y cells. For the SHSY-5Y cells, treatment with MPP$^+$ or H$_2$O$_2$ significantly increased the ratio of Bax/Bcl-2 mRNA expression (MPP$^+$, 1.59 ± 0.02; H$_2$O$_2$, 1.55 ± 0.04). Yet, the ratio of Bax/Bcl-2 mRNA expression in MPP$^+$- and H$_2$O$_2$-stimulated GPR4-OE cells was significantly higher (MPP$^+$, 1.92 ± 0.03; H$_2$O$_2$, 1.88 ± 0.07) than the neurotoxin-treated SH-SY5Y and non-treated GPR4-OE cells. Meanwhile, the ratio of Bax/Bcl-2 mRNA expression in both MPP$^+$- and H$_2$O$_2$-stimulated GPR4-KO cells was significantly lower (MPP$^+$, 1.38 ± 0.02; H$_2$O$_2$, 1.25 ± 0.02) than both the neurotoxin-treated SH-SY5Y and GPR4-OE cells.

2.4. Effect of a GPR4 Antagonist on the Cellular Morphology and GPR4 mRNA Expression of SH-SY5Y Cells

To investigate the effect of the pharmacological inhibition of GPR4, we adopted a GPR4 antagonist, NE52-QQ57. At the physiological pH, NE 52-QQ57 has been reported to effectively block the cAMP that is released by GPR4 activation (IC$_{50}$ 26.8 nM) in HEK293 cells [3]. In this study, the impact of GPR4 antagonist, NE52-QQ57, on the cellular morphology and mRNA expression of GPR4 in MPP$^+$- (1 mM) treated, 24 h serum-starved SH-SY5Y cells was assessed (Figure 4). The SH-SY5Y cells were treated with NE52-QQ57 (1 µM) at pH 7.4 and incubated for 24 h in serum-free culture media, to evaluate the effect of NE52-QQ57 on cell morphology and viability. However, no morphological alteration or cellular toxicity was observed in the control SH-SY5Y, GPR4-OE, or GPR4-KO cells (Figure 4A).

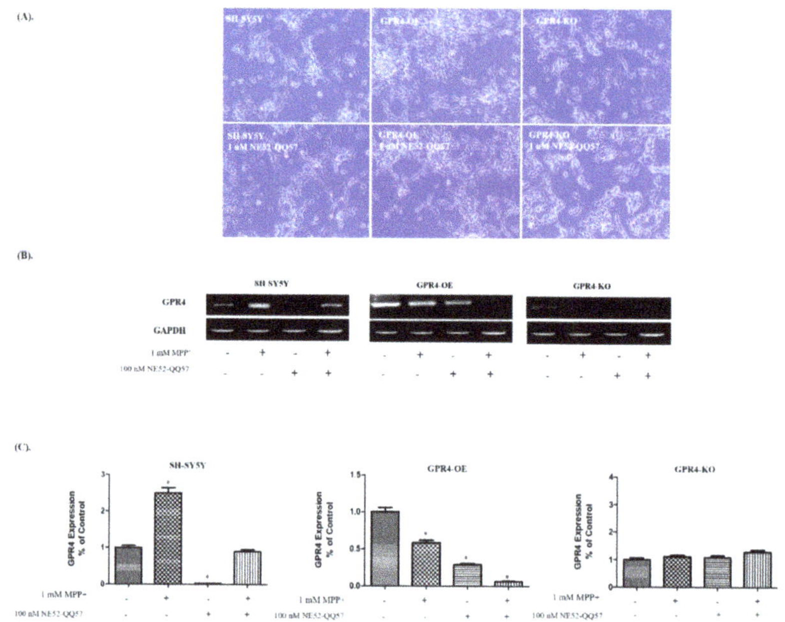

Figure 4. The effect of GPR4 antagonist, NE52-QQ57, on the cellular morphology and on GPR4 mRNA expression in MPP$^+$-treated SH-SY5Y cells that were stably GPR4-OE or GPR4-KO. 24 h serum-starved SH-SY5Y cells were treated for 1 h with NE52-QQ57 followed by either a 24 h incubation or MPP$^+$ (1 mM) stimulation for 24 h in serum-free culture media. (**A**) The cell viability and morphology of SH-SY5Y, GPR4-OE, and GPR4-KO cells that were pre-treated for 1 h with NE52-QQ57. (**B**) An RT-PCR highlighting the mRNA expression of GPR4 and GAPDH in SH-SY5Y, GPR4-OE, and GPR4-KO cells pre-treated for 1 h with NE52-QQ57. (**C**) A semi-quantification of GPR4 mRNA expression relative to GAPDH. This semi-quantification of the respective mRNA expression levels was performed on ImageJ software; GAPDH was utilised as an internal control. Mean ± SEM ($n = 3$) was employed to express the data. Tukey's multiple comparison test was performed using a one-way ANOVA. Each * $p < 0.05$ refers to the sample concentration compared with the same group of non-treated cells.

For an RT-PCR, RNA was isolated from the cells pre-treated with NE52-QQ57 (1 µM) at pH 7.4; this was incubated for an additional 24 h in serum-free media, with or without MPP$^+$ (1 mM) stimulation. The resulting levels of GPR4 mRNA expression illustrated that the NE52-QQ5 (100 nM) had effectively blocked the expression of GPR4 on the control SH-SY5Y and the GPR4-OE cells, similar to genetically GPR4-KO cells (Figure 4B).

A semi-quantitative analysis (Figure 4C) of the RT-PCR bands demonstrated that the NE52-QQ5 significantly reduced GPR4 expression in the SH-SY5Y (0.02 ± 0.001 folds) and the GPR4-OE cells (0.29 ± 0.02 folds), compared with the nontreated SH-SY5Y and GPR4-OE cells, respectively. Moreover, treatment with NE52-QQ5 (100 nM) decreased GPR4 mRNA expression in the MPP$^+$ (1 mM) treated SH-SY5Y (0.9 ± 0.06 folds) and GPR4-OE cells (0.06 ± 0.004 folds), compared with the MPP$^+$-stimulated SH-SY5Y and GPR4-OE cells, respectively.

2.5. Knockout of GPR4 Decreases the Bax/Bcl-2 Protein Ratio and the Cleavage of PARP Expression in Neurotoxin-Stimulated SH-SY5Y Cells

MPP$^+$-treated SH-SY5Y cells were assessed with immunoblotting to evaluate the effect of GPR4 antagonist, NE52-QQ57, on GPR4; the pro-apoptotic proteins, Bax and Bcl-2; and cleaved PARP expression (Figure 5). The SH-SY5Y cells were pre-treated with NE52-QQ57 (100 nM) at pH 7.4. This was followed by 24 h incubation, or MPP$^+$ (1 mM) stimulation for 24 h in serum-free media.

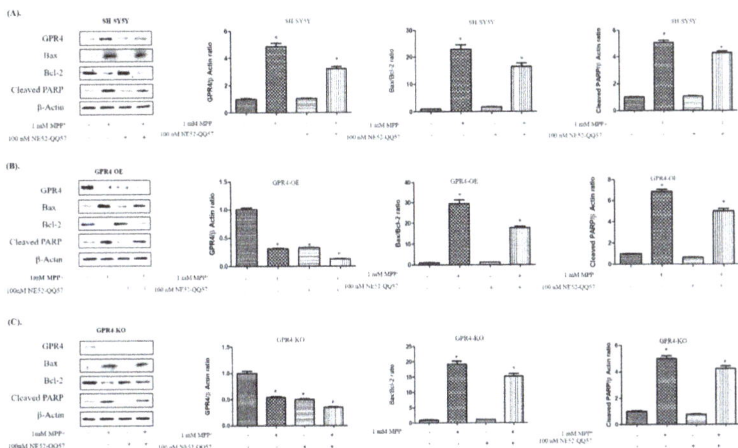

Figure 5. The effect of GPR4 antagonist, NE52-QQ57, on the GPR4 and pro-apoptotic protein expressions in MPP$^+$-treated SH-SY5Y cells that were stably GPR4-OE or GPR4-KO. 24 h serum-starved SH-SY5Y cells were pre-treated for 1 h with NE52-QQ57 (100 μM); this was followed by MPP$^+$ (1 mM) stimulation for 24 h in serum-free culture media. (**A**) An immunoblot and semi-quantification of the respective protein expressions of GPR4, Bax, Bcl-2, Cleaved PARP, and β-Actin in SH-SY5Y cells. (**B**) An immunoblot and semi-quantification of the respective protein expressions of GPR4, Bax, Bcl-2, cleaved PARP, and β-Actin in GPR4-OE cells. (**C**) An immunoblot and semi-quantification of the respective protein expressions of GPR4, Bax, Bcl-2, cleaved PARP, and β-Actin in GPR4-KO cells. β-Actin was utilised as an internal control. Mean ± SEM (n = 3) was employed to express the data. Tukey's multiple comparison test was performed using a one-way ANOVA. Each * p < 0.05 refers to the sample concentration compared with the same group of non-treated cells.

For the control SH-SY5Y cells, MPP$^+$ stimulation significantly increased the Bax/Bcl-2 ratio (22.94 ± 2.02 folds) and the cleaved PARP (5.09 ± 0.18 folds), in comparison with the non-treated control SH-SY5Y cells. Pre-treatment with the NE52-QQ57 (100 nM) significantly lowered the GPR4 expression (3.21 ± 0.18 folds), Bax/Bcl-2 ratio (16.47 ± 1.45 folds), and cleavage of PARP (4.29 ± 0.15 folds) in the MPP$^+$-stimulated cells, in comparison with the SH-SY5Y cells that were only MPP$^+$-treated (Figure 5A). For GPR4-OE cells, MPP$^+$ stimulation significantly increased the Bax/Bcl-2 ratio (29.49 ± 2.06 folds) and the cleaved PARP (6.85 ± 0.23 folds), in comparison with the non-treated GPR4-OE cells. Similar to the RT-PCR results, pre-treatment with NE52-QQ57 (100 nM) significantly lowered the GPR4 expression (0.13 ± 0.003 folds), Bax/Bcl-2 ratio (17.81 ± 0.86 folds), and cleavage of PARP (4.98 ± 0.26 folds) in the MPP$^+$- and NE52-QQ57-treated cells, in comparison with the GPR4-OE cells that were only treated with MPP$^+$ (Figure 5B). In the GPR4-KO cells, MPP$^+$ increased the Bax/Bcl-2 ratio (19.15 ± 1.45 folds) and cleaved PARP (4.99 ± 0.27 folds), in comparison with non-treated GPR4-KO cells. In contrast, MPP$^+$-stimulated GPR4-KO cells that were pre-treated with NE52-QQ57 (100 nM) demonstrated only a minor increase in Bax/Bcl-2 ratio (15.24 ± 0.91 folds) and cleavage of PARP (4.23 ± 0.22 folds), in comparison with GPR4-KO cells that were only MPP$^+$-treated (Figure 5C). Overall, the GPR4-KO cells displayed a lesser increase in the Bax/Bcl-2 ratio (19.15 ± 1.45 folds) and a decrease in the PARP cleavage (4.99 ± 0.27 folds), in comparison with both the MPP$^+$-stimulated SHSY-5Y (Bax/Bcl-2, 22.94 ± 2.02 folds; cleaved PARP, 5.09 ± 0.18 folds) and GPR4-OE cells (Bax/Bcl-2, 29.49 ± 2.06 folds; cleaved PARP, 6.85 ± 0.23 folds). Thus, less apoptotic cell death was induced by MPP$^+$.

2.6. Knockout of GPR4 Decreases the Caspase-3 Activity and Lowers the ROS Generation in Neurotoxin-Stimulated SH-SY5Y Cells

Caspases are important factors that trigger apoptosis. Caspase-3, in particular, is a crucial biomarker and executor of neuronal apoptosis [31]. To evaluate Caspase-3 activity, immunoblotting and a caspase activity assay were performed (Figure 6). SH-SY5Y cells were treated with MPP$^+$ (1 mM) for 24 h in serum-free media, while cell lysates were analysed through a western blot and a caspase activity assay. MPP$^+$-treated GPR4-OE cells demonstrated a significant increase in their cleaved Caspase-3 protein levels, whereas knockout of GPR4 prevented an MPP$^+$ stimulated increase in the level of cleaved Caspase-3 (Figure 6A).

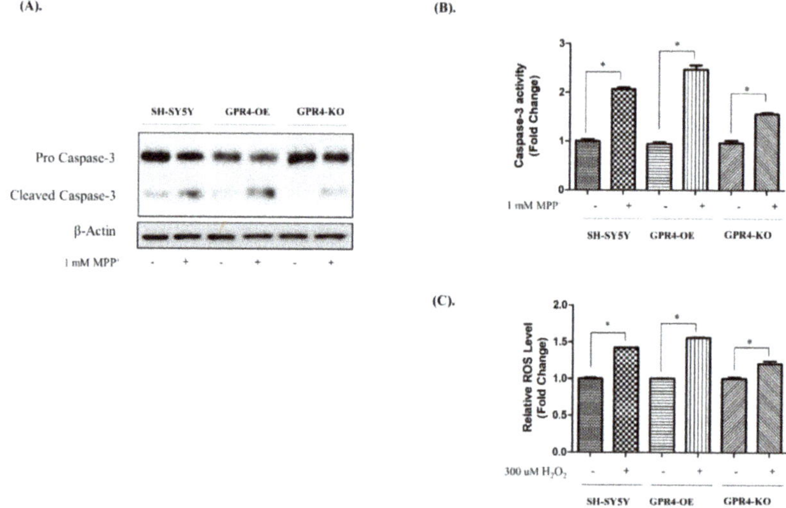

Figure 6. The Caspase-3 activity and intracellular ROS generation in MPP$^+$- and H$_2$O$_2$-treated SH-SY5Y cells that were stably GPR4-OE or GPR4-KO. 24 h serum-starved SH-SY5Y cells were treated with MPP$^+$ (1 mM) for 24 h in serum-free culture media for an immunoblot and a caspase activity assay. SH-SY5Y, GPR4-OE, and GPR4-KO cells were treated with H$_2$O$_2$ (300 μM) for 1 h in serum-free culture media for a 2′,7′-dichlorofluorescein diacetate (DCFDA) assay. (**A**) An immunoblot of the pro caspase & cleaved Caspase-3 and β-Actin. (**B**) Caspase-3 activity was measured using a colorimetric assay kit (Sigma, CAS No. CASP-3-C) in MPP$^+$-induced apoptotic cells. (**C**) The relative intracellular ROS level after 1 h of H$_2$O$_2$ (300 μM) treatment of 24 h serum-starved SH-SY5Y, GPR4-OE, and GPR4-KO cells. β-Actin was utilised as an internal control. Mean ± SEM (n = 3) was employed to express the data. Tukey's multiple comparison test was performed using a one-way ANOVA. Each * $p < 0.05$ refers to the sample concentration compared with the same group of non-treated cells.

The Caspase-3 activity assay results were in complete agreement with the Caspase-3 activity demonstrated in the immunoblot data (Figure 6B). MPP$^+$-treated GPR4-KO cells presented a lower level of caspase activity (1.55 ± 0.03 folds) than the MPP$^+$-treated SHSY-5Y (2.06 ± 0.04 folds) and GPR4-OE cells (2.47 ± 0.011 folds; Figure 6B).

We evaluated the effects of GPR4 overexpression and knockout on H_2O_2-induced intracellular ROS generation in SH-SY5Y cells [19]. In our study, the protective effect of the knockout of GPR4 against H_2O_2 resulted in lower intracellular ROS levels measured in the SH-SY5Y cells. DCFDA, a fluorescent dye that in the presence of ROS is oxidised to fluorescent DCF, was utilised for the detection of intracellular ROS levels. Treatment of the SH-SY5Y cells with H_2O_2 (300 μM) for 1 h led to a marked increase in their intracellular ROS levels. In H_2O_2-treated GPR4-OE cells, the level of intracellular ROS generation was 1.56 ± 0.01 folds higher than that of the non-treated GPR4-OE cells, whereas the H_2O_2-treated GPR4-KO cells demonstrated a lower level of ROS generation (1.20 ± 0.04 folds) in comparison with both the H_2O_2-treated control SH-SY5Y (1.42 ± 0.001 folds) and the H_2O_2-treated GPR4-OE cells (1.56 ± 0.01 folds; Figure 6B).

2.7. Knockout of GPR4 Increases the Mitochondrial Membrane Potential (MMP) in Neurotoxin-Stimulated SH-SY5Y Cells

Excess intracellular ROS leads to swelling of the mitochondrial matrix and rupture of the outer membrane, which opens up the mitochondrial permeability transition pores (mPTPs). As a result, the mitochondrial membrane potential (MMP) is disrupted and mitochondrial oxidative stress-mediated apoptosis is initiated [4].

In this study, to measure the MMP, 24 h serum-starved SH-SY5Y cells were treated with MPP^+ (1 mM) for 24 h in serum-free culture media. MMP was then determined through a JC-10 fluorescence quantitative assay. Similarly, in a separate experiment in a 6-well plate, cells were utilised for JC-10 fluorescence microscopy to visualise the red and green fluorescence.

Aggregated JC-10 is an indicator of MMP; the greater the ratio of red/green fluorescence, the higher the level of MMP. In a quantitative JC-10 fluorescence microplate assay, the MPP^+-treated SHSY-5Y cells presented a lower red/green fluorescence ratio (44.75 ± 0.82%) than the untreated SH-SY5Y cells (Figure 7A). The MPP^+-treated GPR4-OE cells, meanwhile, displayed the lowest red/green fluorescence ratio of all the samples (39.44 ± 0.39%), indicating a loss of MMP. In contrast, GPR4 knockout prevented the loss of MMP for the MPP^+-treated GPR4-KO cells, as indicated by a higher red/green fluorescence ratio (65.44 ± 0.99%). Therefore, the MPP^+-stimulated GPR4-KO cells demonstrated a higher level of MMP than either the MPP^+-stimulated SH-SY5Y (44.75 ± 0.82%) or the GPR-OE (39.44 ± 0.39%) cells.

In this study, the aggregated JC-10 created red fluorescence in the polarised mitochondrial membrane. When the MMP collapsed in apoptotic cells, the JC-10 retained its monomeric form, which is characterised by green fluorescence. An increase in the red/green fluorescence intensity ratio indicated intact mitochondria. In a separate experiment, to visualise the JC-10 fluorescence dye in the MPP^+-treated cells, JC-10 fluorescence microscopy was employed. In the control SHSY-5Y cells, both red and green fluorescence was observed, with a high level of red fluorescence and low level of green fluorescence. In contrast, the level of green fluorescence was higher and the red fluorescence remarkably lower in both the MPP^+-treated SHSY-5Y and GPR4-OE cells, when compared with the control SHSY-5Y cells. In the MPP^+-treated GPR4-KO cells, the red fluorescence was restored close to that of the control SHSY-5Y cells, while the level of green fluorescence was decreased (Figure 7B). These results suggest that GPR4 knockout restores the MPP^+-induced a loss of MMP in dopaminergic neurons.

2.8. Knockout of GPR4 Decreases the Intracellular Calcium in Neurotoxin-Stimulated SH-SY5Y Cells

Increases in intracellular Ca^{2+} in association with MPP^+- or H_2O_2-mediated apoptotic cell death have been previously reported [32]. Several studies have suggested that an increase in the intracellular Ca^{2+} released from the ER store by the inositol trisphosphate receptor (IP_3R) is directly responsible for mitochondrial Ca^{2+} overload [33,34]. However, the exact mechanism by which MPP^+ or H_2O_2 stimulation increases the intracellular calcium is not clearly understood. Interestingly, several studies have demonstrated that H_2O_2-/MPP^+-mediated mitochondrial oxidative stress is associated with an intracellular Ca^{2+} spike, which increases the Bax/Bcl-2 ratio, the release of cytochrome C, mitochondrial depolarisation, and the Caspase-3 activity in neuronal cells [19,32].

Previous reports have suggested that many G protein coupled-receptors (GPCRs), such as GPR4, which releases $G_{\beta\gamma}$ and activates G_i, are capable of Ca^{2+} signalling. Few GPCRs, however, harness $G_{\beta\gamma}$-dependent activation of PLC_β to release ER-stored Ca^{2+} into the cytoplasm through PIP_2 degradation [35,36]. In this study, the MPP^+-treated GPR4-OE cells demonstrated an increased proteolytic degradation of PIP_2, in comparison with the SH-SY5Y cells treated with MPP^+. Contrastingly, the MPP^+-stimulated GPR4-KO cells presented a particularly low degradation of PIP_2 compared with both the MPP^+-stimulated SH-SY5Y and GPR4-OE cells (Figure 8A).

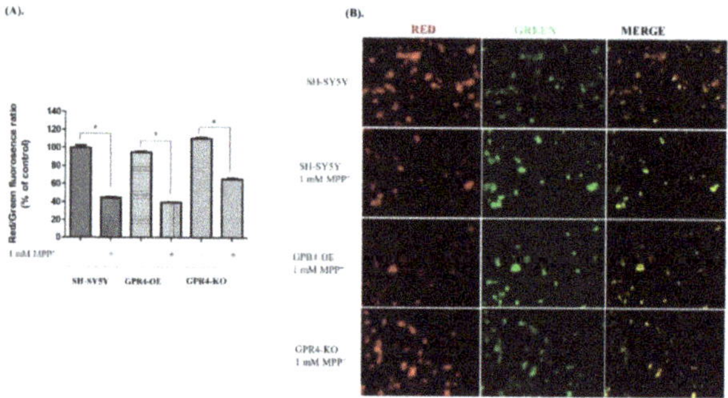

Figure 7. The measurement of mitochondrial membrane potential (MMP) in MPP^+-treated SH-SY5Y cells that were stably GPR4-OE or GPR4-GPR4-KO. 24 h serum-starved SH-SY5Y cells were treated with MPP^+ (1 mM) for 24 h in serum-free culture media; a JC-10 fluorescence quantitative assay (according to the manufacturer's instructions) and fluorescence microscopy were then employed to measure their MMP. (**A**) The percentage ratios of J-aggregates (red) and J-monomers (green). (**B**) MMP changes were monitored with a JC-10 dye that is detectable through fluorescence microscopy. Mean ± SEM ($n = 3$) was employed to express the data. Tukey's multiple comparison test was performed using a one-way ANOVA. Each * $p < 0.05$ refers to the sample concentration compared with the same group of non-treated cells.

To evaluate whether GPR4 overexpression increased intracellular calcium through $G_{\beta\gamma}$ modulation of the PLC_β-PIP_2 pathway, SH-SY5Y cells were treated with MPP^+ (1 mM) for 24 h in serum-free media. Cell lysates were analysed through western blotting to determine the degradation of PIP_2. The SH-SY5Y cells were treated with H_2O_2 (200 µM) for 2 h 30 min to determine their relative intracellular Ca^{2+}, utilising a Fluo-4 AM calcium indicator in a fluorescence microplate assay. Similarly, in a separate 6-well plate, cells stained with a Fluo-4 AM calcium indicator were observed under a fluorescence microscope.

The quantitative analysis of intracellular Ca^{2+}, as indicated by the Fluo-4 AM microplate assay, found levels of intracellular Ca^{2+} for the H_2O_2-treated GPR4-OE cells that were 369.58 ± 24.75% higher than those for the non-treated GPR4-OE cells. In contrast, the H_2O_2-treated GPR4-KO cells presented significantly lower levels of intracellular Ca^{2+} (181.28 ± 0.85%), in comparison with the H_2O_2-treated SH-SY5Y (243.25 ± 7.81%) and H_2O_2-treated GPR4-OE cells (369.58 ± 24.75%; Figure 8B).

In a separate experiment to visualise intracellular Ca^{2+} levels in the H_2O_2-treated cells, Fluo-4 AM fluorescence microscopy was employed. This round of microscopy demonstrated similar results to those obtained from the quantitative microplate assay. H_2O_2-treated GPR4-OE cells displayed the highest levels of green Fluo-4 AM fluorescence, while H_2O_2-treated GPR4-KO cells produced lower levels of green Fluo-4 AM fluorescence than both the H_2O_2-treated SH-SY5Y and GPR4-KO cells (Figure 8C). Overall, these data suggest that the increase in intracellular calcium associated with H_2O_2- or MPP^+-mediated mitochondrial oxidative stress is exaggerated by GPR4 overexpression, whereas GPR4 knockout prevents an increase in intracellular Ca^{2+} through the decrease of PIP_2 degradation, and thus restricts the release of Ca^{2+} from the ER by preventing the degradation of

PIP$_2$. Therefore, GPR4-PLC$_\beta$-PIP$_2$ signalling may act as a key factor through which GPR4 increases intracellular calcium and potentiates mitochondrial oxidative stress-mediated apoptosis.

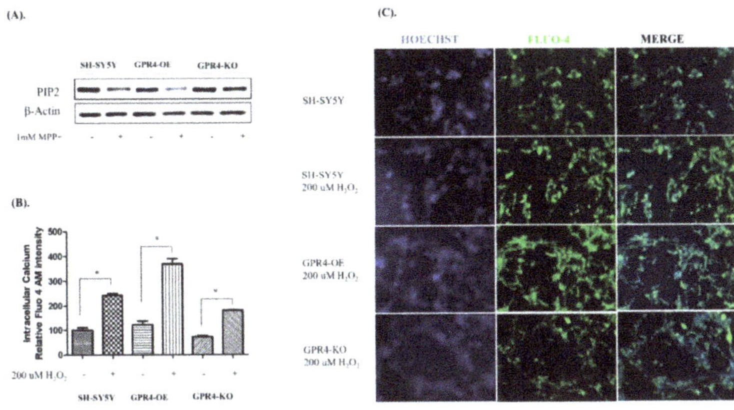

Figure 8. Phosphatidylinositol biphosphate (PIP$_2$) calcium signalling and intracellular calcium levels in MPP$^+$-treated SH-SY5Y cells that were stably GPR4-OE or GPR4-KO. 24 h serum-starved SH-SY5Y cells were treated with MPP$^+$ (1 mM) for 24 h in serum-free culture media for the purpose of immunoblotting. The SH-SY5Y cells were then treated with H$_2$O$_2$ (200 μM) for 2 h 30 min in serum-free culture media and subjected to a Fluo-4 AM fluorescence assay and fluorescence microscopy. Detection of the Fluo-4 AM fluorescence intensity and related imaging were carried out according to the manufacturer's instructions. Cells were counter-stained with Hoechst dye. (**A**) An immunoblot of the PIP$_2$ and β-Actin. (**B**) A quantitative analysis of the Fluo-4 AM fluorescence intensity in H$_2$O$_2$- (200 μM) treated cells. (**C**) Fluo-4 AM calcium imaging of the intracellular calcium level. Mean ± SEM (n = 3) was employed to express the data. Tukey's multiple comparison test was performed using a one-way ANOVA. Each * $p < 0.05$ refers to the sample concentration compared with the same group of non-treated cells.

3. Discussion

In this study, we investigated the roles of GPR4 overexpression, pharmacological inhibition, and genetic knockout in the mitochondrial oxidative stress-induced apoptotic cell death that is associated with PD. Although many studies have reported the activation of GPR4 at the physiological pH range (7.0–7.4), overexpression of GPR4 showed relatively high GPR4 activity at neutral pH 7.4 [37]. In transiently GPR4-overexpressing HEK293 cells, GPR4 is inactive at pHs higher than 8.0, whereas it is highly active at the physiological pH, 7.4, and substantially less active at pHs down to 6.8 (plausible in the range of physiological acidification) [38]. The pH sensitivity of GPR4 has been reported to vary for different cells, though potentially due to the methods employed in different laboratories [25]. In the natively GPR4-expressing cell, HUVEC, pHs from 7.4 to 7.0 have been shown to result in a 1.5-fold activation of GPR4 [25]. In this study, we found an increase in GPR4 mRNA expression at pH 7.4 in both SH-SY5Y and stably GPR4-OE cells in serum-starved media (data not added). A very slight increase in the expression of GPR4 was observed at pH 6.4. Therefore, to maintain consistency, we conducted all the experiments at a pH ~7.4. This was also the pH of the culture media that we employed.

Human-derived neuroblastoma SH-SY5Y cells are widely used in neuroscientific research as an in vitro model for the investigation of neuronal differentiation and neuroprotective events. Stimulation with several neurotoxins, such as MPP$^+$, MPTP, rotenone, 6-OHDA, and H$_2$O$_2$, has been utilised to induce oxidative stress-mediated apoptotic death, thereby mimicking neurodegenerative diseases, including PD and aging [39–41]. To determine the final concentration of H$_2$O$_2$ and MPP$^+$, SH-SY5Y cells were treated with H$_2$O$_2$ at different concentrations, ranging from 75 μM to 125 μM, for 24 h, as well as with MPP$^+$, ranging from 250 µM to 1 mM. H$_2$O$_2$ and MPP$^+$ both decreased the cell viability in a concentration-dependent manner, with optimum cytotoxicity being observed at concentrations of 125 µM for H$_2$O$_2$ and 1 mM for MPP$^+$; these concentrations were selected for further experiments to determine their cytotoxicity in the serum-free SH-SY5Y cell line. In our study, GPR4 mRNA and protein expressions were increased in a time-dependent manner for 24 h in both MPP$^+$- and H$_2$O$_2$-treated SH-SY5Y cells. Hence, GPR4 is directly linked with MPP$^+$- and H$_2$O$_2$-induced apoptotic cell death.

Both the pro-apoptotic protein, Bax, and the anti-apoptotic protein, Bcl-2, are members of the Bcl-2 family and are directly involved in apoptotic cell death. The balance between these two proteins of the Bcl-2 family, or an increase in the Bax/Bcl-2 ratio, indicates the early phases of an apoptotic cascade [29,30]. Significant increases in ROS, or the Bax/Bcl-2 ratio, result in the collapse of the mitochondrial membrane potential, the release of cytochrome C, the activation of Caspase-3, the cleavage of PARP, and, subsequently, apoptotic cell death [6,7]. Both MPP$^+$- and H$_2$O$_2$-induced apoptotic cell deaths bear the characteristic hallmarks of an increase in the Bax/Bcl-2 ratio, the release of cytochrome-C, and the activation of the proteolytic enzyme, Caspase-3, which cleaves PARP and induces apoptotic cell death [7,8]. In our study, the overexpression of GPR4 in SH-SY5Y cells significantly increased the effect of either MPP$^+$ or H$_2$O$_2$ and increased the Bax/Bcl-2 ratio, as was seen in both the immunoblot and RT-PCR. As a result, this significantly increased the protein level of the cleaved Caspase-3, the Caspase-3 mediated cleavage of PARP, and the Caspase-3 activity. On the contrary, the CRISPR/Cas9 knockout of GPR4 was found to result in a lesser increase in the Bax/Bcl-2 mRNA and protein ratio in both MPP$^+$- and H$_2$O$_2$-treated cells. Knockout of GPR4 was also shown to reduce the cleavage of PARP after MPP$^+$ treatment. NE52-QQ57, a selective antagonist of GPR4, demonstrated a similar level of the inhibition of GPR4 expression, as was determined through both our immunoblots and RT-PCR.

We further investigated the effect of GPR4 on mitochondrial oxidative stress-induced increases in intracellular ROS generation and MMP. Surprisingly, GPR4-OE was found to significantly increase tricellular ROS generation in SH-SY5Y cells, whereas GPR4-KO generated a lower level of intracellular ROS accumulation, after a high concentration of H$_2$O$_2$ treatment. Similarly, through both a JC-10 assay and fluorescence microscopy, the knockout of GPR4 was found to decrease mitochondrial membrane depolarisation. In JC-10-tagged fluorescence microscopy, knockout of GPR4 was seen to prevent MPP$^+$ stimulated decrease red fluorescence and increase green fluorescence. The latter was highly increased in the case of GPR4-OE as membrane depolarisation occurs 24 h after MPP$^+$ treatment.

Besides mitochondrial dysfunction, abnormal protein aggregation and dysregulated Ca^{2+} homeostasis are other factors that may be involved in the neurodegeneration observed in individuals with PD [42]. Recent findings suggest that increases in cytosolic Ca^{2+} occur at both early and late stages of the apoptotic pathway. In both cases, ER Ca^{2+} channels are linked with the release of Ca^{2+} to the cytoplasm [33,34]. However, the exact mechanism by which intracellular Ca^{2+} modulates mitochondrial oxidative stress-mediated apoptosis remains elusive. Many studies have suggested that MPP$^+$- and H$_2$O$_2$-induced apoptosis are associated with an increase in intracellular calcium levels [43]. For example, Sing et al. (2016) demonstrated that the administration of Nimodipine, an L-type calcium channel blocker, protected from MPTP-induced dopaminergic neuronal death in an animal model of PD. More importantly, providing evidence for Nimodipine as a means to improve mitochondrial integrity and function. In the study, Nimodipine attenuated the MPTP-induced loss of tyrosine hydroxylase-positive dopaminergic neurons in the SN. It also improved mitochondrial oxygen consumption and inhibited ROS production, as well as improving mitochondrial integrity and function in striatal mitochondria [43]. These findings provide evidence in support of the notion that calcium signalling is linked with neurotoxin-induced mitochondrial dysfunction and neurodegeneration. GPR4 is a G$_s$-coupled receptor that signals through adenylate cyclase and also via G proteins G$_{13}$ and G$_{q/11}$. GPR4 is well known for its ability to recognise phospholipase C β (PLC$_\beta$) as its canonical target [44]. G$_q$ class α subunits, or G$_{\beta\gamma}$ released by GPCR, activate Ca^{2+} signalling through G$_{\beta\gamma}$-dependent activation of PLC$_\beta$. Upon activation, PLC$_\beta$ hydrolyses PIP$_2$ to generate IP$_3$. IP$_3$ binds to the ER-resident IP_3 receptors, which act as Ca^{2+} release channels to release ER-stored Ca^{2+} into the cytoplasm [35]. In this study, overexpression of GPR4 significantly increased the intracellular calcium level in both MPP^+- and H_2O_2-treated cells (MPP^+ data not given), whereas knockout generated very little change in the intracellular calcium. These findings were also observed in our study when employing the Fluo-4 AM indicator. To determine how GPR4 can modulate the intracellular calcium level, we investigated GPCR-mediated calcium signalling. We found that GPR4 knockout decreases the breakdown of PIP_2, which is a critical step in Ca^{2+} release from the ER to the cytoplasm. Therefore, decreased intracellular Ca^{2+} may be responsible for GPR4-mediated neuroprotection against MPP^+- or H_2O_2-induced apoptotic cell death. Our study, for the first time, demonstrated that the knockout of GPR4 protects SH-SY5Y cells from both MPP^+- and H_2O_2-stimulated mitochondrial apoptotic cell death, in association with a decrease in intracellular Ca^{2+}.

In summary, our study suggests that overexpression of GPR4 potentiates neurotoxin-induced mitochondrial oxidative stress, whereas a knockout or pharmacological inhibition of GPR4 improves the neurotoxin-induced, caspase-dependent mitochondrial apoptosis of dopaminergic neuronal cells. This study has also found that GPR4 can increase intracellular Ca^{2+} through the degradation of PIP_2. Further investigation is required to determine how GPR4-mediated calcium signalling can mitigate the neuronal cell death seen in neurodegenerative disorders, including PD.

4. Materials and Methods

4.1. Reagents and Antibodies

1-methyl-4-phenylpyridinium ion MPP^+, H_2O_2, and 3-(3,4-dimehylthiazol-2-yl)-2,5-diphenyl-tetrazolium bromide (MTT) were obtained from Sigma-Aldrich (St. Louis, MO, USA). 96-well tissue culture plates, along with six-well and 100 mm culture dishes, were obtained from Nunc Inc. (North Aurora Road, Naperville, IL, USA). Foetal bovine serum (FBS) and Dulbecco's modified Eagle's medium (DMEM/F12) were purchased from Gibco-BRL Technologies (Gaithersburg, MD, USA). RIPA buffer (10×) was purchased from Millipore (Milford, MA, USA). Tween 80 was obtained from Calbiochem (Gibbstown, NJ, USA). All other chemicals utilised in this research were of analytical grade and were purchased, unless otherwise noted, from Sigma-Aldrich.

4.2. Cell Culture and Transfection

The human dopaminergic neuroblastoma SH-SY5Y cell line was acquired from the American Type Culture Collection (ATCC; Manassas, VA, USA). SH-SY5Y cells were cultured in DMEM/F12 with or without phenol red and HEPES, supplemented with 100 U/mL penicillin/streptomycin and 10% (v/v) inactivated foetal bovine serum. The SH-SY5Y cells were maintained in a 5% CO_2 and 95% humidified air incubator at 37 °C for the time indicated in the experiments. MPP^+ and H_2O_2 were dissolved in three-times distilled water (3DW).

To overexpress and knockout the human GPR4 gene, a GPR4 lentiviral vector (CMV; pLenti-GIII-CMV-C-term-HA) and a GPR4 sgRNA CRISPR/Cas9 lentiviral vector (pLenti-U6-sgRNA-SFFV-Cas9-2A-Puro) were designed, to generate stable GPR4-overexpressing (GPR4-OE) and stable GPR4-knockout (GPR4-KO) SH-SY5Y lines. Briefly, the SH-SY5Y cells were transferred into 60 mm plates at a density of 5×10^4 cells/mL to achieve ~70% confluence at the time of transfection. The cells were transfected using the Lipofectamine® 3000 transfection reagent (ThermoFisher, Langenselbold Germany; #L3000015), according to the manufacturer's protocol. Both the GPR4-overexpression and knockout-silencing genes were designed to carry puromycin-resistance genes and were produced using the Lentivector Expression System (Applied Biological Materials Inc. (ABM), Canada). Stable single clones were selected following 3–5 weeks of puromycin treatment (1 µg/µL). GPR4 overexpression and knockout in the stably infected clones were assessed through RT-PCR and western blotting. A sequence analysis of the GPR4 insert was also employed.

4.3. Measurement of Cell Viability

The cytotoxicity of the MPP$^+$- and H$_2$O$_2$-treated SH-SY5Y cells was measured with an MTT assay, involving the reduction of formazan crystals [41]. SH-SY5Y cells (2.2 × 10^4 cells/mL) were pre-treated in 24-well plates with NE 52-QQ57 (100 nM) and left in serum-free cell culture media for 1 h; this was followed by stimulation with or without MPP$^+$ (1 mM) or H$_2$O$_2$ for 24 h. After MPP$^+$ or H$_2$O$_2$ stimulation, the medium was replaced with 0.5 mg/mL MTT solution, before the plates were incubated for 3 h at 37 °C. The supernatant was carefully removed and the formazan crystals were dissolved in dimethyl sulfoxide (DMSO) by gentle shaking for 10 min. A microplate reader (Molecular device, Sunnyvale, CA, USA) was utilised to measure the absorbance at 550 nm.

4.4. Total RNA Isolation for RT-PCR

SH-SY5Y cells (2.2 × 10^4 cells/mL) were pre-treated in 60 mm cell culture dishes with NE 52-QQ57 (100 nM), then left in serum-free cell culture media for 1 h, followed by stimulation with or without MPP$^+$ (1 mM) or H$_2$O$_2$ for 18 h, once again in serum-free media. TRIzol (Invitrogen; Burlington, ON, Canada) was employed to extract the total RNA from the cells. 2.5 μg total RNA from each group was reverse-transcribed using a first-strand cDNA synthesis kit (Invitrogen). The following primers were utilised for PCR: GPR4 sense, 5′-CCGTTGTCAAGACCGGGG-3′; GPR4 anti-sense, 5′-TCCTAGGACCCCCAGAAAGCA-3′; Bax sense, 5′-CACCAAGGTGCCGGAACTGA-3′; Bax anti-sense, 5′-AATGCCCATGTCCCCCAATC-3′; Bcl-2 sense, 5′-ACGACTTCTCCCGCCGCTAC-3′; Bcl-2 anti-sense, 5′-CCCAGCCTCCGTTATCCTGG-3′; GAPDH sense, 5′-GCAGTGGCAAAGTGGAGATTG-3′; and GAPDH anti-sense, 5′-TGCAGGATGCATTGCTGACA-3′. Then, adopting the previously mentioned primers, the cDNA was amplified through PCR [45]. GAPDH was employed as an internal control to evaluate the relative levels of expression of other genes. PCR products were analysed on 1.0–1.2% agarose gels stained with GelRed (Sigma-Aldrich; St. Louis, MO, USA). The gels were photographed and, utilising ImageJ (NIH) software, the pixel intensity for each band in the photographs was measured and normalised to the band intensity of the GAPDH mRNA, to quantify its relative expression.

4.5. Immunoblot Analysis

SH-SY5Y cells (2.2 × 10^4 cells/mL) were pre-treated in 60 mm cell culture dishes with NE 52-QQ57 (100 nM) and left in serum-free media for 1 h. They were then stimulated with or without MPP$^+$ (1 mM) or H$_2$O$_2$ for 24 h, again in a serum-free media. Next, the cells were washed two times with PBS and lysed for 10 min at 4 °C using an RIPA lysis buffer (with protease and phosphatase inhibitors). Supernatants were collected for further investigation after the cell lysates were centrifuged at 14,000 rpm, at 4 °C. The protein concentration of each sample was measured and normalised using a DC Protein Assay kit (Bio-Rad). Equal amounts of proteins (20–30 μg) were loaded and separated electrophoretically in 8, 10, and 12% sodium dodecyl sulphate-polyacrylamide gels; these were then transferred to polyvinylidene difluoride membranes (Millipore; Bedford, MA, USA). The membranes were incubated overnight at 4 °C, with corresponding primary antibodies, GPR4 (1:500) from Novus Biologicals (Centennial, CO, USA), BCL-2 (1:1000), Caspase-3 (1:1000), cleaved Caspase-3 (1:1000), cleaved PARP (1:1000) from Santa Cruz Biotechnology (Santa Cruz, CA, USA), Bax (1:1000) from Cell Signaling Co. (Boston, MA, USA), PIP$_2$ (1:500) from Abcam (Cambridge, United Kingdom), and β-Actin (1:2000) from Sigma-Aldrich (St. Louis, MO, USA), followed by 1 h incubation with horseradish peroxidase (HRP)-conjugated secondary antibodies (1:2000; Cell signalling, MA, USA). The blots were visualised using a Biorad-ECL (Bio-Rad Laboratories; Hercules, CA, USA) and photographed. Using ImageJ (NIH) software, the pixel intensity for each band in the photographs was measured and normalised to the band intensity of β-Actin, to quantify its relative expression.

4.6. Detection of Intracellular ROS

The ROS-sensitive fluorescent dye, 2′,7′-dichlorofluorescein diacetate (DCFDA; Sigma-Aldrich), was utilised to measure the intracellular ROS levels. SH-SY5Y cells (2.2×10^4 cells/mL) were cultured in black 96-well plates in DMEM/F12 without phenol red. Then, 60–70% confluence cells were stimulated with H_2O_2 (300 µM) for 1 h in serum-free media and then washed twice with PBS, followed by a 30 min incubation with DCFDA (10 µM) in PBS. The cells were then rinsed with PBS twice. Finally, 200 µL PBS was added, and fluorescence was measured using 485 nm excitation and 535 nm in a fluorescence microplate reader (Molecular Device; Sunnyvale, CA, USA).

4.7. Assessment of Caspase-3 Activity

Caspase-3 activity was measured using a Colorimetric Caspase-3 Assay Kit (Sigma-Aldrich; St. Louis, MO, USA), as described previously [41]. The reaction mixture (total volume, 200 µL) was distributed in 96-well plates and incubated at 37 °C for 90 min. Absorbance values were measured at wavelengths of 405 nm in a Tecan Microplate Reader (Meilen; Zurich, Switzerland).

4.8. Assessment of Mitochondrial Membrane Potential (MMP)

A JC-10-based Mitochondrial Membrane Potential Assay Kit (Abcam; Cambridge, United Kingdom) was employed to assess MMP, while a fluorescence microscope was utilised to visualise the JC-10 staining, according to the manufacturer's instructions. In brief, SH-SY5Y cells (2.2×10^4 cells/mL) were cultured in black, 96-well plates for quantification and in 6-well plates for imaging in DMEM/F12 without phenol red. Then, 60–70% confluence cells were stimulated with MPP^+ (1 mM) for 24 h in serum-free media. The cells were incubated with a JC-10 dye loading solution at 37 °C for 1 h and protected from the light. For the 96-well plates, their fluorescence intensities ($\lambda_{ex} = 490/\lambda_{em} = 525$ nm) and ($\lambda_{ex} = 540/\lambda_{em} = 590$ nm) the red-green fluorescence ratios were measured using a fluorescence microplate reader (Molecular Device; Sunnyvale, CA, USA), while confocal images were acquired with a Nikon Eclipse Ts2-FL diascopic and epi-fluorescence illumination microscope.

4.9. Detection of Intracellular Calcium

Intracellular calcium was assessed with a Fluo-4 AM dye (Abcam; Cambridge, United Kingdom) and confocal microscopy, following the manufacturer's instructions. Fluo-4 AM was diluted in DMSO containing 2 mM probenecid and 0.02% pluronic F-127. In brief, the SH-SY5Y cells (2.2×10^4 cells/mL) were cultured in black, 96-well plates for quantification and in 6-well plates for imaging in DMEM/F12 without phenol red. Then, 60–70% confluence cells were stimulated with H_2O_2 (200 µM) for 2 h 30 min in serum-free media. The cells were washed with PBS containing probenecid (2 mM) at room temperature. The cells were incubated with the Fluo-4 AM (2 µM) dye loading solution at 37 °C for 30 min and protected from light, then washed with PBS containing probenecid (2 mM) at room temperature for 30 min. For the 96-well plates, the fluorescence intensities ($\lambda_{ex} = 488/\lambda_{em} = 515$ nm) were measured using a fluorescence microplate reader (Molecular Device; Sunnyvale, CA, USA), and confocal images were acquired with a Nikon Eclipse Ts2-FL diascopic and epi-fluorescence illumination microscope.

4.10. Statistical Analyses

Statistical analyses were performed using GraphPad Prism software, version 5 (GraphPad, La Jolla, CA, USA). Data are expressed as means ± standard error (SEM) of at least three independent experiments. One-way analysis of variance (ANOVA) followed by Tukey's post hoc analysis were performed to determine the significant differences between the groups. p-values < 0.05 were considered statistically significant.

Author Contributions: M.E.H., D.-K.C., and I.-S.K. conceptualized and designed the study; I.-S.K. supervised and corresponded; M.E.H. conducted the experiments and wrote the manuscript; M.E.H. and M.A. analyzed the

Int. J. Mol. Sci. **2020**, 21, 7517

data; M.A. and S.A. have reviewed the literature and proof read the manuscript. All authors have read and agreed to the published version of the manuscript.

Funding: This study was supported by the Basic Science Research Program through the National Research Foundation of Korea (NRF) funded by the Ministry of Science and ICT (NRF-2018R1C1B6005129). All authors have read and agree to the published version of the manuscript.

Acknowledgments: Authors especially like to thank Md Jakaria for his support during the whole research.

Conflicts of Interest: The authors declare no conflict of interest.

Abbreviations

MPTP	1-Methyl-4-Phenyl-1,2,3,6-Tetrahydropyridine
MPP$^+$	1-Methyl-4-Phenylpyridinium Ion
MTT	3-(3,4-Dimehylthiazol-2-Yl)-2,5-Diphenyl-Tetrazolium Bromide
DCF-DA	7'-Dichlorofluorescein Diacetate
DMEM	Dulbecco's Modified Eagle's Medium
ER	Endoplasmic Reticulum
FBS	Fetal Bovine Serum
HUVEC	Human Umbilical Vein Endothelial Cells
MMP	Mitochondrial Membrane Potential
Mptp	Mitochondrial Permeability Transition Pore
PIP2	Phosphatidylinositol Biphosphate
PARP	Poly (ADP-Ribose) Polymerase
SN	Substantia Nigra
TDAG8	T-Cell Death-Associated Gene 8
OGR1	The Ovarian Cancer G Protein-Coupled Receptor 1

References

1. Zhang, Q.; Hu, C.; Huang, J.; Liu, W.; Lai, W.; Leng, F.; Tang, Q.; Liu, Y.; Wang, Q.; Zhou, M.; et al. ROCK1 induces dopaminergic nerve cell apoptosis via the activation of Drp1-mediated aberrant mitochondrial fission in Parkinson's disease. *Exp. Mol. Med.* **2019**, *51*, 1–13. [CrossRef] [PubMed]
2. Henchcliffe, C.; Beal, M.F. Mitochondrial biology and oxidative stress in Parkinson disease pathogenesis. *Nat. Clin. Pract. Neurol.* **2008**, *4*, 600–609. [CrossRef]
3. Mattson, M.P. Apoptosis in neurodegenerative disorders. *Nat. Rev. Mol. Cell Biol.* **2000**, *1*, 120–130. [CrossRef]
4. Ly, J.D.; Grubb, D.R.; Lawen, A.J.A. The mitochondrial membrane potential ($\Delta\psi$ m) in apoptosis; an update. *Apoptosis* **2003**, *8*, 115–128. [PubMed]
5. Prenek, L.; Boldizsár, F.; Kugyelka, R.; Ugor, E.; Berta, G.; Németh, P.; Berki, T. The regulation of the mitochondrial apoptotic pathway by glucocorticoid receptor in collaboration with Bcl-2 family proteins in developing T cells. *Apoptosis* **2017**, *22*, 239–253. [CrossRef] [PubMed]
6. Gerace, E.; Masi, A.; Resta, F.; Felici, R.; Landucci, E.; Mello, T.; Pellegrini-Giampietro, D.; Mannaioni, G.; Moroni, F. PARP-1 activation causes neuronal death in the hippocampal CA1 region by increasing the expression of Ca^{2+}-permeable AMPA receptors. *Neurobiol. Dis.* **2014**, *70*, 43–52. [CrossRef] [PubMed]
7. Chaitanya, G.V.; Alexander, J.S.; Babu, P.P. PARP-1 cleavage fragments: Signatures of cell-death proteases in neurodegeneration. *Cell Commun. Signal.* **2010**, *8*, 31. [CrossRef]
8. More, S.V.; Choi, D.-K. Atractylenolide-I Protects Human SH-SY5Y Cells from 1-Methyl-4-Phenylpyridinium-Induced Apoptotic Cell Death. *Int. J. Mol. Sci.* **2017**, *18*, 1012. [CrossRef]
9. Kirichok, Y.; Krapivinsky, G.; Clapham, D.E. The mitochondrial calcium uniporter is a highly selective ion channel. *Nat. Cell Biol.* **2004**, *427*, 360–364. [CrossRef]
10. Surmeier, D.J.; Guzman, J.N.; Sanchez-Pandila, J.; Schumacker, P.T. The role of calcium and mitochondrial oxidant stress in the loss of substantia nigra pars compacta dopaminergic neurons in Parkinson's disease. *Neuroscience* **2011**, *198*, 221–231.
11. Gunter, T.E.; Gunter, K.K. Uptake of Calcium by Mitochondria: Transport and Possible Function. *IUBMB Life* **2001**, *52*, 197–204. [CrossRef]

12. Haupt, S.; Raghu, D.; Haupt, Y. p53 Calls upon CIA (Calcium Induced Apoptosis) to Counter Stress. *Front. Oncol.* **2015**, *5*, 57. [CrossRef] [PubMed]

13. Mosharov, E.V.; Larsen, K.E.; Kanter, E.; Phillips, K.A.; Wilson, K.; Schmitz, Y.; Krantz, D.E.; Kobayashi, K.; Edwards, R.H.; Sulzer, D. Interplay between Cytosolic Dopamine, Calcium, and α-Synuclein Causes Selective Death of Substantia Nigra Neurons. *Neuron* **2009**, *62*, 218–229. [CrossRef] [PubMed]

14. Sheehan, J.P.; Swerdlow, R.H.; Parker, W.D.; Miller, S.W.; Davis, R.E.; Tuttle, J.B. Altered calcium homeostasis in cells transformed by mitochondria from individuals with Parkinson's disease. *J. Neurochem.* **1997**, *68*, 1221–1233. [CrossRef] [PubMed]

15. Singer, T.P.; Ramsay, R.R. Mechanism of the neurotoxicity of MPTP. *FEBS Lett.* **1990**, *274*, 1–8. [CrossRef]

16. Przedborski, S.; Jackson-Lewis, V. Mechanisms of MPTP toxicity. *Mov. Disord.* **1998**, *13*, 35–38.

17. Dumont, A.; Hehner, S.P.; Hofmann, T.G.; Ueffing, M.; Dröge, W.; Schmitz, M.L. Hydrogen peroxide-induced apoptosis is CD95-independent, requires the release of mitochondria-derived reactive oxygen species and the activation of NF-κB. *Oncogene* **1999**, *18*, 747–757. [CrossRef]

18. Wang, C.; Youle, R.J. The role of mitochondria in apoptosis. *Annu. Rev. Genet.* **2009**, *43*, 95–118. [CrossRef]

19. Uğuz, A.C.; Öz, A.; Nazıroğlu, M. Curcumin inhibits apoptosis by regulating intracellular calcium release, reactive oxygen species and mitochondrial depolarization levels in SH-SY5Y neuronal cells. *J. Recept. Signal Transduct.* **2016**, *36*, 395–401. [CrossRef]

20. Knaryan, V.H.; Samantaray, S.; Park, S.; Azuma, M.; Inoue, J.; Banik, N.L.; Sookyoung, P. SNJ-1945, a calpain inhibitor, protects SH-SY5Y cells against MPP$^+$ and rotenone. *J. Neurochem.* **2014**, *130*, 280–290. [CrossRef]

21. Meyer, T.N.; Gloy, J.; Hug, M.J.; Greger, R.; Schollmeyer, P.; Pavenstädt, H. Hydrogen peroxide increases the intracellular calcium activity in rat mesangial cells in primary culture. *Kidney Int.* **1996**, *49*, 388–395. [CrossRef] [PubMed]

22. Ludwig, M.-G.; Vanek, M.; Guerini, D.; Gasser, J.A.; Jones, C.E.; Junker, U.; Hofstetter, H.; Wolf, R.M.; Seuwen, K. Proton-sensing G-protein-coupled receptors. *Nat. Cell Biol.* **2003**, *425*, 93–98. [CrossRef] [PubMed]

23. Ishii, S.; Kihara, Y.; Shimizu, T. Identification of T Cell Death-associated Gene 8 (TDAG8) as a Novel Acid Sensing G-protein-coupled Receptor. *J. Biol. Chem.* **2005**, *280*, 9083–9087. [CrossRef] [PubMed]

24. Justus, C.R.; Dong, L.; Yang, L.V. Acidic tumor microenvironment and pH-sensing G protein-coupled receptors. *Front. Physiol.* **2013**, *4*, 354. [CrossRef]

25. Hosford, P.; Mosienko, V.; Kishi, K.; Jurisic, G.; Seuwen, K.; Kinzel, B.; Ludwig, M.; Wells, J.; Christie, I.; Koolen, L.; et al. CNS distribution, signalling properties and central effects of G-protein coupled receptor 4. *Neuropharmacology* **2018**, *138*, 381–392. [CrossRef]

26. Dong, B.; Zhang, X.; Fan, Y.; Cao, S.; Zhang, Y. Acidosis promotes cell apoptosis through the G protein-coupled receptor 4/CCAAT/enhancer-binding protein homologous protein pathway. *Oncol. Lett.* **2018**, *16*, 6735–6741. [CrossRef]

27. Krewson, E.A.; Sanderlin, E.J.; Marie, M.A.; Akhtar, S.N.; Velcicky, J.; Loetscher, P.; Yang, L.V. The Proton-Sensing GPR4 Receptor Regulates Paracellular Gap Formation and Permeability of Vascular Endothelial Cells. *iScience* **2020**, *23*, 100848. [CrossRef]

28. Dong, B.; Zhou, H.; Han, C.; Yao, J.; Xu, L.; Zhang, M.; Fu, Y.; Xia, Q. Ischemia/Reperfusion-Induced CHOP Expression Promotes Apoptosis and Impairs Renal Function Recovery: The Role of Acidosis and GPR4. *PLoS ONE* **2014**, *9*, e110944. [CrossRef]

29. Mortadza, S.S.; Sim, J.A.; Stacey, M.; Jiang, L.-H. Signalling mechanisms mediating Zn^{2+}-induced TRPM2 channel activation and cell death in microglial cells. *Sci. Rep.* **2017**, *7*, srep45032. [CrossRef]

30. Gross, A.; McDonnell, J.M.; Korsmeyer, S.J. BCL-2 family members and the mitochondria in apoptosis. *Genes Dev.* **1999**, *13*, 1899–1911. [CrossRef]

31. Hartmann, A.; Hunot, S.; Michel, P.P.; Muriel, M.-P.; Vyas, S.; Faucheux, B.A.; Mouatt-Prigent, A.; Turmel, H.; Srinivasan, A.; Ruberg, M.; et al. Caspase-3: A vulnerability factor and final effector in apoptotic death of dopaminergic neurons in Parkinson's disease. *Proc. Natl. Acad. Sci. USA* **2000**, *97*, 2875–2880. [CrossRef] [PubMed]

32. Jung, S.; Chung, Y.; Lee, Y.; Lee, Y.; Cho, J.W.; Shin, E.-J.; Kim, H.-C.; Oh, Y.J. Buffering of cytosolic calcium plays a neuroprotective role by preserving the autophagy-lysosome pathway during MPP$^+$-induced neuronal death. *Cell Death Discov.* **2019**, *5*, 130. [CrossRef] [PubMed]

33. Kruman, I.; Guo, Q.; Mattson, M.P. Calcium and reactive oxygen species mediate staurosporine-induced mitochondrial dysfunction and apoptosis in PC12 cells. *J. Neurosci. Res.* **1998**, *51*, 293–308. [CrossRef]

34. Lynch, K.; Fernández, G.; Pappalardo, A.; Peluso, J.J. Basic Fibroblast Growth Factor Inhibits Apoptosis of Spontaneously Immortalized Granulosa Cells by Regulating Intracellular Free Calcium Levels through a Protein Kinase Cδ-Dependent Pathway. *Endocrinol.* **2000**, *141*, 4209–4217. [CrossRef]

35. Kiselyov, K.; Shin, D.M.; Muallem, S. Signalling specificity in GPCR-dependent Ca^{2+} signalling. *Cell. Signal.* **2003**, *15*, 243–253. [CrossRef]

36. Frick, K.K.; Krieger, N.S.; Nehrke, K.; Bushinsky, A.D. Metabolic Acidosis Increases Intracellular Calcium in Bone Cells Through Activation of the Proton Receptor OGR1. *J. Bone Miner. Res.* **2009**, *24*, 305–313. [CrossRef] [PubMed]

37. Tobo, M.; Tomura, H.; Mogi, C.; Wang, J.-Q.; Liu, J.-P.; Komachi, M.; Damirin, A.; Kimura, T.; Murata, N.; Kurose, H.; et al. Previously postulated "ligand-independent" signaling of GPR4 is mediated through proton-sensing mechanisms. *Cell. Signal.* **2007**, *19*, 1745–1753. [CrossRef] [PubMed]

38. Kumar, N.N.; Velic, A.; Soliz, J.; Shi, Y.; Li, K.; Wang, S.; Weaver, J.L.; Sen, J.; Abbott, S.B.; Lazarenko, R.M.; et al. Regulation of breathing by CO_2 requires the proton-activated receptor GPR4 in retrotrapezoid nucleus neurons. *Science* **2015**, *348*, 1255–1260. [CrossRef]

39. Wang, X.; Wang, W.; Li, L.; Perry, G.; Lee, H.-G.; Zhu, X. Oxidative stress and mitochondrial dysfunction in Alzheimer's disease. *Biochim. Biophys. Acta (BBA) Mol. Basis Dis.* **2014**, *1842*, 1240–1247. [CrossRef]

40. Fall, C.P.; Bennett, J.P. Characterization and time course of MPP^+ -induced apoptosis in human SH-SY5Y neuroblastoma cells. *J. Neurosci. Res.* **1999**, *55*, 620–628. [CrossRef]

41. Kim, I.-S.; Ganesan, P.; Choi, D.-K. Cx43 Mediates Resistance against MPP^+-Induced Apoptosis in SH-SY5Y Neuroblastoma Cells via Modulating the Mitochondrial Apoptosis Pathway. *Int. J. Mol. Sci.* **2016**, *17*, 1819. [CrossRef] [PubMed]

42. Dauer, W.; Przedborski, S.J.N. Parkinson's disease: Mechanisms and models. *Neuron* **2003**, *39*, 889–909.

43. Singh, A.; Verma, P.; Balaji, G.; Samantaray, S.; Mohanakumar, K.P. Nimodipine, an L-type calcium channel blocker attenuates mitochondrial dysfunctions to protect against 1-methyl-4-phenyl-1,2,3,6-tetrahydropyridine-induced Parkinsonism in mice. *Neurochem. Int.* **2016**, *99*, 221–232. [CrossRef] [PubMed]

44. Liu, J.-P.; Nakakura, T.; Tomura, H.; Tobo, M.; Mogi, C.; Wang, J.-Q.; He, X.-D.; Takano, M.; Damirin, A.; Komachi, M.; et al. Each one of certain histidine residues in G-protein-coupled receptor GPR4 is critical for extracellular proton-induced stimulation of multiple G-protein-signaling pathways. *Pharmacol. Res.* **2010**, *61*, 499–505. [CrossRef]

45. Karthivashan, G.; Park, S.-Y.; Kweon, M.-H.; Kim, J.; Haque, E.; Cho, D.-Y.; Kim, I.-S.; Cho, E.-A.; Ganesan, P.; Choi, D.-K. Ameliorative potential of desalted Salicornia europaea *L. extract* in multifaceted Alzheimer's-like scopolamine-induced amnesic mice model. *Sci. Rep.* **2018**, *8*, 7174. [CrossRef] [PubMed]

MDPI

St. Alban-Anlage 66

4052 Basel

Switzerland

Tel. +41 61 683 77 34

Fax +41 61 302 89 18

www.mdpi.com

International Journal of Molecular Sciences Editorial Office

E-mail: ijms@mdpi.com

www.mdpi.com/journal/ijms